Annotated Instructor's Edition

BASIC COLLEGE MATHEMATICS

K. Elayn Martin-Gay

University of New Orleans

PRENTICE HALL
Upper Saddle River, New Jersey 07458

*To Jewett B. Gay, and to the memory
of her husband, Jack Gay*

Acquisitions Editor: Karin E. Wagner
Editor-in-Chief: Jerome Grant
Editorial Director: Tim Bozik
Associate Editor-in-Chief, Development: Carol Trueheart
Senior Managing Editor: Linda Mihatov Behrens
Executive Managing Editor: Kathleen Schiaparelli
Assistant Vice President of Production and Manufacturing: David W. Riccardi
Marketing Manager: Jolene Howard
Manufacturing Buyer: Alan Fischer
Manufacturing Manager: Trudy Pisciotti
Editorial Assistant/Supplements Editor: Kate Marks
Associate Editor, Math/Statistics Media: Audra J. Walsh
Art Director/Cover Designer: Maureen Eide
Associate Creative Director: Amy Rosen
Director of Creative Services: Paula Maylahn
Assistant to Art Director: John Christiana
Art Manager: Gus Vibal
Art Editor: Grace Hazeldine
Cover image: Roland Seitre/Peter Arnold, Inc.
Text Design and Project Management: Elm Street Publishing Services, Inc.
Photo Researcher: Beth Boyd
Photo Research Administrator: Melinda Reo
Art Studio: Academy Artworks

Printed in the United States of America
10 9 8 7 6 5 4 3 2 1

ISBN: 0-13-080985-3   (Student Edition 0-13-376823-6)

Prentice-Hall International (UK) Limited, *London*
Prentice-Hall of Australia Pty. Limited, *Sydney*
Prentice-Hall Canada Inc., *Toronto*
Prentice-Hall Hispanoamericana, S.A., *Mexico*
Prentice-Hall of India Private Limited, *New Delhi*
Prentice-Hall of Japan, Inc., *Tokyo*
Simon & Schuster Asia Pte. Ltd., *Singapore*
Editora Prentice-Hall do Brasil, Ltda., *Rio de Janeiro*

Contents

Preface

About This Book

This worktext was written to provide a solid foundation in the basics of college mathematics, including the topics of whole numbers, fractions, decimals, ratio and proportion, percent, and measurement as well as introductions to geometry, statistics and probability, and algebra topics. Specific care was taken to ensure that students have the most up-to-date relevant text preparation for their next mathematics course or for nonmathematical courses that require an understanding of basic mathematical concepts. I have tried to achieve this by writing a user-friendly text that is keyed to objectives and contains many worked-out examples. As suggested by the AMATYC Crossroads Document and the NCTM Standards (plus Addenda), real-life and real-data applications, data interpretation, conceptual understanding, problem solving, writing, cooperative learning, appropriate use of technology, mental mathematics, number sense, estimation, critical thinking, and geometric concepts are emphasized and integrated throughout the book.

Key Pedagogical Features

Readability and Connections   I have tried to make the writing style as clear as possible while still retaining the mathematical integrity of the content. When a new topic is presented, an effort has been made to relate the new ideas to those that students may already know. Constant reinforcement and connections within problem-solving strategies, data interpretation, geometry, patterns, graphs, and situations from everyday life can help students gradually master both new and old information. In addition, each section begins with a list of objectives covered in the section. Clear organization of section material based on objectives further enhances readability.

Problem-Solving Process   This is formally introduced in Chapter 1 with a four-step process that is integrated throughout the text. The four steps are Understand, Translate, Solve, and Interpret. The repeated use of these steps in a variety of examples shows their wide applicability. Reinforcing the steps can increase students' comfort level and confidence in tackling problems.

Applications and Connections   Every effort was made to include as many interesting and relevant real-life applications as possible throughout the text in both worked-out examples and exercise sets. The applications help to motivate students and strengthen their understanding of mathematics in the real world. They show connections to a wide range of fields including agriculture, allied health, astronomy, automotive ownership and maintenance, aviation, biology, business, chemistry, communication, computer technology, construction, consumer affairs, cooking, demographics, earth science, education, entertainment, environmental issues, finance, food service, geography, government, history, hobbies, labor and career issues, medicine, music, nutrition, physics, recreation, sports, transportation, travel, weather, and zoology. Many of the applications are based on recent real data. Sources for data include newspapers, magazines, publicly held companies, government agencies, special-interest groups, research organizations, and reference books. Opportunities for obtaining your own real data are also included.

Exercise Sets   Each text section ends with an Exercise Set. Each exercise in the set is keyed to one of the objectives of the section. Wherever possible, a specific example is also given. Throughout the text exercises there is an emphasis on *Data and Graphical Interpretation* via tables and graphs. The ability to interpret data and read and create a variety of types of graphs is developed gradually so students become comfortable with it. In addition, Chapter 9 gives a concise introduction to statistical graphs and descriptive statistics.

In addition to the approximately 4000 exercises in end-of-section exercise sets, exercises may also be found in the Pretests, Integrated Reviews, Chapter Reviews, Chapter Tests, and Cumulative Reviews. Each Exercise Set contains one or more of the following features:

Mental Math   Found at the beginning of an exercise set, these mental warmups reinforce concepts found in the accompanying section and increase students' confidence before they tackle an exercise set. By relying on their own mental skills, students increase not only their confidence in themselves but also their number sense and estimation ability.

Review and Preview   These exercises occur in each exercise set (except for those in Chapter 1) after the exercises keyed to the objectives of the section. Review and Preview problems are keyed to earlier sections and review concepts learned earlier in the text that are needed in the next section or in the next chapter. These exercises show the links between earlier topics and later material.

Combining Concepts   These exercises are found at the end of each exercise set after the Review and Preview exercises. Combining Concepts exercises require students to combine several concepts from that section or to take the concepts of the section a step further by combining them with concepts learned in previous sections. For instance, sometimes students are required to combine the concepts of a section with the problem-solving process they learned in Chapter 1 to try their hand at solving an application problem.

Internet Excursions   These exercises occur once per chapter. Internet Excursions require students to use the Internet as a source of information or a data-collection tool to complete the exercises, allowing students first-hand experience with manipulating and working with real data.

Conceptual and Writing Exercises   These exercises occur in almost every exercise set and are marked with the icon ◲. They require students to show an understanding of a concept learned in the corresponding section. This is accomplished by asking students questions that require them to use two or more concepts together. Some require students to stop, think, and explain in their own words the concept(s) used in the exercises they have just completed. Guidelines recommended by the American Mathematical Association of Two Year Colleges (AMATYC) and other professional groups recommend incorporating writing in mathematics courses to reinforce concepts.

Practice Problems   Throughout the text, each worked-out example has a parallel Practice Problem placed next to the example in the margin. These invite students to be actively involved in the learning process before beginning the end-of-section exercise set. Practice Problems immediately reinforce a skill after it is developed.

Concept Checks   These margin exercises are appropriately placed in many sections of the text. They allow students to gauge their grasp of an idea as it is being explained in the text. Concept Checks stress conceptual understanding at point of use and help suppress misconceived notions before they start.

Integrated Reviews   These "mid-chapter reviews" are appropriately placed once per chapter. Integrated Reviews allow students to review and assimilate the many different skills learned separately over several sections before moving on to related material in the chapter.

Helpful Hints   Helpful Hints contain practical advice on applying mathematical concepts. These are found throughout the text and strategically placed where students are most likely to need immediate reinforcement. Helpful Hints are highlighted for quick reference.

Focus On   Appropriately placed throughout each chapter, these are divided into Focus on Study Skills, Focus on Mathematical Connections, Focus on Business and Career, Focus on the Real World, and Focus on History. They are written to help students develop effective habits for studying mathematics, engage in investigations of other branches of mathematics, understand the importance of mathematics in various careers and in the world of business, and see the relevance of mathematics in both the present and past through critical thinking exercises and group activities.

Calculator Explorations   These optional explorations offer point-of-use instruction, through examples and exercises, on the proper use of calculators as tools in the mathematical problem-solving process. Placed appropriately throughout the text, Calculator Explorations also reinforce concepts learned in the corresponding section.

Additional exercises building on the skill developed in the Explorations may be found in exercise sets throughout the text. Exercises requiring a calculator are marked with the 🖩 icon. The inside back cover of the text includes a brief description of selected keys on a scientific calculator for reference as desired.

Chapter Activity   These features occur once per chapter at the end of the chapter, often serving as a chapter wrap-up. For individual or group completion, the Chapter Activity, usually hands-on or data-based, complements and extends the concepts of the chapter, allowing students to make decisions and interpretations and to think and write about mathematics.

Visual Reinforcement of Concepts   The text contains a wealth of graphics, models, photographs, and illustrations to visually clarify and reinforce concepts. These include bar graphs, line graphs, calculator screens, application illustrations, and geometric figures.

Pretests   Each chapter begins with a pretest that is designed to help students identify areas where they need to pay special attention in the upcoming chapter.

Chapter Highlights   Found at the end of each chapter, these contain key definitions, concepts, and examples to help students understand and retain what they have learned.

Chapter Review and Test   The end of each chapter contains a review of topics introduced in the chapter. The Chapter Review offers exercises that are keyed to sections of the chapter. The Chapter Test is a practice test and is not keyed to sections of the chapter.

Cumulative Review   These are found at the end of each chapter (except Chapter 1). Each problem contained in the cumulative review is actually an earlier worked example in the text that is referenced in the back of the book along with the answer. Students who need to see a complete worked-out solution, with explanation, can do so by turning to the appropriate example in the text.

Student Resource Icons   At the beginning of each section, videotape, software, and solutions manual icons are displayed. These icons help reinforce that these learning aids are available should students wish to use them to help them review concepts and skills at their own pace. These items have direct correlation to the text and emphasize the text's methods of solution.

Functional Use of Color and Design   Elements of the text are highlighted with color or design to make it easier for students to read and study. Color is also used to clarify the problem-solving process in worked examples.

SUPPLEMENTS FOR THE INSTRUCTOR

Printed Supplements

Annotated Instructor's Edition (0-13-080985-3)

▲ Answers to all exercises printed on the same text page.
▲ Teaching Tips throughout the text placed at key points in the margin.

Instructor's Solution Manual (0-13-752528-1)

▲ Solutions to even-numbered section exercises.
▲ Solutions to every (even and odd) Mental Math exercise.
▲ Solutions to every (even and odd) Practice Problem (margin exercise).
▲ Solutions to every (even and odd) exercise found in the Chapter Pretests, Integrated Reviews (mid-chapter reviews), Chapter Reviews, Chapter Tests, Cumulative Reviews.

Instructor's Resource Manual with Tests (0-13-752536-2)

▲ Notes to the Instructor that includes an introduction to Interactive Learning, Interpreting Graphs and Data, Alternative Assessment, Using Technology and Helping Students Succeed.
▲ Two free-response Pretests per chapter.
▲ Eight Chapter Tests per chapter (3 multiple-choice, 5 free-response).
▲ Two Cumulative Review Tests (one multiple-choice, one free-response) every two chapters (after chapters 2, 4, 6, 9).
▲ Eight Final Exams (3 multiple-choice, 5 free-response).
▲ Twenty additional exercises per section for added test exercises if needed.

Media Supplements

TestPro Computerized Testing

▲ Algorithmically driven, text-specific testing program.
▲ Networkable for administering tests and capturing grades on-line.
▲ Edit and add your own questions—create nearly unlimited number of tests and drill worksheets.

Companion Web site
▲ www.prenhall.com/martin-gay
▲ Links related to the Internet Excursions in each chapter allow you to collect data to solve specific internet exercises.
▲ Additional links to helpful, generic sites include Fun Math and For Additional Help.

SUPPLEMENTS FOR THE STUDENT
Printed Supplements
Student Solution's Manual (0-13-752544-3)
▲ Solutions to odd-numbered section exercises.
▲ Solutions to every (even and odd) Mental Math exercise.
▲ Solutions to every (even and odd) Practice Problem (margin exercise).
▲ Solutions to every (even and odd) exercise found in the Chapter Pretests, Integrated Reviews (mid-chapter reviews), Chapter Reviews, Chapter Tests, Cumulative Reviews.

New York Times *Themes of the Times*
▲ Have your instructor contact the local Prentice Hall sales representative.

How to Study Mathematics
▲ Have your instructor contact the local Prentice Hall sales representative.

Internet Guide
▲ Have your instructor contact the local Prentice Hall sales representative.

Media Supplements
MathPro4 computerized tutorial
▲ Keyed to each section of the text for text-specific tutorial exercises and instruction.
▲ Includes Warm-up exercises and graded Practice Problems.
▲ Algorithmically driven and fully networkable.
▲ Have your instructor contact the local Prentice Hall sales representative—also available for purchase for home use.

Videotape Series (0-13-752650-4)
▲ Written and presented by Elayn Martin-Gay.
▲ Keyed to each section of the text.
▲ Step-by-step solutions to exercises from each section of the text. Exercises that are worked in the videos are marked with a video icon ( ).

Companion Web site
▲ www.prenhall.com/martin-gay
▲ Links related to the Internet Excursions in each chapter allow you to collect data to solve specific internet exercises.
▲ Additional links to generic sites include Fun Math, For Additional Help . . .

ACKNOWLEDGMENTS
First, as usual, I would like to thank my husband, Clayton, for his constant encouragement. I would also like to thank my children, Eric and Bryan, for their sense of humor and especially for their suggestion of letting Dad cook the bacon that I always used to burn.

I would also like to thank my extended family for their invaluable help and also their sense of humor. Their contributions are too numerous to list. They are Rod and Karen Pasch; Michael, Christopher, Matthew, and Jessica Callac; Stuart, Earline, Melissa, Mandy and Bailey Martin; Mark, Sabrina, and Madison Martin; Leo and Barbara Miller; and Jewett Gay.

I would like to thank the following reviewers for their input and suggestions:

Fatemah Bicksler, *Delgado Community College*
Beverly Broomell, *Suffolk County Community College*
Karla Childs, *Pittsburg State University*
Dorothy French, *Community College of Philadelphia*
Mary Ellen Gallegos, *Santa Fe Community College*
James Harris, *John A. Logan College*
Brian Hayes, *Triton College*
Rosa Kavanaugh, *Ozarks Technical Community College*
Deanna Li, *North Seattle Community College*
Hubert McClure, *Tri County Technical College*
Christopher McNally, *Tallahassee Community College*
Julie Miller, *Daytona Beach Community College*
Ellen O'Connell, *Triton College*
Maureen O'Grady, *Suffolk County Community College*
Marilyn Platt, *Gaston College*
Pat Roux, *Delgado Community College*

There were many people who helped me develop this text and I will attempt to thank some of them here. Penny Korn did an excellent job of providing answers. Cheryl Roberts Cantwell was invaluable for contributing to the overall accuracy of this text. Emily Keaton was also invaluable for her many suggestions and contributions during the development and writing of this first edition. Ingrid Mount at Elm Street Publishing Services provided guidance throughout the production process. I thank Terri Bittner, Cindy Trimble, Jeff Rector, and Teri Lovelace at Laurel Technical Services for all their work on the supplements, text and thorough accuracy check. Lastly, a special thank you to my editor, Karin Wagner, for her support and assistance throughout the development and production of this text and to all the staff at Prentice Hall: Linda Behrens, Alan Fischer, Maureen Eide, Grace Hazeldine, Gus Vibal, Kate Marks, Jolene Howard, John Tweeddale, Jerome Grant, and Tim Bozik.

K. Elayn Martin-Gay

ABOUT THE AUTHOR

K. Elayn Martin-Gay has taught mathematics at the University of New Orleans for more than 19 years. Her numerous teaching awards include the local University Alumni Association's Award for Excellence in Teaching, and Outstanding Developmental Educator at University of New Orleans, presented by the Louisiana Association of Developmental Educators.

Elayn is the author of an entire product line of highly successful textbooks. The Martin-Gay library includes the exciting new worktext series: *Basic College Mathematics*, *Introductory Algebra* and *Intermediate Algebra*, the successful hardbound series; *Beginning Algebra* 2e, *Intermediate Algebra* 2e, *Intermediate Algebra A Graphing Approach* (co-authored with Margaret Greene), *Prealgebra* 2e, and *Introductory and Intermediate Algebra*, a combined algebra text.

Prior to writing textbooks, Elayn developed an acclaimed series of lecture videos to support developmental mathematics students in their quest for success. These highly successful videos originally served as the foundation material for her texts. Today the tapes specifically support each book in the Martin-Gay series.

Index of Applications

How to Use This Worktext: A Guide for Students

Basic College Mathematics has been written and designed to help you succeed in this course. The goals that the author will help you achieve include:

▲ Exposure to real-world applications you will encounter in life and other courses

▲ An organized, integrated system of learning that combines the text with a comprehensive set of print and media tools

▲ Better preparation relevant to the *next* course you will take in mathematics

Take a few moments now to acquaint yourself with some of the features that have been built into *Basic College Mathematics* to help you excel.

Adding and Subtracting Fractions

CHAPTER 3

Having learned what fractions are and how to multiply and divide them in Chapter 2, we are ready to continue our study of fractions. In this chapter, we will learn how to add and subtract fractions and mixed numbers. We then conclude this chapter with comparing fractions and reviewing the order of operations.

3.1 Adding and Subtracting Like Fractions

3.2 Least Common Multiple

3.3 Adding and Subtracting Unlike Fractions

Integrated Review—Adding and Subtracting Fractions

3.4 Adding and Subtracting Mixed Numbers

3.5 Order, Exponents, and the Order of Operations

The American Lobster, *Homarus americanus*, is big business in New England. Lobster fishers caught more than 70 million pounds of lobsters in 1996, and more than $\frac{4}{5}$ of the lobster catch was made off the coasts of Maine, Massachusetts, and Rhode Island. Most American lobsters are a dark blue-green or grey color when alive. There are rare instances of blue, yellow, and white lobsters. However, the only red lobsters are ones that have been cooked. In Example 8 on pages 189–190 and in the Chapter Activity, we will see how fractions are used in the lobster fishing industry.

161

> The photo applications at the opening of every chapter, and applications throughout, introduce you to real world situations that are applicable to the mathematics you will learn in the upcoming chapter. These applications are often discussed later in the chapter as well.

Page 161

Connect the Concepts

Learning and succeeding in a math course requires practice and a broader understanding of how everything works together. As you study, make connections using this text's organization to help you. All of these features have been included to enhance your understanding of mathematical concepts.

Concept Checks are special margin exercises found in most sections. Work these to gauge your grasp of the idea being explained in the text.

✓ CONCEPT CHECK

True or false: Two different numbers can have exactly the same prime factorization. Explain your answer.

Page 116

COMBINING CONCEPTS

The table gives the leading bowling averages for the Professional Bowlers Association Tour for each of the years listed. Use the table to answer Exercises 55–57.

LEADING PBA AVERAGES BY YEAR		
Year	Bowler	Average Score
1988	Mark Roth	216.081
1989	Pete Weber	218.036
1990	Amleto Monacelli	215.432
1991	Norm Duke	218.158
1992	Dave Ferraro	219.702
1993	Walter Ray Williams, Jr.	222.980
1994	Norm Duke	222.830
1995	Mike Aulby	225.490
1996	Walter Ray Williams, Jr.	225.370
1997	Walter Ray Williams, Jr.	222.000

(*Source:* Professional Bowlers Association)

55. What is the highest average score on the list? Which bowler achieved that average?

56. What is the lowest average score on the list? Which bowler achieved that average?

Look for **Combining Concepts** exercises at the end of each exercise set. Solving these exercises will expose you to the way mathematical ideas build upon each other.

Page 231

Practice Problem 4

Using the figure below, estimate how much farther it is between Dodge City and Pratt than it is between Garden City and Dodge City by rounding each given distance to the nearest mile.

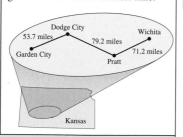

Example 4 Estimating Distance

Use the figure in the margin to estimate the distance in miles between Garden City, Kansas, and Wichita, Kansas, by rounding each given distance to the nearest ten.

Solution:

Calculated Distance		Estimate
53.7	rounds to	50
79.2	rounds to	80
+71.2	rounds to	+70
		200

The distance between Garden City and Wichita is approximately 200 miles. (The calculated distance is 204.1 miles.)

Practice Problems occur in the margins next to every Example. Work these problems after you read an Example to reinforce your understanding.

Page 262

Test Yourself and Check Your Understanding

Good exercise sets are an essential ingredient for a solid basic college mathematics textbook. The exercises you will find in this worktext are intended to help you build skills and understand concepts as well as motivate and challenge you. Note that the features like Chapter Highlights, Tests, and Cumulative Reviews are found at the end of each chapter to help you study and organize your notes.

CHAPTER 2 PRETEST

1. Use a fraction to represent the shaded area of the figure.

2. Write each mixed number as an improper fraction.
   a. $3\frac{4}{7}$ b. $9\frac{5}{8}$

3. Write each improper fraction as a mixed number or a whole number.
   a. $\frac{52}{13}$ b. $\frac{71}{6}$

4. List all the factors of 36.

Identify each number as prime or composite.
5. 19   6. 91   7. Find the prime factorization of 80.

Pretests open each chapter. Take a **Pretest** to evaluate where you need help the most before beginning a new chapter.

Page 100

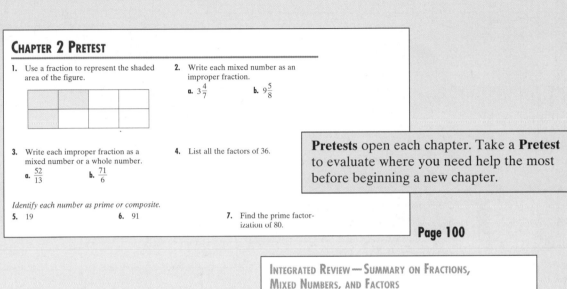

INTEGRATED REVIEW—SUMMARY ON FRACTIONS, MIXED NUMBERS, AND FACTORS

Use a fraction to represent the shaded area of each figure or figure group.
1.   2.

4. In a survey, 73 people out of 85 get less than 8 hours of sleep each night. What fraction of people in the survey get less than 8 hours of sleep?

Write each mixed number as an improper fraction.
5. $3\frac{1}{8}$ 6. $5\frac{3}{5}$ 7. $9\frac{6}{7}$

Integrated Reviews serve as a mid-chapter review and ask you to assimilate the new skills you have learned separately over several sections.

Page 127

MENTAL MATH

State whether the fractions in each list are like or unlike fractions.
1. $\frac{7}{8}, \frac{7}{10}$ 2. $\frac{2}{3}, \frac{4}{9}$ 3. $\frac{9}{10}, \frac{1}{10}$ 4. $\frac{8}{11}, \frac{2}{11}$

5. $\frac{2}{31}, \frac{30}{31}, \frac{19}{31}$ 6. $\frac{3}{10}, \frac{3}{11}, \frac{3}{13}$ 7. $\frac{5}{12}, \frac{7}{12}, \frac{12}{11}$ 8. $\frac{1}{5}, \frac{2}{5}, \frac{4}{5}$

Confidence-building **Mental Math** problems are in most sections.

Page 167

REVIEW AND PREVIEW

Add or subtract as indicated. See Section 3.1.
53. $\frac{7}{10} - \frac{2}{10}$ 54. $\frac{8}{13} - \frac{3}{13}$ 55. $\frac{1}{5} + \frac{1}{5}$ 56. $\frac{1}{8} + \frac{3}{8}$

57. $\frac{23}{18} - \frac{15}{18}$ 58. $\frac{36}{30} - \frac{12}{30}$ 59. $\frac{2}{9} + \frac{1}{9} + \frac{6}{9}$ 60. $\frac{2}{12} + \frac{7}{12} + \frac{3}{12}$

Review and Preview

Page 176

Get Involved!

Discover how mathematics relates to and appears in the world around you. Evaluate and interpret real data in graphs, tables, and charts—make an educated guess on the answers and outcomes! Knowing how to use data and graphs to estimate and predict is a valuable skill in the workplace as well as in other courses.

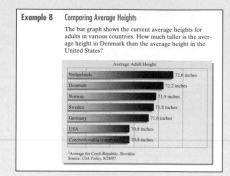

Example 8 Comparing Average Heights

The bar graph shows the current average heights for adults in various countries. How much taller is the average height in Denmark than the average height in the United States?

Average Adult Height

Netherlands	72.6 inches
Denmark	72.2 inches
Norway	71.9 inches
Sweden	71.8 inches
Germany	71.6 inches
USA	70.8 inches
Czechoslovakia (combined)†	70.8 inches

†Average for Czech Republic, Slovakia
Source: *USA Today*, 8/28/97

The following graph is called a circle graph or pie chart. Each sector (shaped like a piece of pie) shows the fractional part of a car's total mileage that falls into a particular category. The whole circle represents a car's total mileage.

Work
$\frac{8}{25}$

Vacation/other $\frac{3}{50}$

$\frac{1}{100}$ Medical

Shopping $\frac{3}{25}$

$\frac{3}{25}$ Visit friends

$\frac{13}{100}$

$\frac{1}{5}$

$\frac{2}{50}$ School/church

Social/recreational

Family business

(*Source:* The American Automobile Manufacturers Assn. and The National Automobile Dealers Assn.)

Real data is integrated throughout the worktext, drawn from current and familiar sources.

Page 137

Page 235

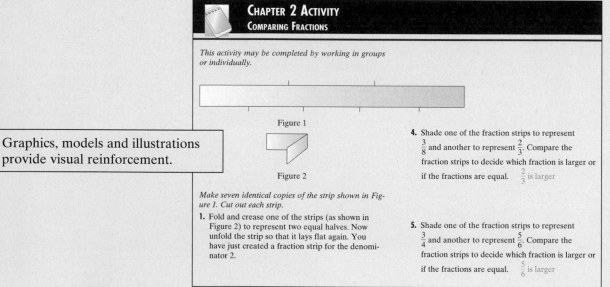

CHAPTER 2 ACTIVITY
COMPARING FRACTIONS

This activity may be completed by working in groups or individually.

Figure 1

Figure 2

Make seven identical copies of the strip shown in Figure 1. Cut out each strip.

1. Fold and crease one of the strips (as shown in Figure 2) to represent two equal halves. Now unfold the strip so that it lays flat again. You have just created a fraction strip for the denominator 2.

4. Shade one of the fraction strips to represent $\frac{3}{8}$ and another to represent $\frac{2}{3}$. Compare the fraction strips to decide which fraction is larger or if the fractions are equal. $\frac{2}{3}$ is larger

5. Shade one of the fraction strips to represent $\frac{3}{4}$ and another to represent $\frac{5}{6}$. Compare the fraction strips to decide which fraction is larger or if the fractions are equal. $\frac{5}{6}$ is larger

Graphics, models and illustrations provide visual reinforcement.

Page 147

CALCULATOR EXPLORATIONS
ENTERING DECIMAL NUMBERS

To enter a decimal number, find the key marked [.]. To enter the number 2.56, for example, press the keys

[2] [.] [5] [6]

The display will read [2.56].

OPERATIONS ON DECIMAL NUMBERS

Operations on decimal numbers are performed in the same way as operations on whole or signed numbers. For example, to find 8.625 − 4.29, press the keys

[8.625] [−] [4.29] [=]

or

[ENTER]

The display will read [4.335].

(Although entering 8.625, for example, requires pressing more than one key, we group numbers together for easier reading.)

...cated operation.

29.68 + 85.902   115.582

5.238 − 0.682   4.556

47.006   499.523
0.17
313.259
+622.65   +139.088

Calculator Explorations and exercises are woven into appropriate sections.

Internet Excursions

Go to http://www.prenhall.com/martin-gay
This World Wide Web site will provide access to a site called the Low Fat Vegetarian Archive, or a related site. It contains more than 2500 recipes for healthy low-fat and fat-free vegetarian dishes. Users can search the recipes by category, such as Mexican foods or desserts, or by ingredient.

81. Visit this site and find a recipe that is interesting to you. Be sure that the ingredient list contains at least two measures that are fractions (for example, $\frac{1}{3}$ cup). Print or copy the recipe. Then rewrite the list of ingredients to show the amounts that would be needed to make $\frac{1}{2}$ of the recipe.

82. Find a different recipe at this site that is interesting to you. Make sure that the recipe lists at least two fractional measures. Print or copy the recipe. Then rewrite the list of ingredients to show the amounts that would be ...

Visit the Internet through the Martin-Gay companion Web site to gather and manipulate data to complete exercises in **Internet Excursions.**

Page 236

Page 138

Focus On boxes found throughout each chapter help you see the relevance of math through critical thinking exercises and group activities. Try these on your own or with a classmate.

Focus on Study Skills

Focus On Study Skills

STUDYING FOR AND TAKING A MATH EXAM

Remember that one of the best ways to start preparing for an exam is to keep current with your assignments as they are made. Make an effort to clear up any confusion on topics as you cover them.

Begin reviewing for your exam a few days in advance. If you find a topic during your review that you still don't understand, you'll have plenty of time to ask your instructor, another student in your class, or a math tutor for help. Don't wait until the last minute to "cram" for the test.

▲ Reread your notes and carefully review the Chapter Highlights at the end of each chapter.
▲ Try solving a few exercises from each section.
▲ Pay special attention to any new terminology or definitions in the chapter. Be sure you can state the meanings of definitions in your own words.
▲ Find a quiet place to take the Chapter Test found at the end of the chapter to be covered. This gives you a chance to practice taking the real exam, so try the Chapter Test

Page 196

Focus on History

Focus On History

MULTICULTURAL FRACTION USE

Many ancient cultures were familiar with the use of fractions and used them in everyday life. Here are some interesting facts about the history of fractions:

▲ Although the ancient Egyptians were comfortable with the idea of fractions, their system of hieroglyphic numbers allowed them to only write fractions with 1 as the numerator, like $\frac{1}{3}, \frac{1}{7},$ or $\frac{1}{10}$.
To write fractions with numerators other than 1, the Egyptians had to write a sum of fractions with 1 as the numerator, and they preferred to never use the same denominator in a sum of fractions twice! For instance, rather than representing $\frac{2}{5}$ as $\frac{1}{5} + \frac{1}{5}$ (which uses the denominator 5 twice), they would use $\frac{1}{3} + \frac{1}{15}$.

Page 242

Focus on Mathematical Connections

Focus On Mathematical Connections

MODELING FRACTIONS

There are several different physical models for representing fractions.

SET MODEL

In this model, a fraction represents the portion of a set of objects that has a certain characteristic. For example, in the set of 10 shapes, 3 are hearts. That is, $\frac{3}{10}$ of the shapes are hearts.

AREA MODEL

In this model, a shape is divided into a number of equal-sized regions. A fraction can be represented by shading some of the regions. For example, both of the following

Page 106

Focus on the Real World

Focus On the Real World

BLOOD AND BLOOD DONATION

Blood is the workhorse of the body. It carries to the body's tissues everything they need, from nutrients to antibodies to heat. Blood also carries away waste products like carbon dioxide. Blood contains three types of cells—red blood cells, white blood cells, and platelets—suspended in clear, watery fluid called plasma. Blood is $\frac{11}{20}$ plasma, and plasma itself is $\frac{9}{10}$ water. In the average healthy adult human, blood accounts for $\frac{1}{11}$ of a person's body weight.

Roughly every 2 seconds someone in the United States needs blood. Although only $\frac{1}{20}$ of eligible donors donate blood, the American Red Cross is still able to collect nearly 6 million volun-

Page 132

Focus on Business and Career

Focus On Business and Career

IN-DEMAND OCCUPATIONS

According to U.S. Bureau of Labor Statistics projections, the careers listed below will have the largest job growth into the next century.

	Occupation	Employment [Numbers in thousands]		
		1994	2005	Change
1.	Cashiers	3005	3567	562
2.	Janitors and cleaners, including maids and housekeeping cleaners	3043	3602	559
3.	Salespersons, retail	3842	4374	532
4.	Waiters and waitresses	1847	2326	479
5.	Registered nurses	1906	2379	473
6.	General managers and top executives	3046	3512	466
7.	Systems analysts	483	928	445
8.	Home health aides	420	848	428
9.	Guards	867	1282	415
10.	Nursing aides, orderlies, and attendants	1265	1652	387
11.	School teachers, secondary	1340	1726	386
12.	Marketing and sales worker supervisors	2293	2673	380
13.	Teacher aides and educational associates	932	1296	364
14.	Receptionists and information clerks	1019	1337	318
15.	Truckdrivers, light and heavy	2565	2837	272

Page 270

Enrich Your Learning

Seek out these additional Student Resources and tools to match your personal learning style.

Text-specific **videos** hosted by the award-winning teacher and author of *Basic College Mathematics* cover each objective in every chapter section as a supplementary review. Many of the examples and exercises worked out in the videos are selected directly from *Basic College Mathematics*.

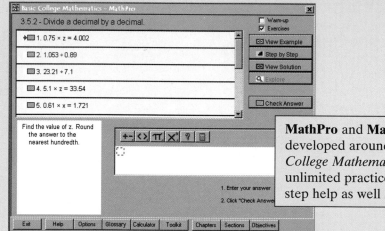

MathPro and **MathPro Explorer** tutorial software is developed around the content and concepts of *Basic College Mathematics*. **MathPro Explorer** provides unlimited practice problems and interactive step-by-step help as well as exploratory activities.

STUDENT SOLUTIONS MANUAL
Laurel Technical Services

Basic College Mathematics

K. Elayn Martin-Gay

Also available:

Ask your instructor or bookstore about these additional study aids.

The Whole Numbers

Mathematics is an important tool for everyday life. Knowing basic mathematical skills can help simplify tasks such as creating a monthly budget. Whole numbers are the basic building blocks of mathematics. The whole numbers answer the question "How many?"

This chapter covers basic operations on whole numbers. Knowledge of these operations provides a good foundation on which to build further mathematical skills.

The American Kennel Club (AKC) was founded in 1884. This nonprofit organization is devoted to the raising and welfare of purebred dogs. The AKC recognizes 140 different dog breeds, ranging in size from the Chihuahua to the Great Dane. In addition to sanctioning dog shows, the AKC maintains a registry of individual purebred dogs. AKC registration means that a dog's parents and ancestors were purebreds. In Exercises 52–55 on page 9, we will see how whole numbers can be used to keep track of and compare American Kennel Club registrations.

Name _____ **Section** _____ **Date** _____

CHAPTER 1 PRETEST

1. Determine the place value of the digit 7 in the whole number 5732.

2. Write the whole number 23,490 in words.

3. Add: 58 + 29

4. Multiply: 413
   $\times$ 9

Subtract. Check by adding.

5.   857
   −231

6.   51
   −19

Solve.

7. Karen Lewis is reading a 329-page novel. If she has just finished reading page 193, how many more pages must she read to finish the novel?

8. Round 9045 to the nearest ten.

9. Round each number to the nearest hundred to find an estimated sum.
   382
   436
   2084
   + 176

10. Use the distributive property to rewrite the following expression.
   $9(3 + 11)$

11. Find the perimeter of the following figure.

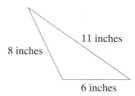

8 inches   11 inches
6 inches

12. Find the area of the following rectangle.

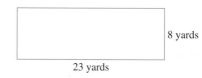

8 yards
23 yards

13. The seats in the history lecture hall are arranged in 32 rows with 18 seats in each row. Find how many seats are in this room.

Divide and then check by multiplying.

14. $2187 \div 9$

15. $\dfrac{5361}{12}$

Solve.

16. Find the average of the following list of numbers. 29, 36, 84, 41, 6, 12, 65

17. Paul Crandall left $29,640 in his will to go to his three favorite nephews. If each boy was to receive the same amount of money, how much did each boy receive?

18. Write $9 \cdot 9 \cdot 9 \cdot 9 \cdot 9 \cdot 9 \cdot 9$ using exponential notation.

19. Evaluate: 7^4

20. Simplify: $36 + 18 \div 6$

2

1.1 PLACE VALUE AND NAMES FOR NUMBERS

The **digits** 0, 1, 2, 3, 4, 5, 6, 7, 8, and 9 can be used to write numbers. For example, the **whole numbers** are

0, 1, 2, 3, 4, 5, 6, 7, 8, 9, 10, 11, . . .

The three dots (. . .) after the 11 means that this list continues indefinitely. That is, there is no largest whole number. The smallest whole number is 0.

A FINDING THE PLACE VALUE OF A DIGIT IN A WHOLE NUMBER

The position of each digit in a number determines its **place value**. A place-value chart is shown next with the whole number 48,337,000 entered.

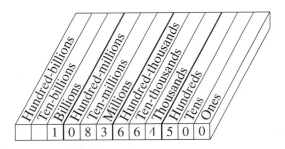

The two 3s in 48,337,000 represent different amounts because of their different placements. The place value of the 3 to the left is hundred-thousands. The place value of the 3 to the right is ten-thousands.

Examples   Find the place value of the digit 4 in each whole number.

1. 48,761	**2.** 249	**3.** 524,007,656
↑	↑	↑
ten-thousands	tens	millions

B WRITING A WHOLE NUMBER IN WORDS AND IN STANDARD FORM

A whole number such as 1,083,664,500 is written in **standard form**. Notice that commas separate the digits into groups of threes, starting from the right. Each group of three digits is called a **period**. The names of the first four periods are shown in red.

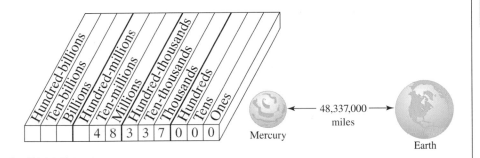

WRITING A WHOLE NUMBER IN WORDS

To write a whole number in words, write the number in each period followed by the name of the period. (The ones period is usually not written.) This same procedure can be used to read a whole number.

Objectives

A Find the place value of a digit in a whole number.

B Write a whole number in words and in standard form.

C Write the expanded form of a whole number.

D Read tables.

SSM   CD-ROM   Video 1.1

TEACHING TIP

Before beginning this lesson, ask students how many digits we use to represent numbers. Make sure students understand that we can write *all* numbers with the ten digits 0 through 9.

Practice Problems 1–3

Find the place value of the digit 7 in each whole number.

1. 72,589,620

2. 67,890

3. 50,722

TEACHING TIP

Provide a list of numbers for students to say aloud. As an extra challenge, consider asking them to say the number which comes immediately before or after the number listed.
Sample List: 63; 25,345; 399,499; 455,699; 7832; 3,000,000; 34,004,000,002; 483

Answers

1. ten-millions, **2.** thousands, **3.** hundreds

For example, we write 1,083,664,500 as

one **billion**,

eighty-three **million**,

six hundred sixty-four **thousand**,

five **hundred**

HELPFUL HINT

The name of the ones period is not used when reading and writing whole numbers. For example,

9,265

is read as

"nine **thousand**, two **hundred** sixty-five."

TEACHING TIP

Have your students practice writing numbers which you say aloud.
Sample List: 48; 21,924; 7,344,899; 236,341; 345; 876,349; 23,000,126,007; 9342

Practice Problems 4–5

Write each number in words.

4. 67

5. 395

Practice Problem 6

Write 321,670,200 in words.

✓ CONCEPT CHECK

True or false? When writing a check for $2600, the word name we write for the dollar amount of the check would be "two thousand sixty." Explain your answer.

Practice Problems 7–10

Write each number in standard form.

7. twenty-nine

8. seven hundred ten

9. twenty-six thousand seventy-one

10. six thousand, five hundred seven

Answers

4. sixty-seven,   **5.** three hundred ninety-five,
6. three hundred twenty-one million,
six hundred seventy thousand, two hundred,
7. 29,   **8.** 710,   **9.** 26,071,   **10.** 6507

✓ Concept Check: False

Examples   Write each number in words.

4. 85   eighty-five
5. 126   one hundred twenty-six

HELPFUL HINT

The word "and" is *not* used when reading and writing whole numbers. It is used when reading and writing mixed numbers and some decimal values, as shown later in this text.

Example 6   Write 106,052,447 in words.

Solution: 106,052,447 is written as

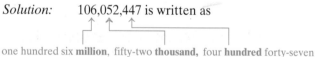

one hundred six **million**, fifty-two **thousand**, four **hundred** forty-seven

---

TRY THE CONCEPT CHECK IN THE MARGIN.

WRITING A WHOLE NUMBER IN STANDARD FORM

To write a whole number in standard form, write the number in each period followed by a comma.

Examples   Write each number in standard form.

7. sixty-one   61
8. eight hundred five   805
9. two million, five hundred sixty-four thousand, three hundred fifty

2,564,350

10. nine thousand, three hundred eighty-six

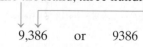

9,386   or   9386

HELPFUL HINT

A comma may or may not be inserted in a four-digit number. For example, both

   9,386 and 9386

are acceptable ways of writing nine thousand, three hundred eighty-six.

C WRITING A WHOLE NUMBER IN EXPANDED FORM

The place value of a digit can be used to write a number in expanded form. The **expanded form** of a number shows each digit of the number with its place value. For example, 5672 is written in expanded form as

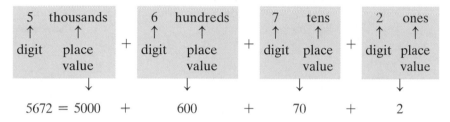

$$5672 = 5000 \ + \ 600 \ + \ 70 \ + \ 2$$

Example 11   Write 706,449 in expanded form.

Solution: $700,000 + 6000 + 400 + 40 + 9$ ▬▬▬

Practice Problem 11

Write 1,047,608 in expanded form.

D READING TABLES

Now that we know about place value and names for whole numbers, we introduce one way that whole number data may be presented. **Tables** are often used to organize and display facts that contain numbers. The table below shows the countries that have won the most medals during the Olympic winter games. (Although the medals are won by athletes from the various countries, for simplicity, we will state that countries have won the medals.)

MOST MEDALS WINTER GAMES (1924–1998)			
	Gold	Silver	Bronze	Total
USSR[1]/Russia	107	77	74	258
Norway	83	85	68	236
U.S.	59	58	40	157
Germany[2]	56	52	44	152
Austria	39	53	53	145
Finland	37	49	48	134
GDR[3]	39	36	35	110
Sweden	39	28	35	102
Switzerland	29	31	32	92
Canada	24	25	28	77

[1]Includes former USSR to 1992, Russia 1994, 1998
[2]Includes West Germany 1952, 1968–1988.
[3]GDR (East Germany) 1968–1988.
Source: Winter Games Net

Answer

11. $1,000,000 + 40,000 + 7000 + 600 + 8$

For example, by reading from left to right along the row marked U.S., we find that the United States has won 59 gold, 58 silver, and 40 bronze medals for the years 1924–1998.

Practice Problem 12

Use the Winter Games table to answer the following questions.

a. How many bronze medals has Austria won during the winter games of the Olympics?

b. Which countries shown have won more than 70 gold medals?

Example 12

Use the Winter Games table on page 5 to answer each question.

a. How many silver medals has Finland won during the winter games of the Olympics?

b. Which country shown has won fewer gold medals than Switzerland?

Solution:

a. Read from left to right across the line marked Finland until the "Silver" column is reached. We find that Finland has won 49 silver medals.

b. Switzerland has won 29 gold medals while Canada has won 24, so Canada has won fewer gold medals than Switzerland.

Answers

12. a. 53,   **b.** USSR/Russia and Norway

Name _____ Section _____ Date _____

Exercise Set 1.1

A *Determine the place value of the digit 5 in each whole number. See Examples 1 through 3.*

1. 352   **2.** 905   **3.** 5890   **4.** 6527

5. 62,500,000   **6.** 79,050,000   **7.** 5,070,099   **8.** 51,682,700

B *Write each whole number in words. See Examples 4 through 6.*

9. 5420   **10.** 3165   📼**11.** 26,990   **12.** 42,009

13. 1,620,000   **14.** 3,204,000   **15.** 53,520,170   **16.** 47,033,107

Write each number in the sentence in words. See Examples 4 through 6.

17. At this writing, the population of Libya is 5,648,359. (*Source: U.N. Statistical Yearbook*)

Libya

18. Liz Harold has the number 16,820,409 showing on her calculator display.

16820409

19. In a recent year, zinc mines in the United States mined 620,000 metric tons of zinc. (*Source:* U.S. Bureau of Mines)

20. The highest point in Montana is at Granite Peak, at an elevation of 12,799 feet. (*Source:* U.S. Geological Survey)

21. In a recent year, there were 3893 patients in the United States waiting for a heart transplant. (*Source:* United Network for Organ Sharing)

22. Each Home Depot store in the United States and Canada stocks at least 40,000 different kinds of building materials, home improvement supplies, and lawn and garden products. (*Source:* The Home Depot, Inc.)

23. 6508

24. 3370

25. 29,900

26. 42,006

27. 6,504,019

28. 10,037,016

29. 3,000,014

30. 7,000,012

31. 1821

32. 93,000,000

33. 63,100,000

34. 1076

35. 500,000,000

36. 34,000

37. 400 + 6

38. 700 + 80 + 9

39. 5000 + 200 + 90

40. 6000 + 40

Write each whole number in standard form. See Examples 7 through 10.

23. Six thousand, five hundred eight

24. Three thousand, three hundred seventy

25. Twenty-nine thousand, nine hundred

26. Forty-two thousand, six

27. Six million, five hundred four thousand, nineteen

28. Ten million, thirty-seven thousand, sixteen

29. Three million, fourteen

30. Seven million, twelve

Write the whole number in each sentence in standard form. See Examples 7 through 10.

31. The world's tallest self-supporting structure is the CN Tower in Toronto, Canada. It is one thousand, eight hundred twenty-one feet tall. (*Source: World Almanac*, 1998)

32. The average distance between Earth and the sun is more than 93 million miles.

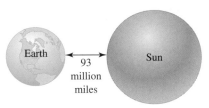

Earth — 93 million miles — Sun

33. Grain storage facilities in Toledo, Ohio, have a capacity of sixty-three million, one hundred thousand bushels. (*Source:* Chicago Board of Trade Market Information Department)

34. Bobby Hurley of Duke University holds the NCAA Men's Division I record for most career assists. He is credited with one thousand, seventy-six assists from 1990 to 1993.

35. In 1996, Kodak diverted five hundred million pounds of materials from landfills in its recycle and reuse programs. (*Source:* Eastman Kodak Company)

36. FedEx employees retrieve packages from over thirty-four thousand FedEx drop boxes around the world each business day. (*Source:* Federal Express Corporation)

C *Write each whole number in expanded form. See Example 11.*

37. 406   **38.** 789   **39.** 5290   **40.** 6040

41. 62,407   42. 20,215   43. 30,680   44. 99,032

45. 39,680,000   46. 47,703,029

D *The table shows the five longest rivers in the world. Use this table to answer Exercises 47–51. See Example 12.*

River	Miles
Chang jiang-Yangtze (China)	3964
Amazon (Brazil)	4000
Tenisei-Angara (Russia)	3442
Mississippi-Missouri (U.S.)	3740
Nile (Egypt)	4145

47. Write the length of the Amazon River in words.

48. Write the length of the Tenisei-Angara River in words.

49. Write the length of the Nile River in expanded form.

50. Write the length of the Mississippi-Missouri River in expanded form.

51. Which river is the longest in the world?

The table shows the top ten breeds of dogs in 1996 according to the American Kennel Club. Use this table to answer Exercises 52–55. See Example 12.

TOP TEN AMERICAN KENNEL CLUB REGISTRATIONS IN 1996	
Breed	Number of Registered Dogs
Beagle	56,946
Cocker Spaniel	45,305
Dachshund	48,426
German Shepherd	79,076
Golden Retriever	68,993
Labrador Retriever	149,505
Pomeranian	39,712
Poodle	56,803
Rottweiler	89,867
Yorkshire Terrier	40,216

(*Source:* American Kennel Club)

52. Which breed has the most American Kennel Club registrations? Write the number of registrations for this breed in words.

53. Which breed has the fewest registrations? Write the number of registered dogs for this breed in words.

54. Which breed has more dogs registered: Dachshund or Poodle?

55. Which breed has fewer dogs registered: Rottweiler or German Shepherd?

---

41. 60,000 + 2000 + 400 + 7

42. 20,000 + 200 + 10 + 5

43. 30,000 + 600 + 80

44. 90,000 + 9000 + 30 + 2

45. 30,000,000 + 9,000,000 + 600,000 + 80,000

46. 40,000,000 + 7,000,000 + 700,000 + 3000 + 20 + 9

47. four thousand

48. three thousand, four hundred forty-two

49. 4000 + 100 + 40 + 5

50. 3000 + 700 + 40

51. Nile

52. Labrador Retriever; one hundred forty-nine thousand, five hundred five

53. Pomeranian; thirty-nine thousand, seven hundred twelve

54. Poodle

55. German Shepherd

9

Name _____

◣ **COMBINING CONCEPTS**

56. Write the largest four-digit number that can be made from the digits 3, 6, 7, and 2 if each digit must be used once. __ __ __ __

57. Write the largest five-digit number that can be made using the digits 4, 5, and 3 if each digit must be used at least once. __ __, __ __ __

58. If a number is given in words, describe the process used to write this number in standard form.

59. If a number is written in standard form, describe the process used to write this number in expanded form.

60. The Pro-Football Hall of Fame was established on September 7, 1963 in this town. Use the information and the diagram below to find the name of the town.

▲ Alliance is East of Massillon.
▲ Dover is between Canton and New Philadelphia.
▲ Massillon is not next to Alliance.
▲ Canton is North of Dover.

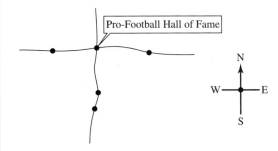

1.2 ADDING WHOLE NUMBERS

A ADDING WHOLE NUMBERS

If one computer in an office has a 2-megabyte memory and a second computer has a 4-megabyte memory, the total memory in the two computers can be found by adding 2 and 4.

$$2 \text{ megabytes} + 4 \text{ megabytes} = 6 \text{ megabytes}$$

The **sum** is 6 megabytes of memory. Each of the numbers 2 and 4 is called an **addend**.

$$\underset{\underset{\boxed{\text{addend}}}{\uparrow}}{2} \quad + \quad \underset{\underset{\boxed{\text{addend}}}{\uparrow}}{4} \quad = \quad \underset{\underset{\boxed{\text{sum}}}{\uparrow}}{6}$$

To add whole numbers, we add the digits in the ones place, then the tens place, then the hundreds place, and so on. For example, let's add $2236 + 160$.

```
  2 2 3 6     Line up numbers vertically so that the place values corespond. Then
+ 1 6 0       add digits in corresponding place values, starting with the ones place.
  2 3 9 6
  ↑ ↑ ↑ ↑
  │ │ │ └── sum of ones
  │ │ └──── sum of tens
  │ └────── sum of hundreds
  └──────── sum of thousands
```

Example 1   Add: $23 + 136$

Solution:
```
   23
 +136
  159
```

When the sum of digits in corresponding place values is more than 9, "carrying" is necessary. For example, to add $365 + 89$, add the ones-place digits first.

```
  1
  3 6 5      5 ones + 9 ones = 14 ones or 1 ten + 4 ones.
+   8 9
      4      Write the 4 ones in the ones place and carry the 1 ten to the tens place.
```

TEACHING TIP

Point out to students that they can and should rewrite problems in a form which makes it easy for them to solve. If an addition problem is given horizontally, they can rewrite it vertically so it is easy to line up numbers with the same place value.

Practice Problem 1

Add: $7235 + 542$

Answer
1. 7777

Next, add the tens-place digits.

$$\begin{array}{r} 1\ 1 \\ 3\ 6\ 5 \\ +\ 8\ 9 \\ \hline 5\ 4 \end{array}$$

1 ten + 6 tens + 8 tens = **15 tens** or **1 hundred** + **5 tens**.
Write the 5 tens in the tens place and carry the 1 hundred to the hundreds place.

Next, add the hundreds-place digits.

$$\begin{array}{r} 1\ 1 \\ 365 \\ +\ 89 \\ \hline 454 \end{array}$$

1 hundred + 3 hundreds = 4 hundreds.
Write the 4 hundreds in the hundreds place.

Example 2   Add: 34,285 + 149,761

Solution:
$$\begin{array}{r} 1\ 1\ 1 \\ 34,285 \\ +149,761 \\ \hline 184,046 \end{array}$$

TRY THE CONCEPT CHECK IN THE MARGIN.

Before we continue adding whole numbers, let's review some properties of addition that you may have already discovered. The first property that we will review is the **addition property of 0**. This property reminds us that the sum of 0 and any number is that same number.

ADDITION PROPERTY OF 0

The sum of 0 and any number is that number. For example,

$$7 + 0 = 7$$
$$0 + 7 = 7$$

Next, notice that we can add any two whole numbers in any order and the sum is the same. For example,

$$4 + 5 = 9 \quad \text{and} \quad 5 + 4 = 9$$

We call this special property of addition the **commutative property of addition**.

COMMUTATIVE PROPERTY OF ADDITION

Changing the **order** of two addends does not change their sum. For example,

$$2 + 3 = 5 \quad \text{and} \quad 3 + 2 = 5$$

Another property that can help us when adding numbers is the **associative property of addition**. This property states that, when adding numbers, the grouping of the numbers can be changed without changing the sum. We use parentheses to group numbers. They indicate what numbers to add first. For example, let's use two different groupings to find the sum of 2 + 1 + 5.

$$2 + \underbrace{(1 + 5)} = 2 + 6 = 8$$

Practice Problem 2

Add: 27,364 + 92,977

✓ CONCEPT CHECK

What is wrong with the following computation?

$$\begin{array}{r} 394 \\ +\ 283 \\ \hline 577 \end{array}$$

Also,

$$\underbrace{(2 + 1)}_{} + 5 = 3 + 5 = 8$$

Both groupings give a sum of 8.

ASSOCIATIVE PROPERTY OF ADDITION

Changing the grouping of addends does not change their sum. For example

$$3 + \underbrace{(5 + 7)}_{} = 3 + 12 = 15 \qquad \text{and} \qquad \underbrace{(3 + 5)}_{} + 7 = 8 + 7 = 15$$

The commutative and associative properties tell us that we can add whole numbers using any order and grouping that we want.

When adding several numbers, it is often helpful to look for two or three numbers whose sum is 10, 20, and so on.

Example 3   Add: $13 + 2 + 7 + 8 + 9$

Solution: $13 + 2 + 7 + 8 + 9 = 39$

$$20 + 10 + 9$$

$$39$$

Example 4   Add: $1647 + 246 + 32 + 85$

Solution:
```
  122
 1647
  246
   32
+  85
 ────
 2010
```

B FINDING THE PERIMETER OF A POLYGON

A special application of addition is finding the perimeter of a polygon. A **polygon** can be described as a flat figure formed by line segments connected at their ends. (For more review, see Appendix C.) Geometric figures such as triangles, squares, and rectangles are called polygons.

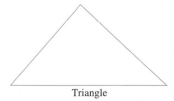
Triangle

Square

Rectangle

The **perimeter** of a polygon is the *distance around* the polygon. This means that the perimeter of a polygon is the sum of the lengths of its sides.

Practice Problem 3

Add: $11 + 7 + 8 + 9 + 13$

Practice Problem 4

Add: $19 + 5042 + 638 + 526$

Answers
3. 48, **4.** 6225

Practice Problem 5

Find the perimeter of the polygon shown. (A centimeter is a unit of length in the metric system.)

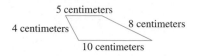

Practice Problem 6

A new shopping mall has its floor plan in the shape of a triangle. Each of the mall's three sides is 532 feet. Find the perimeter of the building.

532 feet

Example 5   Find the perimeter of the polygon shown.

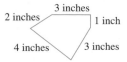

Solution: To find the perimeter (distance around), we add the lengths of the sides.

2 in. + 3 in. + 1 in. + 3 in. + 4 in. = 13 in.

The perimeter is 13 inches.

Example 6   ## Calculating the Perimeter of a Building

The largest commercial building in the world under one roof is the flower auction building of the cooperative VBA in Aalsmeer, Netherlands. The floor plan is a rectangle that measures 776 meters by 639 meters. Find the perimeter of this building. (A meter is a unit of length in the metric system.) (*Source: The Handy Science Answer Book*, Visible Ink Press, 1994)

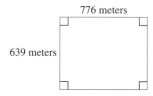

Solution: Recall that opposite sides of a rectangle have the same lengths. To find the perimeter of this building, we add the lengths of the sides. The sum of the lengths of its sides is

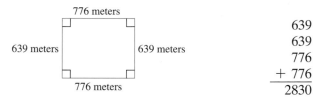

$$\begin{array}{r} 639 \\ 639 \\ 776 \\ + 776 \\ \hline 2830 \end{array}$$

The perimeter of the building is 2830 meters.

C SOLVING PROBLEMS BY ADDING

Often, real-life problems occur that can be solved by writing an addition statement. The first step to solving any word problem is to *understand* the problem by reading it carefully. Descriptions of problems solved through addition *may* include any of these key words or phrases:

Key Words or Phrases	Example	Symbols
added to	5 added to 7	7 + 5
plus	0 plus 78	0 + 78
increased by	12 increased by 6	12 + 6
more than	11 more than 25	25 + 11
total	the total of 8 and 1	8 + 1
sum	the sum of 4 and 133	4 + 133

Answers

5. 27 centimeters,   **6.** 1596 feet

To solve a word problem that involves addition, we first use the facts described to write an addition statement. Then we write the corresponding solution of the real-life problem. It is sometimes helpful to write the statement in words (brief phrases) and then translate to numbers.

Example 7   Finding a Salary

Ivy Vine is station manager of Smoky Hills Public Television. Her monthly salary of $3460 has been increased by a raise of $427. What is her new salary?

Solution: The key phrase here is "increased by" and suggests that we add. To find Ivy's new salary, we add her old salary, $3460, to the increase, $427.

In Words		Translate to Numbers
old salary	→	3460
+ increase	→	+ 427
new salary	→	3887

Her new salary is $3887 per month.   ▬▬▬

Example 8   Determining the Number of Baseball Cards in a Collection

Alan Mayfield collects baseball cards. He has 109 cards for New York Yankees, 96 for Chicago White Sox, 79 for Kansas City Royals, 42 for Seattle Mariners, 67 for Oakland Athletics, and 52 for California Angels. How many cards does he have in total?

Solution: The key word here is "total." To find the total number of Alan's baseball cards, we find the sum of the quantities from each team.

In Words		Translate to Numbers
		33
New York Yankee cards	→	109
Chicago White Sox cards	→	96
Kansas City Royal cards	→	79
Seattle Mariner cards	→	42
Oakland Athletic cards	→	67
+ California Angel cards	→	+ 52
Total cards	→	445

Alan has a total of 445 baseball cards.   ▬▬▬

Practice Problem 7

A new Janda Spirit-L motorbike costs $2431. The Spirit-X costs $486 more than the Spirit-L. How much does a Spirit-X motorbike cost?

Practice Problem 8

Elham Abo-Zahrah collects thimbles. She has 42 glass thimbles, 17 steel thimbles, 37 porcelain thimbles, 9 silver thimbles, and 15 plastic thimbles. How many thimbles are in her collection?

Answers
7. $2917, **8.** 120 thimbles

CALCULATOR EXPLORATIONS
ADDING NUMBERS

To add numbers on a calculator, find the keys marked $\boxed{+}$ and $\boxed{=}$.

or

$\boxed{\text{ENTER}}$

For example, to add 5 and 7 on a calculator, press the keys $\boxed{5}$ $\boxed{+}$ $\boxed{7}$ $\boxed{=}$.

or

$\boxed{\text{ENTER}}$

The display will read $\boxed{12}$.

Thus, $5 + 7 = 12$.

To add 687 and 981 on a calculator, press the keys $\boxed{687}$ $\boxed{+}$ $\boxed{981}$ $\boxed{=}$.

or

$\boxed{\text{ENTER}}$

The display will read $\boxed{1668}$.

Thus, $687 + 981 = 1668$. (Although entering 687, for example, requires pressing more than one key, numbers are grouped together for easier reading.)

Use a calculator to add.

1. $89 + 45$ 134

2. $76 + 97$ 173

3. $285 + 55$ 340

4. $8773 + 652$ 9425

5.
$$\begin{array}{r} 985 \\ 1210 \\ 562 \\ + 77 \\ \hline 2834 \end{array}$$

6.
$$\begin{array}{r} 465 \\ 9888 \\ 620 \\ + 1550 \\ \hline 12{,}523 \end{array}$$

Name _____ **Section** _____ **Date** _____

MENTAL MATH

Find each sum.

1. $5 + 7$ 2. $20 + 30$ 3. $5000 + 4000$

4. $4300 + 26$ 5. $1620 + 0$ 6. $6 + 126 + 4$

EXERCISE SET 1.2

A *Add. See Examples 1 through 4.*

1. $\begin{array}{r} 14 \\ +22 \\ \hline \end{array}$ 2. $\begin{array}{r} 27 \\ +31 \\ \hline \end{array}$ 3. $\begin{array}{r} 62 \\ +30 \\ \hline \end{array}$ 4. $\begin{array}{r} 37 \\ +42 \\ \hline \end{array}$

5. $\begin{array}{r} 12 \\ 13 \\ +24 \\ \hline \end{array}$ 6. $\begin{array}{r} 23 \\ 45 \\ +30 \\ \hline \end{array}$ 7. $\begin{array}{r} 5267 \\ +\ 132 \\ \hline \end{array}$ 8. $\begin{array}{r} 236 \\ +6243 \\ \hline \end{array}$

9. $53 + 64$ 10. $41 + 74$ 11. $22 + 49$ 12. $35 + 47$

13. $38 + 79$ 14. $92 + 37$ 15. $\begin{array}{r} 8 \\ 9 \\ 2 \\ 5 \\ +1 \\ \hline \end{array}$ 16. $\begin{array}{r} 3 \\ 5 \\ 8 \\ 5 \\ +7 \\ \hline \end{array}$

17. $\begin{array}{r} 6 \\ 21 \\ 14 \\ 9 \\ +12 \\ \hline \end{array}$ 18. $\begin{array}{r} 12 \\ 4 \\ 8 \\ 26 \\ +10 \\ \hline \end{array}$ 19. $\begin{array}{r} 81 \\ 17 \\ 23 \\ 79 \\ +12 \\ \hline \end{array}$ 20. $\begin{array}{r} 64 \\ 28 \\ 56 \\ 25 \\ +32 \\ \hline \end{array}$

21. $62 + 18 + 14$ 22. $23 + 49 + 18$ 23. $40 + 800 + 70$

24. $30 + 900 + 20$ 25. $7542 + 49 + 682$ 26. $1624 + 1832 + 1976$

27. $24 + 9006 + 489 + 2407$ 28. $16 + 748 + 1056 + 770$

29. $\begin{array}{r} 627 \\ 628 \\ +629 \\ \hline \end{array}$ 30. $\begin{array}{r} 427 \\ 383 \\ +\ 229 \\ \hline \end{array}$ 31. $\begin{array}{r} 6820 \\ 4271 \\ +5626 \\ \hline \end{array}$ 32. $\begin{array}{r} 6789 \\ 4321 \\ +5555 \\ \hline \end{array}$

Name _____

33.	**34.**	**35.**	**36.**
507	864	4200	5000
593	733	2107	400
+ 10	+356	+2692	+3021

37.	**38.**	**39.**	**40.**
49	26	121,742	504,218
628	582	57,279	321,920
5762	4763	6586	38,507
+29,462	+62,511	+426,782	+594,687

B *Find the perimeter of each figure. See Examples 5 and 6.*

41.
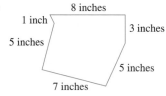
8 inches
1 inch
3 inches
5 inches
5 inches
7 inches

42.
3 kilometers   3 kilometers
5 kilometers   5 kilometers

43.
7 feet   8 feet
10 feet

44.
3 centimeters
5 centimeters   4 centimeters

45.
4 inches
Rectangle   8 inches

46.
8 miles
Rectangle   4 miles

47.
2 yards
2 yards   Square

48.
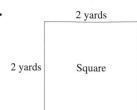
6 inches
5 inches   5 inches
7 inches   7 inches
3 inches
4 inches

18

Name _____

49. 383 mi

50. 29,028 ft

51. 340 ft

52. 210 ft

53. 123,100 people

54. 9544 people

55. 6117 performances

56. 1957

57. 11,949

C *Solve. See Examples 7 and 8.*

49. The distance from Kansas City, Kansas, to Hays, Kansas, is 285 miles. Colby, Kansas, is 98 miles farther from Kansas City than Hays. Find how far it is from Kansas City to Colby.

50. The highest point in Kansas is Mt. Sunflower at 4039 feet above sea level. The highest mountain in the world is Mt. Everest in Asia. Its peak is 24,989 feet higher than Mt. Sunflower. Find how high Mt. Everest is. (*Sources:* U.S. Geological Survey and National Geographic Society)

51. Leo Callier is installing an invisible fence in his back yard. How many feet of wiring is needed to enclose the yard below?

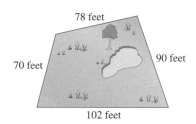

78 feet
70 feet
90 feet
102 feet

52. A homeowner is considering adding gutters around her home. Find the perimeter of her rectangular home.

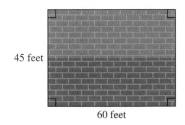

45 feet
60 feet

53. The Limited, Inc., had 105,600 employees worldwide in 1995. In 1997, there were 17,500 more Limited employees than in 1995. How many people were employed by the Limited, Inc., in 1997? (*Source:* The Limited, Inc.)

54. Papa John's is the fourth largest pizza chain in the United States. In 1995, Papa John's employed 6762 people. In 1996, employment increased by 2782 employees. How many people did Papa John's employ in 1996? (*Source:* Papa John's International Inc.)

55. As of June 1, 1997, the Broadway show *Cats* had staged 1934 more performances than the Broadway show *Les Miserables*. At that time, *Les Miserables* had staged 4183 performances. How many performances had *Cats* staged by June 1, 1997? (*Source:* The League of American Theatres and Producers)

56. Charles Schulz, the creator of the Peanuts comic strip, was born in 1922. Scott Adams, the creator of the Dilbert comic strip, was born 35 years later. In what year was Scott Adams born? (*Source: The World Almanac,* 1998)

57. There were 11,099 kidney transplants and 850 kidney-pancreas transplants performed in the United States in 1996. What was the total number of organ transplants performed in 1996 that involved a kidney? (*Source:* United Network for Organ Sharing)

58. There were 2342 heart transplants and 39 heart-lung transplants performed in the United States in 1996. What was the total number of organ transplants performed in 1996 that involved a heart? (*Source:* United Network for Organ Sharing)

20

Name _____

The table shows the number of Wal-Mart stores in 14 states. Use this table to answer Exercises 59–64.

THE TOP STATES FOR WAL-MART STORES IN 1997	
State	Number of Stores
Alabama	84
Arkansas	80
California	119
Florida	163
Georgia	102
Illinois	130
Indiana	87
Louisiana	84
Missouri	119
North Carolina	99
Ohio	99
Oklahoma	84
Tennessee	98
Texas	293

(*Source:* Wal-Mart Stores, Inc.)

59. Which state has the most Wal-Mart stores?

60. The mid-South can be defined as Arkansas, Louisiana, Texas, and Oklahoma. What is the total number of Wal-Mart stores in the mid-South?

61. What is the total number of Wal-Mart stores located in the three states with the most Wal-Mart stores?

62. Which pair of neighboring states has more Wal-Mart stores: Indiana and Illinois or Florida and Georgia?

63. How many Wal-Mart stores are located in the 14 states given in the table? Use a calculator to check your total.

64. There are 1099 Wal-Mart stores in the remaining 36 states. What is the total number of Wal-Mart stores in the United States?

65. In your own words, explain the commutative property of addition.

66. In your own words, explain the associative property of addition.

67. Add: $78,962 + 129,968,350 + 36,462,880$

68. Add: $56,468,980 + 1,236,785 + 986,768,000$

1.3 SUBTRACTING WHOLE NUMBERS

A SUBTRACTING WHOLE NUMBERS

If you have $5 and someone gives you $3, you have a total of $8, since $5 + 3 = 8$. Similarly, if you have $8, and then someone borrows $3, you have $5 left. **Subtraction** is finding the **difference** of two numbers.

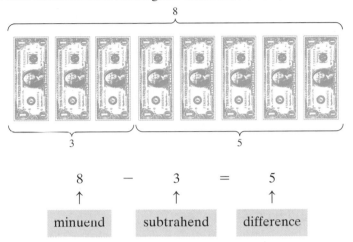

$$8 \quad - \quad 3 \quad = \quad 5$$

| minuend | subtrahend | difference |

Notice that addition and subtraction are very closely related. In fact, subtraction is defined in terms of addition.

$$8 - 3 = 5 \text{ because } 5 + 3 = 8$$

This means that subtraction can be *checked* by addition, and we say that addition and subtraction are reverse operations.

Example 1   Subtract. Check each answer by adding.

a. $12 - 9$ **b.** $11 - 6$ **c.** $5 - 5$ **d.** $7 - 0$

Solution:
a. $12 - 9 = 3$ because $3 + 9 = 12$
b. $11 - 6 = 5$ because $5 + 6 = 11$
c. $5 - 5 = 0$ because $0 + 5 - 5$
d. $7 - 0 = 7$ because $7 + 0 = 7$

Look again at Examples 1(c) and 1(d).

1(c) $5 - 5 = 0$ 1(d) $7 - 0 = 7$

| same number | difference is 0 | a number minus 0 | difference is the same number |

These two examples illustrate the subtraction properties of 0.

SUBTRACTION PROPERTIES OF 0

The difference of any number and that same number is 0. For example,
$$11 - 11 = 0$$
The difference of any number and 0 is that same number. For example,
$$45 - 0 = 45$$

When subtraction involves numbers of two or more digits, it is more convenient to subtract vertically. For example, to subtract $893 - 52$,

Objectives

A Subtract whole numbers.
B Subtract whole numbers when borrowing is necessary.
C Solve problems by subtracting whole numbers.

SSM   CD-ROM   Video 1.3

TEACHING TIP

Ask students to verbalize all the different ways they could state that $8 - 3 = 5$.
For example:
Eight minus 3 equals 5.
Eight take away 3 equals 5.
The difference of 8 and 3 is 5.
Eight decreased by 3 is 5.
Three subtracted from 8 is 5.
Subtracting 3 from 8 results in 5.
Eight less 3 is 5.
Three less than 8 is 5.
Point out the similarity in the wording in the last two sentences but the difference in the placement of the numbers.

Practice Problem 1

Subtract. Check each answer by adding.

a. $14 \quad 9$

b. $9 - 9$

c. $4 - 0$

Answers
1. a. 5,   **b.** 0,   **c.** 4

$$
\begin{array}{r}
8\ 9\ 3 \quad \leftarrow \text{minuend} \\
-\ \ \ 5\ 2 \quad \leftarrow \text{subtrahend} \\
\hline
8\ 4\ 1 \quad \leftarrow \text{difference}
\end{array}
$$

Line up numbers vertically so that the minuend is on top and the place values correspond. Subtract in corresponding places, starting with the ones place.

$$
\begin{array}{l}
3 - 2 \\
9 - 5 \\
8 - 0
\end{array}
$$

To check, add.

$$
\begin{array}{rr}
\text{difference} \quad \text{or} & 841 \\
+\ \text{subtrahend} & +\ 52 \\
\hline
\text{minuend} & 893 \quad \leftarrow
\end{array}
$$

Since this is the original minuend, the problem checks.

Practice Problem 2

Subtract. Check by adding.

a. $4689 - 253$

b. $981 - 630$

TEACHING TIP

Some students may find it easy to solve subtraction problems by thinking "what do I add to the bottom number (subtrahend) to get the top number (minuend)." For example,

$$
\begin{array}{r}
739 \\
-\ 24 \\
\hline
\end{array}
$$

ones: What do I add to 4 to get 9? (5)

tens: What do I add to 2 to get 3? (1)

hundreds: What do I add to 0 to get 7? (7)

Answer: 715

Practice Problem 3

Subtract. Check by adding.

a. $\begin{array}{r} 227 \\ -\ 175 \\ \hline \end{array}$

b. $\begin{array}{r} 1136 \\ -\ 914 \\ \hline \end{array}$

c. $\begin{array}{r} 8627 \\ -\ 4119 \\ \hline \end{array}$

Answers

2. **a.** 4436, **b.** 351, **3. a.** 52, **b.** 222, **c.** 4508

Example 2   Subtract: $7826 - 505$. Check by adding.

Solution:

$$
\begin{array}{r}
7826 \\
-\ 505 \\
\hline
7321
\end{array}
$$

Check:

$$
\begin{array}{r}
7321 \\
+\ 505 \\
\hline
7826
\end{array}
$$

B SUBTRACTING WITH BORROWING

When a digit in the second number (subtrahend) is larger than the corresponding digit in the first number (minuend), **borrowing** is necessary. For example, consider

$$
\begin{array}{r}
81 \\
-\ 63 \\
\hline
\end{array}
$$

Since 3 in the ones place of 63 is larger than 1 in the ones place of 81, borrowing is necessary. We borrow 1 ten from the tens place and add it to the ones places.

Borrowing

$$
\underset{\text{tens}}{8} - \underset{\text{ten}}{1} = \underset{\text{tens}}{7} \rightarrow 7 \ 11 \quad \leftarrow 1 \text{ ten} + 1 \text{ one} = 11 \text{ ones}
$$

$$
\begin{array}{r}
\cancel{8}\ \cancel{1} \\
-\ 6\ 3 \\
\hline
\end{array}
$$

Now we subtract the ones-place digits and then the tens-place digits.

$$
\begin{array}{r}
7 \quad 11 \\
\cancel{8} \quad \cancel{1} \\
-\ 6 \quad 3 \\
\hline
1 \quad 8 \quad \leftarrow 11 - 3 = 8 \\
 \quad \quad 7 - 6 = 1
\end{array}
$$

Check:

$$
\begin{array}{r}
18 \\
+\ 63 \\
\hline
81
\end{array}
$$

The original minuend

Example 3   Subtract $43 - 29$. Check by adding.

Solution:

$$
\begin{array}{r}
3 \ 13 \\
\cancel{4}\ \cancel{3} \\
-\ 2\ 9 \\
\hline
1\ 4
\end{array}
$$

Check:

$$
\begin{array}{r}
14 \\
+\ 29 \\
\hline
43
\end{array}
$$

Sometimes we may have to borrow from more than one place. For example, to subtract $7631 - 152$, we first borrow from the tens place.

$$
\begin{array}{r}
{\scriptstyle 2\ \ 11} \\
7\ 6\ \cancel{3}\ \cancel{1} \\
-\ \ \ 1\ 5\ 2 \\
\hline
9
\end{array}
\quad \leftarrow 11 - 2 = 9
$$

In the tens place, 5 is greater than 2, so we borrow again. This time we borrow from the hundreds place.

$$
\begin{array}{r}
\text{6 hundreds} - \textbf{1 hundred} = \textbf{5 hundreds} \\
12 \leftarrow \begin{cases} \textbf{1 hundred} + \text{2 tens or} \\ \text{10 tens} + \text{2 tens} = \text{12 tens} \end{cases}
\end{array}
$$

$$
\begin{array}{r}
{\scriptstyle 5\ \ \cancel{2}\ \ 11} \\
7\ \cancel{6}\ \cancel{3}\ \cancel{1} \\
-\ \ \ 1\ 5\ 2 \\
\hline
7\ 4\ 7\ 9
\end{array}
$$

Check:

$$
\begin{array}{r}
7479 \\
+\ \ 152 \\
\hline
7631
\end{array}
\quad \text{The original minuend}
$$

Example 4

Subtract: $900 - 174$. Check by adding.

Solution: In the ones place, 4 is larger than 0, so we borrow from the tens place. But the tens place of 900 is 0, so to borrow from the tens place we must first borrow from the hundreds place.

$$
\begin{array}{r}
{\scriptstyle 8\ \ 10} \\
\cancel{9}\ \cancel{0}\ 0 \\
-1\ 7\ 4
\end{array}
$$

Now borrow from the tens place.

$$
\begin{array}{r}
{\scriptstyle 9} \\
{\scriptstyle 8\ \ \cancel{10}\ \ 10} \\
\cancel{9}\ \cancel{0}\ \cancel{0} \\
-1\ 7\ 4 \\
\hline
7\ 2\ 6
\end{array}
$$

Check:

$$
\begin{array}{r}
{\scriptstyle 1\ 1} \\
726 \\
+174 \\
\hline
900
\end{array}
$$

C SOLVING PROBLEMS BY SUBTRACTING

Descriptions of real-life problems that suggest solving by subtracting include these key words or phrases:

Key Words or Phrases	Examples	Symbols
subtract	subtract 5 from 8	$8 - 5$
difference	the difference of 10 and 2	$10 - 2$
less	17 less 3	$17 - 3$
take away	14 take away 9	$14 - 9$
decreased by	7 decreased by 5	$7 - 5$
subtracted from	9 subtracted from 12	$12 - 9$

Practice Problem 4

Subtract. Check by adding.

a.
$$
\begin{array}{r}
400 \\
-\ 164
\end{array}
$$

b.
$$
\begin{array}{r}
200 \\
-\ \ 45
\end{array}
$$

c.
$$
\begin{array}{r}
1000 \\
-\ \ 762
\end{array}
$$

TEACHING TIP

Point out to students that the order of the numbers in a subtraction translation depends on which key word or phrase is used.

Answers

4. a. 236, **b.** 155, **c.** 238

✓ Concept Check

In each of the following problems, identify which number is the minuend and which number is the subtrahend.
a. What is the result when 9 is subtracted from 20?
b. What is the difference of 15 and 8?
c. Find a number that is 15 fewer than 23.

Practice Problem 5

The radius of Earth is 6378 kilometers. The radius of Mars is 2981 kilometers less than the radius of Earth. What is the radius of Mars? (*Source:* National Space Science Data Center)

Earth

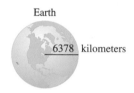

6378 kilometers

Mars

?

Practice Problem 6

A new suit originally priced at $92 is now on sale for $47. How much money was taken off the original price?

Try the Concept Check in the margin.

Example 5   Finding the Radius of a Planet

The radius of Venus is 6052 kilometers. The radius of Mercury is 3612 kilometers less than the radius of Venus. Find the radius of Mercury. (*Source:* National Space Science Data Center)

Venus

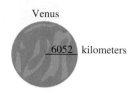

6052 kilometers

Mercury

?

Solution:

Words		Translate to Numbers
radius of Venus	⟶	6052
less 3612	⟶	− 3612
radius of Mercury	⟶	2440

The radius of Mercury is 2440 kilometers.

Example 6   Calculating Miles Per Gallon

A subcompact car gets 42 miles per gallon of gas. A full-size car gets 17 miles per gallon of gas. How many more miles per gallon does the subcompact car get than the full-size car?

Solution:

Words		Translate to Numbers
		3 12
subcompact miles per gallon	⟶	4̸2̸
− full–size miles per gallon	⟶	− 1 7
more miles per gallon		2 5

The subcompact car gets 25 more miles per gallon than the full-size car.

> **HELPFUL HINT**
>
> Since subtraction and addition are reverse operations, don't forget that a subtraction problem can be checked by adding.

Graphs can be used to visualize data. The graph shown next is called a **bar graph**.

Example 7   Reading a Bar Graph

A telephone survey was taken to identify favorite sport activities, and the results from the five most popular activities are shown in the form of a bar graph. In this particular graph, each bar represents a different sport activity, and the height of each bar represents the num-

Answers

5. 3397 kilometers,   **6.** $45

✓ Concept Check

a. minuend: 20, subtrahend: 9,
b. minuend: 15, subtrahend: 8,
c. minuend: 23, subtrahend: 15

ber of people who responded that the particular sport was their favorite activity. Use this graph to answer the questions.

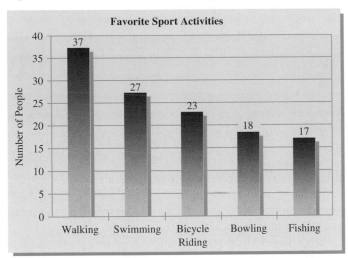

Favorite Sport Activities

a. What activity was preferred by the most people?
b. How many more people responded that swimming rather than fishing is their favorite activity?

Solution: a. The activity preferred by the most people is the one with the highest bar, which is walking.
b. The number of people who responded with swimming is 27. The number of people who responded with fishing is 17. To find how many more responded with swimming than with fishing, we find the difference:

$$27 - 17 = 10$$

Ten more people responded with swimming than with fishing.

Practice Problem 7

Use the graph of Example 7 to answer the following.

a. Find the total number of people who responded to the five favorite sport activities. (Assume that each person could respond to only one activity.)

b. How many more people responded that walking rather than bowling is their favorite sport?

CALCULATOR EXPLORATION
Subtracting Numbers

To subtract numbers on a calculator, find the keys marked $\boxed{-}$ and $\boxed{=}$.

or

$\boxed{\text{ENTER}}$

For example, to find $83 - 49$ on a calculator, press the keys $\boxed{83}\ \boxed{-}\ \boxed{49}\ \boxed{=}$.

or

$\boxed{\text{ENTER}}$

The display will read $\boxed{34}$. Thus, $83 - 49 = 34$.

Use a calculator to subtract.

1. $865 - 95$ 770
2. $76 - 27$ 49
3. $147 - 38$ 109
4. $366 - 87$ 279
5. $9625 - 647$ 8978
6. $10,711 - 8925$ 1786

Answers

7. a. 122 people,   **b.** 19 people

Focus On Study Skills

STUDY TIPS

Have you wondered what you can do to be successful in your math course? If so, that may well be your first step to success! Here are some tips on how to use this text and how to study mathematics in general.

Using This Text:

1. Each example in the section has a parallel Practice Problem. As you read a section, try each Practice Problem after you've finished the corresponding example. This "learn-by-doing" approach will help you grasp ideas before you move on to other concepts.
2. The main section of exercises in an exercise set are referenced by an objective, such as **A** or **B** and also an example(s). Use this referencing if you have trouble completing an assignment from the exercise set.
3. If you need extra help in a particular section, check at the beginning of the section to see what videotapes and software are available.
4. Integrated Reviews in each chapter offer you a chance to practice—in one place—the many concepts that you have learned separately over several sections.
5. There are many opportunities at the end of each chapter to help you understand the concepts of the chapter.

   ▲ **Highlights** contains chapter summaries with examples.
   ▲ **Chapter Review** contains review problems organized by section.
   ▲ **Chapter Test** is a sample test to help you prepare for an exam.
   ▲ **Cumulative Review** is a review consisting of material from the beginning of the book to the end of the particular chapter.

General Tips:

1. Choose to attend all class periods. If possible, sit near the front of the classroom. This way, you will see, hear, and focus on the presentation better. It may be easier for you to participate in classroom activities.
2. Do your homework. You've probably heard the phrase "practice makes perfect" in relation to music and sports. It also applies to mathematics. You will find that the more time you spend solving mathematics problems, the easier the process becomes. Be sure to block out enough time in your schedule to complete your assignments.
3. Check your work. Review the steps you made while working a problem. Learn to check your answers in the original problems. You can also compare your answers to the answers to selected exercises listed in the back of the book. If you have made a mistake, figure out what went wrong. Then correct your mistake.
4. Learn from your mistakes. Everyone, even your instructors, make mistakes. You can use your mistakes to become a better math student. The key is finding and understanding your mistakes. Was your mistake a careless mistake? If so, you can try to work more slowly and make a conscious effort to carefully check your work. Did you make a mistake because you don't understand a concept? If so, take the time to review the concept or ask questions to better understand the concept.
5. Know how to get help if you need it. It's OK to ask for help. In fact, it's a good idea to ask for help whenever there is something that you don't understand. Make sure you know when your instructor has office hours and how to find his or her office. Find out if math tutoring services are available on your campus and check out the hours, location, and requirements of the tutoring service. You might also want to find another student in your class that you can call to discuss your assignment.

Name _____ Section _____ Date _____

MENTAL MATH

Find each difference.

1. $9 - 2$
2. $6 - 6$
3. $5 - 0$
4. $44 - 22$
5. $93 - 93$
6. $700 - 400$
7. $700 - 300$
8. $700 - 700$
9. $600 - 100$
10. $600 - 0$

EXERCISE SET 1.3

A *Subtract. Check by adding. See Examples 1 and 2.*

1.
```
  67
-23
```
2.
```
  72
-41
```
3.
```
  82
-22
```
4.
```
  27
-10
```

5.
```
  389
-124
```
6.
```
  572
-321
```
7.
```
  677
-423
```
8.
```
  766
-324
```

9.
```
  998
-453
```
10.
```
  912
-610
```
11.
```
  749
-149
```
12.
```
  257
-257
```

B *Subtract. Check by adding. See Examples 1 through 4.*

13.
```
  62
-37
```
14.
```
  55
-29
```
15.
```
  70
-25
```
16.
```
  80
-37
```

17.
```
  938
-792
```
18.
```
  436
-275
```
19.
```
  922
-634
```
20.
```
  674
-299
```

21.
```
  600
-432
```
22.
```
  300
-149
```
23.
```
  42
-36
```
24.
```
  73
-29
```

25.
```
  923
-476
```
26.
```
  813
-227
```
27.
```
  6283
-  560
```
28.
```
  5349
-  720
```

29.
```
  533
-  29
```
30.
```
  724
-  16
```
31.
```
  200
-111
```
32.
```
  300
-211
```

28

Name _____

33. 1983 −1904	**34.** 1983 −1914	**35.** 56,422 −16,508	**36.** 76,652 −29,498

37. 50,000 − 17,289 **38.** 40,000 − 23,582 **39.** 7020 − 1979

40. 6050 − 1878 📼 **41.** 51,111 − 19,898 **42.** 62,222 − 39,898

43. Subtract 5 from 9. **44.** Subtract 9 from 21.

45. Find the difference of 41 and 21. **46.** Find the difference of 16 and 5.

47. Subtract 56 from 63. **48.** Subtract 41 from 59.

C *Solve. See Examples 5 through 7.*

49. Michelle and Ronnie Brower entered a crawfish eating contest. Michelle ate 63 crawfish, which was 7 more than Ronnie ate. Find how many crawfish Ronnie ate.

50. A stock worth $135 per share on October 10 dropped to $78 per share on October 30 of the same year. Find how much it lost in value from October 10th to the 30th.

51. Dyllis King is reading a 503-page book. If she has just finished reading page 239, how many more pages must she read to finish the book?

52. When Lou and Judy Zawislak began a trip, the odometer read 55,492. When the trip was over, the odometer read 59,320. How many miles did they drive on their trip?

53. The peak of Mt. McKinley in Alaska is 20,320 feet above sea level. The peak of Long's Peak in Colorado is 14,255 feet above sea level. How much higher is the peak of Mt. McKinley than Long's Peak? (*Source:* U.S. Geological Survey)

54. On one day in May the temperature in Paddin, Indiana, dropped 27 degrees from 2 p.m. to 4 p.m. If the temperature at 2 p.m. was 73° Fahrenheit, what was the temperature at 4 p.m.?

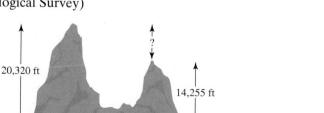

20,320 ft   14,255 ft

73°F
27 degrees
?

55. Buhler Gomez has a total of $539 in his checking account. If he writes a check for each of the items below, how much money will be left in his account?

South Central Bell	$27
Central LA Electric Co.	$101
Mellon Finance	$236

56. Pat Salanki's blood cholesterol level is 243. The doctor tells him it should be decreased to 185. How much of a decrease is this?

57. The distance from Kansas City to Denver is 645 miles. Hays, Kansas, lies on the road between the two and is 287 miles from Kansas City. What is the distance between Hays and Denver?

58. Alan Little is trading his car in on a new car. The new car costs $15,425. His car is worth $7998. How much more money does he need to buy the new car?

59. A new VCR with remote control costs $525. Prunella Pasch has $914 in her savings account. How much will she have left in her savings account after she buys the VCR?

60. Christopher Columbus landed in America in 1492. The Revolutionary War started in 1776. How many years after Columbus landed did the war begin?

61. A stereo that sells regularly for $547 is discounted by $99 in a sale. What is the sale price?

62. A meeting of the board of governors of the Mathematical Association of America was attended by 72 governors. Forty-six of the governors were men. How many were women?

63. The numbers of women enrolled in mathematics classes at FHSU during the fall semester were 78 in Basic Mathematics, 185 in College Algebra, and 23 in Calculus. The total number of students enrolled in these mathematics classes was 459. How many men were enrolled in these classes?

64. The number of cable TV systems operating in the United States in 1995 was 11,215. In 1990, there were 9575 cable TV systems in operation. How many more cable systems were operating in 1995 than in 1990? (*Source: Television and Cable Factbook*, Warren Publishing, Inc.)

55. $175 _____

56. 58 _____

57. 358 miles _____

58. $7427 _____

59. $389 _____

60. 284 years _____

61. $448 _____

62. 26 women _____

63. 173 men _____

64. 1640 cable systems _____

29

65. Jo Keen and Trudy Waterbury were candidates for student government president. Who won the election if the votes were cast as follows? By how many votes did the winner win?

Class	Candidate	
	Jo	Trudy
Freshman	276	295
Sophomore	362	122
Junior	201	312
Senior	179	182

66. Two students submitted advertising budgets for a student government fund-raiser.

	Student A	Student B
Radio ads	$600	$300
Newspaper ads	$200	$400
Posters	$150	$240
Hand bills	$120	$170

If $1200 is available for advertising, how much excess would each budget have?

67. Until recently, the world's largest permanent maze was located in Ruurlo, Netherlands. This maze of beech hedges covers 94,080 square feet. A new hedge maze using hibiscus bushes at the Dole Plantation in Wahiawa, Hawaii, covers 100,000 square feet. How much larger is the Dole Plantation maze than the Ruurlo maze? (*Source: The 1998 Guinness Book of Records*)

68. There were only 27 California condors in the entire world in 1987. By 1997, the number of California condors had increased to 119. How much of an increase was this? (*Source:* California Department of Fish and Game)

69. In 1997, the size of North Korea's armed forces was estimated at 1,130,000 soldiers. South Korea's fighting force was estimated at 633,000. How much bigger is the fighting force of North Korea than South Korea? (*Source: The Top 10 of Everything 1997* by Russell Ash)

70. The Gap, Inc., had a net income of $453,000,000 in 1997. In 1994 the Gap's net income was only $258,000,000. How much did the Gap's net income increase from 1994 to 1997? (*Source:* The Gap, Inc.)

Name _____

The bar graph shows the number of aircraft departures in 1995 for the top five airports in the United States. Use this graph to answer Exercises 71–74. See Example 7.

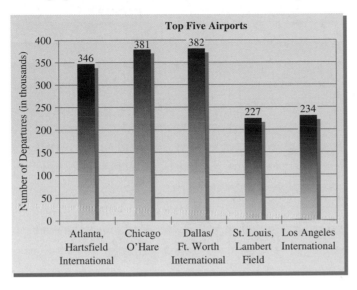

71. Which airport is busiest?

72. Which airports have more than 300 thousand departures per year?

73. How many more departures per year does Dallas/Ft. Worth International Airport have than Los Angeles International Airport?

74. How many more departures per year does Chicago O'Hare Airport have than Atlanta, Hartsfield International Airport?

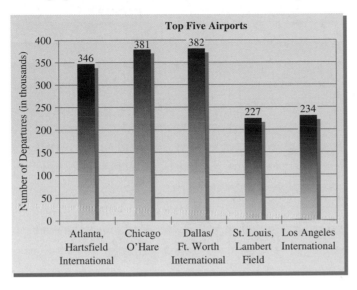

◆ COMBINING CONCEPTS

The table shows the top ten leading United States advertisers in 1996 and the amount of money spent in that year on ads. Use this table to answer Exercises 75–79. (Source: Competitive Media Reporting and Publishers Information Bureau)

General Motors	$1,712,102,000
Procter & Gamble	$1,493,456,000
Philip Morris	$1,236,978,000
Chrysler	$1,119,851,000
Ford Motor	$897,666,000
Johnson & Johnson	$840,483,000
Walt Disney	$775,621,000
Pepsico	$767,978,000
Time Warner	$747,320,000
AT&T	$659,750,000

75. Which companies spent more than $1 billion on ads?

76. Which companies spent less than $700 million on ads?

71. Dallas/Ft. Worth International.

72. Atlanta, Hartsfield International; Chicago O'Hare; Dallas/Ft. Worth International

73. 148 thousand or 148,000

74. 35 thousand or 35,000

75. Procter & Gamble, General Motors, Philip Morris, Chrysler

76. AT&T

77. $461,357,000

78. $72,505,000

79. $10,251,205,000

80. no, 1089 more pages

81.
  5269
 −2385
  2884

82.
  10,244
 − 8 534
  1 710

77. How much more money did Philip Morris spend on ads than Walt Disney?

78. How much more money did Johnson & Johnson spend on ads than Pepsico?

79. Find the total amount of money spent by these ten companies on ads.

80. The local college library is having a Million Pages of Reading promotion. The freshmen have read a total of 289,462 pages, the sophomores have read a total of 369,477 pages, the juniors have read a total of 218,287 pages, and the seniors have read a total of 121,685 pages. Have they reached a goal of a million pages? If not, how many more pages need to be read?

Fill in the missing digits in each problem.

81.
  5 26_
 −2 _85
  2 8_4

82.
  10,_4_
 −8 5_4
  _ 710

1.4 ROUNDING AND ESTIMATING

A ROUNDING WHOLE NUMBERS

Rounding a whole number means approximating it. A rounded whole number is often easier to use, understand, or remember than the precise whole number. For example, instead of trying to remember the Iowa state population as 2,851,792, it is much easier to remember it rounded to the nearest million: 3 million people.

To understand rounding, let's look at the following illustrations. The whole number 36 is closer to 40 than 30, so 36 rounded to the nearst ten is 40.

The whole number 52 rounded to the nearest ten is 50, because 52 is closer to 50 than to 60.

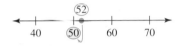

Trying to round 25 to the nearest ten, we see that 25 is halfway between 20 and 30. It is not closer to either number. In such a case, we round to the larger ten, that is, to 30.

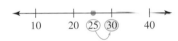

To round a whole number without using a number line, follow these steps.

ROUNDING WHOLE NUMBERS TO A GIVEN PLACE VALUE

Step 1. Locate the digit to the right of the given place value.

Step 2. If this digit is 5 or greater, add 1 to the digit in the given place value and replace each digit to its right by 0.

Step 3. If this digit is less than 5, replace it and each digit to its right by 0.

Example 1 Round 568 to the nearest ten.

Solution: 5 6 (8)   The digit to the right of the tens place is the ones place, which is circled.
↑
tens place

5 6 (8)   Since the circled digit is 5 or greater, add 1 to the 6 in
↑ ↖   the tens place and replace the digit to the right by 0.
Add 1. Replace
with 0.

We find that 568 rounded to the nearest ten is 570.

Objectives

A Round whole numbers.
B Use rounding to estimate sums and differences.
C Solve problems by estimating.

SSM   CD-ROM   Video
1.4

Practice Problem 1

Round to the nearest ten.

a. 46

b. 731

c. 125

Answers

1. a. 50, **b.** 730, **c.** 130

Practice Problem 2

Round to the nearest thousand.

a. 56,702

b. 7444

c. 291,500

Practice Problem 3

Round to the nearest hundred.

a. 2777

b. 38,152

c. 762,955

✓ CONCEPT CHECK

Round each of the following numbers to the nearest *hundred*. Explain your reasoning.
a. 79
b. 33

Practice Problem 4

Round each number to the nearest ten to find an estimated sum:

$$
\begin{array}{r}
79 \\
35 \\
42 \\
21 \\
+\ 98 \\
\end{array}
$$

Example 2   Round 278,362 to the nearest thousand.

Solution:

Thousands place

↓ ┌─ 3 is less than 5.

278,③62

↑   ↑

Do not   Replace with

add 1.   zeros.

The number 278,362 rounded to the nearest thousand is 278,000.

Example 3   Round 248,982 to the nearest hundred.

Solution:

Hundreds place

↓ ┌─ 8 is greater than or equal to 5.

248,9⑧2

↑

Add 1. $9 + 1 = 10$, so replace the digit 9 by 0 and carry 1 to the place value to the left.

$$
\begin{array}{ccccccc}
& & 8+1 & 0 & & & \\
2 & 4 & \not{8}, & \not{9} & 8 & 2 & \\
& & \uparrow & & \multicolumn{2}{c}{\underbrace{}} & \\
& & \text{Add 1.} & & \multicolumn{2}{l}{\text{Replace}} & \\
& & & & \multicolumn{2}{l}{\text{with zeros.}} & \\
\end{array}
$$

The number 248,982 rounded to the nearest hundred is 249,000.

TRY THE CONCEPT CHECK IN THE MARGIN.

B ESTIMATING SUMS AND DIFFERENCES

By rounding addends, we can estimate sums. An estimated sum is appropriate when an exact sum is not necessary. To estimate the sum shown, round each number to the nearest hundred and then add.

$$
\begin{array}{rll}
768 & \text{rounds to} & 800 \\
1952 & \text{rounds to} & 2000 \\
225 & \text{rounds to} & 200 \\
+\ 149 & \text{rounds to} & +\ 100 \\
\hline
& & 3100 \\
\end{array}
$$

The estimated sum is 3100, which is close to the exact sum of 3094.

Example 4   Round each number to the nearest hundred to find an estimated sum:

$$
\begin{array}{r}
294 \\
625 \\
1071 \\
+\ 349 \\
\end{array}
$$

Answers

2. a. 57,000, **b.** 7000, **c.** 292,000, **3. a.** 2800,
b. 38,200, **c.** 763,000, **4.** 280

✓ Concept Check

a. 100, **b.** 0

Solution:

294	rounds to	300
625	rounds to	600
1071	rounds to	1100
+ 349	rounds to	+ 300
		2300

The estimated sum is 2300. (The exact sum is 2339.)

Example 5   Round each number to the nearest hundred to find an estimated difference:

$$4725$$
$$-2879$$

Solution:

4725	rounds to	4700
−2879	rounds to	−2900
		1800

The estimated difference is 1800. (The exact difference is 1846.)

C SOLVING PROBLEMS BY ESTIMATING

Making estimates is often the quickest way to solve real-life problems when their solutions do not need to be exact.

Example 6   **Estimating Distances**

Jose Guillermo is trying to estimate quickly the distance from Temple, Texas, to Brenham, Texas. Round each distance given on the map to the nearest ten to estimate the total distance.

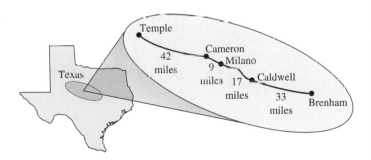

Solution:

Distance		**Estimation**
42	rounds to	40
9	rounds to	10
17	rounds to	20
+33	rounds to	+30
		100

It is approximately 100 miles from Temple to Brenham. (The exact distance is 101 miles.)

TEACHING TIP

Check students' understanding of rounding by asking them to give the range of values which would round to a certain number. For example, if a number rounded to the nearest hundred is 2400, the smallest the original number could be is 2350 and the largest it could be is 2449.

Practice Problem 5

Round each number to the nearest thousand to find an estimated difference:

$$4725$$
$$- 2879$$

Practice Problem 6

Tasha Kilbey is trying to estimate how far it is from Grove, Kansas, to Hays, Kansas. Round each given distance on the map to the nearest ten to estimate the total distance.

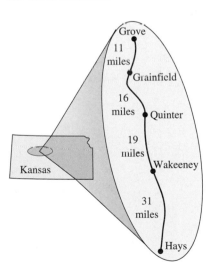

Answers

5. 2000,   **6.** 80 miles

Practice Problem 7

In a recent year, there were 120,624 reported cases of chicken pox, 22,866 reported cases of tuberculosis, and 45,970 reported cases of salmonellosis in the United States. Round each number to the nearest ten-thousand to estimate the total number of cases reported for these diseases. (*Source:* Centers for Disease Control and Prevention)

Example 7   Estimating Data

In three recent years the numbers of reported cases of mumps in the United States were 906, 1537, and 1692. Round each number to the nearest hundred to estimate the total number of cases reported over this period. (*Source:* Centers for Disease Control and Prevention)

Solution:

Number of Cases		Estimation
906	rounds to	900
1537	rounds to	1500
+1692	rounds to	+1700
		4100

The approximate number of cases reported over this period is 4100.

Focus On Study Skills

WHAT IS CRITICAL THINKING?

Although exact definitions often vary, critical thinking usually refers to evaluating, analyzing, and interpreting information in order to make a decision, draw a conclusion, reach a goal, make a prediction, or form an opinion. It often involves problem solving, communication, and reasoning skills. Critical thinking is more than a technique that helps you pass your courses—critical thinking skills are life skills. Developing these skills can help you solve problems in your workplace and in everyday life. For instance, well-developed critical thinking skills would be useful in the following situation:

> Suppose you work as a medical lab technician. Your lab supervisor has decided that some lab equipment should be replaced. She asks you to collect information on several different models from equipment manufacturers. Your assignment is to study the data and then make a recommendation on which model the lab should buy.

HOW CAN CRITICAL THINKING BE DEVELOPED?

Just as physical exercise can help to develop and strengthen certain muscles of the body, mental exercise can help to develop critical thinking skills. Mathematics is ideal for helping to develop such skills because it requires using logic and reasoning, recognizing patterns, making conjectures and educated guesses, and drawing conclusions. You will find many opportunities to build your critical thinking skills throughout *Basic College Mathematics*:

▲ In real-life application problems (see Exercise 17 in Section 2.1)
▲ In conceptual and writing exercises marked with the ✎ icon (see Exercise 55 in Section 3.3)
▲ In the Combining Concepts subsection of the exercise sets (see Exercise 56 in Section 1.1)
▲ In the Chapter Activities (see the Chapter 1 Activity)
▲ In the Critical Thinking and Group Activities found in Focus On features like this one throughout the book (see pages 42 and 106)

Answer

7. 190,000

EXERCISE SET 1.4

A *Round each whole number to the given place. See Examples 1 through 3.*

1. 632 to the nearest ten

2. 273 to the nearest ten

3. 635 to the nearest ten

4. 275 to the nearest ten

5. 792 to the nearest ten

6. 394 to the nearest ten

7. 395 to the nearest ten

8. 582 to the nearest ten

9. 1096 to the nearest ten

10. 2198 to the nearest ten

11. 42,682 to the nearest thousand

12. 42,682 to the nearest ten-thousand

13. 248,695 to the nearest hundred

14. 179,406 to the nearest hundred

15. 36,499 to the nearest thousand

16. 96,501 to the nearest thousand

17. 99,995 to the nearest ten

18. 39,994 to the nearest ten

19. 59,725,642 to the nearest ten-million

20. 39,523,698 to the nearest million

1. 630

2. 270

3. 640

4. 280

5. 790

6. 390

7. 400

8. 580

9. 1100

10. 2200

11. 43,000

12. 40,000

13. 248,700

14. 179,400

15. 36,000

16. 97,000

17. 100,000

18. 39,990

19. 60,000,000

20. 40,000,000

38

Name _____

Complete the table by estimating the given number to the given place value.

		Ten	Hundred	Thousand
21.	5281	5280	5300	5000
22.	7619	7620	7600	8000
23.	9444	9440	9400	9000
24.	7777	7780	7800	8000
25.	14,876	14,880	14,900	15,000
26.	85,049	85,050	85,000	85,000

27. Estimate to the nearest thousand the 1994–1995 enrollment of East Tennessee State University: 11,512 (*Source:* East Tennessee State University)

28. Estimate to the nearest hundred the hourly cost of operating a B747-400 aircraft: $6592 (*Source:* Air Transport Association of America)

Round each number to the indicated place.

29. In 1998, Superbowl XXXII was won by the Denver Broncos over the Green Bay Packers. Attendance at the game was 68,912. Round the attendance figure to the nearest thousand. (*Source:* National Football League)

30. It takes 90,465 days for Pluto to make a complete orbit around the sun. Round this number to the nearest hundred. (*Source:* National Space Science Data Center)

31. In 1996, U.S. farms produced 155,225,000 bushels of oats. Round the oat production figure to the nearest ten million. (*Source:* U.S. Department of Agriculture)

32. Revenues collected for Florida public schools during the 1996–1997 school year totaled $13,896,990,000. Round the revenue figure to the nearest million. (*Source:* National Education Association)

33. In 1996, JCPenney spent $305,718,800 on advertising. Round the advertising figure to the nearest hundred-thousand. (*Source:* Competitive Media Reporting and Publishers Information Bureau)

34. In May 1997, there were 487,297 U.S. Army personnel on active duty. Round this personnel figure to the nearest ten-thousand. (*Source:* U.S. Department of Defense)

B *Estimate the sum or difference by rounding each number to the nearest ten. See Examples 4 and 5.*

35.	**36.**	**37.**	**38.**
29	62	649	555
35	72	−272	−235
42	15		
+16	+19		

Name _____

Estimate the sum or difference by rounding each number to the nearest hundred. See Examples 4 and 5.

39. 1812
1776
+1945

40. 2010
2001
+1984

41. 1774
−1492

42. 1989
−1870

43. 2995
1649
+3940

44. 799
1655
+271

Estimation is useful to check for incorrect answers when using a calculator. For example, pressing a key too hard may result in a double digit while pressing a key too softly may result in the number not appearing in the display.

Two of the given calculator answers below are incorrect. Find them by estimating each sum.

45. 362 + 419   781

46. 522 + 785   1307

47. 432 + 679 + 198   1139

48. 229 + 443 + 606   1278

49. 7806 + 5150   12,956

50. 5233 + 4988   9011

51. 31,439 + 18,781   50,220

52. 68,721 + 52,335   121,056

C *Solve each problem by estimating. See Examples 6 and 7.*

53. Campo Appliance Store advertises three refrigerators on sale at $799, $1299, and $999. Round each cost to the nearest hundred to estimate the total cost.

54. Jared Nuss scored 89, 92, 100, 67, 75, and 79 on his calculus tests. Round each score to the nearest ten to estimate his total score.

39. 5500

40. 6000

41. 300

42. 100

43. 8500

44. 2800

45. correct

46. correct

47. incorrect

48. correct

49. correct

50. incorrect

51. correct

52. correct

53. $3100

54. $90 + 90 + 100 + 70 + 80 + 80 = 510$

39

55. 80 miles

56. $400

57. 25,000 feet

58. 2400 miles

59. 5,700,000

60. 900 miles

61. 14,000,000 votes

62. 2000 credit hours

63. 211,000 children

64. 7400 associates

55. Arlene Neville wants to estimate quickly the distance from Stockton to LaCrosse. Round each distance given on the map to the nearest ten miles to estimate the total distance.

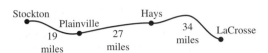

Stockton Plainville Hays LaCrosse
19 miles 27 miles 34 miles

56. Carmelita Watkins is pricing new stereo systems. One system sells for $1895 and another system sells for $1524. Round each price to the nearest hundred dollars to estimate the difference in price of these systems.

57. The peak of Mt. Everest, in Asia, is 29,028 feet above sea level. The top of Mt. Sunflower, in Kansas, is 4039 feet above sea level. Round each height to the nearest thousand to estimate the difference in elevation of these two peaks. (*Sources:* U.S. Geological Survey and National Geographic Society)

58. The Gonzales family took a trip and traveled 458, 489, 377, 243, 69, and 702 miles on six consecutive days. Round each distance to the nearest hundred to estimate the distance they traveled.

59. In 1990 the population of New York City was 7,322,564 and the population of Houston, Texas, was 1,629,902. Round each population to the nearest hundred-thousand to estimate how much larger New York was than Houston. (*Source:* U.S. Bureau of the Census, 1990 census)

60. The distance from Kansas City to Boston is 1429 miles and from Kansas City to Chicago is 530 miles. Round each distance to the nearest hundred to estimate how much farther Boston is from Kansas City than Chicago is.

61. In the 1964 presidential election, Lyndon Johnson received 41,126,233 votes and Barry Goldwater received 27,174,898 votes. Round each number of votes to the nearest million to estimate the number of votes by which Johnson won the election.

62. Enrollment figures at Normal State University showed an increase from 49,713 credit hours in 1988 to 51,746 credit hours in 1989. Round each number to the nearest thousand to estimate the increase.

63. Head Start is a national program that provides developmental and social services for America's low-income, preschool children ages three to five. Enrollment figures in Head Start programs showed an increase from 540,930 children in 1990 to 752,077 in 1996. Round each number of children to the nearest thousand to estimate this increase. (*Source:* Head Start Bureau)

64. In 1986, Rubbermaid employed 6509 associates on average. By 1996, the average number of Rubbermaid associates had increased to 13,861. Round each number of associates to the nearest hundred to estimate this increase. (*Source:* Rubbermaid, Inc.)

40

The following table (from the previous section) shows the top ten leading United States advertisers in 1996 and the amount of money spent in that year on ads. Use this table to answer Exercises 65–68. (Source: Competitive Media Reporting and Publishers Information Bureau)

General Motors	$1,712,102,000
Procter & Gamble	$1,493,456,000
Philip Morris	$1,236,978,000
Chrysler	$1,119,851,000
Ford Motor	$897,666,000
Johnson & Johnson	$840,483,000
Walt Disney	$775,621,000
Pepsico	$767,978,000
Time Warner	$747,320,000
AT&T	$659,750,000

65. Approximate the amount of money spent by Pepsico to the nearest hundred-million.

66. Approximate the amount of money spent by General Motors to the nearest hundred-million.

67. Approximate the amount of money spent by Chrysler to the nearest million.

68. Approximate the amount of money spent by Ford Motor to the nearest million.

COMBINING CONCEPTS

A number rounded to the nearest hundred is 8600.

69. Determine the smallest possible number.

70. Determine the largest possible number.

71. On August 23, 1989, it was estimated that one million five hundred thousand people joined hands in a human chain stretching 370 miles to protest the fiftieth anniversary of the pact that allowed the then Soviet Union to annex the Baltic nations in 1939. If the estimate of the number of people is to the nearest one hundred-thousand, determine the largest possible number of people in the chain.

72. In your own words, explain how to round a number to the nearest thousand.

65. $800,000,000

66. $1,700,000,000

67. $1,120,000,000

68. $898,000,000

69. 8550

70. 8649

71. 1,549,999

72. answers may vary

41

Focus On the Real World

JUDGING DISTANCES

Do you know how to estimate a distance without using a tape measure? One easy way to do this is to use the length of your own stride. First, measure the length of your stride in inches. You can do this with the following steps:

- ▲ Lay a yardstick on the floor.
- ▲ Stand next to the yardstick with feet together so that both heels line up with the 0-mark on the yardstick.
- ▲ Take a normal-sized step forward.
- ▲ Find the whole-inch mark nearest the toe of the foot farthest from the 0-mark. This is roughly the length of your stride.

To judge a distance, simply pace it off using normal-sized strides. Multiply the number of strides by the length of your stride to get a rough estimate of the distance.

Suppose you need to measure a distance that can't easily be paced, such as a pond or a busy street. You can easily "transfer" the distance to a more easily paced area by using a baseball cap.

- ▲ Stand at the edge of the pond or street while wearing a baseball cap.
- ▲ Bend your head until your chin rests on your chest.
- ▲ Pull the bill of the cap up or down until it appears to touch the other side of the pond or street.
- ▲ Without moving your head or the cap, pivot your body to the right until you have a clear path straight ahead for walking.
- ▲ Notice where the bill seems to be touching the ground now. The distance to this point is the same as the distance across the pond or street you are measuring.
- ▲ Pace off the distance to this point and find an estimate of the distance as before.

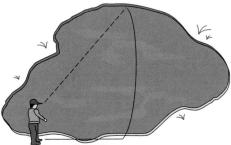

Pace off this distance

Suppose a distance estimate using this method is 1302 inches. To write the distance in terms of feet, divide the estimate by 12. The quotient is the number of whole feet and the remainder is the number of inches. A distance of 1302 inches is the same as 108 feet 6 inches.

$$
\begin{array}{r}
108 \text{R} 6 \\
12\overline{)1302} \\
\underline{12} \\
10 \\
\underline{0} \\
102 \\
\underline{96} \\
6
\end{array}
$$

GROUP ACTIVITY

Materials: yardstick, baseball cap

Use the baseball cap procedure to estimate the distance across a river, stream, pond, or busy road on or near your campus. Have each person in your group estimate the same distance using the length of his or her own stride. Compare your results. Write a brief report summarizing your findings. Be sure to include what distance your group estimated, each member's stride length, the number of strides each member paced off, and each member's distance estimate. Conclude by discussing reasons for any differences in estimates.

1.5 MULTIPLYING WHOLE NUMBERS

Suppose that we wish to count the number of desks in a classroom. The desks are arranged in 5 rows and each row has 6 desks.

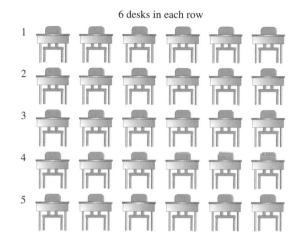

6 desks in each row

Adding 5 sixes gives the total number of desks: $6 + 6 + 6 + 6 + 6 = 30$ desks. When each addend is the same, we refer to this as **repeated addition**. **Multiplication** is repeated addition but with different notation.

$$\underbrace{6 + 6 + 6 + 6 + 6}_{\text{5 sixes}} = \underset{\text{factor}}{5} \times \underset{\text{factor}}{6} = \underset{\text{product}}{30}$$

The $\times$ is called a multiplication sign. The numbers 5 and 6 are called **factors**. The number 30 is called the **product**. The notation 5×6 is read as "five times six." The symbols $\cdot$ and $(\ )$ can also be used to indicate multiplication.

$$5 \times 6 = 30, \quad 5 \cdot 6 = 30, \quad (5)(6) = 30, \quad \text{and} \quad 5(6) = 30$$

TRY THE CONCEPT CHECK IN THE MARGIN.

A USING THE PROPERTIES OF MULTIPLICATION

As for addition, we memorize products of one-digit whole numbers and then use certain properties of multiplication to multiply larger numbers. (If necessary, review the multiplication of one-digit numbers in Appendix B.) Notice in the appendix that when any number is multiplied by 0, the result is always 0. This is called the **multiplication property of 0**.

> **MULTIPLICATION PROPERTY OF 0**
>
> The product of 0 and any number is 0. For example,
>
> $$5 \cdot 0 = 0$$
> $$0 \cdot 8 = 0$$

Also notice in the appendix that when any number is multiplied by 1, the result is always the original number. We call this result the **multiplication property of 1**.

Objectives

A Use the properties of multiplication.

B Multiply whole numbers.

C Find the area of a rectangle.

D Solve problems by multiplying whole numbers.

SSM   CD-ROM   Video 1.5

TEACHING TIP

Show students how to use rectangles to visualize multiplication. For instance, 3×2 can be represented by the problem of finding how many 1 inch squares are in a rectangle which is 3 inches by 2 inches.

✓ CONCEPT CHECK

a. Rewrite $4 + 4 + 4 + 4 + 4 + 4 + 4$ using multiplication.

b. Rewrite 3×16 as repeated addition. Is there more than one way to do this? If so, show all ways.

TEACHING TIP

Help your students build number flexibility by asking them to notice patterns in the multiplication table in the appendix. For instance, they may notice repeating patterns in the ones digit when multiplying by 2, 4, 5, 6, or 8.

Answers

✓ Concept Check

a. $7 \times 4 = 28$, **b.** $16 + 16 + 16 = 48$; yes, $3 + 3 + 3 + 3 + 3 + 3 + 3 + 3 + 3 + 3 + 3 + 3 + 3 + 3 + 3 + 3 = 48$

> ## MULTIPLICATION PROPERTY OF 1
>
> The product of 1 and any number is that same number. For example,
>
> $$1 \cdot 9 = 9$$
> $$6 \cdot 1 = 6$$

Practice Problem 1

Multiply.

a. 3×0

b. $4(1)$

c. $(0)(34)$

d. $1 \cdot 76$

Example 1   Multiply.

      a. 6×1 **b.** $0(8)$ **c.** $1 \cdot 45$ **d.** $(75)(0)$

Solution: **a.** $6 \times 1 = 6$ **b.** $0(8) = 0$
              c. $1 \cdot 45 = 45$ **d.** $(75)(0) = 0$ ▬▬▬

    Like addition, multiplication is commutative and associative. Notice that, when multiplying two numbers, the order of these numbers can be changed without changing the product. For example,

$$3 \cdot 5 = 15 \quad \text{and} \quad 5 \cdot 3 = 15$$

This property is the **commutative property of multiplication**.

> ## COMMUTATIVE PROPERTY OF MULTIPLICATION
>
> Changing the **order** of two factors does not change their product. For example,
>
> $$9 \cdot 2 = 18 \quad \text{and} \quad 2 \cdot 9 = 18$$

    Another property that can help us when multiplying is the **associative property of multiplication**. This property states that, when multiplying numbers, the grouping of the numbers can be changed without changing the product. For example,

$$2 \cdot \underbrace{(3 \cdot 4)} = 2 \cdot 12 = 24$$

Also,

$$\underbrace{(2 \cdot 3)} \cdot 4 = 6 \cdot 4 = 24$$

Both groupings give a product of 24.

> ## ASSOCIATIVE PROPERTY OF MULTIPLICATION
>
> Changing the **grouping** of factors does not change their product. For example,
>
> $$5 \cdot (3 \cdot 2) = (5 \cdot 3) \cdot 2$$
> $$5 \cdot \underbrace{(3 \cdot 2)} = 5 \cdot 6 = 30 \quad \text{and} \quad \underbrace{(5 \cdot 3)} \cdot 2 = 15 \cdot 2 = 30$$

    With these properties, along with the **distributive property**, we can find the product of any whole numbers. The distributive property says that multiplication **distributes** over addition. For example, notice that $3(2 + 5)$ is the same as $3 \cdot 2 + 3 \cdot 5$.

Answers

1. a. 0, **b.** 4, **c.** 0, **d.** 76

$$3(2 + 5) = 3(7) = 21$$

$$3 \cdot 2 + 3 \cdot 5 = 6 + 15 = 21$$

Notice in $3(2 + 5) = 3 \cdot 2 + 3 \cdot 5$ that each number inside the parentheses is multiplied by 3.

DISTRIBUTIVE PROPERTY

Multiplication distributes over addition. For example,

$$2(3 + 4) = 2 \cdot 3 + 2 \cdot 4$$

Example 2   Rewrite each using the distributive property.

a. $3(4 + 5)$ **b.** $10(6 + 8)$ **c.** $2(7 + 3)$

Solution: Using the distributive property, we have

a. $3(4 + 5) = 3 \cdot 4 + 3 \cdot 5$
b. $10(6 + 8) = 10 \cdot 6 + 10 \cdot 8$
c. $2(7 + 3) = 2 \cdot 7 + 2 \cdot 3$

B   MULTIPLYING WHOLE NUMBERS

Let's use the distributive property to multiply $7(48)$. To do so, we begin by writing the expanded form of 48 and then applying the distributive property.

$$7(48) = 7(40 + 8)$$
$$= 7 \cdot 40 + 7 \cdot 8 \qquad \text{Apply the distributive property.}$$
$$= 280 + 56 \qquad \text{Multiply.}$$
$$= 336 \qquad \text{Add.}$$

This is how we multiply whole numbers. When multiplying whole numbers, we will use the following notation.

$$\begin{array}{r} 5 \\ 48 \\ \times\ 7 \\ \hline 336 \end{array}$$

$\leftarrow 7 \cdot 8 = 56$ Write 6 in the ones place and carry 5 to the tens place.

$7 \cdot 4 = 28$ and $28 + 5 = 33$

Example 3   Multiply: $\begin{array}{r} 25 \\ \times\ 8 \end{array}$

Solution: $\begin{array}{r} 4 \\ 25 \\ \times\ 8 \\ \hline 200 \end{array}$

Practice Problem 2

Rewrite each using the distributive property.

a. $5(2 + 3)$

b. $9(8 + 7)$

c. $3(6 + 1)$

TEACHING TIP

Show students how to rewrite problems using the associative property.
$$7 \times 40 = 7(4 \times 10)$$
$$= (7 \times 4)(10)$$
$$= 28 \times 10$$
$$= 280$$

Practice Problem 3

Multiply.

a. $\begin{array}{r} 36 \\ \times\ 4 \end{array}$ b. $\begin{array}{r} 92 \\ \times\ 9 \end{array}$

Answers

2. a. $5(2 + 3) = 5 \cdot 2 + 5 \cdot 3$,
b. $9(8 + 7) = 9 \cdot 8 + 9 \cdot 7$,
c. $3(6 + 1) = 3 \cdot 6 + 3 \cdot 1$,   **3. a.** 144,
b. 828

Copyright 1999 Prentice-Hall, Inc.

TEACHING TIP

It may be helpful to show students how they can use generic rectangles to multiply 236×86.

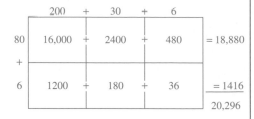

Practice Problem 4

Multiply.

a. $\begin{array}{r} 594 \\ \times\ 72 \\ \hline \end{array}$ b. $\begin{array}{r} 306 \\ \times\ 81 \\ \hline \end{array}$

Practice Problem 5

Multiply.

a. $\begin{array}{r} 726 \\ \times\ 142 \\ \hline \end{array}$ b. $\begin{array}{r} 4 \\ \times\ 288 \\ \hline \end{array}$

✓ CONCEPT CHECK

Find and explain the error in the following multiplication problem.

$$\begin{array}{r} 102 \\ \times\ 33 \\ \hline 306 \\ 306 \\ \hline 612 \end{array}$$

Answers

4. a. 42,768, **b.** 24,786, **5. a.** 103,092,
b. 1152
✓ **Concept Check:** $\begin{array}{r} 102 \\ \times\ 33 \\ \hline 306 \\ 3060 \\ \hline 3366 \end{array}$

To multiply larger whole numbers, use the following similar notation. Multiply 89×52.

Step 1

$$\begin{array}{r} 1 \\ 89 \\ \times\ 52 \\ \hline 178 \end{array} \leftarrow \text{Multiply } 89 \times 2.$$

Step 2

$$\begin{array}{r} 4 \\ 89 \\ \times\ 52 \\ \hline 178 \\ 4450 \end{array} \leftarrow \text{Multiply } 89 \times 50$$

Step 3

$$\begin{array}{r} 89 \\ \times\ 52 \\ \hline 178 \\ 4450 \\ \hline 4628 \end{array} \text{ Add.}$$

The numbers 178 and 4450 are called **partial products**. The sum of the partial products, 4628, is the product of 89 and 52.

Example 4   Multiply: 236×86

Solution:

$$\begin{array}{r} 236 \\ \times\ 86 \\ \hline 1416 \\ 18{,}880 \\ \hline 20{,}296 \end{array}$$

$\leftarrow 6(236)$
$\leftarrow 80(236)$
Add.

Example 5   Multiply: 631×125

Solution:

$$\begin{array}{r} 631 \\ \times\ 125 \\ \hline 3155 \\ 12{,}620 \\ 63{,}100 \\ \hline 78{,}875 \end{array}$$

$\leftarrow 5(631)$
$\leftarrow 20(631)$
$\leftarrow 100(631)$
Add.

TRY THE CONCEPT CHECK IN THE MARGIN.

C   FINDING THE AREA OF A RECTANGLE

A special application of multiplication is finding the area of a region. Area measures the amount of surface of a region. For example, we measure a plot of land or the living space of a home by area. The figures show two examples of units of area measure. (A centimeter is a unit of length in the metric system.)

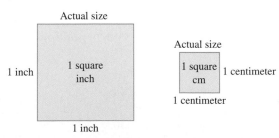

To measure the area of a geometric figure such as the rectangle shown, count the number of square units that cover the region.

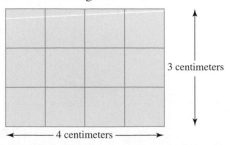

This rectangular region contains 12 square units, each 1 square centimeter. Thus, the area is 12 square centimeters. This total number of squares can be found by counting or by multiplying **4 · 3** (length · width).

$$\begin{aligned} \text{Area of a rectangle} &= \text{length} \cdot \text{width} \\ &= (4 \text{ centimeters})(3 \text{ centimeters}) \\ &= 12 \text{ square centimeters} \end{aligned}$$

In this section, we find the areas of rectangles only. In later sections, we find the areas of other geometric regions.

Example 6   Finding the Area of a State

The state of Colorado is in the shape of a rectangle whose length is 380 miles and whose width is 280 miles. Find its area.

Solution: The area of a rectangle is the product of its length and its width.

$$\begin{aligned} \text{Area} &= \text{length} \cdot \text{width} \\ &= (380 \text{ miles})(280 \text{ miles}) \\ &= 106{,}400 \text{ square miles} \end{aligned}$$

The area of Colorado is 106,400 square miles.

D   SOLVING PROBLEMS BY MULTIPLYING

There are several words or phrases that indicate the operation of multiplication. Some of these are as follows:

Key Words or Phrases	Example	Symbols
multiply	multiply 5 by 7	5 · 7
product	the product of 3 and 2	3 · 2
times	10 times 13	10 · 13

Many key words or phrases describing real-life problems that suggest addition might be better solved by multiplication instead. For example, to find the **total** cost of 8 shirts, each selling for $27, we can either add 27 + 27 + 27 + 27 + 27 + 27 + 27 + 27, or we can multiply 8(27).

Example 7   Finding Disk Space

A single-density computer disk can hold 370 thousand bytes of information. How many total bytes can 42 such disks hold?

Solution: Forty-two disks will hold 42 × 370 thousand bytes.

Words		Translate to Numbers
bytes per disk	→	370
× disks	→	× 42
		740
		14,800
total bytes		15,540

Forty-two disks will hold 15,540 thousand bytes.

Practice Problem 6

The state of Wyoming is in the shape of a rectangle whose length is 360 miles and whose width is 280 miles. Find its area.

Practice Problem 7

A computer printer can print 240 characters per second in draft mode. How many total characters can it print in 15 seconds?

Answers

6. 100,800 square miles,   **7.** 3600 characters

Practice Problem 8

Softball T-shirts come in two styles: plain at $6 each and striped at $7 each. The team orders 4 plain shirts and 5 striped shirts. Find the total cost of the order.

Example 8   Budgeting Money

Earline Martin agrees to take her children and their cousins to the San Antonio Zoo. The ticket price for each child is $4 and for each adult, $6. If 8 children and 1 adult plan to go, how much money is needed for admission?

Solution: If the price of one child's ticket is $4, the price for 8 children is $8 \cdot 4 = \$32$. The price of one adult ticket is $6, so the total cost is

Words		Translate to Numbers
price of 8 children	$\rightarrow$	32
+ price of adult	$\rightarrow$	+ 6
total cost		38

The total cost is $38.

Practice Problem 9

If an average page in a book contains 259 words, estimate, rounding to the nearest hundred, the total number of words contained on 195 pages.

Example 9   Estimating Word Count

The average page of a book contains 259 words. Estimate, rounding to the nearest ten, the total number of words contained on 22 pages.

Solution: The exact number of words is 259×22. Estimate this product by rounding each factor to the nearest ten.

$$
\begin{array}{r}
\overset{1}{259} \text{ rounds to } 260 \\
\underline{\times\ 22} \text{ rounds to } \underline{\times\ 20} \\
5200
\end{array}
$$

There are approximately 5200 words contained on 22 pages.

CALCULATOR EXPLORATION
MULTIPLYING NUMBERS

To multiply numbers on a calculator, find the key marked $\boxed{\times}$ and $\boxed{=}$. For example, to find $31 \cdot 66$ on a calculator, press the

or
$\boxed{\text{ENTER}}$

keys $\boxed{31}\ \boxed{\times}\ \boxed{66}\ \boxed{=}$. The display will read $\boxed{2046}$. Thus,

or
$\boxed{\text{ENTER}}$

$31 \cdot 66 = 2046$.

Use a calculator to multiply.

1. 72×48 3456		**2.** 81×92 7452
3. $163 \cdot 94$ 15,322		**4.** $285 \cdot 144$ 41,040
5. $983(277)$ 272,291		**6.** $1562(843)$ 1,316,766

Answers

8. $59,   **9.** 60,000 words

Name _____ **Section** _____ **Date** _____

MENTAL MATH

Multiply. See Example 1.

1. $1 \cdot 24$ **2.** $55 \cdot 1$ **3.** $0 \cdot 19$ **4.** $27 \cdot 0$

5. $8 \cdot 0 \cdot 9$ **6.** $7 \cdot 6 \cdot 0$ **7.** $87 \cdot 1$ **8.** $1 \cdot 41$

EXERCISE SET 1.5

A *Use the distributive property to rewrite each expression. See Example 2.*

1. $4(3 + 9)$ **2.** $5(8 + 2)$ **3.** $2(4 + 6)$

4. $6(1 + 4)$ **5.** $10(11 + 7)$ **6.** $12(12 + 3)$

B *Multiply. See Example 3.*

7. 42   **8.** 79   **9.** 624   **10.** 638
  $\times\ 6$ $\times\ 3$ $\times\ 3$ $\times\ 5$

11. 277   **12.** 882   **13.** 1062   **14.** 9021
  $\times\ 6$ $\times\ 2$ $\times\ 5$ $\times\ 3$

Multiply. See Examples 4 and 5.

15. 298   **16.** 591   **17.** 231   **18.** 526
  $\times\ 14$ $\times\ 72$ $\times\ 47$ $\times\ 23$

19. 809   **20.** 307   **21.** $(620)(40)$ **22.** $(720)(80)$
  $\times\ 14$ $\times\ 16$

23. $(998)(12)(0)$ **24.** $(593)(47)(0)$ **25.** $(590)(1)(10)$ **26.** $(240)(1)(20)$

27. 1234   **28.** 1357   **29.** 609   **30.** 505
  $\times\ 48$ $\times\ 79$ $\times\ 234$ $\times\ 127$

Name _____

31. 5621
$\times$ 324

32. 1234
$\times$ 567

33. 1941
$\times$ 235

34. 1876
$\times$ 437

35. 589
$\times$ 110

36. 426
$\times$ 110

Estimate the products by rounding each factor to the nearest hundred. See Example 9.

37. 576×354

38. 982×650

39. 604×451

40. 111×999

C *Find the area of each rectangle. See Example 6.*

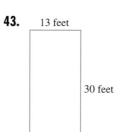 **41.**
9 meters

7 meters

42.
4 inches

12 inches

43.
13 feet

30 feet

44.
25 centimeters

20 centimeters

C D *Solve. See Examples 6 through 9.*

45. One tablespoon of olive oil contains 125 calories. How many calories are in 3 tablespoons of olive oil? (*Source: Home and Garden Bulletin No. 72*, U.S. Department of Agriculture)

46. One ounce of hulled sunflower seeds contains 14 grams of fat. How many grams of fat are in 6 ounces of hulled sunflower seeds? (*Source: Home and Garden Bulletin No. 72*, U.S. Department of Agriculture)

Name _____

47. $1295

47. The textbook for a course in Civil War history costs $37. There are 35 students in the class. Find the total cost of the history books for the class.

48. The seats in the mathematics lecture hall are arranged in 12 rows with 6 seats in each row. Find how many seats are in this room.

48. 72 seats

49. A case of canned peas has *two layers* of cans. In each layer are 8 rows with 12 cans in each row. Find how many cans are in a case.

50. An apartment building has *three floors*. Each floor has five rows of apartments with four apartments in each row. Find how many apartments there are.

49. 192 cans

51. A plot of land measures 90 feet by 110 feet. Find its area.

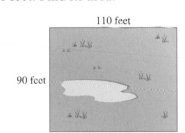

110 feet

90 feet

52. A house measures 45 feet by 60 feet. Find the floor area of the house.

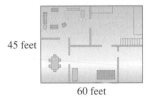

45 feet

60 feet

50. 60 apartments

51. 9900 square feet

52. 2700 square feet

53. Recall from an earlier section that the largest commercial building in the world under one roof is the flower auction building of the cooperative VBA in Aalsmeer, Netherlands. The floor plan is a rectangle that measures 776 meters by 639 meters. Find the area of this building. (A meter is a unit of length in the metric system.) (*Source: The Handy Science Answer Book*, Visible Ink Press, 1994)

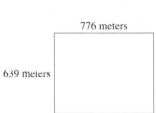

776 meters

639 meters

54. The largest lobby can be found at the Hyatt Regency in San Francisco, CA. It is in the shape of a rectangle that measures 350 feet by 160 feet. Find its area.

53. 495,864 square meters

54. 56,000 square feet

51

52

Name _____

55. A pixel is a rectangular dot on a graphing calculator screen. If a graphing calculator screen contains 62 pixels in a row and 94 pixels in a column, find the total number of pixels on a screen.

56. A high-density computer disk can hold 1510 thousand bytes of information. Find how many bytcs of information 17 disks can hold.

57. A line of print on a computer contains 60 characters (letters, spaces, punctuation marks). Find how many characters there are in 25 lines.

58. An average cow eats 3 pounds of grain per day. Find how much grain a cow eats in a year. (Assume 365 days in 1 year.)

59. One ounce of Planters® Dry Roasted Peanuts has 160 calories. How many calories are in 8 ounces? (*Source:* RJR Nabisco, Inc.)

60. One ounce of Planters® Dry Roasted Peanuts has 13 grams of fat. How many grams of fat are in 8 ounces? (*Source:* RJR Nabisco, Inc.)

61. The diameter of the planet Saturn is 9 times as great as the diameter of Earth. The diameter of Earth is 7927 miles. Find the diameter of Saturn.

Earth   Saturn

62. The planet Uranus orbits the sun every 84 Earth years. Find how many Earth days two orbits take. (Assume 365 days in 1 year.)

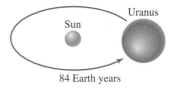

Sun   Uranus

84 Earth years

Name _____

63. 22,464 acres

63. Suppose a wild elk needs 39 acres of natural grazing area to survive. If there are 576 elk in a certain herd, find how much grazing area the herd needs.

64. A window washer in New York City is bidding for a contract to wash the windows of a 23-story building. To write a bid, the number of windows in the building is needed. If there are 7 windows in each row of windows on 2 sides of the building and 4 windows per row on the other 2 sides of the building, find the total number of windows.

64. 506 windows

65. 21,700,000 qts

65. Hershey's main chocolate factory in Hershey, Pennsylvania, uses 700,000 quarts of milk each day. How many quarts of milk would be used during the month of March, assuming that chocolate is made at the factory every day of the month? (*Source:* Hershey Foods Corp.)

66. Among older Americans (age 65 years and older), there are 5 times as many widows as widowers. There were 1,700,000 widowers in 1995. How many widows were there in 1995? (*Source:* Administration on Aging)

66. 8,500,000 widows

67. In 1996, the average cost of the Head Start program was $4571 per child. That year 752,077 children participated in the program. Round each number to the nearest thousand and estimate the total cost of the Head Start program in 1996. (*Source:* Head Start Bureau)

68. Approximately 5570 people celebrated their 65th birthday *each day* of 1996 in the United States. Round each number to the nearest hundred and estimate the total number of people who turned 65 *throughout the year* in 1996. (*Source:* Administration on Aging)

67. $3,760,000,000

◢ **COMBINING CONCEPTS**

68. 2,240,000 people

In a survey, college students were asked to name their favorite fruit. The results are shown in the picture graph. Use this graph to answer Exercises 69–72.

Pick Your Favorite Fruit Survey

Apple	🍎🍎🍎🍎🍎🍎🍎🍎🍎
Orange	🍊🍊🍊🍊🍊🍊🍊
Banana	🍌🍌🍌🍌
Grapes	🍇🍇🍇🍇🍇

Each 🍎 Represents 10 Students

69. 50 students

69. How many students chose grapes as their favorite fruit?

70. How many students chose the banana as their favorite fruit?

70. 40 students

71. apple and orange

71. Which two fruits were the most popular?

72. Which two fruits were chosen by a total of 110 students?

72. orange and banana

Fill in the missing digits in each problem.

73.
```
     4_
  ×  _3
    126
   3780
   3906
```

74.
```
     _7
  ×  6_
    171
   3420
   3591
```

73. 2, 9

75. Explain how to multiply two 2-digit numbers using partial products.

76. A slice of enriched white bread has 65 calories. One tablespoon of jam has 55 calories, and one tablespoon of peanut butter has 95 calories. Suppose a peanut butter and jelly sandwich is made with two slices of white bread, one tablespoon of jam, and one tablespoon of peanut butter. How many calories are in two such peanut butter and jelly sandwiches? (*Source: Home and Garden Bulletin No. 72*, U.S. Department of Agriculture)

74. 5, 3

75. answers may vary

77. In 1997 the WNBA's top-scorer was Cynthia Cooper of the Houston Comets. She made 172 free throws, 124 two-point field goals, and 67 three-point field goals. (Free throws are worth one point.) How many points did Cynthia Cooper score during the 1997 season? (*Source:* Women's National Basketball Association)

76. 560 calories

77. 621 points

1.6 DIVIDING WHOLE NUMBERS

Suppose three people pooled their money and bought a raffle ticket at a local fund-raiser. Their ticket was the winner and they won a $60 cash prize. They then divided the prize into three equal parts so that each person received $20.

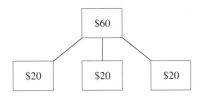

A DIVIDING WHOLE NUMBERS

The process of separating the quantity into these three equal parts is called **division**. Division can be symbolized by several notations.

$$\overset{\text{quotient}}{3)\overline{60}} \leftarrow \text{dividend} \qquad \underset{\text{divisor}}{\frac{60}{3}} = 20 \leftarrow \text{quotient} \qquad \underset{\text{dividend}}{60} \div 3 = \overset{\text{quotient}}{20}$$
divisor ↑   divisor

(In the notation $\frac{60}{3}$, the bar separating 60 and 3 is called a **fraction bar**.) Just as subtraction is the reverse of addition, division is the reverse of multiplication. This means that division can be checked by multiplication.

$$3)\overline{60}^{20} \qquad \text{because} \qquad 20 \cdot 3 = 60$$

Since multiplication and division are related in this way, you can use the multiplication table in Appendix B to review quotients of one-digit divisors, if necessary.

Example 1 Find each quotient. Check by multiplying.

a. $42 \div 7$ **b.** $\frac{81}{9}$ **c.** $4)\overline{24}$

Solution: **a.** $42 \div 7 = 6$, because $6 \cdot 7 = 42$.

b. $\frac{81}{9} = 9$, because $9 \cdot 9 = 81$.

c. $4)\overline{24}^{6}$, because $6 \cdot 4 = 24$.

Example 2 Find each quotient. Check by multiplying.

a. $1)\overline{8}$ **b.** $11 \div 1$ **c.** $\frac{9}{9}$ **d.** $7 \div 7$ **e.** $\frac{10}{1}$ **f.** $6)\overline{6}$

Solution: **a.** $1)\overline{8}^{8}$, because $8 \cdot 1 = 8$.
b. $11 \div 1 = 11$, because $11 \cdot 1 = 11$.
c. $\frac{9}{9} = 1$, because $1 \cdot 9 = 9$.
d. $7 \div 7 = 1$, because $1 \cdot 7 = 7$.

Objectives

A Divide whole numbers.
B Perform long division.
C Solve problems that require dividing by whole numbers.
D Find the average of a list of numbers.

SSM   CD-ROM   Video 1.6

TEACHING TIP

Ask students to verbalize all the different ways they could state the equation, $60 \div 3 = 20$. For example, 60 divided by 3 equals 20. 3 goes into 60 twenty times. The quotient of 60 and 3 is 20.

Practice Problem 1

Find each quotient. Check by multiplying.

a. $8)\overline{48}$

b. $35 \div 5$

c. $\frac{49}{7}$

Practice Problem 2

Find each quotient. Check by multiplying.

a. $\frac{8}{8}$ b. $3 \div 1$ c. $1)\overline{12}$

d. $2 \div 1$ e. $\frac{5}{1}$

Answers

1. a. 6, **b.** 7, **c.** 7, **2. a.** 1, **b.** 3, **c.** 12, **d.** 2, **e.** 5

e. $\frac{10}{1} = 10$, because $10 \cdot 1 = 10$.

f. $6\overline{)6}$ with quotient 1, because $1 \cdot 6 = 6$

Example 2 illustrates important properties of division as described next.

> **DIVISION PROPERTIES OF 1**
>
> The quotient of any number and that same number is 1. For example,
>
> $$8 \div 8 = 1 \qquad \frac{7}{7} = 1 \qquad 4\overline{)4} \text{ with quotient } 1$$
>
> The quotient of any number and 1 is that same number. For example,
>
> $$9 \div 1 = 9 \qquad \frac{6}{1} = 6 \qquad 1\overline{)3} \text{ with quotient } 3 \qquad \frac{0}{1} = 0$$

Practice Problem 3

Find each quotient. Check by multiplying.

a. $\frac{0}{7}$ 　　　 **b.** $5\overline{)0}$ 　　　 **c.** $9 \div 0$

d. $0 \div 6$

⬛ **Example 3**   Find each quotient. Check by multiplying.

a. $9\overline{)0}$ 　 **b.** $0 \div 12$ 　 **c.** $\frac{0}{5}$ 　 **d.** $\frac{3}{0}$

Solution: 　 **a.** $9\overline{)0}$ with quotient 0, because $0 \cdot 9 = 0$.

b. $0 \div 12 = 0$, because $0 \cdot 12 = 0$.

c. $\frac{0}{5} = 0$, because $0 \cdot 5 = 0$.

d. If $\frac{3}{0} = a$ **number**, then the **number** $\cdot 0 = 3$. Recall that any number multiplied by 0 is 0 and not 3. We say then that $\frac{3}{0}$ is **undefined**.

Example 3 illustrates important division properties of 0.

> **DIVISION PROPERTIES OF 0**
>
> The quotient of 0 and any number (except 0) is 0. For example,
>
> $$0 \div 9 = 0 \qquad \frac{0}{5} = 0 \qquad 14\overline{)0} \text{ with quotient } 0$$
>
> The quotient of any number and 0 is not a number. We say that
>
> $$\frac{3}{0}, \qquad 0\overline{)3}, \qquad \text{and} \qquad 3 \div 0$$
>
> are **undefined**.

B PERFORMING LONG DIVISION

When dividends are larger, the quotient can be found by a process called **long division**. For example, let's divide 2541 by 3.

$$3\overline{)\,2541}$$

We can't divide 3 into 2, so try dividing 3 into the first two digits.

Answers

3. a. 0,　**b.** 0,　**c.** undefined,　**d.** 0

$$\begin{array}{r} 8 \\ 3\overline{)2541} \end{array}$$ $25 \div 3 = 8$ with 1 left, so our best estimate is 8. We place 8 over the 5 in 25.

Next, multiply 8 and 3 and subtract this product from 25. Make sure that this difference is less than the divisor.

$$\begin{array}{r} 8 \\ 3\overline{)2541} \\ -24 \\ \hline 1 \end{array}$$ $8(3) = 24$

$25 - 24 = 1$ and 1 is less than the divisor 3.

Next, bring down the next digit and start the process again.

$$\begin{array}{r} 84 \\ 3\overline{)2541} \\ -24\downarrow \\ \hline 14 \\ -12 \\ \hline 2 \end{array}$$ $14 \div 3 = 4$ with 2 left

$4(3) = 12$

$14 - 12 = 2$

Again, bring down the next digit and start the process again.

$$\begin{array}{r} 847 \\ 3\overline{)2541} \\ -24 \\ \hline 14 \\ -12\downarrow \\ \hline 21 \\ -21 \\ \hline 0 \end{array}$$ $21 : 3 = 7$

$7(3) = 21$

$21 - 21 = 0$

The quotient is 847. To check, see that $847 \times 3 - 2541$.

Example 4   Divide: $3705 \div 5$. Check by multiplying.

Solution:

$$\begin{array}{r} 7 \\ 5\overline{)3705} \\ -35\downarrow \\ \hline 20 \end{array}$$ $37 \div 5 = 7$ with 2 left. Place this estimate 7 over the 7 in 37.

$7(5) = 35$

$37 - 35 = 2$ and 2 is less than the divisor 5.

↑—— Bring down the 0.

$$\begin{array}{r} 74 \\ 5\overline{)3705} \\ -35 \\ \hline 20 \\ -20\downarrow \\ \hline 05 \end{array}$$ $20 \div 5 = 4$

$4(5) = 20$

$20 - 20 = 0$ and 0 is less than the divisor 5.

↑— Bring down the 5.

$$\begin{array}{r} 741 \\ 5\overline{)3705} \\ -35 \\ \hline 20 \\ -20\downarrow \\ \hline 5 \\ -5 \\ \hline 0 \end{array}$$ $5 \div 5 = 1$

$1(5) = 5$

$5 - 5 = 0$

Practice Problem 4

Divide. Check by multiplying.

a. $6\overline{)5382}$

b. $4\overline{)2212}$

Check:
$$\begin{array}{r} 741 \\ \times\quad 5 \\ \hline 3705 \end{array}$$

HELPFUL HINT

Since division and multiplication are reverse operations, don't forget that a division problem can be checked by multiplying.

Practice Problem 5

Divide and check.

a. $3\overline{)2397}$

b. $7\overline{)2520}$

Example 5   Divide and check: $1872 \div 9$

Solution:

$$
\begin{array}{r}
208 \\
9\overline{)1872} \\
-18\downarrow \\
\hline
07 \\
-0\downarrow \\
\hline
72 \\
-72 \\
\hline
0
\end{array}
$$

$2(9) = 18$
$18 - 18 = 0$, bring down the 7.
$0(9) = 0$
$7 - 0 = 7$, bring down the 2.
$8(9) = 72$
$72 - 72 = 0$

Check: $208 \cdot 9 = 1872$

Naturally, quotients don't always "come out even." Making 4 rows out of 26 chairs, for example, isn't possible if each row is supposed to have exactly the same number of chairs. Each of 4 rows can have 6 chairs, but 2 chairs are still left over.

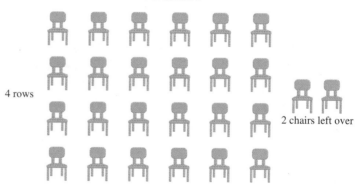

6 chairs in each row

4 rows

2 chairs left over

We signify "leftovers" or *remainders* in this way:

$$4\overline{)26}^{\,6\,R\,2}$$

The **whole number part of the quotient** is 6; the **remainder part of the quotient** is 2. Checking by multiplying,

whole number part	·	divisor	+	remainder part	=	dividend
↓		↓		↓		↓
6	·	4	+	2		
		24	+	2	=	26

Answers

5. a. 799, **b.** 360

Example 6   Divide and check: 2557 ÷ 7

Solution:

$$\begin{array}{r} 365 \text{ R } 2 \\ 7)\overline{2557} \end{array}$$

$-21\!\downarrow$ $3(7) = 21$
45   $25 - 21 = 4$, bring down the 5.
$-42\!\downarrow$ $6(7) = 42$
37   $45 - 42 = 3$, bring down the 7.
-35 $5(7) = 35$
2   $37 - 35 = 2$ The remainder is 2.

Check: 365 · 7 + 2 = 2557

↑ whole number part   · ↑ divisor   + ↑ remainder part   = ↑ dividend

Practice Problem 6

Divide and check.

a. $5)\overline{949}$

b. $6)\overline{4399}$

Example 7   Divide and check: 56,717 ÷ 8

Solution:

$$\begin{array}{r} 7089 \text{ R } 5 \\ 8)\overline{56717} \end{array}$$

$-56\!\downarrow\downarrow$ $7(8) = 56$
7   Subtract and bring down the 7.
$-0\!\downarrow$ $0(8) = 0$
71   Subtract and bring down the 1.
$-64\!\downarrow$ $8(8) = 64$
77   Subtract and bring down the 7.
-72 $9(8) = 63$
5   Subtract. The remainder is 5.

Check: 7089 · 8 + 5 = 56,717

↑ whole number part   · ↑ divisor   + ↑ remainder part   = ↑ dividend

Practice Problem 7

Divide and check.

a. $5)\overline{40,841}$

b. $7)\overline{22,430}$

When the divisor has more than one digit, the same pattern applies. For example, let's find 1358 ÷ 23.

$$\begin{array}{r} 5 \\ 23)\overline{1358} \\ 115\!\downarrow \\ \overline{208} \end{array}$$ $135 \div 23 = 5$ with 20 left over. Our estimate is 5.
   $5(23) = 115$
   $135 - 115 = 20$. Bring down the 8.

Now we continue estimating.

$$\begin{array}{r} 59 \text{ R } 1 \\ 23)\overline{1358} \\ -115 \\ \overline{208} \\ -207 \\ \overline{1} \end{array}$$ $208 \div 23 = 9$ with 1 left over.
   $9(23) = 207$
   $208 - 207 = 1$. The remainder is 1.

To check, see that 59 · 23 + 1 = 1358.

Answers

6. a. 189 R 4,   **b.** 733 R 1,   **7. a.** 8168 R 1,
b. 3204 R 2

Practice Problem 8

Divide: $5740 \div 19$

Practice Problem 9

Divide: $16,589 \div 247$

✓ CONCEPT CHECK

Which of the following is the correct way to represent "the quotient of 20 and 5"? Or are both correct? Explain your answer.

a. $5 \div 20$

b. $20 \div 5$

Practice Problem 10

Marina, Manual, and Min bought 10 dozen high-density computer diskettes to share equally. How many diskettes did each person get?

Example 8

Divide: $6819 \div 17$

Solution:

$$
\begin{array}{r}
401 \text{ R } 2 \\
17\overline{)6819} \\
-68 \\
\hline
01 \\
-0 \\
\hline
19 \\
-17 \\
\hline
2
\end{array}
$$

$4(17) = 68$
Subtract and bring down the 1.
$0(17) = 0$
Subtract and bring down the 9.
$1(17) = 17$
Subtract. The remainder is 2.

To check, see that $401 \cdot 17 + 2 = 6819$.

Example 9

Divide: $51,600 \div 403$

Solution:

$$
\begin{array}{r}
128 \text{ R } 16 \\
403\overline{)51600} \\
-403 \\
\hline
1130 \\
-806 \\
\hline
3240 \\
-3224 \\
\hline
16
\end{array}
$$

$1(403) = 403$
Subtract and bring down the 0.
$2(403) = 806$
Subtract and bring down the 0.
$8(403) = 3224$
Subtract. The remainder is 16.

To check, see that $128 \cdot 403 + 16 = 51,600$.

C SOLVING PROBLEMS BY DIVIDING

Below are some key words and phrases that indicate the operation of multiplication.

Key Words or Phrases	Examples	Symbols
divide	divide 10 by 5	$10 \div 5$ or $\frac{10}{5}$
quotient	the quotient of 64 and 4	$64 \div 4$ or $\frac{64}{4}$
divided by	9 divided by 3	$9 \div 3$ or $\frac{9}{3}$
divided or shared equally among	$100 divided equally among five people	$100 \div 5$ or $\frac{100}{5}$

TRY THE CONCEPT CHECK IN THE MARGIN.

Example 10 Finding Shared Earnings

Zachary, Tyler, and Stephanie McMillan shared a paper route to earn money for college expenses. The total in their fund after expenses was $2895. How much is each person's equal share?

Solution:

Each person's equal share is (total) $\div$ (number of people) or
$$2895 \div 3$$

Answers

8. 302 R 2, **9.** 67 R 40, **10.** 40 diskettes

✓ Concept Check

a. incorrect, **b.** correct

Then
$$
\begin{array}{r}
965 \\
3\overline{)2895} \\
-27 \\ \hline
19 \\
-18 \\ \hline
15 \\
-15 \\ \hline
0
\end{array}
$$

Each person's share is $965.

Example 11   Calculating Shipping Needs

How many boxes are needed to ship 56 pairs of Nikes to a shoe store in Texarkana if 9 pairs of shoes will fit in each shipping box?

Solution:

number of boxes	=	total pairs of shoes	÷	how many pairs in a box
↓		↓		↓
number of boxes	=	56	÷	9

$$
\begin{array}{r}
6\,R\,2 \\
9\overline{)56} \\
-54 \\ \hline
2
\end{array}
$$

There are 6 full boxes with 2 pairs of shoes left over, so 7 boxes will be needed.

Practice Problem 11

Peanut butter and cheese cracker sandwiches are packaged 6 sandwiches to a package. How many full packages are formed from 195 sandwiches?

Example 12   Dividing Holiday Favors Among Students

Mary Schultz has 48 kindergarten students. She buys 260 stickers as Thanksgiving Day favors for her students. Can she divide the stickers up equally among her students? If not, how many stickers will be left over?

Solution:

number of stickers	÷	number of students
↓		↓
260	÷	48

$$
\begin{array}{r}
5\,R\,20 \\
48\overline{)260} \\
-240 \\ \hline
20
\end{array}
$$

No, the stickers cannot be divided equally among her students since there is a nonzero remainder. There will be 20 stickers left over.

Practice Problem 12

Calculators can be packed 24 to a box. If 497 calculators are to be packed, but only full boxes are shipped, how many full boxes will be shipped? How many calculators are left over and are not shipped?

Answers

11. 32 full packages,   **12.** 20 full boxes; 17 calculators leftover

Consider leading a discussion about where averages are used in real life and what these averages mean.

TEACHING TIP

Practice Problem 13

To compute a safe time to wait after allergy shots are administered, a lab technician is given a list of elapsed times between administered shots and reactions. Find the average of the times: 5 minutes, 7 minutes, 20 minutes, 6 minutes, 9 minutes, 3 minutes, and 48 minutes.

D FINDING AVERAGES

A special application of division (and addition) is finding the average of a list of numbers. The **average** of a list of numbers is the sum of the numbers divided by the *number* of numbers.

$$\text{average} = \frac{\text{sum of numbers}}{\textit{number} \text{ of numbers}}$$

Example 13   Averaging Scores

Liam Reilly's scores in his mathematics class so far are 93, 86, 71, and 82. Find his average score.

Solution: To find his average score, we find the sum of his scores and divide by 4, the number of scores.

$$
\begin{array}{r}
93\\
86\\
71\\
+\ 82\\
\hline
332
\end{array}
\text{ sum}
\qquad
\text{average} = \frac{332}{4} = 83
\qquad
\begin{array}{r}
83\\
4\overline{)\,332}\\
-32\\
\hline
12\\
-12\\
\hline
0
\end{array}
$$

His average score is 83.

CALCULATOR EXPLORATION
DIVIDING NUMBERS

To divide numbers on a calculator, find the keys marked $\div$ and $=$. For example, to find $435 \div 5$ on a calculator, press the

or

ENTER

keys 435 $\div$ 5 $=$. The display will read 87. Thus, $435 \div 5 = 87$.   or

ENTER

Use a calculator to divide.

1. $848 \div 16$ 53
2. $564 \div 12$ 47
3. $95\overline{)\,5890}$ 62
4. $27\overline{)\,1053}$ 39
5. $\dfrac{32,886}{126}$ 261
6. $\dfrac{143,088}{264}$ 542
7. $0 \div 315$ 0
8. $315 \div 0$ Error

Answer

13. 14 minutes

Name _____ Section _____ Date _____

MENTAL MATH ANSWERS
1. 5
2. 8
3. 9
4. 8
5. 0
6. 0
7. 9
8. 12
9. 1
10. 1
11. 5
12. 5
13. undefined
14. undefined
15. 7
16. 1
17. 0
18. undefined
19. 8
20. 6

MENTAL MATH

A *Find each quotient. See Examples 1 through 3.*

1. $40 \div 8$
2. $72 \div 9$
3. $45 \div 5$
4. $24 \div 3$

5. $0 \div 5$
6. $0 \div 8$
7. $9 \div 1$
8. $12 \div 1$

9. $\dfrac{16}{16}$
10. $\dfrac{49}{49}$
11. $\dfrac{25}{5}$
12. $\dfrac{45}{9}$

13. $6 \div 0$
14. $\dfrac{12}{0}$
15. $7 \div 1$
16. $6 \div 6$

17. $0 \div 4$
18. $7 \div 0$
19. $16 \div 2$
20. $18 \div 3$

EXERCISE SET 1.6

B *Divide and then check by multiplying. See Examples 4 and 5.*

1. $9\overline{)108}$
2. $5\overline{)85}$
3. $6\overline{)222}$

4. $8\overline{)640}$
5. $3\overline{)1014}$
6. $4\overline{)504}$

Divide and then check by multiplying. See Examples 6 and 7.

7. $6\overline{)98}$
8. $7\overline{)422}$
9. $2\overline{)1127}$

10. $3\overline{)1240}$
11. $186 \div 5$
12. $167 \div 3$

13. $2121 \div 8$
14. $333 \div 4$

Divide and then check by multiplying. See Examples 8 and 9.

15. $23\overline{)1127}$
16. $42\overline{)2016}$
▣**17.** $55\overline{)715}$
18. $32\overline{)1856}$

19. $97\overline{)9449}$
20. $1938 \div 44$
21. $3708 \div 18$
22. $7224 \div 12$

23. $6578 \div 13$
24. $5670 \div 14$
▣**25.** $9299 \div 46$
26. $2539 \div 64$

27. $\dfrac{10,620}{236}$
28. $\dfrac{5781}{123}$
29. $\dfrac{10,194}{103}$
30. $\dfrac{23,048}{240}$

ANSWERS
1. 12
2. 17
3. 37
4. 80
5. 338
6. 126
7. 16 R 2
8. 60 R 2
9. 563 R 1
10. 413 R 1
11. 37 R 1
12. 55 R 2
13. 265 R 1
14. 83 R 1
15. 49
16. 48
17. 13
18. 58
19. 97 R 40
20. 44 R 2
21. 206
22. 602
23. 506
24. 405
25. 202 R 7
26. 39 R 43
27. 45
28. 47
29. 98 R 100
30. 96 R 8

31. 202 R 15

32. 201

33. 202

34. 303 R 63

35. 58 students

36. 26 students

37. $252,000

38. 165 lb

39. 415 bushels

40. 105 stripes

41. 88 bridges

42. 310 yd

43. Yes, she needs 176 feet; she has 9 feet left over.

64

Name _____

31. 20,619 ÷ 102　　**32.** 40,803 ÷ 203　　**33.** 45,046 ÷ 223　　**34.** 164,592 ÷ 543

🌙 *Solve. See Examples 10 through 12.*

35. Kathy Gomez teaches Spanish lessons for $85 per student for a 5-week session. From one group of students, she collects $4930. Find how many students are in the group.

36. Martin Thieme teaches American Sign Language classes for $55 per student for a 7-week session. He collects $1430 from the group of students. Find how many students are in the group.

37. Twenty-one people pooled their money and bought lottery tickets. One ticket won a prize of $5,292,000. Find how many dollars each person receives.

38. The gravity of Jupiter is 318 times as strong as the gravity of Earth, so objects on Jupiter weigh 318 times as much as they weigh on Earth. If a person would weigh 52,470 pounds on Jupiter, find how much the person weighs on Earth.

39. A truck hauls wheat to a storage granary. It carries a total of 5810 bushels of wheat in 14 trips. How much does the truck haul each trip, if each trip it hauls the same amount?

40. The white stripes dividing the lanes on a highway are 25 feet long and the spaces between them are 25 feet long. Find how many whole stripes there are in 1 mile of highway. (A mile is 5280 feet.)

41. There is a bridge over highway I-35 every three miles. Find how many bridges there are over 265 miles of I-35.

42. An 18-hole golf course is 5580 yards long. If the distance to each hole is the same, find the distance between holes.

43. Wendy Holladay has a piece of rope 185 feet long that she cuts into pieces for an experiment in her second-grade class. Each piece of rope is to be 8 feet long. Determine whether she has enough rope for her 22-student class. Determine the amount extra or the amount short.

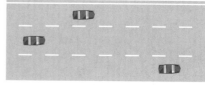

Name _____

44. Jesse White is in the requisitions department of Central Electric Lighting Company. Light poles along a highway are placed 492 feet apart. Find how many poles he should order for a 1-mile strip of highway. (A mile is 5280 feet.)

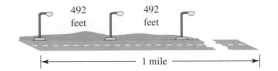

45. Karim Abdul-Jabbar of the Miami Dolphins led the NFL in touchdowns during the 1997 football season. Abdul-Jabbar scored a total of 96 points as touchdowns during the season. If a touchdown is worth 6 points, how many touchdowns did Abdul-Jabbar make during 1997? (*Source: National Football League*)

46. Hershey's main chocolate factory in Hershey, Pennsylvania, uses 700,000 quarts of milk each day. There are 4 quarts in a gallon. How many gallons of milk are used each day? (*Source: Hershey Foods Corp.*)

47. Find how many yards there are in 1 mile. (A mile is 5280 feet; a yard is 3 feet.)

1 foot	1 foot	1 foot	1 foot	1 foot	1 foot	5280 feet ? yards

1 yard   1 yard

48. Find how many whole feet there are in 1 rod. (A mile is 5280 feet; 1 mile is 320 rods.)

D *Find the average of each list of numbers. See Example 13.*

49. 14, 22, 45, 18, 30, 27

50. 38, 26, 15, 29, 29, 51, 22

51. 204, 968, 552, 268

52. 121, 200, 185, 176, 163

53. 86, 79, 81, 69, 80

54. 92, 96, 90, 85, 92, 79

The monthly normal temperature in degrees Fahrenheit for Minneapolis, Minnesota, is given in the table. Use this table to answer Exercises 55 and 56. (Source: National Climatic Data Center)

January	12°	July	74°
February	18°	August	71°
March	31°	Sept.	61°
April	46°	Oct.	49°
May	59°	Nov.	33°
June	68°	Dec.	18°

44. 10 light poles

45. 16 touchdowns

46. 175,000 gallons

47. 1760 yd

48. 16 ft

49. 26

50. 30

51. 498

52. 169

53. 79

54. 89

55. Find the average temperature for December, January, and February.

56. Find the average temperature for the entire year.

COMBINING CONCEPTS

The following table (from previous sections) shows the top ten leading United States advertisers in 1996 and the amount of money spent in that year on ads. Use this table to answer Exercises 57 and 58.

General Motors	$1,712,102,000
Procter & Gamble	$1,493,456,000
Philip Morris	$1,236,978,000
Chrysler	$1,119,851,000
Ford Motor	$897,666,000
Johnson & Johnson	$840,483,000
Walt Disney	$775,621,000
Pepsico	$767,978,000
Time Warner	$747,320,000
AT&T	$659,750,000

57. Find the average amount of money spent on ads for the year by the top four companies.

58. Find the average amount of money spent on ads for the year by Walt Disney, Pepsico, Time Warner, and AT&T.

In Example 13 in this section, we found that the average of 93, 86, 71, and 82 is 83. Use this information to answer Exercises 59 and 60.

59. If the number 71 is removed from the list of numbers, does the average increase or decrease?

60. If the number 93 is removed from the list of numbers, does the average increase or decrease?

61. Without computing it, tell whether the average of 126, 135, 198, 113 is 86. Explain why it is or why it is not.

62. If the area of a rectangle is 30 square feet and its width is 3 feet, what is its length?

```
┌──────────────┐
│  30 square   │  3 feet
│    feet      │
└──────────────┘
      ?
```

Name _____ **Section** _____ **Date** _____

INTEGRATED REVIEW — OPERATIONS ON WHOLE NUMBERS

Perform each indicated operation.

1. 23
46
+79

2. 7006
− 451

3. 36
× 45

4. $8\overline{)4496}$

5. $1 \cdot 79$

6. $\dfrac{36}{0}$

7. $9 \div 1$

8. $9 \div 9$

9. $0 \cdot 13$

10. $7 \cdot 0 \cdot 8$

11. $0 \div 2$

12. $12 \div 4$

13. 4219
−1786

14. 1861
+7965

15. $5\overline{)1068}$

16. 1259
× 63

17. $3 \cdot 9$

18. $45 \div 5$

19. 207
− 69

20. 207
+ 69

21. $32\overline{)21{,}222}$

22. $65\overline{)70{,}000}$

1. 148
2. 6555
3. 1620
4. 562
5. 79
6. undefined
7. 9
8. 1
9. 0
10. 0
11. 0
12. 3
13. 2433
14. 9826
15. 213 R 3
16. 79,317
17. 27
18. 9
19. 138
20. 276
21. 663 R 6
22. 1076 R 60

67

Focus On Mathematical Connections

MODELING SUBTRACTION OF WHOLE NUMBERS

A mathematical concept can be represented or modeled in many different ways. For instance, subtraction can be represented by the following symbolic model.

$$11 - 4$$

The following verbal models can also represent subtraction of these same quantities.

"Four subtracted from eleven"   or   "Eleven take away four"

Physical models can also represent mathematical concepts. In these models, a number is represented by that many objects. For example, the number 5 could be represented by five pennies, squares, paper clips, tiles, or bottle caps.

A physical representation of the number 5

Take Away Model for Subtraction: $11 - 4$

▲ Start with 11 objects.

▲ Take 4 objects away.

▲ How many objects remain?

Comparison Model for Subtraction: $11 - 4$

▲ Start with a set of 11 of one type of object and a set of 4 of another type of object.

▲ Make as many pairs including one object of each type as possible.

▲ How many more objects are in the larger set?

Missing Addend Model for Subtraction: $11 - 4$

▲ Start with 4 objects.

▲ Continue adding objects until a total of 11 is reached:

▲ How many more objects were needed to give a total of 11?

CRITICAL THINKING

Use an appropriate physical model for subtraction to solve each of the following problems. Explain your reasoning for choosing each model.

1. Sneha has assembled 12 computer components so far this shift. If his quota is 20 components, how many more components must he assemble to reach his quota?   8 components

2. Yuko wants to plant 14 daffodil bulbs in her yard. She planted 5 bulbs in the front yard. How many bulbs does she have left for planting in the backyard?   9 bulbs

3. Todd is 19 years old and his sister Tanya is 13 years old. How much older is Todd than Tanya?   6 years

1.7   AN INTRODUCTION TO PROBLEM SOLVING

A   SOLVING PROBLEMS INVOLVING ADDITION, SUBTRACTION, MULTIPLICATION, OR DIVISION

In this section, we will decide which operation to perform in order to solve a problem. Don't forget that key words and phrases help indicate which operation to use. Some of these are listed below and were introduced earlier in the chapter. Also included are several words and phrases that translate to the symbol "=".

Addition (+)	Subtraction (−)	Multiplication (·)	Division (÷)	Equality (=)
sum	difference	product	quotient	equals
plus	minus	times	divide	is equal to
added to	subtract	multiply	shared equally	is/was
more than	less than		among	
increased by	decreased by		divided by	
total	less			

The following problem-solving steps may be helpful to you.

PROBLEM-SOLVING STEPS

1. UNDERSTAND the problem. Some ways of doing this are to read and reread the problem, construct a drawing, look for key words to identify an operation, and estimate a solution.
2. TRANSLATE the problem. That is, write the problem in short form using words, then numbers and symbols.
3. SOLVE the problem. Carry out the indicated operation from Step 2.
4. INTERPRET the results. *Check* the proposed solution in the stated problem and *state* your conclusions. Write your results with correct units attached.

Example 1   Calculating the Length of a River

The Hudson River in New York State is 306 miles long. The Snake River in the northwestern United States is 732 miles longer than the Hudson River. How long is the Snake River? (*Source:* U.S. Department of the Interior)

Solution: 1. UNDERSTAND. Read and reread the problem, and then draw a picture. Notice that we are told that Snake River is 732 miles longer than the Hudson River. The phrase "longer than" means that we add.

306 miles
Hudson River |◄── 732 miles ──►|
Snake River

2. TRANSLATE.

In words:

Snake River	is	732 miles	longer than	the Hudson River
↓	↓	↓	↓	↓

Translate: Snake River =   732   +   306

TEACHING TIP

After going over Example 1, ask students to write a word problem that can be translated as "Answer = 245 + 96."

Practice Problem 1

The Bank of America Building is the second-tallest building in San Francisco, California, at 779 feet. The tallest building in San Francisco is the Transamerica Pyramid, which is 74 feet taller than the Bank of America Building. How tall is the Transamerica Pyramid? (*Source: The 1998 World Almanac*)

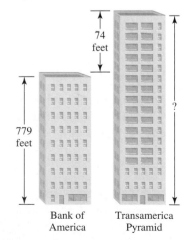

74 feet

779 feet

?

Bank of America   Transamerica Pyramid

Answer
1. 853 ft

3. Solve: 732
           + 306
           ‾‾‾‾‾
           1038

4. INTERPRET. *Check* your work. *State* your conclusion: The Snake River is 1038 miles long. ▬▬

Practice Problem 2

Four friends bought a lottery ticket and won $65,000. If each person is to receive the same amount of money, how much did each person receive?

Example 2   Filling a Shipping Order

How many cases can be filled with 9900 cans of jalapenos if each case holds 48 cans? How many cans will be left over? Will there be enough cases to fill an order for 200 cases?

Solution:

1. UNDERSTAND. Read and reread the problem. Draw a picture to help visualize the situation.

TEACHING TIP

After going over Example 2, ask students to write a word problem that can be translated as "Answer = 294 ÷ 6."

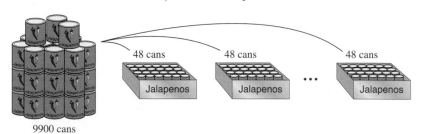

9900 cans

Since each case holds 48 cans, we want to know how many 48s there are in 9900. We find this by dividing.

2. TRANSLATE.

In words:	Number of cases	is	9900	divided by	48
	↓	↓	↓	↓	↓
Translate:	Number of cases	=	9900	÷	48.

3. SOLVE.

```
       206 R 12
   48) 9900
      −96
      ‾‾‾‾
       300
      −288
      ‾‾‾‾
        12
```

4. INTERPRET. *Check* your work. *State* your conclusion: 206 cases will be filled, with 12 cans left over. There will be enough cases to fill an order for 200 cases. ▬▬

Practice Problem 3

The director of the learning lab also needs to include a budget line for 425 blank tapes at a cost of $4 each. What is this total cost?

Example 3   Calculating Budget Costs

The director of a learning lab at a local community college is working on next year's budget. Thirty-three new video players are needed at a cost of $540 each. What is the total cost of these video players?

Solution:

1. UNDERSTAND. Read and reread the problem, and then draw a diagram.

33 Video Players

$540 $540 $540

Answers

2. $16,250,   **3.** $1700

From the phrase "total cost," we might decide to solve this problem by adding. This would work, but repeated addition, or multiplication, would save time.

2. TRANSLATE.

In words:

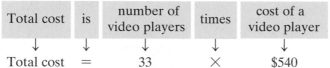

Total cost	is	number of video players	times	cost of a video player
↓	↓	↓	↓	↓

Translate: Total cost   =   33   × $540

3. SOLVE.

$$
\begin{array}{r}
1 \\
540 \\
\times\ \ 33 \\
\hline
1620 \\
1620 \\
\hline
17,820
\end{array}
$$

4. INTERPRET. *Check* your work. *State* your conclusion: The total cost of the video players is $17,820.

Example 4   Calculating a Public School Teacher's Salary

In 1996, the average salary of a public school teacher in California was $43,100. For the same year, the average salary for a public school teacher in Louisiana was $16,300 less than this. What was the average public school teacher's salary in Louisiana? (*Source:* National Education Association)

Solution: **1.** UNDERSTAND. Read and reread the problem. Notice that we are told that the Louisiana salary is $16,300 less than the California salary. The phrase "less than" indicates subtraction.

2. TRANSLATE. Remember that order matters when subtracting, so be careful when translating.

In words:

Louisiana salary	is	California salary	minus	$16,300
↓	↓	↓	↓	↓

Translate: Louisiana salary =   43,100   −   16,300

3. SOLVE:

$$
\begin{array}{r}
43,100 \\
-16,300 \\
\hline
26,800
\end{array}
$$

4. INTERPRET. *Check* your work. *State* your conclusion: The average Louisiana teacher's salary in 1996 was $26,800.

B SOLVING PROBLEMS THAT REQUIRE MORE THAN ONE OPERATION

We must sometimes use more than one operation to solve a problem.

Example 5   Planting a New Garden

A gardener bought enough plants to fill a rectangular garden with length 30 feet and width 20 feet. Because of shading problems with a nearby tree, the gardener

---

Practice Problem 4

In 1996, the average salary of a public school teacher in Idaho was $30.900. For the same year, the average salary for a public school teacher in Mississippi was $3200 less than this. What was the average public school teacher's salary in Mississippi? (*Source:* National Education Association)

TEACHING TIP

After going over Examples 1–4, have your students work in groups of 4 to write an addition, subtraction, multiplication, and division word problem for another group to solve.

Answer
4. $27,700

Practice Problem 5

A gardener is trying to decide how much fertilizer to buy for his yard. He knows that his lot is in the shape of a rectangle that measures 90 feet by 120 feet. He also knows that the floor of his house is in the shape of a rectangle that measures 45 feet by 65 feet. How much area of the lot is not covered by the house?

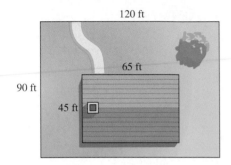

Solution:

changed the width of the garden to 15 feet. If the area is to remain the same, what is the new length of the garden?

1. **UNDERSTAND.** Read and reread the problem. Then draw a picture to help visualize the problem.

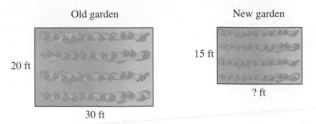

2. **TRANSLATE.** First we find the area of the old garden. Recall from Section 1.5 that area of a rectangle is its length times its width.

In words:

Area	=	length	times	width
↓	↓	↓	↓	↓

Translate: Area   =   30   ×   20

3. **SOLVE.** $30(20) = 600$, so the area is 600 square feet. Since the area of the new garden is to be 600 square feet, we need to see how many 15s there are in 600. This means division.

$$
\begin{array}{r}
40 \\
15\overline{)600} \\
-60 \\
\hline
00
\end{array}
$$

4. **INTERPRET.** *Check* your work. *State* your conclusion: The length of the new garden is 40 feet. ■

Answer

5. 7875 sq. ft

Name _____ **Section** _____ **Date** _____

EXERCISE SET 1.7

A *Solve. See Examples 1 through 4.*

1. 41 increased by 8 is what number?

2. What is the product of 12 and 9?

3. What is the quotient of 1185 and 5?

4. 78 decreased by 12 is what number?

5. What is the total of 35 and 7?

6. What is the difference of 48 and 8?

7. 60 times 10 is what number?

8. 60 divided by 10 is what number?

9. A vacant lot in the shape of a rectangle measures 120 feet by 80 feet. What is the area of the lot?

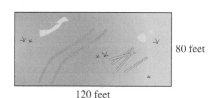

80 feet

120 feet

10. A parking lot in the shape of a rectangle measures 100 feet by 150 feet. What is the area of the parking lot?

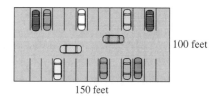

100 feet

150 feet

11. The Henrick family bought a house for $85,700 and later sold the house for $101,200. How much money did they make by selling the house?

12. Three people dream of equally sharing a $147 million lottery. How much would each person receive?

13. There are 24 hours in a day. How many hours are in a week?

14. In 1997, Notre Dame had the most alumni playing in the National Football League with 44 players. They had 8 more players than Penn State. How many players were Penn State alumni? (*Source:* NFL)

ANSWERS

1. 49

2. 108

3. 237

4. 66

5. 42

6. 40

7. 600

8. 6

9. 9600 sq. ft

10. 15,000 sq. ft

11. $15,500

12. $49 million

13. 168 hours

14. 36 players

15. $800

16. 247,800 cubic feet

17. $1,149,000,000

18. 2,109,000 workers

19. $40,992

15. In the game of Monopoly, a player must own all properties in a color group before building houses. The yellow color-group properties are Atlantic Avenue, Ventnor Avenue, and Marvin Gardens. These cost $260, $260 and $280, respectively, when purchased from the bank. What total amount must a player pay to the bank before houses can be built on the yellow properties? (*Source:* Hasbro, Inc.)

16. The Goodyear Tire & Rubber Company maintains a fleet of four blimps. The *Stars and Stripes* can hold 202,700 cubic feet of helium. Its larger sister, the *Spirit of Akron*, can hold 45,100 cubic feet more helium than *Stars and Stripes*. How much helium can *Spirit of Akron* hold? (*Source:* Goodyear Tire & Rubber Company)

17. PepsiCo, Inc., markets beverages (including Pepsi and Mountain Dew) and snack foods (including Fritos and Ruffles). In 1995, PepsiCo's net income was $1,606,000,000. In 1996, PepsiCo's net income had decreased by $457,000,000. What was PepsiCo's net income in 1996? (*Source:* PepsiCo, Inc.)

18. In December 1996, there were 135,060,000 Americans in the civilian labor force. By December 1997, the civilian labor force had grown to 137,169,000 workers. How much bigger was the civilian labor force in December 1997 than in December 1996? (*Source:* U.S. Bureau of Labor Statistics)

19. The average annual cost of tuition at private four-year colleges in the United States was $13,664 during the 1997–1998 academic year. Suppose a family had three children enrolled in private four-year colleges at that time. What total amount would you expect the family to have paid for tuition during the 1997–1998 year? (*Source:* National Center for Education Statistics)

20. In November 1996, the average hourly pay for a production worker in the United States was $12 per hour. How much would a production worker have earned working a 36-hour week at that time? (*Source:* U.S. Bureau of Labor Statistics)

20. $432

21. Federated Department Stores, Inc., operates Macy's, Bloomingdale's, and other department stores across the country. It also operates 25 Stern's stores in New Jersey and New York. In 1997, Stern's had sales of $889,200,000. What is the average amount of sales made by each Stern's store? (*Source:* Federated Department Stores, Inc.)

22. In 1996, the United States Postal Service delivered approximately 900,000,000 packages shipped Parcel Post. The total weight of all Parcel Post packages shipped that year was approximately 2,700,000,000 pounds. What was the average weight of a package shipped Parcel Post during 1996? (*Source:* United States Postal Service)

23. In 1996, the average weekly pay for an aircraft engine mechanic in the United States was $720. If a mechanic works 40 hours in one week, what is his or her hourly pay? (*Source:* U.S. Bureau of Labor Statistics)

24. In 1996, the average weekly pay for a clinical lab technician in the United States was $520. If a technician works 40 hours in one week, what is his or her hourly pay? (*Source:* U.S. Bureau of Labor Statistics)

25. Three ounces of canned tuna in oil has 165 calories. How many calories does 1 ounce have? (*Source: Home and Garden Bulletin No. 72*, U.S. Department of Agriculture)

26. A whole cheesecake has 3360 calories. If the cheesecake is cut into 12 pieces, how many calories will each piece have? (*Source: Home and Garden Bulletin No. 72*, U.S. Department of Agriculture)

27. An average of 14,616 tickets per Major League Soccer (MLS) game were sold in 1997. A total of 160 games were played during the 1997 MLS season. What was the total number of MLS tickets sold during 1997? (*Source:* Major League Soccer)

28. Bridgette Gordon of the WNBA's Sacramento Monarchs scored an average of 13 points per basketball game during the 1997 regular season. A total of 28 games were played during the season. What was the total number of points scored by Bridgette Gordon during 1997? (*Source:* Women's National Basketball Association)

29. The enrollment of all students in elementary and secondary schools in the United States in 2001 is projected to be 54,807,000. Of these students, 15,395,000 are expected to be enrolled in secondary schools. How many students are expected to be enrolled in elementary schools in 2001? (*Source:* National Center for Education Statistics)

30. Kroger is the largest retail food company in the United States. In 1996, Kroger operated a total of 2187 grocery and convenience stores. Of this total, 831 were convenience stores. How many grocery stores did Kroger operate in 1996? (*Source:* The Kroger Company)

31. Marcel Rockett receives a paycheck every four weeks. Find how many paychecks he receives in a year. (A year has 52 weeks.)

32. A loan of $6240 is to be paid in 48 equal payments. How much is each payment?

21. $35,568,000

22. 3 lb

23. $18/hr

24. $13/hr

25. 55 calories

26. 280 calories

27. 2,338,560 tickets

28. 364 points

29. 39,412,000 students

30. 1356 grocery stores

31. 13 paychecks

32. $130

Name _____

B *Solve. See Example 5.*

33. Find the total cost of 3 sweaters at $38 each and 5 shirts at $25 each.

34. Find the total cost of 10 computers at $2100 each and 7 boxes of diskettes at $12 each.

35. A college student has $950 in an account. She spends $205 from the account on books and then deposits $300 in the account. How much money is now in the account?

36. The Meish's yard is in the shape of a rectangle and measures 50 feet by 75 feet. In their yard, they have a rectangular swimming pool that measures 15 feet by 25 feet. How much of their yard is not part of the swimming pool?

The table shows the menu from Corky's, a concession stand at the county fair. Use this menu to answer Exercises 37–39.

37. A couple orders the following items: 1 hot dog, 1 hamburger, 1 order of onion rings, and 2 sodas. How much will their order cost?

CORKY'S CONCESSION STAND MENU	
Item	Price
Hot dog	$3
Hamburger	$4
Soda	$1
Onion rings	$3
French fries	$2
Apple	$1
Candy bar	$2

38. A hungry college student is debating between the following two orders:
a. a hamburger, an order of onion rings, a candy bar, and a soda.

b. a hot dog, an apple, an order of french fries, and a soda.
Which order will be cheaper? By how much?

39. A family of four is debating between the following two orders:
a. 6 hot dogs, 4 orders of onion rings, and 4 sodas.

b. 4 hamburgers, 4 orders of french fries, 2 apples, and 4 sodas.
Will the family save any money by ordering (b) instead of (a)? If so, how much?

76

40. 24,000 sq. ft

COMBINING CONCEPTS

40. The United States Postal Service owns 6947 buildings with a total area of 168,000,000 square feet. Round the number of buildings to the nearest thousand to estimate the average area of each building. (*Source:* United States Postal Service)

41. In 1997, there were 284 Fairfield Inn properties in the United States with a total of 27,300 guestrooms. Round the number of properties to the nearest hundred to estimate the average number of guestrooms per hotel. (*Source:* Marriott International, Inc.)

41. 91 guest rooms

42. IBM

Use the table to answer Exercises 42–48.

42. Which company received the most patents in 1996?

SELECTED CORPORATIONS RECEIVING U.S. PATENTS IN 1996		
Company	Country	Number of Patents
AT&T	U.S.	510
Canon	Japan	1541
Eastman Kodak	U.S.	768
General Electric	U.S.	819
Hitachi	Japan	963
IBM	U.S.	1867
Minnesota Mining & Manufacturing	U.S.	537
Mitsubishi	Japan	934
Motorola	U.S.	1064
NEC Corp.	Japan	1043
Sony	Japan	855
Texas Instruments	U.S.	600
Toshiba	Japan	914
Xerox	U.S.	703

(*Source:* U.S. Patent and Trademark Office)

43. Which company received the least patents in 1996?

43. AT&T

44. How many more patents did the company with the most patents receive than the company with the least patents?

44. 1357 patents

45. How many more patents did Motorola receive than Eastman Kodak?

45. 296 patents

46. 838 patents

46. How many more patents did Canon receive than Xerox?

47. Which company received more patents: Toshiba or Minnesota Mining & Manufacturing? How many more patents did it receive?

47. Toshiba by 377 patents

48. Did the U.S. or Japanese companies listed receive more patents overall? How many more patents did they receive?

48. U.S. by 618 patents

Name _____

Internet Excursions

Go to http://www.prenhall.com/martin-gay
This World Wide Web site provides access to a site that allows the user to find the distance, as the crow flies, between two cities. It also gives population and elevation figures for each city. Visit this site and answer the following questions.

49. Choose three cities in your state: (A) _____ , (B) _____ , and (C) _____ .

50. Find the distance between city A and city B. Then find the distance from city B to city C. If a crow were to fly directly from city A to city B and then continue on to city C, how far would it fly?

51. List the population for each city. Which city has the greatest population? How many more people does that city have than the city with the least population?

1.8 EXPONENTS AND ORDER OF OPERATIONS

A USING EXPONENTIAL NOTATION

An **exponent** is a shorthand notation for repeated multiplication. When the same number is a factor a number of times, an exponent may be used. In the product

$$\underbrace{2 \cdot 2 \cdot 2 \cdot 2 \cdot 2}_{\text{2 is a factor 5 times.}}$$

Using an exponent, this product can be written as

$$\overset{\text{exponent}}{\underset{\text{base}}{2^5}} \qquad \text{Read as "two to the fifth power."}$$

Thus,

$$2 \cdot 2 \cdot 2 \cdot 2 \cdot 2 = 2^5$$

This is called **exponential notation**. The **exponent**, 5, indicates how many times the **base**, 2, is a factor.

Certain expressions are used when reading exponential notation.

$5 = 5^1$ is read as "five to the first power."

$5 \cdot 5 = 5^2$ is read as "five to the second power" or "five squared."

$5 \cdot 5 \cdot 5 = 5^3$ is read as "five to the third power" or "five cubed."

$5 \cdot 5 \cdot 5 \cdot 5 = 5^4$ is read as "five to the fourth power."

Usually, an exponent of 1 is not written, so when no exponent appears, we assume that the exponent is 1. For example, $2 = 2^1$ and $7 = 7^1$.

Examples   Write using exponential notation.

1. $4 \cdot 4 \cdot 4 = 4^3$
2. $7 \cdot 7 = 7^2$
3. $5 \cdot 5 \cdot 5 \cdot 5 = 5^4$
4. $6 \cdot 6 \cdot 6 \cdot 8 \cdot 8 \cdot 8 \cdot 8 \cdot 8 = 6^3 \cdot 8^5$ ▬▬▬

B EVALUATING EXPONENTIAL EXPRESSIONS

To **evaluate** an exponential expression, we write the expression as a product and then find the value of the product.

Examples   Evaluate.

5. $8^2 = 8 \cdot 8 = 64$
6. $7^1 = 7$
7. $2^5 = 2 \cdot 2 \cdot 2 \cdot 2 \cdot 2 = 32$
8. $5 \cdot 6^2 = 5 \cdot 6 \cdot 6 = 180$ ▬▬▬

Example 8 illustrates an important property: An exponent applies only to its base. The exponent 2, in $5 \cdot 6^2$ applies only to its base, 6.

┌───
HELPFUL HINT

An exponent applies only to its base. For example, $4 \cdot 2^3$ means $4 \cdot 2 \cdot 2 \cdot 2$.
└───

Objectives

A Write repeated factors using exponential notation.

B Evaluate expressions containing exponents.

C Use the order of operations.

D Find the area of a square.

SSM   CD-ROM   Video 1.8

TEACHING TIP

Using the multiplication table in the Appendix, consider having students highlight the squared numbers which fall on the diagonal from the upper left corner to the lower right corner. Encourage them to become familiar with these squared numbers. Ask if they notice any pattern in the numbers. Be sure that they see that the difference between consecutive squared numbers is the sequence of odd numbers beginning with 3.

Practice Problems 1–4

Write using exponential notation.

1. $2 \cdot 2 \cdot 2$

2. $3 \cdot 3$

3. $10 \cdot 10 \cdot 10 \cdot 10 \cdot 10 \cdot 10$

4. $5 \cdot 5 \cdot 4 \cdot 4 \cdot 4$

Practice Problems 5–8

Evaluate.

5. 2^3

6. 5^2

7. 10^1

8. $4 \cdot 5^2$

Answers

1. 2^3, **2.** 3^2, **3.** 10^6, **4.** $5^2 \cdot 4^3$, **5.** 8, **6.** 25, **7.** 10, **8.** 100

> **HELPFUL HINT**
>
> Don't forget that 2^4, for example, is *not* $2 \cdot 4$. 2^4 means repeated multiplication of the same factor.
>
> $$2^4 = 2 \cdot 2 \cdot 2 \cdot 2 = 16 \quad \text{whereas} \quad 2 \cdot 4 = 8$$

✓ CONCEPT CHECK

Which of the following statements is correct?

a. 3^6 is the same as $6 \cdot 6 \cdot 6$.

b. "Eight to the fourth power" is the same as 8^4.

c. "Ten squared" is the same as 10^3.

d. 11^2 is the same as $11 \cdot 2$.

TRY THE CONCEPT CHECK IN THE MARGIN.

C   USING THE ORDER OF OPERATIONS

Suppose that you are in charge of taking inventory at a local bookstore. An employee has given you the number of a certain book in stock as the expression

$$3 + 2 \cdot 10$$

To calculate the value of this expression, do you add first or multiply first? If you add first, the answer is 50. If you multiply first, the answer is 23.

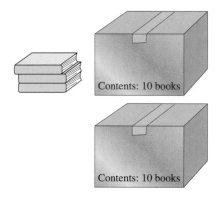

Contents: 10 books

Contents: 10 books

Mathematical symbols wouldn't be very useful if two values were possible for one expression like this. Thus, mathematicians have agreed that, given a choice, we multiply first.

$$3 + 2 \cdot 10 = 3 + 20 \quad \text{Multiply.}$$
$$= 23 \quad \text{Add.}$$

This agreement is one of several **order of operations** agreements.

> **ORDER OF OPERATIONS**
>
> 1. Do all operations within grouping symbols such as parentheses or brackets.
> 2. Evaluate any expressions with exponents.
> 3. Multiply or divide in order from left to right.
> 4. Add or subtract in order from left to right.

For example, using order of operations, let's evaluate $2^3 \cdot 4 - (10 \div 5)$.

$$2^3 \cdot 4 - (10 \div 5) = 2^3 \cdot 4 - 2 \quad \text{Simplify inside parentheses.}$$
$$= 8 \cdot 4 - 2 \quad \text{Write } 2^3 \text{ as 8.}$$
$$= 32 - 2 \quad \text{Multiply } 8 \cdot 4.$$
$$= 30 \quad \text{Subtract.}$$

Answer

✓ Concept Check: b

Example 9   Simplify: $2 \cdot 4 - 3 \div 3$

Solution: There are no parentheses and no exponents, so we start by multiplying and dividing, from left to right.

$$2 \cdot 4 - 3 \div 3 = 8 - 3 \div 3 \qquad \text{Multiply.}$$
$$= 8 - 1 \qquad \text{Divide.}$$
$$= 7 \qquad \text{Subtract.} \qquad \blacksquare$$

Example 10   Simplify: $(8 - 6)^2 + 2^3 \cdot 3$

Solution: $(8 - 6)^2 + 2^3 \cdot 3 = 2^2 + 2^3 \cdot 3 \qquad$ Simplify inside parentheses.

$$= 4 + 8 \cdot 3 \qquad \text{Write } 2^2 \text{ as 4 and } 2^3 \text{ as 8.}$$
$$= 4 + 24 \qquad \text{Multiply.}$$
$$= 28 \qquad \text{Add.} \qquad \blacksquare$$

Example 11   Simplify: $4^3 + [3^2 - (10 \div 2)] - 7 \cdot 3$

Solution: Here we begin within the innermost set of parentheses.

$$4^3 + [3^2 - (10 \div 2)] - 7 \cdot 3 = 4^3 + [3^2 - 5] - 7 \cdot 3 \qquad \begin{matrix}\text{Simplify inside} \\ \text{parentheses.} \\ \text{Write } 3^2 \text{ as 9.}\end{matrix}$$
$$= 4^3 + [9 - 5] - 7 \cdot 3$$

$$= 4^3 + 4 - 7 \cdot 3 \qquad \begin{matrix}\text{Simplify inside} \\ \text{brackets.} \\ \text{Write } 4^3 \text{ as 64.}\end{matrix}$$
$$= 64 + 4 - 7 \cdot 3$$

$$= 64 + 4 - 21 \qquad \text{Multiply.}$$
$$= 47 \qquad \begin{matrix}\text{Add and sub-} \\ \text{tract from left to} \\ \text{right.}\end{matrix} \qquad \blacksquare$$

Example 12   Simplify: $\dfrac{7 - 2 \cdot 3 + 3^2}{2^2 + 1}$

Solution: Here, the fraction bar is like a grouping symbol. We simplify above and below the fraction bar separately.

$$\frac{7 - 2 \cdot 3 + 3^2}{2^2 + 1} = \frac{7 - 2 \cdot 3 + 9}{4 + 1} \qquad \text{Evaluate } 3^2 \text{ and } 2^2.$$

$$= \frac{7 - 6 + 9}{5} \qquad \begin{matrix}\text{Multiply } 2 \cdot 3 \text{ in the numerator and add 4 and} \\ \text{1 in the denominator.}\end{matrix}$$

$$= \frac{10}{5} \qquad \text{Add and subtract from left to right.}$$

$$= 2 \qquad \text{Divide.} \qquad \blacksquare$$

D FINDING THE AREA OF A SQUARE

Since a square is a special rectangle, we can find its area by finding the product of its length and its width.

Area of a rectangle = length · width

By recalling that each side of a square has the same measurement, we can use the following to find its area.

Practice Problem 9

Simplify: $16 \div 4 - 2$

Practice Problem 10

Simplify: $(9 - 8)^3 + 3 \cdot 2^4$

Practice Problem 11

Simplify:
$24 \div [20 - (3 \cdot 4)] + 2^3 - 5$

TEACHING TIP

If your students are working in groups, ask each group to make a list in 10 minutes of expressions which equal the numbers from 1 to 100 using the basic operations (add, subtract, multiply, divide, and raising to exponents) and a digit of their choice. Their digit must be used four times in each expression. For example, if their digit is 3:

$$1 = \frac{(3 + 3)}{(3 + 3)}$$
$$2 = \frac{3}{3} + \frac{3}{3}$$
$$\vdots$$
$$28 = 3^3 + \frac{3}{3}$$

Practice Problem 12

Simplify: $\dfrac{60 - 5^2 + 1}{3(1 + 1)}$

$$\text{Area of a square} = \text{length} \cdot \text{width}$$
$$= \text{side} \cdot \text{side}$$
$$= (\text{side})^2$$

Square

Side

Side

Practice Problem 13

Find the area of the square whose side measures 11 centimeters.

11 centimeters

Example 13   Find the area of the square whose side measures 5 inches.

Solution: $\text{Area of a square} = (\text{side})^2$
$$= (5 \text{ inches})^2$$
$$= 25 \text{ square inches}$$

5 inches

The area of the square is 25 square inches.

CALCULATOR EXPLORATION

EXPONENTS

To evaluate an exponent such as 4^7 on a calculator, find the key marked $\boxed{y^x}$ and $\boxed{=}$. To evaluate 4^7, press

or

$\boxed{\text{ENTER}}$

the keys $\boxed{4}$ $\boxed{y^x}$ $\boxed{7}$ $\boxed{=}$. The display will read $\boxed{16384}$. Thus, $4^7 = 16{,}384$.

or

$\boxed{\text{ENTER}}$

Use a calculator to evaluate.

1. 3^6 729   **2.** 5^6 15,625
3. 4^5 1024   **4.** 7^6 117,649
5. 2^{11} 2048   **6.** 6^8 1,679,616

ORDER OF OPERATIONS

To see whether your calculator has order of operations built in, evaluate $5 + 2 \cdot 3$ by pressing the keys

$\boxed{5}$ $\boxed{+}$ $\boxed{2}$ $\boxed{\times}$ $\boxed{3}$ $\boxed{=}$. If the display reads 11, your calculator does have order of operations built in. This

or

$\boxed{\text{ENTER}}$

means that most of the time you can key in a problem exactly as it is written and the calculator will perform operations in the proper order. When evaluating an expression containing parentheses, key in the parentheses. (If an expression contains brackets, key in parentheses.) For example, to evaluate $2[25 - (8 + 4)] - 11$,

press the keys $\boxed{2}$ $\boxed{\times}$ $\boxed{(}$ $\boxed{25}$ $\boxed{-}$ $\boxed{(}$ $\boxed{8}$ $\boxed{+}$ $\boxed{4}$ $\boxed{)}$ $\boxed{)}$ $\boxed{-}$ $\boxed{11}$ $\boxed{=}$

The display will read $\boxed{15}$.   or

$\boxed{\text{ENTER}}$

Use a calculator to evaluate.

7. $7^4 + 5^3$ 2526   **8.** $12^4 - 8^4$ 16,640
9. $63 \cdot 75 - 43 \cdot 10$ 4295   **10.** $8 \cdot 22 + 7 \cdot 16$ 288
11. $4(15 \div 3 + 2) - 10 \cdot 2$ 8   **12.** $155 - 2(17 + 3) + 185$ 300

Answer

13. 121 square centimeters

EXERCISE SET 1.8

A *Write using exponential notation. See Examples 1 through 4.*

1. $3 \cdot 3 \cdot 3 \cdot 3$ **2.** $5 \cdot 5 \cdot 5$

3. $7 \cdot 7 \cdot 7 \cdot 7 \cdot 7 \cdot 7 \cdot 7 \cdot 7$ **4.** $6 \cdot 6 \cdot 6 \cdot 6 \cdot 6$

5. $12 \cdot 12 \cdot 12$ **6.** $10 \cdot 10$

7. $6 \cdot 6 \cdot 5 \cdot 5 \cdot 5$ **8.** $4 \cdot 4 \cdot 4 \cdot 3 \cdot 3$

9. $9 \cdot 9 \cdot 9 \cdot 8$ **10.** $7 \cdot 7 \cdot 7 \cdot 4$

11. $3 \cdot 2 \cdot 2 \cdot 2 \cdot 2 \cdot 2$ **12.** $4 \cdot 6 \cdot 6 \cdot 6 \cdot 6$

13. $3 \cdot 2 \cdot 2 \cdot 5 \cdot 5 \cdot 5$ **14.** $6 \cdot 6 \cdot 2 \cdot 9 \cdot 9 \cdot 9 \cdot 9$

B *Evaluate. See Examples 5 through 8.*

15. 5^2 **16.** 6^2 **17.** 5^3 **18.** 6^3

19. 2^6 **20.** 2^7 **21.** 2^{10} **22.** 1^{12}

23. 7^1 **24.** 8^1 **25.** 3^5 **26.** 5^4

27. 2^8 **28.** 3^3 **29.** 4^3 **30.** 4^4

31. 9^2 **32.** 8^2 **33.** 9^3 **34.** 8^3

35.	100
36.	1000
37.	10,000
38.	100,000
39.	10
40.	14
41.	1920
42.	6849
43.	729
44.	1024
45.	21
46.	42
47.	8
48.	5
49.	29
50.	22
51.	4
52.	0
53.	17
54.	36
55.	46
56.	21
57.	28
58.	42
59.	10
60.	9
61.	7
62.	5
63.	4
64.	6
65.	14
66.	7
67.	72
68.	5
69.	2
70.	2
71.	35
72.	60
73.	4
74.	3
75.	undefined
76.	undefined

84

Name _____

35. 10^2 **36.** 10^3 **37.** 10^4 **38.** 10^5

39. 10^1 **40.** 14^1 **41.** 1920^1 **42.** 6849^1

43. 3^6 **44.** 4^5

C *Simplify. See Examples 9 through 12.*

45. $15 + 3 \cdot 2$ **46.** $24 + 6 \cdot 3$ **47.** $20 - 4 \cdot 3$ **48.** $17 - 2 \cdot 6$

49. $5 \cdot 9 - 16$ **50.** $8 \cdot 4 - 10$ **51.** $28 \div 4 - 3$ **52.** $42 \div 7 - 6$

53. $14 + \dfrac{24}{8}$ **54.** $32 + \dfrac{8}{2}$ **55.** $6 \cdot 5 + 8 \cdot 2$ **56.** $3 \cdot 4 + 9 \cdot 1$

57. $0 \div 6 + 4 \cdot 7$ **58.** $0 \div 8 + 7 \cdot 6$ **59.** $6 + 8 \div 2$ **60.** $6 + 9 \div 3$

61. $(6 + 8) \div 2$ **62.** $(6 + 9) \div 3$ **63.** $(6^2 - 4) \div 8$ **64.** $(7^2 - 7) \div 7$

65. $(3 + 5^2) \div 2$ **66.** $(13 + 6^2) \div 7$ **67.** $6^2 \cdot (10 - 8)$ **68.** $5^3 \div (10 + 15)$

69. $\dfrac{18 + 6}{2^4 - 4}$ **70.** $\dfrac{15 + 17}{5^2 - 3^2}$ **71.** $(2 + 5) \cdot (8 - 3)$

72. $(9 - 7) \cdot (12 + 18)$ **73.** $\dfrac{7(9 - 6) + 3}{3^2 - 3}$ **74.** $\dfrac{5(12 - 7) - 4}{5^2 - 2^3 - 10}$

75. $5 \div 0 + 24$ **76.** $18 - 7 \div 0$

Name _____

77. $3^4 - [35 - (12 - 6)]$

78. $[40 - (8 - 2)] - 2^5$

**79.** $(7 \cdot 5) + [9 \div (3 \div 3)]$

80. $(18 \div 6) + [(3 + 5) \cdot 2]$

81. $8 \cdot [4 + (6 - 1) \cdot 2] - 50 \cdot 2$

82. $35 \div [3^2 + (9 - 7) - 2^2] + 10 \cdot 3$

83. $7^2 - \{18 - [40 \div (4 \cdot 2) + 2] + 5^2\}$

84. $29 - \{5 + 3[8 \cdot (10 - 8)] - 50\}$

D *Find the area of each square. See Example 13.*

85.

20 miles

86.

4 meters

87.

8 centimeters

88.

31 feet

89. The Eiffel Tower stands on a square base measuring 100 meters on each side. Find the area of the base.

90. A square lawn that measures 72 feet on each side is to be fertilized. If 5 bags of fertilizer are available and each bag can fertilize 1000 square feet, is there enough fertilizer to cover the lawn?

78. 2

79. 44

80. 19

81. 12

82. 35

83. 13

84. 26

85. 400 square miles

86. 16 square meters

87. 64 square centimeters

88. 961 square feet

89. 10,000 square meters

90. no

91. $(2 + 3) \cdot 6 - 2$

92. $(2 + 3) \cdot (6 - 2)$

93. $\begin{array}{c}24 \div \\ (3 \cdot 2) + 2 \cdot 5\end{array}$

94. $\begin{array}{c}24 \div \\ (3 \cdot 2 + 2) \cdot 5\end{array}$

95. 1260 feet

96. 781,250

97. 6,384,814

Name _____

Copyright 1999 Prentice-Hall, Inc.

◆ **COMBINING CONCEPTS**

Insert grouping symbols (parentheses) so that each given expression evaluates to the given number.

91. $2 + 3 \cdot 6 - 2$; Evaluate to 28

92. $2 + 3 \cdot 6 - 2$; Evaluate to 20

93. $24 \div 3 \cdot 2 + 2 \cdot 5$; Evaluate to 14

94. $24 \div 3 \cdot 2 + 2 \cdot 5$; Evaluate to 15

95. A building contractor is bidding on a contract to install gutters on seven homes in a retirement community all in the shape shown. To estimate his cost of materials, he needs to know the total perimeter of all seven homes. Find the total perimeter.

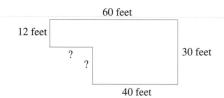

Simplify.

🖩 **96.** $25^3 \cdot (45 - 7 \cdot 5) \cdot 5$

🖩 **97.** $(7 + 2^4)^5 - (3^5 - 2^4)^2$

98. Explain why $2 \cdot 3^2$ is not the same as $(2 \cdot 3)^2$.

CHAPTER 1 ACTIVITY
INVESTIGATING ENDANGERED AND THREATENED SPECIES

This activity may be completed by working in groups or individually.

An **endangered** species is one that is thought to be in danger of becoming extinct throughout all or a major part of its range. A **threatened** species is one that is thought to be in danger of becoming endangered. The Division of Endangered Species at the U.S. Fish and Wildlife Service keeps close tabs on the state of threatened and endangered wildlife in the United States and around the world. The table below was compiled on February 28, 1998. The "Total Species" column gives the total number of endangered and threatened species for each group.

1. Round each number of *endangered animal* species to the nearest ten to estimate the animal total.   880 endangered animal species

2. Round each number of *endangered plant* species to the nearest ten to estimate the plant total.
560 endangered plant species

3. Find the exact numbers of endangered animal and plant species, and record these numbers in the Endangered Species column in the table as

"Animal Subtotal" and "Plant Subtotal," respectively. Then find the total number of endangered species (animals and plants combined) and record this number in the table as the "Grand Total" of the Endangered Species column.

4. Find the Animal Subtotal, Plant Subtotal, and Grand Total for the Total Species column. Record these values in the table.

5. Use the data in the table to complete the Threatened Species column.

6. Write a paragraph discussing the conclusions that can be drawn from the table.   answers may vary

7. (Optional) Visit the current U.S. Fish and Wildlife Service box score of endangered and threatened species at http://www.fws.gov/~r9endspp/boxscore.html on the World Wide Web. How have the figures changed since February 28, 1998? Write a paragraph summarizing the changes in the numbers of endangered and threatened animals and plants.   answers may vary

ENDANGERED AND THREATENED SPECIES WORLDWIDE				
	Group	Endangered Species	Threatened Species	Total Species
ANIMALS	Mammals	309	22	331
	Birds	253	21	274
	Reptiles	80	34	114
	Amphibians	17	8	25
	Fishes	78	41	119
	Snails	16	7	23
	Clams	58	6	64
	Crustaceans	16	3	19
	Insects	32	9	41
	Arachnids	5	0	5
	Animal Total	864	151	1015
PLANTS	Flowering plants	526	113	639
	Conifers	2	2	4
	Ferns and other	26	2	28
	Plant Total	554	117	671
	Grand Total	1418	268	1686

(*Source:* U.S. Fish and Wildlife Service, Division of Endangered Species)

CHAPTER 1 HIGHLIGHTS

DEFINITIONS AND CONCEPTS	EXAMPLES

SECTION 1.1   PLACE VALUE AND NAMES FOR NUMBERS

The **whole numbers** are 0, 1, 2, 3, 4, 5, . . .
The position of each digit in a number determines its **place value**. A place-value chart is shown next with the names of the periods shown.

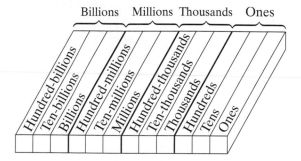

0,   14,   968,   5,268,619

To write a whole number in words, write the number in the period followed by the name of the period. The name of the ones period is not included.

9,078,651,002 is written as nine billion, seventy-eight million, six hundred fifty-one thousand, two.

SECTION 1.2   ADDING WHOLE NUMBERS

To add whole numbers, add the digits in the ones place, then the tens place, then the hundreds place, and so on, carrying when necessary.

Find the sum:

```
  2 1 1
  2689   ←  addend
  1735   ←  addend
+  662   ←  addend
  5086   ←  sum
```

The **perimeter** of a polygon is its distance around or the sum of the lengths of its sides.

Find the perimeter of the polygon shown.

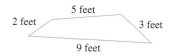

The perimeter is 5 feet + 3 feet + 9 feet + 2 feet = 19 feet.

SECTION 1.3   SUBTRACTING WHOLE NUMBERS

To subtract whole numbers, subtract the digits in the ones place, then the tens place, then the hundreds place, and so on, borrowing when necessary.

Subtract:

```
  8 15
  7954   ←  minuend
 −5673   ←  subtrahend
  2281   ←  difference
```

SECTION 1.4 ROUNDING AND ESTIMATING

ROUNDING WHOLE NUMBERS TO A GIVEN PLACE VALUE

Step 1. Locate the digit to the right of the given place value.

Step 2. If this digit is 5 or greater, add 1 to the digit in the given place value and replace each digit to its right with 0.

Step 3. If this digit is less than 5, replace it and each digit to its right with 0.

Round 15,721 to the nearest thousand.

15,⑦21   Since the circled digit is 5 or
Add 1 ⌐ Replace   greater, add 1 to the given
         with   place value and replace digits
         zeros   to its right with zeros.

15,721 rounded to the nearest thousand is 16,000.

SECTION 1.5 MULTIPLYING WHOLE NUMBERS

To multiply, for example, 73 and 58, multiply 73 and 8, then 73 and 50. The sum of these partial products is the product of 73 and 58. Use the notation to the right.

To find the **area** of a rectangle, multiply length and width.

$$
\begin{array}{r}
73 \\
\times\ 58 \\
\hline
584 \\
3650 \\
\hline
4234
\end{array}
$$

← factor
← factor
← 73 × 8
← 73 × 50
← product

Find the area of the rectangle shown.

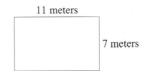

11 meters

7 meters

area of a rectangle = length · width
= (11 meters)(7 meters)
= 77 square meters

SECTION 1.6 DIVIDING WHOLE NUMBERS

To divide larger whole numbers, use the process called **long division** as shown to the right.

$$
\begin{array}{r}
507\ \text{R}\ 2 \\
14\overline{)7100} \\
-70 \\
\hline
10 \\
-\ 0 \\
\hline
100 \\
98 \\
\hline
2
\end{array}
$$

← quotient
divisor → ← dividend
5(14) = 70
Subtract and bring down the 0
0(14) = 0
Subtract and bring down the 0
7(14) = 98
Subtract. The remainder is 2.

To check, see that 507 · 14 + 2 = 7100

The **average** of a list of numbers is

$$\text{average} = \frac{\text{sum of numbers}}{\text{number of numbers}}$$

Find the average of 23, 35, and 38.

$$\text{average} = \frac{23 + 35 + 38}{3} = \frac{96}{3} = 32$$

SECTION 1.7   AN INTRODUCTION TO PROBLEM SOLVING

PROBLEM-SOLVING STEPS

1. UNDERSTAND the problem.

2. TRANSLATE the problem.

3. SOLVE the problem.

4. INTERPRET the results.

Suppose that 225 tickets are sold for each performance of a play. How many tickets are sold for 5 performances?

1. UNDERSTAND. Read and reread the problem. Since we want the number of tickets for 5 performances, we multiply.

2. TRANSLATE

number of tickets	is	number of performances	times	tickets per performance
↓	↓	↓	↓	↓

$$\text{Number of tickets} = 5 \cdot 225$$

3. SOLVE
$$\begin{array}{r} {\scriptstyle 12} \\ 225 \\ \times\ \ 5 \\ \hline 1125 \end{array}$$

4. INTERPRET. **Check** your work and **state** your conclusions. There are 1125 tickets sold for 5 performances.

SECTION 1.8   EXPONENTS AND ORDER OF OPERATIONS

An **exponent** is a shorthand notation for repeated multiplication of the same factor.

ORDER OF OPERATIONS

Simplify expressions using the following order. If grouping symbols such as parentheses ( ) or brackets [ ] are present, simplify expressions within those first, starting with the innermost set. If fraction bars are present, simplify above and below the fraction bar separately.

1. Simplify any expressions with exponents.

2. Perform multiplications or divisions in order from left to right.

3. Perform additions or subtractions in order from left to right.

The **area of a square** is $(\text{side})^2$.

$$3^4 = \underbrace{3 \cdot 3 \cdot 3 \cdot 3}_{\substack{4 \text{ factors} \\ \text{of } 3}} = 81$$

base ↑   exponent ↗

Simplify: $\dfrac{5 + 3^2}{2(7 - 6)}$

Simplify above and below the fraction bar separately.

$$\frac{5 + 3^2}{2(7 - 6)} = \frac{5 + 9}{2(1)} \qquad \text{Evaluate } 3^2.$$
$$\text{Subtract: } 7 - 6.$$
$$= \frac{14}{2} \qquad \text{Add.}$$
$$\text{Multiply.}$$
$$= 7 \qquad \text{Divide.}$$

Find the area of the square shown.

9 inches

$$\begin{aligned} \text{Area of the square} &= (\text{side})^2 \\ &= (9 \text{ inches})^2 \\ &= 81 \text{ square inches} \end{aligned}$$

CHAPTER 1 REVIEW

(1.1) *Determine the place value of the digit 4 in each whole number.*

1. 5480   hundreds

2. 46,200,120   ten-millions

Write each whole number in words.

3. 5480   five thousand, four hundred eighty

4. 46,200,120   forty-six million, two hundred thousand, one hundred twenty

Write each whole number in expanded form.

5. 6279   6000 + 200 + 70 + 9

6. 403,225,000   400,000,000 + 3,000,000 + 200,000 + 20,000 + 5000

Write each whole number in standard form.

7. Fifty-nine thousand eight hundred   59,800

8. Six billion, three hundred four million   6,304,000,000

The following table shows populations of the ten largest cities in the United States. Use this table to answer Exercises 9–12.

Rank	City	1990	1980	1970
1	New York, NY	7,322,564	7,071,639	7,895,563
2	Los Angeles, CA	3,485,557	2,968,528	2,811,801
3	Chicago, IL	2,783,726	3,005,072	3,369,357
4	Houston, TX	1,629,902	1,595,138	1,233,535
5	Philadelphia, PA	1,585,577	1,688,210	1,949,996
6	San Diego, CA	1,110,554	875,538	697,471
7	Detroit, MI	1,027,974	1,203,368	1,514,063
8	Dallas, TX	1,007,618	904,599	844,401
9	Phoenix, AZ	983,403	789,704	584,303
10	San Antonio, TX	935,393	785,940	654,153

(*Source:* The World Almanac, 1998.)

9. Find the population of Houston, Texas, in 1980.   1,595,138

10. Find the population of Los Angeles, California, in 1970.   2,811,801

11. Find the increase in population for Phoenix, Arizona, from 1980 to 1990.   193,699

12. Find the decrease in population for Detroit, Michigan, from 1980 to 1990.   175,394

(1.2) *Add.*

13. 7 + 6   13

14. 8 + 9   17

15. 3 + 0   3

16. 0 + 10   10

17. 25 + 8 + 5   38

18. 27 + 41   68

19. 32 + 24   56

20. 19 + 21   40

21. 47 + 63   110

22. 77 + 43   120

23. 567 + 383   950

24. 463 + 787   1250

25. 591 + 623 + 497   1711

26. 5982 + 1647 + 2238   9867

27. Sean Cruise earned salaries of $62,589, $65,340, and $69,770 during the years of 1987, 1988, and 1989, respectively. Find his total earnings during those three years. $197,699

28. The distance from Chicago to New York City is 714 miles. The distance from New York City to New Delhi, India, is 7,318 miles. Find the total distance from Chicago to New Delhi if traveling through New York City.   8032 miles

Find the perimeter of each figure.

29.

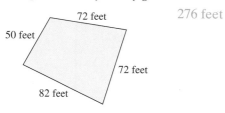

72 feet

50 feet

72 feet

82 feet

276 feet

30.

11 kilometers   20 kilometers

35 kilometers

66 kilometers

(1.3) *Subtract and then check.*

31. 42 − 9   33

32. 67 − 24   43

33. 93 − 79   14

34. 60 − 27   33

35. 599 − 237   362

36. 462 − 397   65

37. 583 − 279   304

38. 600 − 124   476

39. 4000 − 1886   2114

40. 4268 − 3947   321

41. Shelly Winters bought a new car listed at $18,425. She received a discount of $1599 and a factory rebate of $1200. Find how much she paid for the car. $15,626

42. Bob Roma is proofreading the Yellow Pages for his county. If he has finished 315 pages of the total 712 pages, how many pages does he have left to proof-read?   397 pages

The following bar graph shows the monthly savings account balance for a freshman attending a local community college.
Use this graph to answer Exercises 47–50.

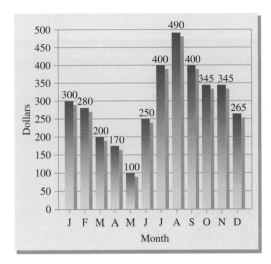

43. During what month was the balance the least?
May

44. During what month was the balance the greatest?
August

45. For what months was the balance greater than
$350?   July, August, September

46. For what months was the balance less than $200?
April, May

(1.4) _Round to the given place._

47. 93 to the nearest ten   90

48. 45 to the nearest ten   50

49. 467 to the nearest ten   470

50. 493 to the nearest hundred   500

51. 4832 to the nearest hundred   4800

52. 57,534 to the nearest thousand   58,000

53. 49,683,712 to the nearest million   50,000,000

54. 768,542 to the nearest hundred thousand   800,000

Estimate the sum or difference by rounding each number to the nearest hundred.

55. 4892 + 647 + 1876   7400

56. 5925 − 1787   4100

57. In 1995, there were 10,160 U.S. commercial radio stations. Round this number to the nearest hundred. (_Source:_ M Street Corporation, New York, NY)
10,200

58. _The Wall Street Journal_ has a circulation of 1,780,442. Round this number to the nearest thousand. (_Source: 1995 Editor and Publisher International Yearbook_)   1,780,000

(1.5) *Multiply.*

59. 6 · 7   42

60. 8 · 3   24

51. 5(0)   0

62. 0(9)   0

63.   47   1410
     × 30

64.   69   2898
     × 42

65. 20(8)(5)   800

66. 25(9 × 4)   900

67.   48   3696
     × 77

68.   77   1694
     × 22

69. 49 · 49 · 0   0

70. 62 · 88 · 0   0

71.   586   16,994
     × 29

72.   242   8954
     × 37

73.   642   113,634
     × 177

74.   347   44,763
     × 129

75.   1026   411,426
     × 401

76.   2107   636,314
     × 302

Estimate each product by rounding each factor to the given place.

77. 49 · 32; tens   1500

78. 586 · 357; hundreds   240,000

79. 5231 · 243; hundreds   1,040,000

80. 7836 · 912; hundreds   7,020,000

81. There were 5283 students enrolled at Weskan State University in the fall semester. Each paid $927 tuition. Find the total tuition collected.
$4,897,341

82. One serving of a particular cake contains 490 calories. If the whole cake contains 8 servings, how many calories are in the whole cake?
3920 calories

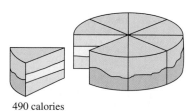

490 calories

Find the area of each rectangle.

83.   12 miles   60 square miles
          5 miles

84.   20 centimeters   500 square centimeters
          25 centimeters

(1.6) *Divide and then check.*

85. 18 ÷ 6   3   **86.** 36 ÷ 9   4   **87.** 42 ÷ 7   6   **88.** 25 ÷ 5   5

89. 27 ÷ 5   5 R 2   **90.** 18 ÷ 4   4 R 2   **91.** 16 ÷ 0   undefined   **92.** 0 ÷ 8   0

93. 9 ÷ 9   1   **94.** 10 ÷ 1   10   **95.** 918 ÷ 0   undefined   **96.** 0 ÷ 668   0

97. 5)75   15   **98.** 8)159   19 R 7   **99.** 26)626   24 R 2   **100.** 6)336   56

101. 32)49   1 R 17   **102.** 19)680   35 R 15   **103.** 20)10,000   500   **104.** 43)909   21 R 6

105. 47)23,782   506   **106.** 30)480   16   **107.** 16)3192   199 R 8   **108.** 25)5000   200

109. One foot is 12 inches. Find how many feet there are in 5496 inches.   458 feet

110. Find the average of the numbers 76, 49, 32, and 47.   51

(1.7) *Solve.*

111. 20 increased by 5 is what number?   25

112. The difference of 20 and 5 is what number?   15

113. The product of 20 and 5 is what number?   100

114. The quotient of 20 and 5 is what number?   4

115. A box can hold 24 cans of corn. How many boxes can be filled with 648 cans of corn?   27 boxes

116. If a ticket to a movie costs $6, how much does 32 tickets cost? $192

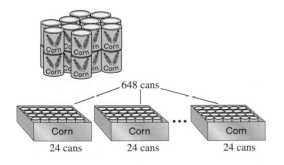

117. The cost to banks for using an ATM (Automatic Teller Machine) is 27¢. The cost to banks when you deposit a check with a teller is 48¢ more. How much is this cost?   75¢

118. Aspirin was 100 years old in 1997 and was the first U.S. drug made in tablet form. Today, people take 11 billion tablets a year for heart disease prevention and 4 billion tablets a year for headaches. How many more tablets are taken a year for heart disease prevention? (*Source:* Bayer Market Research)   7 billion

(1.8) *Write using exponential notation.*

119. $7 \cdot 7 \cdot 7 \cdot 7$ 7^4

120. $6 \cdot 6 \cdot 3 \cdot 3 \cdot 3$ $6^2 \cdot 3^3$

121. $4 \cdot 2 \cdot 2 \cdot 2 \cdot 3 \cdot 3$ $4 \cdot 2^3 \cdot 3^2$

122. $5 \cdot 5 \cdot 7 \cdot 7 \cdot 7 \cdot 2 \cdot 2$ $5^2 \cdot 7^3 \cdot 2^2$

Simplify.

123. 7^2 49

124. 2^6 64

125. $5^3 \cdot 3^2$ 1125

126. $4^1 \cdot 10^2 \cdot 7^2$ 19,600

127. $18 \div 3 + 7$ 13

128. $12 - 8 \div 4$ 10

129. $\dfrac{(6^2 - 3)}{3^2 + 2}$ 3

130. $\dfrac{16 - 8}{2^3}$ 1

131. $2 + 3[1 + (20 - 17) \cdot 3]$ 32

132. $21 - [2^4 - (7 - 5) - 10] + 8 \cdot 2$ 33

Find the area of each square.

133.

7 meters
49 square meters

134.
3 inches
9 square inches

Name _____ **Section** _____ **Date** _____

CHAPTER 1 TEST

Evaluate.

1. $59 + 82$

2. $600 - 487$

3.
$$\begin{array}{r} 496 \\ \times\ 30 \\ \hline \end{array}$$

4. $52,896 \div 69$

5. $2^3 \cdot 5^2$

6. $6^1 \cdot 2^3$

7. $98 \div 1$

8. $0 \div 49$

9. $62 \div 0$

10. $(2^4 - 5) \cdot 3$

11. $16 + 9 \div 3 \cdot 4 - 7$

12. $2\left[(6 - 4)^2 + (22 - 19)^2\right] + 10$

13. Round 52,369 to the nearest thousand.

Estimate each sum or difference by rounding each number to the nearest hundred.

14. $6289 + 5403 + 1957$

15. $4267 - 2738$

16. Twenty-nine cans of Sherwin Williams paint cost $493. How much was each can?

17. Admission to a movie costs $7 per ticket. The Math Club has 17 members who are going to a movie together. What is the total cost of their tickets?

18. Jo McElory is looking at two new refrigerators for her apartment. One costs $599 and the other costs $725.

How much more expensive is the higher-priced one?

Find the perimeter and the area of each figure.

19.

Square | 5 centimeters

20.
20 yards
Rectangle | 10 yards

ANSWERS

1. 141

2. 113

3. 14,880

4. 766 R 42

5. 200

6. 48

7. 98

8. 0

9. undefined

10. 33

11. 21

12. 36

13. 52,000

14. 13,700

15. 1600

16. $17

17. $119

18. $126

19. 20 centimeters, 25 square centimeters

20. 60 yards, 200 square yards

Name

The following table shows SAT average scores by state for some selected states. Use this table to answer Exercises 21–22.

SAT AVERAGE SCORES BY STATE					
	1995		1996		1997	
	Verbal	Math	Verbal	Math	Verbal	Math
Louisiana	560	552	559	550	560	553
Maine	504	497	504	498	507	504
Maryland	506	503	507	504	507	507
Massachusetts	505	502	507	504	508	508
Michigan	559	565	557	565	557	566
Minnesota	580	591	582	593	582	592
Mississippi	572	557	569	557	567	551
Missouri	569	566	570	569	567	568
Montana	549	553	546	547	545	548
Nebraska	568	570	567	568	562	564

Source: College Entrance Examination Board

21. Find the average 1997 SAT verbal score for students from the state of Massachusetts.

22. Find the *increase* in average SAT math scores for students from the state of Maine for the years 1995 and 1997.

Multiplying and Dividing Fractions

Fractions are numbers, and like whole numbers, they can be added, subtracted, multiplied, and divided. Fractions are very useful and appear frequently in everyday language, as in common phrases like "half an hour," "quarter of a pound," and "third of a cup." This chapter introduces the concept of fractions, presents some basic vocabulary, and demonstrates how to multiply and divide fractions.

Established in 1872, Yellowstone National Park was the United States' first national park. Today the National Park Service operates a total of 376 areas, ranging from 54 national parks such as Yellowstone and the Grand Canyon to 73 national monuments such as Devil's Tower in Wyoming to 74 national historic sites such as Abraham Lincoln Birthplace in Kentucky. The largest park in the system is Wrangell–St. Elias National Park and Preserve in Alaska with 13,200,000 acres of land. The smallest park in the system is Thaddeus Kosciuszko National Memorial in Pennsylvania, covering only $\frac{1}{50}$ of an acre of land. In Example 9 and Exercise 52 on pages 122 and 125, we will find the fraction of national parks that can be found in Alaska and in Washington state.

Name _____ **Section** _____ **Date** _____

CHAPTER 2 PRETEST

1. Use a fraction to represent the shaded area of the figure.

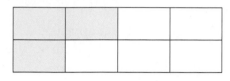

2. Write each mixed number as an improper fraction.

   a. $3\frac{4}{7}$ b. $9\frac{5}{8}$

3. Write each improper fraction as a mixed number or a whole number.

   a. $\frac{52}{13}$ b. $\frac{71}{6}$

4. List all the factors of 36.

Identify each number as prime or composite.

5. 19 6. 91 7. Find the prime factorization of 80.

Write each fraction in simplest form.

8. $\frac{42}{72}$ 9. $\frac{13}{91}$

Determine whether each pair of fractions are equal.

10. $\frac{6}{108}$ and $\frac{1}{18}$ 11. $\frac{30}{134}$ and $\frac{5}{22}$

12. There are 12 inches in 1 foot. What fraction of a foot is 8 inches?

Multiply. Write each answer in simplest form.

13. $\frac{3}{5} \cdot \frac{2}{7}$ 14. $\frac{4}{9} \cdot \frac{3}{28}$ 15. $1\frac{3}{5} \cdot 1\frac{1}{4}$

16. Pam Wiseman has a take home pay of $800 a month. $\frac{2}{5}$ of the monthly income is spent on her car payment. How much is her monthly car payment?

17. Find the reciprocal of $\frac{12}{19}$.

Divide. Write all answers in simplest form.

18. $\frac{4}{5} \div \frac{2}{15}$ 19. $5\frac{5}{8} \div 3$

20. Jillian Grant used $1\frac{1}{2}$ gallons of gas to drive $34\frac{1}{2}$ miles. How many miles could she drive using one gallon of gas?

2.1 INTRODUCTION TO FRACTIONS AND MIXED NUMBERS

A IDENTIFYING NUMERATORS AND DENOMINATORS

Whole numbers are used to count whole things or units, such as cars, ball games, horses, dollars, and people. To refer to a part of a whole, fractions are used. For example, a whole baseball game is divided into nine parts called innings. If a player pitched 5 complete innings, the fraction $\frac{5}{9}$ can be used to show the part of a whole game he or she pitched. The 9 in the fraction $\frac{5}{9}$ is called the **denominator**, and it refers to the total number of equal parts (innings) in the whole game. The 5 in the fraction $\frac{5}{9}$ is called the **numerator**, and it tells how many of those equal parts (innings) the pitcher pitched.

$$\frac{5}{9} \quad \begin{array}{l} \leftarrow \text{ how many of the parts being considered} \\ \leftarrow \text{ number of equal parts in the whole} \end{array}$$

Examples Identify the numerator and the denominator of each fraction.

1. $\dfrac{3}{7}$ $\begin{array}{l} \leftarrow \text{ numerator} \\ \leftarrow \text{ denominator} \end{array}$

2. $\dfrac{13}{5}$ $\begin{array}{l} \leftarrow \text{ numerator} \\ \leftarrow \text{ denominator} \end{array}$

> **HELPFUL HINT**
>
> $\dfrac{3}{7}$ $\leftarrow$ Remember that the bar in a fraction means division. Since division by 0 is undefined, a fraction with a denominator of 0 is undefined.

B WRITING FRACTIONS TO REPRESENT SHADED AREAS OF FIGURES

One way to become familiar with the concept of fractions is to visualize fractions with shaded figures. We can then write a fraction to represent the shaded area of the figure.

Examples Write a fraction to represent the shaded area of each figure.

3.

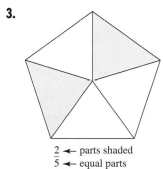

$\dfrac{2}{5}$ $\begin{array}{l} \leftarrow \text{ parts shaded} \\ \leftarrow \text{ equal parts} \end{array}$

The figure is divided into 5 equal parts and 2 of them are shaded.

Thus, $\frac{2}{5}$ of the figure is shaded.

Practice Problems 1–2

Identify the numerator and the denominator of each fraction.

1. $\dfrac{9}{2}$ 　　　　 2. $\dfrac{10}{17}$

Practice Problems 3–4

Write a fraction to represent the shaded area of each figure.

3.

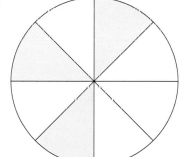

4.

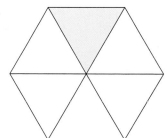

Answers

1. numerator = 9, denominator = 2,
2. numerator = 10, denominator = 17,
3. $\dfrac{3}{8}$, 4. $\dfrac{1}{6}$

Copyright 1999 Prentice-Hall, Inc.

TEACHING TIP

Ask students to use fractions to represent the unshaded area of the figure. Then ask them if they notice any relationship between the fraction for the shaded area and the fraction for the unshaded area. Be sure they see that the sum of the numerators equals the denominator.

Practice Problem 5

Of the nine planets in our solar system, seven are further from the sun than Venus. What fraction of the planets are further from the sun than Venus?

TEACHING TIP

Point out to students that any shape can represent 1 whole. Regular polygons and circles are often used because they can be divided into equal parts more easily than irregular shapes. You may want to have students work in groups to create interesting shaded pictures of 1/4. Finally have the groups share their different pictures of 1/4 with the rest of the class.

4.

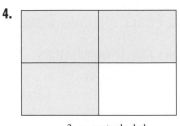

The figure is divided into 4 equal parts and 3 of them are shaded. Thus, $\frac{3}{4}$ of the figure is shaded.

$\frac{3}{4}$ ◄ parts shaded
     ◄ equal parts

Example 5   Writing Fractions from Real-Life Data

Of the nine planets in our solar system, two are closer to the sun than the Earth. What fraction of the planets are closer to the sun than the Earth?

Solution: The fraction closer to the sun than the Earth is:

$\frac{2}{9}$ ← number of planets considered
    ← number of planets in our solar system

Thus, $\frac{2}{9}$ of our solar system planets are closer to the sun than the Earth.

C   IDENTIFYING PROPER FRACTIONS, IMPROPER FRACTIONS, AND MIXED NUMBERS

A **proper fraction** is a fraction whose numerator is less than its denominator. Proper fractions are less than 1. The shaded portion of the triangle's area is represented by $\frac{2}{3}$.

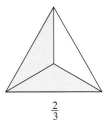

$\frac{2}{3}$

An **improper fraction** is a fraction whose numerator is greater than or equal to its denominator. Improper fractions are greater than or equal to 1. The shaded part of the group of circles' area is $\frac{9}{4}$. The shaded part of the rectangle's area is $\frac{6}{6}$.

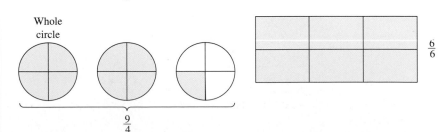

Whole circle

$\frac{9}{4}$

$\frac{6}{6}$

Answer

5. $\frac{7}{9}$

A **mixed number** contains a whole number and a fraction. Mixed numbers are greater than 1. The shaded area of the group of circles' area is $2\frac{1}{4}$. (Read "two and one fourth.")

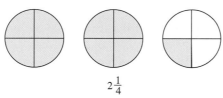

$$2\frac{1}{4}$$

TRY THE CONCEPT CHECK IN THE MARGIN.

> **HELPFUL HINT**
>
> The mixed number $2\frac{1}{4}$ represents $2 + \frac{1}{4}$.

Examples   Represent the shaded part of each figure group's area as both an improper fraction and a mixed number.

6.   Whole object

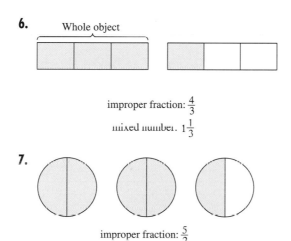

improper fraction: $\frac{4}{3}$

mixed number: $1\frac{1}{3}$

7.

improper fraction: $\frac{5}{2}$

mixed number: $2\frac{1}{2}$

TRY THE CONCEPT CHECK IN THE MARGIN.

D WRITING MIXED NUMBERS AS IMPROPER FRACTIONS

Notice from Examples 6 and 7 that mixed numbers and improper fractions were both used to represent the shaded area of the figure groups. For example,

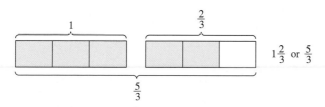

$1\frac{2}{3}$ or $\frac{5}{3}$

The following steps may be used to write a mixed number as an improper fraction.

Copyright 1999 Prentice-Hall, Inc.

TEACHING TIP

Tell students that the skill of converting mixed numbers to improper fractions will be useful to them when they begin multiplying mixed numbers.

Practice Problem 8

Write each as an improper fraction.

a. $2\frac{5}{7}$ b. $5\frac{1}{3}$

c. $9\frac{3}{10}$ d. $1\frac{1}{5}$

TEACHING TIP

After discussing converting mixed numbers to improper fractions, use the same technique to ask students to convert whole numbers to improper fractions with various denominators. For instance, convert 5 to an improper fraction with denominator 2 as $5 = 5\frac{0}{2} = \frac{2 \cdot 5 + 0}{2} = \frac{10}{2}$.

WRITING A MIXED NUMBER AS AN IMPROPER FRACTION

To write a mixed number as an improper fraction,

Step 1. Multiply the whole number by the denominator of the fraction.

Step 2. Add the numerator of the fraction to the product from Step 1.

Step 3. Write the sum from Step 2 as the numerator of the improper fraction over the original denominator.

For example,

$$1\frac{2}{3} = \frac{3 \cdot 1 + 2}{3} = \frac{3 + 2}{3} = \frac{5}{3}$$

Example 8   Write each as an improper fraction.

a. $4\frac{2}{9}$ b. $1\frac{8}{11}$

Solution: a. $4\frac{2}{9} = \frac{9 \cdot 4 + 2}{9} = \frac{36 + 2}{9} = \frac{38}{9}$

b. $1\frac{8}{11} = \frac{11 \cdot 1 + 8}{11} = \frac{11 + 8}{11} = \frac{19}{11}$

E WRITING IMPROPER FRACTIONS AS MIXED NUMBERS OR WHOLE NUMBERS

Just as there are times when an improper fraction is preferred, sometimes a mixed or a whole number better suits a situation. To write improper fractions as mixed or whole numbers, just remember that the fraction bar means division.

WRITING AN IMPROPER FRACTION AS A MIXED NUMBER OR A WHOLE NUMBER

To write an improper fraction as a mixed number or a whole number,

1. Divide the denominator into the numerator.
2. The whole number part of the mixed number is the quotient. The fraction part of the mixed number is the remainder over the original denominator.

$$\text{quotient} \frac{\text{remainder}}{\text{original denominator}}$$

For example,

$$\frac{5}{3} = 3\overline{)5} \; = 1\frac{2}{3} \quad \leftarrow \text{remainder}$$
$$\qquad \frac{3}{2} \qquad \qquad \leftarrow \text{original denominator}$$
$$\qquad \qquad \uparrow$$
$$\qquad \qquad \text{quotient}$$

Answers

8. a. $\frac{19}{7}$, b. $\frac{16}{3}$, c. $\frac{93}{10}$, d. $\frac{6}{5}$

Example 9   Write each as a mixed number or a whole number.

a. $\dfrac{30}{7}$ b. $\dfrac{16}{15}$ c. $\dfrac{84}{6}$

Solution:

a. $7\overline{)30}$ with quotient 4, $\dfrac{28}{2}$ $\dfrac{30}{7} = 4\dfrac{2}{7}$

b. $15\overline{)16}$ with quotient 1, $\dfrac{15}{1}$ $\dfrac{16}{15} = 1\dfrac{1}{15}$

c. $6\overline{)84}$ quotient 14, $\dfrac{6}{24}$, $\dfrac{24}{0}$ $\dfrac{84}{6} = 14$

Since the remainder is 0, the result is the whole number 14.

HELPFUL HINT

When the remainder is 0, the improper fraction is a whole number. For example, $\dfrac{92}{4} = 23$.

$4\overline{)92}$, $\dfrac{8}{12}$, $\dfrac{12}{0}$ quotient 23

Practice Problem 9

Write each as a mixed number or a whole number.

a. $\dfrac{8}{5}$ b. $\dfrac{17}{6}$

c. $\dfrac{48}{4}$ d. $\dfrac{35}{4}$

e. $\dfrac{51}{7}$ f. $\dfrac{21}{20}$

Answers

9. a. $1\dfrac{3}{5}$, **b.** $2\dfrac{5}{6}$, **c.** 12, **d.** $8\dfrac{3}{4}$, **e.** $7\dfrac{2}{7}$,

f. $1\dfrac{1}{20}$

Focus On Mathematical Connections

MODELING FRACTIONS

There are several different physical models for representing fractions.

SET MODEL

In this model, a fraction represents the portion of a set of objects that has a certain characteristic. For example, in the set of 10 shapes, 3 are hearts. That is, $\frac{3}{10}$ of the shapes are hearts.

AREA MODEL

In this model, a shape is divided into a number of equal-sized regions. A fraction can be represented by shading some of the regions. For example, both of the following models represent the fraction $\frac{7}{8}$.

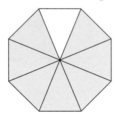 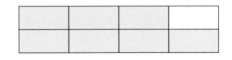

NUMBER LINE MODEL

For this model, draw a line, label a point 0 and a point to its right 1. Now subdivide this distance from 0 to 1 depending on the denominator of fraction that is to be represented. For example, the fraction $\frac{3}{4}$ is graphed by subdividing the portion of the number line between 0 and 1 into 4 equal lengths and placing a dot at the mark that represents $\frac{3}{4}$ of the distance between 0 and 1.

CRITICAL THINKING

1. **a.** Represent the fraction $\frac{5}{6}$ in two different ways using the set model.

   b. Represent the fraction $\frac{5}{6}$ in two different ways using the area model.

   c. Represent the fraction $\frac{5}{6}$ using the number line model.

2. Which model do you prefer? Why?

3. Which model do you think would be most useful for representing multiplication of fractions? Explain how this could be done.

Name _____ **Section** _____ **Date** _____

MENTAL MATH

A *Identify the numerator and the denominator of each fraction. See Examples 1 and 2.*

1. $\frac{1}{2}$ **2.** $\frac{1}{4}$ **3.** $\frac{10}{3}$ **4.** $\frac{53}{21}$

5. $\frac{3}{7}$ **6.** $\frac{11}{15}$

EXERCISE SET 2.1

B *Write a fraction to represent the shaded area of each figure. See Examples 3 and 4.*

1.

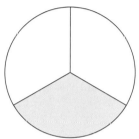

2.

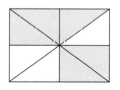

3.

4.

5.

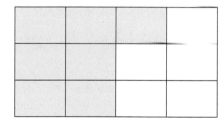

6.

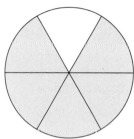

7.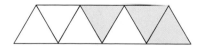

8.

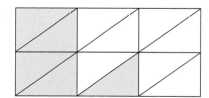

9. $\frac{4}{9}$

10. $\frac{7}{8}$

11. $\frac{42}{131}$

12. $\frac{61}{78}$

13. $89; \frac{89}{131}$

14. $17; \frac{17}{78}$

15. $\frac{4}{10}$

16. $\frac{1}{24}$

17. $\frac{8}{42}$

18. $\frac{4}{9}$

9.

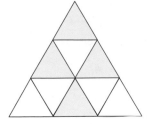

10.

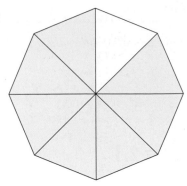

Write each fraction. See Example 5.

📼 **11.** Of the 131 students at a small private school, 42 are freshmen. What fraction of the students are freshmen?

12. Of the 78 executives at a private accounting firm, 61 are men. What fraction of the executives are men?

13. From Exercise 11, how many students are *not* freshmen? What fraction of the students are *not* freshmen?

14. From Exercise 12, how many of the executives are women? What fraction of the executives are women?

15. According to a recent study, four out of ten visits to U.S. hospital emergency rooms were for an injury. What fraction of emergency room visits are injury-related? (*Source:* National Center for Health Statistics)

16. The average American driver spends about 1 hour each day in a car or truck. What fraction of a day does an average American driver spend in a car or truck? (*Hint:* How many hours are in a day?) (*Source:* U.S. Department of Transportation— Federal Highway Administration)

17. As of 1998, the United States has had 42 different presidents. A total of eight U.S. presidents were born in the state of Virginia, more than any other state. What fraction of U.S. presidents were born in Virginia? (*Source: 1998 World Almanac and Book of Facts*)

Eight U.S. Presidents

18. Of the nine planets in our solar system, four have days that are longer than the 24-hour Earth day. What fraction of the planets have longer days than Earth has? (*Source:* National Space Science Data Center)

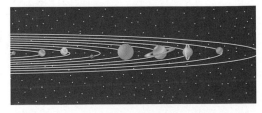

Name _____

C *Write the shaded area in each figure group as (a) a mixed number and (b) an improper fraction. See Examples 6 and 7.*

19.

20.

21.

22.

23.

24.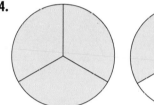

25.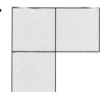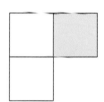

26.

27.

28. a. $1\frac{2}{5}$

b. $\frac{7}{5}$

29. a. $1\frac{5}{9}$

b. $\frac{14}{9}$

30. a. $2\frac{7}{9}$

b. $\frac{25}{9}$

31. $\frac{7}{3}$

32. $\frac{27}{4}$

33. $\frac{11}{3}$

34. $\frac{27}{8}$

35. $\frac{21}{8}$

36. $\frac{63}{5}$

37. $\frac{41}{15}$

38. $\frac{39}{8}$

39. $\frac{83}{7}$

40. $\frac{38}{3}$

41. $\frac{53}{8}$

42. $\frac{89}{10}$

43. $\frac{18}{5}$

44. $\frac{211}{12}$

45. $\frac{109}{24}$

46. $\frac{284}{27}$

47. $\frac{84}{13}$

48. $\frac{142}{25}$

49. $\frac{187}{20}$

50. $\frac{187}{15}$

110

Name _____

28.

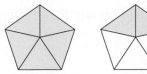

29.

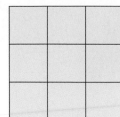 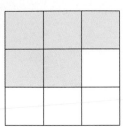

30.

D *Write each mixed number as an improper fraction. See Example 8.*

📼 **31.** $2\frac{1}{3}$ **32.** $6\frac{3}{4}$ **33.** $3\frac{2}{3}$ **34.** $3\frac{3}{8}$

35. $2\frac{5}{8}$ **36.** $12\frac{3}{5}$ **37.** $2\frac{11}{15}$ **38.** $4\frac{7}{8}$

39. $11\frac{6}{7}$ **40.** $12\frac{2}{3}$ **41.** $6\frac{5}{8}$ **42.** $8\frac{9}{10}$

📼 **43.** $3\frac{3}{5}$ **44.** $17\frac{7}{12}$ **45.** $4\frac{13}{24}$ **46.** $10\frac{14}{27}$

47. $6\frac{6}{13}$ **48.** $5\frac{17}{25}$ 📼 **49.** $9\frac{7}{20}$ **50.** $12\frac{7}{15}$

Name _____

E *Write each improper fraction as a mixed number or a whole number. See Example 9.*

51. $\dfrac{17}{5}$ **52.** $\dfrac{13}{7}$ **53.** $\dfrac{42}{13}$ **54.** $\dfrac{39}{3}$

55. $\dfrac{47}{15}$ **56.** $\dfrac{65}{12}$ **57.** $\dfrac{198}{6}$ **58.** $\dfrac{112}{7}$

59. $\dfrac{225}{15}$ **60.** $\dfrac{23}{5}$ **61.** $\dfrac{37}{8}$ **62.** $\dfrac{46}{21}$

63. $\dfrac{18}{17}$ **64.** $\dfrac{149}{143}$ **65.** $\dfrac{247}{23}$ **66.** $\dfrac{437}{53}$

67. $\dfrac{46}{11}$ **68.** $\dfrac{67}{17}$

REVIEW AND PREVIEW

Simplify. See Section 1.8.

69. 3^2 **70.** 4^3 **71.** 5^3 **72.** 3^4

73. 7^2 **74.** 5^4 **75.** $2^3 \cdot 3$ **76.** $4^2 \cdot 5$

COMBINING CONCEPTS

77. Write $38\dfrac{41}{79}$ as an improper fraction.   **78.** Write $\dfrac{2178}{31}$ as a mixed number.

---

51. $3\dfrac{2}{5}$

52. $1\dfrac{6}{7}$

53. $3\dfrac{3}{13}$

54. 13

55. $3\dfrac{2}{15}$

56. $5\dfrac{5}{12}$

57. 33

58. 16

59. 15

60. $4\dfrac{3}{5}$

61. $4\dfrac{5}{8}$

62. $2\dfrac{4}{21}$

63. $1\dfrac{1}{17}$

64. $1\dfrac{6}{143}$

65. $10\dfrac{17}{23}$

66. $8\dfrac{13}{53}$

67. $4\dfrac{2}{11}$

68. $3\dfrac{16}{17}$

69. 9

70. 64

71. 125

72. 81

73. 49

74. 625

75. 24

76. 80

77. $\dfrac{3043}{79}$

78. $70\dfrac{8}{31}$

Name _____

79. Habitat for Humanity is a nonprofit organization that helps provide affordable housing to families in need. Habitat for Humanity does its work of building and renovating houses through 1300 local affiliates in the United States and 285 international affiliates. What fraction of the total Habitat for Humanity affiliates are located in the United States? (*Source:* Habitat for Humanity International)

80. The United States Marine Corps (USMC) has five principal training centers in California, three in North Carolina, two in South Carolina, one in Arizona, one in Hawaii, and one in Virginia. What fraction of the total USMC principal training centers are located in California? (*Source:* U.S. Department of Defense)

81. The Public Broadcasting Service (PBS) provides programming to the noncommercial public TV stations of the United States. The table shows a breakdown of the public television licensees by type. Each licensee operates one or more PBS member TV stations. What fraction of the public television licensees are universities or colleges?

82. The table shows the number of Wendy's restaurants operated in various regions. What fraction of Wendy's restaurants are located in the United States?

WENDY'S RESTAURANTS	
Region	Number
United States	4377
Canada	229
Asia/Pacific	201
Latin America	68
Europe & other	58

(*Source:* Wendy's International, Inc.)

PUBLIC TELEVISION LICENSEES	
Type	Number
Community organizations	87
Universities/colleges	56
State authorities	21
Local education/ municipal authorities	8

(*Source:* The Public Broadcast Service)

2.2 FACTORS AND PRIME FACTORIZATION

To perform many operations with fractions it is necessary to be able to factor a number. In this section, only the **natural numbers**—1, 2, 3, 4, 5, and so on—will be considered.

TRY THE CONCEPT CHECK IN THE MARGIN.

A FACTORING NUMBERS

Recall that when numbers are multiplied to form a product, each number is called a factor. Since $3 \cdot 4 = 12$, both 3 and 4 are **factors** of 12 and $3 \cdot 4$ is called a **factorization** of 12.

The two-number factorizations of 12 are

$$1 \cdot 12 \qquad 2 \cdot 6 \qquad 3 \cdot 4$$

Thus, we say that the factors of 12 are 1, 2, 3, 4, 6, and 12.

HELPFUL HINT

A **factor** of a number divides the number evenly (with a remainder of 0). For example,

$$\frac{12}{1)\overline{12}} \qquad \frac{6}{2)\overline{12}} \qquad \frac{4}{3)\overline{12}} \qquad \frac{3}{4)\overline{12}} \qquad \frac{2}{6)\overline{12}} \qquad \frac{1}{12)\overline{12}}$$

Example 1
Find all the factors of 20.

Solution: First we write all the two-number factorizations of 20.

$$1 \cdot 20 = 20$$
$$2 \cdot 10 = 20$$
$$4 \cdot 5 = 20$$

The factors of 20 are 1, 2, 4, 5, 10, and 20.

B IDENTIFYING PRIME AND COMPOSITE NUMBERS

Of all the ways to factor a number, one special way is called the prime factorization. To help us write prime factorizations, we first review prime and composite numbers.

PRIME NUMBERS

A **prime number** is a natural number that has exactly two different factors, 1 and itself.

Example 2
Determine whether each number is prime. Explain your answers.

3, 9, 11, 17, 26

Solution: The number 3 is prime. Its only factors are 1 and 3.
The number 9 is not prime. It has more than two factors: 1, 3, and 9.
The number 11 is prime. Its only factors are 1 and 11.
The number 17 is prime. Its only factors are 1 and 17.
The number 26 is not prime. Its factors are 1, 2, 13, and 26.

Objectives

A Find the factors of a number.
B Identify prime and composite numbers.
C Find the prime factorization of a number.

SSM   CD-ROM   Video 2.2

✓ CONCEPT CHECK

How are the natural numbers and the whole numbers alike? How are they different?

Practice Problem 1

Find all the factors of each number.

a. 15   b. 7

Practice Problem 2

Determine whether each number is prime. Explain your answers.

15, 11, 24, 29, 39

Answers

✓ **Concept Check:** answers may vary
1. a. 1, 3, 5, 15, **b.** 1, 7, **2.** 11, 29 are prime. 15, 24, and 39 are not prime.

The first ten prime numbers are

2, 3, 5, 7, 11, 13, 17, 19, 23, 29

It would be helpful to memorize these.

If a natural number, other than 1, is not a prime number, it is called a **composite number**.

COMPOSITE NUMBERS

A **composite number** is any natural number, other than 1, that is not prime.

HELPFUL HINT

The natural number 1 is neither prime nor composite.

C FINDING PRIME FACTORIZATIONS

Now we are ready to find **prime factorizations** of numbers.

PRIME FACTORIZATIONS

The **prime factorization** of a number is a factorization in which all the factors are prime numbers.

For example, the prime factorization of 12 is $2 \cdot 2 \cdot 3$ because

$$12 = \underbrace{2 \cdot 2 \cdot 3}$$

This product is 12 and each number is a prime number.

There is only one prime factorization for any given number. In other words, the prime factorization of a number is unique.

Practice Problem 3

Find the prime factorization of 28.

Example 3   Find the prime factorization of 45.

Solution: The first prime, 2, does not divide 45 evenly (with a remainder of 0). The second prime, 3, does, so we divide 45 by 3.

$$\begin{array}{r} 15 \\ 3\overline{)45} \end{array}$$

Because 15 is not prime and 3 also divides 15 evenly, we divide by 3 again.

$$\begin{array}{r} 5 \\ 3\overline{)15} \\ 3\overline{)45} \end{array}$$

The quotient, 5, is a prime number, so we are finished. The prime factorization of 45 is

$$45 = 3 \cdot 3 \cdot 5 \quad \text{or} \quad 45 = 3^2 \cdot 5, \quad \text{using exponents.}$$

Answer

3. $2^2 \cdot 7$

There are a few quick **divisibility tests** to determine if a number is divisible by the primes 2, 3, or 5. (A number is divisible by 2, for example, if 2 divides it evenly.)

DIVISIBILITY TESTS

A number is divisible by
▲ **2** if the ones digit is 0, 2, 4, 6, or 8.
    ↓
    132 is divisible by 2 since the ones digit is a 2.

▲ **3** if the sum of the digits is divisible by 3.
    144 is divisible by 3 since $1 + 4 + 4 = 9$ is divisible by 3.

▲ **5** if the ones digit is 0 or 5.
    ↓
    1115 is divisible by 5 since the ones digit is a 5.

Example 4   Find the prime factorization of 180.

Solution: We divide 180 by 2 and continue dividing until the quotient is no longer divisible by 2. We then divide by the next largest prime, 3, until the quotient is no longer divisible by 3. We continue this process until the quotient is a prime number.

$$\begin{array}{r} 5 \\ 3\overline{)15} \\ 3\overline{)45} \\ 2\overline{)90} \\ 2\overline{)180} \end{array}$$

Thus the prime factorization of 180 is

$180 = 2 \cdot 2 \cdot 3 \cdot 3 \cdot 5$ or $180 = 2^2 \cdot 3^2 \cdot 5$, using exponents.

Example 5   Find the prime factorization of 80.

Solution:
$$\begin{array}{r} 5 \\ 2\overline{)10} \\ 2\overline{)20} \\ 2\overline{)40} \\ 2\overline{)80} \end{array}$$

The prime factorization of 80 is

$80 = 2 \cdot 2 \cdot 2 \cdot 2 \cdot 5$ or $80 = 2^4 \cdot 5$, using exponents.

Another way to find the prime factorization is to use a factor tree, as shown in the next example.

Example 6   Use a factor tree to find the prime factorization of 18.

Solution: We begin by writing 18 as a product of two numbers, say, $2 \cdot 9$.

Practice Problem 4

Find the prime factorization of 378.

Practice Problem 5

Find the prime factorization of 75.

Practice Problem 6

Use a factor tree to find the prime factorization of 70.

Answers
4. $2 \cdot 3^3 \cdot 7$,   **5.** $3 \cdot 5^2$,   **6.** $2 \cdot 5 \cdot 7$

$$18$$
$$2 \cdot 9$$

The number 2 is prime, but 9 is not. So we write 9 as $3 \cdot 3$.

$$18$$
$$2 \cdot 9$$
$$2 \cdot 3 \cdot 3$$

Each factor is now prime, so the prime factorization is

$$18 = 2 \cdot 3 \cdot 3 \quad \text{or} \quad 18 = 2 \cdot 3^2 \quad \text{using exponents.}$$

Don't forget that multiplication is commutative, so that $2 \cdot 3 \cdot 3$ can also be written as $3 \cdot 3 \cdot 2$ or $3 \cdot 2 \cdot 3$. Any of these can be called the one prime factorization of 18.

Example 7   Use a factor tree to find the prime factorization of 24.

Solution:

$$24$$
$$4 \cdot 6$$
$$2 \cdot 2 \cdot 2 \cdot 3$$

The prime factorization of 24 is

$$24 = 2 \cdot 2 \cdot 2 \cdot 3 \quad \text{or} \quad 2^3 \cdot 3, \quad \text{using exponents.}$$

TRY THE CONCEPT CHECK IN THE MARGIN.

Practice Problem 7

Use a factor tree to find the prime factorization of each number.

a. 10

b. 30

c. 72

✓ **CONCEPT CHECK**

True or false: Two different numbers can have exactly the same prime factorization. Explain your answer.

Answers

7. a. $2 \cdot 5$, **b.** $2 \cdot 3 \cdot 5$, **c.** $2^3 \cdot 3^2$

✓ **Concept Check:**   False

EXERCISE SET 2.2

A *List all the factors of each number. See Example 1.*

1. 8   **2.** 6   🔲 **3.** 25   **4.** 30

5. 4   **6.** 9   **7.** 18   **8.** 24

9. 7   **10.** 5   **11.** 80   **12.** 100

🔲 **13.** 12   **14.** 20   **15.** 34   **16.** 26

B *Identify each number as prime or composite. See Example 2.*

17. 7   **18.** 5   🔲 **19.** 4   **20.** 49

🔲 **21.** 10   **22.** 13   **23.** 29   **24.** 45

25. 6   **26.** 2   **27.** 15   **28.** 21

29. 31   **30.** 27   **31.** 33   **32.** 51

C *Find the prime factorization of each number. Write any repeated factors using exponents. See Examples 3 through 7.*

33. 12   **34.** 20   🔲 **35.** 15   **36.** 21

37. 40   **38.** 63   🔲 **39.** 36   **40.** 64

ANSWERS

1. 1, 2, 4, 8
2. 1, 2, 3, 6
3. 1, 5, 25
4. 1, 2, 3, 5, 6, 10, 15, 30
5. 1, 2, 4
6. 1, 3, 9
7. 1, 2, 3, 6, 9, 18
8. 1, 2, 3, 4, 6, 8, 12, 24
9. 1, 7
10. 1, 5
11. 1, 2, 4, 5, 8, 10, 16, 20, 40, 80
12. 1, 2, 4, 5, 10, 20, 25, 50, 100
13. 1, 2, 3, 4, 6, 12
14. 1, 2, 4, 5, 10, 20
15. 1, 2, 17, 34
16. 1, 2, 13, 26
17. prime
18. prime
19. composite
20. composite
21. composite
22. prime
23. prime
24. composite
25. composite
26. prime
27. composite
28. composite
29. prime
30. composite
31. composite
32. composite
33. $2^2 \cdot 3$
34. $2^2 \cdot 5$
35. $3 \cdot 5$
36. $3 \cdot 7$
37. $2^3 \cdot 5$
38. $3^2 \cdot 7$
39. $2^2 \cdot 3^2$
40. 2^6

Name _____

41. 39	**42.** 33	**43.** 48	**44.** 28
45. 54	**46.** 56	**47.** 60	**48.** 100
49. 110	**50.** 140	**51.** 88	**52.** 93
53. 128	**54.** 81	**55.** 150	**56.** 175
57. 300	**58.** 360	**59.** 240	**60.** 400
61. 945	**62.** 504	**63.** 700	**64.** 1000

REVIEW AND PREVIEW

Round each whole number to the indicated place value. See Section 1.4.

65. 4267   hundreds   **66.** 7,658,240   ten-thousands

67. 4,286,340   tens   **68.** 19,764   thousands

69. 55,342   hundreds   **70.** 10,292,876   millions

71. 3499   tens   **72.** 2,437,831   hundred-thousands

73. 1247   thousands   **74.** 9485   tens

COMBINING CONCEPTS

75. Find the prime factorization of each number.
    a. 3600   **b.** 1350

76. In your own words, define a prime number.

77. The number 2 is a prime number. All other even natural numbers are composite numbers. Explain why.

2.3 SIMPLEST FORM OF A FRACTION

A WRITING FRACTIONS IN SIMPLEST FORM

Fractions that represent the same portion of a whole are called **equivalent fractions**.

$$\frac{1}{3} \qquad \frac{2}{6} \qquad \frac{4}{12}$$

For example, $\frac{1}{3}$, $\frac{2}{6}$ and $\frac{4}{12}$ all represent the same shaded portion of the rectangle's area, so they are equivalent fractions.

$$\frac{1}{3} = \frac{2}{6} = \frac{4}{12}$$

A special form of a fraction is called **simplest form**.

SIMPLEST FORM OF A FRACTION

A fraction is written in **simplest form** or **lowest terms** when the numerator and the denominator have no common factors other than 1.

For example, the fraction $\frac{2}{6}$ *is not* in simplest form because 2 and 6 both have a factor of 2. That is, 2 is a common factor of 2 and 6. The fraction $\frac{1}{3}$ *is* in simplest form because 1 and 3 have no common factor other than 1. The process of writing a fraction in simplest form is called **simplifying** the fraction.

We can use the prime factorization of a number to help us write a fraction in simplest form.

WRITING A FRACTION IN SIMPLEST FORM

To write a fraction in simplest form, write the prime factorization of the numerator and the denominator and then divide both by all common factors.

For example,

$$\frac{2}{6} = \frac{2}{2 \cdot 3} = \frac{2 \div 2}{2 \cdot 3 \div 2} = \frac{1}{3}$$ Divide the numerator and the denominator by 2.

In the future, we will use the following notation to show dividing the numerator and denominator by common factors.

$$\frac{2}{6} = \frac{\overset{1}{\cancel{2}}}{\underset{1}{\cancel{2}} \cdot 3} = \frac{1}{1 \cdot 3} = \frac{1}{3}$$

Example 1   Write in simplest form: $\frac{12}{20}$

Objectives

A Write a fraction in simplest form or lowest terms.

B Determine whether two fractions are equivalent.

C Solve problems by writing fractions in simplest form.

SSM   CD-ROM   Video 2.3

TEACHING TIP

Before discussing writing fractions in lowest terms, have students help you generate fractions equivalent to 1/2. Get started by asking, "If half of a class of 36 students is girls, what fraction with denominator 36 describes the fraction of girls in the class?" Also, "If half a wedding cake which originally had 250 pieces was eaten, what fraction with denominator 250 was eaten?" Once you have a list of 8 fractions equivalent to 1/2, ask students how many of these kinds of fractions could be generated. Which of these forms is easiest to understand? Why?

Practice Problem 1

Write in simplest form: $\frac{30}{45}$

Answer

1. $\frac{2}{3}$

Solution: First, we write the prime factorization of the numerator and the denominator.

$$\frac{12}{20} = \frac{2 \cdot 2 \cdot 3}{2 \cdot 2 \cdot 5}$$

Next, we divide the numerator and the denominator by all common factors.

$$\frac{12}{20} = \frac{\overset{1}{\cancel{2}} \cdot \overset{1}{\cancel{2}} \cdot 3}{\underset{1}{\cancel{2}} \cdot \underset{1}{\cancel{2}} \cdot 5} = \frac{3}{5}$$

Practice Problem 2

Write in simplest form: $\frac{39}{51}$

Example 2   Write in simplest form: $\frac{42}{66}$

Solution:

$$\frac{42}{66} = \frac{\overset{1}{\cancel{2}} \cdot \overset{1}{\cancel{3}} \cdot 7}{\underset{1}{\cancel{2}} \cdot \underset{1}{\cancel{3}} \cdot 11} = \frac{7}{11}$$

Practice Problem 3

Write in simplest form: $\frac{9}{50}$

Example 3   Write in simplest form: $\frac{10}{27}$

Solution: $\frac{10}{27} = \frac{2 \cdot 5}{3 \cdot 3 \cdot 3}$

Since 10 and 27 have no common factors, $\frac{10}{27}$ is already in simplest form.

Practice Problem 4

Write in simplest form: $\frac{49}{63}$

Example 4   Write in simplest form: $\frac{30}{108}$

Solution:

$$\frac{30}{108} = \frac{\overset{1}{\cancel{2}} \cdot \overset{1}{\cancel{3}} \cdot 5}{2 \cdot \underset{1}{\cancel{2}} \cdot \underset{1}{\cancel{3}} \cdot 3 \cdot 3} = \frac{5}{18}$$

Practice Problem 5

Write in simplest form: $\frac{24}{20}$

Example 5   Write in simplest form: $\frac{72}{26}$

Solution:

$$\frac{72}{26} = \frac{\overset{1}{\cancel{2}} \cdot 2 \cdot 2 \cdot 3 \cdot 3}{\underset{1}{\cancel{2}} \cdot 13} = \frac{36}{13}$$

which can also be written as

$$2\frac{10}{13}$$

✓ CONCEPT CHECK

Which is the correct way to simplify the fraction $\frac{15}{25}$? Or are both correct? Explain.

a. $\frac{15}{25} = \frac{3 \cdot \overset{1}{\cancel{5}}}{5 \cdot \underset{1}{\cancel{5}}} = \frac{3}{5}$

b. $\frac{1\cancel{5}}{2\cancel{5}} = \frac{11}{21}$

TRY THE CONCEPT CHECK IN THE MARGIN.

Example 6   Write in simplest form: $\frac{6}{60}$

Solution:

$$\frac{6}{60} = \frac{\overset{1}{\cancel{2}} \cdot \overset{1}{\cancel{3}}}{\underset{1}{\cancel{2}} \cdot 2 \cdot \underset{1}{\cancel{3}} \cdot 5} = \frac{1}{10}$$

Practice Problem 6

Write in simplest form: $\frac{8}{56}$

Answers

2. $\frac{13}{17}$,   **3.** $\frac{9}{50}$,   **4.** $\frac{7}{9}$,   **5.** $\frac{6}{5}$ or $1\frac{1}{5}$,   **6.** $\frac{1}{7}$

✓ **Concept Check**

a. correct,   **b.** incorrect

HELPFUL HINT

Be careful when all factors of the numerator or denominator are divided out. In Example 6, the numerator was $1 \cdot 1 = 1$, so the final result was $\dfrac{1}{10}$.

In the fraction of Example 6, $\dfrac{6}{60}$ you may immediately notice that the largest common factor of 6 and 60 is 6. If so, you may simply divide out that common factor.

$$\frac{6}{60} = \frac{\overset{1}{\cancel{6}}}{\underset{1}{\cancel{6}} \cdot 10} = \frac{1}{10}$$ Divide out a common factor of 6.

Notice the result $\dfrac{1}{10}$ is in simplest form. If it were not, we would repeat the same procedure until the result was in simplest form.

B DETERMINING WHETHER TWO FRACTIONS ARE EQUIVALENT

How can we check to see whether a simplified fraction is equivalent to an original fraction? Any two fractions are equivalent if their **cross products** are equal. This test for equality is shown in the next example.

Example 7　Determine whether $\dfrac{6}{60}$ and $\dfrac{1}{10}$ are equivalent.

Solution:

$$\frac{6}{60} \times \frac{1}{10}$$

$60 \cdot 1 = 60$　These cross products are
$6 \cdot 10 = 60$　equal, so the fractions are equivalent.

$$\frac{6}{60} = \frac{1}{10}$$

Example 8　Determine whether $\dfrac{8}{11}$ and $\dfrac{19}{26}$ are equivalent.

Solution:

$$\frac{8}{11} \times \frac{19}{26}$$

$11 \cdot 19 = 209$　These cross products are *not*
$8 \cdot 26 = 208$　equal, so the fractions are not equivalent.

$$\frac{8}{11} \neq \frac{19}{26}$$

HELPFUL HINT

Not equal to symbol.

C SOLVING PROBLEMS BY WRITING FRACTIONS IN SIMPLEST FORM

Many real-life problems can be solved by writing fractions. To make the answers more clear, these fractions should be written in simplest form.

TEACHING TIP

After discussing cross products, have students use cross products to check if $\dfrac{6}{15} = \dfrac{34}{85}$. Then have them rewrite the fractions by writing the prime factorization for each part of the fraction: $\dfrac{2 \cdot 3}{3 \cdot 5} = \dfrac{2 \cdot 17}{5 \cdot 17}$. Help students see that if you eliminate common factors from the numerator and denominator, both sides are equal to $\dfrac{2}{5}$.

Practice Problem 7

Determine whether $\dfrac{7}{9}$ and $\dfrac{21}{27}$ are equivalent.

Practice Problem 8

Determine whether $\dfrac{4}{13}$ and $\dfrac{5}{18}$ are equivalent.

Answers

7. are equivalent,　**8.** are not equivalent

Practice Problem 9

Eighty pigs were used in a recent study of olestra, a calorie-free fat substitute. A group of 12 of these pigs were fed a diet high in fat. What fraction of the pigs were fed the high-fat diet in this study? Write your answer in simplest form. (*Source*: from a study conducted by the Procter & Gamble Company)

Example 9   Calculating Fraction of Parks in Washington State

As of 1998, there were 54 national parks in the United States. Three of these parks are located in the state of Washington. What fraction of the United States' national parks can be found in Washington state? Write the fraction in simplest form. (*Source*: National Park Service)

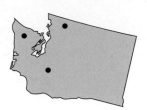

Solution: First we determine the fraction of parks found in Washington state.

$$\dfrac{3}{54} \quad \leftarrow \text{national parks in Washington}$$
$$\phantom{\dfrac{3}{54}} \quad \leftarrow \text{total national parks}$$

Next we simplify the fraction.

$$\frac{3}{54} = \frac{\overset{1}{\cancel{3}}}{\underset{1}{\cancel{3}} \cdot 18} = \frac{1}{18}$$

Thus, $\dfrac{1}{18}$ of the United States' national parks are in Washington.   ▬▬▬▬

Answer

9. $\dfrac{3}{20}$

EXERCISE SET 2.3

A *Write each fraction in simplest form. See Examples 1 through 6.*

1. $\frac{3}{12}$

2. $\frac{5}{20}$

3. $\frac{7}{35}$

4. $\frac{9}{48}$

5. $\frac{14}{16}$

6. $\frac{18}{34}$

7. $\frac{24}{30}$

8. $\frac{70}{80}$

9. $\frac{35}{42}$

10. $\frac{25}{55}$

11. $\frac{63}{72}$

12. $\frac{56}{64}$

13. $\frac{21}{49}$

14. $\frac{14}{35}$

15. $\frac{24}{40}$

16. $\frac{36}{54}$

17. $\frac{36}{63}$

18. $\frac{39}{42}$

19. $\frac{12}{15}$

20. $\frac{18}{24}$

21. $\frac{25}{40}$

22. $\frac{36}{42}$

23. $\frac{27}{90}$

24. $\frac{60}{150}$

25. $\frac{36}{24}$

26. $\frac{60}{36}$

27. $\frac{40}{64}$

28. $\frac{28}{60}$

29. $\frac{70}{196}$

30. $\frac{98}{126}$

31. $\frac{66}{308}$

32. $\frac{65}{234}$

33. equivalent

34. equivalent

35. not equivalent

36. not equivalent

37. not equivalent

38. not equivalent

39. equivalent

40. equivalent

41. equivalent

42. not equivalent

43. not equivalent

44. equivalent

45. $\frac{3}{4}$ of a shift

46. $\frac{1}{10}$

47. $\frac{1}{2}$ mile

48. $\frac{1}{5}$ meter

49. $\frac{47}{74}$ of individuals

50. $\frac{59}{201}$ of Hallmark employees

124

Name _____

B *Determine whether each pair of fractions are equivalent. See Examples 7 and 8.*

33. $\frac{10}{15}$ and $\frac{6}{9}$

34. $\frac{9}{12}$ and $\frac{15}{20}$

35. $\frac{7}{11}$ and $\frac{5}{8}$

36. $\frac{6}{7}$ and $\frac{7}{8}$

37. $\frac{10}{13}$ and $\frac{12}{15}$

38. $\frac{2}{3}$ and $\frac{29}{43}$

39. $\frac{3}{9}$ and $\frac{6}{18}$

40. $\frac{3}{11}$ and $\frac{33}{121}$

41. $\frac{4}{10}$ and $\frac{6}{15}$

42. $\frac{2}{5}$ and $\frac{4}{11}$

43. $\frac{1}{7}$ and $\frac{2}{8}$

44. $\frac{2}{8}$ and $\frac{5}{20}$

C *Solve. Write each fraction in simplest form. See Example 9.*

45. A work shift for an employee at McDonald's consists of 8 hours. What fraction of the employee's work shift is represented by 6 hours?

46. Two thousand baseball caps were sold one year at the U.S. Open Golf Tournament. What fractional part of this total does 200 caps represent?

47. There are 5280 feet in a mile. What fraction of a mile is represented by 2640 feet?

48. There are 100 centimeters in 1 meter. What fraction of a meter is 20 centimeters?

49. As of the beginning of 1997, a total of 370 individuals from around the world had flown in space. Of these, 235 were citizens of the United States. What fraction of individuals who have flown in space were Americans? (*Source:* Congressional Research Service)

50. Hallmark Cards employs 20,100 full-time employees worldwide. About 5900 employees work at the Hallmark headquarters in Kansas City, Missouri. What fraction of Hallmark employees work in Kansas City? (*Source:* Hallmark Cards, Inc.)

Name _____

51. Fifteen states in the United States have Ritz-Carlton hotels. (*Source:* Marriott International)

a. What fraction of states can claim at least one Ritz-Carlton hotel?

b. How many states do not have a Ritz-Carlton hotel?

c. Write the fraction of states without a Ritz-Carlton hotel.

52. As of 1998, there were 54 national parks in the United States. Eight of these parks are located in Alaska. (*Source:* National Park Service)

a. What fraction of the United States' national parks can be found in Alaska?

b. How many of the United States' national parks are found outside Alaska?

c. Write the fraction of national parks found in states other than Alaska.

REVIEW AND PREVIEW

Multiply. See Section 1.5.

53.
$$\begin{array}{r} 91 \\ \times\ 4 \\ \hline \end{array}$$

54.
$$\begin{array}{r} 73 \\ \times\ 8 \\ \hline \end{array}$$

55.
$$\begin{array}{r} 387 \\ \times\ 6 \\ \hline \end{array}$$

56.
$$\begin{array}{r} 562 \\ \times\ 9 \\ \hline \end{array}$$

57.
$$\begin{array}{r} 72 \\ \times\ 35 \\ \hline \end{array}$$

58.
$$\begin{array}{r} 238 \\ \times\ 26 \\ \hline \end{array}$$

COMBINING CONCEPTS

Determine whether each is true or false. If false, explain why.

59. $\dfrac{14}{42} = \dfrac{\overset{1}{\cancel{2}} \cdot \overset{1}{\cancel{7}}}{\underset{1}{\cancel{2}} \cdot 3 \cdot \underset{1}{\cancel{7}}} = \dfrac{0}{3}$

60. A proper fraction cannot be equivalent to an improper fraction.

Write each fraction in simplest form.

61. $\dfrac{372}{620}$

62. $\dfrac{9506}{12,222}$

---

51. a. $\dfrac{3}{10}$

b. 35 states

c. $\dfrac{7}{10}$

52. a. $\dfrac{4}{27}$

b. 46 parks

c. $\dfrac{23}{27}$

53. 364

54. 584

55. 2322

56. 5058

57. 2520

58. 6188

59. false; $1 \cdot 1 = 1$

60. true

61. $\dfrac{3}{5}$

62. $\dfrac{7}{9}$

125

63. $\frac{9}{25}$

64. $\frac{11}{25}$

65. $\frac{1}{25}$

66. $\frac{1}{10}$

67. $\frac{3}{20}$

Name _____

There are generally considered to be eight basic blood types. The table shows the number of people with the various blood types in a typical group of 100 blood donors. Use the table to answer Exercises 63–67. Write each answer in simplest form.

DISTRIBUTION OF BLOOD TYPES IN BLOOD DONORS	
Blood type	Number of People
O Rh-positive	37
O Rh-negative	7
A Rh-positive	36
A Rh-negative	6
B Rh-positive	9
B Rh-negative	1
AB Rh-positive	3
AB Rh-negative	1

(*Source:* American Red Cross Biomedical Services)

63. What fraction of blood donors have blood type A Rh-positive?

64. What fraction of blood donors have an O blood type?

65. What fraction of blood donors have an AB blood type?

66. What fraction of blood donors have a B blood type?

67. What fraction of blood donors have the negative Rh-factor?

126

Integrated Review—Summary on Fractions, Mixed Numbers, and Factors

Use a fraction to represent the shaded area of each figure or figure group.

1.

2.

3.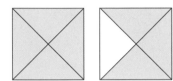

4. In a survey, 73 people out of 85 get less than 8 hours of sleep each night. What fraction of people in the survey get less than 8 hours of sleep?

Write each mixed number as an improper fraction.

5. $3\frac{1}{8}$ **6.** $5\frac{3}{5}$ **7.** $9\frac{6}{7}$

Write each improper fraction as a mixed number or a whole number.

8. $\frac{20}{7}$ **9.** $\frac{55}{11}$ **10.** $\frac{39}{8}$

List the factors of each number.

11. 35   **12.** 40   **13.** 72

Write the prime factorization of each number. Write any repeated factors using exponents.

14. 6   **15.** 70   **16.** 252

ANSWERS

1. $\frac{1}{4}$

2. $\frac{3}{6}$

3. $\frac{7}{4}$

4. $\frac{73}{85}$

5. $\frac{25}{8}$

6. $\frac{28}{5}$

7. $\frac{69}{7}$

8. $2\frac{6}{7}$

9. 5

10. $4\frac{7}{8}$

11. 1, 5, 7, 35

12. 1, 2, 4, 5, 8, 10, 20, 40

13. 1, 2, 3, 4, 6, 8, 9, 12, 18, 24, 36, 72

14. $2 \cdot 3$

15. $2 \cdot 5 \cdot 7$

16. $2^2 \cdot 3^2 \cdot 7$

17. $\frac{1}{7}$

18. $\frac{5}{6}$

19. $\frac{9}{19}$

20. $\frac{21}{55}$

21. $\frac{1}{2}$

22. $\frac{9}{10}$

23. not equivalent

24. equivalent

25. $\frac{1}{25}$

128

Write each fraction in simplest form.

17. $\frac{2}{14}$ **18.** $\frac{20}{24}$ **19.** $\frac{18}{38}$

20. $\frac{42}{110}$ **21.** $\frac{32}{64}$ **22.** $\frac{72}{80}$

Determine whether each pair of fractions are equivalent.

23. $\frac{7}{8}$ and $\frac{9}{10}$ **24.** $\frac{10}{12}$ and $\frac{15}{18}$

25. Of the 50 United States, 2 states are not adjacent to any other states. What fraction of the states are not adjacent to other states? Write the fraction in simplest form.

2.4 MULTIPLYING FRACTIONS

A MULTIPLYING FRACTIONS

Let's use a diagram to discover how fractions are multiplied. For example, to multiply $\frac{1}{2}$ and $\frac{3}{4}$, we find $\frac{1}{2}$ of $\frac{3}{4}$. To do this, we begin with a diagram showing $\frac{3}{4}$ of rectangle's area shaded.

 $\frac{3}{4}$ of the rectangle's area is shaded.

To find $\frac{1}{2}$ of $\frac{3}{4}$, we heavily shade $\frac{1}{2}$ of the part that is already shaded.

By counting smaller rectangles, we see that $\frac{3}{8}$ of the larger rectangle is now heavily shaded, so that $\frac{1}{2}$ of $\frac{3}{4}$ is $\frac{3}{8}$. This means that

$$\frac{1}{2} \cdot \frac{3}{4} = \frac{3}{8}$$ Notice that $\frac{1}{2} \cdot \frac{3}{4} = \frac{1 \cdot 3}{2 \cdot 4} = \frac{3}{8}$.

> **MULTIPLYING FRACTIONS**
>
> To multiply two fractions, multiply the numerators and multiply the denominators.

Examples   Multiply.

1. $\dfrac{2}{3} \cdot \dfrac{5}{11} = \dfrac{2 \cdot 5}{3 \cdot 11} = \dfrac{10}{33}$

2. $\dfrac{1}{4} \cdot \dfrac{1}{2} = \dfrac{1 \cdot 1}{4 \cdot 2} = \dfrac{1}{8}$

Example 3   Multiply and simplify: $\dfrac{6}{7} \cdot \dfrac{14}{27}$

Solution: $\dfrac{6}{7} \cdot \dfrac{14}{27} = \dfrac{6 \cdot 14}{7 \cdot 27}$

We can simplify by prime factoring and dividing out common factors.

$$\frac{6 \cdot 14}{7 \cdot 27} = \frac{2 \cdot \overset{1}{\cancel{3}} \cdot 2 \cdot \overset{1}{\cancel{7}}}{\underset{1}{\cancel{7}} \cdot \underset{1}{\cancel{3}} \cdot 3 \cdot 3}$$

$$= \frac{4}{9}$$

Objectives

A Multiply fractions.

B Multiply fractions and mixed numbers or whole numbers.

C Solve problems by multiplying fractions.

SSM   CD-ROM   Video 2.4

TEACHING TIP

Before multiplying two fractions, consider asking students to write a fraction as a multiplication problem:

$\frac{3}{4} = 3 \cdot \frac{1}{4}$

Then have them write words to describe

$2 \cdot \frac{3}{4}$

Two sets of $\frac{3}{4}$ which is

Two sets of three sets of $\frac{1}{4}$ which is

Six sets of $\frac{1}{4}$ which is

Six fourths

Finally have them convert their words to math symbols:

$2 \cdot \frac{3}{4} = (2 \cdot 3) \cdot \frac{1}{4} = 6 \cdot \frac{1}{4} = \frac{6}{4}$

Practice Problems 1–2

Multiply.

1. $\dfrac{3}{8} \cdot \dfrac{5}{7}$ 2. $\dfrac{1}{3} \cdot \dfrac{1}{6}$

Practice Problem 3

Multiply and simplify: $\dfrac{6}{11} \cdot \dfrac{5}{8}$

TEACHING TIP

Before discussing multiplying fractions, help students develop an intuitive understanding of why denominators are multiplied. Ask, "If you were to eat $\frac{1}{2}$ of $\frac{1}{4}$ of a pie, how much of the pie would you eat?" Allow students to answer $\frac{1}{8}$ and show them that this is equivalent to $\frac{1}{2} \cdot \frac{1}{4} = \frac{1}{8}$. Ask what happened to the denominators during multiplication.

Answers

1. $\dfrac{15}{56}$, 2. $\dfrac{1}{18}$, 3. $\dfrac{15}{44}$

HELPFUL HINT

In simplifying a product, it may be possible to identify common factors without actually writing the prime factorization. For example,

$$\frac{10}{11} \cdot \frac{1}{20} = \frac{10 \cdot 1}{11 \cdot 20} = \frac{\overset{1}{\cancel{10}} \cdot 1}{11 \cdot \cancel{10} \cdot 2} = \frac{1}{22}$$

Practice Problem 4

Multiply and simplify: $\frac{4}{15} \cdot \frac{3}{8}$

Example 4 Multiply and simplify: $\frac{23}{32} \cdot \frac{4}{7}$

Solution: Notice that 4 and 32 have a common factor of 4.

$$\frac{23}{32} \cdot \frac{4}{7} = \frac{23 \cdot 4}{32 \cdot 7} = \frac{23 \cdot \overset{1}{\cancel{4}}}{\underset{1}{\cancel{4}} \cdot 8 \cdot 7} = \frac{23}{56}$$

After multiplying two fractions, always check to see whether the product can be simplified.

Practice Problems 5–6

Multiply.

5. $\frac{2}{5} \cdot \frac{15}{17}$

6. $\frac{4}{11} \cdot \frac{33}{16}$

Examples Multiply.

5. $\frac{3}{4} \cdot \frac{8}{5} = \frac{3 \cdot 8}{4 \cdot 5} = \frac{3 \cdot \overset{1}{\cancel{4}} \cdot 2}{\underset{1}{\cancel{4}} \cdot 5} = \frac{6}{5}$

6. $\frac{6}{13} \cdot \frac{26}{30} = \frac{6 \cdot 26}{13 \cdot 30} = \frac{\overset{1}{\cancel{6}} \cdot \overset{1}{\cancel{13}} \cdot 2}{\underset{1}{\cancel{13}} \cdot \underset{1}{\cancel{6}} \cdot 5} = \frac{2}{5}$

B **MULTIPLYING FRACTIONS AND MIXED NUMBERS OR WHOLE NUMBERS**

When multiplying a fraction and a mixed or a whole number, remember that mixed and whole numbers can be written as fractions.

> **MULTIPLYING FRACTIONS AND MIXED NUMBERS OR WHOLE NUMBERS**
>
> To multiply a fraction and a mixed number or a whole number, first write the mixed or whole number as a fraction and then multiply as usual.

Practice Problem 7

Multiply and simplify: $2\frac{1}{2} \cdot \frac{8}{15}$

Example 7 Multiply: $3\frac{1}{3} \cdot \frac{7}{8}$

Solution: The mixed number $3\frac{1}{3}$ can be written as the fraction $\frac{10}{3}$. Then,

$$3\frac{1}{3} \cdot \frac{7}{8} = \frac{10}{3} \cdot \frac{7}{8} = \frac{\overset{1}{\cancel{2}} \cdot 5 \cdot 7}{3 \cdot \underset{1}{\cancel{2}} \cdot 4} = \frac{35}{12} \quad \text{or} \quad 2\frac{11}{12}$$

Don't forget that a whole number can be written as a fraction by writing the whole number over 1. For example,

$$20 = \frac{20}{1} \quad \text{and} \quad 7 = \frac{7}{1}$$

Answers

4. $\frac{1}{10}$, **5.** $\frac{6}{17}$, **6.** $\frac{3}{4}$, **7.** $\frac{4}{3}$ or $1\frac{1}{3}$

Examples   Multiply.

8. $1\frac{2}{3} \cdot 2\frac{1}{4} = \frac{5}{3} \cdot \frac{9}{4} = \frac{5 \cdot 9}{3 \cdot 4} = \frac{5 \cdot \overset{1}{\cancel{3}} \cdot 3}{\underset{1}{\cancel{3}} \cdot 4} = \frac{15}{4}$ or $3\frac{3}{4}$

9. $\frac{3}{4} \cdot 20 = \frac{3}{4} \cdot \frac{20}{1} = \frac{3 \cdot 20}{4 \cdot 1} = \frac{3 \cdot \overset{1}{\cancel{4}} \cdot 5}{\underset{1}{\cancel{4}} \cdot 1} = \frac{15}{1}$ or 15

TRY THE CONCEPT CHECK IN THE MARGIN.

C SOLVING PROBLEMS BY MULTIPLYING FRACTIONS

To solve real-life problems that involve multiplying fractions, we will use our four problem-solving steps from Chapter 1. In Example 10, a new key word that implies multiplication is used. That key word is "of."

Example 10   Finding Number of Roller Coasters in an Amusement Park

Cedar Point is an amusement park located in Sandusky, Ohio. Its collection of 60 rides is the largest in the world. One-fifth of Cedar Point's rides are roller coasters. How many roller coasters are in Cedar Point's collection of rides? (*Source:* Cedar Fair, L.P.)

Solution: **1.** UNDERSTAND the problem. To do so, read and reread the problem. We are told that $\frac{1}{5}$ of Cedar Point's rides are roller coasters. The word "of" here means multiplication.
2. TRANSLATE.

In words:	Number of roller coasters	is	$\frac{1}{5}$	of	total rides at Cedar Point
	↓	↓	↓	↓	↓
Translate:	Number of roller coasters	=	$\frac{1}{5}$	·	60

3. SOLVE.

$\frac{1}{5} \cdot 60 = \frac{1}{5} \cdot \frac{60}{1} = \frac{1 \cdot 60}{5 \cdot 1} = \frac{1 \cdot \overset{1}{\cancel{5}} \cdot 12}{\underset{1}{\cancel{5}} \cdot 1} = \frac{12}{1}$ or 12

4. INTERPRET. *Check* your work. *State* your conclusion: The number of roller coasters at Cedar Point is 12.

Practice Problems 8–9

Multiply.

8. $3\frac{1}{5} \cdot 2\frac{3}{4}$

9. $\frac{2}{3} \cdot 18$

✓ CONCEPT CHECK

a. Find the error.
$2\frac{1}{4} \cdot \frac{1}{2} = 2\frac{1 \cdot 1}{4 \cdot 2} = 2\frac{1}{8}$

b. How could you estimate the product $3\frac{1}{7} \cdot 4\frac{5}{6}$?

Practice Problem 10

About $\frac{1}{3}$ of all plant and animal species in the United States are at risk of becoming extinct. There are 20,439 known species of plants and animals in the United States. How many species are at risk of extinction? (*Source:* The Nature Conservancy)

TEACHING TIP   Classroom Activity

Ask students to work in groups to make up their own word problems involving the multiplication of fractions. Have students write 4 word problems: one which involves a fraction and a whole number, one which involves two fractions, one which involves a fraction and a mixed number, and one which involves two mixed numbers. Then have them give their problems to another group to solve.

Answers

8. $\frac{44}{5}$ or $8\frac{4}{5}$, **9.** 12, **10.** 6813 species

✓ **Concept Check**

a. forgot to change mixed number to fraction,
b. round mixed numbers to closest whole numbers and multiply

Focus On the Real World

BLOOD AND BLOOD DONATION

Blood is the workhorse of the body. It carries to the body's tissues everything they need, from nutrients to antibodies to heat. Blood also carries away waste products like carbon dioxide. Blood contains three types of cells—red blood cells, white blood cells, and platelets—suspended in clear, watery fluid called plasma. Blood is $\frac{11}{20}$ plasma, and plasma itself is $\frac{9}{10}$ water. In the average healthy adult human, blood accounts for $\frac{1}{11}$ of a person's body weight.

Roughly every 2 seconds someone in the United States needs blood. Although only $\frac{1}{20}$ of eligible donors donate blood, the American Red Cross is still able to collect nearly 6 million volunteer donations of blood each year. This volume makes Red Cross Biomedical Services the largest blood supplier for blood transfusions in the United States.

The modern Red Cross blood donation program has its roots in World War II. In 1940, Dr. Charles Drew headed the Red Cross-sponsored American blood collection efforts for bombing victims in Great Britain. During that time, Dr. Drew developed techniques for separating plasma from blood cells that allowed mass production of plasma. Dr. Drew had discovered that plasma could be preserved longer than whole blood. He also found that dried plasma could be stored longer than its liquid form. By 1941, Dr. Drew had become the first medical director of the first American Red Cross Blood Bank in the United States. Plasma collected through this program saved the lives of many wounded civilians and Allied soldiers by reducing the high rate of death from shock.

GROUP ACTIVITY

Contact your local Red Cross blood service office. Find out how many people donated blood in your area in the past two months. Ask if it is possible to get a breakdown of the blood donations by blood type. (For more on blood type, see Exercises 63–67 in Section 2.3.)

1. Research the population of the area served by your local Red Cross Blood Service office. Write the fraction of the local population who gave blood in the past two months.
2. Use the breakdown by blood type to write the fraction of donors giving each type of blood.

EXERCISE SET 2.4

A *Multiply. Write each answer in simplest form. See Examples 1 through 6.*

1. $\frac{1}{3} \cdot \frac{2}{5}$ **2.** $\frac{2}{3} \cdot \frac{4}{7}$ 📼**3.** $\frac{6}{5} \cdot \frac{1}{7}$ **4.** $\frac{7}{3} \cdot \frac{2}{3}$

5. $\frac{3}{10} \cdot \frac{3}{8}$ **6.** $\frac{2}{5} \cdot \frac{7}{11}$ **7.** $\frac{7}{8} \cdot \frac{2}{3}$ **8.** $\frac{5}{9} \cdot \frac{7}{4}$

📼**9.** $\frac{2}{7} \cdot \frac{5}{8}$ **10.** $\frac{5}{8} \cdot \frac{1}{3}$ **11.** $\frac{1}{2} \cdot \frac{2}{15}$ **12.** $\frac{3}{8} \cdot \frac{5}{12}$

13. $\frac{5}{8} \cdot \frac{9}{4}$ **14.** $\frac{8}{15} \cdot \frac{5}{32}$ 📼**15.** $\frac{5}{28} \cdot \frac{2}{25}$ **16.** $\frac{4}{25} \cdot \frac{5}{8}$

17. $\frac{18}{20} \cdot \frac{36}{99}$ **18.** $\frac{5}{32} \cdot \frac{64}{100}$ **19.** $\frac{3}{2} \cdot \frac{7}{3}$ **20.** $\frac{15}{2} \cdot \frac{3}{5}$

21. $\frac{14}{21} \cdot \frac{15}{16}$ **22.** $\frac{15}{20} \cdot \frac{2}{7}$ **23.** $\frac{1}{10} \cdot \frac{1}{11}$ **24.** $\frac{1}{9} \cdot \frac{1}{4}$

25. $\frac{3}{8} \cdot \frac{9}{10}$ **26.** $\frac{4}{5} \cdot \frac{8}{25}$ **27.** $\frac{6}{5} \cdot \frac{5}{6}$ **28.** $\frac{9}{10} \cdot \frac{10}{9}$

29. $\frac{7}{72} \cdot \frac{9}{49}$ **30.** $\frac{3}{80} \cdot \frac{2}{9}$

B *Multiply. Write each answer in simplest form. See Examples 7 through 9.*

31. $3 \cdot \frac{1}{4}$ **32.** $\frac{2}{3} \cdot 6$ 📼**33.** $\frac{5}{8} \cdot 4$ **34.** $3 \cdot \frac{7}{8}$

35. $\frac{1}{5}$

36. $\frac{66}{35}$ or $1\frac{31}{35}$

37. $\frac{5}{3}$ or $1\frac{2}{3}$

38. $\frac{1}{2}$

39. $\frac{2}{3}$

40. $\frac{20}{9}$ or $2\frac{2}{9}$

41. $\frac{77}{10}$ or $7\frac{7}{10}$

42. $\frac{513}{32}$ or $16\frac{1}{32}$

43. $\frac{836}{35}$ or $23\frac{31}{35}$

44. 42

45. 12

46. 21

47. $\frac{25}{2}$ or $12\frac{1}{2}$

48. 20

49. 15

50. $\frac{42}{5}$ or $8\frac{2}{5}$

51. $\frac{3}{2}$ in. or $1\frac{1}{2}$ in.

52. 30 gallons

53. $\frac{55}{2}$ miles or $27\frac{1}{2}$ miles

54. \$400/month

Name _____

35. $1\frac{1}{4} \cdot \frac{4}{25}$ **36.** $2\frac{1}{5} \cdot \frac{6}{7}$ **37.** $\frac{2}{5} \cdot 4\frac{1}{6}$ **38.** $\frac{3}{22} \cdot 3\frac{2}{3}$

39. $\frac{2}{3} \cdot 1$ **40.** $4 \cdot \frac{5}{9}$ 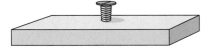 **41.** $2\frac{1}{5} \cdot 3\frac{1}{2}$ **42.** $2\frac{1}{4} \cdot 7\frac{1}{8}$

43. $3\frac{4}{5} \cdot 6\frac{2}{7}$ **44.** $5\frac{5}{6} \cdot 7\frac{1}{5}$ **45.** $\frac{3}{4} \cdot 16$ **46.** $\frac{7}{8} \cdot 24$

47. $5 \cdot 2\frac{1}{2}$ **48.** $6 \cdot 3\frac{1}{3}$ **49.** $1\frac{1}{5} \cdot 12\frac{1}{2}$ **50.** $1\frac{1}{6} \cdot 7\frac{1}{5}$

C *Solve. Write each answer in simplest form. See Example 10.*

51. Each turn of a screw sinks it $\frac{3}{16}$ of an inch deeper into a piece of wood. Find how deep the screw is after 8 turns.

52. A veterinarian's dipping vat holds 36 gallons of liquid. She normally fills it $\frac{5}{6}$ full of a medicated flea dip solution. Find how many gallons of solution are normally in the vat.

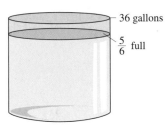

36 gallons

$\frac{5}{6}$ full

53. Holly Hanson drives $5\frac{1}{2}$ miles a day to and from the Star Five television station where she is the anchor woman for the evening news 5 days a week. How far does she drive in a week?

54. The Braybendre family has a take home pay of \$1400 a month. They spend $\frac{2}{7}$ of their monthly income on their house payment. How much is their house payment?

55. An estimate for the measure of an adult's wrist is $\frac{1}{4}$ of the waist size. If Jorge has a 34-inch waist, estimate the size of his wrist.

56. The market value of a house is $\frac{3}{4}$ of its appraised value. If the appraised value of Pauline's house is $60,000, find its market value.

57. The plans for a deck call for $\frac{2}{5}$ of a 4-foot post to be underground. Find the length of the post that is to be buried.

58. A patient was told that no more than $\frac{1}{5}$ of his calories should come from fat. If his diet consists of 3000 calories a day, how many of these calories can come from fat?

59. The radius of a circle is one-half of its diameter as shown. If the diameter of a circle is $\frac{3}{8}$ of an inch, what is its radius?

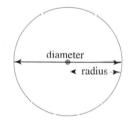

60. A recipe calls for $\frac{1}{3}$ of a cup of flour. How much flour should be used if only $\frac{1}{2}$ of the recipe is being made?

61. A special on a cruise to the Bahamas is advertised to be $\frac{2}{3}$ of the regular price. If the regular price is $2757, what is the sale price?

62. The Gonzales' recently sold their house for $102,000, but $\frac{3}{50}$ of this amount goes to the real estate companies that helped them sell their house. How much money do the Gonzales' pay to the real estate companies?

55. $\frac{17}{2}$ in. or $8\frac{1}{2}$ in.

56. $45,000

57. $\frac{8}{5}$ ft or $1\frac{3}{5}$ ft

58. 600 calories

59. $\frac{3}{16}$ of an in.

60. $\frac{1}{6}$ of a cup

61. $1838

62. $6120

135

63. $\dfrac{39}{2}$ or $19\dfrac{1}{2}$ in.

64. $\dfrac{45}{2}$ or $22\dfrac{1}{2}$ g

65. $\dfrac{1}{14}$ sq. ft

66. $\dfrac{3}{16}$ sq. mi

67. $\dfrac{7}{2}$ or $3\dfrac{1}{2}$ sq. yd

68. $\dfrac{35}{2}$ or $17\dfrac{1}{2}$ sq. in.

69. $\dfrac{121}{4}$ or $30\dfrac{1}{4}$ sq. in.

70. $\dfrac{57}{2}$ or $28\dfrac{1}{2}$ sq. yd

136

63. A sidewalk is built 6 bricks wide by laying each brick side by side. How many inches wide is the sidewalk if each brick measures $3\dfrac{1}{4}$ inches wide?

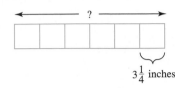

64. The nutrition label on a can of crushed pineapple shows there are 9 grams of carbohydrates for each cup of pineapple. How many grams of carbohydrates are in a $2\dfrac{1}{2}$ cup can?

Find the area of each rectangle. Recall that area $= (length) \cdot (width)$.

65.

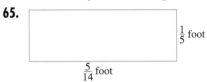

$\dfrac{1}{5}$ foot

$\dfrac{5}{14}$ foot

66.

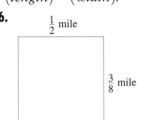

$\dfrac{1}{2}$ mile

$\dfrac{3}{8}$ mile

67.

$1\dfrac{3}{4}$ yards

2 yards

68.

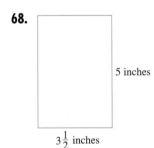

5 inches

$3\dfrac{1}{2}$ inches

69. A model for a proposed computer chip measures $5\dfrac{1}{2}$ inches by $5\dfrac{1}{2}$ inches. Find its area.

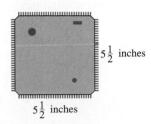

$5\dfrac{1}{2}$ inches

$5\dfrac{1}{2}$ inches

70. The Saltalamachio's are planning to build a deck that measures $4\dfrac{1}{2}$ yards by $6\dfrac{1}{3}$ yards. Find the area of their proposed deck.

The following graph is called a circle graph or pie chart. Each sector (shaped like a piece of pie) shows the fractional part of a car's total mileage that falls into a particular category. The whole circle represents a car's total mileage.

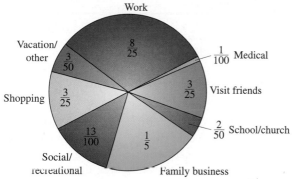

(*Source:* The American Automobile Manufacturers Assn. and The National Automobile Dealers Assn.)

In one year, the Rodriguez family drove 12,000 miles in the family car. Use the circle graph to determine how many of these miles might be expected to fall in the categories shown in Exercises 71–74.

71. Work   **72.** Shopping   **73.** Family business   **74.** Medical

REVIEW AND PREVIEW

Divide. See Section 1.6.

75. $8\overline{)1648}$ **76.** $7\overline{)3920}$ **77.** $23\overline{)1300}$ **78.** $31\overline{)2500}$

COMBINING CONCEPTS

79. In 1996 there were 33,861,000 Americans aged 65 years or older. About $\frac{2}{5}$ of these older Americans had annual incomes under $10,000. How many older Americans had incomes of less than $10,000? (*Source:* U.S. Bureau of the Census)

80. Approximately $\frac{4}{5}$ of the funding for public television comes from sources other than the federal government, such as individuals, businesses, and state governments. In 1996, public television stations' total income was $1,490,000,000. About how much of this income came from nonfederal sources? (*Source:* Corporation for Public Broadcasting)

71. 3840 mi

72. 1440 mi

73. 2400 mi

74. 120 mi

75. 206

76. 560

77. 56 R 12

78. 80 R 20

79. 13,544,400

80. $1,192,000,000

Name _____

Internet Excursions

Go to http://www.prenhall.com/martin-gay
This World Wide Web site will provide access to a site called the Low Fat Vegetarian Archive, or a related site. It contains more than 2500 recipes for healthy low-fat and fat-free vegetarian dishes. Users can search the recipes by category, such as Mexican foods or desserts, or by ingredient.

81. Visit this site and find a recipe that is interesting to you. Be sure that the ingredient list contains at least two measures that are fractions (for example, $\frac{1}{3}$ cup). Print or copy the recipe. Then rewrite the list of ingredients to show the amounts that would be needed to make $\frac{1}{2}$ of the recipe.

82. Find a different recipe at this site that is interesting to you. Make sure that the recipe lists at least two fractional measures. Print or copy the recipe. Then rewrite the list of ingredients to show the amounts that would be needed to triple the recipe.

2.5   DIVIDING FRACTIONS

A   FINDING RECIPROCALS OF FRACTIONS

Before we can divide fractions, we need to know how to find the **reciprocal** of a fraction.

> **RECIPROCAL OF A FRACTION**
>
> Two numbers are **reciprocals** of each other if their product is 1.

For example,

$$\frac{2}{3} \cdot \frac{3}{2} = \frac{2 \cdot 3}{3 \cdot 2} = \frac{6}{6} = 1 \qquad \text{so } \tfrac{2}{3} \text{ and } \tfrac{3}{2} \text{ are reciprocals.}$$

$$4 \cdot \frac{1}{4} = \frac{4}{1} \cdot \frac{1}{4} = \frac{4 \cdot 1}{1 \cdot 4} = \frac{4}{4} = 1 \qquad \text{so } 4 \text{ and } \tfrac{1}{4} \text{ are reciprocals.}$$

> **FINDING THE RECIPROCAL OF A FRACTION**
>
> To find the reciprocal of a fraction, interchange its numerator and denominator.

For example, the reciprocal of $\frac{6}{11}$ is $\frac{11}{6}$.

Examples   Find the reciprocal of each fraction.

1. The reciprocal of $\frac{5}{6}$ is $\frac{6}{5}$. $\frac{5}{6} \cdot \frac{6}{5} = \frac{5 \cdot 6}{6 \cdot 5} = \frac{30}{30} = 1$

2. The reciprocal of $\frac{11}{8}$ is $\frac{8}{11}$. $\frac{11}{8} \cdot \frac{8}{11} = \frac{11 \cdot 8}{8 \cdot 11} = \frac{88}{88} = 1$

3. The reciprocal of $\frac{1}{3}$ is $\frac{3}{1}$ or 3. $\frac{1}{3} \cdot \frac{3}{1} = \frac{1 \cdot 3}{3 \cdot 1} = \frac{3}{3} = 1$

4. The reciprocal of 5, or $\frac{5}{1}$, is $\frac{1}{5}$. $\frac{5}{1} \cdot \frac{1}{5} = \frac{5 \cdot 1}{1 \cdot 5} = \frac{5}{5} = 1$

Practice Problems 1–4

Find the reciprocal of each number.

1. $\frac{4}{9}$ 2. $\frac{15}{7}$

3. 7   4. $\frac{1}{8}$

> **HELPFUL HINT**
>
> Every number has a reciprocal except 0. The number 0 has no reciprocal because there is no number that when multiplied by 0 gives a result of 1.

Answers

1. $\frac{9}{4}$,   **2.** $\frac{7}{15}$,   **3.** $\frac{1}{7}$,   **4.** 8

Copyright 1990 Prentice-Hall Inc.

TEACHING TIP

Ask students to rewrite $10 \div 5$ as an equivalent fraction $\left(\dfrac{10}{5}\right)$ and then as an equivalent multiplication problem $10 \div \dfrac{1}{5}$. Then have them write and fill in the blanks for "Dividing by 5 is the same as multiplying by the _____ of five."

B DIVIDING FRACTIONS

Division of fractions has the same meaning as division of whole numbers. For example,

$10 \div 5$ means: How many 5s are there in 10?

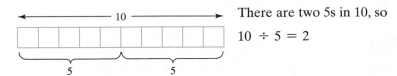

There are two 5s in 10, so

$$10 \div 5 = 2$$

$\dfrac{3}{4} \div \dfrac{1}{8}$ means: How many $\dfrac{1}{8}$s are there in $\dfrac{3}{4}$?

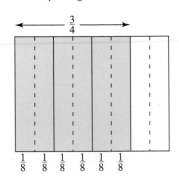

There are six $\dfrac{1}{8}$s in $\dfrac{3}{4}$, so

$$\frac{3}{4} \div \frac{1}{8} = 6$$

We can use reciprocals to divide fractions.

DIVIDING FRACTIONS

To divide two fractions, multiply the first fraction by the reciprocal of the second fraction.

For example,

multiply by reciprocal

$$\frac{3}{4} \div \frac{1}{8} = \frac{3}{4} \cdot \frac{8}{1} = \frac{3 \cdot 8}{4 \cdot 1} = \frac{3 \cdot 2 \cdot \overset{1}{\cancel{4}}}{\underset{1}{\cancel{4}} \cdot 1} = \frac{6}{1} \quad \text{or} \quad 6$$

Just as when you are multiplying fractions, always check to see whether your answer can be simplified when you divide fractions.

Practice Problems 5–7

Divide and simplify.

5. $\dfrac{3}{2} \div \dfrac{14}{5}$

6. $\dfrac{8}{7} \div \dfrac{2}{9}$

7. $\dfrac{4}{9} \div \dfrac{1}{2}$

Examples

Divide and simplify.

5. $\dfrac{7}{8} \div \dfrac{2}{9} = \dfrac{7}{8} \cdot \dfrac{9}{2} = \dfrac{7 \cdot 9}{8 \cdot 2} = \dfrac{63}{16}$

6. $\dfrac{5}{16} \div \dfrac{3}{4} = \dfrac{5}{16} \cdot \dfrac{4}{3} = \dfrac{5 \cdot 4}{16 \cdot 3} = \dfrac{5 \cdot \overset{1}{\cancel{4}}}{\underset{1}{\cancel{4}} \cdot 4 \cdot 3} = \dfrac{5}{12}$

7. $\dfrac{2}{5} \div \dfrac{1}{2} = \dfrac{2}{5} \cdot \dfrac{2}{1} = \dfrac{2 \cdot 2}{5 \cdot 1} = \dfrac{4}{5}$

After dividing two fractions, always check to see whether the product can be simplified.

Answers

5. $\dfrac{15}{28}$,   **6.** $\dfrac{36}{7}$ or $5\dfrac{1}{7}$,   **7.** $\dfrac{8}{9}$

HELPFUL HINT

When dividing fractions, do *not* look for common factors to divide out until you rewrite the division as multiplication.

Do not try to divide out these two 2s.

$$\frac{1}{2} \div \frac{2}{3} = \frac{1}{2} \cdot \frac{3}{2} = \frac{3}{4}$$

C **DIVIDING FRACTIONS AND MIXED NUMBERS OR WHOLE NUMBERS**

Just as with multiplying, remember that mixed or whole numbers are written as fractions before you divide them.

> **DIVIDING FRACTIONS AND MIXED NUMBERS OR WHOLE NUMBERS**
>
> To divide with a mixed number or a whole number, first write the mixed or whole number as a fraction and then divide as usual.

Examples   Divide.

8. $\dfrac{3}{4} \div 5 = \dfrac{3}{4} \cdot \dfrac{1}{5} = \dfrac{3 \cdot 1}{4 \cdot 5} = \dfrac{3}{20}$

9. $\dfrac{11}{18} \div 2\dfrac{5}{6} = \dfrac{11}{18} \div \dfrac{17}{6} = \dfrac{11}{18} \cdot \dfrac{6}{17} = \dfrac{11 \cdot 6}{18 \cdot 17} = \dfrac{11 \cdot \overset{1}{\cancel{6}}}{\underset{1}{\cancel{6}} \cdot 3 \cdot 17} = \dfrac{11}{51}$

10. $5\dfrac{2}{3} \div 2\dfrac{5}{9} = \dfrac{17}{3} \div \dfrac{23}{9} = \dfrac{17}{3} \cdot \dfrac{9}{23} = \dfrac{17 \cdot 9}{3 \cdot 23} = \dfrac{17 \cdot \overset{1}{\cancel{3}} \cdot 3}{\underset{1}{\cancel{3}} \cdot 23} = \dfrac{51}{23}$

    or $\;2\dfrac{5}{23}$

TRY THE CONCEPT CHECK IN THE MARGIN.

D **SOLVING PROBLEMS BY DIVIDING FRACTIONS**

To solve real-life problems that involve dividing fractions, we will continue to use our four problem-solving steps.

Example 11   **Calculating Manufacturing Materials Needed**

In a manufacturing process, a metal-cutting machine cuts strips $1\dfrac{3}{5}$ inches long from a piece of metal stock. How many such strips can be cut from a 48-inch piece of stock?

Practice Problems 8–10

Divide.

8. $\dfrac{4}{9} \div 5$ **9.** $\dfrac{8}{15} \div 3\dfrac{4}{5}$

10. $3\dfrac{2}{5} \div 2\dfrac{2}{15}$

✓ CONCEPT CHECK

a. Which of the following is the correct way to divide $\dfrac{2}{5}$ by $\dfrac{3}{4}$? Or are both correct? Explain.

1. $\dfrac{5}{2} \cdot \dfrac{3}{4}$ 2. $\dfrac{2}{5} \cdot \dfrac{4}{3}$

b. How could you estimate the quotient $8\dfrac{6}{7} \div 2\dfrac{4}{5}$?

Practice Problem 11

A designer of women's clothing designs a woman's dress that requires $2\dfrac{1}{7}$ yards of material. How many dresses can be made from a 30-yard bolt of material?

Answers

8. $\dfrac{4}{45}$,   **9.** $\dfrac{8}{57}$,   **10.** $\dfrac{51}{32}$ or $1\dfrac{19}{32}$,   **11.** 14 dresses

✓ Concept Check

a. 2 is correct,   **b.** Round the mixed numbers to the nearest whole numbers and divide.

Ask students to work in groups to make up word problems involving the division of fractions. Have them write six word problems involving division of a fraction and a whole number, a whole number and fraction, two fractions, a fraction and a mixed number, a mixed number and a fraction, two mixed numbers. Have them give their problems to another group to solve.

Solution:

1. UNDERSTAND the problem. To do so, read and reread the problem. Then draw a diagram.

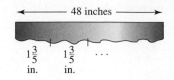

We want to know how many $1\frac{3}{5}$s there are in 48.

2. TRANSLATE.

In words:

Number of strips	is	48	divided by	$1\frac{3}{5}$
↓	↓	↓	↓	↓

Translate: $\text{Number of strips}\ =\ 48\ \div\ 1\frac{3}{5}$

3. SOLVE.

$$48 \div 1\frac{3}{5} = 48 \div \frac{8}{5} = \frac{48}{1} \cdot \frac{5}{8} = \frac{48 \cdot 5}{1 \cdot 8} = \frac{\overset{6}{\cancel{8}} \cdot 6 \cdot 5}{1 \cdot \underset{1}{\cancel{8}}} = \frac{30}{1} \quad \text{or} \quad 30$$

4. INTERPRET. *Check* your work. *State* your conclusion: Thirty strips can be cut from the 48-inch piece of stock.

Name _____ **Section** _____ **Date** _____

EXERCISE SET 2.5

A *Find the reciprocal of each fraction. See Examples 1 through 4.*

1. $\dfrac{4}{7}$

2. $\dfrac{9}{10}$

3. $\dfrac{1}{11}$

4. $\dfrac{1}{20}$

5. 15

6. 13

7. $\dfrac{12}{7}$

8. $\dfrac{10}{3}$

B *Divide. Write each answer in simplest form. See Examples 5 through 7.*

9. $\dfrac{2}{3} \div \dfrac{5}{6}$

10. $\dfrac{5}{8} \div \dfrac{2}{3}$

11. $\dfrac{6}{15} \div \dfrac{12}{5}$

12. $\dfrac{4}{15} \div \dfrac{8}{3}$

13. $\dfrac{8}{9} \div \dfrac{1}{2}$

14. $\dfrac{10}{11} \div \dfrac{4}{5}$

15. $\dfrac{3}{7} \div \dfrac{5}{6}$

16. $\dfrac{16}{27} \div \dfrac{8}{15}$

17. $\dfrac{3}{5} \div \dfrac{4}{5}$

18. $\dfrac{11}{16} \div \dfrac{13}{16}$

19. $\dfrac{11}{20} \div \dfrac{3}{11}$

20. $\dfrac{9}{20} \div \dfrac{2}{9}$

21. $\dfrac{1}{10} \div \dfrac{10}{1}$

22. $\dfrac{3}{13} \div \dfrac{13}{3}$

23. $\dfrac{7}{9} \div \dfrac{7}{3}$

24. $\dfrac{6}{11} : \dfrac{6}{5}$

25. $\dfrac{3}{7} \div \dfrac{4}{7}$

26. $\dfrac{5}{8} \div \dfrac{3}{8}$

27. $\dfrac{7}{8} \div \dfrac{5}{6}$

28. $\dfrac{3}{8} \div \dfrac{5}{8}$

29. $\dfrac{7}{45} \div \dfrac{4}{25}$

30. $\dfrac{14}{52} \div \dfrac{1}{13}$

31. $\dfrac{2}{37} \div \dfrac{1}{7}$

32. $\dfrac{100}{158} \div \dfrac{10}{79}$

33. $\dfrac{3}{25} \div \dfrac{27}{40}$

34. $\dfrac{6}{15} \div \dfrac{7}{10}$

ANSWERS

1. $\dfrac{7}{4}$
2. $\dfrac{10}{9}$
3. 11
4. 20
5. $\dfrac{1}{15}$
6. $\dfrac{1}{13}$
7. $\dfrac{7}{12}$
8. $\dfrac{3}{10}$
9. $\dfrac{4}{5}$
10. $\dfrac{15}{16}$
11. $\dfrac{1}{6}$
12. $\dfrac{1}{10}$
13. $\dfrac{16}{9}$ or $1\dfrac{7}{9}$
14. $\dfrac{25}{22}$ or $1\dfrac{3}{22}$
15. $\dfrac{18}{35}$
16. $\dfrac{10}{9}$ or $1\dfrac{1}{9}$
17. $\dfrac{3}{4}$
18. $\dfrac{11}{13}$
19. $\dfrac{121}{60}$ or $2\dfrac{1}{60}$
20. $\dfrac{81}{40}$ or $2\dfrac{1}{40}$
21. $\dfrac{1}{100}$
22. $\dfrac{9}{169}$
23. $\dfrac{1}{3}$
24. $\dfrac{5}{11}$
25. $\dfrac{3}{4}$
26. $\dfrac{5}{3}$ or $1\dfrac{2}{3}$
27. $\dfrac{21}{20}$ or $1\dfrac{1}{20}$
28. $\dfrac{3}{5}$
29. $\dfrac{35}{36}$
30. $\dfrac{7}{2}$ or $3\dfrac{1}{2}$
31. $\dfrac{14}{37}$
32. 5
33. $\dfrac{8}{45}$
34. $\dfrac{4}{7}$

35. $\frac{1}{6}$

36. $\frac{1}{12}$

37. $\frac{40}{3}$ or $13\frac{1}{3}$

38. $\frac{77}{2}$ or $38\frac{1}{2}$

39. 5

40. $\frac{35}{3}$ or $11\frac{2}{3}$

41. $\frac{5}{28}$

42. $\frac{8}{75}$

43. $\frac{36}{35}$ or $1\frac{1}{35}$

44. $\frac{7}{12}$

45. 96

46. 54

47. $\frac{35}{11}$ or $3\frac{2}{11}$

48. $\frac{29}{11}$ or $2\frac{7}{11}$

49. $\frac{5}{6}$ Tbsp

50. $\frac{11}{12}$ ft

51. $\frac{235}{34}$ or $6\frac{31}{34}$ gal

52. 111 quarter-pound hamburgers

144

Name _____

C *Divide. Write each answer in simplest form. See Examples 8 through 10.*

35. $\frac{2}{3} \div 4$ **36.** $\frac{5}{6} \div 10$ **37.** $8 \div \frac{3}{5}$ **38.** $7 \div \frac{2}{11}$

39. $2\frac{1}{2} \div \frac{1}{2}$ **40.** $4\frac{2}{3} \div \frac{2}{5}$ **41.** $\frac{5}{12} \div 2\frac{1}{3}$ **42.** $\frac{4}{15} \div 2\frac{1}{2}$

43. $3\frac{3}{7} \div 3\frac{1}{3}$ **44.** $2\frac{5}{6} \div 4\frac{6}{7}$ **45.** $12 \div \frac{1}{8}$ **46.** $9 \div \frac{1}{6}$

47. $4\frac{5}{11} \div 1\frac{2}{5}$ **48.** $8\frac{2}{7} \div 3\frac{1}{7}$

D *Solve. Write each answer in simplest form. See Example 11.*

49. Sharon Bollen is to take $3\frac{1}{3}$ tablespoons of medicine per day in 4 equally divided doses. How much medicine is to be taken in each dose?

50. Vonshay Bartlet wants to build a $5\frac{1}{2}$-foot tall bookcase with 6 equally spaced shelves. How far apart should he space the shelves?

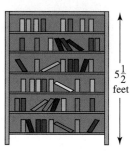

$5\frac{1}{2}$ feet

51. A small airplane used $58\frac{3}{4}$ gallons of fuel to fly an $8\frac{1}{2}$ hour trip. How many gallons of fuel were used for each hour?

52. Kesha Jonston is planning a Memorial Day barbeque. She has $27\frac{3}{4}$ pounds of hamburger. How many quarter-pound hamburgers can she make?

Name _____

53. The record for rainfall during a 24-hour period in Alaska is $15\frac{1}{5}$ inches. This record was set in Angoon, Alaska, in October 1982. How much rain fell per hour on average? (*Source*: National Climatic Data Center)

54. On March 19, 1998, the closing price for Kellogg Company stock was $43\frac{3}{4}$ per share. Severo Guiterrez spent $525 buying shares of Kellogg stock at the close of the stock market on that day. How many shares did he purchase? (*Source*: based on data from Standard & Poor's ComStock)

55. An order for 125 custom-made candle stands was placed with Mr. Levi, the manager of Just For You, Inc. The worker available to assign to the job can produce $2\frac{3}{5}$ candle stands per hour. Using this worker, how many work hours will be required to complete the order?

56. Yoko's Fine Jewelery sells a $\frac{3}{4}$-carat gem for $450. At this price, what is the cost of one carat?

Review and Preview

Perform each indicated operation. See Sections 1.2 and 1.3.

57.
$$\begin{array}{r} 27 \\ 76 \\ +98 \\ \hline \end{array}$$

58.
$$\begin{array}{r} 811 \\ 42 \\ +\ 69 \\ \hline \end{array}$$

59.
$$\begin{array}{r} 968 \\ -772 \\ \hline \end{array}$$

60.
$$\begin{array}{r} 882 \\ -773 \\ \hline \end{array}$$

61.
$$\begin{array}{r} 2000 \\ -\ 431 \\ \hline \end{array}$$

62.
$$\begin{array}{r} 500 \\ -\ 92 \\ \hline \end{array}$$

53. $\frac{19}{30}$ in.

54. 12 shares

55. $\frac{625}{13}$ or $48\frac{1}{13}$ hours

56. $600

57. 201

58. 922

59. 196

60. 109

61. 1569

62. 408

Name _____

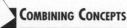
COMBINING CONCEPTS

63. One-third of all native flowering plant species in the United States are at risk of becoming extinct. That translates into 5144 at-risk flowering plant species. Based on this data, how many flowering plant species are native to the United States overall? (*Source*: The Nature Conservancy) (Hint: How many $\frac{1}{3}$s are in 5144?)

64. The FedEx fleet of aircraft includes 264 Cessnas. These Cessnas make up $\frac{66}{149}$ of the FedEx fleet. What is the size of the entire FedEx fleet of aircraft? (*Source*: Federal Express Corp.)

65. During the 1997 NFL regular season, the Denver Broncos lost four games. These losses represented $\frac{1}{4}$ of the regular season games they played. How many regular season games did the Broncos play? (*Source*: National Football League)

CHAPTER 2 ACTIVITY
COMPARING FRACTIONS

This activity may be completed by working in groups or individually.

Figure 1

Figure 2

Make seven identical copies of the strip shown in Figure 1. Cut out each strip.

1. Fold and crease one of the strips (as shown in Figure 2) to represent two equal halves. Now unfold the strip so that it lays flat again. You have just created a fraction strip for the denominator 2.

2. Make similar fraction strips for the denominators 3, 4, 5, 6, 8, and 12 by folding. (Note: The markings along the top of Figure 1 will help you gauge equal thirds. The markings along the bottom of Figure 1 will help you gauge equal fifths.)

3. Shade one of the fraction strips to represent $\frac{1}{2}$ and another to represent $\frac{2}{5}$. Compare the fraction strips to decide which fraction is larger or if the fractions are equal. $\frac{1}{2}$ is larger

4. Shade one of the fraction strips to represent $\frac{3}{8}$ and another to represent $\frac{2}{3}$. Compare the fraction strips to decide which fraction is larger or if the fractions are equal. $\frac{2}{3}$ is larger

5. Shade one of the fraction strips to represent $\frac{3}{4}$ and another to represent $\frac{5}{6}$. Compare the fraction strips to decide which fraction is larger or if the fractions are equal. $\frac{5}{6}$ is larger

6. Shade one of the fraction strips to represent $\frac{2}{8}$ and another to represent $\frac{3}{12}$. (To reuse a fraction strip, just flip it over and use the blank side.) Compare the fraction strips to decide which fraction is larger or if the fractions are equal. fractions are equal

7. Shade one of the fraction strips to represent $\frac{4}{5}$ and another to represent $\frac{7}{12}$. (To reuse a fraction strip, just flip it over and use the blank side.) Compare the fraction strips to decide which fraction is larger or if the fractions are equal. $\frac{4}{5}$ is larger

CHAPTER 2 HIGHLIGHTS

DEFINITIONS AND CONCEPTS	EXAMPLES

SECTION 2.1   INTRODUCTION TO FRACTIONS AND MIXED NUMBERS

A **fraction** is of the form

$$\frac{\text{numerator}}{\text{denominator}} \begin{array}{l} \leftarrow \text{number of parts considered} \\ \leftarrow \text{number of equal parts in the whole} \end{array}$$

Write a fraction to represent the shaded part of the figure.

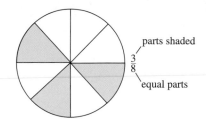

A fraction is called a **proper fraction** if its numerator is less than its denominator.

$$\frac{1}{3}, \quad \frac{2}{5}, \quad \frac{7}{8}, \quad \frac{100}{101}$$

A fraction is called an **improper fraction** if its numerator is greater than or equal to its denominator.

$$\frac{5}{4}, \quad \frac{2}{2}, \quad \frac{9}{7}, \quad \frac{101}{100}$$

A **mixed number** contains a whole number and a fraction.

$$1\frac{1}{2}, \quad 5\frac{7}{8}, \quad 25\frac{9}{10}$$

TO WRITE A MIXED NUMBER AS AN IMPROPER FRACTION

1. Multiply the whole number part of the mixed number by the denominator of the fraction.
2. Add the numerator of the fraction to the product from Step 1.
3. Write this sum as the numerator of the improper fraction over the original denominator.

$$5\frac{2}{7} = \frac{5 \cdot 7 + 2}{7} = \frac{35 + 2}{7} = \frac{37}{7}$$

TO WRITE AN IMPROPER FRACTION AS A MIXED NUMBER OR A WHOLE NUMBER

1. Divide the denominator into the numerator.
2. The whole number part of the mixed number is the quotient. The fraction is the remainder over the original denominator.

$$\text{quotient} \; \frac{\text{remainder}}{\text{original denominator}}$$

$$\frac{17}{3} = 5\frac{2}{3}$$

$$\begin{array}{r} 5 \\ 3\overline{)17} \\ \underline{15} \\ 2 \end{array}$$

<div align="center">**SECTION 2.2** **FACTORS AND PRIME FACTORIZATION**</div>	
A **prime number** is a natural number that has exactly two different factors, 1 and itself.	$2, 3, 5, 7, 11, 13, 17, \ldots$
A **composite number** is any natural number other than 1 that is not prime.	$4, 6, 8, 9, 10, 12, 14, 15, 16, \ldots$
The prime factorization of a number is unique.	Write the prime factorization of 60. $$60 = 6 \cdot 10$$ $$= 2 \cdot 3 \cdot 2 \cdot 5$$

<div align="center">**SECTION 2.3** **SIMPLEST FORM OF A FRACTION**</div>	
Fractions that represent the same portion of a whole are called **equivalent fractions**.	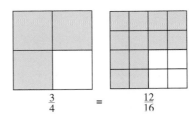 <div align="center">$\dfrac{3}{4} \quad = \quad \dfrac{12}{16}$</div>
A fraction is in **simplest form** or **lowest terms** when the numerator and the denominator have no common factors other than 1.	Write in simplest form $\dfrac{30}{36}$: $$\frac{30}{36} = \frac{\overset{1}{\cancel{2}} \cdot \overset{1}{\cancel{3}} \cdot 5}{\underset{1}{\cancel{2}} \cdot 2 \cdot \underset{1}{\cancel{3}} \cdot 3} = \frac{5}{6}$$
Two fractions are equivalent if their cross products are equal.	Determine whether $\dfrac{7}{8}$ and $\dfrac{21}{24}$ are equivalent. $\left. \begin{aligned} 8 \cdot 21 &= 168 \\ 7 \cdot 24 &= 168 \end{aligned} \right\}$ Since the cross products are equal, the fractions are equivalent. <div align="center">$\dfrac{7}{8} = \dfrac{21}{24}$</div>

<div align="center">**SECTION 2.4** **MULTIPLYING FRACTIONS**</div>	
To multiply two fractions, multiply the numerators and multiply the denominators.	Multiply: $$\frac{7}{8} \cdot \frac{3}{5} = \frac{7 \cdot 3}{8 \cdot 5} = \frac{21}{40}$$ $$\frac{3}{4} \cdot \frac{1}{6} = \frac{3 \cdot 1}{4 \cdot 6} = \frac{\overset{1}{\cancel{3}} \cdot 1}{4 \cdot \underset{1}{\cancel{3}} \cdot 2} = \frac{1}{8}$$ $$2\frac{1}{3} \cdot \frac{1}{9} = \frac{7}{3} \cdot \frac{1}{9} = \frac{7 \cdot 1}{3 \cdot 9} = \frac{7}{27}$$

SECTION 2.5   DIVIDING FRACTIONS

Two numbers are **reciprocals** of each other if their product is 1.

To divide two fractions, multiply the first fraction by the reciprocal of the second fraction.

The reciprocal of $\frac{3}{5}$ is $\frac{5}{3}$.

Divide:

$$\frac{3}{10} \div \frac{7}{9} = \frac{3}{10} \cdot \frac{9}{7} = \frac{3 \cdot 9}{10 \cdot 7} = \frac{27}{70}$$

CHAPTER 2 REVIEW

(2.1) *Determine whether each number is an improper fraction, a proper fraction or a mixed number.*

1. $\frac{11}{23}$ proper

2. $\frac{9}{8}$ improper

3. $\frac{1}{2}$ proper

4. $2\frac{1}{4}$ mixed number

5. $\frac{0}{3}$ proper

6. $5\frac{6}{7}$ mixed number

Write a fraction to represent the shaded area.

7. 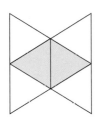 $\frac{2}{6}$

8.  $\frac{4}{7}$

9. 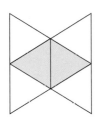 $\frac{7}{3}$

10.  $\frac{9}{4}$

11. A basketball player made 11 free throws out of 12 during a game. What fraction of free throws did he make? $\frac{11}{12}$

12. A new-car lot contained 23 blue cars out of a total of 131 cars. What fraction of cars on the lot are blue? $\frac{23}{131}$

Write each improper fraction as a mixed number or a whole number.

13. $\frac{15}{4}$ $3\frac{3}{4}$

14. $\frac{39}{13}$ 3

15. $\frac{275}{6}$ $45\frac{5}{6}$

16. $\frac{125}{4}$ $31\frac{1}{4}$

Write each mixed number as an improper fraction.

17. $1\frac{1}{5}$ $\frac{6}{5}$

18. $2\frac{8}{9}$ $\frac{26}{9}$

19. $3\frac{11}{12}$ $\frac{47}{12}$

20. $5\frac{10}{17}$ $\frac{95}{17}$

(2.2) *Identify each number as prime or composite.*
21. 51   composite   **22.** 17   prime   **23.** 27   composite   **24.** 21   composite

List all factors of each number.
25. 42   1, 2, 3, 6, 7, 14, 21, 42   **26.** 30   1, 2, 3, 5, 6, 10, 15, 30

Find the prime factorization of each number.
27. 68 $2^2 \cdot 17$ **28.** 90 $2 \cdot 3^2 \cdot 5$ **29.** 785 $5 \cdot 157$ **30.** 255 $3 \cdot 5 \cdot 17$

(2.3) *Write each fraction in simplest form.*
31. $\frac{12}{28}$ $\frac{3}{7}$ **32.** $\frac{15}{27}$ $\frac{5}{9}$ **33.** $\frac{25}{75}$ $\frac{1}{3}$ **34.** $\frac{36}{72}$ $\frac{1}{2}$

35. $\frac{29}{32}$ $\frac{29}{32}$ **36.** $\frac{18}{23}$ $\frac{18}{23}$ **37.** $\frac{45}{27}$ $\frac{5}{3}$ or $1\frac{2}{3}$ **38.** $\frac{42}{30}$ $\frac{7}{5}$ or $1\frac{2}{5}$

39. $\frac{48}{6}$ 8   **40.** $\frac{54}{9}$ 6   **41.** $\frac{140}{150}$ $\frac{14}{15}$ **42.** $\frac{84}{140}$ $\frac{3}{5}$

(2.4) *Multiply. Write each answer in simplest form.*
43. $\frac{3}{5} \cdot \frac{1}{2}$ $\frac{3}{10}$ **44.** $\frac{6}{7} \cdot \frac{5}{12}$ $\frac{5}{14}$ **45.** $\frac{7}{8} \cdot \frac{2}{3}$ $\frac{7}{12}$ **46.** $\frac{6}{15} \cdot \frac{5}{8}$ $\frac{1}{4}$

47. $\frac{24}{5} \cdot \frac{15}{8}$ 9   **48.** $\frac{27}{21} \cdot \frac{7}{18}$ $\frac{1}{2}$ **49.** $5 \cdot \frac{7}{8}$ $\frac{35}{8}$ or $4\frac{3}{8}$ **50.** $6 \cdot \frac{5}{12}$ $\frac{5}{2}$ or $2\frac{1}{2}$

51. $\frac{39}{3} \cdot \frac{7}{13}$ 7   **52.** $\frac{42}{5} \cdot \frac{15}{6}$ 21   **53.** $1\frac{5}{8} \cdot \frac{2}{3}$ $\frac{13}{12}$ or $1\frac{1}{12}$ **54.** $3\frac{6}{11} \cdot \frac{5}{13}$ $\frac{15}{11}$ or $1\frac{4}{11}$

55. $4\frac{1}{6} \cdot 2\frac{2}{5}$ 10   **56.** $5\frac{2}{3} \cdot 2\frac{1}{4}$ $\frac{51}{4}$ or $12\frac{3}{4}$

57. There are 58 calories in 1 ounce of turkey. How many calories are there in a $3\frac{1}{2}$ ounce serving of turkey?   203 calories

58. There are $3\frac{1}{3}$ grams of fat in each ounce of lean hamburger. How many grams of fat are in a 4-ounce hamburger patty? $\frac{40}{3}$ or $13\frac{1}{3}$ grams

59. Find the area of the rectangle.

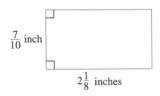

$\frac{7}{10}$ inch

$2\frac{1}{8}$ inches $\frac{119}{80}$ or $1\frac{39}{80}$ sq. in.

(2.5) *Find the reciprocal of each fraction.*

60. 7 $\frac{1}{7}$

61. $\frac{1}{8}$ 8

62. $\frac{14}{23}$ $\frac{23}{14}$

63. $\frac{17}{5}$ $\frac{5}{17}$

Divide. Write each answer in simplest form.

64. $\frac{3}{4} \div \frac{3}{8}$ 2

65. $\frac{21}{4} \div \frac{7}{5}$ $\frac{15}{4}$ or $3\frac{3}{4}$

66. $\frac{18}{5} \div \frac{2}{5}$ 9

67. $\frac{9}{2} \div \frac{1}{3}$ $\frac{27}{2}$ or $13\frac{1}{2}$

68. $\frac{5}{3} \div 2$ $\frac{5}{6}$

69. $5 \div \frac{15}{8}$ $\frac{8}{3}$ or $2\frac{2}{3}$

70. $6\frac{3}{4} \div 1\frac{2}{7}$ $\frac{21}{4}$ or $5\frac{1}{4}$

71. $5\frac{1}{2} \div 2\frac{1}{11}$ $\frac{121}{46}$ or $2\frac{29}{46}$

72. $\frac{7}{2} \div 1\frac{1}{2}$ $\frac{7}{3}$ or $2\frac{1}{3}$

73. $1\frac{3}{5} \div \frac{1}{4}$ $\frac{32}{5}$ or $6\frac{2}{5}$

74. A suit is on sale for $120, which is $\frac{3}{5}$ of the regular price. What is the regular price of the suit? $200

75. Herman Heltznutt walks 5 days a week for a total distance of $5\frac{1}{4}$ miles per week. If he walks the same distance each day, find the distance he walks each day. $\frac{21}{20}$ or $1\frac{1}{20}$ mi

Focus On Study Skills

TIME MANAGEMENT

As a college student, you know the demands that classes, homework, work, campus life, and family place on your time. Some days you probably wonder how you'll ever get everything done! One key to managing your time is developing a schedule. Here are some hints for making a schedule:

1. Make a list of all the weekly commitments you have for the term. Be sure to include classes, work, regular meetings, extracurricular activities, etc. You may also find it helpful to list such things as regular workouts, doing laundry, and grocery shopping.

2. Next, estimate the time needed for each item on the list. Also make a note of how often you will need to do each item. Don't forget to include time estimates for the reading, studying, and homework you do outside your classes. Some instructors suggest spending a minimum of 2 hours studying outside class for every hour you spend in class. You might want to check with your instructor to see if this estimate is realistic. You may also need to consider commuting time.

3. Block out a typical week on a schedule grid like the one below. Start with items with fixed time slots, like classes and work. You might want to shade or draw an X through the boxes on the grid to indicate that these time slots are filled.

4. Add in the items on your list with flexible time slots. You'll need to think carefully about how best to schedule some items such as study time.

5. Don't fill up every time slot on the schedule. Remember: you need to allow time for eating, sleeping, and relaxing! You should also leave a little cushion in case things take longer than planned.

6. You may find it helpful to make a copy of your typical weekly schedule for each week of the term. Then you can add to it or adjust it for the specific events going on each week.

7. If you find that your basic weekly schedule is too full for you to handle, you may need to make some changes in your work load, class load, or in other areas of your life. You may want to talk your advisor, manager or supervisor at work, or someone in your college's academic counseling center for help with such decisions.

	Monday	Tuesday	Wednesday	Thursday	Friday	Saturday	Sunday
7:00 a.m.							
8:00 a.m.							
9:00 a.m.							
10:00 a.m.							
11:00 a.m.							
12:00 p.m.							
1:00 p.m.							
2:00 p.m.							
3:00 p.m.							
4:00 p.m.							
5:00 p.m.							
6:00 p.m.							
7:00 p.m.							
8:00 p.m.							
9:00 p.m.							

Name _____ **Section** _____ **Date** _____

CHAPTER 2 TEST

Perform each indicated operation. Write each answer in simplest form.

1. $\dfrac{4}{4} \div \dfrac{3}{4}$

2. $\dfrac{4}{3} \cdot \dfrac{4}{4}$

3. $2 \cdot \dfrac{1}{8}$

4. $\dfrac{2}{3} \cdot \dfrac{8}{15}$

5. $8 \div \dfrac{1}{2}$

6. $13\dfrac{1}{2} \div 3$

7. $\dfrac{3}{8} \cdot \dfrac{16}{6}$

8. $5\dfrac{1}{4} \div \dfrac{7}{12}$

9. $\dfrac{16}{3} \div \dfrac{3}{12}$

10. $3\dfrac{1}{3} \cdot 6\dfrac{3}{4}$

11. $12 \div 3\dfrac{1}{3}$

12. $\dfrac{14}{5} \cdot \dfrac{25}{21}$

Write each mixed number as an improper fraction.

13. $7\dfrac{2}{3}$

14. $3\dfrac{6}{11}$

Write each improper fraction as a mixed number or a whole number.

15. $\dfrac{23}{5}$

16. $\dfrac{75}{4}$

17. Find the area of the figure.

$\dfrac{2}{3}$ mile [rectangle] $1\dfrac{8}{9}$ miles

18. During a 258-mile trip, a car used $10\dfrac{3}{4}$ gallons of gas. How many miles would we expect the car to travel on one gallon of gas?

Write each fraction in simplest form.

19. $\dfrac{24}{210}$

20. $\dfrac{42}{70}$

1. $\dfrac{4}{3}$ or $1\dfrac{1}{3}$

2. $\dfrac{4}{3}$ or $1\dfrac{1}{3}$

3. $\dfrac{1}{4}$

4. $\dfrac{16}{45}$

5. 16

6. $\dfrac{9}{2}$ or $4\dfrac{1}{2}$

7. 1

8. 9

9. $\dfrac{64}{3}$ or $21\dfrac{1}{3}$

10. $\dfrac{45}{2}$ or $22\dfrac{1}{2}$

11. $\dfrac{18}{5}$ or $3\dfrac{3}{5}$

12. $\dfrac{10}{3}$ or $3\dfrac{1}{3}$

13. $\dfrac{23}{3}$

14. $\dfrac{39}{11}$

15. $4\dfrac{3}{5}$

16. $18\dfrac{3}{4}$

17. $\dfrac{34}{27}$ or $1\dfrac{7}{27}$ sq. mi

18. 24 miles

19. $\dfrac{4}{35}$

20. $\dfrac{3}{5}$

21. 8800 sq. yds

21. How many square yards of artificial turf are necessary to cover a football field, including the end zones and 10 yards beyond the side lines? (*Hint:* A football field measures $100 \times 53\frac{1}{3}$ yards and the end zones are 10 yards deep.)

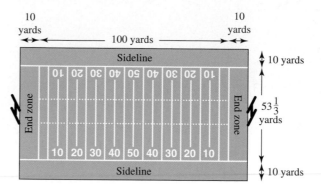

22. $2^3 \cdot 5 \cdot 7$

Find the prime factorization of each natural number.

22. 280

23. 84

23. $2^2 \cdot 3 \cdot 7$

24. Prior to an oil spill, the stock in an oil company sold for $120 per share. As a result of the liability the company incurred from the spill, the price per share fell to $\frac{3}{4}$ of the price before the spill. What did the stock sell for after the spill?

24. $90 per share

Name _____ **Section** _____ **Date** _____

CUMULATIVE REVIEW

1. Find the place value of the digit 4 in the whole number 48,761.

2. Write the number, eight hundred five, in standard form.

3. Add: 34,285 + 149,761

4. Find the perimeter of the polygon shown.

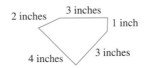

5. Ivy is a station manager of Smoky Hills Public Television. Her monthly salary of $3460 is increased by a raise of $427. What is her new salary?

6. Subtract: 7826 − 505. Check by adding.

7. A telephone survey was taken to identify favorite sport activities, and the results from the five most popular activities are shown below in the form of a bar graph. In this particular graph, each bar represents a different sport activity, and the height of each bar represents the number of people who responded that the particular sport was their favorite activity. Use this graph to answer the questions below.

Favorite Sport Activities

a. What activity was prefered by the most people?

b. How many more people responded that swimming rather than fishing is their favorite activity?

8. Round 568 to the nearest ten.

9. Round each number to the nearest hundred to find an estimated difference.

$$4725$$
$$-2879$$

10. Multiply.
   a. 6×1

   b. $0(8)$

   c. $1 \cdot 45$

   d. $(75)(0)$

11. Rewrite each using the distributive property.
   a. $3(4 + 5)$

   b. $10(6 + 8)$

   c. $2(7 + 3)$

12. Find each quotient, and then check the answer by multiplying.
   a. $9\overline{)0}$

   b. $0 \div 12$

   c. $\dfrac{0}{5}$

   d. $\dfrac{3}{0}$

13. Divide and check: $1872 \div 9$

14. How many boxes are needed to ship 56 pairs of Nikes to a shoe store in Texarkana if 9 pairs of shoes will fit in each shipping box?

15. A gardener bought enough plants to fill a rectangular garden with length 30 feet and width 20 feet. Because of shading problems with a nearby tree, the gardener changed the width of the garden to 15 feet. If the area is to remain the same, what is the new length of the garden?

Name _____

Write using exponential notation.

16. $4 \cdot 4 \cdot 4$

17. $6 \cdot 6 \cdot 6 \cdot 8 \cdot 8 \cdot 8 \cdot 8 \cdot 8$

18. Simplify: $2 \cdot 4 - 3 \div 3$

19. Use a fraction to represent the shaded part of the figure.

20. Write as improper fractions.

   a. $4\frac{2}{9}$

   b. $1\frac{8}{11}$

21. Find all the factors of 20.

22. Write $\frac{42}{66}$ in simplest form.

23. Multiply: $3\frac{1}{3} \cdot \frac{7}{8}$

24. Find the reciprocal of $\frac{1}{3}$.

25. Divide and simplify: $\frac{5}{16} \div \frac{3}{4}$

16. 4^3; (Sec. 1.8, Ex. 1)

17. $6^3 \cdot 8^5$; (Sec. 1.8, Ex. 4)

18. 7; (Sec. 1.8, Ex. 9)

19. $\frac{3}{4}$; (Sec. 2.1, Ex. 4)

20. a. $\frac{38}{9}$

   b. $\frac{19}{11}$; (Sec. 2.1, Ex. 8)

21. 1, 2, 4, 5, 10, 20; (Sec. 2.2, Ex. 1)

22. $\frac{7}{11}$; (Sec. 2.3, Ex. 2)

23. $\frac{35}{12}$ or $2\frac{11}{12}$; (Sec. 2.4, Ex. 7)

24. $\frac{3}{1}$ or 3; (Sec. 2.5, Ex. 3)

25. $\frac{5}{12}$; (Sec. 2.5, Ex. 6)

Adding and Subtracting Fractions

Having learned what fractions are and how to multiply and divide them in Chapter 2, we are ready to continue our study of fractions. In this chapter, we will learn how to add and subtract fractions and mixed numbers. We then conclude this chapter with comparing fractions and reviewing the order of operations.

The American Lobster, *Homarus americanus*, is big business in New England. Lobster fishers caught more than 70 million pounds of lobsters in 1996, and more than $\frac{4}{5}$ of the lobster catch was made off the coasts of Maine, Massachusetts, and Rhode Island. Most American lobsters are a dark blue-green or grey color when alive. There are rare instances of blue, yellow, and white lobsters. However, the only red lobsters are ones that have been cooked. In Example 8 on pages 189–190 and in the Chapter Activity, we will see how fractions are used in the lobster fishing industry.

CHAPTER 3 PRETEST

Add and simplify.

1. $\dfrac{1}{9} + \dfrac{7}{9}$

2. $\dfrac{5}{12} + \dfrac{7}{12}$

3. $\dfrac{1}{6} + \dfrac{3}{4}$

4. $\dfrac{3}{20} + \dfrac{4}{15}$

5. $\quad 5\dfrac{4}{9}$

$\quad +3\dfrac{3}{4}$

Subtract and simplify.

6. $\dfrac{8}{17} - \dfrac{3}{17}$

7. $\dfrac{7}{18} - \dfrac{5}{18}$

8. $\dfrac{11}{14} - \dfrac{2}{21}$

9. $\dfrac{4}{5} - \dfrac{2}{3}$

10. $\quad 10\dfrac{2}{5}$

$\quad -\ 4\dfrac{5}{6}$

Find the LCM of the following lists of numbers.

11. 6, 15

12. 20, 24, 45

Write each fraction as an equivalent fraction using the given denominator.

13. $\dfrac{2}{5} = \dfrac{}{35}$

14. $\dfrac{13}{9} = \dfrac{}{27}$

15. Insert $<$ or $>$ to make a true statement.

$\dfrac{4}{7} \qquad \dfrac{3}{5}$

16. Evaluate: $\left(\dfrac{3}{5}\right)^3$

17. Use order of operations to simplify the following expression.

$\dfrac{2}{9} + \dfrac{1}{4} \cdot \dfrac{2}{3}$

18. Find the average of the following list of numbers.

$\dfrac{1}{2}, \dfrac{2}{3},$ and $\dfrac{1}{30}$

19. Peggy Brown mixes $\dfrac{1}{4}$ pound of cashews with $\dfrac{5}{8}$ pound of peanuts. How much does the mixture weigh?

20. A plumber cuts a pipe $2\dfrac{5}{8}$ feet long from a pipe 8 feet long. How long is the remaining piece?

3.1   ADDING AND SUBTRACTING LIKE FRACTIONS

Fractions with the same denominators are called **like fractions**. Fractions that have different denominators are called **unlike fractions**.

Like Fractions

$\frac{2}{5}$ and $\frac{3}{5}$

$\frac{5}{21}, \frac{16}{21},$ and $\frac{7}{21}$

Unlike Fractions

$\frac{2}{5}$ and $\frac{3}{4}$

$\frac{5}{7}$ and $\frac{5}{9}$

A   ADDING LIKE FRACTIONS

To see how we add like fractions (fractions with the same denominator), study the figures below.

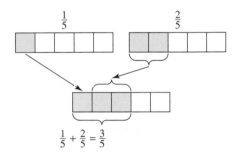

$$\frac{1}{5} + \frac{2}{5} = \frac{3}{5}$$

ADDING LIKE FRACTIONS

To add like fractions, add the numerators and write the sum over the common denominator.

For example, $\frac{1}{4} + \frac{2}{4} = \frac{1+2}{4} = \frac{3}{4}$ Add the numerators.
Keep the denominator.

⌐HELPFUL HINT

As usual, don't forget to write all answers in simplest form.

Examples   Add and simplify.

1. $\frac{2}{7} + \frac{3}{7} = \frac{2+3}{7} = \frac{5}{7}$ ← Add the numerators.
← Keep the common denominator.

2. $\frac{3}{16} + \frac{7}{16} = \frac{3+7}{16} = \frac{10}{16} = \frac{\cancel{2} \cdot 5}{\cancel{2} \cdot 8} = \frac{5}{8}$

3. $\frac{7}{8} + \frac{6}{8} + \frac{3}{8} = \frac{7+6+3}{8} = \frac{16}{8} = 2$

TRY THE CONCEPT CHECK IN THE MARGIN.

B   SUBTRACTING LIKE FRACTIONS

To see how we subtract like fractions (fractions with the same denominator), study the following figure.

TEACHING TIP

Begin this lesson by asking students, "How would you calculate the *value* of 3 quarters and 4 dimes?" Then ask why wouldn't you just add 3 and 4? (because quarters and dimes are not the same denomination or type of coin).

Practice Problems 1–3

Add and simplify.

1. $\frac{5}{9} + \frac{2}{9}$ 2. $\frac{5}{8} + \frac{1}{8}$

3. $\frac{10}{11} + \frac{1}{11} + \frac{7}{11}$

✓ **CONCEPT CHECK**

Find and correct the error in the following.

$\frac{1}{5} + \frac{1}{5} = \frac{2}{10}$

Answers

1. $\frac{7}{9}$,   2. $\frac{3}{4}$,   3. $\frac{18}{11} = 1\frac{7}{11}$

✓ **Concept Check:**   Don't add denominators together. Correct solution $\frac{1}{5} + \frac{1}{5} = \frac{2}{5}$.

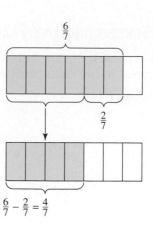

$$\frac{6}{7} - \frac{2}{7} = \frac{4}{7}$$

SUBTRACTING LIKE FRACTIONS

To subtract like fractions, subtract the numerators and write the difference over the common denominator.

For example,

$$\frac{4}{5} - \frac{2}{5} = \frac{4 - 2}{5} = \frac{2}{5}$$ Subtract the numerators.
                 Keep the denominator.

Practice Problems 4–5

Subtract and simplify.

4. $\dfrac{7}{12} - \dfrac{2}{12}$

5. $\dfrac{9}{10} - \dfrac{1}{10}$

Examples   Subtract and simplify.

4. $\dfrac{8}{9} - \dfrac{1}{9} = \dfrac{8 - 1}{9} = \dfrac{7}{9}$ ← Subtract the numerators.
                           ← Keep the common denominator.

5. $\dfrac{7}{8} - \dfrac{5}{8} = \dfrac{7 - 5}{8} = \dfrac{2}{8} = \dfrac{\overset{1}{\cancel{2}}}{\underset{1}{\cancel{2}} \cdot 4} = \dfrac{1}{4}$

$\blacksquare$

⬛ SOLVING PROBLEMS BY ADDING AND SUBTRACTING LIKE FRACTIONS

Many real-life problems involve finding the perimeters of square or rectangular areas, such as pastures, swimming pools, and so on. We can use our knowledge of adding fractions to find perimeters.

Practice Problem 6

Find the perimeter of the square.

$\frac{1}{7}$ mile

Example 6   Find the perimeter of the rectangle.

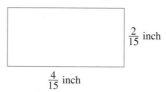

$\frac{2}{15}$ inch

$\frac{4}{15}$ inch

Solution: Recall that perimeter means distance around and that opposite sides of a rectangle are the same length.

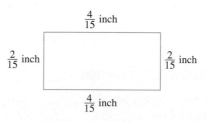

$\frac{4}{15}$ inch

$\frac{2}{15}$ inch $\frac{2}{15}$ inch

$\frac{4}{15}$ inch

Answers

4. $\dfrac{5}{12}$,   5. $\dfrac{4}{5}$,   6. $\dfrac{4}{7}$ mile

$$\text{Perimeter} = \frac{2}{15} + \frac{4}{15} + \frac{2}{15} + \frac{4}{15} = \frac{2+4+2+4}{15}$$

$$= \frac{12}{15} = \frac{\cancel{3} \cdot 4}{\cancel{3} \cdot 5} = \frac{4}{5}$$

The perimeter of the rectangle is $\frac{4}{5}$ inch.

We can combine our skills in adding and subtracting fractions with our four problem-solving steps from Chapter 1 to solve many kinds of real-life problems.

Example 7   Total Amount of an Ingredient in a Recipe

A recipe calls for $\frac{1}{3}$ of a cup of flour at the beginning and $\frac{2}{3}$ of a cup of flour later. How much total flour is needed to make that recipe?

$\frac{1}{3}$ cup $\frac{1}{2}$ cup

Solution:

1. UNDERSTAND the problem. To do so, read and reread the problem. Since we are finding total flour, we will add.

2. TRANSLATE.

In words:	total flour	is	flour at the beginning	added to	flour later
	↓	↓	↓	↓	↓
Translate:	total flour	=	$\frac{1}{3}$	+	$\frac{2}{3}$

3. SOLVE. $\frac{1}{3} + \frac{2}{3} = \frac{1+2}{3} = \frac{\cancel{3}}{\cancel{3}} = 1$

4. INTERPRET. *Check* your work. *State* your conclusion: The total flour needed for the recipe is 1 cup.

Example 8   Calculating Distance

The distance from home to the World Gym is $\frac{7}{8}$ of a mile and from home to the post office is $\frac{5}{8}$ of a mile. How much farther is it from home to the World Gym than from home to the post office?

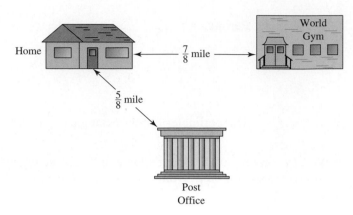

Solution:

1. UNDERSTAND. Read and reread the problem. The phrase "How much farther" tells us to subtract distances.

2. TRANSLATE.

In words:

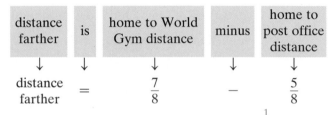

distance farther	is	home to World Gym distance	minus	home to post office distance
↓	↓	↓	↓	↓

Translate:

$$\text{distance farther} = \frac{7}{8} - \frac{5}{8}$$

3. SOLVE: $\dfrac{7}{8} - \dfrac{5}{8} = \dfrac{7-5}{8} = \dfrac{2}{8} = \dfrac{\overset{1}{\cancel{2}}}{\underset{1}{\cancel{2}} \cdot 4} = \dfrac{1}{4}$

4. INTERPRET. *Check* your work. *State* your conclusion: The distance from home to the World Gym is $\frac{1}{4}$ mile farther than from home to the post office. ■

Name _____ **Section** _____ **Date** _____

MENTAL MATH

State whether the fractions in each list are like or unlike fractions.

1. $\dfrac{7}{8}, \dfrac{7}{10}$

2. $\dfrac{2}{3}, \dfrac{4}{9}$

3. $\dfrac{9}{10}, \dfrac{1}{10}$

4. $\dfrac{8}{11}, \dfrac{2}{11}$

5. $\dfrac{2}{31}, \dfrac{30}{31}, \dfrac{19}{31}$

6. $\dfrac{3}{10}, \dfrac{3}{11}, \dfrac{3}{13}$

7. $\dfrac{5}{12}, \dfrac{7}{12}, \dfrac{12}{11}$

8. $\dfrac{1}{5}, \dfrac{2}{5}, \dfrac{4}{5}$

EXERCISE SET 3.1

A Add and simplify. See Examples 1 through 3.

1. $\dfrac{1}{7} + \dfrac{2}{7}$

2. $\dfrac{5}{9} + \dfrac{2}{9}$

3. $\dfrac{1}{10} + \dfrac{1}{10}$

4. $\dfrac{1}{4} + \dfrac{1}{4}$

5. $\dfrac{2}{9} + \dfrac{4}{9}$

6. $\dfrac{3}{10} + \dfrac{2}{10}$

7. $\dfrac{6}{20} + \dfrac{1}{20}$

8. $\dfrac{5}{24} + \dfrac{7}{24}$

9. $\dfrac{3}{14} + \dfrac{4}{14}$

10. $\dfrac{2}{8} + \dfrac{3}{8}$

11. $\dfrac{10}{11} + \dfrac{3}{11}$

12. $\dfrac{4}{17} + \dfrac{10}{17}$

13. $\dfrac{4}{13} + \dfrac{2}{13} + \dfrac{1}{13}$

14. $\dfrac{5}{11} + \dfrac{1}{11} + \dfrac{2}{11}$

15. $\dfrac{7}{18} + \dfrac{3}{18} + \dfrac{2}{18}$

16. $\dfrac{7}{15} + \dfrac{4}{15} + \dfrac{1}{15}$

B Subtract and simplify. See Examples 4 and 5.

17. $\dfrac{10}{11} - \dfrac{4}{11}$

18. $\dfrac{9}{13} - \dfrac{5}{13}$

19. $\dfrac{4}{5} - \dfrac{1}{5}$

20. $\dfrac{7}{8} - \dfrac{4}{8}$

21. $\dfrac{7}{4} - \dfrac{3}{4}$

22. $\dfrac{18}{5} - \dfrac{3}{5}$

23. $\dfrac{7}{8} - \dfrac{1}{8}$

24. $\dfrac{5}{6} - \dfrac{1}{6}$

MENTAL MATH ANSWERS

1. unlike
2. unlike
3. like
4. like
5. like
6. unlike
7. unlike
8. like

ANSWERS

1. $\dfrac{3}{7}$

2. $\dfrac{7}{9}$

3. $\dfrac{1}{5}$

4. $\dfrac{1}{2}$

5. $\dfrac{2}{3}$

6. $\dfrac{1}{2}$

7. $\dfrac{7}{20}$

8. $\dfrac{1}{2}$

9. $\dfrac{1}{2}$

10. $\dfrac{5}{8}$

11. $\dfrac{13}{11} = 1\dfrac{2}{11}$

12. $\dfrac{14}{17}$

13. $\dfrac{7}{13}$

14. $\dfrac{8}{11}$

15. $\dfrac{2}{3}$

16. $\dfrac{4}{5}$

17. $\dfrac{6}{11}$

18. $\dfrac{4}{13}$

19. $\dfrac{3}{5}$

20. $\dfrac{3}{8}$

21. 1

22. 3

23. $\dfrac{3}{4}$

24. $\dfrac{2}{3}$

Name _____

25. $\frac{25}{12} - \frac{15}{12}$ **26.** $\frac{30}{20} - \frac{15}{20}$ **27.** $\frac{11}{10} - \frac{3}{10}$ **28.** $\frac{14}{15} - \frac{4}{15}$

29. $\frac{27}{33} - \frac{8}{33}$ **30.** $\frac{37}{45} - \frac{18}{45}$

C *Find the perimeter of each figure. See Example 6.*

31.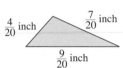
$\frac{4}{20}$ inch $\frac{7}{20}$ inch $\frac{9}{20}$ inch

32. Square $\frac{1}{6}$ centimeter

33. $\frac{5}{12}$ meter Rectangle $\frac{7}{12}$ meter

34. $\frac{3}{13}$ foot $\frac{2}{13}$ foot $\frac{6}{13}$ foot $\frac{3}{13}$ foot $\frac{4}{13}$ foot

Solve. Write each answer in simplest form. See Examples 7 and 8.

35. Emil Vasquez, a body builder, worked out $\frac{7}{8}$ of an hour one morning before school and $\frac{5}{8}$ of an hour that evening. How long did he work out that day?

36. A recipe for Heavenly Hash cake calls for $\frac{3}{4}$ cup of flour and later $\frac{1}{4}$ cup of flour. How much flour is needed to make the recipe?

37. Jenny Joyce, a railroad inspector, must inspect $\frac{19}{20}$ of a mile of railroad track. If she has already inspected $\frac{5}{20}$ of a mile, how much more does she need to inspect?

38. Scott Davis has run $\frac{11}{8}$ miles already and plans to complete $\frac{16}{8}$ miles. To do this, how much farther must he run?

39. The Commonwealth of Massachusetts gets $\frac{48}{100}$ of its operating revenue from taxes and $\frac{13}{100}$ of its revenue from lottery sales. What fraction of Massachusetts' operating revenue comes from lottery sales and taxes? (*Source:* Massachusetts State Lottery Commission)

40. In the United States, about $\frac{27}{98}$ of all households owning at least one television set use broadcast TV services only. Approximately $\frac{65}{98}$ of all households owning at least one television set subscribe to cable TV services. What fraction of television-owning households use either broadcast or cable services? (*Source:* Carmel Group)

41. As of October 1997, the fraction of states in the United States with maximum interstate highway speed limits up to and including 70 mph was $\frac{39}{50}$. The fraction of states with 70 mph speed limits was $\frac{16}{50}$. What fraction of states had speed limits that were less than 70 mph? (*Source:* National Motorists Association)

42. When people take aspirin, $\frac{31}{50}$ of the time it is used to treat some type of pain. Approximately $\frac{7}{50}$ of all aspirin use is for treating headaches. What fraction of aspirin use is for treating pain other than headaches? (*Source:* Bayer Market Research)

REVIEW AND PREVIEW

Write the prime factorization of each number. See Section 2.2.

43. 10   **44.** 12   **45.** 8

46. 20   **47.** 55   **48.** 28

COMBINING CONCEPTS

Perform each indicated operation.

49. $\frac{3}{8} + \frac{7}{8} - \frac{5}{8}$ **50.** $\frac{12}{20} - \frac{1}{20} - \frac{3}{20}$

51. $\frac{4}{11} + \frac{5}{11} - \frac{3}{11} + \frac{2}{11}$ **52.** $\frac{9}{12} + \frac{1}{12} - \frac{3}{12} - \frac{5}{12}$

39. $\frac{61}{100}$

40. $\frac{46}{49}$

41. $\frac{23}{50}$

42. $\frac{12}{25}$

43. $2 \cdot 5$

44. $2^2 \cdot 3$

45. 2^3

46. $2^2 \cdot 5$

47. $5 \cdot 11$

48. $2^2 \cdot 7$

49. $\frac{5}{8}$

50. $\frac{2}{5}$

51. $\frac{8}{11}$

52. $\frac{1}{6}$

Solve. Write each answer in simplest form.

53. Mike Cannon jogged $\frac{3}{8}$ of a mile from home and then rested. Then he continued jogging for another $\frac{3}{8}$ of a mile until he discovered his watch had fallen off. He walked back along the same path for $\frac{4}{8}$ of a mile until he found his watch. Find how far he was from his *starting point*.

54. A person's marital status is either single (never married), married, widowed, or divorced. Of American men over the age of 65, $\frac{38}{50}$ are married and $\frac{7}{50}$ are widowed. What fraction of American men over age 65 are either single or divorced? (*Source:* based on data from the U.S. Bureau of the Census)

55. In your own words, explain how to add like fractions.

56. In your own words, explain how to subtract like fractions.

3.2 Least Common Multiple

A Finding the Least Common Multiple Using Multiples

A multiple of a number is the product of that number and a natural number. For example, multiples of 5 are

$$\underbrace{5 \cdot 1}_{\downarrow} \quad \underbrace{5 \cdot 2}_{\downarrow} \quad \underbrace{5 \cdot 3}_{\downarrow} \quad \underbrace{5 \cdot 4}_{\downarrow} \quad \underbrace{5 \cdot 5}_{\downarrow} \quad \underbrace{5 \cdot 6}_{\downarrow} \quad \underbrace{5 \cdot 7}_{\downarrow} \quad \underbrace{5 \cdot 8}_{\downarrow}$$
$$5, \quad 10, \quad 15, \quad 20, \quad 25, \quad 30, \quad 35, \quad 40, \ldots$$

Multiples of 4 are

$$4, 8, 12, 16, 20, 24, 28, 32, 36, 40, 44, \ldots$$

Common multiples of both 4 and 5 are numbers that are found in both lists above. If we study the lists of multiples and extend them we have:

Common multiples of 4 and 5: 20, 40, 60, 80, . . .

We call the smallest number in the list of common multiples the **least common multiple** (**LCM**). From the list of common multiples of 4 and 5, we see that the LCM of 4 and 5 is 20.

Example 1   Find the LCM of 6 and 8.

Solution: Multiples of 6: 6, 12, 18, ㉔, 30, 36, 42, ㊽, . . .

Multiples of 8: 8, 16, ㉔, 32, 40, ㊽, 56, . . .

The common multiples are 24, 48, ... The least common multiple (LCM) is 24.   ▬

Listing all multiples of every number in a list can be cumbersome and tedious. We can condense the procedure shown in Example 1 with the following steps.

Method 1: Finding the LCM of a List of Numbers Using Multiples of the Largest Number

Step 1. Write multiples of the largest number (starting with the number itself) until you find one that is a multiple of the other numbers.

Step 2. The multiple found in Step 1 is the LCM.

Example 2   Find the LCM of 6 and 9.

Solution: We write the multiples of 9 until we find a number that is also a multiple of 6.

$9 \cdot 1 = 9$ Not a multiple of 6

$9 \cdot 2 = 18$ A multiple of 6

The LCM of 6 and 9 is 18.   ▬

Example 3   Find the LCM of 7 and 14.

Solution: We write multiples of 14 until we find one that is also a multiple of 7.

$14 \cdot 1 = 14$ A multiple of 7

The LCM of 7 and 14 is 14.   ▬

Objectives

A Find the least common multiple (LCM) of a list of numbers using multiples.

B Find the LCM using prime factorization.

C Write an equivalent fraction with a given denominator.

SSM   CD-ROM   Video 3.2

Practice Problem 1

Find the LCM of 15 and 25.

TEACHING TIPS
Begin the lesson by having the first person in a row select a natural number and then have the row count by multiples of that number. Next, ask the class what number the 50th or 80th person would say. Point out that this can be found by multiplying and that all of the numbers named are multiples of the first number. Repeat the process with several rows.

After discussing the first method for finding the LCM, consider leading a discussion on why it is a better strategy to use the multiples of the larger number rather than the multiples of the smaller number.

Practice Problem 2

Find the LCM of 8 and 10.

Practice Problem 3

Find the LCM of 8 and 16.

Answers

1. 75,   **2.** 40,   **3.** 16

Practice Problem 4

Find the LCM of 25 and 30.

Practice Problem 5

Find the LCM of 40 and 108.

Example 4   Find the LCM of 12 and 20.

Solution: We write multiples of 20 until we find one that is also a multiple of 12.

$20 \cdot 1 = 20$ Not a multiple of 12

$20 \cdot 2 = 40$ Not a multiple of 12

$20 \cdot 3 = 60$ A multiple of 12

The LCM of 12 and 20 is 60.

B FINDING THE LCM USING PRIME FACTORIZATION

Method 1 for finding multiples works fine for smaller numbers, but may get tedious for larger numbers. A second method that uses prime factorization may be easier to use for larger numbers.

> **METHOD 2: FINDING THE LCM OF A LIST OF NUMBERS USING PRIME FACTORIZATION**
>
> **Step 1.**   Write the prime factorization of each number.
> **Step 2.**   For each different factor from the prime factors in Step 1, circle the largest number of factors found in any one factorization.
> **Step 3.**   The LCM is the product of the circled factors.

Example 5   Find the LCM of 72 and 60.

Solution: First we write the prime factorization of each number.

$72 = 2 \cdot 2 \cdot 2 \cdot 3 \cdot 3$

$60 = 2 \cdot 2 \cdot 3 \cdot 5$

For the prime factors shown, we circle the largest number of factors found in either factorization.

$72 = \boxed{2 \cdot 2 \cdot 2} \cdot \boxed{3 \cdot 3}$

$60 = 2 \cdot 2 \cdot 3 \cdot \boxed{5}$

The LCM is the product of the circled factors.

$LCM = 2 \cdot 2 \cdot 2 \cdot 3 \cdot 3 \cdot 5 = 360$

The LCM is 360.

> **HELPFUL HINT**
>
> If the number of factors of a prime number are equal, circle either one, but not both. For example,
>
> $12 = \boxed{2 \cdot 2} \cdot \boxed{3}$
>
> $15 = 3 \cdot \boxed{5}$ Circle either 3, but not both.
>
> The LCM is $2 \cdot 2 \cdot 3 \cdot 5 = 60$.

Answers
4. 150,   **5.** 1080

Example 6   Find the LCM of 15, 18, and 54.

Solution: $15 = 3 \cdot ⑤$

$18 = ② \cdot 3 \cdot 3$

$54 = 2 \cdot ③ \cdot 3 \cdot 3$

The LCM is $2 \cdot 3 \cdot 3 \cdot 3 \cdot 5$ or 270.

Example 7   Find the LCM of 11 and 33.

Solution: $11 = ⑪$

$33 = ③ \cdot 11$

The LCM is $3 \cdot 11$ or 33.

C WRITING EQUIVALENT FRACTIONS

To add or subtract unlike fractions in the next section, we will first write equivalent fractions with the LCM as the denominator. Recall that fractions that represent the same portion of a whole are called equivalent fractions.

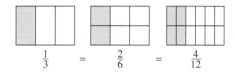

$$\frac{1}{3} = \frac{2}{6} = \frac{4}{12}$$

To write $\frac{1}{3}$ as an equivalent fraction with a denominator of 12, we multiply by 1.

$$\frac{1}{3} = \frac{1}{3} \cdot 1 = \frac{1}{3} \cdot \frac{4}{4} = \frac{1 \cdot 4}{3 \cdot 4} = \frac{4}{12}$$

Recall that
$$\frac{4}{4} = 1$$

So $\frac{1}{3} = \frac{4}{12}$. We will use the following notation to write equivalent fractions.

$$\frac{1}{3} = \frac{1 \cdot 4}{3 \cdot 4} = \frac{4}{12}$$

TRY THE CONCEPT CHECK IN THE MARGIN.

Example 8   Write an equivalent fraction with the indicated denominator.

$$\frac{3}{4} = \frac{}{20}$$

Solution: $\frac{3}{4} = \frac{}{20}$

$4 \cdot 5 = 20$

Practice Problem 6

Find the LCM of 20, 24, and 120.

Practice Problem 7

Find the LCM of 7 and 21.

TEACHING TIP

Before showing students mathematically that $1/3 = 4/12$, use the following illustration. Suppose you have two pies of equal size. One pie is cut into 3 equal pieces while the other is cut into 12 equal pieces. Notice that there are 4 times as many pieces of pie in the second pie as in the first pie. Suppose a friend takes a piece of pie from the first pie. How many pieces must you take from the second pie to get an equal amount of pie?

✓ **CONCEPT CHECK**

Which of the following is not equivalent to $\frac{3}{4}$?

a. $\frac{6}{8}$ b. $\frac{18}{24}$ c. $\frac{9}{14}$ d. $\frac{30}{40}$

Practice Problem 8

Write an equivalent fraction with the indicated denominator. $\frac{7}{8} = \frac{}{56}$

Answers

6. 120,   **7.** 21,   **8.** $\frac{49}{56}$

✓ Concept Check: c

Since $4 \cdot 5 = 20$, we multiply the numerator and the denominator of $\frac{3}{4}$ by 5.

$$\frac{3}{4} = \frac{3 \cdot 5}{4 \cdot 5} = \frac{15}{20}$$ Multiply the numerator and the denominator by 5.

HELPFUL HINT

To check Example 8, write $\frac{15}{20}$ in simplest form.

$$\frac{15}{20} = \frac{3 \cdot \overset{1}{\cancel{5}}}{4 \cdot \underset{1}{\cancel{5}}} = \frac{3}{4}, \quad \text{the original fraction.}$$

If the original fraction is in lowest terms we can check our work by writing the new equivalent fraction in simplest form. This form should be the original fraction.

Practice Problem 9

Write an equivalent fraction with the indicated denominator.

$$\frac{3}{5} = \frac{}{15}$$

TEACHING TIP

Remind students that they can check their solutions using cross products.

✓ CONCEPT CHECK

True or false? When the fraction $\frac{2}{9}$ is rewritten as an equivalent fraction with 27 as the denominator, the result is $\frac{2}{27}$.

Example 9 Write an equivalent fraction with the indicated denominator.

$$\frac{1}{2} = \frac{}{14}$$

Solution: Since $2 \cdot 7 = 14$, we multiply the numerator and the denominator by 7.

$$\frac{1}{2} = \frac{1 \cdot 7}{2 \cdot 7} = \frac{7}{14}$$

Thus, $\frac{1}{2} = \frac{7}{14}$.

TRY THE CONCEPT CHECK IN THE MARGIN.

Answers

9. $\frac{9}{15}$

✓ **Concept Check:** false; the correct result would be $\frac{6}{27}$

EXERCISE SET 3.2

A B *Find the LCM of each list of numbers. See Examples 1 through 7 .*

1. 3, 4   **2.** 4, 6   ▭ **3.** 9, 15   **4.** 12, 20

5. 12, 18   **6.** 12, 15   **7.** 24, 36   **8.** 42, 70

9. 18, 21   **10.** 24, 45   **11.** 15, 25   **12.** 21, 14

▭ **13.** 8, 24   **14.** 15, 90   **15.** 6, 7   **16.** 15, 8

▭ **17.** 25, 15, 6   **18.** 4, 6, 18   **19.** 34, 68   **20.** 25, 175

21. 84, 294   **22.** 48, 54   ▭ **23.** 30, 36, 50   **24.** 5, 10, 25

25. 3, 21, 51   **26.** 70, 98, 100   **27.** 11, 33, 121   **28.** 10, 15, 100

29. 8, 6, 27   **30.** 6, 25, 10   **31.** 4, 6, 10, 15   **32.** 25, 3, 15, 10

C *Write each fraction as an equivalent fraction with the given denominator. See Examples 8 and 9.*

▭ **33.** $\dfrac{4}{7} = \dfrac{}{35}$ **34.** $\dfrac{3}{5} = \dfrac{}{20}$ **35.** $\dfrac{2}{3} = \dfrac{}{21}$ **36.** $\dfrac{1}{6} = \dfrac{}{24}$

37. $\dfrac{2}{5} = \dfrac{}{25}$ **38.** $\dfrac{9}{10} = \dfrac{}{70}$ **39.** $\dfrac{1}{2} = \dfrac{}{30}$ **40.** $\dfrac{1}{3} = \dfrac{}{30}$

ANSWERS

1. 12
2. 12
3. 45
4. 60
5. 36
6. 60
7. 72
8. 210
9. 126
10. 360
11. 75
12. 42
13. 24
14. 90
15. 42
16. 120
17. 150
18. 36
19. 68
20. 175
21. 588
22. 432
23. 900
24. 50
25. 357
26. 4900
27. 363
28. 300
29. 216
30. 150
31. 60
32. 150
33. $\dfrac{20}{35}$
34. $\dfrac{12}{20}$
35. $\dfrac{14}{21}$
36. $\dfrac{4}{24}$
37. $\dfrac{10}{25}$
38. $\dfrac{63}{70}$
39. $\dfrac{15}{30}$
40. $\dfrac{10}{30}$

41. $\dfrac{30}{21}$

42. $\dfrac{35}{21}$

43. $\dfrac{21}{28}$

44. $\dfrac{36}{45}$

45. $\dfrac{30}{45}$

46. $\dfrac{50}{75}$

47. $\dfrac{36}{81}$

48. $\dfrac{40}{88}$

49. $\dfrac{12}{9}$

50. $\dfrac{6}{4}$

51. $\dfrac{18}{10}$

52. $\dfrac{21}{12}$

53. $\dfrac{5}{10} = \dfrac{1}{2}$

54. $\dfrac{5}{13}$

55. $\dfrac{2}{5}$

56. $\dfrac{4}{8} = \dfrac{1}{2}$

57. $\dfrac{8}{18} = \dfrac{4}{9}$

58. $\dfrac{24}{30} = \dfrac{4}{5}$

59. $\dfrac{9}{9} = 1$

60. $\dfrac{12}{12} = 1$

61. $\dfrac{814}{3630}$

62. $\dfrac{2052}{4085}$

63. answers may vary

64. answers may vary

176

Name _____

41. $\dfrac{10}{7} = \dfrac{}{21}$ **42.** $\dfrac{5}{3} = \dfrac{}{21}$ **43.** $\dfrac{3}{4} = \dfrac{}{28}$ **44.** $\dfrac{4}{5} = \dfrac{}{45}$

45. $\dfrac{2}{3} = \dfrac{}{45}$ **46.** $\dfrac{2}{3} = \dfrac{}{75}$ 📼 **47.** $\dfrac{4}{9} = \dfrac{}{81}$ **48.** $\dfrac{5}{11} = \dfrac{}{88}$

49. $\dfrac{4}{3} = \dfrac{}{9}$ **50.** $\dfrac{3}{2} = \dfrac{}{4}$ **51.** $\dfrac{9}{5} = \dfrac{}{10}$ **52.** $\dfrac{7}{4} = \dfrac{}{12}$

REVIEW AND PREVIEW

Add or subtract as indicated. See Section 3.1.

53. $\dfrac{7}{10} - \dfrac{2}{10}$ **54.** $\dfrac{8}{13} - \dfrac{3}{13}$ **55.** $\dfrac{1}{5} + \dfrac{1}{5}$ **56.** $\dfrac{1}{8} + \dfrac{3}{8}$

57. $\dfrac{23}{18} - \dfrac{15}{18}$ **58.** $\dfrac{36}{30} - \dfrac{12}{30}$ **59.** $\dfrac{2}{9} + \dfrac{1}{9} + \dfrac{6}{9}$ **60.** $\dfrac{2}{12} + \dfrac{7}{12} + \dfrac{3}{12}$

COMBINING CONCEPTS

Write each fraction as an equivalent fraction with the indicated denominator.

61. $\dfrac{37}{165} = \dfrac{}{3630}$ **62.** $\dfrac{108}{215} = \dfrac{}{4085}$

63. In your own words, explain how to find the LCM of two numbers.

64. In your own words, explain how to write a fraction as an equivalent fraction with a given denominator.

3.3 ADDING AND SUBTRACTING UNLIKE FRACTIONS

A ADDING UNLIKE FRACTIONS

In this section we add and subtract fractions with unlike denominators. To add or subtract these unlike fractions, we first write the fractions as equivalent fractions with a common denominator and then add or subtract the like fractions. The common denominator that we will use is the LCM of the denominators. This denominator is called the **least common denominator (LCD).**

To begin, let's add the unlike fractions $\frac{3}{4} + \frac{1}{6}$. The LCM of denominators 4 and 6 is 12. This means that the LCD for the denominators 4 and 6 is 12. So we write each fraction as an equivalent fraction with a denominator of 12.

$$\frac{3}{4} = \frac{3 \cdot 3}{4 \cdot 3} = \frac{9}{12} \text{ and } \frac{1}{6} = \frac{1 \cdot 2}{6 \cdot 2} = \frac{2}{12}$$

Then

$$\frac{3}{4} + \frac{1}{6} = \frac{9}{12} + \frac{2}{12} = \frac{11}{12}$$

ADDING OR SUBTRACTING UNLIKE FRACTIONS

Step 1. Find the LCD of the denominators of the fractions.

Step 2. Write each fraction as an equivalent fraction whose denominator is the LCD.

Step 3. Add or subtract the like fractions.

Step 4. Write the sum or difference in simplest form.

Example 1   Add: $\frac{2}{5} + \frac{4}{15}$

Solution: **Step 1.** The LCD for the denominators 5 and 15 is 15.

Step 2. $\frac{2}{5} = \frac{2 \cdot 3}{5 \cdot 3} = \frac{6}{15}$, $\frac{4}{15} = \frac{4}{15}$ ← This fraction already has a denominator of 15.

Step 3.

$$\frac{2}{5} + \frac{4}{15} = \frac{6}{15} + \frac{4}{15} = \frac{10}{15}$$

Step 4. Write in simplest form.

$$\frac{10}{15} = \frac{2 \cdot \overset{1}{\cancel{5}}}{3 \cdot \underset{1}{\cancel{5}}} = \frac{2}{3}$$

Example 2   Add: $\frac{2}{15} + \frac{3}{10}$

Solution: **Step 1.** The LCD for 15 and 10 is 30.

Step 2. $\frac{2}{15} = \frac{2 \cdot 2}{15 \cdot 2} = \frac{4}{30}$ $\frac{3}{10} = \frac{3 \cdot 3}{10 \cdot 3} = \frac{9}{30}$

Objectives

A Add unlike fractions.
B Subtract unlike fractions.
C Solve problems by adding or subtracting unlike fractions.

SSM   CD-ROM   Video
                              3.3

Practice Problem 1

Add: $\frac{4}{7} + \frac{3}{14}$

Practice Problem 2

Add: $\frac{5}{6} + \frac{2}{9}$

Answers

1. $\frac{11}{14}$, **2.** $\frac{19}{18} = 1\frac{1}{18}$

Step 3. $\dfrac{2}{15} + \dfrac{3}{10} = \dfrac{4}{30} + \dfrac{9}{30} = \dfrac{13}{30}$

Step 4. $\dfrac{13}{30}$ is in simplest form.

Practice Problem 3

Add: $\dfrac{2}{5} + \dfrac{4}{9}$

Example 3   Add: $\dfrac{2}{3} + \dfrac{1}{7}$

Solution: The LCD for 3 and 7 is 21.

$$\dfrac{2}{3} + \dfrac{1}{7} = \dfrac{2 \cdot 7}{3 \cdot 7} + \dfrac{1 \cdot 3}{7 \cdot 3}$$

$$= \dfrac{14}{21} + \dfrac{3}{21}$$

$$= \dfrac{17}{21}$$

Practice Problem 4

Add: $\dfrac{1}{4} + \dfrac{1}{5} + \dfrac{1}{10}$

Example 4   Add: $\dfrac{1}{2} + \dfrac{1}{3} + \dfrac{1}{6}$

Solution: The LCD for 2, 3, and 6 is 6.

$$\dfrac{1}{2} + \dfrac{1}{3} + \dfrac{1}{6} = \dfrac{1 \cdot 3}{2 \cdot 3} + \dfrac{1 \cdot 2}{3 \cdot 2} + \dfrac{1}{6}$$

$$= \dfrac{3}{6} + \dfrac{2}{6} + \dfrac{1}{6}$$

$$= \dfrac{6}{6} = 1$$

TRY THE CONCEPT CHECK IN THE MARGIN.

✓ CONCEPT CHECK

Find and correct the error in the following: $\dfrac{2}{9} + \dfrac{4}{11} = \dfrac{6}{20} = \dfrac{3}{10}$

B SUBTRACTING UNLIKE FRACTIONS

As indicated in the box on page 177, we follow the same steps when subtracting unlike fractions as when adding them.

Example 5   Subtract: $\dfrac{2}{5} - \dfrac{3}{20}$

Practice Problem 5

Subtract: $\dfrac{7}{12} - \dfrac{5}{24}$

Solution: **Step 1.** The LCD for the denominators 5 and 20 is 20.

Step 2. $\dfrac{2}{5} = \dfrac{2 \cdot 4}{5 \cdot 4} = \dfrac{8}{20}$ $\dfrac{3}{20} = \dfrac{3}{20}$ ← The fraction already has a denominator of 20.

Practice Problem 6

Subtract: $\dfrac{9}{10} - \dfrac{3}{7}$

Step 3. $\dfrac{2}{5} - \dfrac{3}{20} = \dfrac{8}{20} - \dfrac{3}{20} = \dfrac{5}{20}$

Step 4. Write in simplest form.

$$\dfrac{5}{20} = \dfrac{\overset{1}{\cancel{5}}}{\underset{1}{\cancel{5}} \cdot 4} = \dfrac{1}{4}$$

Answers

3. $\dfrac{38}{45}$,   **4.** $\dfrac{11}{20}$,   **5.** $\dfrac{3}{8}$,   **6.** $\dfrac{33}{70}$

✓ **Concept Check**

When adding unlike fractions you don't add the denominators. Correct solution:
$\dfrac{2}{9} + \dfrac{4}{11} = \dfrac{22}{99} + \dfrac{36}{99} = \dfrac{58}{99}$

Example 6   Subtract: $\dfrac{10}{11} - \dfrac{2}{3}$

Solution: **Step 1.** The LCD for 11 and 3 is 33.

Step 2. $\dfrac{10}{11} = \dfrac{10 \cdot 3}{11 \cdot 3} = \dfrac{30}{33}$ $\dfrac{2}{3} = \dfrac{2 \cdot 11}{3 \cdot 11} = \dfrac{22}{33}$

Step 3. $\dfrac{10}{11} - \dfrac{2}{3} = \dfrac{30}{33} - \dfrac{22}{33} = \dfrac{8}{33}$

Step 4. $\dfrac{8}{33}$ is in simplest form.

Example 7   Subtract: $\dfrac{11}{12} - \dfrac{2}{9}$

Solution: The LCD for 12 and 9 is 36.

$$\dfrac{11}{12} - \dfrac{2}{9} = \dfrac{11 \cdot 3}{12 \cdot 3} - \dfrac{2 \cdot 4}{9 \cdot 4}$$

$$= \dfrac{33}{36} - \dfrac{8}{36}$$

$$= \dfrac{25}{36}$$

C SOLVING PROBLEMS BY ADDING OR SUBTRACTING UNLIKE FRACTIONS

Very often, real-world problems involve adding or subtracting unlike fractions.

Example 8   Finding Total Weight

A freight truck has $\dfrac{1}{4}$ ton of computers, $\dfrac{1}{3}$ ton of televisions, and $\dfrac{3}{8}$ ton of small appliances. Find the total weight of its load.

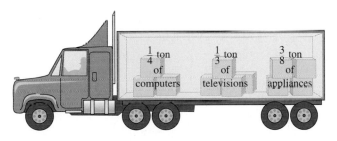

Solution: **1.** UNDERSTAND. Read and reread the problem. The phrase "total weight" tells us to add.
2. TRANSLATE.

In words:

total weight	is	weight of computers	plus	weight of televisions	plus	weight of appliances

Translate:

total weight	=	$\dfrac{1}{4}$	+	$\dfrac{1}{3}$	+	$\dfrac{3}{8}$

3. SOLVE. The LCD is 24.

$$\dfrac{1}{4} + \dfrac{1}{3} + \dfrac{3}{8} = \dfrac{1 \cdot 6}{4 \cdot 6} + \dfrac{1 \cdot 8}{3 \cdot 8} + \dfrac{3 \cdot 3}{8 \cdot 3}$$

$$= \dfrac{6}{24} + \dfrac{8}{24} + \dfrac{9}{24}$$

$$= \dfrac{23}{24}$$

Practice Problem 7

Subtract: $\dfrac{7}{8} - \dfrac{5}{6}$

Practice Problem 8

To repair her sidewalk, a homeowner must pour small amounts of cement in three different locations. She needs $\dfrac{3}{5}$ of a cubic yard, $\dfrac{2}{10}$ of a cubic yard, and $\dfrac{2}{15}$ of a cubic yard for these locations. Find the total amount of concrete the homeowner needs. If she bought enough cement to mix 1 cubic yard, did she buy enough?

Answers

7. $\dfrac{1}{24}$, **8.** $\dfrac{14}{15}$ cubic yard; Yes, she bought enough.

Practice Problem 9

Find the difference in length of two boards if one board is $\frac{4}{5}$ of a foot long and the other is $\frac{2}{3}$ of a foot long.

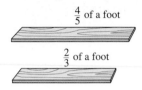
$\frac{4}{5}$ of a foot

$\frac{2}{3}$ of a foot

TEACHING TIP   Classroom Activity

Ask students to work in groups to make up their own word problems involving the addition and subtraction of fractions with unlike denominators. Then have them give their problems to another group to solve.

4. INTERPRET. *Check* the solution. *State* your conclusion: The total weight of the truck's load is $\frac{23}{24}$ ton.

Example 9   Calculating Flight Time

A flight from Tucson to Phoenix, Arizona, requires $\frac{5}{12}$ of an hour. If the plane has been flying $\frac{1}{4}$ of an hour, find how much time remains before landing.

Solution:

1. UNDERSTAND. Read and reread the problem. The phrase "how much time remains" tells us to subtract.

2. TRANSLATE.

In words:

time remaining	is	flight time from Tucson to Phoenix	minus	flight time already passed
↓	↓	↓	↓	↓

Translate:

time remaining	=	$\frac{5}{12}$	−	$\frac{1}{4}$

3. SOLVE. The LCD is 12.

$$\frac{5}{12} - \frac{1}{4} = \frac{5}{12} - \frac{1 \cdot 3}{4 \cdot 3}$$

$$= \frac{5}{12} - \frac{3}{12}$$

$$= \frac{2}{12} = \frac{\cancel{2}^{1}}{\cancel{2}_{1} \cdot 6} = \frac{1}{6}$$

4. INTERPRET. *Check* the solution. *State* your conclusion: The flight time remaining is $\frac{1}{6}$ of an hour.

Answer

9. $\frac{2}{15}$ feet

EXERCISE SET 3.3

A *Add and simplify. See Examples 1 through 4.*

1. $\dfrac{2}{3} + \dfrac{1}{6}$ **2.** $\dfrac{5}{6} + \dfrac{1}{12}$ **3.** $\dfrac{1}{2} + \dfrac{1}{3}$ **4.** $\dfrac{2}{3} + \dfrac{1}{4}$

5. $\dfrac{2}{11} + \dfrac{2}{33}$ **6.** $\dfrac{5}{9} + \dfrac{1}{3}$ **7.** $\dfrac{3}{14} + \dfrac{3}{7}$ **8.** $\dfrac{2}{5} + \dfrac{2}{15}$

9. $\dfrac{11}{35} + \dfrac{2}{7}$ **10.** $\dfrac{2}{5} + \dfrac{3}{25}$ **11.** $\dfrac{5}{12} + \dfrac{1}{9}$ **12.** $\dfrac{7}{12} + \dfrac{5}{18}$

13. $\dfrac{7}{15} + \dfrac{5}{12}$ **14.** $\dfrac{5}{8} + \dfrac{3}{20}$ **15.** $\dfrac{2}{28} + \dfrac{2}{21}$ **16.** $\dfrac{6}{25} + \dfrac{7}{35}$

17. $\dfrac{5}{7} + \dfrac{1}{8} + \dfrac{1}{2}$ **18.** $\dfrac{10}{13} + \dfrac{7}{10} + \dfrac{1}{5}$ **19.** $\dfrac{5}{11} + \dfrac{3}{9} + \dfrac{2}{3}$ **20.** $\dfrac{7}{18} + \dfrac{2}{9} + \dfrac{5}{6}$

B *Subtract and simplify. See Examples 5 through 7.*

21. $\dfrac{7}{8} - \dfrac{3}{16}$ **22.** $\dfrac{5}{13} - \dfrac{3}{26}$ **23.** $\dfrac{5}{6} - \dfrac{3}{7}$ **24.** $\dfrac{3}{4} - \dfrac{1}{7}$

25. $\dfrac{5}{7} - \dfrac{1}{8}$ **26.** $\dfrac{10}{13} - \dfrac{7}{10}$ **27.** $\dfrac{5}{11} - \dfrac{3}{9}$ **28.** $\dfrac{7}{18} - \dfrac{2}{9}$

ANSWERS

1. $\dfrac{5}{6}$
2. $\dfrac{11}{12}$
3. $\dfrac{5}{6}$
4. $\dfrac{11}{12}$
5. $\dfrac{8}{33}$
6. $\dfrac{8}{9}$
7. $\dfrac{9}{14}$
8. $\dfrac{8}{15}$
9. $\dfrac{3}{5}$
10. $\dfrac{13}{25}$
11. $\dfrac{19}{36}$
12. $\dfrac{31}{36}$
13. $\dfrac{53}{60}$
14. $\dfrac{31}{40}$
15. $\dfrac{1}{6}$
16. $\dfrac{11}{25}$
17. $\dfrac{75}{56} = 1\dfrac{19}{56}$
18. $\dfrac{217}{130} = 1\dfrac{87}{130}$
19. $\dfrac{16}{11} = 1\dfrac{5}{11}$
20. $\dfrac{13}{9} - 1\dfrac{4}{9}$
21. $\dfrac{11}{16}$
22. $\dfrac{7}{26}$
23. $\dfrac{17}{42}$
24. $\dfrac{17}{28}$
25. $\dfrac{33}{56}$
26. $\dfrac{9}{130}$
27. $\dfrac{4}{33}$
28. $\dfrac{1}{6}$

181

Name _____

29. $\frac{11}{35} - \frac{2}{7}$ **30.** $\frac{2}{5} - \frac{3}{25}$ **31.** $\frac{5}{12} - \frac{1}{9}$ **32.** $\frac{7}{12} - \frac{5}{18}$

33. $\frac{7}{15} - \frac{5}{12}$ **34.** $\frac{5}{8} - \frac{3}{20}$ **35.** $\frac{3}{28} - \frac{2}{21}$ **36.** $\frac{6}{25} - \frac{7}{35}$

C *Find the perimeter of each geometric figure. (Hint: Recall that perimeter means distance around.)*

37.

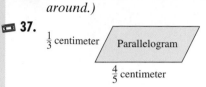
$\frac{1}{3}$ centimeter Parallelogram
$\frac{4}{5}$ centimeter

38.

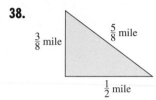
$\frac{3}{8}$ mile $\frac{5}{8}$ mile
$\frac{1}{2}$ mile

39.

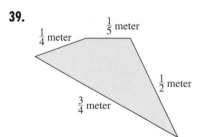
$\frac{1}{4}$ meter $\frac{1}{5}$ meter
$\frac{1}{2}$ meter
$\frac{3}{4}$ meter

40.

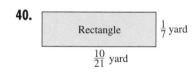
Rectangle $\frac{1}{7}$ yard
$\frac{10}{21}$ yard

Solve. See Examples 8 and 9.

41. Martell Wilson, a sales clerk at a Godiva candy shop, mixes $\frac{3}{4}$ pound of chocolate covered almonds with $\frac{5}{8}$ pound of coconut creams in a gift container. How much does the mixture weigh?

42. Jillian Frize has $\frac{5}{8}$ of a box of computer paper. If she places $\frac{1}{4}$ of the contents of the box on the printer stand, what portion remains in the box?

43. About $\frac{13}{20}$ of American students ages 10 to 17 name math, science, or art as their favorite subject in school. Art is the favorite subject for about $\frac{4}{25}$ of these students. For what fraction of students this age is math or science their favorite subject? (*Source:* Peter D. Hart Research Associates for the National Science Foundation)

44. Together, the United States' and Japan's postal services handle $\frac{49}{100}$ of the world's mail volume. Japan's postal service alone handles $\frac{3}{50}$ of the world's mail. What fraction of the world's mail is handled by the postal service of the United States? (*Source:* United States Postal Service)

Name _____

The table gives the fraction of Americans who eat pasta at various intervals. Use this table for Exercises 45 and 46.

How Often Americans Eat Pasta	
Frequency	Fraction
3 times per week	$\dfrac{31}{100}$
1 or 2 times per week	$\dfrac{23}{50}$
1 or 2 times per month	$\dfrac{17}{100}$
Less often	$\dfrac{3}{50}$

(*Source:* Princton Survey Research)

45. What fraction of Americans eat pasta 1, 2, or 3 times a week?

46. What fraction of Americans eat pasta 1 or 2 times a month or less often?

REVIEW AND PREVIEW

Multiply or divide as indicated. See Sections 2.4 and 2.5.

47. $1\dfrac{1}{2} \cdot 3\dfrac{1}{3}$

48. $2\dfrac{5}{6} \div 5$

49. $4 \div 7\dfrac{1}{4}$

50. $4\dfrac{3}{4} \cdot 5\dfrac{1}{5}$

51. $3 \cdot 2\dfrac{1}{9}$

52. $6\dfrac{2}{7} \cdot 14$

COMBINING CONCEPTS

Perform each indicated operation.

53. $\dfrac{30}{55} + \dfrac{1000}{1760}$;
the LCD for 55 and 1760 is 1760.

54. $\dfrac{19}{26} - \dfrac{968}{1352}$;
the LCD for 26 and 1352 is 1352.

55. In your own words, describe how to add two fractions with different denominators.

---

45. $\dfrac{77}{100}$ of Americans

46. $\dfrac{23}{100}$ of Americans

47. 5

48. $\dfrac{17}{30}$

49. $\dfrac{16}{29}$

50. $\dfrac{247}{10} = 24\dfrac{7}{10}$

51. $\dfrac{19}{3} = 6\dfrac{1}{3}$

52. 88

53. $\dfrac{49}{44} = 1\dfrac{5}{44}$

54. $\dfrac{5}{338}$

55. answers may vary

56. $\frac{163}{579}$

57. $\frac{212}{579}$

58. 16,300,000 square miles

59. 21,200,000 square miles

60. $\frac{175}{193}$

The table shows the fraction of the world's land area taken up by each continent. Use the table to answer Exercises 56–60.

Continent	Fraction of World's Land Area
Africa	$\frac{39}{193}$
Antarctica	$\frac{18}{193}$
Asia	$\frac{58}{193}$
Australia	$\frac{11}{193}$
Europe	$\frac{38}{579}$
North America	$\frac{94}{579}$
South America	$\frac{23}{193}$

(Source: 1998 World Almanac)

56. What fraction of the world's land area is accounted for by North and South America?

57. What fraction of the world's land area is accounted for by Asia and Europe?

58. If the total land area of Earth's surface is 57,900,000 square miles, what is the combined land area of the North and South American continents?

59. If the total land area of Earth's surface is 57,900,000 square miles, what is the combined land area of the European and Asian continents?

60. Antarctica is generally considered to be uninhabited. What fraction of the world's land area is accounted for by inhabited continents?

INTEGRATED REVIEW — ADDING AND SUBTRACTING FRACTIONS

Find the LCM of each list of numbers.

1. 5, 6

2. 3, 7

3. 2, 14

4. 5, 25

5. 4, 20, 25

6. 6, 18, 30

Write each fraction as an equivalent fraction with the indicated denominator.

7. $\dfrac{3}{8} = \dfrac{}{24}$

8. $\dfrac{7}{9} = \dfrac{}{36}$

9. $\dfrac{1}{4} = \dfrac{}{40}$

10. $\dfrac{2}{5} = \dfrac{}{30}$

11. $\dfrac{11}{15} = \dfrac{}{75}$

12. $\dfrac{5}{6} = \dfrac{}{48}$

Add or subtract as indicated. Simplify if necessary.

13. $\dfrac{3}{8} + \dfrac{1}{8}$

14. $\dfrac{7}{10} - \dfrac{3}{10}$

15. $\dfrac{4}{15} + \dfrac{9}{15}$

16. $\dfrac{17}{24} - \dfrac{3}{24}$

17. $\dfrac{1}{4} + \dfrac{1}{2}$

18. $\dfrac{1}{3} - \dfrac{1}{5}$

19. $\dfrac{7}{9} - \dfrac{2}{5}$

20. $\dfrac{3}{10} + \dfrac{2}{25}$

21. $\dfrac{7}{8} + \dfrac{1}{20}$

22. $\dfrac{5}{12} - \dfrac{2}{18}$

23. $\dfrac{1}{11} + \dfrac{1}{11}$

24. $\dfrac{3}{17} - \dfrac{2}{17}$

25. $\dfrac{9}{11} - \dfrac{2}{3}$ | 26. $\dfrac{1}{6} - \dfrac{1}{7}$ | 27. $\dfrac{2}{9} + \dfrac{1}{18}$

28. $\dfrac{4}{13} + \dfrac{2}{26}$ | 25. $\dfrac{2}{9} + \dfrac{1}{18} + \dfrac{1}{3}$ | 30. $\dfrac{3}{10} + \dfrac{1}{5} + \dfrac{6}{25}$

25. $\dfrac{5}{33}$ _____

26. $\dfrac{1}{42}$ _____

27. $\dfrac{5}{18}$ _____

28. $\dfrac{5}{13}$ _____

29. $\dfrac{11}{18}$ _____

30. $\dfrac{37}{50}$ _____

3.4 ADDING AND SUBTRACTING MIXED NUMBERS

A ADDING MIXED NUMBERS

Recall that a mixed number has a whole number part and a fraction part.

$$2\frac{3}{8} \text{ means } 2 + \frac{3}{8}$$

with "whole number" labeling the 2, and "fraction" labeling the $\frac{3}{8}$.

TRY THE CONCEPT CHECK IN THE MARGIN.

> **ADDING OR SUBTRACTING MIXED NUMBERS**
>
> To add or subtract mixed numbers, add or subtract the fractions and then add or subtract the whole numbers.

For example,

$$\begin{array}{r} 2\frac{2}{7} \\ +6\frac{3}{7} \\ \hline 8\frac{5}{7} \end{array}$$ ← Add the fractions,

then add the whole numbers.

Example 1   Add: $2\frac{1}{3} + 5\frac{3}{8}$

Solution: The LCD is 24.

$$\begin{array}{r} 2\dfrac{1 \cdot 8}{3 \cdot 8} = 2\dfrac{8}{24} \\ +5\dfrac{3 \cdot 3}{8 \cdot 3} = 5\dfrac{9}{24} \\ \hline 7\dfrac{17}{24} \end{array}$$ ← Add the fractions.

Add the whole numbers.

Example 2   Add: $3\frac{4}{5} + 1\frac{4}{15}$

Solution: The LCD of 5 and 15 is 15.

$$\begin{array}{r} 3\dfrac{4}{5} = 3\dfrac{12}{15} \\ +1\dfrac{4}{15} = 1\dfrac{4}{15} \\ \hline 4\dfrac{16}{15} \end{array}$$

Add the fractions, then add the whole numbers.

Notice that the fractional part is improper.

Since $\frac{16}{15}$ is $1\frac{1}{15}$, we can write the sum as

$$4\frac{16}{15} = 4 + 1\frac{1}{15} = 5\frac{1}{15}$$

Objectives

A Add mixed numbers.
B Subtract mixed numbers.
C Solve problems by adding or subtracting mixed numbers.

SSM   CD-ROM   Video 3.4

✓ CONCEPT CHECK

Which of the following are equivalent to 7?

a. $6\frac{5}{5}$ b. $6\frac{7}{7}$

c. $5\frac{8}{4}$ d. $6\frac{17}{17}$

e. all of these

Practice Problem 1

Add: $4\frac{2}{5} + 5\frac{3}{10}$

Practice Problem 2

Add: $2\frac{5}{14} + 5\frac{6}{7}$

TEACHING TIP

Ask students how they can use estimation to check operations on mixed numbers.

Answers

1. $9\frac{7}{10}$,   **2.** $8\frac{3}{14}$

✓ Concept Check: e

Practice Problem 3

Add: $10 + 2\frac{1}{7} + 3\frac{1}{5}$

✓ CONCEPT CHECK

Explain how you could estimate the sum: $5\frac{1}{9} + 14\frac{10}{11}$.

Practice Problem 4

Subtract: $29\frac{7}{8} - 13\frac{3}{16}$

Practice Problem 5

Subtract: $9\frac{7}{15} - 5\frac{4}{5}$

Answers

3. $15\frac{12}{35}$, **4.** $16\frac{11}{16}$, **5.** $3\frac{2}{3}$

✓ Concept Check

Round each mixed number to the nearest whole number and add. $5\frac{1}{9}$ rounds to 5 and $14\frac{10}{11}$ rounds to 15, and their sum is $5 + 15 = 20$.

Example 3   Add: $1\frac{4}{5} + 4 + 2\frac{1}{2}$

Solution: The LCD of 5 and 2 is 10.

$$1\frac{4}{5} = 1\frac{8}{10}$$
$$4\phantom{\frac{4}{5}} = 4$$
$$+2\frac{1}{2} = 2\frac{5}{10}$$
$$\overline{\phantom{+2\frac{1}{2}}}$$
$$7\frac{13}{10} = 7 + 1\frac{3}{10} = 8\frac{3}{10}$$

TRY THE CONCEPT CHECK IN THE MARGIN.

B SUBTRACTING MIXED NUMBERS

Example 4   Subtract: $9\frac{3}{7} - 5\frac{4}{21}$

Solution: The LCD of 7 and 21 is 21.

$$9\frac{3}{7} = 9\frac{9}{21} \quad \leftarrow \text{The LCD of 7 and 21 is 21.}$$
$$-5\frac{4}{21} = -5\frac{4}{21}$$
$$\overline{\phantom{-5\frac{4}{21}}}$$
$$4\frac{5}{21} \quad \leftarrow \text{Subtract the fractions.}$$
$$\uparrow$$
$$\text{Subtract the whole numbers.}$$

When subtracting mixed numbers, borrowing may be needed, as shown in the next example.

Example 5   Subtract: $7\frac{3}{14} - 3\frac{6}{7}$

Solution: The LCD of 7 and 14 is 14.

$$7\frac{3}{14} = 7\frac{3}{14} \quad \text{Notice that we cannot subtract } \frac{12}{14} \text{ from } \frac{3}{14},$$
$$-3\frac{6}{7} = -3\frac{12}{14} \quad \text{so we borrow from the whole number 7.}$$
$$\overline{\phantom{-3\frac{6}{7}}}$$

borrow 1 from 7

$$7\frac{3}{14} = 6 + 1\frac{3}{14} = 6 + \frac{17}{14} \text{ or } 6\frac{17}{14}$$
$$\text{Now subtract.}$$

$$7\frac{3}{14} = 7\frac{3}{14} = 6\frac{17}{14}$$
$$-3\frac{6}{7} = -3\frac{12}{14} = -3\frac{12}{14}$$
$$\overline{\phantom{-3\frac{12}{14}}}$$
$$3\frac{5}{14} \quad \leftarrow \text{Subtract the fractions.}$$
$$\uparrow$$
$$\text{Subtract the whole numbers.}$$

TRY THE CONCEPT CHECK IN THE MARGIN.

Example 6   Subtract: $12 - 8\frac{3}{7}$

Solution:

$$12 = 11\frac{7}{7} \quad \text{Borrow 1 from 12 and write it as } \frac{7}{7}.$$

$$-8\frac{3}{7} = -8\frac{3}{7}$$

$$\overline{\qquad\qquad\qquad 3\frac{4}{7}} \leftarrow \text{Subtract the fractions.}$$

$\uparrow$
Subtract the whole numbers.

C SOLVING PROBLEMS BY ADDING OR SUBTRACTING MIXED NUMBERS

Now that we know how to add and subtract mixed numbers we can solve real-life problems.

Example 7   Calculating Total Weight

Sarah Grahamm purchases two packages of ground round. One package weighs $2\frac{3}{8}$ pounds and the other $1\frac{4}{5}$ pounds. What is the combined weight of the ground round?

Solution: **1.** UNDERSTAND. Read and reread the problem. The phrase "combined weight" tells us to add.

2. TRANSLATE.

In words:

combined weight	is	weight of one package	plus	weight of second package

Translate:

$$\text{combined weight} = 2\frac{3}{8} + 1\frac{4}{5}$$

3. Solve.

$$2\frac{3}{8} = 2\frac{15}{40}$$
$$+1\frac{4}{5} = 1\frac{32}{40}$$
$$\overline{\qquad\qquad 3\frac{47}{40}} = 4\frac{7}{40}$$

4. INTERPRET. *Check* your work. *State* your conclusion: The combined weight of the ground round is $4\frac{7}{40}$ pounds.

Example 8   Finding Legal Lobster Size

Lobster fisherman must measure the upper body shells of the lobsters they catch. Lobsters that are too small are thrown back into the ocean. Each state has its own size standard for lobsters to help control the breeding stock. In 1988, Massachusetts increased its legal lobster size

Practice Problem 6

Subtract: $25 - 10\frac{2}{9}$

Practice Problem 7

Two rainbow trout weigh $2\frac{1}{2}$ pounds and $3\frac{2}{3}$ pounds. What is the total weight of the two trout?

TEACHING TIP   Classroom Activity

Ask students to work in groups to make up their own word problems involving the addition and subtraction of mixed numbers. Then have them give their problems to another group to solve.

Answers

6. $14\frac{7}{9}$,   **7.** $6\frac{1}{6}$ pounds

✓ Concept Check

Rewrite $5\frac{1}{4}$ as $4\frac{5}{4}$ by borrowing from the 5.

Practice Problem 8

The measurement around the trunk of a tree just below shoulder height is called its girth. The largest known American Beech tree in the United States has a girth of $23\frac{1}{4}$ feet. The largest known Sugar Maple tree in the United States has a girth of $19\frac{5}{12}$ feet.

How much larger is the girth of the largest known American Beech tree than the girth of the largest known Sugar Maple tree? (*Source: American Forests*)

← girth

from $3\frac{3}{16}$ inches to $3\frac{7}{32}$ inches. How much of an increase was this? (*Source:* Peabody Essex Museum, Salem, Massachusetts)

Solution:

1. UNDERSTAND. Read and reread the problem carefully. The word "increase" found in the problem might make you think that we add to solve the problem. But the phrase "how much of an increase" tells us to subtract to find the increase.

2. TRANSLATE.

In words:

increase	is	new lobster size	minus	old lobster size
↓	↓	↓	↓	↓

Translate: increase   = $3\frac{7}{32}$ − $3\frac{3}{16}$

3. Solve. $3\frac{7}{32} = 3\frac{7}{32}$

$$-3\frac{3}{16} = 3\frac{6}{32}$$
$$\frac{1}{32}$$

4. INTERPRET. *Check* your work. *State* your conclusion: The increase in lobster size is $\frac{1}{32}$ of an inch.

Answer

8. $3\frac{5}{6}$ feet

EXERCISE SET 3.4

A *Add. See Examples 1 through 3.*

1. $4\frac{7}{10}$
$+2\frac{1}{10}$

2. $7\frac{4}{9}$
$+3\frac{2}{9}$

3. $10\frac{3}{14}$
$+\ 3\frac{4}{7}$

4. $12\frac{5}{12}$
$+\ 4\frac{1}{6}$

5. $9\frac{1}{5}$
$+8\frac{2}{25}$

6. $6\frac{2}{13}$
$+8\frac{7}{26}$

7. $1\frac{5}{6}$
$+5\frac{3}{8}$

8. $2\frac{5}{12}$
$+\ 1\frac{5}{8}$

9. $3\frac{1}{2}$
$+4\frac{1}{8}$

10. $9\frac{3}{4}$
$+2\frac{1}{8}$

11. $8\frac{2}{5}$
$+11\frac{2}{3}$

12. $7\frac{3}{7}$
$+3\frac{1}{5}$

13. $15\frac{1}{6}$
$+13\frac{5}{12}$

14. $21\frac{3}{10}$
$+\ 11\frac{3}{5}$

15. $3\frac{5}{8}$
$2\frac{1}{6}$
$+7\frac{3}{4}$

16. $4\frac{1}{3}$
$9\frac{2}{5}$
$+3\frac{1}{6}$

B *Subtract. See Examples 4 through 6.*

17. $4\frac{7}{10}$
$-2\frac{1}{10}$

18. $7\frac{4}{9}$
$-3\frac{2}{9}$

19. $10\frac{13}{14}$
$+\ 3\frac{4}{7}$

20. $12\frac{5}{12}$
$+\ 4\frac{1}{6}$

21. $9\frac{1}{5}$
$-8\frac{6}{25}$

22. $6\frac{2}{13}$
$-4\frac{7}{26}$

23. $15\frac{4}{7}$
$-9\frac{11}{14}$

24. $23\frac{3}{5}$
$-8\frac{8}{15}$

ANSWERS

1. $6\frac{4}{5}$

2. $10\frac{2}{3}$

3. $13\frac{11}{14}$

4. $16\frac{7}{12}$

5. $17\frac{7}{25}$

6. $14\frac{11}{26}$

7. $7\frac{5}{24}$

8. $4\frac{1}{24}$

9. $7\frac{5}{8}$

10. $11\frac{7}{8}$

11. $20\frac{1}{15}$

12. $10\frac{22}{35}$

13. $28\frac{7}{12}$

14. $32\frac{9}{10}$

15. $13\frac{13}{24}$

16. $16\frac{9}{10}$

17. $2\frac{3}{5}$

18. $4\frac{2}{9}$

19. $7\frac{5}{14}$

20. $8\frac{1}{4}$

21. $\frac{24}{25}$

22. $1\frac{23}{26}$

23. $5\frac{11}{14}$

24. $15\frac{1}{15}$

25. $2\frac{1}{2}$ _____

26. $2\frac{9}{16}$ _____

27. $23\frac{5}{8}$ _____

28. $\frac{11}{14}$ _____

29. $1\frac{4}{5}$ _____

30. $5\frac{1}{4}$ _____

31. $1\frac{13}{15}$ _____

32. $2\frac{3}{10}$ _____

33. $3\frac{5}{9}$ _____

34. $6\frac{3}{10}$ _____

35. $15\frac{3}{4}$ _____

36. $58\frac{5}{6}$ _____

37. $2\frac{3}{8}$ hours _____

38. $2\frac{5}{8}$ feet _____

39. $10\frac{1}{4}$ hours _____

40. $8\frac{23}{48}$ hours _____

41. $7\frac{13}{20}$ inches _____

42. $1\frac{3}{8}$ inches _____

192

25. $5\frac{2}{3} - 3\frac{1}{6}$ **26.** $5\frac{3}{8} - 2\frac{13}{16}$ **27.** $47\frac{5}{12} - 23\frac{19}{24}$ **28.** $6\frac{4}{7} - 5\frac{11}{14}$

29. 10 **30.** 23 **31.** $11\frac{3}{5}$ **32.** $9\frac{2}{5}$

$\quad\underline{-8\frac{1}{5}}$ $\quad\underline{-17\frac{3}{4}}$ $\quad\underline{-9\frac{11}{15}}$ $\quad\underline{-7\frac{1}{10}}$

33. 6 **34.** 8 **35.** $63\frac{1}{6}$ **36.** $86\frac{2}{15}$

$\quad\underline{-2\frac{4}{9}}$ $\quad\underline{-1\frac{7}{10}}$ $\quad\underline{-47\frac{5}{12}}$ $\quad\underline{-27\frac{3}{10}}$

C *Solve. See Examples 7 and 8.*

37. Jerald Divis, a tax consultant, takes $3\frac{1}{2}$ hours to prepare a personal tax return and $5\frac{7}{8}$ hours to prepare a business return. How much longer does it take him to prepare the business return?

38. Shamalika Corning, a trim carpenter, cuts a board $3\frac{3}{8}$ feet long from one 6 feet long. How long is the remaining piece?

39. On four consecutive days, Jose Gonzalez, a concert pianist, practiced for $2\frac{1}{2}$ hours, $1\frac{2}{3}$ hours, $2\frac{1}{4}$ hours, and $3\frac{5}{6}$ hours. Find his total practice time.

40. Coach Marlene Lawfield was preparing her team for a tennis tournament and enforced this practice schedule: Monday, $2\frac{1}{2}$ hours; Tuesday, $2\frac{2}{3}$ hours; Wednesday, $1\frac{3}{4}$ hours; and Thursday, $1\frac{9}{16}$ hours. How long did the team practice that week before Friday's tournament?

41. If Tucson's average rainfall is $11\frac{1}{4}$ inches and Yuma's is $3\frac{3}{5}$ inches, how much more rain, on the average, does Tucson get than Yuma?

42. A pair of crutches needs adjustment. One crutch is 43 inches and the other is $41\frac{5}{8}$ inches. Find how much the short crutch should be lengthened to make both crutches the same length.

43. Charlotte Dowlin has $15\frac{2}{3}$ feet of plastic pipe. She cuts off a $2\frac{1}{2}$-foot length and then a $3\frac{1}{4}$-foot length. If she now needs a 10-foot piece of pipe, will the remaining piece do?

44. Jessica Callac takes $2\frac{3}{4}$ hours to clean her room. Her brother Matthew takes $1\frac{1}{3}$ hours to clean his room. If they start at the same time, how long does Matthew have to wait for Jessica to finish?

45. If the total weight allowable without overweight charges is 50 pounds and the traveler's luggage weighs $60\frac{5}{8}$ pounds, on how many pounds will the traveler's overweight charges be based?

46. What is the difference between interest rates of $11\frac{1}{2}\%$ and $9\frac{3}{4}\%$?

47. The longest floating pontoon bridge in the United States is the Evergreen Point Bridge in Seattle, Washington. It is 2526 yards long. The second-longest pontoon bridge in the United States is the Hood Canal Bridge in Point Gamble, Washington. The second-longest pontoon bridge is $2173\frac{2}{3}$ yards long. How much longer is the Evergreen Point Bridge than the Hood Canal Bridge? (*Source:* Federal Highway Administration)

48. The record for largest rainbow trout ever caught is $42\frac{1}{8}$ pounds and was set in Alaska in 1970. The record for largest tiger trout ever caught is $20\frac{13}{16}$ pounds and was set in Michigan in 1978. How much more did the record-setting rainbow trout weigh than the record-setting tiger trout? (*Source:* International Game Fish Association)

49. Located on an island in New York City's harbor, the Statue of Liberty is one of the largest statues in the world. The copper figure is $152\frac{1}{6}$ feet tall from feet to tip of torch. The figure stands on a pedestal that is $154\frac{1}{2}$ feet tall. What is the overall height of the Statue of Liberty from the base of the pedestal to the tip of the torch? (*Source: Microsoft Encarta 97 Encyclopedia*)

43. no, she will be $\frac{1}{12}$ of a foot short

44. $1\frac{5}{12}$ hours

45. $10\frac{5}{8}$ pounds

46. $1\frac{3}{4}\%$

47. $352\frac{1}{3}$ yards

48. $21\frac{5}{16}$ pounds

49. $306\frac{2}{3}$ feet

194

During the first decade of the 21st century, there will be three total eclipses of the sun visible from the Atlantic Ocean. The duration of each eclipse is listed in the table. Use the table to answer Exercises 50–52.

TOTAL SOLAR ECLIPSES VISIBLE FROM THE ATLANTIC OCEAN	
Date of Eclipse	Duration (in minutes)
June 21, 2001	$4\frac{14}{15}$
March 29, 2006	$4\frac{7}{60}$
July 11, 2010	$5\frac{1}{3}$

(Source: 1988 World Almanc)

50. What is the total duration for the three eclipses?

51. How much longer is the July 11, 2010 eclipse than the June 21, 2001 eclipse?

52. How much longer is the June 21, 2001 eclipse than the March 29, 2006 eclipse?

Find the perimeter of each figure.

53.

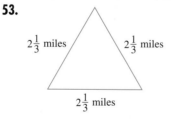

$2\frac{1}{3}$ miles $\quad$ $2\frac{1}{3}$ miles

$2\frac{1}{3}$ miles

54.

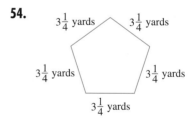

$3\frac{1}{4}$ yards $\quad$ $3\frac{1}{4}$ yards

$3\frac{1}{4}$ yards $\quad$ $3\frac{1}{4}$ yards

$3\frac{1}{4}$ yards

55.

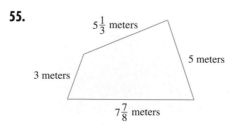

$5\frac{1}{3}$ meters

5 meters

3 meters

$7\frac{7}{8}$ meters

56.

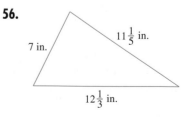

$11\frac{1}{5}$ in.

7 in.

$12\frac{1}{3}$ in.

Name _____

REVIEW AND PREVIEW

Evaluate each expression. See Section 1.8.

57. 2^3 **58.** 3^2 **59.** 5^2 **60.** 2^5

61. 3^4 **62.** 1^{10} **63.** 4^3 **64.** 9^2

COMBINING CONCEPTS

Solve.

65. Carmen's Candy Clutch is famous for its "Nutstuff," a special blend of nuts and candy. A Supreme box of Nutstuff has $2\frac{1}{4}$ pounds of nuts and $3\frac{1}{2}$ pounds of candy. A Deluxe box has $1\frac{3}{8}$ pounds of nuts and $4\frac{1}{4}$ pounds of candy. Which box is heavier and by how much?

66. Willie Cassidie purchased three Supreme boxes and two Deluxe boxes of Nutstuff from Carmen's Candy Clutch. What is the total weight of his purchase?

67. Explain in your own words why $9\frac{13}{9}$ is equal to $10\frac{4}{9}$.

68. In your own words, explain how to borrow when subtracting mixed numbers.

Internet Excursions

Go to http://www.prenhall.com/martin-gay
Did you know that a 2 x 4 board doesn't actually measure 2 inches by 4 inches? Visit this Web site to find the actual sizes of all kinds of pieces of lumber.

69. Visit this Web site to find the actual sizes of a nominal (or so-called) 2 x 4 board, a 6 x 8 timber, and a 1 x 12 common board. How much smaller is the actual width and thickness compared to the nominal width and thickness of each piece of lumber?

70. Visit this Web site to find the actual sizes of a nominal 1 x 6 common board and a nominal 6 x 10 timber. Compare the actual widths and thicknesses of these two pieces of lumber. How much wider and thicker is the timber than the board?

58. 9

59. 25

60. 32

61. 81

62. 1

63. 64

64. 81

65. Supreme is heavier by $\frac{1}{8}$ pound

66. $28\frac{1}{2}$ pounds

67. answers may vary

68. answers may vary

69. 2 x 4: width $= \frac{1}{2}$ inch, thickness $- \frac{1}{2}$ inch; 6 x 8: width $= \frac{3}{4}$ inch, thickness $= \frac{1}{2}$ inch; 1 x 12: width $= \frac{3}{4}$ inch, thickness $= \frac{1}{4}$ inch

70. Timber is $3\frac{3}{4}$ inches wider and $4\frac{3}{4}$ inches thicker than the board.

Focus On Study Skills

STUDYING FOR AND TAKING A MATH EXAM

Remember that one of the best ways to start preparing for an exam is to keep current with your assignments as they are made. Make an effort to clear up any confusion on topics as you cover them.

Begin reviewing for your exam a few days in advance. If you find a topic during your review that you still don't understand, you'll have plenty of time to ask your instructor, another student in your class, or a math tutor for help. Don't wait until the last minute to "cram" for the test.

▲ Reread your notes and carefully review the Chapter Highlights at the end of each chapter.
▲ Try solving a few exercises from each section.
▲ Pay special attention to any new terminology or definitions in the chapter. Be sure you can state the meanings of definitions in your own words.
▲ Find a quiet place to take the Chapter Test found at the end of the chapter to be covered. This gives you a chance to practice taking the real exam, so try the Chapter Test without referring to your notes or looking up anything in your book. Give yourself the same amount of time to take the Chapter Test as you will have to take the exam for which you are preparing. If your exam covers more than one chapter, you should try taking the Chapter Tests for each chapter covered. You may also find working through the Cumulative Reviews helpful when preparing for a multi-chapter test.
▲ When you have finished taking the Chapter Test, check your answers in the back of the book. Redo any of the problems you missed. Then spend extra time solving similar problems.
▲ If you tend to get anxious while taking an exam, try to visualize yourself taking the exam in advance. Picture yourself being calm, clearheaded, and successful. Picture yourself remembering concepts and definitions with no trouble. When you are well prepared for an exam, a lot of nervousness can be avoided through positive thinking.
▲ Get lots of rest the night before the exam. It's hard to show how well you know the material if your brain is foggy from lack of sleep.

When it's time to take your exam, remember these hints:
▲ Make sure you have all the tools you will need to take the exam, including an extra pencil and eraser, paper (if needed), and calculator (if allowed).
▲ Try to relax. Taking a few deep breaths, inhaling and then exhaling slowly before you begin, might help.
▲ Are there any special definitions or solution steps that you'll need to remember during the exam? As soon as you get your exam, write these down at the top, bottom, or on the back of your paper.
▲ Scan the entire test to get an idea of what questions are being asked.
▲ Start with the questions that are easiest for you. This will help build your confidence. Then return to the harder ones.
▲ Read all directions carefully. Make sure that your final result answers the question being asked.
▲ Show all of your work. Try to work neatly.
▲ Don't spend too much time on a single problem. If you get stuck, try moving on to other problems so you can increase your chances of finishing the test. If you have time, you can return to the problem giving you trouble.
▲ Before turning in your exam, check your work carefully if time allows. Be on the lookout for careless mistakes.

3.5 ORDER, EXPONENTS, AND THE ORDER OF OPERATIONS

A COMPARING FRACTIONS

Recall that whole numbers can be shown on a number line using equally spaced distances.

From the number line, we can see order of numbers. For example, we can see that 3 is less than 5 because 3 is to the left of 5.

For any two numbers on a number line, the number to the left is always the smaller number and the number to the right is always the larger number.

We can use **inequality symbols** $<$ or $>$ to write the order of numbers.

INEQUALITY SYMBOLS

$<$ means *is less than*
$>$ means *is greater than*

For example,

$\underbrace{\text{3 is less than 5}}$ or $\underbrace{\text{5 is greater than 3}}$

$\qquad 3 < 5 \qquad\qquad\qquad 5 > 3$

We can compare fractions the same way. To see fractions on a number line, divide spaces between whole numbers into equal parts.

For example, let's compare $\dfrac{2}{5}$ and $\dfrac{4}{5}$.

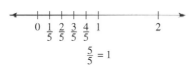

$$\frac{5}{5} = 1$$

Since $\dfrac{4}{5}$ is to the right of $\dfrac{2}{5}$,

$$\frac{2}{5} < \frac{4}{5}. \qquad \text{Notice that } 2 < 4 \text{ also.}$$

COMPARING FRACTIONS

To determine which of two fractions is greater,

Step 1. Write the fractions as like fractions.
Step 2. The fraction with the greater numerator is the greater fraction.

Example 1   Insert $<$ or $>$ to form a true statement.

$$\frac{3}{10} \qquad \frac{2}{7}$$

Solution: The LCD is 70.

$$\frac{3}{10} = \frac{3 \cdot 7}{10 \cdot 7} = \frac{21}{70} \qquad \frac{2}{7} = \frac{2 \cdot 10}{7 \cdot 10} = \frac{20}{70}$$

Practice Problem 1

Insert $<$ or $>$ to form a true statement.

$$\frac{8}{9} \qquad \frac{9}{10}$$

Answer
1. $<$

Since $21 > 20$, then $\dfrac{21}{70} > \dfrac{20}{70}$ or

$$\dfrac{3}{10} > \dfrac{2}{7}$$

Practice Problem 2

Insert $<$ or $>$ to form a true statement.

$\dfrac{4}{7} \quad \dfrac{3}{5}$

Example 2   Insert $<$ or $>$ to form a true statement.

$\dfrac{9}{10} \quad \dfrac{11}{12}$

Solution: The LCD is 60.

$$\dfrac{9}{10} = \dfrac{9 \cdot 6}{10 \cdot 6} = \dfrac{54}{60} \qquad \dfrac{11}{12} = \dfrac{11 \cdot 5}{12 \cdot 5} = \dfrac{55}{60}$$

Since $54 < 55$, then $\dfrac{54}{60} < \dfrac{55}{60}$ or

$$\dfrac{9}{10} < \dfrac{11}{12}$$

TEACHING TIP

After discussing objective A, have students do cross product calculations on all the examples and practice problems in this section to see if they can use the cross product to determine which fraction is larger. Then have them write a conjecture about their observations and a convincing argument to support their conclusion.

HELPFUL HINT

If we think of $<$ and $>$ as arrowheads, a true statement is always formed when the arrow points to the smaller number.

$$\underset{\uparrow}{\dfrac{2}{3}} > \dfrac{1}{3} \qquad\qquad \dfrac{5}{6} < \underset{\uparrow}{\dfrac{7}{6}}$$

points to smaller number   points to smaller number

B EVALUATING FRACTIONS RAISED TO POWERS

Recall from Section 1.8 that exponents indicate repeated multiplication.

$$\overset{\text{exponent}}{\underset{\underset{\text{base}}{\uparrow}}{5^3}} = \underbrace{5 \cdot 5 \cdot 5}_{\text{3 factors of 5}} = 125$$

Exponents mean the same when the base is a fraction. For example,

$$\left(\dfrac{1}{3}\right)^4 = \dfrac{1}{3} \cdot \dfrac{1}{3} \cdot \dfrac{1}{3} \cdot \dfrac{1}{3} = \dfrac{1}{81}$$

Practice Problems 3–5

Evaluate each expression.

3. $\left(\dfrac{1}{5}\right)^2$ \qquad 4. $\left(\dfrac{2}{3}\right)^3$

5. $\left(\dfrac{1}{4}\right)^2\left(\dfrac{2}{3}\right)^3$

Examples   Evaluate each expression.

3. $\left(\dfrac{1}{4}\right)^2 = \dfrac{1}{4} \cdot \dfrac{1}{4} = \dfrac{1}{16}$

4. $\left(\dfrac{3}{5}\right)^3 = \dfrac{3}{5} \cdot \dfrac{3}{5} \cdot \dfrac{3}{5} = \dfrac{27}{125}$

5. $\left(\dfrac{1}{6}\right)^2 \cdot \left(\dfrac{3}{4}\right)^3 = \left(\dfrac{1}{6} \cdot \dfrac{1}{6}\right) \cdot \left(\dfrac{3}{4} \cdot \dfrac{3}{4} \cdot \dfrac{3}{4}\right) = \dfrac{1 \cdot 1 \cdot \overset{1}{\cancel{3}} \cdot \overset{1}{\cancel{3}} \cdot 3}{2 \cdot \underset{1}{\cancel{3}} \cdot 2 \cdot \underset{1}{\cancel{3}} \cdot 4 \cdot 4 \cdot 4} = \dfrac{3}{256}$

TEACHING TIP

Ask students if $\dfrac{3^2}{5} = \left(\dfrac{3}{5}\right)^2$?

C REVIEWING OPERATIONS ON FRACTIONS

To get ready to use the order of operations with fractions, let's first review operations on fractions that we have learned.

Answers

2. $<$, 3. $\dfrac{1}{25}$, 4. $\dfrac{8}{27}$, 5. $\dfrac{1}{54}$

Examples   Perform each indicated operation.

6. $\dfrac{1}{2} \div \dfrac{8}{7} = \dfrac{1}{2} \cdot \dfrac{7}{8} = \dfrac{1 \cdot 7}{2 \cdot 8} = \dfrac{7}{16}$

7. $\dfrac{3}{5} + \dfrac{7}{10} = \dfrac{3 \cdot 2}{5 \cdot 2} + \dfrac{7}{10} = \dfrac{6}{10} + \dfrac{7}{10} = \dfrac{13}{10} = 1\dfrac{3}{10}$ The LCD is 10.

8. $\dfrac{2}{9} \cdot \dfrac{3}{11} = \dfrac{2 \cdot 3}{9 \cdot 11} = \dfrac{2 \cdot \overset{1}{\cancel{3}}}{\cancel{3} \cdot 3 \cdot 11} = \dfrac{2}{33}$

9. $\dfrac{6}{7} - \dfrac{1}{3} = \dfrac{6 \cdot 3}{7 \cdot 3} - \dfrac{1 \cdot 7}{3 \cdot 7} = \dfrac{18}{21} - \dfrac{7}{21} = \dfrac{11}{21}$ The LCD is 21.

D USING THE ORDER OF OPERATIONS

The order of operations that we use on whole numbers applies to expressions containing fractions also.

> **ORDER OF OPERATIONS**
>
> 1. Do all operations within parentheses or brackets.
> 2. Evaluate any expressions with exponents.
> 3. Multiply or divide in order from left to right.
> 4. Add or subtract in order from left to right.

Example 10   Simplify: $\dfrac{1}{5} + \dfrac{2}{3} \cdot \dfrac{4}{5}$

Solution: $\dfrac{1}{5} + \dfrac{2}{3} \cdot \dfrac{4}{5} = \dfrac{1}{5} + \dfrac{8}{15}$ Multiply first.

$\qquad\qquad = \dfrac{1 \cdot 3}{5 \cdot 3} + \dfrac{8}{15}$ The LCD is 15.

$\qquad\qquad = \dfrac{3}{15} + \dfrac{8}{15}$

$\qquad\qquad = \dfrac{11}{15}$ Add.

Example 11   Simplify: $\left(\dfrac{2}{3}\right)^2 \div \left(\dfrac{8}{27} + \dfrac{2}{3}\right)$

Solution:

$\left(\dfrac{2}{3}\right)^2 \div \left(\dfrac{8}{27} + \dfrac{2}{3}\right) = \left(\dfrac{2}{3}\right)^2 \div \left(\dfrac{8}{27} + \dfrac{18}{27}\right)$ Write $\dfrac{2}{3}$ as $\dfrac{18}{27}$.

$\qquad\qquad = \left(\dfrac{2}{3}\right)^2 \div \dfrac{26}{27}$ Simplify inside the parentheses.

$\qquad\qquad = \dfrac{4}{9} \div \dfrac{26}{27}$ Write $\left(\dfrac{2}{3}\right)^2$ as $\dfrac{4}{9}$.

$\qquad\qquad = \dfrac{4}{9} \cdot \dfrac{27}{26}$

$\qquad\qquad = \dfrac{\overset{1}{\cancel{2}} \cdot 2 \cdot 3 \cdot \overset{1}{\cancel{9}}}{\underset{1}{\cancel{9}} \cdot \underset{1}{\cancel{2}} \cdot 13}$

$\qquad\qquad = \dfrac{6}{13}$

Practice Problems 6–9

Perform each indicated operation.

6. $\dfrac{3}{7} \div \dfrac{2}{11}$

7. $\dfrac{5}{12} + \dfrac{1}{6}$

8. $\dfrac{2}{3} \cdot \dfrac{9}{10}$

9. $\dfrac{11}{12} - \dfrac{2}{5}$

Practice Problem 10

Simplify: $\dfrac{2}{9} + \dfrac{3}{8} \cdot \dfrac{4}{9}$

Practice Problem 11

Simplify: $\left(\dfrac{2}{5}\right)^2 \div \left(\dfrac{3}{5} - \dfrac{11}{25}\right)$

Answers

6. $\dfrac{33}{14} = 2\dfrac{5}{14}$, **7.** $\dfrac{7}{12}$, **8.** $\dfrac{3}{5}$, **9.** $\dfrac{31}{60}$, **10.** $\dfrac{7}{18}$,

11. 1

Recall that the average of a list of numbers is their sum divided by the number of numbers in the list.

Practice Problem 12

Find the average of $\frac{1}{2}$, $\frac{3}{8}$, and $\frac{7}{24}$.

Example 12   Find the average of $\frac{1}{3}$, $\frac{2}{5}$, and $\frac{2}{9}$.

Solution: The average is their sum, divided by 3.

$$\left(\frac{1}{3} + \frac{2}{5} + \frac{2}{9}\right) \div 3 = \left(\frac{15}{45} + \frac{18}{45} + \frac{10}{45}\right) \div 3 \quad \text{The LCD is 45.}$$

$$= \frac{43}{45} \div 3 \qquad\qquad \text{Add.}$$

$$= \frac{43}{45} \cdot \frac{1}{3}$$

$$= \frac{43}{135} \qquad\qquad \text{Multiply.}$$

TRY THE CONCEPT CHECK IN THE MARGIN.

✓ CONCEPT CHECK

What should be done first to simplify

$$3\left[\left(\frac{1}{4}\right)^2 + \frac{3}{2}\left(\frac{6}{7} - \frac{1}{3}\right)\right]?$$

Answers

12. $\frac{7}{18}$

✓ **Concept Check:** $\frac{6}{7} - \frac{1}{3}$

EXERCISE SET 3.5

A *Insert* $<$ *or* $>$ *to form a true statement. See Examples 1 and 2.*

1. $\dfrac{7}{9}$ $\dfrac{6}{9}$
2. $\dfrac{12}{17}$ $\dfrac{13}{17}$
3. $\dfrac{3}{3}$ $\dfrac{5}{3}$
4. $\dfrac{3}{23}$ $\dfrac{4}{23}$

5. $\dfrac{9}{42}$ $\dfrac{5}{21}$
6. $\dfrac{17}{20}$ $\dfrac{5}{6}$
7. $\dfrac{9}{8}$ $\dfrac{17}{16}$
8. $\dfrac{3}{8}$ $\dfrac{14}{40}$

9. $\dfrac{3}{4}$ $\dfrac{2}{3}$
10. $\dfrac{5}{7}$ $\dfrac{16}{21}$
11. $\dfrac{3}{5}$ $\dfrac{9}{14}$
12. $\dfrac{3}{10}$ $\dfrac{7}{25}$

13. $\dfrac{27}{100}$ $\dfrac{7}{25}$
14. $\dfrac{10}{17}$ $\dfrac{19}{34}$
15. $\dfrac{1}{10}$ $\dfrac{1}{11}$
16. $\dfrac{2}{5}$ $\dfrac{1}{3}$

B *Evaluate each expression. See Examples 3 through 5.*

17. $\left(\dfrac{1}{2}\right)^{4}$
18. $\left(\dfrac{1}{7}\right)^{2}$
19. $\left(\dfrac{2}{5}\right)^{3}$
20. $\left(\dfrac{3}{4}\right)^{3}$

21. $\left(\dfrac{4}{7}\right)^{3}$
22. $\left(\dfrac{2}{3}\right)^{4}$
23. $\left(\dfrac{2}{9}\right)^{2}$
24. $\left(\dfrac{7}{11}\right)^{2}$

25. $\left(\dfrac{3}{4}\right)^{2} \cdot \left(\dfrac{2}{3}\right)^{3}$
26. $\left(\dfrac{1}{6}\right)^{2} \cdot \left(\dfrac{9}{10}\right)^{2}$

ANSWERS

1. $>$
2. $<$
3. $<$
4. $<$
5. $<$
6. $>$
7. $>$
8. $>$
9. $>$
10. $<$
11. $<$
12. $>$
13. $<$
14. $>$
15. $>$
16. $>$
17. $\dfrac{1}{16}$
18. $\dfrac{1}{49}$
19. $\dfrac{8}{125}$
20. $\dfrac{27}{64}$
21. $\dfrac{64}{343}$
22. $\dfrac{16}{81}$
23. $\dfrac{4}{81}$
24. $\dfrac{49}{121}$
25. $\dfrac{1}{6}$
26. $\dfrac{9}{400}$

27. $\frac{11}{15}$

28. $\frac{27}{20} = 1\frac{7}{20}$

29. $\frac{3}{35}$

30. $\frac{5}{4} = 1\frac{1}{4}$

31. $\frac{1}{12}$

32. $\frac{3}{88}$

33. $\frac{9}{11}$

34. $\frac{29}{50}$

35. $\frac{2}{5}$

36. $\frac{61}{60} = 1\frac{1}{60}$

37. $\frac{2}{77}$

38. $\frac{6}{5} = 1\frac{1}{5}$

39. $\frac{17}{60}$

40. $\frac{5}{9}$

41. $\frac{5}{8}$

42. $\frac{1}{2}$

43. $\frac{1}{2}$

44. 2

45. $\frac{29}{10} = 2\frac{9}{10}$

46. $\frac{4}{5}$

47. $\frac{27}{32}$

48. $\frac{16}{27}$

49. $\frac{1}{81}$

50. $\frac{27}{125}$

51. $\frac{3}{4}$

52. $\frac{15}{28}$

53. $\frac{1}{4}$

54. $\frac{13}{60}$

Name _____

C *Perform each indicated operation. See Examples 6 through 9.*

27. $\frac{2}{15} + \frac{3}{5}$ 28. $\frac{9}{10} \div \frac{2}{3}$ 29. $\frac{3}{7} \cdot \frac{1}{5}$ 30. $\frac{5}{12} + \frac{5}{6}$

31. $\frac{5}{6} - \frac{3}{4}$ 32. $\frac{3}{8} \cdot \frac{1}{11}$ 33. $\frac{6}{11} \div \frac{2}{3}$ 34. $\frac{7}{10} - \frac{3}{25}$

35. $\frac{20}{35} \cdot \frac{7}{10}$ 36. $\frac{11}{20} + \frac{7}{15}$ 37. $\frac{4}{7} - \frac{6}{11}$ 38. $\frac{18}{25} \div \frac{3}{5}$

D *Use the order of operations to simplify each expression. See Examples 10 and 11.*

39. $\frac{1}{5} + \frac{1}{3} \cdot \frac{1}{4}$ 40. $\frac{1}{2} + \frac{1}{6} \cdot \frac{1}{3}$ 41. $\frac{5}{6} \div \frac{1}{3} \cdot \frac{1}{4}$ 42. $\frac{7}{8} \div \frac{1}{4} \cdot \frac{1}{7}$

43. $\frac{1}{5} \cdot \left(2\frac{5}{6} - \frac{1}{3}\right)$ 44. $\frac{4}{7} \cdot \left(6 - 2\frac{1}{2}\right)$ 45. $2 \cdot \left(\frac{1}{4} + \frac{1}{5}\right) + 2$

46. $\frac{2}{5} \cdot \left(5 - \frac{1}{2}\right) - 1$ 47. $\left(\frac{3}{4}\right)^2 \div \left(\frac{3}{4} - \frac{1}{12}\right)$ 48. $\left(\frac{8}{9}\right)^2 \div \left(2 - \frac{2}{3}\right)$

49. $\left(\frac{2}{3} - \frac{5}{9}\right)^2$ 50. $\left(1 - \frac{2}{5}\right)^3$

Find the average of each list of numbers. See Example 12.

51. $\frac{5}{6}$ and $\frac{2}{3}$ 52. $\frac{1}{2}$ and $\frac{4}{7}$ 53. $\frac{1}{3}, \frac{1}{4},$ and $\frac{1}{6}$ 54. $\frac{1}{5}, \frac{3}{10},$ and $\frac{3}{20}$

Name _____

REVIEW AND PREVIEW

Simplify each fraction. See Section 2.3.

55. $\dfrac{10}{12}$

56. $\dfrac{8}{12}$

57. $\dfrac{20}{24}$

58. $\dfrac{22}{24}$

59. $\dfrac{50}{75}$

60. $\dfrac{30}{65}$

COMBINING CONCEPTS

Solve.

61. In 1997, about $\dfrac{22}{125}$ of the total weight of mail delivered by the United States Postal Service was first class mail. That same year, about $\dfrac{93}{500}$ of the total weight of mail delivered by the United States Postal Service were periodicals. Which of these two categories accounted for a greater portion of the mail handled by weight? (*Source*: United States Postal Service)

62. The National Park System (NPS) in the United States includes a wide variety of park types. National historic parks account for $\dfrac{19}{188}$ of all NPS parks, and $\dfrac{2}{47}$ of NPS parks are classified as national preserves. Which category, national historic park or national preserve, is bigger? (*Source*: National Park Service)

63. Approximately $\dfrac{7}{10}$ of U.S. adults have a savings account. About $\dfrac{11}{25}$ of U.S. adults have a non-interest bearing checking account. Which type of banking service, savings account or non-interest checking account, do adults in the United States use more? (*Source*: Scarborough Research/ USData.com, Inc.)

64. About $\dfrac{127}{500}$ of U.S. adults rent one or two videos per month. Approximately $\dfrac{31}{200}$ of U.S. adults rent three or four videos per month. Which video rental category, 1–2 videos or 3–4 videos per month, is bigger? (*Source*: Telenation/Market Facts, Inc.)

Name _____

Focus On Mathematical Connections

INDUCTIVE REASONING

Inductive reasoning is the process of drawing a general conclusion from just a few observations. Many times in mathematics, we observe similarities among numbers or calculations and notice a pattern emerging. Identifying a pattern in this way uses inductive reasoning.

For example, look at the following list of numbers. The dots at the end of the list indicate that the list continues indefinitely. What do you notice?

$$2, 4, 6, 8, 10, 12, \ldots$$

You probably noticed the pattern that each number in the list is 2 greater than the previous number and that all of the numbers are even numbers. If we were asked to guess the next number in the list, we could be confident that "14" would be a good response.

Let's try finding another pattern in a list of numbers.

$$10, 13, 18, 25, 34, 45, \ldots$$

What do you notice? You might find it useful to find the difference between successive numbers in the list. Using the differences found below, we can see that each successive difference is 2 greater than the previous difference. We can guess that the next difference will be 13, so the next number in the list is probably $45 + 13 = 58$.

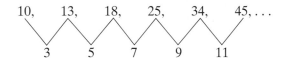

CRITICAL THINKING

Give the next two numbers in each list.

1. 5, 8, 11, 14, 17, 20, . . .   **2.** 100, 95, 90, 85, 80, . . .

3. $\frac{1}{2}, \frac{1}{3}, \frac{1}{4}, \frac{1}{5}, \ldots$ **4.** 5, 1, 6, 1, 1, 7, 1, 1, 1, 8, 1, 1, 1, . . .

5. 3, 4, 6, 9, 13, 18, . . .   **6.** 2, 4, 8, 16, 32, . . .

204

CHAPTER 3 ACTIVITY
USING FRACTIONS

This activity may be completed by working in groups or individually.

Lobsters are normally classified by weight. Use the weight classification table to answer the questions in this activity.

CLASSIFICATION OF LOBSTERS	
Class	Weight (in pounds)
Chicken	1 to $1\frac{1}{8}$
Quarter	$1\frac{1}{4}$
Half	$1\frac{1}{2}$ to $1\frac{3}{4}$
Select	$1\frac{3}{4}$ to $2\frac{1}{2}$
Large Select	$2\frac{1}{2}$ to $3\frac{1}{2}$
Jumbo	Over $3\frac{1}{2}$

(*Source:* The Maine Lobster Promotion Council)

1. A lobster fisher has kept four lobsters from a lobster trap. Classify each lobster if they have the following weights:

a. $1\frac{7}{8}$ pounds   Select

b. $1\frac{9}{16}$ pounds   Half

c. $2\frac{3}{4}$ pounds   Large Select

d. $2\frac{3}{8}$ pounds   Select

2. A recipe requires 5 pounds of lobster. Using the minimum weight for each class, decide whether a chicken, half, and select lobster will be enough for the recipe, and explain your reasoning. If not, suggest a better choice of lobsters to meet the recipe requirements.   no, you will be $\frac{3}{4}$ pound short; answers may vary

3. A lobster market customer has selected two chickens, a select, and a large select. What is the most that these four lobsters could weigh? What is the least that these four lobsters could weigh? $8\frac{1}{4}$ pounds; $6\frac{1}{4}$ pounds

4. A lobster market customer wishes to buy three quarters. If lobsters sell for $7 per pound, how much will the customer owe for her purchase? $26.25

5. Why do you think there is no classification for lobsters weighing under 1 pound?   answers may vary

CHAPTER 3 HIGHLIGHTS

DEFINITIONS AND CONCEPTS	EXAMPLES

SECTION 3.1 ADDING AND SUBTRACTING LIKE FRACTIONS

Fractions that have the same denominator are called **like fractions**.

To add or subtract like fractions, combine the numerators and place the sum or difference over the common denominator.

$\frac{1}{3}$ and $\frac{2}{3}$; $\frac{5}{7}$ and $\frac{6}{7}$

$\frac{2}{7} + \frac{3}{7} = \frac{5}{7}$ ← Add the numerators.
← Keep the common denominator.

$\frac{7}{8} - \frac{4}{8} = \frac{3}{8}$ ← Subtract the numerators.
← Keep the common denominator.

SECTION 3.2 LEAST COMMON MULTIPLE

The **least common multiple (LCM)** is the smallest number that is a multiple of all numbers in a list of numbers.

METHOD 1 FOR FINDING THE LCM OF A LIST OF NUMBERS USING MULTIPLES

Step 1. Write the multiples of the largest number until a multiple common to all numbers in the list is found.
Step 2. The multiple found in Step 1 is the LCM.

METHOD 2 FOR FINDING THE LCM OF A LIST OF NUMBERS USING PRIME FACTORIZATION

Step 1. Write the prime factorization of each number.
Step 2. For each different factor from the prime factors in Step 1, circle the largest number of factors found in any one factorization.
Step 3. The LCM is the product of the circled factors.

Equivalent fractions represent the same portion of a whole.

The LCM of 2 and 6 is 6 because 6 is the smallest number that is a multiple of both 2 and 6.

Find the LCM of 4 and 6 using Method 1.

$6 \cdot 1 = 6$ Not a multiple of 4
$6 \cdot 2 = 12$ A multiple of 4

The LCM is 12.

Find the LCM of 6 and 20 using Method 2.

$6 = 2 \cdot ③$
$20 = ②·②·⑤$

The LCM is

$2 \cdot 2 \cdot 3 \cdot 5 = 60$

Write an equivalent fraction with the indicated denominator.

$\frac{2}{8} = \frac{}{16}$

$\frac{2 \cdot 2}{8 \cdot 2} = \frac{4}{16}$

SECTION 3.3	ADDING AND SUBTRACTING UNLIKE FRACTIONS

To Add or Subtract Fractions with Unlike Denominators	Add: $\dfrac{3}{20} + \dfrac{2}{5}$
Step 1. Find the LCD.	Step 1. The LCD is 20.
Step 2. Write equivalent fractions with the LCD as denominator.	Step 2. $\dfrac{2}{5} = \dfrac{2 \cdot 4}{5 \cdot 4} = \dfrac{8}{20}.$
Step 3. Add or subtract the like fractions.	Step 3. $\dfrac{3}{20} + \dfrac{2}{5} = \dfrac{3}{20} + \dfrac{8}{20} = \dfrac{11}{20}.$
Step 4. Write the result in simplest form.	Step 4. $\dfrac{11}{20}$ is in simplest form.

SECTION 3.4	ADDING AND SUBTRACTING MIXED NUMBERS

To add or subtract mixed numbers, add or subtract the fractions and then add or subtract the whole numbers.	Add: $2\dfrac{1}{2} + 5\dfrac{7}{8}$

$$2\dfrac{1}{2} = 2\dfrac{4}{8}$$
$$+5\dfrac{7}{8} = 5\dfrac{7}{8}$$
$$\overline{\qquad\qquad} $$
$$7\dfrac{11}{8} = 7 + 1\dfrac{3}{8} = 8\dfrac{3}{8}$$

SECTION 3.5	ORDER, EXPONENTS, AND THE ORDER OF OPERATIONS

To compare like fractions, compare the numerators. The order of the fractions is the same as the order of the numerators.	Compare $\dfrac{3}{10}$ and $\dfrac{4}{10}.$
	$\dfrac{3}{10} < \dfrac{4}{10}$ since $3 < 4$
To compare unlike fractions, first write them with a common denominator, then compare the resulting like fractions.	Compare $\dfrac{2}{5}$ and $\dfrac{3}{7}$
	$\dfrac{2}{5} = \dfrac{2 \cdot 7}{5 \cdot 7} = \dfrac{14}{35} \qquad \dfrac{3}{7} = \dfrac{3 \cdot 5}{7 \cdot 5} = \dfrac{15}{35}$
	Since $14 < 15$, then
	$\dfrac{14}{35} < \dfrac{15}{35}$ or $\dfrac{2}{5} < \dfrac{3}{7}$
Exponents mean repeated multiplication when the base is a whole number or a fraction.	$\left(\dfrac{1}{2}\right)^3 = \dfrac{1}{2} \cdot \dfrac{1}{2} \cdot \dfrac{1}{2} = \dfrac{1}{8}$

SECTION 3.5   (CONTINUED)

ORDER OF OPERATIONS

1. Simplify inside parentheses first.
2. Simplify any expressions with exponents.
3. Multiply or divide in order from left to right.
4. Add or subtract in order from left to right.

Peform each indicated operation.

$$\frac{1}{2} + \frac{2}{3} \cdot \frac{1}{5} = \frac{1}{2} + \frac{2}{15} \qquad \text{Multiply.}$$

$$= \frac{1 \cdot 15}{2 \cdot 15} + \frac{2 \cdot 2}{15 \cdot 2} \qquad \text{The LCD is 30.}$$

$$= \frac{15}{30} + \frac{4}{30}$$

$$= \frac{19}{30} \qquad \text{Add.}$$

CHAPTER 3 REVIEW

(3.1) *Add or subtract as indicated. Simplify your answers.*

1. $\dfrac{7}{11} + \dfrac{3}{11}$ $\dfrac{10}{11}$

2. $\dfrac{4}{9} + \dfrac{2}{9}$ $\dfrac{2}{3}$

3. $\dfrac{5}{12} - \dfrac{3}{12}$ $\dfrac{1}{6}$

4. $\dfrac{3}{10} - \dfrac{1}{10}$ $\dfrac{1}{5}$

5. $\dfrac{11}{15} - \dfrac{1}{15}$ $\dfrac{2}{3}$

6. $\dfrac{4}{21} - \dfrac{1}{21}$ $\dfrac{1}{7}$

7. $\dfrac{4}{15} + \dfrac{3}{15} + \dfrac{2}{15}$ $\dfrac{3}{5}$

8. $\dfrac{3}{20} + \dfrac{7}{20} + \dfrac{2}{20}$ $\dfrac{3}{5}$

9. $\dfrac{1}{12} + \dfrac{11}{12}$ 1

10. $\dfrac{3}{4} + \dfrac{1}{4}$ 1

11. $\dfrac{11}{25} + \dfrac{6}{25} + \dfrac{2}{25}$ $\dfrac{19}{25}$

12. $\dfrac{4}{21} + \dfrac{1}{21} + \dfrac{11}{21}$ $\dfrac{16}{21}$

Solve.

13. Gregor Krowsky studied math for $\dfrac{3}{8}$ of an hour and geography for $\dfrac{1}{8}$ of an hour. How long did he study?
$\dfrac{1}{2}$ hour

14. Beryl Goldstein mixed $\dfrac{5}{8}$ of a gallon of water with $\dfrac{1}{8}$ of a gallon of punch concentrate. Then she and her friends drank $\dfrac{3}{8}$ of a gallon of the punch. How much was left? $\dfrac{3}{8}$ of a gallon

15. One evening Mark Alorenzo did $\dfrac{3}{8}$ of his homework before supper, another $\dfrac{2}{8}$ of it while his children did their homework, and $\dfrac{1}{8}$ after his children went to bed. What part of his homework did he do that evening? $\dfrac{3}{4}$ of his homework

16. The Simpson's will be fencing in their land. In order to do this, they need to find its perimeter. Find the perimeter of their land. $\dfrac{3}{2}$ miles $= 1\dfrac{1}{2}$ miles

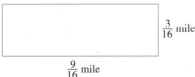

$\dfrac{3}{16}$ mile

$\dfrac{9}{16}$ mile

(3.2) *Find the LCM of each list of numbers.*

17. 5, 11   55

18. 20, 30   60

19. 20, 24   120

20. 12, 21   84

21. 12, 21, 63   252

22. 6, 8, 18   72

Write each fraction as an equivalent fraction with the given denominator.

23. $\dfrac{7}{8} = \dfrac{}{64}$ $\dfrac{56}{64}$

24. $\dfrac{2}{3} = \dfrac{}{30}$ $\dfrac{20}{30}$

25. $\dfrac{7}{11} = \dfrac{}{33}$ $\dfrac{21}{33}$

26. $\dfrac{10}{13} = \dfrac{}{26}$ $\dfrac{20}{26}$

27. $\dfrac{4}{15} = \dfrac{}{60}$ $\dfrac{16}{60}$

28. $\dfrac{5}{12} = \dfrac{}{60}$ $\dfrac{25}{60}$

(3.3) *Add or subtract as indicated. Simplify your answers.*

29. $\dfrac{7}{18} + \dfrac{2}{9}$ $\dfrac{11}{18}$

30. $\dfrac{4}{13} - \dfrac{1}{26}$ $\dfrac{7}{26}$

31. $\dfrac{1}{3} + \dfrac{1}{4}$ $\dfrac{7}{12}$

32. $\dfrac{2}{3} + \dfrac{1}{4}$ $\dfrac{11}{12}$

33. $\dfrac{5}{11} + \dfrac{2}{55}$ $\dfrac{27}{55}$

34. $\dfrac{4}{15} + \dfrac{1}{5}$ $\dfrac{7}{15}$

35. $\dfrac{7}{12} - \dfrac{1}{9}$ $\dfrac{17}{36}$

36. $\dfrac{7}{18} - \dfrac{2}{9}$ $\dfrac{1}{6}$

37. $\dfrac{4}{9} + \dfrac{5}{6}$ $\dfrac{23}{18} = 1\dfrac{5}{18}$

38. $\dfrac{9}{14} - \dfrac{3}{7}$ $\dfrac{3}{14}$

Find the perimeter of each figure.

39.
$\dfrac{2}{9}$ meter | Rectangle | ⌐
$\dfrac{5}{6}$ meter $2\dfrac{1}{9}$ meters

40.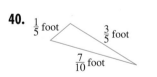
$\dfrac{1}{5}$ foot $\dfrac{3}{5}$ foot $\dfrac{7}{10}$ foot $1\dfrac{1}{2}$ feet

41. Find the difference in length of two scarves if one scarf is $\dfrac{5}{12}$ of a yard long and the other is $\dfrac{2}{3}$ of a yard long. $\dfrac{1}{4}$ of a yard

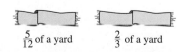
$\dfrac{5}{12}$ of a yard $\dfrac{2}{3}$ of a yard

42. Truman Kalzote cleaned $\dfrac{3}{5}$ of his house yesterday and $\dfrac{1}{10}$ of it today. How much of the house has been cleaned? $\dfrac{7}{10}$ has been cleaned

Name _____

(3.4) *Add or subtract as indicated. Simplify your answers.*

43. $31\frac{2}{7}$

$+14\frac{10}{21}$ $45\frac{16}{21}$

44. $24\frac{4}{5}$

$+35\frac{1}{5}$ 60

45. $69\frac{5}{22}$

$-36\frac{7}{11}$ $32\frac{13}{22}$

46. $36\frac{3}{20}$

$-32\frac{5}{6}$ $3\frac{19}{60}$

47. $29\frac{2}{9}$

$27\frac{7}{18}$

$+54\frac{2}{3}$ $111\frac{5}{18}$

48. $7\frac{3}{8}$

$9\frac{5}{6}$

$+3\frac{1}{12}$ $20\frac{7}{24}$

49. $9\frac{3}{5}$

$-4\frac{1}{7}$ $5\frac{16}{35}$

50. $8\frac{3}{11}$

$-5\frac{1}{5}$ $3\frac{4}{55}$

Solve.

51. Two packages of soup bones weigh $3\frac{3}{4}$ pounds and $2\frac{3}{5}$ pounds. Find their combined weight.

$6\frac{7}{20}$ pounds

52. A ribbon $5\frac{1}{2}$ yards long is cut from a reel of ribbon with 50 yards on it. Find the length of the piece remaining on the reel. $44\frac{1}{2}$ yards

53. The average annual snowfall at a certain ski resort is $62\frac{3}{10}$ inches. Last year it had $54\frac{1}{2}$ inches. How many inches below average was last year's snowfall?

$7\frac{4}{5}$ inches

54. Find the perimeter of a rectangular sheet of gift wrap that is $2\frac{1}{4}$ feet by $3\frac{1}{3}$ feet. $11\frac{1}{6}$ feet

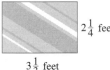

$2\frac{1}{4}$ feet

$3\frac{1}{3}$ feet

55. Find the perimeter of a sheet of shelf paper needed to fit exactly a square drawer $1\frac{1}{4}$ feet long on each side.   5 feet

$1\frac{1}{4}$ feet

56. Dinah's canned peaches contain $15\frac{3}{5}$ ounces per can. A can of Amy's brand contains $15\frac{5}{8}$ ounces per can. Amy's brand weighs how much more than Dinah's? $\frac{1}{40}$ ounce

(3.5) *Insert* $<$ *or* $>$ *to form a true statement.*

57. $\dfrac{5}{11}$ $\dfrac{6}{11}$ $\quad <$

58. $\dfrac{4}{35}$ $\dfrac{3}{35}$ $\quad >$

59. $\dfrac{5}{14}$ $\dfrac{16}{42}$ $\quad <$

60. $\dfrac{6}{35}$ $\dfrac{17}{105}$ $\quad >$

61. $\dfrac{7}{8}$ $\dfrac{6}{7}$ $\quad >$

62. $\dfrac{7}{10}$ $\dfrac{2}{3}$ $\quad >$

Evaluate each expression. Use the order of operations to simplify.

63. $\left(\dfrac{3}{7}\right)^2$ $\quad \dfrac{9}{49}$

64. $\left(\dfrac{4}{5}\right)^3$ $\quad \dfrac{64}{125}$

65. $\left(\dfrac{1}{2}\right)^4 \cdot \left(\dfrac{3}{5}\right)^2$ $\quad \dfrac{9}{400}$

66. $\left(\dfrac{1}{3}\right)^2 \cdot \left(\dfrac{9}{10}\right)^2$ $\quad \dfrac{9}{100}$

67. $\left(\dfrac{6}{7} - \dfrac{3}{14}\right)^2$ $\quad \dfrac{81}{196}$

68. $\dfrac{2}{7} \cdot \left(\dfrac{1}{5} + \dfrac{3}{10}\right)$ $\quad \dfrac{1}{7}$

69. $\dfrac{2}{5} + \left(\dfrac{2}{5}\right)^2 - \dfrac{3}{25}$ $\quad \dfrac{11}{25}$

70. $\dfrac{1}{4} + \left(\dfrac{1}{2}\right)^2 - \dfrac{3}{8}$ $\quad \dfrac{1}{8}$

71. $\left(\dfrac{1}{3}\right)^2 - \dfrac{2}{27}$ $\quad \dfrac{1}{27}$

72. $\dfrac{9}{10} \div \left(\dfrac{1}{5} + \dfrac{1}{20}\right)$ $\quad 3\dfrac{3}{5}$

73. $\left(\dfrac{3}{4} + \dfrac{1}{2}\right) \div \left(\dfrac{4}{9} + \dfrac{1}{3}\right)$ $\quad 1\dfrac{17}{28}$

74. $\left(\dfrac{3}{8} - \dfrac{1}{16}\right) \div \left(\dfrac{1}{2} - \dfrac{1}{8}\right)$ $\quad \dfrac{5}{6}$

Name _____ **Section** _____ **Date** _____

CHAPTER 3 TEST

1. Find the LCM of 4 and 15.

2. Find the LCM of 8, 9, and 12.

Insert < or > to form a true statement.

3. $\dfrac{5}{6}$ _____ $\dfrac{26}{30}$

4. $\dfrac{7}{8}$ _____ $\dfrac{8}{9}$

Perform each indicated operation. Simplify your answers.

5. $\dfrac{7}{9} + \dfrac{1}{9}$

6. $\dfrac{8}{15} - \dfrac{2}{15}$

7. $\dfrac{9}{10} + \dfrac{2}{5}$

8. $\dfrac{1}{6} + \dfrac{3}{14}$

9. $\dfrac{7}{8}$ _____ $\dfrac{1}{3}$

10. $\dfrac{6}{21} - \dfrac{1}{7}$

11. $\dfrac{9}{20} + \dfrac{2}{3}$

12. $\dfrac{16}{25} - \dfrac{1}{2}$

13. $\dfrac{11}{12} + \dfrac{3}{8} + \dfrac{5}{24}$

14. $\begin{aligned} 3\tfrac{7}{8} \\ 7\tfrac{2}{5} \\ +2\tfrac{3}{4} \\ \hline \end{aligned}$

15. $\begin{aligned} 5\tfrac{1}{6} \\ -3\tfrac{7}{8} \\ \hline \end{aligned}$

16. $\dfrac{2}{7} \cdot \left(6 - \dfrac{1}{6}\right)$

17. $\left(\dfrac{2}{3}\right)^4$

18. $\left(\dfrac{1}{2} + \dfrac{1}{3}\right) \div \left(\dfrac{1}{2}\right)^2$

19. $\left(\dfrac{4}{5}\right)^2 + \left(\dfrac{1}{2}\right)^3$

20. $\left(\dfrac{3}{4}\right)^2 \div \left(\dfrac{2}{3} + \dfrac{5}{6}\right)$

ANSWERS

1. 60

2. 72

3. <

4. <

5. $\dfrac{8}{9}$

6. $\dfrac{2}{5}$

7. $\dfrac{13}{10} = 1\dfrac{3}{10}$

8. $\dfrac{8}{21}$

9. $\dfrac{13}{24}$

10. $\dfrac{1}{7}$

11. $\dfrac{67}{60} = 1\dfrac{7}{60}$

12. $\dfrac{7}{50}$

13. $\dfrac{3}{2} = 1\dfrac{1}{2}$

14. $14\dfrac{1}{40}$

15. $1\dfrac{7}{24}$

16. $\dfrac{5}{3} = 1\dfrac{2}{3}$

17. $\dfrac{16}{81}$

18. $\dfrac{10}{3} = 3\dfrac{1}{3}$

19. $\dfrac{153}{200}$

20. $\dfrac{3}{8}$

213

21. $3\frac{3}{4}$ feet

21. A carpenter cuts a piece $2\frac{3}{4}$ feet long from a cedar plank that is $6\frac{1}{2}$ feet long. How long is the remaining piece?

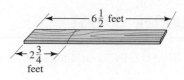

As shown in the circle graph, the market for backpacks is divided among five companies. For instance, Wilderness, Inc.'s backpack accounts for $\frac{1}{4}$ of all backpack sales. Use the graph to answer Questions 22 and 23.

22. $\frac{5}{16}$

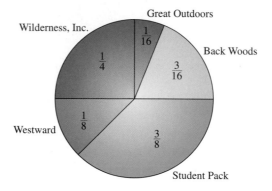

22. What fraction of backpack sales goes to Back Woods and Westward combined?

23. If a total of 500,000 backpacks are sold each year, how many backpacks does Wilderness, Inc. sell?

23. 125,000 backpacks

Name _____ Section _____ Date _____

CUMULATIVE REVIEW

Write each number in words.

1. 85

2. 126

3. Add: 23 + 136

4. Subtract 43 − 29. Then check by adding.

5. Round 278,362 to the nearest thousand.

6. Multiply: 236 × 86

7. Find each quotient and then check the answer by multiplying.

a. 1)8

b. 11 ÷ 1

c. $\frac{9}{9}$

d. 7 ÷ 7

e. $\frac{10}{1}$

f. 6)6

8. The Hudson River in New York State is 306 miles long. The Snake River, in the northwestern U.S., is 732 miles longer than the Hudson River. How long is the Snake River? (*Source:* U.S. Department of the Interior)

1. eighty-five (Sec. 1.1, Ex. 4)

2. one hundred twenty-six (Sec. 1.1, Ex. 5)

3. 159 (Sec. 1.2, Ex. 1)

4. 14 (Sec. 1.3, Ex. 3)

5. 278,000 (Sec. 1.4, Ex. 2)

6. 20,296 (Sec. 1.5, Ex. 4)

7. a. 8

b. 11

c. 1

d. 1

e. 10

f. 1 (Sec. 1.6, Ex. 2)

8. 1038 miles (Sec. 1.7, Ex. 1)

215

Name _____

Evaluate.

9. 8^2

10. 2^5

Write the shaded part as an improper fraction and a mixed number.

11.

12.

13. Of the numbers 3, 9, 11, 17, 26, which are prime?

14. Find the prime factorization of 180.

15. Write $\frac{72}{26}$ in simplest form.

16. Determine whether $\frac{8}{11}$ and $\frac{19}{26}$ are equivalent.

Multiply.

17. $\frac{2}{3} \cdot \frac{5}{11}$

18. $\frac{1}{4} \cdot \frac{1}{2}$

Divide.

19. $\frac{11}{18} \div 2\frac{5}{6}$

20. $5\frac{2}{3} \div 2\frac{5}{9}$

21. Add and simplify: $\frac{3}{16} + \frac{7}{16}$

22. Find the LCM of 6 and 8.

23. Add: $\frac{1}{2} + \frac{1}{3} + \frac{1}{6}$

24. Subtract: $9\frac{3}{7} - 5\frac{4}{21}$

25. Simplify: $\left(\frac{2}{3}\right)^2 \div \left(\frac{8}{27} + \frac{2}{3}\right)$

Decimals

Decimal numbers represent parts of a whole, just like fractions. In this chapter, we learn to perform arithmetic operations using decimals and analyze the relationship between fractions and decimals. We also learn how decimals are used in the real world. For instance, we use decimal numbers in our money system. One penny is 0.01 dollar and one dime is 0.10 dollar. Among many other uses, decimals also express averages, such as a batting average. A baseball player with a 0.333 batting average is a pretty good batter.

Did you know that eating chocolate dates as far back as 200 B.C.? The ancient Mayans and Aztecs of Central America ate mashed cacao beans for centuries before cocoa, the powdered product of the cacao bean, was introduced to Europe by Spanish explorers. Europeans then developed cocoa into the chocolate confection we know today. However, American Milton S. Hershey became the first to mass produce milk chocolate at an affordable price. He founded the Hershey Chocolate Company in 1894 and introduced the Hershey's milk chocolate bar in 1900. In 1903 he established a community (Hershey, Pennsylvania) around the building of what is today the largest chocolate factory in the world. Soon after, Hershey's Kisses and Hershey's milk chocolate bars with almonds were introduced. Today, Hershey plants in the United States can produce 12 billion Kisses per year and are the largest single users of almonds in the country. In Exercise 29 on page 247, we will see how a decimal number can be used to describe Hershey's milk chocolate bar production.

Name _____ Section _____ Date _____

CHAPTER 4 PRETEST

1. Write the decimal five and four hundredths in standard form.

2. Write the given decimal as a fraction or a mixed number, in lowest terms. 0.68

3. Write the given fraction as a decimal. $\dfrac{81}{1000}$

4. Insert $<$, $>$, or $=$ to form a true statement. 0.213   0.2421

5. Round 364.6551 to the nearest tenth.

6. Add: 13.165 + 59.08

7. Subtract: 91.307 − 43.59

Multiply.

8. 5.4 × 0.26

9. 27.01 × 0.001

10. Find the circumference of a circle whose radius is 11 inches. Then use the approximation 3.14 for π to approximate the circumference.

Divide.

11. 15 ÷ 0.003

12. 1.26 ÷ 0.28

13. $\dfrac{17.23}{1000}$

Perform the indicated operation. Estimate to see whether each proposed result is reasonable.

14. 5.8 + 3.4 + 7.9

15. 4 × 129.21

Perform the indicated operation.

16. 3.2(7 − 5.6)

17. $(1.4)^2 - 0.82$

18. $\dfrac{3}{5} - (0.3)(0.15)$

19. Write the following numbers in order from smallest to largest. $\dfrac{7}{18}$, $\dfrac{1}{3}$, 0.37

20. Jennifer Kline was hired to babysit for 7.6 hours. Her hourly rate is $3.25. How much money did Jennifer make on this job?

4.1   INTRODUCTION TO DECIMALS

A   DECIMAL NOTATION AND WRITING DECIMALS IN WORDS

Like fractional notation, decimal notation is used to denote a part of a whole. Numbers written in decimal notation are called **decimal numbers**, or simply **decimals**. The decimal 17.758 has three parts.

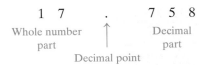

In Section 1.1, we introduced place value for whole numbers. Place names and place values for the whole-number part of a decimal number are exactly the same, as shown next. Place names and place values for the decimal part are also shown.

PLACE-VALUE CHART

hundreds	tens	ones	decimal	tenths	hundredths	thousandths	ten-thousandths	hundred-thousandths
	1	7	.	7	5	8		
100	10	1		$\frac{1}{10}$	$\frac{1}{100}$	$\frac{1}{1000}$	$\frac{1}{10,000}$	$\frac{1}{100,000}$

Notice that the value of each place is $\frac{1}{10}$ of the value of the place to its left. For example,

$$1 \cdot \frac{1}{10} = \frac{1}{10}$$
ones   tenths

$$\frac{1}{10} \cdot \frac{1}{10} = \frac{1}{100}$$
tenths   hundredths

The decimal number 17.758 means

1 ten	+	7 ones	+	7 tenths	+	5 hundredths	+	8 thousandths

$$\text{or } 1 \cdot 10 + 7 \cdot 1 + 7 \cdot \frac{1}{10} + 5 \cdot \frac{1}{100} + 8 \cdot \frac{1}{1000}$$

$$\text{or } 10 + 7 + \frac{7}{10} + \frac{5}{100} + \frac{8}{1000}$$

WRITING (OR READING) A DECIMAL IN WORDS

Step 1.   Write the whole-number part in words.

Step 2.   Write "and" for the decimal point.

Step 3.   Write the decimal part in words as though it were a whole number, followed by the place value of the last digit.

Objectives

A Know the meaning of place value for a decimal number and write decimals in words.

B Write decimals in standard form.

C Write decimals as fractions.

D Write fractions as decimals.

SSM   CD-ROM   Video 4.1

TEACHING TIP

Have students compare the names for the place values which are symmetric to the ones column. For instance, compare "tens" with "tenths."

TEACHING TIP

After having students write 17.758 in expanded form, have them write it as a sum of fractions with a common denominator of 1000, as a single fraction, in words, and then as a decimal number.

$$1 \times 10 + 7 \times 1 + 7 \times \frac{1}{10}$$
$$+ 5 \times \frac{1}{100} + 8 \times \frac{1}{1000}$$
$$= 1 \times \frac{10,000}{1000} + 7 \times \frac{1000}{1000}$$
$$+ 7 \times \frac{100}{1000} + 5 \times \frac{10}{1000} + 8 \times \frac{1}{1000}$$
$$= \frac{10,000 + 7000 + 700 + 50 + 8}{1000}$$
$$= \frac{17,758}{1000}$$

seventeen thousand seven hundred fifty-eight thousandths
= 17.758

Then ask, "How many thousandths are in 17.758?"

Practice Problem 1

Write the decimal 8.7 in words.

Practice Problem 2

Write the decimal 97.28 in words.

Example 1   Write the decimal 1.3 in words.

Solution: one and three tenths

Example 2   Write the decimal in the sentence in words. The Golden Jubilee Diamond is a 545.67 carat cut diamond. (*Source: The Guinness Book of Records*, 1998)

Solution: five hundred forty-five and sixty-seven hundredths

Practice Problem 3

Write the decimal 302.1056 in words.

Example 3   Write the decimal 19.5023 in words.

Solution: nineteen and five thousand twenty-three ten-thousandths

Practice Problem 4

Write the decimal 72.1085 in words.

Example 4   Write the decimal in the sentence in words. The oldest known fragments of the Earth's crust are Zircon crystals. They were discovered in Australia and thought to be 4.276 billion years old. (*Source: The Guinness Book of Records*, 1998)

Solution: four and two hundred seventy-six thousandths

B WRITING DECIMALS IN STANDARD FORM

A decimal written in words can be written in standard form by reversing the above procedure.

Practice Problems 5–6

Write each decimal in standard form.

5. Three hundred and ninety-six hundredths

6. Thirty-nine and forty-two thousandths

Examples   Write each decimal in standard form.

5. Forty-eight and twenty-six hundredths is

48.26

↑ hundredths place

6. Six and ninety-five thousandths is

6.095

↑ thousandths place

Answers

1. eight and seven tenths,   **2.** ninety-seven and twenty-eight hundredths,   **3.** three hundred two and one thousand fifty-six ten-thousandths,   **4.** seventy-two and one thousand eighty-five ten-thousandths,   **5.** 300.96,   **6.** 39.042

┌ HELPFUL HINT

When writing a decimal from words to decimal notation, make sure the last digit is in the correct place by inserting 0s if necessary. For example,

Two and thirty-eight thousandths is 2.038

thousandths place

TEACHING TIP

After discussing Example 6, ask students to write 6.950 in words using hundredths and using thousandths.

C WRITING DECIMALS AS FRACTIONS

Once you master reading and writing decimals, writing a decimal as a fraction follows naturally.

Decimal	In Words	Fraction
0.7	seven tenths	$\frac{7}{10}$
0.51	fifty-one hundredths	$\frac{51}{100}$
0.009	nine thousandths	$\frac{9}{1000}$

Notice that the number of decimal places in a decimal number is the same as the number of zeros in the denominator of the equivalent fraction. We can use this fact to write decimals as fractions.

$$0.51 = \frac{51}{100}$$

↑ ↑
2 decimal places 2 zeros

$$0.009 = \frac{9}{1000}$$

↑ ↑
3 decimal places 3 zeros

Example 7   Write 0.43 as a fraction.

Solution:

$$0.43 = \frac{43}{100}$$

↑ ↑
2 2
decimal zeros
places

Example 8   Write 5.6 as a mixed number.

Solution:

$$5.6 = 5\frac{6}{10} = 5\frac{3}{5} \quad \text{in simplest form}$$

↑ ↑
1 1
decimal zero
place

Examples   Write each decimal as a fraction or mixed number. Write your answer in simplest form.

9. $0.125 = \frac{125}{1000} = \frac{1}{8}$

10. $23.5 = 23\frac{5}{10} = 23\frac{1}{2}$

11. $105.083 = 105\frac{83}{1000}$

Practice Problem 7

Write 0.037 as a fraction.

Practice Problem 8

Write 14.97 as a mixed number.

Practice Problems 9–11

Write each decimal as a fraction or mixed number. Write your answer in simplest form.

  9. 0.12

10. 57.8

11. 209.986

Answers

7. $\frac{37}{1000}$,   **8.** $14\frac{97}{100}$,   **9.** $\frac{3}{25}$,   **10.** $57\frac{4}{5}$,

11. $209\frac{493}{500}$

Practice Problems 12–15

Write each fraction as a decimal.

12. $\dfrac{12}{100}$

13. $\dfrac{59}{100}$

14. $\dfrac{9}{1000}$

15. $\dfrac{172}{10}$

D WRITING FRACTIONS AS DECIMALS

If the denominator of a fraction is a power of 10, we can write it as a decimal by reversing the procedure above.

Examples   Write each fraction as a decimal.

12. $\dfrac{8}{10} = 0.8$

  ↑   ↑
1 zero   1 decimal place

13. $\dfrac{87}{10} = 8.7$

  ↑   ↑
1 zero   1 decimal place

14. $\dfrac{18}{1000} = 0.018$

  ↑   ↑
3 zeros   3 decimal places

15. $\dfrac{507}{100} = 5.07$

  ↑   ↑
2 zeros   2 decimal places

Answers
12. 0.12,   **13.** 0.59,   **14.** 0.009,   **15.** 17.2

Name _____ **Section** _____ **Date** _____

MENTAL MATH

Determine the place value for the digit 7 in each number.

1. 70　　　　**2.** 700　　　　**3.** 0.7　　　　**4.** 0.07

EXERCISE SET 4.1

A _Write each decimal number in words. See Examples 1 through 4._

1. 6.52　　　**2.** 7.59　　　📼 **3.** 16.23　　　**4.** 47.65

5. 0.205　　　**6.** 0.495　　　📼 **7.** 167.009　　　**8.** 233.056

9. The English Channel Tunnel is 31.04 miles long. (_Source: Railway Directory & Year Book_)

10. Saturn makes a complete orbit of the sun every 29.48 years. (_Source: National Space Science Data Center_)

11. The recommended daily allowance of riboflavin for teenage boys between the ages of 15 and 18 is 1.8 milligrams. (_Source:_ Food and Nutrition Board of the Institute of Medicine, National Academy of Sciences)

12. The top-rated television series for the 1996–1997 viewing season was E.R., which received a rating of 21.2. (_Source:_ Nielsen Media Research)

B _Write each decimal number in standard form. See Examples 5 and 6._

13. Six and five tenths

14. Three and nine tenths

📼 **15.** Nine and eight hundredths

16. Twelve and six hundredths

17. Five and six hundred twenty-five thousandths

18. Four and three hundred ninety-nine thousandths

19. Sixty-four ten-thousandths

20. Thirty-eight ten-thousandths

21. The record rainfall amount for a 24-hour period in Alabama is twenty and thirty-three hundredths inches. This record was set in Axis, Alabama, in 1955. (_Source:_ National Climatic Data Center)

22. The United States Postal Service vehicle fleet averages nine and sixty-two hundredths miles per gallon of fuel. (_Source:_ United States Postal Service)

23. In 1996, there was an average of twelve and nine tenths on-the-job illnesses and injuries for every 100 workers in the dairy farm industry. (_Source:_ Bureau of Labor Statistics)

24. Cynthia Cooper of the WNBA's Houston Comets scored an average of twenty-two and two tenths points per basketball game during the 1997 regular season. (_Source:_ Women's National Basketball Association)

C _Write each decimal as a fraction or a mixed number. Write your answer in simplest form. See Examples 7 through 11._

25. 0.3　　　**26.** 0.7　　　📼 **27.** 0.27　　　**28.** 0.39

29. $5\frac{47}{100}$

30. $6\frac{3}{10}$

31. $\frac{6}{125}$

32. $\frac{41}{500}$

33. $7\frac{7}{100}$

34. $9\frac{9}{100}$

35. $15\frac{401}{500}$

36. $11\frac{203}{500}$

37. $\frac{601}{2000}$

38. $\frac{1003}{5000}$

39. $487\frac{8}{25}$

40. $298\frac{31}{50}$

41. 0.6

42. 0.3

43. 0.45

44. 0.75

45. 3.7

46. 2.8

47. 0.268

48. 0.709

49. 0.09

50. 0.07

51. 4.026

52. 3.601

53. 0.028

54. 0.063

55. 56.3

56. 20.6

57. 47,260

58. 47,300

59. 47,000

60. 50,000

61. answers may vary

twenty-six million, eight hundred forty-nine thousand, five hundred seventy-six

62. hundred-billionths

63. 7.12

64. 17.268

Name _____

29. 5.47   **30.** 6.3   **31.** 0.048   **32.** 0.082

33. 7.07   **34.** 9.09   **35.** 15.802   **36.** 11.406

37. 0.3005   **38.** 0.2006   **39.** 487.32   **40.** 298.62

D *Write each fraction as a decimal. See Examples 12 through 15.*

41. $\frac{6}{10}$ **42.** $\frac{3}{10}$ **43.** $\frac{45}{100}$ **44.** $\frac{75}{100}$

45. $\frac{37}{10}$ **46.** $\frac{28}{10}$ **47.** $\frac{268}{1000}$ **48.** $\frac{709}{1000}$

49. $\frac{9}{100}$ **50.** $\frac{7}{100}$ **51.** $\frac{4026}{1000}$ **52.** $\frac{3601}{1000}$

53. $\frac{28}{1000}$ **54.** $\frac{63}{1000}$ **55.** $\frac{563}{10}$ **56.** $\frac{206}{10}$

REVIEW AND PREVIEW

Round 47,261 to the indicated place value. See Section 1.1.

57. tens   **58.** hundreds   **59.** thousands   **60.** ten-thousands

COMBINING CONCEPTS

61. In your own words, describe how to write a decimal as a fraction or mixed number.

62. Write 0.00026849576 in words.

63. Write $7\frac{12}{100}$ as a decimal.

64. Write $17\frac{268}{1000}$ as a decimal.

4.2 ORDER AND ROUNDING

A COMPARING DECIMALS

One way to compare decimals is to compare their graphs on a number line. Recall that for any two numbers on a number line, the number to the left is smaller and the number to the right is larger. The decimals 0.5 and 0.8 are graphed as follows:

Comparing decimals by comparing their graphs on a number line can be time consuming. Another way to compare the size of decimals is by comparing digits in corresponding places.

COMPARING TWO DECIMALS

Compare digits in the same places from left to right. When two digits are not equal, the number with the larger digit is the larger decimal. If necessary, insert 0s after the last digit to the right of the decimal point to continue comparing.

Compare hundredths place digits

28.253   28.263
  ↑   ↑
  5   <   6
so 28.253   <   28.263

HELPFUL HINT

For any decimal, inserting 0s after the last digit to the right of the decimal point does not change the value of the number.

    7.6 = 7.60 = 7.600, and so on

When a whole number is written as a decimal, the decimal point is placed to the right of the ones digit.

    25 = 25.0 = 25.00, and so on

Example 1   Insert < , > , or = to form a true statement.

    0.378   0.368

Solution: 0.378   0.368   The tenths places are the same.

    0.378   0.368   The hundredths places are different.

Since 7 > 6, then 0.378 > 0.368.

Example 2   Insert < , > , or = to form a true statement.

    0.052   0.236

Solution: 0.052 < 0.236   0 is smaller than 2 in the tenths place.

Practice Problem 1

Insert < , > , or = to form a true statement.

13.208   13.28

Practice Problem 2

Insert < , > , or = to form a true statement.

0.12   0.086

Answers
1. < , 2. >

B ROUNDING DECIMALS

We **round the decimal part** of a decimal number in nearly the same way as we round whole numbers. The only difference is that we drop digits to the right of the rounding place, instead of replacing these digits by 0s. For example,

24.954   rounded to the nearest hundredth is   24.95
    ↑

> **ROUNDING DECIMALS TO A PLACE VALUE TO THE RIGHT OF THE DECIMAL POINT**
>
> **Step 1.** Locate the digit to the right of the given place value.
> **Step 2.** If this digit is 5 or greater, add 1 to the digit in the given place value and drop all digits to its right. If this digit is less than 5, drop all digits to the right of the given place.

Practice Problem 3

Round 123.7817 to the nearest thousandth.

Example 3   Round 736.2359 to the nearest tenth.

Solution: **Step 1.** We locate the digit to the right of the tenths place.

$$736.2\,③\,59$$

tenths place / digit to the right

Step 2. Since this digit to the right is less than 5, we drop it and all digits to its right.

Thus, 736.2359 rounded to the nearest tenth is 736.2.

Practice Problem 4

Round 123.7817 to the nearest tenth.

Example 4   Round 736.2359 to the nearest hundredth.

Solution: **Step 1.** We locate the digit to the right of the hundredths place.

$$736.23\,⑤\,9$$

hundredths place / digit to the right

Step 2. Since this digit to the right is 5, we add 1 to the digit in the hundredths place and drop all digits to the right of the hundredths place.

Thus, 736.2359 rounded to the nearest hundredth is 736.24.

Rounding often occurs with money amounts. Since there are 100 cents in a dollar, each cent is $\frac{1}{100}$ of a dollar. This means that if we want to round to the nearest cent, we round to the nearest hundredth of a dollar.

Practice Problem 5

In Cititown, the price of a gallon of gasoline is $1.0789. Round this to the nearest cent.

Example 5   The price of a gallon of gasoline in Aimsville is currently $1.0279. Round this to the nearest cent.

Solution:

$$\$1.02\,⑦\,9$$

hundredths place / digit to the right

Answers

3. 123.782,   **4.** 123.8,   **5.** $1.08

Since the digit to the right is greater than 5, we add one to the hundredths digit and drop all digits to the right of the hundredths digit. Thus, $1.0279 rounded to the nearest cent is $1.03.

Example 6   Round $0.098 to the nearest cent.

Solution:

$0.09 ⑧

— hundredths place

⟶ digit to the right

Since the digit to the right is greater than 5, we add 1 to the hundredths digit and drop all digits to the right of the hundredths digit.

$$\begin{array}{r} 0.09 \\ +0.01 \\ \hline 0.10 \end{array}$$ Add 1 hundredth to 9 hundredths.

Thus, $0.098 rounded to the nearest cent is $0.10.

Try the Concept Check in the margin.

Example 7   Determining State Taxable Income

A high school teacher's taxable income is $31,567.72. The tax tables in the teacher's state use amounts to the nearest dollar. Round the teacher's income to the nearest whole dollar.

Solution: Rounding to the nearest whole dollar means rounding to the nearest ones place.

$31,567. 7 2

— ones place

⟶ digit to the right

Since the digit to the right is 5 or greater, we add 1 to the ones place and drop all digits to the right of the ones place.

The teacher's income rounded to the nearest dollar is $31,568.

Practice Problem 6

Round $1.095 to the nearest cent.

✓ **Concept Check**

1756.0894 rounded to the nearest ten is
a. 1756.1   b. 1760.0894

c. 1760

Practice Problem 7

Water bills in Gotham City are always rounded to the nearest dime. Lois's water bill was $24.43. Round her bill to the nearest dime (tenth).

Answers

6. $1.10,   **7.** $24.40

✓ Concept Check: c

Focus On Mathematical Connections

MODELING MULTIPLICATION WITH DECIMAL NUMBERS

We can use an area model to represent multiplying decimal numbers. For instance, we can think of the 10 × 10 grid shown below on the left as representing 1, or a whole. We can show the product 0.5 × 0.2 by shading five tenths of the grid along one side of square and shading two tenths of the grid along the other side.

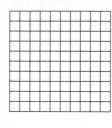 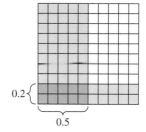

CRITICAL THINKING

1. What does the region where the shading overlaps represent? See if you can figure it out by comparing to the product 0.5 × 0.2 found numerically. (Remember that one square within the grid represents one hundredth.)

2. How could you use this type of area model to represent the product of a decimal number (such as 0.6) and a whole number (such as 2)?

3. Use the model to show each of the following products.
   a. 0.4 × 0.4
   b. 0.9 × 0.1
   c. 0.8 × 2
   d. 0.7 × 1.5

EXERCISE SET 4.2

A *Insert* $<$, $>$, *or* $=$ *to form a true statement. See Examples 1 and 2.*

1. 0.15   0.16

2. 0.12   0.15

▱ 3. 0.57   0.54

4. 0.59   0.52

5. 0.098   0.1

6. 0.0756   0.2

7. 0.54900   0.549

8. 0.98400   0.984

▱ 9. 167.908   167.980

10. 519.3405   519.3054

11. 420,000   0.000042

12. 0.000987   987,000

B *Round each decimal to the given place value. See Examples 3 and 4.*

13. 0.57, nearest tenth

14. 0.54, nearest tenth

▱ 15. 0.234, nearest hundredth

16. 0.452, nearest hundredth

17. 0.5942, nearest thousandth

18. 63.4523, nearest thousandth

19. 98,207.23, nearest ten

20. 68,934.543, nearest ten

21. 12.342, nearest tenth

22. 42.9878, nearest thousandth

▱ 23. 17.667, nearest hundredth

24. 0.766, nearest hundredth

25. 0.501, nearest tenth

26. 0.602, nearest tenth

27. 0.1295, nearest thousandth

28. 0.8295, nearest thousandth

29. 3829.34, nearest ten

30. 4520.876, nearest hundred

ANSWERS

1. $<$
2. $<$
3. $>$
4. $>$
5. $<$
6. $<$
7. $=$
8. $=$
9. $<$
10. $>$
11. $>$
12. $<$
13. 0.6
14. 0.5
15. 0.23
16. 0.45
17. 0.594
18. 63.452
19. 98,210
20. 68,930
21. 12.3
22. 42.988
23. 17.67
24. 0.77
25. 0.5
26. 0.6
27. 0.130
28. 0.830
29. 3830
30. 4500

Name _____

Round each money amount to the nearest cent or dollar as indicated. See Examples 5 through 7.

31. $0.067, nearest cent

32. $0.025, nearest cent

33. $42,650.14, nearest dollar

34. $768.95, nearest dollar

35. $26.95, nearest dollar

36. $14,769.52, nearest dollar

37. $0.1992, nearest cent

38. $0.7633, nearest cent

39. Which number(s) rounds to 0.26?
0.26559   0.26499   0.25786   0.25186

40. Which number(s) rounds to 0.06?
0.0612   0.066   0.0586   0.0506

Round each number to the given place value.

41. The attendance at a Mets baseball game was reported to be 39,867. Round this number to the nearest thousand.

42. A used office desk is advertised at $19.95 by Drawley's Office Furniture. Round this price to the nearest dollar.

43. During the 1997 Boston Marathon, Fatuma Roba of Ethiopia was the first woman to cross the finish line. Her time was 2.43972 hours. Round this time to the nearest hundredth. (*Source: 1998 World Almanac*)

44. The population density of the state of Ohio is roughly 271.0961 people per square mile. Round this population density to the nearest tenth. (*Source: U.S. Bureau of the Census*)

Ohio

45. The length of a day on Mars is 24.6229 hours. Round this figure to the nearest thousandth. (*Source: National Space Science Date Center*)

46. Venus makes a complete orbit around the sun every 224.695 days. Round this figure to the nearest whole day. (*Source: National Space Science Data Center*)

230

Name _____

47. During the 1997 NFL season, the average length of a Denver Broncos' punt was 43.3 yards. Round this figure to the nearest whole yard. (*Source:* National Football League)

48. Raptor is a roller coaster at Cedar Point, an amusement park in Sandusky, Ohio. It is one of the world's tallest, fastest, and steepest inverted roller coasters. A ride on Raptor lasts about 2.267 minutes. Round this figure to the nearest tenth. (*Source:* Cedar Fair, L.P.)

REVIEW AND PREVIEW

Perform each indicated operation. See Sections 1.2 and 1.3.

49. 3452 + 2314

50. 8945 + 4536

51. 94 − 23

52. 82 − 47

53. 482 − 239

54. 4002 − 3897

COMBINING CONCEPTS

The table gives the leading bowling averages for the Professional Bowlers Association Tour for each of the years listed. Use the table to answer Exercises 55–57.

LEADING PBA AVERAGES BY YEAR		
Year	Bowler	Average Score
1988	Mark Roth	216.081
1989	Pete Weber	218.036
1990	Amleto Monacelli	215.432
1991	Norm Duke	218.158
1992	Dave Ferraro	219.702
1993	Walter Ray Williams, Jr.	222.980
1994	Norm Duke	222.830
1995	Mike Aulby	225.490
1996	Walter Ray Williams, Jr.	225.370
1997	Walter Ray Williams, Jr.	222.000

(*Source:* Professional Bowlers Association)

55. What is the highest average score on the list? Which bowler achieved that average?

56. What is the lowest average score on the list? Which bowler achieved that average?

57. Make a list of the leading averages in order from greatest to least for the years shown in the table.

47. 43 yards

48. 2.3 minutes

49. 5766

50. 13,481

51. 71

52. 35

53. 243

54. 105

55. 225.490; Mike Aulby

56. 215.432; Amleto Monacelli

57. 225.490; 225.370; 222.980; 222.830; 222.000; 219.702; 218.158; 218.036; 216.081; 215.432

58. answers may vary

58. Write a 4-digit number that rounds to 26.3.

59. Write a 5-digit number that rounds to 1.7.

59. answers may vary

60. Explain how to identify the value of the 9 in the decimal 486.3297.

60. answers may vary

Internet Excursions

Go to http://www.prenhall.com/martin-gay
This World Wide Web site will direct you to the National Weather Service's Interactive Weather Information Network, or a related site. The table gives up-to-date weather summaries and forecasts for selected cities across the nation. The weather summary includes the previous day's high and low temperatures as well as precipitation amounts in inches, if any.

61. Record the date and time that the table was created. Scan the list of selected cities. How many cities received precipitation? Which city received the most precipitation?

62. Find the five cities that received the most precipitation. List the cities along with their precipitation amounts in order from most to least precipitation.

61. answers may vary

62. answers may vary

4.3 ADDING AND SUBTRACTING DECIMALS

A ADDING DECIMALS

Adding decimals is similar to adding whole numbers. We add digits in corresponding place values from right to left, carrying if necessary. To make sure that digits in corresponding place values are added, we line up the decimal points vertically.

ADDING OR SUBTRACTING DECIMALS

Step 1.   Write the decimals so that the decimal points line up vertically.

Step 2.   Add or subtract as for whole numbers.

Step 3.   Place the decimal point in the sum or difference so that it lines up vertically with the decimal points in the problem.

Example 1   Add: 23.85 + 1.604

Solution: First we line up the decimal points vertically.

$$\begin{array}{r} 23.850 \\ + \ 1.604 \\ \hline \end{array} \quad \text{Write one 0.}$$

↑
Line up decimal points

Then we add the digits from right to left as for whole numbers.

$$\begin{array}{r} 1 \\ 23.850 \\ + \ 1.604 \\ \hline 25.454 \end{array}$$

↑_____ Place the decimal point in the sum so that all decimal points line up.

⌐ HELPFUL HINT

Recall that 0's may be placed after the last digit to the right of the decimal point without changing the value of the decimal. This may be used to help line up place values when adding decimals.

$$\begin{array}{r} 3.2 \\ 15.567 \\ + \ 0.11 \\ \hline \end{array} \quad \text{becomes} \quad \begin{array}{r} 3.200 \\ 15.567 \\ + \ 0.110 \\ \hline 18.877 \end{array} \quad \begin{array}{l} \text{Insert two 0s.} \\ \\ \text{Insert one 0.} \\ \text{Add.} \end{array}$$

Example 2   Add: 763.7651 + 22.001 + 43.89

Solution: First we line up the decimal points.

$$\begin{array}{r} 763.7651 \\ 22.0010 \\ + \ 43.8900 \\ \hline 829.6561 \end{array} \quad \begin{array}{l} \\ \text{Write one 0.} \\ \text{Write two 0s.} \\ \text{Add.} \end{array}$$

Objectives

A Add decimals.
B Subtract decimals.
C Solve problems that involve adding or subtracting decimals.

SSM   CD-ROM   Video
4.3

Practice Problem 3

Add: 26.072 + 119

✓ CONCEPT CHECK

What is wrong with the following calculation of the sum of 7.03, 2.008, 19.16, and 3.1415?

$$
\begin{array}{r}
7.03 \\
2.008 \\
19.16 \\
+3.1415 \\
\hline
3.6042
\end{array}
$$

Practice Problem 4

Subtract. Check your answers.

a. 82.75 − 15.9

b. 126.032 − 95.71

TEACHING TIP

Show students how to check their subtraction by adding from bottom to top rather than rewriting.

$$
\begin{array}{r}
\text{Check:} \quad 35.218 \\
-23.650 \\
\hline
11.568 \uparrow + \\
1\,1 \leftarrow \text{carry}
\end{array}
$$

Practice Problem 5

Subtract. Check your answers.

a. 5.8 − 3.92

b. 9.72 − 4.068

Practice Problem 6

Subtract. Check your answers.

a. 53 − 29.31

b. 120 − 68.22

Answers

3. 145.072,   **4. a.** 66.85,   **b.** 30.322,   **5. a.** 1.88, **b.** 5.652,   **6. a.** 23.69,   **b.** 51.78

✓ **Concept Check:**   The decimal places are not lined up properly.

┌─ **HELPFUL HINT**
Don't forget that the decimal point on a whole number is after the last digit.
─┘

Example 3   Add: 45 + 2.06

Solution:

$$
\begin{array}{r}
45.00 \qquad \text{Write the decimal point and two 0s.} \\
+\ 2.06 \qquad \text{Line up decimal points.} \\
\hline
47.06 \qquad \text{Add.}
\end{array}
$$

TRY THE CONCEPT CHECK IN THE MARGIN.

B SUBTRACTING DECIMALS

Subtracting decimals is similar to subtracting whole numbers. We line up digits and subtract from right to left, borrowing when needed.

Example 4   Subtract: 35.218 − 23.65. Check your answer.

Solution: First we line up the decimal points.

$$
\begin{array}{r}
{\scriptstyle 4\ \ 11\,11} \\
3\,\cancel{5}.\cancel{2}\cancel{1}8 \\
-2\,3.6\,5\,0 \qquad \text{Write one 0.} \\
\hline
1\,1.5\,6\,8 \qquad \text{Subtract.}
\end{array}
$$

Recall that we can check a subtraction problem by adding.

$$
\begin{array}{r}
{\scriptstyle 1\ \ 1} \\
11.568 \qquad \text{Difference} \\
+23.650 \qquad \text{Subtrahend} \\
\hline
35.218 \qquad \text{Minuend}
\end{array}
$$

Example 5   Subtract: 3.5 − 0.068. Check your answer.

Solution:

$$
\begin{array}{r}
{\scriptstyle 9} \\
{\scriptstyle 4\ \cancel{10}\,10} \\
3.\cancel{5}\cancel{0}\cancel{0} \qquad \text{Write two 0s.} \\
-0.0\,6\,8 \qquad \text{Line up decimal points.} \\
\hline
3.4\,3\,2 \qquad \text{Subtract.}
\end{array}
$$

Check:

$$
\begin{array}{r}
3.432 \qquad \text{Difference} \\
+0.068 \qquad \text{Subtrahend} \\
\hline
3.500 \qquad \text{Minuend}
\end{array}
$$

Example 6   Subtract: 85 − 17.31. Check your answer.

Solution:

$$
\begin{array}{r}
{\scriptstyle 9} \\
{\scriptstyle 7\ 14\ \cancel{10}\,10} \\
\cancel{8}\cancel{5}.\cancel{0}\cancel{0} \\
-1\,7.3\,1 \\
\hline
6\,7.6\,9
\end{array}
$$

$$
\begin{array}{r}
\text{Check:} \quad 67.69 \qquad \text{Difference} \\
+17.31 \qquad \text{Subtrahend} \\
\hline
85.00 \qquad \text{Minuend}
\end{array}
$$

C SOLVING PROBLEMS BY ADDING OR SUBTRACTING DECIMALS

Decimals are very common in real-life problems.

Example 7   Calculating the Cost of Owning an Automobile

Find the total monthly cost of owning and operating a certain automobile given the expenses shown.

Monthly car payment: $256.63
Monthly insurance cost: $47.52
Average gasoline bill per month: $95.33

Solution: **1.** UNDERSTAND. Read and reread the problem. The phrase "total monthly cost" tells us to add.
2. TRANSLATE.

In words:

total monthly cost	is	car payment	plus	insurance	plus	gasoline bill
↓	↓	↓	↓	↓	↓	↓

Translate:

$$\text{total monthly cost} = \$256.63 + \$47.52 + \$95.33$$

3. SOLVE.

$$\begin{array}{r} \overset{1}{2}56.63 \\ 47.52 \\ + \ 95.33 \\ \hline \$399.48 \end{array}$$

4. INTERPRET. *Check* your work. *State* your conclusion: The total monthly cost is $399.48.

Example 8   Comparing Average Heights

The bar graph shows the current average heights for adults in various countries. How much taller is the average height in Denmark than the average height in the United States?

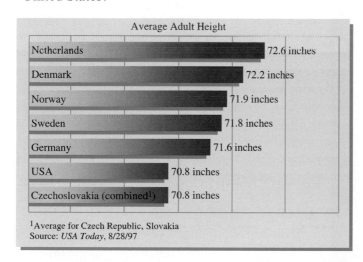

Average Adult Height

Netherlands	72.6 inches
Denmark	72.2 inches
Norway	71.9 inches
Sweden	71.8 inches
Germany	71.6 inches
USA	70.8 inches
Czechoslovakia (combined[1])	70.8 inches

[1]Average for Czech Republic, Slovakia
Source: *USA Today*, 8/28/97

Solution: **1.** UNDERSTAND. Read and reread the problem. Since we want to know "how much taller," we subtract.
2. TRANSLATE.

Practice Problem 7

Find the total monthly cost of owning and operating a certain automobile given the expenses shown.

Monthly car payment:	$536.50
Monthly insurance cost:	$52.70
Average gasoline bill per month:	$87.50

Practice Problem 8

Use the bar graph for Example 8. How much taller is the average height in the Netherlands than the average height in Czechoslovakia?

TEACHING TIP   Classroom Activity

Ask students to work in groups to make up their own word problems involving the addition and subtraction of decimals. Then have them give their problems to another group to solve.

Answers

7. $676.70, **8.** 1.8 inches

In words:

How much taller?	is	Denmark's average height	minus	U.S. average height
↓	↓	↓	↓	↓

Translate: How much taller = 72.2 − 70.8

3. SOLVE.

$$
\begin{array}{r}
{\scriptstyle 1\ \ 12} \\
7\,\cancel{2}\,.\,\cancel{2} \\
-\,7\,0\,.\,8 \\
\hline
1\,.\,4
\end{array}
$$

4. INTERPRET. *Check* your work. *State* your conclusion: The average height in Denmark is 1.4 inches more than the average U.S. height.

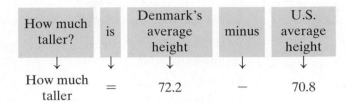

CALCULATOR EXPLORATIONS
ENTERING DECIMAL NUMBERS

To enter a decimal number, find the key marked $\boxed{\cdot}$. To enter the number 2.56, for example, press the keys

$\boxed{2}\ \boxed{\cdot}\ \boxed{5}\ \boxed{6}$

The display will read $\boxed{\qquad 2.56}$.

OPERATIONS ON DECIMAL NUMBERS

Operations on decimal numbers are performed in the same way as operations on whole or signed numbers. For example, to find $8.625 - 4.29$, press the keys

$\boxed{8.625}\ \boxed{-}\ \boxed{4.29}\ \boxed{=}$

or

$\boxed{\text{ENTER}}$

The display will read $\boxed{\qquad 4.335}$.
(Although entering 8.625, for example, requires pressing more than one key, we group numbers together for easier reading.)

Use a calculator to perform each indicated operation.

1. $315.782 + 12.96$ 328.742   **2.** $29.68 + 85.902$ 115.582

3. $6.249 - 1.0076$ 5.2414   **4.** $5.238 - 0.682$ 4.556

5.
$$
\begin{array}{r}
12.555 \\
224.987 \\
5.2 \\
+\,622.65 \\
\end{array}
$$
865.392

6.
$$
\begin{array}{r}
47.006 \\
0.17 \\
313.259 \\
+\,139.088 \\
\end{array}
$$
499.523

Name _____ **Section** _____ **Date** _____

MENTAL MATH

State the sum.

1. 0.3
      +0.2

2. 0.4
      +0.5

3. 1.00
      +0.26

4. 3.00
      +0.19

5. 7.6
      +1.3

6. 4.5
      +3.2

7. 0.9
      −0.3

8. 0.6
      −0.2

EXERCISE SET 4.3

A *Add. See Examples 1 through 3.*

1. $1.3 + 2.2$

2. $2.5 + 4.1$

3. $5.7 + 1.13$

4. $2.31 + 6.4$

5. $0.003 + 0.091$

6. $0.004 + 0.085$

7. $19.23 + 602.782$

8. $47.14 + 409.567$

9. $490 + 93.09$

10. $600 + 83.0062$

11. 234.89
       +230.67

12. 734.89
       +640.56

13. 100.009
         6.08
       + 9.034

14. 200.89
         7.49
       + 62.83

15. $24.6 + 2.39 + 0.0678$

16. $32.4 + 1.58 + 0.0934$

17. 45.023
         3.006
       + 8.403

18. 65.0028
         5.0903
       + 6.9003

B *Subtract and check. See Examples 4 through 6.*

19. $8.8 − 2.3$

20. $7.6 − 2.1$

21. $18 − 2.7$

22. $28 − 3.3$

23. 654.9
       − 56.67

24. 863.23
       − 39.453

25. $5.9 − 4.07$

26. $6.4 − 3.04$

27. $923.5 − 61.9$

28. $845.93 − 45.8$

MENTAL MATH ANSWERS

1. 0.5
2. 0.9
3. 1.26
4. 3.19
5. 8.9
6. 7.7
7. 0.6
8. 0.4

ANSWERS

1. 3.5
2. 6.6
3. 6.83
4. 8.71
5. 0.094
6. 0.089
7. 622.012
8. 456.707
9. 583.09
10. 683.0062
11. 465.56
12. 1375.45
13. 115.123
14. 271.21
15. 27.0578
16. 34.0734
17. 56.432
18. 76.9934
19. 6.5
20. 5.5
21. 15.3
22. 24.7
23. 598.23
24. 823.777
25. 1.83
26. 3.36
27. 861.6
28. 800.13

237

29. 376.89

30. 136.76

31. 876.6

32. 1672.53

33. 194.4

34. 791.1

35. 2.9988

36. 6.903

37. 16.3

38. 35.8

39. $454.71

40. $506.42

41. $0.06

42. $250.66

43. $7.52

44. $1.74

45. 197.8 lb

46. $0.45 per hour

Name _____

29. 500.34 − 123.45   **30.** 600.74 − 463.98   **31.**   1000
                                                                    − 123.4

32.   2000   **33.** 200 − 5.6   **34.** 800 − 8.9
        − 327.47

35. 3 − 0.0012   **36.** 7 − 0.097

37. Subtract 6.7 from 23   **38.** Subtract 9.2 from 45

C *Solve. See Examples 7 and 8.*

39. Find the total monthly cost of owning and maintaining a car given the information shown.

Monthly car payment:	$275.36
Monthly insurance cost:	$83.00
Average cost of gasoline per month:	$81.60
Average maintenance cost per month:	$14.75

40. Find the total monthly cost of owning and maintaining a car given the information shown.

Monthly car payment:	$306.42
Monthly insurance cost:	$53.50
Average cost of gasoline per month:	$123.00
Average maintenance cost per month:	$23.50

41. Gasoline was $1.039 per gallon on one day and $0.979 per gallon the next day. By how much did the price change?

42. A pair of eyeglasses costs a total of $347.89. The frames of the glasses are $97.23. How much do the lenses of the eyeglasses cost?

43. Ann-Margaret Tober bought a book for $32.48. If she paid with two $20 bills, what was her change?

44. Phillip Guillot bought a car part for $8.26. If he paid with a $10 bill, what was his change?

45. Swaziland and Singapore are the number one and number two sugar-consuming countries in the world, respectively. Swaziland's citizens consume an annual average of 373.7 pounds of sugar per person. Singapore's citizens consume an annual average of 175.9 pounds of sugar per person. How much more sugar does the average person in Swaziland consume than the average person in Singapore in a year? (*Source: The Top 10 of Everything, 1997,* by Russell Ash)

46. In December 1996, the average wage for U.S. workers was $12.03 per hour. By December 1997, the average wage for U.S. workers had climbed to $12.48 per hour. How much of an increase was this? (*Source:* Bureau of Labor Statistics)

Name _____

47. The average annual rainfall in Houston, Texas, is 46.07 inches. The average annual rainfall in New Orleans, Louisiana, is 61.88 inches. On average, how much more rain does New Orleans receive annually than Houston? (*Source:* National Climatic Data Center)

48. The average wind speed at the weather station on Mt. Washington in New Hampshire is 35.3 miles per hour. The average wind speed in Chicago, Illinois, is 10.4 miles per hour. How much faster is the average wind speed on Mt. Washington than in Chicago? (*Source:* National Climatic Data Center)

49. In October 1997, Andy Green set a new one-mile land speed record. This record was 129.567 miles per hour faster than a previous record of 633.468 set in 1983. What was Green's record-setting speed? (*Source:* United States Auto Club)

50. It costs $2.56 to send a 2-pound package locally via parcel post at a U.S. Post Office. A 5-pound package costs $2.77 to send locally via parcel post. What is the total cost of sending a 2-pound and a 5-pound package locally via parcel post? (*Source:* United States Postal Service)

51. The snowiest city in the United States is Blue Canyon, California, which receives an average of 111.6 more inches of snow than the second snowiest city. The second snowiest city in the United States is Marquette, Michigan. Marquette receives an average of 129.2 inches of snow annually. How much snow does Blue Canyon receive on average each year? (*Source:* National Climatic Data Center)

52. The driest city in the world is Aswan, Egypt, which receives an average of only 0.02 inches of rain per year. Yuma, Arizona, is the driest city in the United States. Yuma receives an average of 2.63 more inches of rain each year than Aswan. What is the average annual rainfall in Yuma? (*Source:* National Climatic Data Center)

53. A landscape architect is planning a border for a flower garden shaped like a triangle. The sides of the garden measure 12.4 feet, 29.34 feet, and 25.7 feet. Find the amount of border material needed.

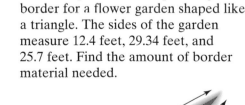
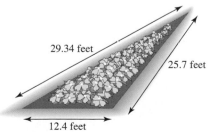
29.34 feet
25.7 feet
12.4 feet

54. A contractor needs to buy railing to completely enclose a newly built deck. Find the amount of railing needed.

15.7 feet
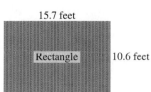
Rectangle
10.6 feet

47. 15.81 inches

48. 24.9 mph

49. 763.035 mph

50. $5.33

51. 240.8 inches

52. 2.65 inches

53. 67.44 feet

54. 52.6 feet

240

Name _____

The table shows the average speeds for the Indianapolis 500 winners for three years. Use this table to answer Exercises 55–56.

INDIANAPOLIS 500 WINNERS		
Year	Winner	Average Speed
1995	Jacques Villeneuve	153.616 mph
1950	Johnnie Parsons	124.002 mph
1911	Ray Harroun	74.602 mph

(*Source:* Indianapolis Motor Speedway)

55. How much faster was the average Indianapolis 500 speed in 1995 than in 1950?

56. How much faster was the average Indianapolis 500 speed in 1995 than in 1911?

The table shows spaceflight information for astronaut James A. Lovell. Use this table to answer Exercises 57–58.

SPACEFLIGHTS OF JAMES A. LOVELL		
Year	Mission	Duration (in hours)
1965	Gemini 6	330.583
1966	Gemini 12	94.567
1968	Apollo 8	147.0
1970	Apollo 13	142.9

(*Source:* NASA)

57. Find the total time spent in spaceflight by astronaut James A. Lovell.

58. Find the total time James A. Lovell spent in spaceflight on all Apollo missions.

The table shows the five top chocolate-consuming nations in the world. Use this table to answer Exercises 59–63.

THE WORLD'S TOP CHOCOLATE-CONSUMING COUNTRIES	
Country	Pounds of Chocolate Per Person
Belgium	13.9
Germany	15.8
Norway	16.0
Switzerland	22.0
United Kingdom	14.5

(*Source:* Hershey Foods Corporation)

59. Which country in the table has the greatest chocolate consumption per person?

60. Which country in the table has the least chocolate consumption per person?

Name _____

61. How much more is the greatest chocolate comsumption than the least chocolate consumption shown in the table?

62. How much more chocolate does the average German consume than the average citizen of the United Kingdom?

63. Make a new chart listing the countries and their corresponding chocolate consumption in order from greatest to least.

REVIEW AND PREVIEW

Multiply. See Sections 1.5 and 2.4.

64. $23 \cdot 2$

65. $46 \cdot 3$

66. $43 \cdot 90$

67. $30 \cdot 32$

68. $\left(\dfrac{2}{3}\right)^2$

69. $\left(\dfrac{1}{5}\right)^3$

70. $\dfrac{12}{7} \cdot \dfrac{14}{3}$

71. $\dfrac{25}{36} \cdot \dfrac{24}{40}$

◤ COMBINING CONCEPTS

72. Laser beams can be used to measure the distance to the moon. One measurement showed the distance to the moon to be 256,435.235 miles. A later measurement showed that the distance is 256,436.012 miles. Find how much farther away the moon is in the second measurement compared to the first.

73. Explain how adding or subtracting decimals is similar to adding or subtracting whole numbers.

---

61. 8.1 pounds

62. 1.3 pounds

Country	Pounds of Chocolate Per Person
Switzerland	22.0
Norway	16.0
Germany	15.8
United Kingdom	14.5
Belgium	13.9

63. _____

64. 46

65. 138

66. 3870

67. 960

68. $\dfrac{4}{9}$

69. $\dfrac{1}{125}$

70. 8

71. $\dfrac{5}{12}$

72. 0.777 mile

73. answers may vary

74. 6.08 in.

Name _____

Find the unknown length in each figure.

74.

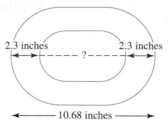

75.

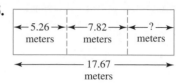

75. 4.59 meters

Focus On History

MULTICULTURAL FRACTION USE

Many ancient cultures were familiar with the use of fractions and used them in everyday life. Here are some interesting facts about the history of fractions:

▲ Although the ancient Egyptians were comfortable with the idea of fractions, their system of hieroglyphic numbers allowed them to only write fractions with 1 as the numerator, like $\frac{1}{3}$, $\frac{1}{7}$, or $\frac{1}{10}$.

To write fractions with numerators other than 1, the Egyptians had to write a sum of fractions with 1 as the numerator, and they preferred to never use the same denominator in a sum of fractions twice! For instance, rather than representing $\frac{2}{5}$ as $\frac{1}{5} + \frac{1}{5}$ (which uses the denominator 5 twice), they would use $\frac{1}{3} + \frac{1}{15}$.

Similarly, $\frac{2}{13}$ would not be represented by $\frac{1}{13} + \frac{1}{13}$ but by $\frac{1}{8} + \frac{1}{52} + \frac{1}{104}$.

▲ The ancient Babylonians also dealt with fractions but in a more direct way then the Egyptians did. For instance, they thought of the fraction $\frac{2}{7}$ as the product of 2 and the reciprocal of 7. They compiled detailed lists of reciprocals and could calculate with fractions very easily and quickly.

▲ The Hindus were the first to write fractions with the numerator above the denominator. Later Arab mathematicians copied with Hindu notation and then improved it by inserting a horizontal bar between the numerator and denominator to separate the two.

4.4 MULTIPLYING DECIMALS

A MULTIPLYING DECIMALS

Multiplying decimals is similar to multiplying whole numbers. The only difference is that we place a decimal point in the product. To discover where a decimal point is placed in the product, let's multiply 0.6 × 0.03. We first write each decimal as an equivalent fraction and then multiply.

$$0.6 \quad \times \quad 0.03 \quad = \frac{6}{10} \times \frac{3}{100} = \frac{18}{1000} = 0.018$$

1 decimal place   2 decimal places   3 decimal places

Now let's multiply 0.03 × 0.002.

$$0.03 \quad \times \quad 0.002 \quad = \frac{3}{100} \times \frac{2}{1000} = \frac{6}{100,000} = 0.00006$$

2 decimal places   3 decimal places   5 decimal places

Instead of writing decimals as fractions each time we want to multiply, we notice a pattern from these examples and state a rule that we can use.

MULTIPLYING DECIMALS

Step 1. Multiply the decimals as though they are whole numbers.

Step 2. The decimal point in the product is placed so the number of decimal places in the product is equal to the *sum* of the number of decimal places in the factors.

Example 1   Multiply: 23.6 × 0.78

Solution:

```
      23.6      1 decimal place
   × 0.78       2 decimal places
    1888
   16520
   18.408       3 decimal places
```

Example 2   Multiply: 0.283 × 0.3

Solution:

```
    0.283       3 decimal places
  ×   0.3       1 decimal place
   0.0849       4 decimal places
```
↑_____ Insert one 0 since the product must have 4 decimal places.

Example 3   Multiply: 0.0531 × 16

Solution:

```
    0.0531      4 decimal places
  ×     16      0 decimal places
    3186
   05310
   0.8496       4 decimal places
```

TRY THE CONCEPT CHECK IN THE MARGIN.

Objectives

A Multiply decimals.
B Multiply by powers of 10.
C Find the circumference of a circle.
D Solve problems by multiplying decimals.

SSM   CD-ROM   Video 4.4

TEACHING TIP

Before discussing the answer to 0.6 × 0.03, point out to students that if they know what 6 × 3 is, they should be able to guess what digits will be in the product. The only other thing they need to solve the problem is where the decimal point is located.

Practice Problem 1

Multiply: 45.9 × 0.42

Practice Problem 2

Multiply: 0.112 × 0.6

Practice Problem 3

Multiply: 0.0721 × 48

✓ CONCEPT CHECK

True or false? The number of decimal places in the product of 0.261 and 0.78 is 6. Explain.

Answers

1. 19.278, **2.** 0.0672, **3.** 3.4608

✓ **Concept Check:** False; 3 decimal places + 2 decimal places is 5 decimal places in the product.

TEACHING TIP

Tell your students that multiplying by powers of ten will be used later when we work with the metric system.

B MULTIPLYING BY POWERS OF 10

There are some patterns that occur when we multiply a number by a power of 10, such as 10, 100, 1000, 10,000, and so on.

$$23.6951 \times 10 = 236.951$$ Move the decimal point *1 place* to the *right*.

1 zero

$$23.6951 \times 100 = 2369.51$$ Move the decimal point *2 places* to the *right*.

2 zeros

$$23.6951 \times 100,000 = 2,369,510.$$ Move the decimal point *5 places* to the *right* (insert a 0).

5 zeros

Notice that we move the decimal point the same number of places as there are zeros in the power of 10.

> **MULTIPLYING DECIMALS BY POWERS OF 10 SUCH AS 10, 100, 1000, 10,000 . . .**
>
> Move the decimal point to the *right* the same number of places as there are *zeros* in the power of 10.

Practice Problems 4–6

Multiply.

4. 23.7×10

5. 203.004×100

6. 1.15×1000

Examples

Multiply.

4. $7.68 \times 10 = 76.8$ 7.68

5. $23.702 \times 100 = 2370.2$ 23.702

6. $76.3 \times 1000 = 76,300$ 76.300

There are also powers of 10 that are less than 1. The decimals 0.1, 0.01, 0.001, 0.0001, and so on, are examples of powers of 10 less than 1. Notice the pattern when we multiply by these powers of 10.

$$569.2 \times 0.1 = 56.92$$ Move the decimal point *1 place* to the *left*.

1 decimal place

$$569.2 \times 0.01 = 5.692$$ Move the decimal point *2 places* to the *left*.

2 decimal places

$$569.2 \times 0.0001 = 0.05692$$ Move the decimal point *4 places* to the *left* (insert one 0).

4 decimal places

> **MULTIPLYING DECIMALS BY POWERS OF 10 SUCH AS 0.1, 0.01, 0.001, 0.0001 . . .**
>
> Move the decimal point to the *left* the same number of places as there are *decimal places* in the power of 10.

Answers

4. 237, **5.** 20,300.4, **6.** 1150

Examples   Multiply.

7. $42.1 \times 0.1 = 4.21$

8. $76,805 \times 0.01 = 768.05$

9. $9.2 \times 0.001 = 0.0092$

Many times we see large numbers written, for example, in the form 267.1 million rather than in the longer standard notation. The next example shows how to interpret these numbers.

Example 10   The population of the United States is 267.1 million. Write this number in standard notation. (*Source: The World Almanac*, 1998)

Solution: 267.1 million = 267.1×1 million

$= 267.1 \times 1,000,000 = 267,100,000.$

C **FINDING THE CIRCUMFERENCE OF A CIRCLE**

Recall that the distance around a polygon is called its perimeter. The distance around a circle is given a special name called the **circumference**, and this distance depends on the radius or the diameter of the circle.

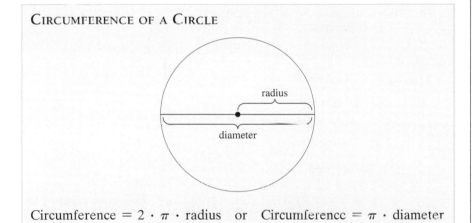

CIRCUMFERENCE OF A CIRCLE

Circumference = $2 \cdot \pi \cdot$ radius   or   Circumference = $\pi \cdot$ diameter

The symbol π is the Greek letter pi, pronounced "pie." It is a number between 3 and 4. The number π rounded to two decimal places is 3.14. Also, a fraction approximation for π is $\frac{22}{7}$.

Example 11   Find the circumference of a circle whose radius is 5 inches. Then use the approximation 3.14 for π to approximate the circumference.

Solution: Circumference = $2 \cdot \pi \cdot$ radius

$= 2 \cdot \pi \cdot 5$ inches

$= 10\pi$ inches

5 inches

Multiply.

7. 7.62×0.1

8. 1.9×0.01

9. 7682×0.001

There are 2064 thousand farms in the United States. Write this number in standard notation. (*Source: The World Almanac*, 1998)

Find the circumference of a circle whose radius is 11 meters. Then use the approximation 3.14 for π to approximate this circumference.

Answers

7. 0.762, **8.** 0.019, **9.** 7.682, **10.** 2,064,000 farms, **11.** 22π meters; 69.08 meters

Copyright 1999 Prentice-Hall, Inc.

Next we replace π with the approximation 3.14.

$$\text{Circumference} = 10\pi \text{ inches}$$

(is approximately) $\rightarrow$ $\approx 10(3.14)$ inches

$$= 31.4 \text{ inches}$$

The *exact* circumference or distance around the circle is 10π inches, which is *approximately* 31.4 inches. ◼

D SOLVING PROBLEMS BY MULTIPLYING DECIMALS

The solutions to many real-life problems are found by multiplying decimals. We continue using our four problem-solving steps to solve such problems.

Example 12   Finding Total Cost of Materials for a Job

A college student is hired to paint a billboard with paint costing $2.49 per quart. If the job requires 3 quarts of paint, what is the total cost of the paint?

Solution: **1.** UNDERSTAND. Read and reread the problem. The phrase "total cost" might make us think addition, but since this is repeated addition, let's multiply.

2. TRANSLATE.

In words:

Total cost	is	cost per quart of paint	times	number of quarts
↓	↓	↓	↓	↓

Translate: Total cost = 2.49 × 3

3. SOLVE.

$$\begin{array}{r} {\scriptstyle 1\ 2} \\ 2.49 \\ \times\quad 3 \\ \hline 7.47 \end{array}$$

4. INTERPRET. *Check* your work. *State* your conclusion: The total cost of the paint is $7.47. ◼

Practice Problem 12

Elaine Rehmann is fertilizing her garden. She uses 5.6 ounces of fertilizer per square yard. The garden measures 60.5 square yards. How much fertilizer does she need?

Answer

12. 338.8 ounces

Exercise Set 4.4

A *Multiply. See Examples 1 through 3.*

1. 0.2
    × 0.6

2. 0.7
    × 0.9

3. 1.2
    × 0.5

4. 6.8
    × 0.3

5. 0.26
    × 5

6. 0.19
    × 6

7. 5.3
    × 4.2

8. 6.2
    × 3.8

9. 5.62
    × 7.7

10. 8.03
    × 5.5

11. 1.0047
    ×   8.2

12. 2.0005
    ×   5.5

13. 490.2
    × 0.023

14. 300.9
    × 0.032

15. 16.003
    × 5.31

16. 31.006
    × 3.71

B *Multiply. See Examples 4 through 9.*

17. 6.5×10

18. 7.2×100

19. 6.5×0.1

20. 7.2×0.01

21. 7.093×100

22. 6.046×1000

23. 0.06×0.01

24. 4.7×0.1

25. 9.1×1000

26. 0.5×10

27. 37.62×0.001

28. 14.3×0.001

Write each number in standard form. See Example 10.

29. The storage silos at the main Hershey chocolate factory in Hershey, Pennsylvania, can hold enough cocoa beans to make 5.5 billion Hershey's milk chocolate bars. (*Source:* Hershey Foods Corporation)

30. In 1998, the restaurant industry in the United States employed 9.5 million people. (*Source:* National Restaurant Association)

247

248

Name _____

31. About 36.4 million American house-holds own at least one dog. (*Source:* American Pet Products Manufacturers Association)

32. The most-visited national park in the United States is the Blue Ridge Park-way in Virginia. An estimated 17.17 million people visit the park each year. (*Source:* National Park Service)

33. Americans lose more than 1.6 million hours each day stuck in traffic. (*Source:* United States Department of Transportation—Federal Highway Administration)

34. The top advertiser in the United States in 1996 was General Motors, who spent $1.71 billion on advertising. (*Source:* Competitive Media Report-ing and Publishers Information Bureau)

C *Find the circumference of each circle. Then use the approximation 3.14 for π and approx-imate each circumference. See Example 11.*

35.

4 meters

36.

8 feet

37.

10 centimeters

38.

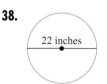

22 inches

39.

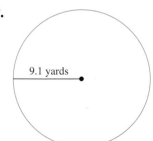

9.1 yards

40.

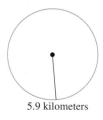

5.9 kilometers

D *Solve. See Example 12.*

41. A 1-ounce serving of cream cheese contains 6.2 grams of saturated fat. How much saturated fat is in 4 ounces of cream cheese? (*Source: Home and Garden Bulletin No. 72*; U.S. Department of Agriculture)

42. A 3.5-ounce serving of lobster meat contains 0.1 gram of saturated fat. How much saturated fat is in 3 servings of lobster meat contain? (*Source:* The National Institute of Health)

43. In 1996, American farmers received an average of $1.90 per bushel for oats. How much did a farmer receive for selling 1000 bushels of oats? (*Source:* National Agricultural Statistics Service)

44. In 1996, American farmers received an average of $6.85 per bushel for soybeans. How much did a farmer receive for selling 10,000 bushels of soybeans? (*Source:* National Agricultural Statistics Service)

45. A meter is a unit of length in the metric system that is approximately equal to 39.37 inches. Sophia Wagner is 1.65 meters tall. Find her approximate height in inches.

46. The doorway to a room is 2.15 meters tall. Approximate this height in inches. (*Hint:* See Exercise 45.)

47. Jose Severos, an electrician for Central Power and Light, worked 40 hours last week. Calculate his pay before taxes for last week if his hourly wage is $13.88.

48. Maribel Chin, an assembly line worker, worked 20 hours last week. Her hourly rate is $8.52 per hour. Calculate Maribel's pay before taxes.

REVIEW AND PREVIEW

Divide. See Sections 1.6 and 2.5.

49. $130 \div 5$

50. $495 \div 27$

51. $2016 \div 56$

52. $1863 \div 69$

53. $2920 \div 365$

54. $2916 \div 6$

55. $\dfrac{24}{7} \div \dfrac{8}{21}$

56. $\dfrac{162}{25} \div \dfrac{9}{75}$

---

41. 24.8 grams

42. 0.3 gram

43. $1900

44. $68,500

45. 64.9605 inches

46. 84.6455 inches

47. $555.20

48. $170.40

49. 26

50. 18 R 9

51. 36

52. 27

53. 8

54. 486

55. 9

56. 54

Name _____

COMBINING CONCEPTS

57. Find how far radio waves travel in 20.6 seconds. (Radio waves travel at a speed of $1.86 \times 100{,}000$ miles per second.)

58. If it takes radio waves approximately 8.3 minutes to travel from the sun to the Earth, find approximately how far it is from the sun to the Earth. (*Hint:* See Exercise 57.)

59. In 1893, the first ride called a ferris wheel was constructed by Washington Gale Ferris. Its diameter was 250 feet. Find its circumference. Give an exact answer and a two-decimal place approximation, using 3.14 for π.

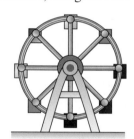

(*Source: The Handy Science Answer Book*, Visible Ink Press, 1994)

60. The radius of the Earth is approximately 3950 miles. Find the distance around the Earth at the equator. Give an exact answer and a two-decimal place approximation, using 3.14 for π. (*Hint:* Find the circumference of a circle with radius 3950 miles.)

3950 miles

61. In your own words, explain how to find the number of decimal places in a product of decimal numbers.

62. In your own words, explain how to multiply by a power of 10.

4.5 Dividing Decimals

A Dividing Decimals

Dividing decimal numbers is similar to dividing whole numbers. The only difference is that we place a decimal point in the quotient. If the divisor is a whole number, place the decimal point in the quotient directly above the decimal point in the dividend, then divide as for whole numbers. Recall that division can be checked by multiplication.

$$
\begin{array}{r}
0.26 \leftarrow \text{quotient} \\
\text{divisor} \rightarrow 32\overline{)8.32} \leftarrow \text{dividend} \\
-6\,4 \\
\hline
192 \\
-192 \\
\hline
0
\end{array}
$$

Check:

$$
\begin{array}{r}
0.26 \quad \text{quotient}\\
\times \ \ 32 \quad \text{divisior}\\
\hline
52 \\
7\,8 \\
\hline
8.32 \quad \text{dividend}
\end{array}
$$

Dividing by a Whole Number

Step 1. Place the decimal point in the quotient directly above the decimal point in the dividend.

Step 2. Divide as for whole numbers.

Example 1
Divide: 270.2 ÷ 7. Check your answer.

Solution:

Write the decimal point.

$$
\begin{array}{r}
38.6 \\
7\overline{)270.2} \\
-21 \\
\hline
60 \\
-56 \\
\hline
4\,2 \\
-4\,2 \\
\hline
0
\end{array}
$$

Check:

$$
\begin{array}{r}
\overset{64}{38.6} \\
\times \ \ \ 7 \\
\hline
270.2
\end{array}
$$

The quotient is 38.6.

Example 2
Divide: 60.24 ÷ 8. Check your answer.

Solution:

Write the decimal point.

$$
\begin{array}{r}
7.53 \\
8\overline{)60.24} \\
-56 \\
\hline
42 \\
-40 \\
\hline
24 \\
-24 \\
\hline
0
\end{array}
$$

Check:

$$
\begin{array}{r}
7.53 \\
\times \ \ 8 \\
\hline
60.24
\end{array}
$$

Sometimes, to continue dividing we may need to insert zeros after the last digit in the dividend.

Objectives

A Divide decimals.
B Divide decimals by powers of 10.
C Solve problems by dividing decimals.

SSM   CD-ROM   Video 4.5

Practice Problem 1

Divide: 517.2 ÷ 6. Check your answer.

Practice Problem 2

Divide: 26.19 ÷ 9. Check your answer.

Answers
1. 86.2, **2.** 2.91

Divide and check.

a. $0.4 \div 8$

b. $13.62 \div 12$

Example 3   Divide: $0.5 \div 4$. Check your answer.

Solution:

$$
\begin{array}{r}
0.125 \\
4\overline{)0.500} \\
\underline{4} \\
10 \\
\underline{-8} \\
20 \\
\underline{-20} \\
0
\end{array}
$$

Insert two 0s to continue dividing.

Check:
$$
\begin{array}{r}
\overset{1\,2}{0.125} \\
\times 4 \\
\hline
0.500
\end{array}
$$

If the divisor is not a whole number, we need to move the decimal point to the right until the divisor is a whole number before we divide.

$$1.5\overline{)64.85}$$

divisor — ↑   ↑— dividend

To understand how this works, let's rewrite

$$1.5\overline{)64.85} \quad \text{as} \quad \frac{64.85}{1.5}$$

and then multiply the numerator and the denominator by 10.

$$\frac{64.85}{1.5} = \frac{64.85 \times 10}{1.5 \times 10} = \frac{648.5}{15}$$

which can be written as $15.\overline{)648.5}$. Notice that

$$1.5\overline{)64.85} \quad \text{is equivalent to} \quad 15.\overline{)648.5}$$

The decimal points in the dividend and the divisor were both moved one place to the right, and the divisor is now a whole number. This procedure is summarized next.

DIVIDING BY A DECIMAL

Step 1. Move the decimal point in the divisor to the right until the divisor is a whole number.

Step 2. Move the decimal point in the dividend to the right the *same number of places* as the decimal point was moved in Step 1.

Step 3. Divide. Place the decimal point in the quotient directly over the moved decimal point in the dividend.

Divide: $166.88 \div 5.6$

Example 4   Divide: $10.764 \div 2.3$

Solution: We move the decimal points in the divisor and the dividend one place to the right so that the divisor is a whole number.

$$2.3\overline{)10.764} \quad \text{becomes} \quad
\begin{array}{r}
4.68 \\
23.\overline{)107.64} \\
\underline{92} \\
15\,6 \\
\underline{13\,8} \\
1\,84 \\
\underline{1\,84} \\
0
\end{array}
$$

Answers

3. a. 0.05,   **b.** 1.135,   **4.** 29.8

Example 5 Divide: $5.264 \div 0.32$

Solution:

$$0.32\overline{)5.264} \quad \text{becomes} \quad 32\overline{)526.40} \quad \text{Insert one 0.}$$

$$
\begin{array}{r}
16.45 \\
32\overline{)526.40} \\
-32 \\
\hline
206 \\
-192 \\
\hline
14\,4 \\
-12\,8 \\
\hline
1\,60 \\
-1\,60 \\
\hline
0
\end{array}
$$

TRY THE CONCEPT CHECK IN THE MARGIN.

Example 6 Divide: $17.5 \div 0.48$. Round the quotient to the nearest hundredth.

Solution: First we move the decimal points in the divisor and the dividend two places. Then we divide and round the quotient to the nearest hundredth.

hundredths place

$$36.458 \approx 36.46$$

$$
\begin{array}{r}
48.\overline{)1750.000} \\
-144 \\
\hline
310 \\
-288 \\
\hline
220 \\
-192 \\
\hline
280 \\
-240 \\
\hline
400 \\
-384 \\
\hline
16
\end{array}
$$

approximately

If rounding to the nearest hundredth, carry the division process out to one more decimal place, the thousandths place.

TRY THE CONCEPT CHECK IN THE MARGIN.

B DIVIDING DECIMALS BY POWERS OF 10

There are patterns that occur when we divide decimals by powers of 10 such as 10, 100, 1000, and so on.

$$\frac{569.2}{10} = 56.92 \quad \text{Move the decimal point } 1 \text{ place to the } left.$$

1 zero

$$\frac{569.2}{10,000} = 0.05692 \quad \text{Move the decimal point } 4 \text{ places to the } left.$$

4 zeros

This pattern suggests the following rule.

DIVIDING DECIMALS BY POWERS OF 10 SUCH AS 10, 100, OR 1000

Move the decimal point of the dividend to the *left* the same number of places as there are *zeros* in the power of 10.

Practice Problem 5

Divide: $1.976 \div 0.16$

✓ **CONCEPT CHECK**

Is it always true that the number of decimal places in a quotient equals the sum of the decimal places in the dividend and divisor?

Practice Problem 6

Divide $23.4 \div 0.57$. Round the quotient to the nearest hundredth.

TEACHING TIP

Encourage students to convert a division problem to a fraction with a whole number in the denominator and then convert this fraction to long division form. For instance, in Example 5: $5.264 \div 0.32 =$

$$\frac{5.264}{0.32} = \frac{526.4}{32} = 32\overline{)526.4}$$

✓ **CONCEPT CHECK**

If a quotient is to be rounded to the nearest thousandth, to what place should the division be carried out?

TEACHING TIP

Consider asking students to fill in the blanks in the following sentences with a fraction and a decimal:
Dividing by 10 is the same as multiplying by ____ or ____.
Dividing by 100 is the same as multiplying by ____ or ____.
Dividing by 1000 is the same as multiplying by ____ or ____.

Answers

5. 12.35, **6.** 41.05

✓ Concept Check: No, it's false.

✓ Concept Check: ten-thousandths place

Practice Problems 7–8

Divide.

7. $\dfrac{28}{1000}$

8. $\dfrac{8.56}{100}$

Practice Problem 9

A bag of fertilizer covers 1250 square feet of lawn. Tim Parker's lawn measures 14,800 square feet. How many bags of fertilizer does he need? If he can buy whole bags of fertilizer only, how many whole bags does he need?

TEACHING TIP   Classroom Activity

Have each person bring 2 circular objects with different diameters. Provide each group with a centimeter and inch measuring tape and a worksheet to record their results. Have the groups measure and record the diameter and circumference of each object in centimeters and inches. Then have them use a calculator to find the circumference/diameter for each object and ask them to make a conclusion from their data. After they have reached a conclusion, have them share their insights with the class. Be sure they see that no matter how large or small a circle is circumference ÷ diameter = π, which is about 3.14.

Examples

Divide.

7. $\dfrac{786}{10,000} = 0.0786$ Move the decimal point *4 places* to the *left*.

   4 zeros

8. $\dfrac{0.12}{10} = 0.012$ Move the decimal point *1 place* to the *left*.

   1 zero

C SOLVING PROBLEMS BY DIVIDING DECIMALS

Many real-life problems involve dividing decimals.

Example 9   Calculating Materials Needed for a Job

A gallon of paint covers a 250-square-foot area. If Betty Adkins wishes to paint a wall that measures 1450 square feet, how many gallons of paint does she need? If she can buy gallon containers of paint only, how many gallon containers does she need?

Solution:

1. UNDERSTAND. Read and reread the problem. To find the number of gallons, we divide 1450 by 250.

2. TRANSLATE.

In words:	number of gallons	is	square feet	divided by	square feet per gallon
	↓	↓	↓	↓	↓
Translate:	number of gallons	=	1450	÷	250

3. SOLVE.

$$
\begin{array}{r}
5.8 \\
250\overline{)1450.0} \\
-1250 \\
\hline
200\,0 \\
-200\,0 \\
\hline
0
\end{array}
$$

4. INTERPRET. *Check* your work. *State* your conclusion: Betty needs 5.8 gallons of paint. If she can buy gallon containers of paint only, she needs 6 gallon containers of paint to complete the job.

Answers

7. 0.028, **8.** 0.0856, **9.** 11.84 bags; 12 bags

Name _____ Section _____ Date _____

EXERCISE SET 4.5

A *Divide. See Examples 1 through 5.*

📼 **1.** $5\overline{)0.47}$ **2.** $2\overline{)11.8}$ **3.** $0.06\overline{)18}$ **4.** $0.04\overline{)20}$

📼 **5.** $0.82\overline{)4.756}$ **6.** $0.92\overline{)3.312}$ **7.** $5.5\overline{)36.3}$ **8.** $2.2\overline{)21.78}$

9. $6.195 \div 15$ **10.** $0.54 \div 12$ **11.** Divide 4.2 by 0.6

12. Divide 3.6 by 0.9   **13.** $0.27\overline{)1.296}$ **14.** $0.34\overline{)2.176}$

15. $0.02\overline{)42}$ **16.** $0.03\overline{)24}$ **17.** $0.6\overline{)18}$ **18.** $0.4\overline{)20}$

19. $0.005\overline{)35}$ **20.** $0.0007\overline{)35}$ **21.** $7.2\overline{)70.56}$ **22.** $6.3\overline{)52.92}$

23. $5.4\overline{)51.84}$ **24.** $7.7\overline{)33.88}$ **25.** $\dfrac{1.215}{0.027}$ **26.** $\dfrac{3.213}{0.051}$

Divide. Round the quotients as indicated. See Example 6.

27. Divide 429.34 by 2.4 and round the quotient to the nearest hundred.

28. Divide 54.8 by 2.6 and round the quotient to the nearest ten.

📼 **29.** Divide 0.549 by 0.023 and round the quotient to the nearest hundredth.

30. Divide 0.0453 by 0.98 and round the quotient to the nearest thousandth.

31. Divide 45.23 by 0.4 and round the quotient to the nearest ten.

32. Divide 983.32 by 0.061 and round the quotient to the nearest thousand.

ANSWERS

1. 0.094
2. 5.9
3. 300
4. 500
5. 5.8
6. 3.6
7. 6.6
8. 9.9
9. 0.413
10. 0.045
11. 7
12. 4
13. 4.8
14. 6.4
15. 2100
16. 800
17. 30
18. 50
19. 7000
20. 50,000
21. 9.8
22. 8.4
23. 9.6
24. 4.4
25. 45
26. 63
27. 200
28. 20
29. 23.87
30. 0.046
31. 110
32. 16,000

Name _____

B *Divide. See Examples 7 and 8.*

33. $\dfrac{54.982}{100}$

34. $\dfrac{342.54}{100}$

35. $\dfrac{12.9}{1000}$

36. $\dfrac{13.49}{10}$

37. $\dfrac{87}{10}$

33. $\dfrac{0.27}{1000}$

C *Solve. See Example 9.*

39. Dorren Schmidt pays $73.86 per month to pay back a loan of $1772.64. In how many months will the loan be paid off?

40. Josef Jones is painting the walls of a room. The walls have a total area of 546 square feet. If a quart of paint covers 52 square feet, how many quarts of paint does he need to buy?

41. The leading money winner in men's professional golf in 1996 was Tom Lehmann. He earned $1,780,159. Suppose he had earned this working 40 hours per week for a year. Determine his hourly wage to the nearest cent. (*Source: The World Almanac*, 1998)

42. Juanita Gomez bought unleaded gasoline for her car at $1.169 per gallon. She wanted to keep a record of how many gallons her car is using but forgot to write down how many gallons she purchased. She wrote a check for $27.71 to pay for it. How many gallons, to the nearest tenth of a gallon, did she buy?

43. A pound of fertilizer covers 39 square feet of lawn. Vivian Bulgakov's lawn measures 7883.5 square feet. How much fertilizer, to the nearest tenth of a pound, does she need to buy?

44. A page of a book contains about 1.5 kilobytes of information. If a computer disk can hold 740 kilobytes of information, how many pages of a book can be stored on one computer disk? Round to the nearest tenth of a page.

45. There are approximately 39.37 inches in 1 meter. How many meters, to the nearest tenth of a meter, are there in 200 inches?

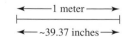

46. There are approximately 2.54 centimeters in 1 inch. How many inches are there in 50 centimeters? Round to the nearest tenth.

47. In 1996, American farmers received an average of $58.70 per 100 pounds of beef cattle. What was the average price per pound for beef cattle? (*Source:* National Agricultural Statistics Service)

48. In 1996, American farmers received an average of $82.20 per 100 pounds of lamb. What was the average price per pound for lambs? (*Source:* National Agricultural Statistics Service)

49. During the 24 Hours of Le Mans endurance auto race in 1997, the winning team of Michele Alboreto, Stefan Johansson, and Tom Kristensen drove a total of 3052.45 miles in 24 hours. What was their average speed in miles per hour? Round to the nearest tenth. (*Source: 1998 World Almanac*)

50. In 1997, Danish runner Wilson Kipketer set a new world record for the 800-meter event. His time for the event was 101.11 seconds. What was his average speed in meters per second? Round to the nearest tenth. (*Source:* International Amateur Athletic Federation)

51. Ruthie Bolton-Holifield of the WNBA's Sacramento Monarchs scored a total of 447 points during the 23 basketball games of the 1997 regular season. What was the average number of points scored by Bolton-Holifield per game? Round to the nearest hundredth. (*Source:* Women's National Basketball Association)

52. During the 1997 Major League Soccer season, DC United was the top scoring team with a total of 70 goals throughout the season. DC United played 32 games. What was the average number of goals scored by DC United per game? Round to the nearest hundredth. (*Source:* Major League Soccer)

REVIEW AND PREVIEW

Round each decimal to the given place value. See Section 4.2.

53. 345.219, nearest hundredth

54. 902.155, nearest hundredth

55. 1000.994, nearest tenth

56. 234.1029, nearest thousandth

Use the order of operations to simplify each expression. See Section 1.8.

57. $2 + 3 \cdot 6$

58. $9 \cdot 2 + 8 \cdot 3$

59. $20 - 10 \div 5$

60. $(20 - 10) \div 5$

47. $0.587 per pound

48. $0.822 per pound

49. 127.2 mph

50. 7.9 meters/second

51. 19.43 points/game

52. 2.19 goals/game

53. 345.22

54. 902.16

55. 1001.0

56. 234.103

57. 20

58. 42

59. 18

60. 2

Name _____

 COMBINING CONCEPTS

Recall from Section 1.6 that the average of a list of numbers is their total divided by how many numbers there are in the list. Use this to find the average of the test scores listed. If necessary, round to the nearest tenth.

61. 86, 78, 91, 85

62. 56, 75, 80

63. In 1997, American manufacturers shipped approximately 753.1 million music CDs to retailers. How many music CDs were shipped per week during 1997 on average? Round to the nearest thousand. (*Source:* Recording Industry Association of America)

64. The area of the rectangle is 38.7 square feet. If its width is 4.5 feet, find its length.

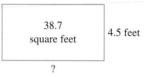

38.7 square feet   4.5 feet

?

65. The perimeter of the square is 180.8 centimeters. Find the length of a side.

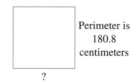

Perimeter is 180.8 centimeters

?

66. Don Larson is building a horse corral shaped like a rectangle with dimensions 24.28 meters by 15.675 meters. (A meter is a unit of length in the metric system.) He plans to make a four-wire fence; that is, he will string four wires around the corral. How much wire will he need?

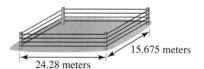

15.675 meters

24.28 meters

Name _____ **Section** _____ **Date** _____

Integrated Review — Operations on Decimals

Perform the indicated operations.

1. $1.6 + 0.97$

2. $3.2 + 0.85$

3. $9.8 - 0.9$

4. $10.2 - 6.7$

5. $\begin{array}{r} 0.8 \\ \times\ 0.2 \\ \hline \end{array}$

6. $\begin{array}{r} 0.6 \\ \times\ 0.4 \\ \hline \end{array}$

7. $8\overline{)2.16}$

8. $6\overline{)3.12}$

9. $\begin{array}{r} 9.6 \\ \times\ 0.5 \\ \hline \end{array}$

10. $\begin{array}{r} 8.7 \\ \times\ 0.7 \\ \hline \end{array}$

11. $\begin{array}{r} 123.6 \\ -\ 48.04 \\ \hline \end{array}$

12. $\begin{array}{r} 325.2 \\ -\ 36.08 \\ \hline \end{array}$

13. $25 + 0.026$

14. $0.125 + 44$

15. $3.4\overline{)29.24}$

16. $1.9\overline{)10.26}$

17. 2.8×100

18. 1.6×1000

19. $\begin{array}{r} 96.21 \\ 7.028 \\ +121.7 \\ \hline \end{array}$

20. $\begin{array}{r} 0.268 \\ 1.93 \\ +142.881 \\ \hline \end{array}$

21. $25.76 \div 46$

22. $27.09 \div 43$

23. $\begin{array}{r} 12.004 \\ \times\ \ \ \ 2.3 \\ \hline \end{array}$

24. $\begin{array}{r} 28.006 \\ \times\ \ \ \ 5.2 \\ \hline \end{array}$

25. Subtract 4.6 from 10

26. Subtract 0.26 from 18

27. 414.44 _____

28. 1295.03 _____

29. 34 _____

30. 28 _____

31. 116.81 _____

32. 18.79 _____

27. 268.19
+146.25

28. 860.18
+ 434.85

29. $\dfrac{2.958}{0.087}$

30. $\dfrac{1.708}{0.061}$

31. $160 - 43.19$

32. $120 - 101.21$

260

4.6 ESTIMATING AND ORDER OF OPERATIONS

A ESTIMATING OPERATIONS ON DECIMALS

Estimating sums, differences, products, and quotients of decimal numbers is an important skill whether you use a calculator or perform decimal operations by hand. When you can estimate results as well as calculate them, you can judge whether the calculations are reasonable. If they aren't, you know you've made an error somewhere along the way.

Example 1   Subtract: $78.62 − $16.85. Then estimate the difference to see whether the proposed result is reasonable by rounding each decimal to the nearest dollar and then subtracting.

Solution:

Given		Estimate
$78.62	rounds to	$79
−$16.85	rounds to	−$17
$61.77		$62

The estimated difference is $62, so $61.77 is reasonable.

TRY THE CONCEPT CHECK IN THE MARGIN.

Example 2   Multiply: 28.06 × 1.95. Then estimate the product to see whether the proposed result is reasonable.

Solution:

Given	Estimate 1	Estimate 2
28.06	28	30
× 1.95	× 2	× 2
14030	56	60
252540		
280600		
54.7170		

The answer 54.7170 is reasonable.

As shown in Example 2, estimated results will vary depending on what estimates are used. Notice that estimating results is a good way to see whether the decimal point has been correctly placed.

Example 3   Divide: 272.356 ÷ 28.4. Then estimate the quotient to see whether the proposed result is reasonable.

Solution:

```
        9.59              9               9
284.)2723.56     30)270    or    300)2700
     2556
     1675
    −1420
     2556
    −2556
        0
```

The estimate is 9, so 9.59 is reasonable.

Objectives

A Estimate operations on decimals.

B Simplify expressions containing decimals using the order of operations.

SSM CD-ROM Video 4.6

Practice Problem 1

Subtract: $65.34 − $14.68. Then estimate the difference to see whether the proposed result is reasonable.

✓ **CONCEPT CHECK**

Why shouldn't the sum 21.98 + 42.36 be estimated as 30 + 50 = 80?

Practice Problem 2

Multiply: 30.26 × 2.98. Then estimate the product to see whether the proposed solution is reasonable.

TEACHING TIP

Point out that when estimating, estimates could be higher or lower than the actual answer. Encourage students to try to determine if the estimate will be too high to too low. For instance, for the second estimate in Example 2, both numbers were rounded up so the estimate will be larger than the actual answer.

Practice Problem 3

Divide: 713.7 ÷ 91.5. Then estimate the quotient to see whether the proposed answer is reasonable.

Answers

1. $50.66, **2.** 90.1748, **3.** 7.8

✓ **Concept Check:** Each number is rounded incorrectly. The estimate is too high.

Practice Problem 4

Using the figure below, estimate how much farther it is between Dodge City and Pratt than it is between Garden City and Dodge City by rounding each given distance to the nearest mile.

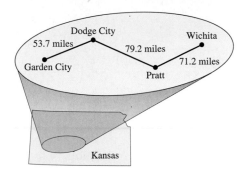

Practice Problem 5

Simplify: $8.69(3.2 - 1.8)$

Practice Problem 6

Simplify: $(0.7)^2$

Practice Problem 7

Simplify: $\dfrac{8.78 - 2.8}{20}$

Practice Problem 8

Simplify: $20.06 - (1.2)^2 \div 10$

Answers

4. 25 miles, **5.** 12.166, **6.** 0.49, **7.** 0.299, **8.** 19.916

Example 4 Estimating Distance

Use the figure in the margin to estimate the distance in miles between Garden City, Kansas, and Wichita, Kansas, by rounding each given distance to the nearest ten.

Solution:

Calculated Distance		Estimate
53.7	rounds to	50
79.2	rounds to	80
+71.2	rounds to	+70
		200

The distance between Garden City and Wichita is approximately 200 miles. (The calculated distance is 204.1 miles.)

B SIMPLIFYING EXPRESSIONS CONTAINING DECIMALS

In the remaining examples, we will review the order of operations by simplifying expressions that contain decimals.

> **ORDER OF OPERATIONS**
>
> 1. Do all operations within grouping symbols such as parentheses or brackets.
> 2. Evaluate any expressions with exponents.
> 3. Multiply or divide in order from left to right.
> 4. Add or subtract in order from left to right.

Example 5 Simplify: $0.5(8.6 - 1.2)$

Solution: According to the order of operations, we simplify inside the parentheses first.

$$0.5(8.6 - 1.2) = 0.5(7.4) \quad \text{Subtract.}$$
$$= 3.7 \quad \text{Multiply.}$$

Example 6 Simplify: $(1.3)^2$

Solution: We recall the meaning of an exponent.

$$(1.3)^2 = (1.3)(1.3) \quad \text{Use the definition of an exponent.}$$
$$= 1.69 \quad \text{Multiply.}$$

Example 7 Simplify: $\dfrac{0.7 + 1.84}{0.4}$

Solution: First we simplify the numerator of the fraction. Then we divide.

$$\frac{0.7 + 1.84}{0.4} = \frac{2.54}{0.4} \quad \text{Simplify the numerator.}$$
$$= 6.35 \quad \text{Divide.}$$

Example 8 Simplify: $5.68 + (0.9)^2 \div 100$

Solution:

$$5.68 + 0.9^2 \div 100 = 5.68 + 0.81 \div 100 \quad \text{Simplify } 0.9^2.$$
$$= 5.68 + 0.0081 \quad \text{Divide.}$$
$$= 5.6881 \quad \text{Add.}$$

Name _____ Section _____ Date _____

EXERCISE SET 4.6

A *Perform each indicated operation. Then estimate to see whether each proposed result is reasonable. See Examples 1 through 3.*

1. $2.1 + 5.8 + 4.1$

2. $6.1 + 5.9 + 3.1$

3. $4.9 - 2.1$

4. $7.3 - 3.8$

📼 **5.** 6×483.11

6. 98.2×405.8

7. $62.16 \div 14.8$

8. $186.75 \div 24.9$

9. $69.2 + 32.1 + 48.5$

10. $179.52 + 221.47 + 397.23$

11. $34.92 - 12.03$

12. $349.87 - 251.32$

Solve. See Example 4.

13. Estimate the perimeter of the rectangle by rounding each measurement to the nearest whole inch.

5.9 inches

12.2 inches

14. Joe Armani wants to put plastic edging around his rectangular flower garden to help control the weeds. The length and width are 16.2 meters and 12.9 meters, respectively. Find how much edging he needs by rounding each measurement to the nearest whole meter.

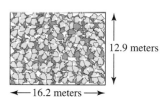

12.9 meters

← 16.2 meters →

15. Estimate the perimeter of the triangle by rounding each measurement to the nearest whole foot.

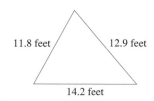

11.8 feet   12.9 feet

14.2 feet

16. The area of a triangle is

$$\text{area} = \frac{1}{2} \cdot \text{base} \cdot \text{height}.$$

Approximate the area of a triangle whose base is 21.9 centimeters and height is 9.9 centimeters. Round each measurement to the nearest whole centimeter.

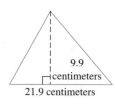

9.9 centimeters

21.9 centimeters

1. 12

2. 15.1

3. 2.8

4. 3.5

5. 2898.66

6. 39,849.56

7. 4.2

8. 7.5

9. 149.8

10. 798.22

11. 22.89

12. 98.55

13. 36 inches

14. 58 meters

15. 39 feet

16. 110 square centimeters

17. 43.96 meters

18. 37.68 centimeters

19. 47 gallons

20. $49.77

21. about $12,000

22. yes

23. about 53 miles

24. 65 feet

Name _____

17. Use 3.14 for π to approximate the circumference of a circle whose radius is 7 meters.

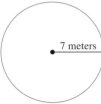

18. Use 3.14 for π to approximate the circumference of a circle whose radius is 6 centimeters.

19. Mike and Sandra Hallahan are taking a trip and plan to travel about 1550 miles. Their car gets 32.8 miles to the gallon. Approximate how many gallons of gasoline their car will use. Round the result to the nearest gallon.

20. Refer to Exercise 19. Suppose gasoline costs $1.059 per gallon. Approximate to the nearest cent how much money Mike and Sandra need for gasoline for their 1550-mile trip.

21. Ken and Wendy Holladay purchased a new car. They pay $198.79 per month on their car loan for 5 years. Approximate how much they will pay for the car altogether by rounding the monthly payment to the nearest hundred.

22. Yoshikazu Sumo is purchasing the following groceries. He has only a ten dollar bill in his pocket. Estimate the cost of each item to see whether he has enough money. Bread, $1.49; milk, $2.09; carrots, $0.97; corn, $0.89; salt, $0.53; and butter, $2.89.

23. Estimate the total distance to the nearest mile between Grove City and Jerome by rounding each distance to the nearest mile.

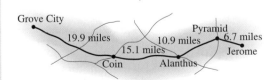

24. Estimate the perimeter of the figure shown by rounding each measurement to the nearest whole foot.

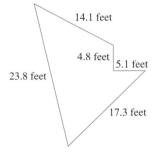

25. The all-time top five movies (that is, those that have earned the most money in the United States) along with the approximate amount of money they have earned are listed in the table. Estimate the total amount of money that these movies have earned.

ALL-TIME TOP FIVE AMERICAN MOVIES	
Movie	Gross Domestic Earnings
Titanic (1997)	$560.6 million
Star Wars (1977)	$461.0 million
E.T. (1982)	$399.8 million
Jurassic Park (1993)	$357.1 million
Forrest Gump (1994)	$329.7 million

(*Source: Variety* magazine)

26. San Marino is one of the smallest countries in the world. In 1997, it had a population density of 1029 people per square mile. The area of San Marino is about 24 square miles. Use this data to estimate the population of San Marino in 1997. (*Source: 1998 World Almanac*)

27. Micronesia is a string of islands in the western Pacific Ocean. In 1997, it had a population density of 470 people per square mile. The area of Micronesia is about 271 square miles. Use this data to estimate the population of Micronesia in 1997. (*Source: 1998 World Almanac*)

28. In 1997, American manufacturers shipped approximately 66.7 million music CD singles to retailers. The value of these shipments was $272.7 million. Use this data to estimate the value of an individual CD single. (*Source:* Recording Industry Association of America)

B *Simplify each expression. See Examples 5 through 8.*

29. $(0.4)^2$

30. $300 - 100 \times 2.3$

31. $\dfrac{1 + 0.8}{0.6}$

32. $(0.09)^2$

33. $1.4(2 - 1.8)$

34. $\dfrac{0.29 + 1.69}{3}$

35. $30.03 + 5.1 \times 9.9$

36. $60 - 6.02 \times 8.97$

37. $7.8 - 4.83 \div 2.1$

38. $90 - 62.1 \div 2.7 + 8.6$

39. $(2.3)^2$

25. $2109 million

26. 20,600 people

27. 126,900 people

28. $3.86

29. 0.16

30. 70

31. 3

32. 0.0081

33. 0.28

34. 0.66

35. 80.52

36. 6.0006

37. 5.5

38. 75.6

39. 5.29

Name _____

40. $(8.2)(100) - (8.2)(10)$ **41.** $(3.1 + 0.7)(2.9 - 0.9)$

42. $\dfrac{3.19 - 0.707}{1.3}$ **43.** $\dfrac{(4.5)^2}{100}$ **44.** $0.9(6.5 - 5.6)$

45. $\dfrac{7 + 0.74}{0.06}$ **46.** $(1.5)^2 + 0.5$

REVIEW AND PREVIEW

Perform each operation. See Sections 2.4, 2.5, and 3.3.

47. $\dfrac{3}{4} \cdot \dfrac{5}{12}$ **48.** $\dfrac{56}{23} \cdot \dfrac{46}{8}$ **49.** $\dfrac{36}{56} \div \dfrac{30}{35}$

50. $\dfrac{5}{7} \div \dfrac{14}{10}$ **51.** $\dfrac{5}{12} - \dfrac{1}{3}$ **52.** $\dfrac{12}{15} + \dfrac{5}{21}$

COMBINING CONCEPTS

Simplify each expression then estimate to see whether the proposed result is reasonable.

53. $1.96(7.852 - 3.147)^2$ **54.** $(6.02)^2 + (2.06)^2$

4.7 FRACTIONS AND DECIMALS

A WRITING FRACTIONS AS DECIMALS

To write a fraction as a decimal, we interpret the fraction bar as division and find the quotient.

> **WRITING FRACTIONS AS DECIMALS**
>
> To write a fraction as a decimal, divide the numerator by the denominator.

Example 1   Write $\frac{1}{4}$ as a decimal.

Solution: $\frac{1}{4} = 1 \div 4$

$$
\begin{array}{r}
0.25 \\
4\overline{)1.00} \\
-8 \\
\hline
20 \\
-20 \\
\hline
0
\end{array}
$$

Thus, $\frac{1}{4}$ written as a decimal is 0.25.

Example 2   Write $\frac{2}{3}$ as a decimal.

Solution:

$$
\begin{array}{r}
0.666\ldots \\
3\overline{)2.000} \\
-1\,8 \\
\hline
20 \\
-18 \\
\hline
20 \\
-18 \\
\hline
2
\end{array}
$$

This pattern will continue so that $\frac{2}{3} = 0.6666\ldots$

A bar can be placed over the digit 6 to indicate that it repeats.

$$\frac{2}{3} = 0.666\ldots = 0.\overline{6}$$

We can also write a decimal approximation for $\frac{2}{3}$. For example, $\frac{2}{3}$ rounded to the nearest hundredth is 0.67.

This can be written as $\frac{2}{3} \approx 0.67$.

Example 3   Write $\frac{22}{7}$ as a decimal. Round to the nearest hundredth.

(The fraction $\frac{22}{7}$ is an approximation for π.)

Objectives

A Write fractions as decimals.
B Compare fractions and decimals.
C Solve area problems containing fractions and decimals.

SSM   CD-ROM   Video
4.7

Practice Problem 1

a. Write $\frac{2}{5}$ as a decimal.

b. Write $\frac{9}{40}$ as a decimal.

Practice Problem 2

a. Write $\frac{5}{6}$ as a decimal.

b. Write $\frac{2}{9}$ as a decimal.

TEACHING TIP
Point out to students that the long division symbol resembles a small multiplication chart where you know one factor and the result and need to find the other factor. Using division of two whole numbers, $20 \div 4$ is the same as asking 4 times what number is 20? Using a fraction, $\frac{3}{4} = 3 \div 4$ is the same as asking 4 times what number is 3?

$$
\begin{array}{c|c}
\times & ? \\
\hline
4 & 20
\end{array}
\quad \rightarrow \quad 4\overline{)20}
$$

$$
\begin{array}{c|c}
\times & ? \\
\hline
4 & 3
\end{array}
\quad \rightarrow \quad 4\overline{)3}
$$

Practice Problem 3

Write $\frac{1}{9}$ as a decimal. Round to the nearest thousandth.

Answers

1. a. 0.4, **b.** 0.225, **2. a.** $0.8\overline{3} \approx 0.83$, **b.** $0.\overline{2} \approx 0.22$, **3.** $0.\overline{1} \approx 0.11$

Solution: $\quad\quad \underset{7\overline{)22.000}}{3.142} \approx 3.14$ <small>Carry the division out to the thousandths place.</small>

$$\begin{array}{r} 3.142 \\ 7\overline{)22.000} \\ -21 \\ \hline 1\,0 \\ -\ 7 \\ \hline 30 \\ -28 \\ \hline 20 \\ -14 \\ \hline 6 \end{array}$$

The fraction $\dfrac{22}{7}$ in decimal form is approximately 3.14.

B COMPARING DECIMALS AND FRACTIONS

Now we can compare decimals and fractions by writing fractions as equivalent decimals.

Practice Problem 4

Insert $<$, $>$, or $=$ to form a true statement.

$\dfrac{1}{5} \quad\quad 0.25$

Example 4   Insert $<$, $>$, or $=$ to form a true statement.

$\dfrac{1}{8} \quad\quad 0.12$

Solution: First we write $\dfrac{1}{8}$ as an equivalent decimal. Then we compare decimal places.

$$\begin{array}{r} 0.125 \\ 8\overline{)1.000} \\ -8 \\ \hline 20 \\ -16 \\ \hline 40 \\ -40 \\ \hline 0 \end{array}$$

Original numbers	$\dfrac{1}{8}$	0.12
Decimals	0.125	0.120
Compare	0.125 $>$ 0.12	
Thus,	$\dfrac{1}{8} > 0.12$	

Practice Problem 5

Insert $<$, $>$, or $=$ to form a true statement.

a. $\dfrac{1}{2} \quad\quad 0.54$ b. $0.\overline{6} \quad\quad \dfrac{2}{3}$

c. $\dfrac{5}{7} \quad\quad 0.72$

Example 5   Insert $<$, $>$, or $=$ to form a true statement.

$0.\overline{7} \quad\quad \dfrac{7}{9}$

Solution: We write $\dfrac{7}{9}$ as a decimal and then compare.

$$\begin{array}{r} 0.77\ldots = 0.\overline{7} \\ 9\overline{)7.00} \\ -6\,3 \\ \hline 70 \\ -63 \\ \hline 7 \end{array}$$

Original numbers	$0.\overline{7}$	$\dfrac{7}{9}$
Decimals	$0.\overline{7}$	$0.\overline{7}$
Compare	$0.\overline{7} = 0.\overline{7}$	
Thus,	$0.\overline{7} = \dfrac{7}{9}$	

Practice Problem 6

Write the numbers in order from smallest to largest.

a. $\dfrac{1}{3}, 0.302, \dfrac{3}{8}$ b. $1.26, 1\dfrac{1}{4}, 1\dfrac{2}{5}$

c. $0.4, 0.41, \dfrac{5}{7}$

Answers

4. $<$,   **5. a.** $<$,   **b.** $=$,   **c.** $<$,

6. a. $0.302, \dfrac{1}{3}, \dfrac{3}{8}$,   **b.** $1\dfrac{1}{4}, 1.26, 1\dfrac{2}{5}$,

c. $0.4, 0.41, \dfrac{5}{7}$

Example 6   Write the numbers in order from smallest to largest.

$\dfrac{9}{20}, \quad \dfrac{4}{9}, \quad 0.456$

Solution:

Original numbers	$\frac{9}{20}$	$\frac{4}{9}$	0.456
Decimals	0.450	0.444 . . .	0.456
Compare in order	2nd	1st	3rd

Then written in order, we have

$$\frac{4}{9}, \frac{9}{20}, 0.456$$

C SOLVING AREA PROBLEMS CONTAINING FRACTIONS AND DECIMALS

Sometimes real-life problems contain both fractions and decimals. In this section, we will solve such problems about area.

Example 7   The area of a triangle is area $= \frac{1}{2} \cdot$ base $\cdot$ height. Find the area of the triangle shown.

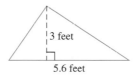

3 feet

5.6 feet

Solution: Area $= \frac{1}{2} \cdot$ base $\cdot$ height

$= \frac{1}{2} \cdot 5.6 \cdot 3$

$= 0.5 \cdot 5.6 \cdot 3$ Write $\frac{1}{2}$ as the decimal 0.5.

$= 8.4$

The area of the triangle is 8.4 square feet.

Practice Problem 7

Find the area of the triangle.

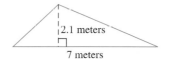

2.1 meters

7 meters

Answer

7. 7.35 square meters

Focus On Business and Career

According to U.S. Bureau of Labor Statistics projections, the careers listed below will have the largest job growth into the next century.

	Occupation	Employment [Numbers in thousands]		
		1994	2005	Change
1.	Cashiers	3005	3567	562
2.	Janitors and cleaners, including maids and housekeeping cleaners	3043	3602	559
3.	Salespersons, retail	3842	4374	532
4.	Waiters and waitresses	1847	2326	479
5.	Registered nurses	1906	2379	473
6.	General managers and top executives	3046	3512	466
7.	Systems analysts	483	928	445
8.	Home health aides	420	848	428
9.	Guards	867	1282	415
10.	Nursing aides, orderlies, and attendants	1265	1652	387
11.	School teachers, secondary	1340	1726	386
12.	Marketing and sales worker supervisors	2293	2673	380
13.	Teacher aides and educational associates	932	1296	364
14.	Receptionists and information clerks	1019	1337	318
15.	Truckdrivers, light and heavy	2565	2837	272
16.	Secretaries, except legal and medical	2842	3109	267
17.	Clerical supervisors and managers	1340	1600	260
18.	Child care workers	757	1005	248
19.	Maintenance repairers, general utility	1273	1505	232
20.	Teachers, elementary	1419	1639	220
21.	Personal and home care aides	179	391	212
22.	Teachers, special education	388	593	205
23.	Licensed practical nurses	702	899	197
24.	Food service and lodging managers	579	771	192
25.	Food preparation workers	1190	1378	188
26.	Social workers	557	744	187
27.	Lawyers	658	839	181
28.	Financial managers	768	950	182
29.	Computer engineers	195	372	177
30.	Hand packers and packagers	942	1102	160

Source: Bureau of Labor Statistics, Office of Employment Projections, November 1995

What do all of these in-demand occupations have in common? They all require a knowledge of math! For some careers like cashiers, salespersons, waiters and waitresses, financial managers, and computer engineers, the ways math is used on the job may be obvious. For other occupations, the use of math may not be quite as apparent. However, tasks common to many jobs like filling in a time sheet, writing up an expense or mileage report, planning a budget, figuring a bill, ordering supplies, completing a packing list, and even making a work schedule all require math.

Critical Thinking

Suppose that your college placement office is planning to publish an occupational handbook on math in popular occupations. Choose one of the occupations from the list above that interests you. Research the occupation. Then write a brief entry for the occupational handbook that describes how a person in that career would use math in his or her job. Include an example if possible.

Exercise Set 4.7

A *Write each fraction as a decimal. See Examples 1 through 3.*

1. $\dfrac{1}{5}$ 2. $\dfrac{2}{5}$ 3. $\dfrac{4}{8}$ 4. $\dfrac{6}{8}$

5. $\dfrac{3}{4}$ 6. $\dfrac{6}{5}$ 7. $\dfrac{2}{25}$ 8. $\dfrac{3}{25}$

9. $\dfrac{3}{8}$ 10. $\dfrac{1}{4}$ 11. $\dfrac{11}{12}$ 12. $\dfrac{19}{25}$

13. $\dfrac{17}{40}$ 14. $\dfrac{5}{12}$ 15. $\dfrac{9}{20}$ 16. $\dfrac{31}{40}$

17. $\dfrac{1}{3}$ 18. $\dfrac{7}{9}$ 19. $\dfrac{7}{16}$ 20. $\dfrac{27}{16}$

21. $\dfrac{2}{9}$ 22. $\dfrac{9}{11}$ 23. $\dfrac{5}{3}$ 24. $\dfrac{4}{11}$

Round each number as indicated.

25. Round your decimal answer to Exercise 17 to the nearest hundredth.

26. Round your decimal answer to Exercise 18 to the nearest hundredth.

27. Round your decimal answer to Exercise 19 to the nearest hundredth.

28. Round your decimal answer to Exercise 20 to the nearest hundredth.

29. Round your decimal answer to Exercise 21 to the nearest tenth.

30. Round your decimal answer to Exercise 22 to the nearest tenth.

31. Round your decimal answer to Exercise 23 to the nearest tenth.

32. Round your decimal answer to Exercise 24 to the nearest tenth.

33. 0.194

34. 0.42

35. 0.44

36. 0.4122807018

37. 0.8

38. 0.284

39. <

40. <

41. >

42. >

43. <

44. <

45. <

46. <

47. <

48. >

49. >

50. <

51. <

52. <

53. <

54. <

272

Name _____

Write each fraction as a decimal.

33. In the United States' National Park System, $\frac{73}{376}$ of all the park units are classified as national monuments. Round your decimal answer to the nearest thousandth. (*Source:* National Park System)

34. About $\frac{21}{50}$ of all blood donors have type A blood. (*Source:* American Red Cross Biomedical Services)

35. The National Athletic Trainers' Association is an organization dedicated to advancing the athletic training profession. Women make up approximately $\frac{11}{25}$ of its membership. (*Source:* National Athletic Trainers' Association)

36. At the beginning of 1997, $\frac{47}{114}$ of all individuals who had flown in space were citizens of the United States. (*Source:* Congressional Research Service)

37. Approximately $\frac{4}{5}$ of the funding for public television comes from sources other than the federal government, such as individuals, businesses, and state governments. (*Source:* Corporation for Public Broadcasting)

38. In 1996, approximately $\frac{71}{250}$ of the civilian labor force had at least a bachelor's degree. (*Source:* U.S. Bureau of the Census)

B *Insert <, >, or = to form a true statement. See Examples 4 and 5.*

39. 0.562   0.569

40. 0.983   0.988

41. 0.823   0.813

42. 0.824   0.821

43. 0.0923   0.0932

44. 0.00536   0.00563

45. $\frac{2}{3}$ $\frac{5}{6}$

46. $\frac{1}{9}$ $\frac{2}{17}$

47. $\frac{5}{9}$ $\frac{51}{91}$

48. $\frac{7}{12}$ $\frac{6}{11}$

49. $\frac{4}{7}$ 0.14

50. $\frac{5}{9}$ 0.557

51. 1.38 $\frac{18}{13}$

52. 0.372 $\frac{22}{59}$

53. 7.123 $\frac{456}{64}$

54. 12.713 $\frac{89}{7}$

Name _____

Write the numbers in order from smallest to largest. See Example 6.

55. 0.34, 0.35, 0.32   **56.** 0.47, 0.42, 0.40   **57.** 0.49, 0.491, 0.498

58. 0.72, 0.727, 0.728   **59.** $\frac{3}{4}$, 0.78, 0.73   **60.** $\frac{2}{5}$, 0.49, 0.42

61. $\frac{4}{7}$, 0.453, 0.412   **62.** $\frac{6}{9}$, 0.663, 0.668   **63.** 5.23, $\frac{42}{8}$, 5.34

64. 7.56, $\frac{67}{9}$, 7.562   **65.** $\frac{12}{5}$, 2.37, $\frac{17}{8}$ **66.** $\frac{29}{16}$, 1.75, $\frac{58}{32}$

C *Find the area of each triangle or rectangle. See Example 7.*

67.

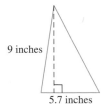

9 inches

5.7 inches

68.

4.4 feet

17 feet

69.

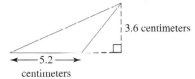

3.6 centimeters

5.2 centimeters

70.

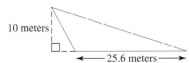

10 meters

25.6 meters

71.

0.62 yard

$\frac{2}{5}$ yard

72.

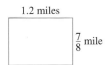

1.2 miles

$\frac{7}{8}$ mile

REVIEW AND PREVIEW

Simplify. See Sections 1.8 and 3.5.

73. 2^3 **74.** 5^4 **75.** $6^2 \cdot 2$ **76.** $4 \cdot 3^4$

55. 0.32, 0.34, 0.35

56. 0.40, 0.42, 0.47

57. 0.49, 0.491, 0.498

58. 0.72, 0.727, 0.728

59. 0.73, $\frac{3}{4}$, 0.78

60. $\frac{2}{5}$, 0.42, 0.49

61. 0.412, 0.453, $\frac{4}{7}$

62. 0.663, $\frac{6}{9}$, 0.668

63. 5.23, $\frac{42}{8}$, 5.34

64. $\frac{67}{9}$, 7.56, 7.562

65. $\frac{17}{8}$, 2.37, $\frac{12}{5}$

66. 1.75, $\frac{29}{16} = \frac{58}{32}$

67. 25.65 square inches

68. 37.4 square feet

69. 9.36 square centimeters

70. 128 square meters

71. 0.248 square yards

72. 1.05 square miles

73. 8

74. 625

75. 72

76. 324

77. $\left(\dfrac{1}{3}\right)^4$ **78.** $\left(\dfrac{4}{5}\right)^3$ **79.** $\left(\dfrac{3}{5}\right)^2$ **80.** $\left(\dfrac{7}{2}\right)^2$

81. $\left(\dfrac{2}{5}\right)\left(\dfrac{5}{2}\right)^2$ **82.** $\left(\dfrac{2}{3}\right)^?\left(\dfrac{3}{2}\right)^3$

COMBINING CONCEPTS

In 1977, there were 10,313 commercial radio stations in the United States. The most popular formats are listed in the table along with their counts for 1997. Use the table to answer Exercises 83–86.

TOP COMMERCIAL RADIO STATION FORMATS IN 1997	
Format	Number of Stations
Country	2502
Adult Contemporary	1521
News/Talk/Business/Sports	1313
Religious	1054
Rock	942

(*Source:* M Street Corporation)

83. Write the fraction of radio stations with a country music format as a decimal. Round to the nearest thousandth.

84. Write the fraction of radio stations with a talk format as a decimal. Round to the nearest hundredth.

85. Estimate the total number of stations with the top five formats in 1997.

86. Use your estimate from Exercise 85 to write the fraction of radio stations accounted for by the top five formats as a decimal. Round to the nearest thousandth.

87. Suppose that you are writing a paper on a word processor. You have been instructed to use $\dfrac{5}{8}$-inch margins. To change margins in the word processing software, you must enter margin widths in inches as a decimal. What number should you enter?

88. Suppose that you are using a word processor to format a table for a presentation. The first column of the table must be $3\dfrac{7}{16}$ inches wide. To adjust the width of the column in the word processing software, you must enter the column width in inches as a decimal. What number should you enter?

89. Describe how to determine the larger of two fractions.

77. $\dfrac{1}{81}$

78. $\dfrac{64}{125}$

79. $\dfrac{9}{25}$

80. $\dfrac{49}{4}$

81. $\dfrac{5}{2}$

82. $\dfrac{3}{2}$

83. 0.243

84. 0.13

85. 7300

86. 0.708

87. 0.625

88. 3.4375

89. answers may vary

CHAPTER 4 ACTIVITY
MAINTAINING A CHECKING ACCOUNT

This activity may be completed by working in groups or individually.

A checking account is a convenient way of handling money and paying bills. To open a checking account, the bank or savings and loan association requires a customer to make a deposit. Then the customer receives a checkbook which contains checks, deposit slips, and a checkbook register for recording checks written and deposits made. It is important to record all payments and deposits that affect the account. It is also important to keep the checkbook balance current by subtracting checks written and adding deposits made.

About once a month checking customers receive a statement from the bank listing all activity that the account has had in the last month. The statement lists a beginning balance, all checks and deposits, any service charges made against the account, and an ending balance. Because it may take several days for checks that a customer has written to clear the banking system, the check register may list checks that do not appear on the monthly bank statement. These checks are called **outstanding checks**. Deposits that are recorded in the check register but do not appear on the statement are called **deposits in transit**. Because of these differences, it is important to balance or reconcile the checkbook against the monthly statement. The steps for doing so are listed at the right.

BALANCING OR RECONCILING A CHECKBOOK

Step 1. Place a check mark in the checkbook register by each check and deposit listed on the monthly bank statement. Any entries in the register without a check mark are outstanding checks or deposits in transit.

Step 2. Find the ending checkbook register balance and add to it any outstanding checks and any interest paid on the account.

Step 3. From the total in Step 2, subtract any deposits in transit and any service charges.

Step 4. Compare the amount found in Step 3 with the ending balance listed on the bank statement. If they are the same, the checkbook balances with the bank statement. Be sure to update the check register with service charges and interest.

Step 5. If the checkbook does not balance, recheck the balancing process. Next, make sure that the running checkbook register balance was calculated correctly. Finally, compare the checkbook register with the statement to make sure that each check was recorded for the correct amount.

For the checkbook register and monthly bank statement given
a. *update the checkbook register*   **b.** *list the outstanding checks and deposits in transit*
c. *balance the checkbook—be sure to update the register with any interest or service fees.*

#	Date	Description	Payment	✓	Deposit	Balance
		Checkbook Register				425.86
114	4/1	Market Basket	30.27	✓		395.59
115	4/3	May's Texaco	8.50	✓		387.09
	4/4	Cash at ATM	50.00	✓		337.09
116	4/6	UNO Bookstore	121.38			215.71
	4/7	Deposit		✓	100.00	315.71
117	4/9	MasterCard	84.16	✓		231.55
118	4/10	Blockbuster	6.12			225.43
119	4/12	Kroger	18.72	✓		206.71
120	4/14	Parking sticker	18.50			188.21
	4/15	Direct deposit		✓	294.36	482.57
121	4/20	Rent	395.00	✓		87.57
122	4/25	Student fees	20.00			67.57
	4/28	Deposit			75.00	142.57

FIRST NATIONAL BANK
Monthly Statement 4/30

BEGINNING BALANCE:		425.86
Date	Number	Amount

CHECKS AND ATM WITHDRAWALS

4/3	114	30.27
4/4	ATM	50.00
4/11	117	84.16
4/13	115	8.50
4/15	119	18.72
4/22	121	395.00

DEPOSITS

4/7		100.00
4/15	Direct deposit	294.36

SERVICE CHARGES

Low balance fee	7.50

INTEREST

Credited 4/30	1.15
ENDING BALANCE:	227.22

Outstanding Checks: #116, #118, #120, #122; Total = $166.00
Deposits in Transit: $75.00
Balance: $142.57 + $166.00 + $1.15 − $75.00 − $7.5 = $227.22

CHAPTER 4 HIGHLIGHTS

DEFINITIONS AND CONCEPTS	EXAMPLES

SECTION 4.1   INTRODUCTION TO DECIMALS

PLACE-VALUE CHART

hundreds	tens	ones	.	tenths	hundredths	thousandths	ten-thousandths	hundred-thousandths
		4	↑	2	6	5		
100	10	1	decimal point	$\frac{1}{10}$	$\frac{1}{100}$	$\frac{1}{1000}$	$\frac{1}{10,000}$	$\frac{1}{100,000}$

4.265 means

$$4 \cdot 1 + 2 \cdot \frac{1}{10} + 6 \cdot \frac{1}{100} + 5 \cdot \frac{1}{1000}$$

or

$$4 + \frac{2}{10} + \frac{6}{100} + \frac{5}{1000}$$

WRITING (OR READING) A DECIMAL IN WORDS

Step 1.   Write the whole-number part in words.
Step 2.   Write "and" for the decimal point.
Step 3.   Write the decimal part in words as though it were a whole number, followed by the place value of the last digit.

A decimal written in words can be written in standard form by reversing the above procedure.

Write 3.08 in words.
Three and eight hundredths

Write "four and twenty-one thousandths" in standard form.

4.021

SECTION 4.2   ORDER AND ROUNDING

To compare decimals, compare digits in the same place from left to right. Where two digits are not equal, the number with the larger digit is the larger decimal.

$$3.0261 < 3.0186 \text{ because}$$
$$\uparrow \qquad \uparrow$$
$$2 \quad < \quad 1$$

TO ROUND DECIMALS TO A PLACE VALUE TO THE RIGHT OF THE DECIMAL POINT.

Step 1.   Locate the digit to the right of the given place value.
Step 2.   If this digit is 5 or greater, add 1 to the digit in the given place value and drop all digits to its right.
Step 3.   If this digit is less than 5, drop all digits to the right of the given place value.

Round 86.1256 to the nearest hundredth.

hundredths place

86.12⑤6

5 or greater

rounds to 86.13

SECTION 4.3   ADDING AND SUBTRACTING DECIMALS

TO ADD OR SUBTRACT DECIMALS

Step 1.   Write the decimals so that the decimal points line up vertically.
Step 2.   Add or subtract as for whole numbers.
Step 3.   Place the decimal point in the sum or difference so that it lines up vertically with the decimal points in the problem.

Add: 4.6 + 0.28

```
  4.60
+ 0.28
------
  4.88
```

Subtract: 2.8 − 1.04

```
    7 10
  2.8̸0̸
− 1.04
------
  1.76
```

SECTION 4.4   MULTIPLYING DECIMALS

TO MULTIPLY DECIMALS

Step 1. Multiply the decimals as though they are whole numbers.
Step 2. The decimal point in the product is placed so that the number of decimal places in the product is equal to the **sum** of the number of decimal places in the factors.

The circumference of a circle is the distance around the circle.

$C = \pi \cdot$ diameter or
$C = 2 \cdot \pi \cdot$ radius

where $\pi \approx 3.14$ or $\pi \approx \frac{22}{7}$.

Multiply 1.48×5.9

$$
\begin{array}{r}
1.4\,8 \quad \leftarrow 2 \text{ decimal places.}\\
\times\ 5.9 \quad \leftarrow 1 \text{ decimal place.}\\
\hline
1\,3\,3\,2\\
7\,4\,0\,0\\
\hline
8.7\,3\,2 \quad \leftarrow 3 \text{ decimal places}
\end{array}
$$

Find the exact circumference and an approximation by using 3.14 for π.

$C = 2 \cdot \pi \cdot$ radius
$= 2 \cdot \pi \cdot 5$
$= 10\pi$
$\approx 10(3.14)$
$= 31.4$

The circumference is exactly 10π miles and approximately 31.4 miles.

SECTION 4.5   DIVIDING DECIMALS

TO DIVIDE DECIMALS

Step 1. Move the decimal point in the divisor to the right until the divisor is a whole number.
Step 2. Move the decimal point in the dividend to the right the **same number of places** as the decimal point was moved in Step 1.
Step 3. Divide. The decimal point in the quotient is directly over the moved decimal point in the dividend.

Divide: $1.118 \div 2.6$

$$
\begin{array}{r}
0.43\\
2.6\overline{)1.118}\\
-1\,04\\
\hline
78\\
-78\\
\hline
0
\end{array}
$$

SECTION 4.6   ESTIMATING AND ORDER OF OPERATIONS

ORDER OF OPERATIONS

1. Do all operations within grouping symbols such as parentheses or brackets.
2. Evaluate any expressions with exponents.
3. Multiply or divide in order from left to right.
4. Add or subtract in order from left to right.

Simplify.

$1.9(12.8 - 4.1) = 1.9(8.7)$ Subtract.
$\qquad\qquad\qquad = 16.53$ Multiply.

SECTION 4.7 FRACTIONS AND DECIMALS	
To write fractions as decimals, divide the numerator by the denominator.	Write $\frac{3}{8}$ as a decimal. $$\begin{array}{r} 0.375 \\ 8\overline{)3.000} \\ -2\,4 \\ \hline 60 \\ -56 \\ \hline 40 \\ -40 \\ \hline 0 \end{array}$$

CHAPTER 4 REVIEW

(4.1) *Determine the place value of the digit 4 in each decimal.*

1. 23.45   tenths

2. 0.000345   hundred-thousandths

Write each decimal in words.

3. 23.45   twenty-three and forty-five hundredths

4. 0.00345   three hundred forty-five hundred-thousandths

5. 109.23   one hundred nine and twenty-three hundredths

6. 200.000032   two hundred and thirty-two millionths

Write each decimal in standard form.

7. Two and fifteen hundredths   2.15

8. Five hundred three and one hundred two thousandths   503.102

9. Sixteen thousand twenty-five and fourteen ten-thousandths   16,025.0014

Write the decimal as a fraction or a mixed number.

10. 0.16 $\dfrac{4}{25}$

11. 12.023 $12\dfrac{23}{1000}$

12. 1.0045 $1\dfrac{9}{2000}$

13. 0.00231 $\dfrac{231}{100,000}$

14. 25.25 $25\dfrac{1}{4}$

Write each fraction as a decimal.

15. $\dfrac{9}{10}$ 0.9

16. $\dfrac{25}{100}$ 0.25

17. $\dfrac{45}{1000}$ 0.045

18. $\dfrac{7}{100}$ 0.07

(4.2) *Insert* $<$, $>$, *or* $=$ *to make a true statement.*

19. 0.49   0.43 $>$

20. 0.973   0.9730 $=$

21. 402.00032   402.000032 $>$

22. 0.230505   0.23505 $<$

Round each decimal to the given place value.

23. 0.623, nearest tenth   0.6

24. 0.9384, nearest hundredth   0.94

25. 42.895, nearest hundredth   42.90

26. 16.34925, nearest thousandth   16.349

Round each money amount to the nearest cent or dollar, as indicated.

27. $0.259, nearest cent $0.26

28. $12.461, nearest cent $12.46

29. $123.46, nearest dollar $123.00

30. $3,645.52, nearest dollar $3,646.00

31. Every day in America an average of 13,490.5 people get married. Round this number to the nearest hundred.   13,500 people

32. A certain kind of chocolate candy bar contains 10.75 teaspoons of sugar. Convert this number to a mixed number. $10\frac{3}{4}$ teaspoons

(4.3) *Add or subtract as indicated.*

33. 2.4 + 7.1   9.5

34. 3.9 + 1.2   5.1

35. 4.9 − 3.2   1.7

36. 5.23 − 2.74   2.49

37. 6.4 + 0.88   7.28

38. 19.02 + 6.98   26

39. 200.49 + 16.82 + 103.002   320.312

40. 0.00236 + 100.45 + 48.29   148.74236

Subtract.

41. 892.1 − 432.4   459.7

42. 100.342 − 0.064   100.278

43. Subtract 34.98 from 100.   65.02

44. Subtract 10.02 from 200.   189.98

45. The price of oil was $17.02 per barrel on October 23. It was $17.46 on October 24. Find by how much the price of oil increased from the 23rd to the 24th. $0.44

46. Find the perimeter.   22.2 inches

6.2 inches

Rectangle   4.9 inches

(4.4) *Multiply.*

47. 7.2 × 10   72

48. 9.345 × 1000   9345

49.   34.02   78.246
        × 2.3

50.   839.02   73,246.446
        × 87.3

51. Find the exact circumference of the circle. Then use the approximation 3.14 for π and approximate the circumference.   14π meters; 43.96 meters

7 meters

52. A kilometer is approximately 0.625 mile. It is 102 kilometers from Hays to Colby. Write 102 kilometers in miles to the nearest tenth mile.
63.8 miles

(4.5) *Divide. Round the quotient to the nearest thousandth if necessary.*

53. 21 ÷ 0.3   70

54. 0.0063 ÷ 0.03   0.21

55. 0.005)‾24.5   4900

56. 2.3)‾54.98   23.904

57. 0.34)‾2.74   8.059

58. 20)‾316.5   15.825

59. $\frac{2.67}{100}$ 0.0267

60. $\frac{93}{10}$ 9.3

61. There are approximately 3.28 feet in 1 meter. Find how many meters there are in 24 feet to the nearest tenth of a meter.   7.3 meters

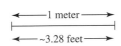

←———1 meter———→
←——~3.28 feet——→

62. George Strait pays $69.71 per month to pay back a loan of $3136.95. In how many months will the loan be paid off?   45 months

(4.6) *Perform each indicated operation. Then estimate to see whether each proposed result is reasonable.*

63. 2.4 + 6.7 + 9.1   18.2

64. 15.9 + 34.1   50

65. 340.03 − 240.98   99.05

66. 100 − 45.9   54.1

67. 6.02 × 5.91   35.5782

68. 0.205 × 1.72   0.3526

69. 62.13 ÷ 1.9   32.7

70. 601.92 ÷ 19.8   30.4

71. Tomaso is going to fertilize his lawn, a rectangle measuring 77.3 feet by 115.9 feet. Approximate the area of the lawn by rounding each measurement to the nearest foot.　8932 square feet

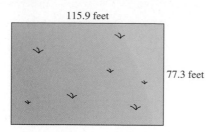

115.9 feet

77.3 feet

72. Estimate the cost of the items to see whether the groceries can be purchased with a $5 bill.　Yes, $5 is enough.

$1.07　　$1.89

3 cans for $0.99

Simplify each expression.

73. $7.6 \times 1.9 + 2.5$　16.94

74. $2.3^2 - 1.4$　3.89

75. $\dfrac{(3.2)^2}{100}$　0.1024

76. $(2.6 + 1.4)(4.5 - 3.6)$　3.6

(4.7) *Write each fraction as a decimal. Round to the nearest thousandth if necessary.*

77. $\dfrac{4}{5}$　0.8

78. $\dfrac{12}{13}$　0.923

79. $\dfrac{3}{7}$　0.429

80. $\dfrac{13}{60}$　$0.21\overline{6}$ or 0.217 rounded

81. $\dfrac{9}{80}$　0.1125 or 0.113 rounded

82. $\dfrac{8935}{175}$　51.057

Insert $<$, $>$, or $=$ to make a true statement.

83. $0.392 \quad 0.392 \quad =$

84. $\dfrac{4}{7} \quad \dfrac{5}{8} \quad <$

85. $0.293 \quad \dfrac{5}{17} \quad <$

86. $\dfrac{6}{11} \quad 0.55 \quad <$

Write the numbers in order from smallest to largest.

87. 0.837, 0.839, 0.832　0.832, 0.837, 0.839

88. $\dfrac{3}{7}, 0.42, 0.43$　$0.42, \dfrac{3}{7}, 0.43$

89. $\dfrac{18}{11}, 1.63, \dfrac{19}{12}$　$\dfrac{19}{12}, 1.63, \dfrac{18}{11}$

90. $\dfrac{6}{7}, \dfrac{8}{9}, \dfrac{3}{4}$　$\dfrac{3}{4}, \dfrac{6}{7}, \dfrac{8}{9}$

Find each area.

91.

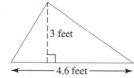

3 feet

4.6 feet

6.9 square feet

92.

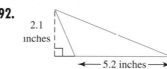

2.1 inches

5.2 inches

5.46 square inches

282

Name _____ **Section** _____ **Date** _____

CHAPTER 4 TEST

Write the decimal as indicated.

1. 45.092, in words

2. Three thousand and fifty-nine thousandths, in standard form.

Round the decimal to the indicated place value.

3. 34.8923, nearest tenth

4. 0.8623, nearest thousandth

Insert $<$, $>$, *or* $=$ *to make a true statement.*

5. 25.0909   25.9090

6. $\dfrac{4}{9}$ 0.445

Write the decimal as a fraction or a mixed number.

7. 0.345

8. 24.73

Write the fraction as a decimal. If necessary, round to the nearest thousandths place.

9. $\dfrac{13}{26}$

10. $\dfrac{16}{17}$

Perform the indicated operations. Round the result to the nearest thousandth if necessary.

11. $2.893 + 4.2 + 10.49$

12. Subtract 8.6 from 20.

13. $\begin{array}{r} 10.2 \\ \times\ 4.3 \\ \hline \end{array}$

14. $0.23\overline{)12.88}$

15. $7\overline{)46.71}$

16. 126.9×100

17. $\dfrac{473}{10}$

283

Name _____

18. $1.57 - (0.6)^2$

19. $\dfrac{0.23 + 1.63}{0.3}$

Find the area.

20.

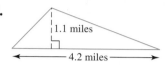

21. Vivian Thomas is going to put insecticide on her lawn to control grubworms. The lawn is a rectangle measuring 123.8 feet by 80 feet. The amount of insecticide required is 0.02 ounces per square foot. Find how much insecticide Vivian needs to purchase.

22. Find the exact circumference of the circle. Then use the approximation 3.14 for π and approximate the circumference.

23. A Superdisk diskette holds 120 megabytes of data. A high-density diskette holds 1.4 megabytes of data. How many high-density diskettes does it take to hold as much data as a Superdisk diskette? Round to the nearest whole diskette.

24. Estimate the total distance from Bayette to Center City by rounding each distance to the nearest mile.

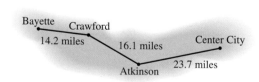

Name _____ **Section** _____ **Date** _____

CUMULATIVE REVIEW

1. Write 106,052,447 in words.

2. Alan Mayfield collects baseball cards. He has 109 cards for New York Yankees, 96 for Chicago White Sox, 79 for Kansas City Royals, 42 for Seattle Mariners, 67 for Oakland Athletics, and 52 for California Angels. How many cards does he have in total?

3. Subtract $900 - 174$. Then check by adding.

4. Round each number to the nearest hundred to find an estimated sum.

$$\begin{array}{r} 294 \\ 625 \\ 1071 \\ + \ 349 \\ \hline \end{array}$$

5. A single-density computer disk can hold 370 thousand bytes of information. How many total bytes can 42 such disks hold?

6. Divide: $6819 \div 17$

7. Simplify: $4^3 + \left[3^2 \quad (10 : 2) \right] - 7 \cdot 3$

8. Identify the numerator and the denominator. $\frac{3}{7}$

9. Write $\frac{6}{60}$ in simplest form.

10. Multiply: $\frac{3}{4} \cdot 20$

11. Divide: $\frac{7}{8} \div \frac{2}{9}$

12. Multiply: $1\frac{2}{3} \cdot 2\frac{1}{4}$

13. Divide: $\frac{3}{4} \div 5$

Subtract and simplify.

14. $\frac{8}{9} - \frac{1}{9}$

15. $\frac{7}{8} - \frac{5}{8}$

Name _____

16. Write an equivalent fraction with the indicated denominator. $\frac{3}{4} = \frac{}{20}$

17. Add: $\frac{2}{15} + \frac{3}{10}$

18. Sarah Grahamm purchases two packages of ground round. One package weighs $2\frac{3}{8}$ pounds and the other $1\frac{4}{5}$ pounds. What is the combined weight of the ground round?

Evaluate each expression.

19. $\left(\frac{1}{4}\right)^2$

20. $\left(\frac{1}{6}\right)^2 \cdot \left(\frac{3}{4}\right)^3$

21. Write 0.43 as a fraction.

22. Insert $<$, $>$, or $=$ to form a true statement.

0.378 ____ 0.368

23. Subtract: $35.218 - 23.65$.

Multiply.

24. 23.702×100

25. $76,805 \times 0.01$

Ratio and Proportion

Having studied fractions in Chapters 2 and 3, we are ready to explore using fractions to express a relationship between two quantities. We often make comparisons between like quantities in real life. When we say there is a 2 to 1 ratio of men to women at a party, we mean that for every two men, there is one woman. When we find gas mileage as miles per gallon, or compare two unlike quantities, we are finding a rate. We can apply both of these concepts when finding a proportion. A proportion compares two equal rates or ratios. Once we know how far a car can go on one gallon of gasoline (a rate), we can use a proportion to compute the number of gallons of gasoline needed for a 200-mile trip.

Lawn care is a more than $25-billion-per-year business. About one in five households in the United States hire professional services to care for and landscape their lawns. The sales of lawn care products alone account for roughly one-third of all money spent on gardening in this country. Why is there so much interest in maintaining lawns? Lawns have many benefits. They are aesthetically pleasing. They provide a cushioned play area for children. They can increase the selling price of a home. However, one of the most important benefits of lawns is their environmental impact. Turf grasses help clean the air with their oxygen production and also help control pollution. In Exercise 27 on page 298 and Exercise 27 on page 316, we will use rates and proportions to compute the effect lawns have on trapping dust and on oxygen production.

1. $\frac{27}{8}$; (5.1A)

2. $\frac{4}{37}$; (5.1A)

3. $\frac{5}{12}$; (5.1B)

4. $\frac{4}{3}$; (5.1B)

5. $\frac{6}{19}$; (5.1B)

6. $\frac{4}{11}$; (5.1B)

7. $\frac{35 \text{ students}}{2 \text{ instructors}}$; (5.2A)

8. $\frac{7 \text{ cups of flour}}{2 \text{ cakes}}$; (5.2A)

9. $\frac{78.75 \text{ kilometers}}{1 \text{ hour}}$ or 78.75 km/hr; (5.2B)

10. $\frac{3 \text{ cookies}}{1 \text{ child}}$ or 3 cookies/child; (5.2B)

11. 12 oz; (5.2C)

12. 12 doughnuts; (5.2C)

13. $\frac{5}{80} = \frac{15}{240}$; (5.3A)

14. false; (5.3B)

15. true; (5.3B)

16. $n = 16$; (5.3C)

17. $n = 4.5$; (5.3C)

18. $n = 133$; (5.3C)

19. $12,000; (5.4A)

20. $472.75; (5.4A)

288

Name _____ Section _____ Date _____

CHAPTER 5 PRETEST

Write each ratio as a fraction in simplest form.

1. 27 to 8

2. 4 to 37

3. $2\frac{1}{2}$ to 6

4. 8.4 to 6.3

5. 12 pints to 38 pints

6. 200 years to 550 years

Write each rate as a fraction in simplest form.

7. 140 students for 8 instructors

8. 21 cups of flour for 6 cakes

Write each phrase as a unit rate.

9. 945 kilometers in 12 hours

10. 780 cookies for 260 children

Compare the unit prices and decide which is the better buy.

11. Coffee:
8 ounces for $3.69
12 ounces for $5.25

12. Doughnuts:
12 doughnuts for $4.20
30 doughnuts for $10.80

13. Write the following sentence as a proportion. 5 incorrect answers is to 80 math problems as 15 incorrect answers is to 240 math problems.

Determine whether the proportions are true or false.

14. $\frac{8}{22} = \frac{24}{72}$

15. $\frac{4.8}{6.4} = \frac{3}{4}$

Find the unknown number n.

16. $\frac{n}{4} = \frac{28}{7}$

17. $\frac{3}{74} = \frac{n}{111}$

18. $\frac{9\frac{1}{2}}{n} = \frac{3\frac{3}{7}}{6}$

Personal property taxes on a vehicle are figured at a rate of $3.05 tax for every $100 of assessed vehicle value.

19. If Tony pays $366 in personal property taxes for his car, find the assessed value of his car.

20. Find the personal property taxes on a truck with an assessed value of $15,500.

5.1 Ratios

A WRITING RATIOS AS FRACTIONS

A **ratio** is the quotient of two quantities. A ratio, in fact, is no different than a fraction, except that a ratio is sometimes written using notation other than fractional notation. For example, the ratio of 1 to 2 can be written as

$$1 \text{ to } 2 \quad \text{or} \quad \frac{1}{2} \quad \text{or} \quad 1 : 2$$

fractional notation   colon notation

These ratios are all read as "the ratio of 1 to 2."

TRY THE CONCEPT CHECK IN THE MARGIN.

In this section, we will write ratios using fractional notation.

> **WRITING A RATIO AS A FRACTION**
>
> The order of the quantities is important when writing ratios. To write a ratio as a fraction, write the *first number* of the ratio as the *numerator* of the fraction and the *second number* as the *denominator*.

For example, the ratio of 6 to 11 is $\frac{6}{11}$, *not* $\frac{11}{6}$.

Example 1

Write the ratio of 12 to 17 using fractional notation.

Solution: The ratio is $\frac{12}{17}$.

> **HELPFUL HINT**
>
> Don't forget that order is important when writing ratios. The ratio $\frac{17}{12}$ is *not* the same as the ratio $\frac{12}{17}$.

Examples

Write each ratio using fractional notation.

2. The ratio of 2.6 to 3.1 is $\frac{2.6}{3.1}$.

3. The ratio of $1\frac{1}{2}$ to $7\frac{3}{4}$ is $\dfrac{1\frac{1}{2}}{7\frac{3}{4}}$.

B SIMPLIFYING RATIOS

To simplify a ratio, we just write the fraction in simplest form. Common factors can be divided out as well as common units.

Example 4

Write the ratio of $15 to $10 as a fraction in simplest form.

Solution:

$$\frac{\$15}{\$10} = \frac{15}{10} = \frac{3 \cdot \cancel{5}}{2 \cdot \cancel{5}} = \frac{3}{2}$$

Objectives

A Write ratios as fractions.
B Write ratios in simplest form.

SSM   CD-ROM   Video
5.1

✓ CONCEPT CHECK

How should each ratio be read aloud?

a. $\frac{8}{5}$ b. $\frac{5}{8}$

TEACHING TIP
While discussing Example 1, have students come up with several situations which would have a ratio of 12 to 7. Use the following situation to get them started. This is the ratio of female students to male students in a class having 12 female students and 7 male students. Then ask, "Are there any other numbers of female students and male students in a class which would have the ratio 12 to 7?"

Practice Problem 1

Write the ratio of 20 to 23 using fractional notation.

TEACHING TIP
While discussing Example 2, try putting the ratio in context for students. Miles are often measured in tenths. This ratio could represent the ratio of miles run to miles walked if I ran 2.6 miles and walked 3.1 miles. What else could it represent?

Practice Problems 2–3

Write each ratio using fractional notation.

2. The ratio of 10.3 to 15.1

3. The ratio of $3\frac{1}{3}$ to $12\frac{1}{5}$

Practice Problem 4

Write the ratio of $8 to $6 as a fraction in simplest form.

Answers

1. $\frac{20}{23}$, 2. $\frac{10.3}{15.1}$, 3. $\dfrac{3\frac{1}{3}}{12\frac{1}{5}}$, 4. $\frac{4}{3}$

✓ **Concept Check**
a. Eight to five, **b.** Five to eight

TEACHING TIP

While discussing Example 3, have students work in groups to come up with several situations which would have this ratio. Then have them share their situations with the other groups in the class. To help them get started, ask them what is commonly measured in mixed numbers with fractional parts in halves or fourths?

Practice Problem 5

Write the ratio of 1.71 to 4.56 as a fraction in simplest form.

Practice Problem 6

Use the circle graph below to write the ratio of work miles to total miles as a fraction in simplest form.

(*Source:* The American Automobile Manufacturers Assn. and The National Automobile Dealers Assn.)

Practice Problem 7

Given the triangle shown:

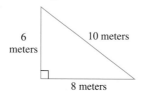

a. Find the ratio of the length of the shortest side to the length of the longest side.

b. Find the ratio of the length of the longest side to the perimeter of the triangle.

✓ CONCEPT CHECK

Explain why the answer $\frac{7}{5}$ would be incorrect for part (a) of Example 7.

Answers

5. $\frac{3}{8}$,   **6.** $\frac{8}{25}$,   **7. a.** $\frac{3}{5}$,   **b.** $\frac{5}{12}$

✓ **Concept Check:** $\frac{7}{5}$ would be the ratio of the rectangle's length to its width.

In the example above, although the fraction $\frac{3}{2}$ is equal to the mixed number $1\frac{1}{2}$, a ratio is a quotient of *two* quantities. For that reason, ratios are not written as mixed numbers.

If a ratio compares two decimal numbers, we will write the ratio as a ratio of whole numbers.

Example 5

Write the ratio of 2.6 to 3.1 as a fraction in simplest form.

Solution: The ratio is

$$\frac{2.6}{3.1}$$

Now let's clear the ratio of decimals.

$$\frac{2.6}{3.1} = \frac{2.6 \cdot 10}{3.1 \cdot 10} = \frac{26}{31} \quad \text{Simplest form}$$

Example 6   Writing a Ratio from a Circle Graph

The circle graph in the margin shows the part of a car's total mileage that falls into a particular category. Write the ratio of medical miles to total miles as a fraction in simplest form.

Solution:

$$\frac{\text{medical miles}}{\text{total miles}} = \frac{150 \text{ miles}}{15,000 \text{ miles}} = \frac{150}{15,000} = \frac{\overset{1}{\cancel{150}}}{\underset{1}{\cancel{150}} \cdot 100} = \frac{1}{100}$$

Example 7

Given the rectangle shown:

a. Find the ratio of its width to its length.

b. Find the ratio of its length to its perimeter.

Solution:

a. The ratio of its width to its length is

$$\frac{\text{width}}{\text{length}} = \frac{5 \text{ feet}}{7 \text{ feet}} = \frac{5}{7}$$

b. Recall that the perimeter of the rectangle is the distance around the rectangle: $7 + 5 + 7 + 5 = 24$ feet. The ratio of its length to its perimeter is

$$\frac{\text{length}}{\text{perimeter}} = \frac{7 \text{ feet}}{24 \text{ feet}} = \frac{7}{24}$$

TRY THE CONCEPT CHECK IN THE MARGIN.

EXERCISE SET 5.1

A *Write each ratio using fractional notation. Do not simplify. See Examples 1 through 3.*

1. 11 to 14   **2.** 7 to 12   ▄ **3.** 23 to 10   **4.** 8 to 5

5. 151 to 201   **6.** 673 to 1000   **7.** 2.8 to 7.6   **8.** 3.9 to 4.2

9. 5 to $7\frac{1}{2}$ **10.** $5\frac{3}{4}$ to 3   **11.** $3\frac{3}{4}$ to $1\frac{2}{3}$ **12.** $2\frac{2}{5}$ to $6\frac{1}{2}$

B *Write each ratio as a ratio of whole numbers using fractional notation. Write the fraction in simplest form. See Examples 4 and 5.*

▄ **13.** 16 to 24   **14.** 25 to 150   ▄ **15.** 7.7 to 10

16. 8.1 to 10   **17.** 4.63 to 8.21   **18.** 9.61 to 7.62

19. 9 inches to 12 inches   **20.** 14 centimeters to 20 centimeters

21. 10 hours to 24 hours   **22.** 18 quarts to 30 quarts

23. $32 to $100   **24.** $46 to $102

▄ **25.** 24 days to 14 days   **26.** 80 miles to 120 miles

ANSWERS

1. $\frac{11}{14}$

2. $\frac{7}{12}$

3. $\frac{23}{10}$

4. $\frac{8}{5}$

5. $\frac{151}{201}$

6. $\frac{673}{1000}$

7. $\frac{2.8}{7.6}$

8. $\frac{3.9}{4.2}$

9. $\frac{5}{7\frac{1}{2}}$

10. $\frac{5\frac{3}{4}}{3}$

11. $\frac{3\frac{3}{4}}{1\frac{2}{3}}$

12. $\frac{2\frac{2}{5}}{6\frac{1}{2}}$

13. $\frac{2}{3}$

14. $\frac{1}{6}$

15. $\frac{77}{100}$

16. $\frac{81}{100}$

17. $\frac{463}{821}$

18. $\frac{961}{762}$

19. $\frac{3}{4}$

20. $\frac{7}{10}$

21. $\frac{5}{12}$

22. $\frac{3}{5}$

23. $\frac{8}{25}$

24. $\frac{23}{51}$

25. $\frac{12}{7}$

26. $\frac{2}{3}$

27. $\frac{16}{23}$ _____

28. $\frac{4}{1}$ _____

29. $\frac{2}{5}$ _____

30. $\frac{9}{2}$ _____

31. $\frac{5}{3}$ _____

32. $\frac{1}{3}$ _____

33. $\frac{17}{40}$ _____

34. $\frac{2}{13}$ _____

35. $\frac{5}{4}$ _____

36. $\frac{4}{9}$ _____

37. $\frac{3}{1}$ _____

38. $\frac{1}{4}$ _____

292

Name _____

27. 32,000 bytes to 46,000 bytes

28. 600 copies to 150 copies

29. 8 inches to 20 inches

30. 9 yards to 2 yards

Find the ratio described in each problem. See Example 7.

31. Find the ratio of the length to the width of the swimming pool.

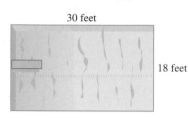

30 feet

18 feet

32. Find the ratio of the base to the height of the triangular mainsail.

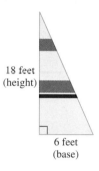

18 feet (height)

6 feet (base)

33. Find the ratio of the longest side to the perimeter of the right-triangular-shaped billboard.

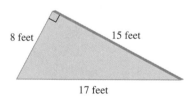

8 feet   15 feet

17 feet

34. Find the ratio of the width to the perimeter of the rectangular vegetable garden.

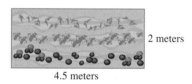

2 meters

4.5 meters

At the Honey Island College Glee Club meeting one night, there were 125 women and 100 men present.

35. Find the ratio of women to men.

36. Find the ratio of men to the total number of people present.

A poll at State University revealed that 4500 students out of 6000 students are single, and the rest are married.

37. Find the ratio of single students to married students.

38. Find the ratio of married students to the total student population.

Name _____

Blood contains three types of cells: red blood cells, white blood cells, and platelets. For approximately every 600 red blood cells in healthy humans, there are 40 platelets and 1 white blood cell. (*Source:* American Red Cross Biomedical Services)

39. Write the ratio of red blood cells to platelet cells.

40. Write the ratio of white blood cells to red blood cells.

41. When the first national minimum wage was enacted in 1938, minimum wage was $0.25 per hour. In 1997, minimum wage was increased to $5.15 per hour. Find the ratio of the 1997 minimum wage to the 1938 minimum wage. (*Source:* U.S. Department of Labor)

42. Citizens of the United States eat an average of 25 pints of ice cream per year. Residents of the New England states eat an average of 39 pints of ice cream per year. Find the ratio of the amount of ice cream eaten by New Englanders to the amount eaten by the average U.S. citizen. (*Source:* International Dairy Foods Association)

43. At the Winter Olympic Games in Nagano, Japan, a total of 205 medals were awarded. Norwegian athletes won a total of 25 medals. Find the ratio of medals won by Norway to the total medals awarded. (*Source:* Organizing Committee for the XVIII Olympic Winter Games, Nagano)

44. At the Winter Olympic Games in Nagano, Japan, a total of 205 medals were awarded. Japanese athletes won a total of 10 medals. Find the ratio of medals won by Japan to the total medals awarded. (*Source:* Organizing Committee for the XVIII Olympic Winter Games, Nagano)

REVIEW AND PREVIEW

Divide. See Section 4.5.

45. $9\overline{)20.7}$ **46.** $7\overline{)60.2}$ **47.** $3.7\overline{)0.555}$ **48.** $4.6\overline{)1.15}$

39. $\dfrac{15}{1}$ _____

40. $\dfrac{1}{600}$ _____

41. $\dfrac{103}{5}$ _____

42. $\dfrac{39}{25}$ _____

43. $\dfrac{5}{41}$ _____

44. $\dfrac{2}{41}$ _____

45. 2.3 _____

46. 8.6 _____

47. 0.15 _____

48. 0.25 _____

49. $\frac{3}{7}$

50. $\frac{17}{33}$

51. $\frac{10}{23}$

52. yes, the machine should be repaired

53. no, the shipment should not be refused

54. $\frac{1}{87}$; $\frac{1}{160}$; answers may vary

55. $\frac{1}{48}$; $\frac{1}{64}$; answers may vary

◆ **COMBINING CONCEPTS**

49. As of March 1988, statewide mandatory bicycle helmet laws had been adopted by 15 states. Find the ratio of states with bicycle helmet laws to states with no such laws. (*Source:* Bicycle Helmet Safety Institute)

50. A total of 17 states have 200 or more public libraries. Find the ratio of states with 200 or more public libraries to states with fewer than 200 public libraries. (*Source:* U.S. Department of Education)

51. Write the ratio $2\frac{1}{2}$ to $5\frac{3}{4}$ as a fraction in simplest form.

52. A pantyhose manufacturing machine will be repaired if the ratio of defective pantyhose to good pantyhose is at least 1 to 20. A quality-control engineer found 10 defective pantyhose in a batch of 200. Determine whether the machine should be repaired.

53. A grocer will refuse a shipment of tomatoes if the ratio of bruised tomatoes to the total batch is at least 1 to 10. A sample is found to contain 3 bruised tomatoes and 30 good tomatoes. Determine whether the shipment should be refused.

Internet Excursions

Go to http://www.prenhall.com/martin-gay

A scale model of an object is one in which there is a particular ratio between each measurement on the model and the corresponding measurement on the object being modeled. For instance, model trains are scale models of actual trains. Model train enthusiasts know that model trains are available in several different scales. By going to the World Wide Web site listed above, you will be directed to the Lionel Glossary in the Lionel model trains Web site, or a related site, where you can look up information to help you answer the questions below.

54. The two most popular model train scales are the HO scale and the N scale. What ratio of measurements is used in HO scale? What ratio of measurements is used in N scale? Describe what each of these means and give an example.

55. Other model train scales include the O scale and the S scale. What ratio of measurements is used in O scale? What ratio of measurements is used in S scale? Describe what each of these means and give an example.

5.2   Rates

A   Writing Rates as Fractions

A special type of ratio is a rate. **Rates** are used to compare *different* kinds of quantities. For example, suppose that a recreational runner can run 3 miles in 33 minutes. If we write this rate as a fraction, we have

$$\frac{3 \text{ miles}}{33 \text{ minutes}} = \frac{1 \text{ mile}}{11 \text{ minutes}} \quad \text{In simplest form}$$

HELPFUL HINT

When comparing quantities with different units, write the units as part of the comparison. They do not divide out.

Same Units: $\dfrac{3 \cancel{\text{ inches}}}{12 \cancel{\text{ inches}}} = \dfrac{1}{4}$

Different Units: $\dfrac{2 \text{ miles}}{20 \text{ minutes}} = \dfrac{1 \text{ mile}}{10 \text{ minutes}}$ Units are still written.

Example 1   Write the rate as a fraction in simplest form: 10 nails every 6 feet

Solution: $\dfrac{10 \text{ nails}}{6 \text{ feet}} = \dfrac{5 \text{ nails}}{3 \text{ feet}}$

Examples   Write each rate as a fraction in simplest form.

2. $2160 for 12 weeks is $\dfrac{2160 \text{ dollars}}{12 \text{ weeks}} = \dfrac{180 \text{ dollars}}{1 \text{ week}}$

3. 360 miles on 16 gallons of gasoline is
$\dfrac{360 \text{ miles}}{16 \text{ gallons}} = \dfrac{45 \text{ miles}}{2 \text{ gallons}}$

TRY THE CONCEPT CHECK IN THE MARGIN.

B   Finding Unit Rates

A **unit rate** is a rate with a denominator of 1. A familiar example of a unit rate is 55 mph, read as "55 **miles per hour**." This means 55 miles per 1 hour or

$$\frac{55 \text{ miles}}{1 \text{ hour}} \quad \text{Denominator of 1}$$

WRITING A RATE AS A UNIT RATE

To write a rate as a unit rate, divide the numerator of the rate by the denominator.

TEACHING TIP
Before beginning this lesson, ask students to estimate each of the following quantities.

    cost of long distance phone call
    cost of movie ticket
    water needed for making pasta
    mileage per gallon of gas
    speed of a car on the highway
    apartment rent
    cost of roses

Then help students see that each estimate can only be described using two units . . . an amount per an amount. Have them convert the estimates to fractional form and find an equivalent fraction for each.

Practice Problem 1

Write the rate as a fraction in simplest form: 12 commercials every 45 minutes

Practice Problems 2–3

Write each rate as a fraction in simplest form.

2. $1680 for 8 weeks

3. 236 miles on 12 gallons of gasoline

✓ Concept Check

True or false: $\dfrac{16 \text{ gallons}}{4 \text{ gallons}}$ is a rate.
Explain.

Answers

1. $\dfrac{4 \text{ commercials}}{15 \text{ minutes}}$,   **2.** $\dfrac{210 \text{ dollars}}{1 \text{ week}}$,

3. $\dfrac{59 \text{ miles}}{3 \text{ gallons}}$

✓ **Concept Check:**   False since a rate compares different kinds of quantities.

Practice Problem 4

Write as a unit rate: 3600 feet every 12 seconds

TEACHING TIP

After discussing Example 4, point out that the unit rate was given in dollars per month. Then ask students if they could write it as a unit rate in dollars per year.

Practice Problem 5

Write as a unit rate: 52 bushels of fruit from 8 trees

TEACHING TIP

Point out that grocery stores often label the shelves with the unit prices so shoppers can easily compare prices.

   As an individual project outside of class, have students make a price comparison for different sizes of a given product. Is there any relationship between the amount of the item being purchased and the unit price?

Practice Problem 6

An automobile rental agency charges $170 for 5 days for a certain model car. What is the unit price in dollars per day?

Practice Problem 7

Approximate each unit price to decide which is the better buy for a bag of nacho chips: 11 ounces for $2.32 or 16 ounces for $3.59.

   As a group project outside of class, have groups make a price comparison of two sizes each of 6 different products at 4 different stores. How do the unit prices compare? Does one store always have the lowest unit price? Does one store always have the highest unit price?

Answers

4. $\frac{300 \text{ feet}}{1 \text{ second}}$ or 300 feet/second,

5. $\frac{6.5 \text{ bushels}}{1 \text{ tree}}$ or 6.5 bushels/tree,

6. $34 per day,   7. 11-ounce bag

Example 4   Write as a unit rate: $27,000 every 6 months

Solution: $\frac{27,000 \text{ dollars}}{6 \text{ months}}$ $6\overline{)27,000}^{\,4,500}$

The unit rate is

$\frac{4500 \text{ dollars}}{1 \text{ month}}$ or   4500 dollars/month   Read as "4500 dollars per month"

Example 5   Write as a unit rate: 318.5 miles every 13 gallons of gas

Solution: $\frac{318.5 \text{ miles}}{13 \text{ gallons}}$ $13\overline{)318.5}^{\,24.5}$

The unit rate is

$\frac{24.5 \text{ miles}}{1 \text{ gallon}}$ or   24.5 miles/gallon   Read as "24.5 miles per gallon"

C FINDING UNIT PRICES

Rates are used extensively in sports, business, medicine, and science applications. One of the most common uses of rates is in consumer economics. When a unit rate is "money per item," it is also called a **unit price**.

Example 6   Finding Unit Price

A store charges $3.36 for a 16-ounce jar of picante sauce. What is the unit price in dollars per ounce?

Solution: $\frac{\text{unit}}{\text{price}} = \frac{\$3.36}{16 \text{ ounces}} = \frac{\$0.21}{1 \text{ ounce}}$ or $0.21 per ounce

Example 7   Finding Best Buy

Approximate each unit price to decide which is the better buy: $0.99 for 4 bars of soap or $1.19 for 5 bars of soap.

Solution:

$\frac{\text{unit}}{\text{price}} = \frac{\$0.99}{4 \text{ bars}} \approx \frac{\$0.25 \text{ per bar}}{\text{of soap}}$ $4\overline{)0.990}^{\,0.247 \approx 0.25}$ ↑___ (is approximately)

$\frac{\text{unit}}{\text{price}} = \frac{\$1.19}{5 \text{ bars}} \approx \frac{\$0.24 \text{ per bar}}{\text{of soap}}$ $5\overline{)1.190}^{\,0.238 \approx 0.24}$

Thus, the 5-bar package is the better buy.

Name _____ Section _____ Date _____

Exercise Set 5.2

A *Write each rate as a fraction in simplest form. See Examples 1 through 3.*

1. 5 shrubs every 15 feet

2. 14 lab tables for 28 students

📼 3. 15 returns for 100 sales

4. 150 graduate students for 8 advisors

5. 8 phone lines for 36 employees

6. 6 laser printers for 28 computers

7. 18 gallons of pesticide for 4 acres of crops

8. 4 inches of rain in 18 hours

9. 6 flight attendants for 200 passengers

10. 240 pounds of grass seed for 9 lawns

11. 355 calories in a 10-fluid ounce chocolate milkshake (*Source: Home and Garden Bulletin No. 72*; U.S. Department of Agriculture)

12. 160 calories in an 8-fluid ounce serving of cream of tomato soup (*Source: Home and Garden Bulletin No. 72*; U.S. Department of Agriculture)

B *Write each rate as a unit rate. See Examples 4 and 5.*

13. 375 riders in 5 subway cars

14. 18 campaign yard signs in 6 blocks

📼 15. 330 calories in a 3-ounce serving

16. 275 miles in 11 hours

17. 144 diapers for 24 babies

18. 420 feet in 3 seconds

19. $1,000,000 lottery paid over 20 years

20. 400,000 library books for 8000 students

21. 600 kilometers in 90 minutes

22. 5000 registered vehicles for 2000 campus parking spaces

23. The state of Tennessee has 2,450,000 registered voters for 2 senators. (*Source*: U.S. Bureau of the Census)

24. The state of Louisiana has 4,320,000 residents for 64 parishes. (*Source*: U.S. Bureau of the Census)

Name _____

25. 12,000 good assemblyline products to 40 defective products

26. 5,000,000 lottery tickets for 4 lottery winners

27. 12 million tons of dust and dirt are trapped by the 25 million acres of lawns in the United States each year. (*Source:* Professional Lawn Care Association of America)

28. Approximately 65,000,000,000 checks are written each year by a total of approximately 260,000,000 Americans. (*Source:* Board of Governors of the Federal Reserve System)

29. The National Zoo in Washington, D.C., has an annual budget of $24,000,000 for its 500 different species. (*Source: 1998 World Almanac*)

30. On average, it costs $26,268 for 4 hours of flight in a B747-100 aircraft. (*Source:* Air Transport Association of America)

31. In Maryland, there is $123,219,000 of funding for 24 public libraries. (*Source:* U.S. Department of Education)

32. The top-grossing concert tour in North America was the 1994 Rolling Stones tour, which grossed $121,200,000 for 60 shows. (*Source:* Pollstar)

33. Greer Krieger can assemble 250 computer boards in an 8-hour shift, while Lamont Williams can assemble 400 computer boards in a 12-hour shift.
a. Find the unit rate of Greer.

b. Find the unit rate of Lamont.

c. Who can assemble computer boards faster, Greer or Lamont?

34. Jerry Stein laid 713 bricks in 46 minutes, while his associate, Bobby Burns, laid 396 bricks in 30 minutes.
a. Find the unit rate of Jerry.

b. Find the unit rate of Bobby.

c. Who is the faster bricklayer?

Name _____

C _Find each unit price. See Example 6._

35. $57.50 for 5 compact disks

36. $0.87 for 3 apples

37. $1.19 for 7 bananas

38. $73.50 for 6 lawn chairs

Find each unit price and decide which is the better buy. Round to 3 decimal places. Assume that we are comparing different sizes of the same brand. See Examples 6 and 7.

39. Crackers:
$1.19 for 8 ounces

$1.59 for 12 ounces

40. Pickles:
$1.89 for 32 ounces

$0.89 for 18 ounces

41. Frozen orange juice:
$1.69 for 16 ounces

$0.69 for 6 ounces

42. Eggs:
$0.69 per dozen

$2.10 for a flat $\left(2\frac{1}{2}\text{ dozen}\right)$

43. Soy sauce:
$2.29 for 12 ounces

$1.49 for 8 ounces

44. Shampoo.
$1.89 for 20 ounces

$3.19 for 32 ounces

45. Napkins:
100 for $0.59

180 for $0.93

46. Crackers:
$2.39 for 20 ounces

$0.99 for 8 ounces

REVIEW AND PREVIEW

Multiply or divide as indicated. See Sections 4.4 and 4.5.

47. 1.7
 $\times$ 6

48. 2.3
 $\times$ 9

49. 3.7
 $\times$1.2

35. $11.50 per compact disk

36. $0.29 per apple

37. $0.17 per banana

38. $12.25 per lawn chair

39. 8 oz: $0.149 per ounce; 12 oz: $0.133 per ounce; 12 ounces

40. 32 oz: $0.059 per ounce; 18 oz: $0.049 per ounce: 18 ounces

41. 16 oz: $0.106 per ounce: 6 oz: $0.115 per ounce; 16 ounces

42. dozen: $0.058 per egg; flat: $0.07 per egg; dozen

43. 12 oz: $0.191 per ounce; 8 oz: $0.186 per ounce; 8 ounces

44. 20 oz: $0.095 per ounce; 32 oz: $0.10 per ounce; 20 ounces

45. 100: $0.006 per napkin; 180: $0.005 per napkin; 180 napkins

46. 20 oz: $0.120 per ounce; 8 oz: $0.124 per ounce; 20 ounces

47. 10.2

48. 20.7

49. 4.44

Name _____

50. 6.6
× 2.5

51. 2.3)‾4.37‾

52. 3.5)‾22.75‾

COMBINING CONCEPTS

53. Fill in the table to calculate miles per gallon.

Beginning Odometer Reading	Ending Odometer Reading	Miles Driven	Gallons of Gas Used	Miles Per Gallon (Round to the nearest tenth)
79,286	79,543		13.4	
79,543	79,895		15.8	
79,895	80,242		16.1	

Find each unit rate.

54. Sammy Joe Wingfield from Arlington, Texas, is the fastest bricklayer on record. On May 20, 1994, he laid 1048 bricks in 60 minutes. Find his unit rate of bricks per minute rounded to the nearest tenth. (*Source: The Guinness Book of Records*, 1996)

55. The longest stairway is the service stairway for the Niesenbahn Cable railway near Spiez, Switzerland. It has 11,674 steps and rises to a height of 7759 feet. Find the unit rate of steps per foot rounded to the nearest tenth of a step. (*Source: The Guinness Book of Records*, 1996)

56. In New Mexico, the total enrollment in public schools was 329,640 in a recent year. At the same time, there were 19,368 public school teachers. Write a unit rate in students per teacher. Round to the nearest whole. (*Source:* National Center for Education Statistics)

57. In Utah, the total number of students enrolled in public schools during a recent year was 477,121. At the same time, there were 20,039 public school teachers. Write a unit rate in students per teacher. Round to the nearest whole. (*Source:* National Center for Education Statistics)

INTEGRATED REVIEW — RATIO AND RATE

Write each ratio as a ratio of whole numbers using fractional notation. Write the fraction in simplest form.

1. 18 to 20

2. 36 to 100

3. 8.6 to 10

4. 1.6 to 4.6

5. 8.65 to 6.95

6. 7.2 to 8.4

7. $3\frac{1}{2}$ to 13

8. $1\frac{2}{3}$ to $2\frac{3}{4}$

9. 8 inches to 12 inches

10. 3 hours to 24 hours

Find the ratio described in each problem.

11. American women between the ages of 25 and 54 watch an average of 3 hours of television between 4:30 P.M. and 7:30 P.M. Monday through Friday. American men between the ages of 25 and 54 watch an average of $2\frac{7}{20}$ hours of television in the same time slot. Find the ratio of the time spent by women watching television to the time spent by men watching television during this time slot. (*Source:* Nielsen Media Research)

12. At the end of 1996, Marriott International had $5075 million in assets and $1010 million in long-term debts. Find the ratio of assets to long-term debt. (*Source:* Marriott International Inc.)

Write each rate as a fraction in simplest form.

13. 5 offices for every 20 graduate assistants

14. 6 lights every 15 feet

15. 100 U.S. Senators for 50 states

16. 5 teachers for every 140 students

Write each rate as a unit rate.

17. 165 miles in 3 hours

18. 560 feet in 4 seconds

19. 63 employees per 3 fax lines

20. 85 phone calls for 5 teenagers

ANSWERS

1. $\frac{9}{10}$

2. $\frac{9}{25}$

3. $\frac{43}{50}$

4. $\frac{8}{23}$

5. $\frac{173}{139}$

6. $\frac{6}{7}$

7. $\frac{7}{26}$

8. $\frac{20}{33}$

9. $\frac{2}{3}$

10. $\frac{1}{8}$

11. $\frac{60}{47}$

12. $\frac{1015}{202}$

13. $\frac{1 \text{ office}}{4 \text{ graduate assistants}}$

14. $\frac{2 \text{ lights}}{5 \text{ feet}}$

15. $\frac{2 \text{ Senators}}{1 \text{ state}}$

16. $\frac{1 \text{ teacher}}{28 \text{ students}}$

17. 55 miles/hour

18. 140 feet/second

19. 21 employees/fax line

20. 17 phone calls/teenager

Write each unit price and decide which is the better buy.

21. Dog Food:
8 pounds for $2.16

18 pounds for $4.99

22. Paper plates:
100 for $1.98

500 for $8.99

23. Microwave popcorn:
3 packs for $2.39

8 packs for $5.99

24. AA Batteries:
4 for $3.69

10 for $9.89

---

21. 8 pounds: $0.27 per pound; 18 pounds: $0.28 per pound; **8 pounds**

22. 100: $0.020 per plate; 500: $0.018 per plate; **500 paper plates**

23. 3 packs: $0.80 per pack; 8 packs: $0.75 per pack; **8 packs**

24. 4: $0.92 per battery; 10: $0.99 per battery; **4 batteries**

5.3 PROPORTIONS

A WRITING PROPORTIONS

A **proportion** is a statement that 2 ratios or rates are equal. For example,

$$\frac{5}{6} = \frac{10}{12}$$

is a proportion. We can read this as "5 is to 6 as 10 is to 12."

Example 1   Write each sentence as a proportion.

a. 12 diamonds is to 15 rubies as 4 diamonds is to 5 rubies.
b. 5 hits is to 9 at-bats as 20 hits is to 36 at-bats.

Solution: **a.** diamonds → $\frac{12}{15} = \frac{4}{5}$ ← diamonds
rubies → ← rubies
b. hits → $\frac{5}{9} = \frac{20}{36}$ ← hits
at bats → ← at bats

┌───┐
│ **HELPFUL HINT**

Notice in the above examples of proportions that numerators contain the same units and denominators contain the same units. In this text, proportions will be written so that this is so.
└───┘

B DETERMINING WHETHER PROPORTIONS ARE TRUE

Like other mathematical statements, a proportion may be either true or false. A proportion is true if its ratios are equal. Since ratios are fractions, one way to determine whether a proportion is true is to write each fraction in simplest form and compare them.

Another way is to compare cross products as we did in Section 2.3.

┌───┐
│ **DETERMINING WHETHER PROPORTIONS ARE TRUE OR FALSE**
│
│ If cross products are *equal*, the proportion is *true*.
│ If cross products are *not equal*, the proportion is *false*.
└───┘

Example 2   Is $\frac{2}{3} = \frac{4}{6}$ a true proportion?

Solution:

$$\frac{2}{3} = \frac{4}{6}$$

$\left. \begin{array}{l} 3 \cdot 4 = 12 \\ 2 \cdot 6 = 12 \end{array} \right\}$ Cross products are equal.

Since the cross products are equal, the proportion is true.

Objectives

A Know the meaning of proportion and write sentences as proportions.

B Determine whether proportions are true.

C Find an unknown number in a proportion.

SSM   CD-ROM   Video 5.3

Practice Problem 1

Write each sentence as a proportion.

a. 24 right is to 6 wrong as 4 right is to 1 wrong.

b. 32 Cubs fans is to 18 Mets fans as 16 Cubs fans is to 9 Mets fans.

TEACHING TIP

Before beginning this lesson, ask students if they ever use the idea of proportions in every day life. If someone volunteers, ask them how they use proportions. If nobody thinks they use proportions, point out that they do any time they double a recipe, or reduce or enlarge a picture or diagram. Then ask what it means to be in the right proportion.

Practice Problem 2

Is $\frac{3}{6} = \frac{4}{8}$ a true proportion?

Answers

1. a. $\frac{24}{6} = \frac{4}{1}$, **b.** $\frac{32}{18} = \frac{16}{9}$, **2.** Yes

Practice Problem 3

Is $\frac{3.6}{6} = \frac{5.4}{8}$ a true proportion?

Practice Problem 4

Is $\frac{4\frac{1}{5}}{2\frac{1}{3}} = \frac{3\frac{3}{10}}{1\frac{5}{6}}$ a true proportion?

✓ CONCEPT CHECK

Write the proportion $\frac{5}{8} = \frac{10}{16}$ in two other ways so that it remains a true proportion.

TEACHING TIP

Before discussing Example 4, provide the following context for students: Suppose a lemonade recipe calls for $1\frac{1}{6}$ cups of sugar and $\frac{1}{2}$ cup of lemon juice and someone told you that you could make a lot more of this recipe by using $10\frac{1}{2}$ cups of sugar and $4\frac{1}{2}$ cups of lemon juice. How could you check that the ratio of the sugar is proportional to the ratio of the lemon juice. Then ask what would happen to the recipe if the ratios were not proportional? How would the recipe taste if you used $10\frac{1}{2}$ cups of sugar and $3\frac{1}{2}$ cups of lemon juice? (weaker, sweeter) How would it taste if you used $10\frac{1}{2}$ cups of sugar and $5\frac{1}{2}$ cups of lemon juice? (stronger, more sour)

Answers

3. No,   **4.** Yes

✓ **Concept Check:**   possible answers: $\frac{8}{5} = \frac{16}{10}$ and $\frac{5}{10} = \frac{8}{16}$

Example 3   Is $\frac{4.1}{7} = \frac{2.9}{5}$ a true proportion?

Solution:

$$\frac{4.1}{7} = \frac{2.9}{5}$$

$$\left.\begin{array}{l} 7 \cdot 2.9 = 20.3 \\ 4.1 \cdot 5 = 20.5 \end{array}\right\} \begin{array}{l}\text{Cross products are} \\ \text{not equal.}\end{array}$$

Since cross products are not equal, $\frac{4.1}{7} \neq \frac{2.9}{5}$. The proportion is not a true proportion.

Example 4   Is $\frac{1\frac{1}{6}}{10\frac{1}{2}} = \frac{\frac{1}{2}}{4\frac{1}{2}}$ a true proportion?

Solution:

$$\frac{1\frac{1}{6}}{10\frac{1}{2}} = \frac{\frac{1}{2}}{4\frac{1}{2}}$$

$$\left.\begin{array}{l} 10\frac{1}{2} \cdot \frac{1}{2} = \frac{21}{2} \cdot \frac{1}{2} = \frac{21}{4} \text{ or } 5\frac{1}{4} \\ 1\frac{1}{6} \cdot 4\frac{1}{2} = \frac{7}{6} \cdot \frac{9}{2} = \frac{63}{12} = \frac{21}{4} \text{ or } 5\frac{1}{4} \end{array}\right\} \begin{array}{l}\text{Cross products} \\ \text{are equal.}\end{array}$$

Since cross products are equal, the proportion is true.

TRY THE CONCEPT CHECK IN THE MARGIN.

C FINDING UNKNOWN NUMBERS IN PROPORTIONS

When one number of a proportion is unknown, we can use cross products to find the unknown number. For example, to find the unknown number, n, in the proportion $\frac{2}{3} = \frac{n}{30}$, we cross multiply.

$$\frac{2}{3} = \frac{n}{30}$$

$$\left.\begin{array}{l} 3 \cdot n \\ 2 \cdot 30 \end{array}\right\} \text{Find the cross products.}$$

If the proportion is true, then cross products are equal.

$3 \cdot n = 2 \cdot 30$ Set the cross products equal to each other.

$3 \cdot n = 60$ Write $2 \cdot 30$ as 60.

To find the unknown number n, we ask ourselves "3 times what number is 60?" The number is 20 and can be found by dividing 60 by 3.

$n = \frac{60}{3}$ Divide 60 by the number multiplied by n.

$n = 20$ Simplify.

Thus, the unknown number is 20.

To *check*, let's replace n with this value, 20, and verify that a true proportion results.

$\frac{2}{3} \stackrel{?}{=} \frac{20}{30}$ ← Let n be 20.

$\frac{2}{3} \overset{?}{=} \frac{20}{30}$ $\left.\begin{array}{l} 3 \cdot 20 = 60 \\ 2 \cdot 30 = 60 \end{array}\right\}$ Cross products are equal.

FINDING AN UNKNOWN VALUE, *n* IN A PROPORTION

Step 1. Find cross products.
Step 2. Set the cross products equal to each other.
Step 3. Divide the number not multiplied by *n* by the number multiplied by *n*.

Example 5   Find the value of the unknown number, *n*.

$$\frac{34}{51} = \frac{n}{3}$$

Solution: **Step 1.**

$\frac{34}{51} = \frac{n}{3}$ $\left.\begin{array}{l} 51 \cdot n \\ 34 \cdot 3 = 102 \end{array}\right\}$ Find the cross products.

Step 2. $51 \cdot n = 102$ Set the cross products equal to each other.

Step 3. $n = \dfrac{102}{51}$ Divide 102 by 51, the number multiplied by *n*.

$n = 2$ Simplify

Check: $\dfrac{34}{51} \overset{?}{=} \dfrac{2}{3}$ Replace *n* with its value, 2.

$\frac{34}{51} \overset{?}{=} \frac{2}{3}$ $\left.\begin{array}{l} 51 \cdot 2 = 102 \\ 34 \cdot 3 = 102 \end{array}\right\}$ Cross products are equal, so proportion is true.

Example 6   Find the unknown number, *n*.

$$\frac{6}{5} = \frac{7}{n}$$

Solution: **Step 1.**

$\frac{6}{5} = \frac{7}{n}$ $\left.\begin{array}{l} 5 \cdot 7 = 35 \\ 6 \cdot n \end{array}\right\}$ Find the cross products.

Step 2. $6 \cdot n = 35$ Set the cross products equal to each other.

Step 3. $n = \dfrac{35}{6}$ Divide 35 by 6, the number multiplied by *n*.

$n = 5\dfrac{5}{6}$

Check to see that $5\frac{5}{6}$ is the unknown number.

Practice Problem 5

Find the value of the unknown number, *n*.

$$\frac{2}{15} = \frac{n}{60}$$

Practice Problem 6

Find the unknown number, *n*.

$$\frac{8}{n} = \frac{5}{9}$$

Answers

5. $n = 8$, **6.** $n = 14\frac{2}{5}$

Practice Problem 7

Find the unknown number, n.

$$\frac{n}{6} = \frac{0.7}{1.2}$$

Example 7

Find the unknown number, n.

$$\frac{n}{3} = \frac{0.8}{1.5}$$

Solution: **Step 1.**

$$\frac{n}{3} = \frac{0.8}{1.5} \qquad \left.\begin{array}{l} 3 \cdot 0.8 = 2.4 \\ n \cdot 1.5 \end{array}\right\} \text{ Find the cross products.}$$

Step 2.

$n \cdot 1.5 = 2.4$ Set the cross products equal to each other.

Step 3.

$n = \dfrac{2.4}{1.5}$ Divide 2.4 by 1.5, the number multiplied by n.

$n = 1.6$ Simplify.

Check to see that 1.6 is the unknown number.   ▬▬▬

Practice Problem 8

Find the unknown number, n.

$$\frac{n}{4\frac{1}{3}} = \frac{4\frac{1}{2}}{1\frac{3}{4}}$$

Example 8

Find the unknown number, n.

$$\frac{1\frac{2}{3}}{3\frac{1}{4}} = \frac{n}{2\frac{3}{5}}$$

Solution: **Step 1.**

$$\frac{1\frac{2}{3}}{3\frac{1}{4}} = \frac{n}{2\frac{3}{5}} \qquad \left.\begin{array}{l} 3\frac{1}{4} \cdot n \\[4pt] 1\frac{2}{3} \cdot 2\frac{3}{5} = \frac{5}{3} \cdot \frac{13}{5} = \frac{\overset{1}{\cancel{5}} \cdot 13}{3 \cdot \underset{1}{\cancel{5}}} = \frac{13}{3} \end{array}\right\} \begin{array}{l}\text{Find the cross}\\\text{products.}\end{array}$$

Step 2.

$3\frac{1}{4} \cdot n = \dfrac{13}{3}$ Set the cross products equal to each other.

$\dfrac{13}{4} \cdot n = \dfrac{13}{3}$ Write $3\frac{1}{4}$ as $\frac{13}{4}$.

Step 3.

$n = \dfrac{13}{3} \div \dfrac{13}{4}$ Divide $\frac{13}{3}$ by $\frac{13}{4}$, the number multiplied by n.

$n = \dfrac{13}{3} \cdot \dfrac{4}{13} = \dfrac{4}{3}$ or $1\frac{1}{3}$ Divide by multiplying by the reciprocal.

Check to see that $1\frac{1}{3}$ is the unknown number.   ▬▬▬

Answers

7. $n = 3.5$,   **8.** $n = 11\frac{1}{7}$

Name _____ Section _____ Date _____

MENTAL MATH ANSWERS
1. true
2. true
3. false
4. false
5. true
6. true

MENTAL MATH

State whether each proportion is true or false.

1. $\dfrac{2}{1} = \dfrac{6}{3}$

2. $\dfrac{3}{1} = \dfrac{15}{5}$

3. $\dfrac{1}{2} = \dfrac{3}{5}$

4. $\dfrac{2}{11} = \dfrac{1}{5}$

5. $\dfrac{2}{3} = \dfrac{4}{6}$

6. $\dfrac{3}{4} = \dfrac{6}{8}$

EXERCISE SET 5.3

A *Write each sentence as a proportion. See Example 1.*

1. 10 diamonds is to 6 opals as 5 diamonds is to 3 opals.

2. 8 books is to 6 courses as 4 books is to 3 courses.

3. 3 printers is to 12 computers as 1 printer is to 4 computers.

4. 4 hit songs is to 16 releases as 1 hit song is to 4 releases.

5. 6 eagles is to 58 sparrows, as 3 eagles is to 29 sparrows.

6. 12 errors is to 8 pages as 1.5 errors is to 1 page.

7. $2\dfrac{1}{4}$ cups of flour is to 24 cookies as $6\dfrac{3}{4}$ cups of flour is to 72 cookies.

8. $1\dfrac{1}{2}$ cups milk is to 10 bagels as $\dfrac{3}{4}$ cup milk is to 5 bagels.

9. 22 vanilla wafers is to 1 cup of cookie crumbs as 55 vanilla wafers is to 2.5 cups of cookie crumbs. (*Source:* based on data from *Family Circle* magazine)

10. 1 cup of instant rice is to 1.5 cups cooked rice as 1.5 cups of instant rice is to 2.25 cups of cooked rice. (*Source:* based on data from *Family Circle* magazine)

B *Determine whether each proportion is a true proportion. See Examples 2 through 4.*

11. $\dfrac{15}{9} = \dfrac{5}{3}$

12. $\dfrac{8}{6} = \dfrac{20}{15}$

13. $\dfrac{8}{6} = \dfrac{9}{7}$

14. $\dfrac{7}{12} = \dfrac{4}{7}$

ANSWERS

1. $\dfrac{10 \text{ diamonds}}{6 \text{ opals}} = \dfrac{5 \text{ diamonds}}{3 \text{ opals}}$

2. $\dfrac{8 \text{ books}}{6 \text{ courses}} = \dfrac{4 \text{ books}}{3 \text{ courses}}$

3. $\dfrac{3 \text{ printers}}{12 \text{ computers}} = \dfrac{1 \text{ printer}}{4 \text{ computers}}$

4. $\dfrac{4 \text{ hits}}{16 \text{ releases}} = \dfrac{1 \text{ hit}}{4 \text{ releases}}$

5. $\dfrac{6 \text{ eagles}}{58 \text{ sparrows}} = \dfrac{3 \text{ eagles}}{29 \text{ sparrows}}$

6. $\dfrac{12 \text{ errors}}{8 \text{ pages}} = \dfrac{1.5 \text{ errors}}{1 \text{ page}}$

7. $\dfrac{2\frac{1}{4} \text{ cups flour}}{24 \text{ cookies}} = \dfrac{6\frac{3}{4} \text{ cups flour}}{72 \text{ cookies}}$

8. $\dfrac{1\frac{1}{2} \text{ cups milk}}{10 \text{ bagels}} = \dfrac{\frac{3}{4} \text{ cup milk}}{5 \text{ bagels}}$

9. $\dfrac{22 \text{ vanilla wafers}}{1 \text{ cup cookie crumbs}} = \dfrac{55 \text{ vanilla wafers}}{2.5 \text{ cups cookie crumbs}}$

10. $\dfrac{1 \text{ cup instant rice}}{1.5 \text{ cups cooked rice}} = \dfrac{1.5 \text{ cups instant rice}}{2.25 \text{ cups cooked rice}}$

11. true

12. true

13. false

14. false

Name _____

15. $\dfrac{9}{36} = \dfrac{2}{8}$

16. $\dfrac{8}{24} = \dfrac{3}{9}$

17. $\dfrac{5}{8} = \dfrac{625}{1000}$

18. $\dfrac{30}{50} = \dfrac{600}{1000}$

19. $\dfrac{0.8}{0.3} = \dfrac{0.2}{0.6}$

20. $\dfrac{0.7}{0.4} = \dfrac{0.3}{0.1}$

21. $\dfrac{4.2}{8.4} = \dfrac{5}{10}$

22. $\dfrac{8}{10} = \dfrac{5.6}{0.7}$

23. $\dfrac{\frac{3}{4}}{\frac{4}{3}} = \dfrac{\frac{1}{2}}{\frac{8}{9}}$

24. $\dfrac{\frac{2}{5}}{\frac{2}{7}} = \dfrac{\frac{1}{10}}{\frac{1}{3}}$

25. $\dfrac{2\frac{2}{5}}{\frac{2}{3}} = \dfrac{\frac{10}{9}}{\frac{1}{4}}$

26. $\dfrac{5\frac{5}{8}}{\frac{5}{3}} = \dfrac{4\frac{1}{2}}{1\frac{1}{5}}$

C *For each proportion, find the unknown number, n. See Examples 5 through 8.*

27. $\dfrac{n}{5} = \dfrac{6}{10}$

28. $\dfrac{n}{3} = \dfrac{12}{9}$

29. $\dfrac{30}{10} = \dfrac{15}{n}$

30. $\dfrac{25}{100} = \dfrac{7}{n}$

31. $\dfrac{n}{8} = \dfrac{50}{100}$

32. $\dfrac{12}{18} = \dfrac{n}{21}$

33. $\dfrac{n}{6} = \dfrac{8}{15}$

34. $\dfrac{24}{n} = \dfrac{60}{96}$

35. $\dfrac{12}{10} = \dfrac{n}{16}$

36. $\dfrac{18}{54} = \dfrac{3}{n}$

37. $\dfrac{\frac{1}{3}}{\frac{3}{8}} = \dfrac{\frac{2}{5}}{n}$

38. $\dfrac{\frac{7}{9}}{\frac{8}{27}} = \dfrac{\frac{1}{4}}{n}$

39. $\dfrac{8}{\frac{1}{3}} = \dfrac{24}{n}$

40. $\dfrac{\frac{3}{4}}{12} = \dfrac{n}{48}$

41. $\dfrac{\frac{2}{3}}{\frac{6}{9}} = \dfrac{12}{n}$

42. $\dfrac{n}{24} = \dfrac{\frac{5}{8}}{3}$

43. $\dfrac{n}{\frac{6}{5}} = \dfrac{4\frac{1}{6}}{6\frac{2}{3}}$

44. $\dfrac{\frac{11}{4}}{\frac{25}{8}} = \dfrac{7\frac{3}{5}}{n}$

45. $\dfrac{n}{0.6} = \dfrac{0.05}{12}$

46. $\dfrac{0.2}{0.7} = \dfrac{8}{n}$

47. $\dfrac{3.5}{12.5} = \dfrac{7}{n}$

48. $\dfrac{7.8}{13} = \dfrac{n}{2.6}$

Name _____

REVIEW AND PREVIEW

Insert < or > to form a true statement. See Sections 2.1 and 4.2.

49. 8.01   8.1

50. 7.26   7.026

51. $2\frac{1}{2}$ $2\frac{1}{3}$

52. $9\frac{1}{5}$ $9\frac{1}{4}$

53. $5\frac{1}{3}$ $6\frac{2}{3}$

54. $1\frac{1}{2}$ $2\frac{1}{2}$

COMBINING CONCEPTS

For each proportion, find the unknown number, n.

55. $\dfrac{n}{7} = \dfrac{0}{8}$

56. $\dfrac{0}{2} = \dfrac{n}{3.5}$

57. $\dfrac{n}{1150} = \dfrac{588}{483}$

58. $\dfrac{585}{n} = \dfrac{117}{474}$

59. $\dfrac{222}{1515} = \dfrac{37}{n}$

60. $\dfrac{1425}{1062} = \dfrac{n}{177}$

61. Explain the difference between a ratio and a proportion.

62. Explain how to find the unknown number in a proportion such as $\dfrac{n}{18} = \dfrac{12}{8}$.

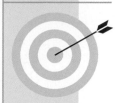

Focus On Real Life

BODY DIMENSIONS

Have you ever noticed how large a baby's head looks compared to the rest of its body? We tend to notice this because we are used to seeing adults with very different "proportions" or dimensions. Try the following group activity to see if there are any patterns in adult body dimensions from person to person.

GROUP ACTIVITY

Have another group member help you measure the height of your head from under your chin to the top of your skull. (Remember that the top of your head is rounded. You may find it helpful to hold a flat surface on top of your head to more accurately locate and measure to the "top"). Use your head measurement to make a measuring stick from a piece of cardboard. You will now use this measuring stick to measure other parts of your body in units equal to the height of your head. With the help of another group member as needed, measure the following distances with your "head" measuring stick:

▲ Your overall height
▲ Your arm length from shoulder to elbow
▲ Your arm length from elbow to longest fingertip
▲ Your leg length from hip to ankle
▲ Your leg length from knee to ankle
▲ The distance from your chin to waistline

(Note: Each group member must make his or her own measurements in relation to the measurement of his or her own head height.)

1. For each of the above measurements made, find the ratio of the measurement to the height measurement of your head.
2. As a group, compile a table summarizing the ratios for each group member by category. (You might list the six measurement categories along the side of the table and group members' names along the top of the table.)
3. Analyze the table. Do you see any patterns? Explain.
4. About how many "heads tall" would you say is a normal adult?

5.4 PROPORTIONS AND PROBLEM SOLVING

A SOLVING PROBLEMS BY WRITING PROPORTIONS

Writing proportions is a powerful tool for solving problems in almost every field, including business, chemistry, biology, health sciences, and engineering, as well as in daily life, too. Given a specified ratio (or rate) of two quantities, a proportion can be used to determine an unknown quantity.

Example 1   Determining Distances from a Map

On a Chamber of Commerce map of Abita Springs, 5 miles corresponds to 2 inches. How many miles correspond to 7 inches?

Solution: **1.** UNDERSTAND. Read and reread the problem. You may want to draw a diagram.

5 miles	5 miles	5 miles	?	= a little over 15 miles
2 inches	2 inches	2 inches	1 inch	= 7 inches

From the diagram we can see that our solution should be a little over 15 miles.

2. TRANSLATE. We will let n represent our unknown number. Since 5 miles corresponds to 2 inches as n miles corresponds to 7 inches, we have the proportion:

$$\begin{array}{ccc} \text{miles} & \rightarrow & \dfrac{5}{2} = \dfrac{n}{7} & \leftarrow & \text{miles} \\ \text{inches} & \rightarrow & & \leftarrow & \text{inches} \end{array}$$

3. SOLVE.

$$\dfrac{5}{2} = \dfrac{n}{7} \qquad \left.\begin{array}{c} 2 \cdot n \\ 5 \cdot 7 = 35 \end{array}\right\} \text{ Find the cross products.}$$

$2 \cdot n = 35$ Set the cross products equal to each other.

$n = \dfrac{35}{2}$ Divide 35 by 2, the number multiplied by n.

$n = 17.5$ Simplify.

4. INTERPRET. *Check* your work. This result is reasonable since it is a little over 15 miles. *State* your conclusion: 7 inches corresponds to 17.5 miles.

Practice Problem 1

On an architect's blueprint, 1 inch corresponds to 12 feet. How long is a wall represented by a $3\frac{1}{2}$-inch line on the blueprint?

Answer
1. 42 feet

> **HELPFUL HINT**
>
> We can also solve Example 1 by writing the proportion
>
> $$\frac{2 \text{ inches}}{5 \text{ miles}} = \frac{7 \text{ inches}}{n \text{ miles}}$$
>
> Although other proportions may be used to solve Example 1, we will solve by writing proportions so that the numerators have the same unit measures and the denominators have the same unit measures.

Practice Problem 2

An auto mechanic recommends that 3 ounces of isopropyl alcohol be mixed with a tankful of gas (14 gallons) to increase the octane of the gasoline for better engine performance. At this rate, how many gallons of gas can be treated with a 16-ounce bottle of alcohol?

Example 2 Finding Medicine Dosage

The standard dose of an antibiotic is 4 cc (cubic centimeters) for every 25 pounds (lb) of body weight. At this rate, find the standard dose for a 140-lb woman.

Solution:

1. UNDERSTAND. Read and reread the problem. You may want to draw a diagram to estimate a reasonable solution.

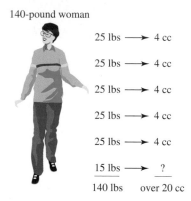

140-pound woman

	25 lbs ⟶	4 cc
	25 lbs ⟶	4 cc
	25 lbs ⟶	4 cc
	25 lbs ⟶	4 cc
	25 lbs ⟶	4 cc
	15 lbs ⟶	?
	140 lbs	over 20 cc

From the diagram, we can see that a reasonable solution is a little over 20 cc.

2. TRANSLATE. We will let n represent the unknown number. From the problem, we know that 4 cc is to 25 pounds as n cc is to 140 pounds, or

cubic centimeters → $\dfrac{4}{25} = \dfrac{n}{140}$ ← cubic centimeters
pounds → $\phantom{\dfrac{4}{25} = \dfrac{n}{140}}$ ← pounds

3. SOLVE.

$$\frac{4}{25} = \frac{n}{140}$$

$$\left.\begin{array}{l} 25 \cdot n \\ 4 \cdot 140 = 560 \end{array}\right\} \text{Find the cross products.}$$

$$25 \cdot n = 560 \qquad \text{Set the cross products equal to each other.}$$

$$n = \frac{560}{25} \qquad \text{Divide 560 by 25, the number multiplied by } n.$$

$$n = 22.4 \qquad \text{Simplify.}$$

Answer

2. $74\frac{2}{3}$ gal

4. INTERPRET. *Check* your work. This result is reasonable since it is a little over 20 cc. *State* your conclusion: The standard dose for a 140-lb woman is 22.4 cc.

Example 3   Calculating Supplies Needed to Fertilize a Lawn

A 50-pound bag of fertilizer covers 2400 square feet of lawn. How many bags of fertilizer are needed to cover a town square containing 15,360 square feet of lawn? Round the answer up to the nearest whole bag.

Solution: **1.** UNDERSTAND. Read and reread the problem. Draw a picture.

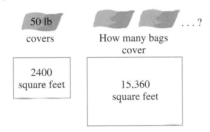

50 lb		How many bags
covers		cover

2400 square feet	15,360 square feet

2. TRANSLATE. We'll let n represent the unknown number. From the problem, we know that 1 bag is to 2400 square feet as n bags is to 15,360 square feet.

$$\begin{array}{ccc} \text{bags} & \rightarrow & \dfrac{1}{2400} = \dfrac{n}{15,360} & \leftarrow \text{bags} \\ \text{square feet} & \rightarrow & & \leftarrow \text{square feet} \end{array}$$

3. SOLVE.

$$\dfrac{1}{2400} = \dfrac{n}{15,360}$$

$$\left. \begin{array}{l} 2400 \cdot n \\ 1 \cdot 15,360 = 15,360 \end{array} \right\} \text{Find the cross products.}$$

$2400 \cdot n = 15,360$ Set the cross products equal to each other.

$n = \dfrac{15,360}{2400}$ Divide 15,360 by 2400, the number multiplied by n.

$n = 6.4$ Simplify.

4. INTERPRET. *Check* that replacing n with 6.4 makes the proportion true. Is the answer reasonable?

Practice Problem 3

If a gallon of paint covers 400 square feet, how many gallons must be bought to paint a retaining wall 260 feet long and 4 feet high? Round the answer up to the nearest whole gallon.

TEACHING TIP   Classroom Activity

Have students work in groups to create 3 word problems involving proportions. Then have them give their problems to another group to solve.

Answer

3. 3 gallons

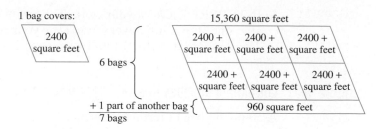

Yes. Since we must buy whole bags of fertilizer, 7 bags are needed. *State* your conclusion: To cover 15,360 square feet of lawn, 7 bags are needed. ▬▬▬

TRY THE CONCEPT CHECK IN THE MARGIN.

✓ **CONCEPT CHECK**

You are told that 12 ounces of ground coffee will brew enough coffee to serve 20 people. How could you estimate how much ground coffee will be needed to serve 95 people?

Answer

✓ **Concept Check:**   Find how much will be needed for 100 people (20×5) by multiplying 12 ounces by 5 which is 60 ounces.

EXERCISE SET 5.4

A *Solve. See Examples 1 through 3.*

The ratio of a quarterback's completed passes to attempted passes is 4 to 9.

1. If he attempted 27 passes, find how many passes he completed.

2. If he completed 20 passes, find how many passes he attempted.

It takes Sandra Hallahan 30 minutes to word process and spell check 4 pages.

3. Find how long it takes her to word process and spell check 22 pages.

4. Find how many pages she can word process and spell check in 4.5 hours.

University Law School accepts 2 out of every 7 applicants.

5. If the school received 630 applications, find how many students were accepted.

6. If the school accepted 150 students, find how many applications were received.

On an architect's blueprint, 1 inch corresponds to 8 feet.

7. Find the length of a wall represented by a line $2\frac{7}{8}$ inches long on the blueprint.

8. If an exterior wall is 42 feet long, find how long the blueprint measurement should be.

A human factors expert recommends that there be at least 9 square feet of floor space in a college classroom for every student in the class.

9. Find the minimum floor space that 30 students require.

10. Due to a space crunch, a university converts a 21′ by 15′ conference room into a classroom. Find the maximum number of students the room can accommodate.

A Honda Civic averages 450 miles on a 12-gallon tank of gas.

11. If Dave Smythe runs out of gas in a Honda Civic and AAA comes to his rescue with $1\frac{1}{2}$ gallons of gas, determine how far he can go. Round to the nearest mile.

12. Find how many gallons of gas Denise Wolcott can expect to burn on a 2000-mile vacation trip in a Honda Civic. Round to the nearest gallon.

The scale on an Italian map states that 1 centimeter corresponds to 30 kilometers (a unit of length in the metric system).

13. Find how far apart Milan and Rome are if their corresponding points on the map are 15 centimeters apart.

14. On the map, a small Italian village is located 0.4 centimeters from the Mediterranean Sea. Find the actual distance.

1. 12 passes

2. 45 passes

3. 165 minutes

4. 36 pages

5. 180 students

6. 525 applications

7. 23 feet

8. $5\frac{1}{4}$ inches

9. 270 square feet

10. 35 students

11. 56 miles

12. 53 gallons

13. 450 kilometers

14. 12 kilometers

Name _____

A drink called Sea Breeze punch is made by mixing 3 parts of grapefruit juice with 4 parts of cranberry juice.

15. Find how much grapefruit juice should be mixed with 32 ounces of cranberry juice.

16. For a party, 6 quarts of grapefruit juice have been purchased to make Sea Breeze punch. Find how much cranberry juice should be purchased.

A bag of Scott fertilizer covers 3000 square feet of lawn.

17. Find how many bags of fertilizer should be purchased to cover a rectangular lawn 260 feet by 180 feet.

18. Find how many bags of fertilizer should be purchased to cover a square lawn measuring 160 feet on each side.

Yearly homeowner property taxes are figured at a rate of $1.45 tax for every $100 of house value.

19. If Janet Blossom, a homeowner, pays $2349 in property taxes, find the value of her home.

20. Find the property taxes on a condominium valued at $72,000.

A Cubs baseball player makes 3 hits in every 8 times at bat.

21. If this Cubs player comes up to bat 40 times in a World Series playoff series, find how many hits he would be expected to make.

22. At this rate, if he made 12 hits, find how many times he batted.

A survey reveals that 2 out of 3 people prefer Coke to Pepsi.

23. In a room of 40 people, how many people are likely to prefer Coke? Round the answer to the nearest person.

24. In a college class of 36 students, find how many students are likely to prefer Pepsi.

An office uses 5 boxes of envelopes every 3 weeks.

25. Find how long a gross of envelope boxes is likely to last. (A gross of boxes is 144 boxes). Round to the nearest week.

26. Find how many boxes should be purchased to last a month. Round to the nearest box.

27. The daily supply of oxygen for one person is provided by 625 square feet of lawn. A total of 3750 square feet of lawn would provide the daily supplies of oxygen for how many people? (*Source:* Professional Lawn Care Association of America)

28. In the United States, approximately 71 million of the 200 million cars and light trucks in service have driver air bags. In a parking lot containing 800 cars and light trucks, how many would be expected to have driver air bags? (*Source:* Insurance Institute for Highway Safety)

Name _____

29. A student would like to estimate the height of the Statue of Liberty in New York City's harbor. According to the *1998 World Almanac*, the length of the Statue of Liberty's right arm is 42 feet. The student's right arm is 2 feet long and her height is $5\frac{1}{3}$ feet.

Use this information to estimate the height of the Statue of Liberty. (The actual height of the Statue of Liberty, from heel to top of the head, is 111 feet, 1 inch. How close is the estimated height to the actual height of the statue?)

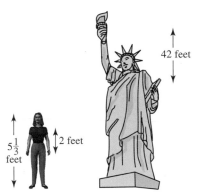

30. There are 72 milligrams of cholesterol in a 3.5 ounce serving of lobster. How much cholesterol is in 5 ounces of lobster? (*Source:* The National Institute of Health)

31. One pound of firmly packed brown sugar yields $2\frac{1}{4}$ cups. How many pounds of brown sugar will be required by a recipe that calls for 6 cups of firmly packed brown sugar? (*Source:* based on data from *Family Circle* magazine)

32. Eleven out of every 25 greeting cards sold in the United States are Hallmark brand cards. If a consumer purchased 75 greeting cards in the past year, how many do you expect would have been Hallmark brand cards? (*Source:* Hallmark Cards, Inc.)

33. Medication is prescribed in 7 out of every 10 hospital emergency room visits that involve an injury. If a large urban hospital had 620 emergency room visits involving an injury in the past month, how many of these visits would you expect included a prescription for medication? (*Source:* National Center for Health Statistics)

34. Currently in the American population of people aged 65 years old and older, there are 145 women for every 100 men. In a nursing home with 280 male residents over the age of 65, how many female residents over the age of 65 would be expected? (*Source:* U.S. Bureau of the Census)

35. $3 \cdot 5$

36. $3 \cdot 7$

37. $2^2 \cdot 5$

38. $2^3 \cdot 3$

39. $2^3 \cdot 5^2$

40. $2^2 \cdot 3 \cdot 5^2$

41. 2^5

42. 3^4

43. $4\frac{2}{3}$ feet

44. $9\frac{3}{5}$ feet

318

REVIEW AND PREVIEW

Find the prime factorization of each number. See Section 2.2.

35. 15   **36.** 21   **37.** 20   **38.** 24

39. 200   **40.** 300   **41.** 32   **42.** 81

COMBINING CONCEPTS

A board such as the one pictured below will balance if the following proportion is true:

$$\frac{\text{first weight}}{\text{second distance}} = \frac{\text{second weight}}{\text{first distance}}$$

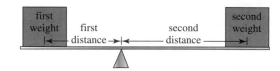

Use this proportion to solve Exercises 43–44.

43. Find the distance *n* that will allow the board to balance.

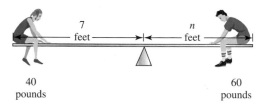

44. Find the length *n* needed to lift the weight below.

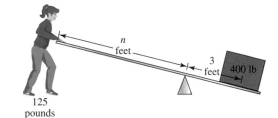

45. Describe a situation in which writing a proportion might solve a problem related to driving a car.

Chapter 5 Activity
Investigating Scale Drawings

Materials:
▲ ruler
▲ tape measure
▲ grid paper (optional)

This activity may be completed by working in groups or individually.

Scale drawings are used by architects, engineers, interior designers, ship builders, and others. In a scale drawing, each unit measurement on the drawing represents a fixed length on the object being drawn. For instance, in an architect's scale drawing, 1 inch on the drawing may represent 10 feet on a building. The scale describes the relationship between the measurements. If the measurements have the same units, the scale can be expressed as a ratio. In this case, the ratio, would be 1:120, representing 1 inch to 120 inches (or 10 feet).

Use a ruler and the scale drawing of an elementary school to answer the following questions.

1. How wide are each of the front doors of the school?   5 feet

2. How long is the front of the school?   50 feet

3. How tall is the front of the school?   25 feet

Now you will draw your own scale floor plan. First choose a room to draw—it can be your math classroom, your living room, your dormitory room, or any room that can be easily measured. Start by using a tape measure to measure the distances around the base of the walls in the room you are drawing.

4. Choose a scale for your floor plan.

5. Convert each measurement in the room you are drawing to the corresponding lengths needed for the scale drawing.

6. Complete your floor plan (you may find it helpful to use grid paper). Mark the locations of doors and windows on your floor plan. Be sure to indicate the scale used in your floor plan on the drawing.

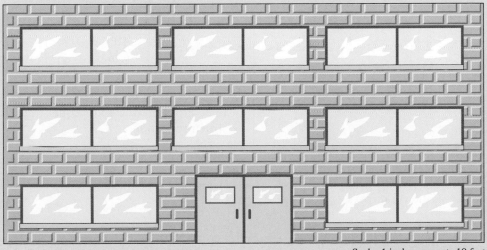

Scale: 1 inch represents 10 feet

CHAPTER 5 HIGHLIGHTS

DEFINITIONS AND CONCEPTS	EXAMPLES

SECTION 5.1   RATIOS

A **ratio** is the quotient of two quantities.	The ratio of 3 to 4 can be written as $$\frac{3}{4} \qquad \text{or} \qquad 3 : 4$$ $\uparrow$ fraction notation $\uparrow$ colon notation

SECTION 5.2   RATES

Rates are used to compare different kinds of quantities.	Write the rate 12 spikes every 8 inches as a fraction in simplest form. $$\frac{12 \text{ spikes}}{8 \text{ inches}} = \frac{3 \text{ spikes}}{2 \text{ inches}}$$
A **unit rate** is a rate with a denominator of 1.	Write as a unit rate: 117 miles on 5 gallons of gas $$\frac{117 \text{ miles}}{5 \text{ gallons}} = \frac{23.4 \text{ miles}}{1 \text{ gallon}}$$ or 23.4 miles per gallon or 23.4 miles/gallon
A **unit price** is a "money per item" unit rate.	Write as a unit price: \$5.88 for 42 ounces of detergent $$\frac{\$5.88}{42 \text{ ounces}} = \frac{\$0.14}{1 \text{ ounce}} = \$0.14 \text{ per ounce}$$

SECTION 5.3   PROPORTIONS

A **proportion** is a statement that two ratios or rates are equal.	$\frac{1}{2} = \frac{4}{8}$ is a proportion.
If cross products are equal, the proportion is true. If cross products are not equal, the proportion is false.	Is $\frac{6}{10} = \frac{9}{15}$ a true proportion? $$\frac{6}{10} = \frac{9}{15} \qquad \left. \begin{array}{l} 10 \cdot 9 = 90 \\ 6 \cdot 15 = 90 \end{array} \right\} \begin{array}{l} \text{Cross products are} \\ \text{equal.} \end{array}$$ Since cross products are equal, the proportion is a true proportion.

TEACHING TIP
As a review of the major concepts in this chapter, have students write a short paragraph comparing and contrasting ratios, rates, unit rates, and proportions. To help them get started have them identify each of the following as a ratio, rate, unit rate, proportion, or any combination of these.
3/4
5/7 = 15/21
\$0.432/lb
63 miles/3 gallons = 21 miles/1 gallon
\$48/12 people
13/7.5
\$17.99/dozen

SECTION 5.3 (CONTINUED)	
FINDING AN UNKNOWN VALUE, *n*, IN A PROPORTION	Find *n*: $\dfrac{n}{7} = \dfrac{5}{8}$
Step 1. Find the cross products.	Step 1. $\dfrac{n}{7} = \dfrac{5}{8}$ $\left.\begin{array}{l} 7 \cdot 5 = 35 \\ n \cdot 8 \end{array}\right\}$ Find the cross products.
Step 2. Set the cross products equal to each other.	Step 2. $n \cdot 8 = 35$ Set the cross products equal to each other.
Step 3. Divide the number not multiplied by *n* by the number multiplied by *n*.	Step 3. $n = \dfrac{35}{8}$ Divide 35 by 8, the number multiplied by *n*. $n = 4\dfrac{3}{8}$

SECTION 5.4 PROPORTIONS AND PROBLEM SOLVING	
Given a specified ratio (or rate) of two quantities, a proportion can be used to determine an unknown quantity.	On a map, 50 miles corresponds to 3 inches. How many miles correspond to 10 inches? **1.** UNDERSTAND. Read and reread the problem. **2.** TRANSLATE. We let *n* represent the unknown number. We are given that 50 miles is to 3 inches as *n* miles is to 10 inches. miles → $\dfrac{50}{3} = \dfrac{n}{10}$ ← miles inches → ← inches **3.** SOLVE. $\dfrac{50}{3} = \dfrac{n}{10}$ $\left.\begin{array}{l} 3 \cdot n \\ 50 \cdot 10 = 500 \end{array}\right\}$ Find the cross products. $3 \cdot n = 500$ Set the cross products equal to each other. $n = \dfrac{500}{3}$ Divide 500 by 3, the number multiplied by *n*. $n = 166\dfrac{2}{3}$ **4.** INTERPRET. *Check* your work. *State* your conclusion: On the map, $166\dfrac{2}{3}$ miles corresponds to 10 inches.

Focus On Business and Career

CONSUMER PRICE INDEX

Do you remember when the regular price of a candy bar was 5¢, 10¢, or 25¢? It is certainly difficult to find a candy bar for that price these days. The reason is inflation: the tendency for the price of a given product to increase over time. Businesses and government agencies use the Consumer Price Index (CPI) to track inflation. The CPI measures the change in prices over time of basic consumer goods and services. It is calculated by the Bureau of Labor Statistics, part of the U.S. Department of Labor.

The CPI is very useful for comparing the prices of fixed items in various years. For instance, suppose an insurance company customer submits a claim for the theft of a fishing boat purchased in 1975. Because the customer's policy includes replacement cost coverage, the insurance company must calculate how much it would cost to replace the boat at today's prices. (Let's assume the theft took place in 1997.) The customer has a receipt for the boat showing that it cost $598 in 1975. The insurance company can use the following proportion to calculate the replacement cost:

$$\frac{\text{Price in earlier year}}{\text{Price in later year}} = \frac{\text{CPI value in earlier year}}{\text{CPI value in later year}}$$

Because the CPI value is 53.8 for 1975 and 160.5 for 1997, the insurance company would use the following proportion for this situation. (We will let n represent the unknown price in 1997.)

$$\frac{\text{Price in 1975}}{\text{Price in 1997}} = \frac{\text{CPI value in 1975}}{\text{CPI value in 1997}}$$

$$\frac{598}{n} = \frac{53.8}{160.5}$$

$$53.8 \cdot n = 598(160.5)$$

$$53.8 \cdot n = 95{,}979$$

$$n = \frac{95{,}979}{53.8}$$

$$n \approx 1784$$

CONSUMER PRICE INDEX	
Year	CPI
1915	10.1
1920	20.0
1925	17.5
1930	16.7
1935	13.7
1940	14.0
1945	18.0
1950	24.1
1955	26.8
1960	29.6
1965	31.5
1970	38.8
1975	53.8
1980	82.4
1985	107.6
1990	130.7
1995	152.4
1997	160.5

(*Source:* Bureau of Labor Statistics, U.S. Dept. of Labor)

The replacement cost of the fishing boat at 1997 prices is $1784.

CRITICAL THINKING

1. What trends do you see in CPI values in the table? Do you think these trends make sense? Explain.   answers may vary

2. A piece of jewelry cost $1225 in 1975. What is its 1997 replacement value? $3654.51
3. In 1997, the cost of a loaf of bread was about $2. What would an equivalent loaf of bread cost in 1950? $0.30
4. Suppose a couple purchased a house for $22,000 in 1920. At what price could they have expected to sell the house in 1990? $143,770
5. An original Ford Model T cost about $850 in 1915. What is the equivalent cost of a Model T in 1995 dollars? $12,825.74

Chapter 5 Review

(5.1) *Write each ratio as a fraction in simplest form.*

1. 6000 people to 4800 people $\dfrac{5}{4}$

2. 121 births to 143 births $\dfrac{11}{13}$

3. $2\dfrac{1}{4}$ days to 10 days $\dfrac{9}{40}$

4. 14 quarters to 5 quarters $\dfrac{14}{5}$

5. 4 weeks to 15 weeks $\dfrac{4}{15}$

6. 4 yards to 8 yards $\dfrac{1}{2}$

7. $3\dfrac{1}{2}$ dollars to 7 dollars $\dfrac{1}{2}$

8. 3.5 centimeters to 75 centimeters $\dfrac{7}{150}$

(5.2) *Write each rate as a fraction in simplest form.*

9. 8 stillborn births to 1000 live births
$\dfrac{1 \text{ stillborn birth}}{125 \text{ live births}}$

10. 6 professors for 20 graduate research assistants
$\dfrac{3 \text{ professors}}{10 \text{ assistants}}$

11. 15 word-processing pages printed in 6 minutes
$\dfrac{5 \text{ pages}}{2 \text{ minutes}}$

12. 8 computers assembled in 6 hours $\dfrac{4 \text{ computers}}{3 \text{ hours}}$

Find each unit rate.

13. 468 miles in 9 hours   52 miles/hour

14. 180 feet in 12 seconds   15 ft/sec

15. $0.93 for 3 pears $0.31/pear

16. $6.96 for 4 diskettes $1.74/diskette

17. 260 kilometers in 4 hours   65 km/hr

18. 8 gallons of pesticide for 6 acres of crops
$1\dfrac{1}{3}$ gal/acre

19. $184 for books for 5 college courses
$36.80/course

20. 52 bushels of fruit from 4 trees   13 bushels/tree

Find each unit price and decide which is the better buy. Assume that we are comparing different sizes of the same brand.

21. Taco sauce: 8 ounces for $0.99 or 12 ounces for $1.69   8-oz size

22. Peanut butter: 18 ounces for $1.49 or 28 ounces for $2.39   18-oz size

23. 2% milk: 16 ounces for $0.59, $\frac{1}{2}$ gallon for $1.69, or 1 gallon for $2.29 (1 gallon = 128 fluid ounces)

1-gallon size

24. Coca-Cola: 12 ounces for $0.59, 16 ounces for $0.79, or 32 ounces for $1.19   32-oz size

(5.3) *Write each sentence as a proportion.*

25. 20 men is to 14 women as 10 men is to 7 women.

$$\frac{20 \text{ men}}{14 \text{ women}} = \frac{10 \text{ men}}{7 \text{ women}}$$

26. 50 tries is to 4 successes as 25 tries is to 2 successes.

$$\frac{50 \text{ tries}}{4 \text{ successes}} = \frac{25 \text{ tries}}{2 \text{ successes}}$$

27. 16 sandwiches is to 8 players as 2 sandwiches is to 1 player.

$$\frac{16 \text{ sandwiches}}{8 \text{ players}} = \frac{2 \text{ sandwiches}}{1 \text{ player}}$$

28. 12 tires is to 3 cars as 4 tires is to 1 car.

$$\frac{12 \text{ tires}}{3 \text{ cars}} = \frac{4 \text{ tires}}{1 \text{ car}}$$

Determine whether each proportion is true.

29. $\frac{21}{8} = \frac{14}{6}$ no

30. $\frac{3}{5} = \frac{60}{100}$ yes

31. $\frac{3.1}{6.2} = \frac{0.8}{0.16}$ no

32. $\frac{3.75}{3} = \frac{7.5}{6}$ yes

Find the unknown number, n, in each proportion.

33. $\dfrac{n}{6} = \dfrac{15}{18}$ 5

34. $\dfrac{n}{9} = \dfrac{5}{3}$ 15

35. $\dfrac{4}{13} = \dfrac{10}{n}$ 32.5

36. $\dfrac{8}{5} = \dfrac{9}{n}$ 5.625

37. $\dfrac{16}{3} = \dfrac{n}{6}$ 32

38. $\dfrac{n}{3} = \dfrac{9}{2}$ 13.5

39. $\dfrac{n}{5} = \dfrac{27}{2\frac{1}{4}}$ 60

40. $\dfrac{2\frac{1}{2}}{6} = \dfrac{3}{n}$ $7\frac{1}{5}$

41. $\dfrac{n}{0.4} = \dfrac{4.7}{2}$ 0.94

42. $\dfrac{6}{0.3} = \dfrac{7.2}{n}$ 0.36

(5.4) *Solve.*

The ratio of a quarterback's completed passes to attempted passes is 3 to 7.

43. If he attempts 32 passes, find how many passes he completed. Round to the nearest whole pass.   14

44. If he completed 15 passes, find how many passes he attempted.   35

One bag of pesticide covers 4000 square feet of crops.

45. Find how many bags of pesticide should be purchased to cover a rectangular garden 180 feet by 175 feet.   8 bags

46. Find how many bags of pesticide should be purchased to cover a square garden 250 feet on each side.   16 bags

An owner of a Ford Escort can drive 420 miles on 11 gallons of gas.

47. If Tom Aloiso runs out of gas in an Escort and AAA comes to his rescue with $1\frac{1}{2}$ gallons of gas, determine whether Tom can then drive to a gas station 65 miles away.   no

48. Find how many gallons of gas Tom can expect to burn on a 3000-mile trip. Round to the nearest gallon.   79 gal

Yearly homeowner property taxes are figured at a rate of $1.15 tax for every $100 of house value.

49. If a homeowner pays $627.90 in property taxes, find the value of his home. $54,600

50. Find the property taxes on a townhouse valued at $89,000. $1023.50

On an architect's blueprint, 1 inch = 12 feet.

51. Find the length of a wall represented by a $3\frac{3}{8}$-inch line on the blueprint. $40\frac{1}{2}$ ft

52. If an exterior wall is 99 feet long, find how long the blueprint measurement should be. $8\frac{1}{4}$ in.

CHAPTER 5 TEST

Write each ratio as a fraction in simplest form.

1. 4500 trees to 6500 trees

2. $75 to $10

3. 28 men to every 4 women

4. 9 inches of rain in 30 days

Find each unit rate.

5. 650 kilometers in 8 hours

6. 8 inches of rain in 12 hours

7. 140 students for 5 teachers

Find each unit price and decide which is the better buy.

8. Steak sauce:
8 ounces for $1.19

12 ounces for $1.89

9. Jelly:
16 ounces for $1.49

24 ounces for $2.39

Determine whether each proportion is true.

10. $\dfrac{28}{16} = \dfrac{14}{8}$

11. $\dfrac{3.6}{2.2} = \dfrac{1.9}{1.2}$

Find the unknown number, n, in each proportion.

12. $\dfrac{n}{3} = \dfrac{15}{9}$

13. $\dfrac{8}{n} = \dfrac{11}{6}$

14. $\dfrac{\frac{15}{12}}{\frac{3}{7}} = \dfrac{n}{\frac{4}{5}}$

15. $\dfrac{1.5}{5} = \dfrac{2.4}{n}$

Solve.

16. On an architect's drawing, 2 inches corresponds to 9 feet. Find the length of a home represented by a line that is 11 inches long.

17. If a car can be driven 80 miles in 3 hours, how long will it take to travel 100 miles?

18. The standard dose of medicine for a dog is 10 grams for every 15 pounds of body weight. What is the standard dose for a dog that weighs 80 pounds?

19. Jerome Grant worked 6 hours and packed 86 cartons of books. At this rate, how many cartons can he pack in 8 hours?

20. Currently in the American adult population, 12 out of 25 of us drink coffee. In a town with a population of 31,000 adults, how many of these adults would you expect to drink coffee? (*Source: Chicago Tribune*, 1/24/98.)

16. $49\frac{1}{2}$ feet

17. $3\frac{3}{4}$ hours

18. $53\frac{1}{3}$ grams

19. $144\frac{2}{3}$ cartons

20. 14,880 adults

| Name _____ | Section _____ Date _____ | **ANSWERS** |

CUMULATIVE REVIEW

1. Subtract. Check each answer by adding.
   a. $12 - 9$

   b. $11 - 6$

   c. $5 - 5$

   d. $7 - 0$

2. Round 248,982 to the nearest hundred.

3. Multiply: $\begin{array}{r} 25 \\ \times\ 8 \\ \hline \end{array}$

4. The director of a learning lab at a local community college is working on next year's budget. Thirty-three new video players are needed at a cost of $540 each. What is the total cost of these video players?

5. Find the prime factorization of 45.

6. Write $\dfrac{12}{20}$ in simplest form.

Multiply.

7. $\dfrac{3}{4} \cdot \dfrac{8}{5}$

8. $\dfrac{6}{13} \cdot \dfrac{26}{30}$

Add and simplify.

9. $\dfrac{2}{7} + \dfrac{3}{7}$

10. $\dfrac{7}{8} + \dfrac{6}{8} + \dfrac{3}{8}$

11. Find the LCM of 6 and 9.

12. Write an equivalent fraction with the indicated denominator.

$$\frac{1}{2} = \frac{}{14}$$

1. a. 3

   b. 5

   c. 0

   d. 7 (Sec. 1.3, Ex. 1)

2. 249,000 (Sec. 1.4, Ex. 3)

3. 200 (Sec. 1.5, Ex. 3)

4. $17,820 (Sec. 1.7, Ex. 3)

5. $3 \cdot 3 \cdot 5$ or $3^2 \cdot 5$ (Sec. 2.2, Ex. 3)

6. $\dfrac{3}{5}$ (Sec. 2.3, Ex. 1)

7. $\dfrac{6}{5}$ (Sec. 2.4, Ex. 5)

8. $\dfrac{2}{5}$ (Sec. 2.4, Ex. 6)

9. $\dfrac{5}{7}$ (Sec. 3.1, Ex. 1)

10. 2 (Sec 3.1, Ex. 3)

11. 18 (Sec. 3.2, Ex. 2)

12. $\dfrac{7}{14}$ (Sec. 3.2, Ex. 9)

13. $\frac{8}{33}$ (Sec. 3.3, Ex. 6)

14. $\frac{1}{6}$ of an hour (Sec. 3.3, Ex. 9)

15. $7\frac{17}{24}$ (Sec. 3.4, Ex. 1)

16. $>$ (Sec. 3.5, Ex. 1)

17. one and three tenths (Sec. 4.1, Ex. 1)

18. 736.2 (Sec. 4.2, Ex. 3)

19. 25.454 (Sec. 4.3, Ex. 1)

20. 0.0849 (Sec. 4.4, Ex. 2)

21. 0.125 (Sec. 4.5, Ex. 3)

22. 3.7 (Sec. 4.6, Ex. 5)

23. $\frac{4}{9}, \frac{9}{20}$, 0.456 (Sec. 4.7, Ex. 6)

24. $\frac{2.6}{3.1}$ (Sec. 5.1, Ex. 2)

25. $\dfrac{1\frac{1}{2}}{7\frac{3}{4}}$ (Sec. 5.1, Ex. 3)

Name _____

13. Subtract: $\dfrac{10}{11} - \dfrac{2}{3}$

14. A flight from Tucson to Phoenix, Arizona, requires $\dfrac{5}{12}$ of an hour. If the plane has been flying $\dfrac{1}{4}$ of an hour, find how much time remains before landing.

15. Add: $2\dfrac{1}{3} + 5\dfrac{3}{8}$

16. Insert $<$ or $>$ to form a true statement.

$$\frac{3}{10} \qquad \frac{2}{7}$$

17. Write the decimal 1.3 in words.

18. Round 736.2359 to the nearest tenth.

19. Add: $23.85 + 1.604$

20. Multiply: 0.283×0.3

21. Divide: $0.5 \div 4$

22. Simplify: $0.5(8.6 - 1.2)$

23. Write the numbers in order from smallest to largest.

$$\frac{9}{20}, \quad \frac{4}{9}, \quad 0.456$$

Write each ratio using fractional notation.

24. The ratio of 2.6 to 3.1

25. The ratio of $1\dfrac{1}{2}$ to $7\dfrac{3}{4}$

Percent

CHAPTER 6

This chapter is devoted to percent, a concept used virtually every day in ordinary and business life. Understanding percent and using it efficiently depends on understanding ratios, because a percent is a ratio whose denominator is 100. We present techniques to write percents as fractions and as decimals and then solve problems relating to sales tax, commission, discounts, interest, and other real-life situations using percents.

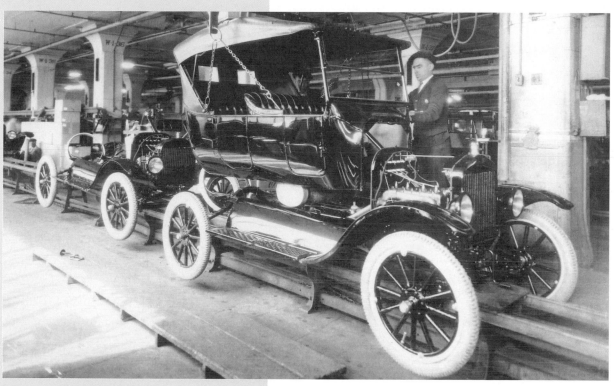

The Model T, developed by Henry Ford in 1908, was the world's first mass-produced automobile. It sold for $850. In 1913, Ford Motor Company introduced interchangeable parts and the moving assembly line, and the automobile industry as we know it today was born. For almost half a century, American automobile manufacturers produced the majority of the world's motor vehicles. However, in the 1970s and 1980s, the sales of foreign auto imports in the United States began to rise. In 1987, 31.1% of all auto sales in the United States were imports. However, American auto producers started to make a comeback, reducing the sales of imported vehicles to just 14.9% of all such sales in 1996. In Exercises 57–58 and 61–62 on page 345, we will see some of the ways percents are used by the automobile manufacturing industry.

Chapter 6 Pretest

1. In a classroom of 100 students, 48 are female. What percent of the students in the classroom are female?

Write each percent as a decimal.

2. 73%

3. 6.8%

Write each decimal as a percent.

4. 0.03

5. 2.1

Write each percent as a fraction in simplest form.

6. 22%

7. 1.5%

Write each fraction or mixed number as a percent.

8. $\dfrac{9}{10}$

9. $2\dfrac{1}{5}$

10. A salesman receives a commission of 3.5% of his total sales. Write 3.5% as a decimal.

11. Translate the following to a percent equation: 4 is what percent of 28?

12. Translate to a proportion: 18% of what number is 70?

Solve.

13. 6 is 15% of what number?

14. What number is 40% of 16?

15. A $230.00 CD player is on sale at 20% off. What is the discount and what is the sale price?

16. In hopes of increasing sales, a local fast food chain decreased the price of its hamburger meal from $2.39 to $1.89. What is the percent decrease? Round to the nearest whole percent.

17. Find the sales tax and the total price on a purchase of a $42.00 dress in a city where the sales tax rate is 6%.

18. A house sold for $125,000 and the real estate agent earned a commission of $5000. Find the rate of commission.

19. Find the simple interest after 3 years on $600 at an interest rate of 8%.

20. Find the monthly payment on a $4000 loan for 4 years. The interest on the 4-year loan is $960.

ANSWERS

1. 48%; (6.1A)

2. 0.73; (6.1B)

3. 0.068; (6.1B)

4. 3%; (6.1C)

5. 210%; (6.1C)

6. $\dfrac{11}{50}$; (6.2A)

7. $\dfrac{3}{200}$; (6.2A)

8. 90%; (6.2B)

9. 220%; (6.2B)

10. 0.035; (6.2C)

11. $4 = n \cdot 28$; (6.3A)

12. $\dfrac{70}{b} = \dfrac{18}{100}$; (6.4A)

13. 40; (6.3B, 6.4B)

14. 6.4; (6.3B, 6.4B)

15. discount: $46; sale price: $184; (6.5A)

16. 21%; (6.5B)

17. sales tax: $2.52; total price: $44.52; (6.6A)

18. 4%; (6.6B)

19. $144; (6.7A)

20. $103.33; (6.7C)

6.1 INTRODUCTION TO PERCENT

A UNDERSTANDING PERCENT

The word **percent** comes from the Latin phrase *per centum*, which means "**per 100**." For example, 53% (percent) means 53 per 100. In the square below, 53 of the 100 squares are shaded. Thus 53% of the figure is shaded.

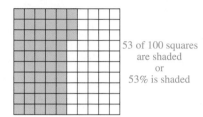

53 of 100 squares
are shaded
or
53% is shaded

Since 53% means 53 per 100, 53% is the ratio of 53 to 100, or $\frac{53}{100}$.

$$53\% = \frac{53}{100}$$

Also,

$$7\% = \frac{7}{100} \quad \text{7 parts per 100 parts}$$

$$73\% = \frac{73}{100} \quad \text{73 parts per 100 parts}$$

$$109\% = \frac{109}{100} \quad \text{109 parts per 100 parts}$$

Percent is used in a variety of everyday situations. For example:

The interest rate is 5.7%.
28% of U.S. homes have Internet access.
The store is having a 25% off sale.
78% of us trust our local fire department.
The federal debt has increased 107% in the last 10 years.

Example 1 In a survey of 100 people, 17 people drive blue cars. What percent of people drive blue cars?

Solution: Since 17 people out of 100 drive blue cars, the fraction is $\frac{17}{100}$. Then

$$\frac{17}{100} = 17\%$$

Example 2 46 out of every 100 college students live at home. What percent of students live at home? (*Source:* Independent Insurance Agents of America)

Solution: $\frac{46}{100} = 46\%$

TEACHING TIP

Ask students to find the percent of unshaded squares. Then ask how the two percents are related.

PERCENT

Percent means **per one hundred**. The "%" symbol is used to denote percent.

TEACHING TIP

You may want to ask students what "109 parts out of 100 parts" means. Illustrate its meaning by having them think of the squares as the total amount earned by the student in an afternoon. Then ask, "Suppose you earned $100 in an afternoon. How much would each square be worth? If I earn 53% of the amount you earn, how much would I earn?" Repeat using all the percents given. When you get to 109%, emphasize "I would earn 109 dollars for every 100 dollars you earned or 109 parts of 100 parts."

Practice Problem 1

Of 100 students in a club, 23 are freshmen. What percent of the students are freshmen?

Practice Problem 2

29 out of 100 executives are in their forties. What percent of executives are in their forties?

Answers
1. 23%, **2.** 29%

WRITING A PERCENT AS A DECIMAL

Drop the percent symbol and move the decimal point two places to the left.

$$43\% = 0.43$$

Practice Problem 3

Write 89% as a decimal.

Practice Problems 4–6

Write each percent as a decimal.

4. 2.7%   5. 150%   6. 0.69%

✓ CONCEPT CHECK

Why is it incorrect to write the percent 0.033% as 3.3 in decimal form?

WRITING A DECIMAL AS A PERCENT

Move the decimal point two places to the right and attach the percent symbol, %.

$$0.27 = 27.\%$$

Practice Problem 7

Write 0.19 as a percent.

Practice Problems 8–10

Write each decimal as a percent.

8. 1.75   9. 0.044   10. 0.7

✓ CONCEPT CHECK

Why is it incorrect to write the decimal 0.0345 as 34.5% in percent form?

Answers

3. 0.89, 4. 0.027, 5. 1.5, 6. 0.0069,
7. 19%, 8. 175%, 9. 4.4%, 10. 70%

✓ **Concept Check:** To write a percent as a decimal, the decimal point should be moved two places to the left, not to the right. So the correct answer is 0.00033.

✓ **Concept Check:** To change a decimal to a percent, the decimal point should be moved *only* two places to the right. So the correct answer is 3.45%.

B WRITING PERCENTS AS DECIMALS

To write a percent as a decimal, we can first write the percent as a fraction.

$$53\% = \frac{53}{100}$$

Now we can write the fraction as a decimal as we did in Section 4.7.

$$\frac{53}{100} = 0.53 \quad \text{(53 hundredths)}$$

Notice that the result is

$$53.\% = 0.53 \quad \text{Drop the percent symbol and move the decimal point two places to the left.}$$

Example 3   Write 23% as a decimal.

Solution: $23\% = 23.\%$ Drop the percent symbol and move the decimal
$= 0.23$ point two places to the left.

Examples   Write each percent as a decimal.

4. $4.6\% = 0.046$ Drop the percent symbol and move the decimal point two places to the left.

5. $190\% = 190.\% = 1.90$ or 1.9

6. $0.74\% = 0.0074$

TRY THE CONCEPT CHECK IN THE MARGIN.

C WRITING DECIMALS AS PERCENTS

To write a decimal as a percent, we can first write the decimal as a fraction.

$$0.38 = \frac{38}{100}$$

Now we can write the fraction as a percent.

$$\frac{38}{100} = 38\%$$

Notice that the result is

$$0.38 = 38.\% \quad \text{Move the decimal point two places to the right and attach a percent symbol.}$$

Example 7   Write 0.65 as a percent.

Solution: $0.65 = 65.\%$ Move the decimal point two places to the right and attach a percent symbol.
$= 65\%$

Examples   Write each decimal as a percent.

8. $1.25 = 125.\%$ or 125%

9. $0.012 = 001.2\%$ or 1.2%

10. $0.6 = 060.\%$ or 60%

TRY THE CONCEPT CHECK IN THE MARGIN.

Name _____ Section _____ Date _____

EXERCISE SET 6.1

A *Solve. See Examples 1 and 2.*

1. A basketball player makes 81 out of 100 attempted free throws. What percent of free throws was made?

2. In a survey of 100 people, 54 preferred chocolate syrup on their ice cream. What percent preferred chocolate syrup?

Adults were asked what type of cookie was their favorite. The circle graph below shows the results for every 100 people. Use this graph to answer Exercises 3–6. See Examples 1 and 2.

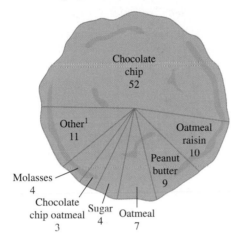

[1]–all 1% or less
Source: USA Today, 2/23/96

3. What percent preferred peanut butter cookies?

4. What percent preferred oatmeal raisin cookies?

5. What type of cookies was preferred by most adults? What percent preferred this type of cookie?

6. What two types of cookies were preferred by the same number of adults? What percent preferred each type?

7. 12 out of 100 adults have watched an entire infomercial. What percent is this? (*Source:* Aragon Consulting Group)

8. 89 out of 100 classrooms for grades K–6 have computers. What percent is this? (*Source:* 1997 Tenth Planet Teachers and Technology Survey)

B *Write each percent as a decimal. See Examples 3 through 6.*

9. 48%   **10.** 64%   **11.** 6%   **12.** 9%

13. 100%   **14.** 136%   **15.** 61.3%   **16.** 52.7%

1. 81%

2. 54%

3. 9%

4. 10%

5. chocolate chip; 52%

6. molasses and sugar; 4%

7. 12%

8. 89%

9. 0.48

10. 0.64

11. 0.06

12. 0.09

13. 1

14. 1.36

15. 0.613

16. 0.527

335

336

17. 2.8%   **18.** 1.7%   **19.** 0.6%   **20.** 0.9%

21. 300%   **22.** 500%   **23.** 32.58%   **24.** 72.18%

Write each percent as a decimal. See Examples 3 through 6.

25. 73.7% of the work force in the United States works 35 hours or more per week. (*Source:* U.S. Bureau of Labor)

26. Approximately 23.6% of new pickup trucks and vans are white, making white the most popular new vehicle color for that class. (*Source:* American Automobile Manufacturers Association)

27. The United States is the largest consumer of energy in the world, using 25% of all energy produced worldwide each year. (*Source:* Energy Information Administration)

28. A health insurance company pays 80% of a person's medical costs.

29. In 1996, 81.8% of all Southwest Airlines' flights were on time, making Southwest the leading airline for on-time arrivals that year. (*Source:* U.S. Department of Transportation)

30. In 1997, 42.5% of all public high schools in the United States owned at least one laser disc player. (*Source:* Quality Education Data, Inc.)

Write each decimal as a percent. See Examples 7 through 10.

31. 0.98   **32.** 0.75   **33.** 3.1   **34.** 4.8

35. 29.00   **36.** 56.00   **37.** 0.003   **38.** 0.006

39. 0.22   **40.** 0.45   **41.** 5.3   **42.** 1.6

Name _____

43. 0.056   **44.** 0.027   **45.** 0.3328   **46.** 0.1115

47. 3.00   **48.** 5.00   **49.** 0.7   **50.** 0.8

Write each decimal as a percent. See Examples 7 through 10.

51. The Munoz family saves 0.10 of their take-home pay.

52. The cost of an item for sale is 0.7 of the sales price.

53. In 1997, 0.702 of all recorded music sales were in the full-length CD format. (*Source:* Recording Industry Association of America)

54. In 1997, 0.522 of all mail delivered by the United States Postal Service was first-class mail. (*Source:* United States Postal Service)

55. People take aspirin for a variety of reasons. The most common use of aspirin is to prevent heart disease, accounting for 0.38 of all aspirin use. (*Source:* Bayer Market Research)

56. The highest income tax rate in the state of Iowa is 0.0998 on taxable income over $50,040. (*Source: CCH State Tax Guide*)

REVIEW AND PREVIEW

Write each fraction as a decimal. See Section 4.7.

57. $\frac{1}{4}$ **58.** $\frac{3}{5}$ **59.** $\frac{13}{20}$

60. $\frac{11}{40}$ **61.** $\frac{9}{10}$ **62.** $\frac{7}{10}$

43. 5.6%

44. 2.7%

45. 33.28%

46. 11.15%

47. 300%

48. 500%

49. 70%

50. 80%

51. 10%

52. 70%

53. 70.2%

54. 52.2%

55. 38%

56. 9.98%

57. 0.25

58. 0.60

59. 0.65

60. 0.275

61. 0.9

62. 0.7

337

63. Home health aides

Human services
64. workers

65. 1.3

66. 1.1

67. answers may vary

68. answers may vary

COMBINING CONCEPTS

The bar graph shows the predicted fastest growing occupations. Use this graph for Exercises 63–66.

Fastest Growing Occupations: 1992 to 2005

Occupation	Percent change
Home health aides	138%
Human services workers	135%
Personal and home care aides	130%
Computer engineers and scientists	111%
Systems analysts	110%
Physical and corrective therapy assistants	92%
Physical therapists	88%
Paralegals	86%
Occupational therapy assistants and aides	78%
Electronic pagination systems workers	77%

Percent change

63. What occupation is predicted to be the fastest growing?

64. What occupation is predicted to be the second fastest growing?

65. Write the percent change for Personal and home care aides as a decimal.

66. Write the percent change for Systems analysts as a decimal.

67. In your own words, explain how to write a percent as a decimal.

68. In your own words, explain how to write a decimal as a percent.

6.2 PERCENTS AND FRACTIONS

A WRITING PERCENTS AS FRACTIONS

When we write a percent as a fraction, we usually then write the fraction in simplest form. For example, recall from the previous section that

$$50\% = \frac{50}{100}$$

Then we write the fraction in simplest form:

$$\frac{50}{100} = \frac{\overset{1}{\cancel{50}}}{2 \cdot \cancel{50}} = \frac{1}{2}$$

> ### WRITING A PERCENT AS A FRACTION
>
> Drop the percent symbol and write the number over 100. Then simplify the fraction if possible.
>
> $$7\% = \frac{7}{100}$$

Examples

Write each percent as a fraction or mixed number in simplest form.

1. $40\% = \frac{40}{100} = \frac{2 \cdot \overset{1}{\cancel{20}}}{5 \cdot \underset{1}{\cancel{20}}} = \frac{2}{5}$

2. $1.9\% = \frac{1.9}{100}$. Next we multiply numerator and denominator by 10.

$$\frac{1.9}{100} = \frac{1.9 \cdot 10}{100 \cdot 10} = \frac{19}{1000}$$

3. $125\% = \frac{125}{100} = \frac{5 \cdot \overset{1}{\cancel{25}}}{4 \cdot \underset{1}{\cancel{25}}} = \frac{5}{4}$ or $1\frac{1}{4}$

4. $33\frac{1}{3}\% = \frac{33\frac{1}{3}}{100} = \frac{\frac{100}{3}}{100}$. Recall that the fraction bar means division. Thus

$$\frac{100}{3} \div 100 = \frac{100}{3} \cdot \frac{1}{100} = \frac{1}{3}$$

5. $100\% = \frac{100}{100} = 1$

B WRITING FRACTIONS AS PERCENTS

Recall that to write a percent as a fraction, we drop the percent symbol and divide by 100. We can reverse these steps to write a fraction as a percent.

Objectives

A Write percents as fractions.
B Write fractions as percents.
C Convert percents, decimals, and fractions.

SSM   CD-ROM   Video 6.2

Practice Problems 1–5

Write each percent as a fraction in simplest form.

1. 25%

2. 2.3%

3. 150%

4. $66\frac{2}{3}\%$

5. 8%

Answers

1. $\frac{1}{4}$, **2.** $\frac{23}{1000}$, **3.** $\frac{3}{2}$, **4.** $\frac{2}{3}$, **5.** $\frac{2}{25}$

WRITING A FRACTION AS A PERCENT

Multiply by 100 and attach a percent symbol. This is the same as multiplying by 100%.

$$\frac{1}{8} = \frac{1}{8} \cdot 100\% = \frac{1}{8} \cdot \frac{100}{1}\% = \frac{100}{8}\% = 12\frac{1}{2}\% \text{ or } 12.5\%$$

HELPFUL HINT

From Example 5, we know that

$$100\% = 1$$

Recall that when we multiply a number by 1, we are not changing the value of that number. This means that when we multiply a number by 100%, we are not changing its value but rather writing the number as an equivalent percent.

Examples   Write each fraction or mixed number as a percent.

6. $\frac{9}{20} = \frac{9}{20} \cdot 100\% = \frac{9}{20} \cdot \frac{100}{1}\% = \frac{900}{20}\% = 45\%$

7. $\frac{2}{3} = \frac{2}{3} \cdot 100\% = \frac{2}{3} \cdot \frac{100}{1}\% = \frac{200}{3}\% = 66\frac{2}{3}\%$

8. $1\frac{1}{2} = \frac{3}{2} \cdot 100\% = \frac{3}{2} \cdot \frac{100}{1}\% = \frac{300}{2}\% = 150\%$

TRY THE CONCEPT CHECK IN THE MARGIN.

Example 9   Write $\frac{1}{12}$ as a percent. Round to the nearest hundredth percent.

Solution:

$$\frac{1}{12} = \frac{1}{12} \cdot 100\% = \frac{1}{12} \cdot \frac{100\%}{1} = \frac{100}{12}\% \approx 8.33\%$$

approximately

$$\begin{array}{r} 8.333 \\ 12\overline{)100.000} \\ -96 \\ \hline 4\,0 \\ -3\,6 \\ \hline 40 \\ -36 \\ \hline 40 \\ -36 \\ \hline 4 \end{array}$$

≈ 8.33

Thus, $\frac{1}{12}$ is approximately 8.33%.

C   CONVERTING PERCENTS, DECIMALS, AND FRACTIONS

Let's summarize what we have learned so far about percents, decimals, and fractions.

TEACHING TIP

For Examples 6 and 8, you may want to point out that students could also write the fraction as a percent by rewriting the fraction with a denominator of 100.

Practice Problems 6–8

Write each fraction or mixed number as a percent.

6. $\frac{1}{2}$ **7.** $\frac{7}{40}$ **8.** $2\frac{1}{4}$

✓ CONCEPT CHECK

Which digit in the percent 76.4582% represents

a. A tenth percent?

b. A thousandth percent?

c. A hundredth percent?

d. A whole percent?

Practice Problem 9

Write $\frac{3}{17}$ as a percent. Round to the nearest hundredth percent.

TEACHING TIP

Before going over the summary, you may want to have students write their own summaries of conversion strategies.

Answers

6. 50%,   **7.** $17\frac{1}{2}\%$,   **8.** 225%,   **9.** 17.65%

✓ Concept Check

a. 4,   **b.** 8,   **c.** 5,   **d.** 6

SUMMARY OF CONVERTING PERCENTS, DECIMALS, AND FRACTIONS

▲ *To write a percent as a decimal*, drop the % symbol and move the decimal point two places to the left.
▲ *To write a decimal as a percent*, move the decimal point two places to the right and attach the % symbol.
▲ *To write a percent as a fraction*, drop the % symbol and write the number over 100.
▲ *To write a fraction as a percent*, multiply the fraction by 100%.

Example 10   17.8% of automobile thefts in the continental United States occur in the Midwest. Write this percent as a decimal. (*Source:* The American Automobile Manufacturers Association)

Solution: $17.8\% = 0.178$. Thus, 17.8% written as a decimal is 0.178.

——

Example 11   An advertisement for a stereo system reads "$\frac{1}{4}$ off." What percent off is this?

Solution: Write $\frac{1}{4}$ as a percent.

$$\frac{1}{4} = \frac{1}{4} \cdot 100\% = \frac{1}{4} \cdot \frac{100\%}{1} = \frac{100}{4}\% = 25\%$$

Then "$\frac{1}{4}$ off" is the same as "25% off."

——

It is helpful to know a few basic percent conversions. Appendix D contains a handy reference of percent, decimal, and fraction equivalences.

Practice Problem 10

A family decides to spend no more than 25% of its monthly income on rent. Write 25% as a decimal.

Practice Problem 11

Provincetown's budget for waste disposal increased by $1\frac{1}{4}$ times over the budget from last year. What percent increase is this?

Answers
10. 0.25,   **11.** 125%

Focus On Real World

MORTGAGES

Buying a house may be one of the most expensive purchases we make. The amount borrowed from a lending institution for real estate is called a **mortgage**. The lending institutions that normally make mortgage loans include banks, savings and loans, credit unions, and mortgage companies.

There are basically three items that define a mortgage loan: the principal, the loan term, and the interest rate. The principal is the dollar amount being borrowed or financed by the home buyers. The loan term is the length of the loan, or how long it will take to pay off the loan. The interest rate, normally expressed as a percent, governs how much must be paid for the privilege of borrowing the money.

Mortgages come in all shapes and sizes. Loan terms can range anywhere from 10 years to 15, 20, 25, 30, or even 40 years. The interest rates on shorter loans are generally lower than the interest rates on longer loans. For instance, the interest rate on a 15-year loan might be 7.25% when the interest rate on a 30-year loan is 7.5%. Interest rates also tend to be lower on loans with a smaller principal as compared to a large principal. For example, many banks offer a jumbo mortgage loan that applies only to principals over a certain limit, generally over $227,150. The interest rates on jumbo mortgages are higher (often by about 0.25%) than on other mortgage programs.

Most lending institutions require a home buyer to make a **down payment** in cash on a home. The size of the down payment usually depends on the buyer's circumstances and the specific mortgage program chosen, but down payments generally range from 3% to 20% of the home's value. A typical down payment on a house is 10% of the purchase price. Once a down payment has been chosen, the mortgage amount can be calculated by subtracting the amount of the down payment from the purchase price:

$$\text{mortgage} = \text{purchase price} - \text{down payment}$$

Besides the down payment, there are a number of initial costs related to the mortgage that must be paid. These costs are called **closing costs**. Two expensive items on the list of closing costs are the **loan origination fee** and **loan discount points**. Both of these items are generally given in "points," where each point is equal to 1% of the mortgage amount. For example, "3 points" means 3% of the mortgage amount. The loan origination fee is the fee charged by the lender to cover the costs of preparing all the loan documents. Loan discount points is prepaid interest paid at closing. Home buyers generally can choose whether or not they will pay loan discount points. Doing so lowers the interest rate on the mortgage.

$$\text{loan origination fee} = \text{mortgage} \cdot \text{points}$$

$$\text{loan discount points} = \text{mortgage} \cdot \text{points}$$

CRITICAL THINKING

Suppose you are considering buying a house with a purchase price of $140,000.

1. Find the amount of the down payment if you plan to make a 10% down payment.
2. Find the mortgage amount.
3. Calculate the loan origination fee of 1 point.
4. To get a lower interest rate on your loan, suppose you choose to pay 2.5 loan discount points at closing. How much will you be paying in loan discount points?

Name _____ **Section** _____ **Date** _____

MENTAL MATH

Write each fraction as a percent.

1. $\dfrac{13}{100}$ 2. $\dfrac{92}{100}$ 3. $\dfrac{87}{100}$

4. $\dfrac{71}{100}$ 5. $\dfrac{1}{100}$ 6. $\dfrac{2}{100}$

EXERCISE SET 6.2

A *Write each percent as a fraction or mixed number in simplest form. See Examples 1 through 5.*

1. 12%   2. 24%   ▭3. 4%   4. 2%

5. 4.5%   6. 7.5%   ▭7. 175%   8. 250%

9. 73%   10. 86%   11. 12.5%   12. 62.5%

13. 6.25%   14. 37.5%   15. 8%   16. 16%

▭17. $10\dfrac{1}{3}$%   18. $7\dfrac{3}{4}$%   19. $22\dfrac{3}{8}$%   20. $15\dfrac{5}{8}$%

B *Write each fraction or mixed number as a percent. See Examples 6 through 8.*

21. $\dfrac{3}{4}$ 22. $\dfrac{1}{2}$ ▭23. $\dfrac{7}{10}$ 24. $\dfrac{3}{10}$

25. $\dfrac{2}{5}$ 26. $\dfrac{4}{5}$ 27. $\dfrac{59}{100}$ 28. $\dfrac{73}{100}$

29. 34%

30. 94%

31. $37\frac{1}{2}\%$

32. $62\frac{1}{2}\%$

33. $31\frac{1}{4}\%$

34. $43\frac{3}{4}\%$

35. 160%

36. 175%

37. $66\frac{2}{3}\%$

38. $33\frac{1}{3}\%$

39. 65%

40. 15%

41. 250%

42. 220%

43. 190%

44. 270%

45. 63.64%

46. 41.67%

47. 26.67%

48. 90.91%

49. 14.29%

50. 11.11%

51. 91.67%

52. 83.33%

53. 0.35, $\frac{7}{20}$; 20%, 0.2; 50%, $\frac{1}{2}$; 0.7, $\frac{7}{10}$; 37.5%, 0.375

54. 52.5%, $\frac{21}{40}$; 75%, 0.75; $0.666\overline{6}$, $\frac{2}{3}$; $83\frac{1}{3}\%$, 0.8333; 1, 1

55. 0.4, $\frac{2}{5}$; $23\frac{1}{2}\%$; $\frac{47}{200}$; 0.3333, $\frac{1}{3}$; 87.5%, 0.875; 0.075, $\frac{3}{40}$

56. 0.5, $\frac{1}{2}$; 40%, 0.4; 25%, $\frac{1}{4}$; 0.125, $\frac{1}{8}$; 62.5%, 0.625; 14%, 0.14

344

Name _____

29. $\frac{17}{50}$

30. $\frac{47}{50}$

31. $\frac{3}{8}$

32. $\frac{5}{8}$

33. $\frac{5}{16}$

34. $\frac{7}{16}$

35. $1\frac{3}{5}$

36. $1\frac{3}{4}$

37. $\frac{2}{3}$

38. $\frac{1}{3}$

39. $\frac{13}{20}$

40. $\frac{3}{20}$

41. $2\frac{1}{2}$

42. $2\frac{1}{5}$

43. $1\frac{9}{10}$

44. $2\frac{7}{10}$

Write each fraction as a percent. Round to the nearest hundredth percent. See Example 9.

45. $\frac{7}{11}$

46. $\frac{5}{12}$

47. $\frac{4}{15}$

48. $\frac{10}{11}$

49. $\frac{1}{7}$

50. $\frac{1}{9}$

51. $\frac{11}{12}$

52. $\frac{5}{6}$

C *Complete each table. See Examples 10 and 11.*

53.

Percent	Decimal	Fraction
35%		
		$\frac{1}{5}$
	0.5	
70%		
		$\frac{3}{8}$

54.

Percent	Decimal	Fraction
	0.525	
		$\frac{3}{4}$
$66\frac{2}{3}\%$		
		$\frac{5}{6}$
100%		

55.

Percent	Decimal	Fraction
40%		
	0.235	
		$\frac{4}{5}$
$33\frac{1}{3}\%$		
		$\frac{7}{8}$
7.5%		

56.

Percent	Decimal	Fraction
50%		
		$\frac{2}{5}$
	0.25	
12.5%		
		$\frac{5}{8}$
		$\frac{7}{50}$

Copyright 1000 Prentice Hall Inc.

57. $\dfrac{47}{250}$ _____

Solve. See Examples 10 and 11.

58. 0.757 _____

59. 0.667 _____

60. $\dfrac{21}{250}$ _____

57. Approximately 18.8% of new full-size cars are dark green, making dark green the most popular new vehicle color for that class. Write this percent as a fraction. (*Source:* American Automobile Manufacturers Association)

58. In 1950, the United States produced 75.7% of all motor vehicles made worldwide. Write this percent as a decimal. (*Source:* American Automobile Manufacturers Association)

59. In 1996, 66.7% of all households with televisions subscribed to a cable television service. Write this percent as a decimal. (*Source:* Nielsen Media Research)

60. In 1996, 8.4% of Wisconsin residents were not covered by some type of health insurance. Write this percent as a fraction. (*Source:* U.S. Bureau of the Census)

61. 8.5% _____

62. 21.2% _____

61. In 1996, $\dfrac{17}{200}$ of all new cars sold in the United States were imported from Japan. Write this fraction as a percent. (*Source:* American Automobile Manufacturers Association)

62. In 1980, $\dfrac{53}{250}$ of all new cars sold in the United States were imported from Japan. Write this fraction as a percent. (*Source:* American Automobile Manufacturers Association)

63. 0.0825 _____

63. The sales tax in Slidell, Louisiana, is 8.25%. Write this percent as a decimal.

64. A real estate agent receives a commission of 3% of the sale price of a house. Write this percent as a decimal.

64. 0.03 _____

65. In 1997, the top-rated prime-time television program was *E.R.*, which had an average audience share of $\dfrac{7}{20}$ of all those watching television during that time slot. Write this fraction as a percent. (*Source:* Nielsen Media Research)

66. Approximately $\dfrac{37}{50}$ of all structure fires take place in personal residences. Write this fraction as a percent. (*Source:* National Fire Protection Association)

65. 35% _____

66. 74% _____

Name _____

REVIEW AND PREVIEW

Find the value of n. See Section 5.3.

67. $3 \cdot n = 45$ **68.** $7 \cdot n = 48$ **69.** $8 \cdot n = 80$

70. $2 \cdot n = 16$ **71.** $6 \cdot n = 72$ **72.** $5 \cdot n = 35$

COMBINING CONCEPTS

Write each fraction as a decimal and then write each decimal as a percent. Round the decimal to three decimal places and the percent to the nearest tenth of a percent.

73. $\dfrac{21}{79}$ **74.** $\dfrac{56}{102}$ **75.** $\dfrac{850}{736}$ **76.** $\dfrac{506}{248}$

Fill in the blanks.

77. A fraction written as a percent is greater than 100% when the numerator is _____ than the denominator.
greater/less

78. A decimal written as a percent is less than 100% when the decimal is _____ than 1.
greater/less

79. In your own words, explain how to write a percent as a fraction.

80. In your own words, explain how to write a fraction as a decimal.

Internet Excursions

Go to http://www.prenhall.com/martin-gay
This World Wide Web address will provide you with access to the official Web site of the Women's National Basketball Association (WNBA), or a related site. You will be able to collect statistics of the league leaders before answering the following questions.

81. Name three WNBA statistics that are reported as percents. Explain how each is calculated and what each means. Field goal percentage, three-point field goal percentage, and free throw percentage. Field goal percentage = total field goals made ÷ total field goals attempted; three-point field goal percentage = total three-point field goals made ÷ total three-point field goals attempted; free throw percentage = total free throws made ÷ total free throws attempted

82. List a real example of each of these three types of statistics. Write each example as a fraction, a decimal, and a percent.

346

6.3   SOLVING PERCENT PROBLEMS WITH EQUATIONS

Throughout this text, we have written mathematical statements such as $3 + 10 = 13$, or area = length · width. These statements are called equations. An equation is simply a statement that contains an equal sign. To solve percent problems, we will translate the problems into such mathematical statements, or equations.

A   WRITING PERCENT PROBLEMS AS EQUATIONS

Recognizing key words in a percent problem is helpful in writing the problem as an equation. Three key words in the statement of a percent problem and their meanings are as follows:

of means **multiplication** (·)
is means **equals** (=)
what (or some equivalent) means **the unknown number**.

In our examples, we will let the letter n stand for the unknown number.

Example 1   Translate to an equation:

5 is what percent of 20?

Solution: 5 is what percent of 20?

$$5 = n \cdot 20$$

HELPFUL HINT

Remember that an equation is simply a mathematical statement that contains an equal sign (=).

$$5 = n \cdot 20$$
↑
equal sign

Example 2   Translate to an equation:

1.2 is 30% of what number?

Solution: 1.2 is 30% of what number?

$$1.2 = 30\% \cdot n$$

Example 3   Translate to an equation:

What number is 25% of 0.008?

Solution: What number is 25% of 0.008?

$$n = 25\% \cdot 0.008$$

Examples   Translate each to an equation.

4. 38% of 200 is what number?

$$38\% \cdot 200 = n$$

Objectives

A Write percent problems as equations.

B Solve percent problems

SSM   CD-ROM   Video
6.3

TEACHING TIP

You may want to begin this lesson by asking students if they recall how to find the unknown number n in an equation such as $7 \cdot n = 105$.

Practice Problem 1

Translate: 6 is what percent of 24?

Practice Problem 2

Translate: 1.8 is 20% of what number?

Practice Problem 3

Translate: What number is 40% of 3.6?

Practice Problems 4–6

Translate each to an equation.

4. 42% of 50 is what number?

5. 15% of what number is 9?

6. What percent of 150 is 90?

Answers

1. $6 = n \cdot 24$, **2.** $1.8 = 20\% \cdot n$,
3. $n = 40\% \cdot 3.6$, **4.** $42\% \cdot 50 = n$,
5. $15\% \cdot n = 9$, **6.** $n \cdot 150 = 90$

5. 40% of what number is 80?

$$40\% \cdot n = 80$$

6. What percent of 85 is 34?

$$n \cdot 85 = 34$$

TRY THE CONCEPT CHECK IN THE MARGIN.

✓ CONCEPT CHECK

In the equation $2 \cdot n = 10$, what step is taken to solve the equation?

B SOLVING PERCENT PROBLEMS

You may have noticed by now that each percent problem has contained three numbers—in our examples, two are known and one is unknown. Each of these numbers is given a special name.

15%   of   60   is   9

| 15% percent | · | 60 base | = | 9 amount |

We call this equation the **percent equation.**

> **PERCENT EQUATION**
>
> percent · base = amount

> **HELPFUL HINT**
>
> Notice that the percent equation given above is a true statement. To see this, simplify the left side as shown.
>
> $$15\% \cdot 60 = 9$$
> $$0.15 \cdot 60 = 9 \quad \text{Write 15\% as 0.15.}$$
> $$9 = 9 \quad \text{Multiply.}$$
>
> The statement $9 = 9$ is true.

Once a percent problem has been written as a percent equation, we can use the equation to find the unknown number. This is called **solving** the equation.

SOLVING PERCENT EQUATIONS FOR THE AMOUNT

Practice Problem 7

What number is 20% of 85?

Answers

7. 17

✓ **Concept Check:** If $2 \cdot n = 10$, then $n = \frac{10}{2}$ or $n = 5$.

Example 7   What number is 35% of 40?

Solution:

$$n = 35\% \cdot 40 \quad \text{Translate to an equation.}$$
$$n = 0.35 \cdot 40 \quad \text{Write 35\% as 0.35.}$$
$$n = 14 \quad \text{Multiply:}$$

$$\begin{array}{r} 40 \\ \underline{0.35} \\ 200 \\ \underline{1200} \\ 14.00 \end{array}$$

Then 14 is 35% of 40.

> **HELPFUL HINT**
>
> When solving a percent equation, write the percent as a decimal or fraction.

Example 8

85% of 300 is what number?

Solution:

$$85\% \cdot 300 = n \qquad \text{Translate to an equation.}$$
$$0.85 \cdot 300 = n \qquad \text{Write 85\% as 0.85.}$$
$$255 = n \qquad \text{Multiply: } 0.85 \cdot 300 = 255.$$

Then 85% of 300 is 255.

Practice Problem 8

90% of 150 is what number?

SOLVING PERCENT EQUATIONS FOR THE BASE

Example 9

12% of what number is 0.6?

Solution:

$$12\% \cdot n = 0.6 \qquad \text{Translate to an equation.}$$
$$0.12 \cdot n = 0.6 \qquad \text{Write 12\% as 0.12.}$$

Recall from Section 5.3 that if "0.12 times some number is 0.6," then the number is 0.6 divided by 0.12.

$$n = \frac{0.6}{0.12} \qquad \text{Divide 0.6 by 0.12, the number multiplied by } n.$$
$$n = 5$$

$$\begin{array}{r} 5. \\ 0.12\overline{)0.60} \\ \underline{60} \\ 0 \end{array}$$

Then 12% of 5 is 0.6.

Practice Problem 9

15% of what number is 1.2?

Example 10

13 is $6\frac{1}{2}$% of what number?

Solution:

$$13 = 6\frac{1}{2}\% \cdot n \qquad \text{Translate to an equation.}$$
$$13 = 0.065 \cdot n \qquad 6\frac{1}{2}\% = 6.5\% = 0.065.$$
$$\frac{13}{0.065} = n \qquad \text{Divide 13 by 0.065, the number multiplied by } n.$$
$$200 = n$$

$$\begin{array}{r} 200. \\ 0.065\overline{)13.000} \\ \underline{130} \\ 0 \end{array}$$

Then 13 is $6\frac{1}{2}$% of 200.

Practice Problem 10

27 is $4\frac{1}{2}$% of what number?

SOLVING PERCENT EQUATIONS FOR PERCENT

Example 11

What percent of 12 is 9?

Solution:

$$n \cdot 12 = 9 \qquad \text{Translate to an equation.}$$
$$n = \frac{9}{12} \qquad \text{Divide 9 by 12, the number multiplied by } n.$$
$$n = 0.75$$

Practice Problem 11

What percent of 80 is 8?

Answers

8. 135, **9.** 8, **10.** 600, **11.** 10%

Next, since we are looking for percent, write 0.75 as a percent.

$n = 75\%$

Then 75% of 12 is 9.

HELPFUL HINT

If your unknown in the percent equation is percent, don't forget to convert your answer to a percent.

Practice Problem 12

35 is what percent of 25?

Example 12   78 is what percent of 65?

Solution: $78 = n \cdot 65$ Translate to an equation.

$\dfrac{78}{65} = n$ Divide 78 by 65, the number multiplied by n.

$1.2 = n$

$120\% = n$ Write 1.2 as a percent.

Then 78 is 120% of 65.

Answer

12. 140%

Name _____ **Section** _____ **Date** _____

MENTAL MATH

Identify the percent, the base, and the amount in each equation. Recall that
percent · base = amount.

1. $42\% \cdot 50 = 21$

2. $30\% \cdot 65 = 19.5$

3. $107.5 = 125\% \cdot 86$

4. $99 = 110\% \cdot 90$

EXERCISE SET 6.3

A Translate each to an equation. Do not solve. See Examples 1 through 6.

1. 15% of 72 is what number?

2. What number is 25% of 55?

3. 30% of what number is 80?

4. 0.5 is 20% of what number?

5. What percent of 90 is 20?

6. 8 is 50% of what number?

7. 1.9 is 40% of what number?

8. 72% of 63 is what number?

9. What number is 9% of 43?

10. 4.5 is what percent of 45?

B Solve. See Examples 7 and 8.

11. 10% of 35 is what number?

12. 25% of 60 is what number?

13. What number is 14% of 52?

14. What number is 30% of 17?

Solve. See Examples 9 and 10.

15. 30 is 5% of what number?

16. 25 is 25% of what number?

17. 1.2 is 12% of what number?

18. 0.22 is 44% of what number?

Solve. See Examples 11 and 12.

19. 66 is what percent of 60?

20. 30 is what percent of 20?

21. 16 is what percent of 50?

22. 27 is what percent of 50?

351

23. 1

24. 10

25. 45

26. 27

27. 500

28. 115.2

29. 400%

30. 800%

31. 25.2

32. 28.8

33. 45%

34. 84%

35. 35

36. 25

37. $n = 30$

38. $n = 25$

39. $n = 3\frac{7}{11}$

40. $n = 1\frac{5}{13}$

41. $\frac{17}{12} = \frac{n}{20}$

42. $\frac{20}{25} = \frac{n}{10}$

43. $\frac{8}{9} = \frac{14}{n}$

44. $\frac{5}{6} = \frac{15}{n}$

45. 686.625

46. 37%

47. 12,285

48. answers may vary

Solve. See Examples 7 through 12.

23. 0.1 is 10% of what number?

24. 0.5 is 5% of what number?

25. 125% of 36 is what number?

26. 200% of 13.5 is what number?

27. 82.5 is $16\frac{1}{2}$% of what number?

28. 7.2 is $6\frac{1}{4}$% of what number?

29. 126 is what percent of 31.5?

30. 264 is what percent of 33?

31. What number is 42% of 60?

32. What number is 36% of 80?

33. What percent of 150 is 67.5?

34. What percent of 105 is 88.2?

35. 120% of what number is 42?

36. 160% of what number is 40?

REVIEW AND PREVIEW

Find the value of n in each proportion. See Section 5.3.

37. $\frac{27}{n} = \frac{9}{10}$

38. $\frac{35}{n} = \frac{7}{5}$

39. $\frac{n}{5} = \frac{8}{11}$

40. $\frac{n}{3} = \frac{6}{13}$

Write each phrase as a proportion.

41. 17 is to 12 as n is to 20

42. 20 is to 25 as n is to 10

43. 8 is to 9 as 14 is to n

44. 5 is to 6 as 15 is to n

COMBINING CONCEPTS

Solve.

45. 1.5% of 45,775 is what number?

46. What percent of 75,528 is 27,945.36?

47. 22,113 is 180% of what number?

48. In your own words, explain how to solve a percent equation.

6.4   SOLVING PERCENT PROBLEMS WITH PROPORTIONS

There is more than one method that can be used to solve percent problems. In the last section, we used the percent equation. In this section, we will use proportions.

A   WRITING PERCENT PROBLEMS AS PROPORTIONS

To understand the proportion method, recall that 70% means the ratio of 70 to 100, or $\frac{70}{100}$.

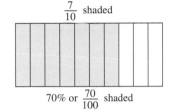

$$70\% = \frac{70}{100} = \frac{7}{10}$$

$\frac{7}{10}$ shaded

70% or $\frac{70}{100}$ shaded

Since the ratio $\frac{70}{100}$ is equal to the ratio $\frac{7}{10}$, we have the proportion

$$\frac{7}{10} = \frac{70}{100}.$$

We call this proportion the percent proportion. In general, we can name the parts of this proportion as follows.

When we translate percent problems to proportions, the **percent** can be identified by looking for the symbol % or the word *percent*. The **base** usually follows the word *of*. The **amount** is the part compared to the whole.

Example 1   Translate to a proportion:

12% of what number is 47?

↓   ↓   ↓

Solution:

percent	base	amount
	It appears after the word *of*.	It is the part compared to the whole.

$$\text{amount} \to \frac{47}{b} = \frac{12}{100} \leftarrow \text{percent}$$
$$\text{base} \to$$

Example 2   Translate to a proportion:

101 is what percent of 200?

↓   ↓   ↓

Solution:

amount	percent	base
It is the part compared to the whole.		It appears after the word *of*.

$$\text{amount} \to \frac{101}{200} = \frac{p}{100} \leftarrow \text{percent}$$
$$\text{base} \to$$

PERCENT PROPORTION

$$\frac{\text{amount}}{\text{base}} = \frac{\text{percent}}{100} \leftarrow \text{always 100}$$

or

$$\begin{array}{c}\text{amount} \to \\ \text{base} \to\end{array} \frac{a}{b} = \frac{p}{100} \leftarrow \text{percent}$$

HELPFUL HINT

Part of Proportion	How It's Identified
Percent	% or percent
Base	Appears after *of*
Amount	Part compared to whole

Practice Problem 1

Translate to a proportion:

15% of what number is 55?

Practice Problem 2

Translate to a proportion:

35 is what percent of 70?

Answers

1. $\frac{15}{100} = \frac{55}{b}$,   2. $\frac{35}{70} = \frac{p}{100}$

Practice Problem 3

Translate to a proportion:

What number is 25% of 68?

Practice Problem 4

Translate to a proportion:

520 is 65% of what number?

Practice Problem 5

Translate to a proportion:

65 is what percent of 50?

Practice Problem 6

Translate to a proportion:
36% of 80 is what number?

✓ CONCEPT CHECK

When solving a percent problem with a proportion, describe how you can check the result.

Practice Problem 7

What number is 8% of 120?

Answers

3. $\dfrac{a}{68} = \dfrac{25}{100}$, 4. $\dfrac{520}{b} = \dfrac{65}{100}$, 5. $\dfrac{65}{50} = \dfrac{p}{100}$,

6. $\dfrac{a}{80} = \dfrac{36}{100}$, 7. 9.6

✓ **Concept Check:** By putting the result into the proportion and checking that the proportion is true

---

Example 3 Translate to a proportion:

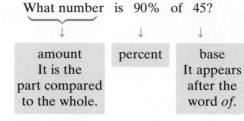

Solution:

$$\underset{\text{base} \rightarrow}{\overset{\text{amount} \rightarrow}{}} \frac{a}{45} = \frac{90}{100} \overset{\leftarrow \text{percent}}{}$$

Example 4 Translate to a proportion:

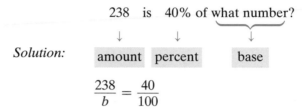

Solution:

$$\frac{238}{b} = \frac{40}{100}$$

Example 5 Translate to a proportion:

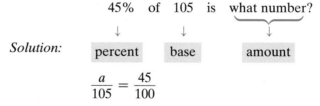

Solution:

$$\frac{75}{30} = \frac{p}{100}$$

Example 6 Translate to a proportion:

45%   of   105   is   what number?

↓   ↓   ↓

| percent | base | amount |

Solution:

$$\frac{a}{105} = \frac{45}{100}$$

TRY THE CONCEPT CHECK IN THE MARGIN.

B SOLVING PERCENT PROBLEMS

The proportions that we have written in this section contain three values that can change: the percent number, the base, and the amount. If any two of these values are known, we can find the third (unknown value). To do this, we write a percent proportion and find the unknown value as we did in Section 5.3.

Example 7 Solving Percent Proportion for the Amount

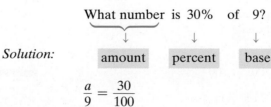

Solution:

$$\frac{a}{9} = \frac{30}{100}$$

To solve, we set cross products equal to each other.

$$\frac{a}{9} = \frac{30}{100}$$

$$9 \cdot 30 = 270$$

$$a \cdot 100$$

$$a \cdot 100 = 270 \quad \text{Set cross products equal.}$$

Recall from Section 5.3 that if "some number times 100 is 270," then the number is 270 divided by 100.

$$a = \frac{270}{100} \quad \text{Divide 270 by 100, the number multiplied by } a.$$

$$a = 2.7 \quad \text{Simplify.}$$

Then 2.7 is 30% of 9.

TRY THE CONCEPT CHECK IN THE MARGIN.

HELPFUL HINT

The proportion in Example 7 contained the ratio $\frac{30}{100}$. A ratio in a proportion may be simplified before solving the proportion. The unknown number in both

$$\frac{a}{9} = \frac{30}{100} \quad \text{and} \quad \frac{a}{9} = \frac{3}{10} \quad \text{is 2.7.}$$

Example 8   Solving Percent Problems for the Base

150% of what number is 30?

Solution:

percent   base   amount

$$\frac{30}{b} = \frac{150}{100} \quad \text{Write the proportion.}$$

$$\frac{30}{b} = \frac{3}{2} \quad \text{Simplify } \frac{150}{100}.$$

$$30 \cdot 2 = b \cdot 3 \quad \text{Set cross products equal.}$$

$$60 = b \cdot 3 \quad \text{Write } 30 \cdot 2 \text{ as 60.}$$

$$\frac{60}{3} = b \quad \text{Divide 60 by 3, the number multiplied by } b.$$

$$20 = b \quad \text{Simplify.}$$

Then 150% of 20 is 30.

TEACHING TIP

After Example 6, you may want to have students translate each of the following proportions into a question of the form "_____ of _____ is _____?" and a question of the form "_____ is _____ of _____?"

a. $\frac{a}{480} = \frac{26}{100}$

b. $\frac{87}{b} = \frac{38}{100}$

c. $\frac{360}{135} = \frac{p}{100}$

✓ CONCEPT CHECK

In the statement "78 is what percent of 350?", which part of the percent proportion is unknown: (a) the amount, (b) the base, or (c) the percent number?

Practice Problem 8

75% of what number is 60?

Answers

8. 80

✓ Concept Check: c

Practice Problem 9

15 is 5% of what number?

Example 9

20.8 is 40% of what number?

Solution:

amount   percent   base

$$\frac{20.8}{b} = \frac{40}{100} \quad \text{or} \quad \frac{20.8}{b} = \frac{2}{5} \qquad \text{Write the proportion and simplify } \frac{40}{100}.$$

$$20.8 \cdot 5 = b \cdot 2 \qquad \text{Set cross products equal.}$$

$$104 = b \cdot 2 \qquad \text{Multiply.}$$

$$\frac{104}{2} = b \qquad \text{Divide 104 by 2, the number multiplied by } b.$$

$$52 = b \qquad \text{Simplify.}$$

Then 20.8 is 40% of 52.

Practice Problem 10

What percent of 40 is 5?

Example 10   Solving Percent Problems for Percent

What percent of 50 is 8?

Solution:

percent   base   amount

$$\frac{8}{50} = \frac{p}{100} \quad \text{or} \quad \frac{4}{25} = \frac{p}{100} \qquad \text{Write the proportion and simplify } \frac{8}{50}.$$

$$4 \cdot 100 = 25 \cdot p \qquad \text{Set cross products equal.}$$

$$400 = 25 \cdot p \qquad \text{Multiply.}$$

$$\frac{400}{25} = p \qquad \text{Divide 400 by 25, the number multiplied by } p.$$

$$16 = p \qquad \text{Simplify.}$$

Then 16% of 50 is 8.

HELPFUL HINT

Recall from our percent proportion that this number already is a percent. Just keep the number as is and attach a % symbol.

Practice Problem 11

What percent of 160 is 336?

TEACHING TIP

After discussing Example 11, you may want to ask students to compare the amount and the base when percent > 100. Then suggest that they check that amount > base when percent > 100, amount < base when percent < 100, and amount = base when percent = 100.

Example 11

504 is what percent of 360?

Solution:

amount   percent   base

$$\frac{504}{360} = \frac{p}{100}$$

Let's choose not to simplify the ratio $\frac{504}{360}$.

$$504 \cdot 100 = 360 \cdot p \qquad \text{Set cross products equal.}$$

$$50,400 = 360 \cdot p \qquad \text{Multiply.}$$

$$\frac{50,400}{360} = p \qquad \text{Divide 50,400 by 360, the number multiplied by } p.$$

$$140 = p \qquad \text{Simplify.}$$

Notice by choosing not to simplify $\frac{504}{360}$, we had larger numbers in our equation. Either way, we find that 504 is 140% of 360.

Answers

9. 300,   **10.** 12.5%,   **11.** 210%

MENTAL MATH

Identify the amount, the base, and the percent in each equation. Recall that $\dfrac{\text{amount}}{\text{base}} = \dfrac{\text{percent}}{100}$.

1. $\dfrac{12.6}{42} = \dfrac{30}{100}$ **2.** $\dfrac{201}{300} = \dfrac{67}{100}$ **3.** $\dfrac{20}{100} = \dfrac{102}{510}$ **4.** $\dfrac{40}{100} = \dfrac{248}{620}$

EXERCISE SET 6.4

A *Translate each to a proportion. Do not solve. See Examples 1 through 6.*

1. 32% of 65 is what number?

2. What number is 5% of 125?

3. 40% of what number is 75?

4. 1.2 is 47% of what number?

5. What percent of 200 is 70?

6. 520 is 85% of what number?

7. 2.3 is 58% of what number?

8. 92% of 30 is what number?

9. What number is 19% of 130?

10. 8.2 is what percent of 82?

B *Solve. See Example 7.*

11. 10% of 55 is what number?

12. 25% of 84 is what number?

13. What number is 18% of 105?

14. What number is 40% of 29?

Solve. See Examples 8 and 9.

15. 60 is 15% of what number?

16. 75 is 75% of what number?

17. 7.8 is 78% of what number?

18. 1.1 is 44% of what number?

Solve. See Examples 10 and 11.

19. 105 is what percent of 84?

20. 77 is what percent of 44?

21. 14 is what percent of 50?

22. 37 is what percent of 50?

23. 29

24. 124

25. 1.92

26. 7.8

27. 1000

28. 500

29. 210%

30. 320%

31. 55.18

32. 68.9

33. 45%

34. 32%

35. 85

36. 130

37. $\frac{7}{8}$

38. $\frac{1}{24}$

39. $3\frac{2}{15}$

40. $7\frac{1}{6}$

41. 0.7

42. 15.29

43. 2.19

44. 13.76

45. 12,011.2

46. 80.0%

47. 7270.6

Name _____

Solve. See Examples 7 through 11.

23. 2.9 is 10% of what number?

24. 6.2 is 5% of what number?

25. 2.4% of 80 is what number?

26. 6.5% of 120 is what number?

27. 160 is 16% of what number?

28. 30 is 6% of what number?

29. 348.6 is what percent of 166?

30. 262.4 is what percent of 82?

31. What number is 89% of 62?

32. What number is 53% of 130?

33. What percent of 8 is 3.6?

34. What percent of 5 is 1.6?

35. 140% of what number is 119?

36. 170% of what number is 221?

REVIEW AND PREVIEW

Add or subtract the fractions. See Sections 3.1, 3.3, and 3.4.

37. $\frac{11}{16} + \frac{3}{16}$

38. $\frac{5}{8} - \frac{7}{12}$

39. $3\frac{1}{2} - \frac{11}{30}$

40. $2\frac{2}{3} + 4\frac{1}{2}$

Add or subtract the decimals. See Section 4.3.

41. 0.41
      $+0.29$

42. 10.78
        4.3
      $+\ 0.21$

43. 2.38
      -0.19

44. 16.37
      $-\ 2.61$

COMBINING CONCEPTS

Solve. Round to the nearest tenth, if necessary.

45. What number is 22.3% of 53,862?

46. What percent of 110,736 is 88,542?

47. 8652 is 119% of what number?

Name _____ **Section** _____ **Date** _____

INTEGRATED REVIEW — PERCENT AND PERCENT PROBLEMS

Write each number as a percent.

1. 0.12   **2.** 0.68   **3.** $\frac{1}{4}$ **4.** $\frac{1}{2}$

5. 5.2   **6.** 7.8   **7.** $\frac{3}{50}$ **8.** $\frac{11}{25}$

9. $2\frac{1}{2}$ **10.** $3\frac{1}{4}$ **11.** 0.03   **12.** 0.05

Write each percent as a decimal.

13. 65%   **14.** 31%   **15.** 8%   **16.** 7%

17. 142%   **18.** 538%   **19.** 2.9%   **20.** 6.6%

Write each percent as a fraction or mixed number in simplest form.

21. 3%   **22.** 8%   **23.** 5.25%   **24.** 12.75%

25. 38%   **26.** 45%   **27.** $12\frac{1}{3}\%$ **28.** $16\frac{2}{3}\%$

Solve each percent problem.

29. 12% of 70 is what number?   **30.** 36 is 36% of what number?

31. 212.5 is 85% of what number?   **32.** 66 is what percent of 55?

33. 28% _____

34. 76 _____

35. 11 _____

36. 130% _____

37. 86% _____

38. 37.8 _____

39. 150 _____

40. 62 _____

33. 23.8 is what percent of 85?

34. 38% of 200 is what number?

35. What number is 25% of 44?

36. What percent of 99 is 128.7?

37. What percent of 250 is 215?

38. What number is 45% of 84?

39. 63 is 42% of what number?

40. 58.9 is 95% of what number?

6.5   APPLICATIONS OF PERCENT

A   SOLVING APPLICATIONS INVOLVING PERCENT

Percent is used in a variety of everyday situations. The next few examples show just a few ways that percent occurs in real-life settings. (Each of these examples shows two ways of solving these problems. If you studied Section 6.3 only, see *Method 1*. If you studied Section 6.4 only, see *Method 2*.)

Example 1   Finding Totals Using Percents

Mr. Buccaran, the principal at Slidell High School, counted 31 freshmen absent during a particular day. If this is 4% of the total number of freshmen, how many freshmen are there at Slidell High School?

Solution: *Method 1*. First we state the problem in words, then we translate.

In words: 31 is 4% of what number?

$$\downarrow \quad \downarrow \quad \downarrow \quad \downarrow \qquad \downarrow$$

Translate: $31 = 4\% \quad \cdot \qquad n$

Next, we solve for n.

$31 = 0.04 \cdot n$ Write 4% as a decimal.

$\dfrac{31}{0.04} = n$ Divide 31 by 0.04, the number multiplied by n.

$775 = n$ Simplify.

There are 775 freshmen at Slidell High School.

Method 2. First we state the problem in words, then we translate.

In words: 31 is 4% of what number?

$$\downarrow \qquad \searrow \qquad \downarrow$$

amount   percent   base

Translate: $\begin{array}{l} \text{amount} \to \\ \text{base} \to \end{array} \dfrac{31}{b} = \dfrac{4}{100} \leftarrow \text{percent}$

Next we solve for b.

$31 \cdot 100 = b \cdot 4$ Set cross products equal.

$3100 = b \cdot 4$ Multiply.

$\dfrac{3100}{4} = b$ Divide 3100 by 4, the number multiplied by b.

$775 = b$ Simplify.

There are 775 freshmen at Slidell High School.   ▬▬▬

Example 2   Finding Percents

Standardized nutrition labeling like the one shown has been on foods since 1994. It is recommended that no

---

Objectives

A Solve applications involving percent.

B Find percent increase and percent decrease.

SSM   CD-ROM   Video 6.5

Practice Problem 1

The freshmen class of 775 students is 31% of all students at Euclid University. How many students go to Euclid University?

Answer

1. 2500

Practice Problem 2

The nutrition label below is from a can of cashews. Find what percent of total calories are from fat. Round to the nearest tenth of a percent.

Nutrition Facts

Serving Size $\frac{1}{4}$ cup (33g)
Servings Per Container About 9

Amount Per Serving

Calories 190 Calories from Fat 130

	% Daily Value
Total Fat 16g	**24%**
Saturated Fat 3g	**16%**
Cholesterol 0mg	**0%**
Sodium 135mg	**6%**
Total Carbohydrate 9g	**3%**
Dietary Fiber 1g	**5%**
Sugars 2g	
Protein 5g	

Vitamin A 0% • Vitamin C 0%
Calcium 0% • Iron 8%

more than 30% of your calorie intake be from fat. Find what percent of the total calories shown are fat.

Nutrition Facts

Serving Size 1 pouch (20g)
Servings Per Container 6

Amount Per Serving

Calories	80
Calories from fat	10

	% Daily Value*
Total Fat 1g	**2%**
Sodium 45mg	**2%**
Total Carbohydrate 17g	**6%**
Sugars 9g	
Protein 0g	

Vitamin C	25%

Not a significant source of saturated fat, cholesterol, dietary fiber, vitamin A, calcium and iron.

*Percent Daily Values are based on a 2,000 calorie diet.

Fruit snacks nutrition label

Solution: *Method 1.*

In words: 10 is what percent of 80?

Translate: $10 = n \cdot 80$

Next we solve for n.

$\dfrac{10}{80} = n$ Divide 10 by 80, the number multiplied by n.

$0.125 = n$ Simplify.

$12.5\% = n$ Write 0.125 as a percent.

This food contains 12.5% of its total calories from fat.

Method 2.

In words: 10 is what percent of 80?

     amount   percent   base

Translate: $\overset{\text{amount} \to}{\underset{\text{base} \to}{}} \dfrac{10}{80} = \dfrac{p}{100} \leftarrow \text{percent}$

Next we solve for p.

$10 \cdot 100 = 80 \cdot p$ Set cross products equal.

$1000 = 80 \cdot p$ Multiply.

$\dfrac{1000}{80} = p$ Divide 1000 by 80, the number multiplied by p.

$12.5 = p$ Simplify.

This food contains 12.5% of its total calories from fat.

Answer

2. 68.4%

B FINDING PERCENT INCREASE AND PERCENT DECREASE

We often use percents to show how much an amount has increased or decreased.

Example 3   Finding an Increase

From 1970 to 1996, the number of U.S. drivers on the road has increased 61%. If the number of drivers on the road in 1970 was 112 million, find the number of drivers on the road in 1996. (*Source:* Federal Highway Administration)

Solution: *Method 1.* First we find the increase in drivers.

In words: What number is 61% of 112?

Translate: n $= 61\% \cdot 112$

Next we solve for n.

$n = 0.61 \cdot 112$ Write 61% as a decimal.
$n = 68.32$ Multiply.

The increase in drivers is 68.32 million. This means that the number of drivers in 1996 was

112 million + 68.32 million = 180.32 million

Method 2. First we find the increase in drivers.

In words: What number is 61% of 112?

| amount | percent | base |

Translate: $\dfrac{\text{amount} \rightarrow\ a}{\text{base} \rightarrow 112} = \dfrac{61}{100} \leftarrow \text{percent}$

Next we solve for a.

$a \cdot 100 = 112 \cdot 61$ Set cross products equal.
$a \cdot 100 = 6832$ Multiply.
$a = \dfrac{6832}{100}$ Divide 6832 by 100, the number multiplied by a.
$a = 68.32$ Simplify.

The increase in drivers is 68.32 million. This means that the number of drivers in 1996 was

112 million + 68.32 million = 180.32 million.   ▬▬▬▬

Suppose that the population of a town is 10,000 people and then it increases by 2000 people. The **percent of increase** is

$\dfrac{\text{amount of increase} \rightarrow\ 2000}{\text{original amount} \rightarrow 10{,}000} = 0.2 = 20\%$

Example 4   Finding Percent Increase

The number of applications for a mathematics scholarship at Yale increased from 34 to 45 in one year. What is

Practice Problem 3

From 1970 to 1996, the number of vehicles on the road has increased 90%. If the number of vehicles on the road in 1970 was 109 million, find the number of vehicles on the orad in 1996, (*Source:* Federal Highway Administration)

PERCENT OF INCREASE

$\dfrac{\text{percent of}}{\text{increase}} = \dfrac{\text{amount of increase}}{\text{original amount}}$

Answer

3. 207.1 million

Practice Problem 4

Saturday's attendance at the play *Peter Pan* increased to 333 people over Friday's attendance of 285 people. What was the percent increase in attendance? Round to the nearest tenth of a percent.

the percent increase? Round to the nearest whole percent.

Solution: First we find the amount of increase by subtracting the original number of applicants from the new number of applicants.

amount of increase $= 45 - 34 = 11$

The amount of increase is 11 applicants. To find the percent of increase,

$$\text{percent of increase} = \frac{\text{amount of increase}}{\text{original amount}} = \frac{11}{34} \approx 0.32 = 32\%$$

> **HELPFUL HINT**
>
> Make sure that this number is the original number and not the new number.

The number of applications increased by about 32%.

✓ CONCEPT CHECK

A student is calculating the percent increase in enrollment from 180 students one year to 200 students the next year. Explain what is wrong with the following calculations.

$$\begin{array}{l}\text{Amount} \\ \text{of increase}\end{array} = 200 - 180 = 20$$

$$\begin{array}{l}\text{Percent of} \\ \text{increase}\end{array} = \frac{20}{200} = 0.1 = 10\%$$

TRY THE CONCEPT CHECK IN THE MARGIN.

Suppose that your income was $300 a week and then it decreased by $30. The percent decrease is

$$\begin{array}{l}\text{amount of decrease} \rightarrow \\ \text{original amount} \rightarrow\end{array} \frac{\$30}{\$300} = 0.1 = 10\%$$

> **PERCENT OF DECREASE**
>
> $$\text{percent of decrease} = \frac{\text{amount of decrease}}{\text{original amount}}$$

Practice Problem 5

A town with a population of 20,145 decreased to 18,430 over a 10-year period. What was the percent decrease? Round to the nearest tenth of a percent.

Example 5   Finding Percent Decrease

In response to a decrease in sales, a company with 1500 employees reduces the number of employees to 1230. What is the percent decrease?

Solution: First we find the amount of decrease by subtracting 1230 from 1500.

amount of decrease $= 1500 - 1230 = 270$

The amount of decrease is 270. To find the percent of decrease,

$$\begin{array}{l}\text{percent of} \\ \text{decrease}\end{array} = \frac{\text{amount of decrease}}{\text{original amount}} = \frac{270}{1500} = 0.18 = 18\%$$

The number of employees decreased by 18%.

Answers

4. 16.8%,   **5.** 8.5%

✓ **Concept Check:**   To find the percent of increase, you have to divide the amount of increase by the original amount $\left(\frac{20}{180}\right)$.

Name _____ Section _____ Date _____

EXERCISE SET 6.5

A *Solve. See Example 1.*

1. An inspector found 24 defective bolts during an inspection. If this is 1.5% of the total number of bolts inspected, how many bolts were inspected?

2. A daycare worker found 28 children absent one day during an epidemic of chicken pox. If this was 35% of the total number of children attending the daycare, how many children attend this daycare?

3. An owner of a repair service company estimates that, for every 40 hours a repairperson is on the job, he can only bill for 75% of the hours. The remaining hours, the repairperson is idle or driving to or from a job. Determine the number of hours per 40-hour week the owner can bill for a repairperson.

4. The Hodder family paid 20% of the purchase price of a $75,000 home as a down payment. Determine the amount of the down payment.

5. Vera Faciane earns $2000 per month and budgets $300 per month for food. What percent of her monthly income is spent on food?

6. Last year, Mai Toberlan bought a share of stock for $83. She was paid a dividend of $4.15. Determine what percent of the stock price is the dividend.

7. A manufacturer of electronic components expects 1.04% of its product to be defective. Determine the number of defective components expected in a batch of 28,350 components. Round to the nearest whole component.

8. 18% of Frank's wages are withheld for income tax. Find the amount withheld from Frank's wages of $3680 per month.

For each food described, find what percent of total calories is from fat. If necessary, round to the nearest tenth of a percent. See Example 2.

9.
Nutrition Facts
Serving Size 18 crackers (29g)
Servings Per Container About 9

Amount Per Serving

Calories 120   Calories from Fat 35

	% Daily Value*
Total Fat 4g	**6%**
Saturated Fat 0.5g	**3%**
Polyunsaturated Fat 0g	
Monounsaturated Fat 1.5g	
Cholesterol 0mg	**0%**
Sodium 220mg	**9%**
Total Carbohydrate 21g	**7%**
Dietary Fiber 2g	**7%**
Sugars 3g	
Protein 2g	

Vitamin A 0% • Vitamin C 0%
Calcium 2% • Iron 4%
Phosphorus 10%

Snack Crackers

10.
Nutrition Facts
Serving Size 28 crackers (31g)
Servings Per Container About 6

Amount Per Serving

Calories 130   Calories from Fat 35

	% Daily Value*
Total Fat 4g	**6%**
Saturated Fat 2g	**10%**
Polyunsaturated Fat 1g	
Monounsaturated Fat 1g	
Cholesterol 0mg	**0%**
Sodium 470mg	**20%**
Total Carbohydrate 23g	**8%**
Dietary Fiber 1g	**4%**
Sugars 4g	
Protein 2g	

Vitamin A 0% • Vitamin C 0%
Calcium 0% • Iron 2%

1. 1600 bolts

2. 80 children

3. 30 hours

4. $15,000

5. 15%

6. 5%

7. 295 components

8. $662.40

9. 29.2%

10. 26.9%

366

Name _____

B *Solve. Round money amounts to the nearest cent and all other amounts to the nearest tenth. See Example 3.*

11. Ace Furniture Company currently produces 6200 chairs per month. If production increases 8%, find the increase and the new number of chairs produced each month.

12. The enrollment at a local college increased 5% over last year's enrollment of 7640. Find the increase in enrollment and the current enrollment.

13. By carefully planning their meals, a family was able to decrease their weekly grocery bill by 20%. Their weekly grocery bill use to be $170. What is their new weekly grocery bill?

14. The profit of Ramone Company last year was $175,000. This year's profit decreased by 11%. Find this year's profit.

15. A car manufacturer announced that next year the price of a certain model car would increase 4.5%. This year the price is $19,286. Find the increase and the new price.

16. A union contract calls for a 6.5% salary increase for all employees. Determine the increase and the new salary that a worker currently making $28,500 under this contract can expect.

17. The nation's Hispanic population is projected to increase 33% during the 10-year period from 1995 to 2005. The number of Hispanics in 1995 was 26.8 million. Find the increase and the projected 2005 population. (*Source:* U.S. Bureau of the Census, *U.S. Census of Population:* 1970)

18. From 1995 to 2004, the number of doctorates awarded to women is projected to increase 15%. The number of women who received doctorates in 1995 was 17,000. Find the predicted number of women to be awarded doctorates in 2004. (*Source:* U.S. National Center for Education Statistics)

Find the amount of increase and the percent increase. See Example 4.

	Original Amount	New Amount	Amount of Increase	Percent Increase
19.	40	50	_____	_____
20.	10	15	_____	_____
21.	85	187	_____	_____
22.	78	351	_____	_____

Find the amount of decrease and the percent decrease. See Example 5.

	Original Amount	New Amount	Amount of Decrease	Percent Decrease
23.	8	6	_____	_____
24.	25	20	_____	_____
25.	160	40	_____	_____
26.	200	162	_____	_____

Name _____

Solve. Round percents to the nearest tenth, if necessary. See Examples 3 through 5.

27. There are 150 calories in a cup of whole milk and only 84 in a cup of skimmed milk. In switching to skimmed milk, find the percent decrease in number of calories.

28. In reaction to a slow economy, the number of employees at a soup company decreased from 530 to 477. What was the percent decrease in employees?

29. By changing his driving routines, Alan Miller increased his car's rate of miles per gallon from 19.5 to 23.7. Find the percent increase.

30. John Smith decided to decrease the number of calories in his diet from 3250 to 2100. Find the percent decrease.

🔲 **31.** The price of a loaf of bread decreased from $1.09 to $0.89. Find the percent decrease.

32. Before taking a typing course, Geoffry Landers could type 32 words per minute. By the end of the course, he was able to type 76 words per minute. Find the percent increase.

33. In 1940, there were approximately 52.1 million sheep on farms in the United States. In 1997, this number had decreased to 7.9 million sheep. What was the percent decrease? (*Source:* National Agricultural Statistics Service)

34. In 1970, there were approximately 12.1 million milk cows on farms in the United States. In 1997, this number had decreased to 9.3 million milk cows. What is the percent decrease? (*Source:* National Agricultural Statistics Service)

27. 44%

28. 10%

29. 21.5%

30. 35.4%

31. 18.3%

32. 137.5%

33. 84.8%

34. 23.1%

368

Name _____

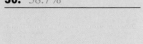 **35.** In 1994, approximately 16,000 occupational therapy assistants and aides were employed in the United States. By 2005, this number is expected to increase to 29,000 assistants and aides. What was the percent increase? (*Source:* Bureau of Labor Statistics)

36. In 1994, approximately 206,000 medical assistants were employed in the United States. By 2005, this number is expected to increase to 327,000 medical assistants. What is the percent increase? (*Source:* Bureau of Labor Statistics)

REVIEW AND PREVIEW

Perform each indicated operation. See Sections 4.3 and 4.4.

37.   0.12
      $\times$ 38

38.   42
      $\times$ 0.7

39. 9.20 + 1.98

40. 46 + 7.89

41. 78 − 19.46

42. 64.80 − 10.72

COMBINING CONCEPTS

Solve. Round percents to the nearest tenth.

43. The population of Tokyo is expected to increase from 26,518 thousand in 1994 to 28,700 thousand in 2015. Find the percent increase. (*Source:* United Nations, Dept. for Economic and Social Information and Policy Analysis)

44. In 1994, approximately 179,000 personal and home care aides were employed in the United States. By 2005, this number is expected to increase to 391,000 aides. What is the percent increase? (*Source:* Bureau of Labor Statistics)

Tokyo

JAPAN

6.6 PERCENT AND PROBLEM SOLVING: SALES TAX, COMMISSION, AND DISCOUNT

A CALCULATING SALES TAX AND TOTAL PRICE

Percents are frequently used in the retail trade. For example, most states charge a tax on certain items when purchased. This tax is called a **sales tax**, and retail stores collect it for the state. Sales tax is almost always stated as a percent of the purchase price.

A 6% sales tax rate on a purchase of a $10.00 item gives a sales tax of

$$\text{sales tax} = 6\% \text{ of } \$10 = 0.06 \cdot \$10.00 = \$0.60$$

The total price to the customer would be

purchase price plus sales tax

$$\$10.00 \quad + \quad \$0.60 = \$10.60$$

This example suggests the following equations.

> **SALES TAX AND TOTAL PRICE**
>
> sales tax = tax rate · purchase price
> total price = purchase price + sales tax

In this section we will round dollar amounts to the nearest cent.

Example 1 Finding Sales Tax and Purchase Price

Find the sales tax and the total price on a purchase of an $85.50 trench coat in a city where the sales tax rate is 7.5%.

$85.50
+ 7.5% tax

Solution: The purchase price is $85.50 and the tax rate is 7.5%.

sales tax = tax rate · purchase price

$$\text{sales tax} = 7.5\% \cdot \$85.50$$
$$= 0.075 \cdot \$85.5 \quad \text{Write 7.5\% as a decimal.}$$
$$\approx \$6.41 \quad \text{Rounded to the nearest cent.}$$

Thus

total price = purchase price + sales tax

$$\text{total price} = \$85.50 + \$6.41$$
$$= \$91.91$$

The sales tax on $85.50 is $6.41 and the total price is $91.91.

Objectives

A Calculate sales tax and total price.
B Calculate commissions.
C Calculate discount and sale price.

SSM   CD-ROM   Video
6.6

TEACHING TIP

You may want to begin this lesson by showing students how easy it is to find 1% and 10% of a price. Then show how they can use these percents to calculate other percents.

Practice Problem 1

If the sales tax rate is 6%, what is the sales tax and the total amount due on a $29.90 Goodgrip tire?

TEACHING TIP

You may want to show students how to quickly calculate tips. Typically people leave a 15% or 20% tip at a restaurant where they received good service. You can quickly estimate a tip by using an estimate of 10% of the bill. A 15% tip can be found by adding on half of the 10% estimate. A 20% tip can be found by doubling the 10% estimate.

Answer

1. tax: $1.79; total: $31.69

✓ CONCEPT CHECK

The purchase price of a textbook is $50 and sales tax is 10%. If you are told by the cashier that the total price is $75, how can you tell that a mistake has been made?

Practice Problem 2

The sales tax on a $13,500 automobile is $1080.00. Find the sales tax rate.

Practice Problem 3

Mr. Olsen is a sales representative for Miko Copiers. Last month he sold $37,632 worth of copy equipment and supplies. What is his commission for the month if he is paid a commission of 6.6% of his total sales for the month?

Answers

2. 8%,   **3.** $2483.71

✓ **Concept Check:**   Since $10\% = \frac{1}{10}$, the sales tax is $\frac{\$50}{10} = \$5$. The total price should have been $55.

TRY THE CONCEPT CHECK IN THE MARGIN.

Example 2   Finding a Sales Tax Rate

The sales tax on a $300 printer is $22.50. Find the sales tax rate.

$300
+ $22.50 sales tax

Solution: Let r by the unknown sales tax rate. Then

$$\boxed{\text{sales tax}} = \boxed{\text{tax rate}} \cdot \boxed{\text{purchase price}}$$

$$\$22.50 = r \cdot \$300$$

$$\frac{22.50}{300} = r \qquad \text{Divide 22.50 by 300, the number multiplied by } r.$$

$$0.075 = r \qquad \text{Simplify.}$$

$$7.5\% = r \qquad \text{Write 0.075 as a percent.}$$

The sales tax rate is 7.5%.

B   CALCULATING COMMISSIONS

A **wage** is payment for performing work. Hourly wage, commissions, and salary are some of the ways wages can be paid. Many people who work in sales are paid a commission. An employee who is paid a **commission** is paid a percent of his or her total sales.

> **COMMISSION**
>
> commission = commission rate · sales

Example 3   Finding a Commission Rate

Sherry Souter, a real estate broker for Wealth Investments, sold a house for $114,000 last week. If her commission is 1.5% of the selling price of the home, find the amount of her commission.

Solution: $$\boxed{\text{commission}} = \boxed{\text{commission rate}} \cdot \boxed{\text{sales}}$$

$$\text{commission} = 1.5\% \cdot \$114,000$$

$$= 0.015 \cdot \$114,000 \qquad \text{Write 1.5\% as 0.015.}$$

$$= \$1710 \qquad \text{Multiply.}$$

Her commission on the house is $1710.

Example 4   Finding a Commission Rate

A salesperson earned $1560 for selling $13,000 worth of television and stereo systems. Find the commission rate.

Solution: Let *r* stand for the unknown commission rate. Then

$$\underbrace{\text{commission}}_{\$1560} = \underbrace{\text{commission rate}}_{r} \cdot \underbrace{\text{sales}}_{\$13,000}$$

$$\$1560 = r \cdot \$13,000$$

$$\frac{1560}{13,000} = r \qquad \text{Divide 1560 by 13,000, the number multiplied by } r.$$

$$0.12 = r \qquad \text{Simplify.}$$

$$12\% = r \qquad \text{Write 0.12 as a percent.}$$

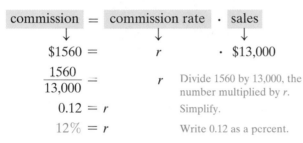

Commission: $1560

The commission rate is 12%.

C CALCULATING DISCOUNT AND SALE PRICE

Suppose that an item that normally sells for $40 is on sale for 25% off. This means that the **original price** of $40 is reduced, or **discounted** by 25% of $40, or $10. The **discount rate** is 25%, the **amount of discount** is $10 and the **sale price** is $40 − $10 or $30.

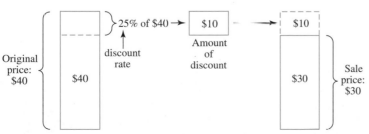

Original price: $40 — $40 — discount rate — 25% of $40 → $10 Amount of discount → $10 — $30 Sale price: $30

Practice Problem 4

A salesperson earns $1290 for selling $8600 worth of appliances. Find the commission rate.

TEACHING TIP   Classroom Activity

Ask students to work in groups to write word problems involving commission. One way to structure this activity is to have them write a word problem for each of the following situations:
Commission
Combination: Fixed Wage and Commission
Then have them give their problems to another group to solve.

TEACHING TIP

You may want to ask students, "If you were given a choice, would you opt for 25% discount on the price of a product or for having the price of a product decreased by $100?"

Answer
4. 15%

To calculate discounts and sale prices, we can use the following equations.

> **DISCOUNT AND SALE PRICE**
>
> amount of discount = discount rate · original price
>
> sale price = original price − amount of discount

Practice Problem 5

A Panasonic TV is advertised on sale for 15% off the regular price of $700. Find the discount and the sale price.

Example 5   Finding a Discount and a Sale Price

A speaker that normally sells for $65.00 is on sale at 25% off. What is the discount and what is the sale price?

25% off

Solution: First we find the discount.

amount of discount	=	discount rate	·	original price

amount of discount = 25% · $65

= 0.25 · $65   Write 25% as 0.25.

= $16.25   Multiply.

The discount is $16.25. Next, find the sale price.

sale price	=	original price	−	discount

sale price = $65 − $16.25

= $48.75   Subtract.

The sale price is $48.75.

TEACHING TIP

You may want to conclude this lesson by having students work in groups to respond to the following question: "When would a given percent discount be a better choice than a markdown of $100? Explain your answer."
Answer:
When:
  0%   Never
 10%   Price > $1000
 25%   Price > $400
 50%   Price > $200
 75%   Price > $133.33
 90%   Price > $111.11
100%   Price > $100

Answer

5. $105; $595

Name _____ Section _____ Date _____

EXERCISE SET 6.6

A *Solve. See Examples 1 and 2.*

1. What is the sales tax on a suit priced at $150.00 if the sales tax rate is 5%?

2. If the sales tax rate is 6%, find the sales tax on a microwave oven priced at $188.

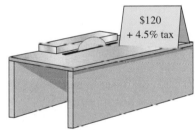 3. The purchase price of a camcorder is $799. What is the total price if the sales tax rate is 7.5%?

4. A stereo system has a purchase price of $426. What is the total price if the sales tax rate is 8%?

5. A chair and ottoman have a purchase price of $600. If the sales tax on this purchase is $54, find the sales tax rate.

6. The sales tax on the purchase of a $2500 computer is $162.50. Find the sales tax rate.

7. A table saw sells for $120. With a sales tax rate of 4.5%, find the total price.

8. A one-half carat diamond ring is priced at $800. The sales tax rate is 6.5%. Find the total price.

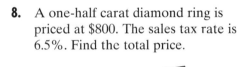

$120
+ 4.5% tax

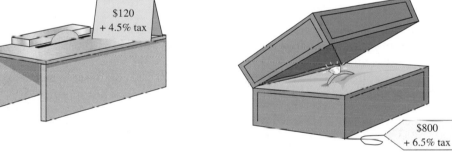

$800
+ 6.5% tax

9. A gold and diamond bracelet sells for $1800. Find the total price if the sales tax rate is 6.5%.

10. The purchase price of a personal computer is $1890.00. If the sales tax rate is 8%, what is the total price?

11. The sales tax on the purchase of a truck is $920. If the tax rate is 8%, find the purchase price of the truck.

12. The sales tax on the purchase of a desk is $27.50. If the tax rate is 5%, find the purchase price of the desk.

1. $7.50

2. $11.28

3. $858.93

4. $460.08

5. 9%

6. 6.5%

7. $125.40

8. $852

9. $1917

10. $2041.20

11. $11,500

12. $550

373

374

Name _____

13. A cordless phone costs $90 and a battery recharger costs $15. What is the total price for purchasing these items if the sales tax rate is 7%?

14. Ms. Warner bought a blouse for $35, a skirt for $55, and a blazer for $95. Find the total price she paid, given a sales tax rate of 6.5%.

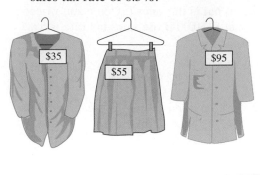

15. The sales tax is $98.70 on a stereo sound system purchase of $1645. Find the sales tax rate.

16. The sales tax is $103.50 on a necklace purchase of $1150. Find the sales tax rate.

B *Solve. See Examples 3 and 4.*

17. Jane Moreschi, a sales representative for a large furniture warehouse, is paid a commission rate of 4%. Find her commission if she sold $1,236,856 worth of furniture last month.

18. Rosie Davis-Smith is a beauty consultant for a home cosmetic business. She is paid a commission rate of 4.8%. Find her commission if she sold $1638 in cosmetics last month.

19. A salesperson earned a commission of $1380.40 for selling $9860.00 worth of paper products. Find the commission rate.

20. A salesperson earned a commission of $3575 for selling $32,500 worth of books to various book stores. Find the commission rate.

21. How much commission will Jack Pruet make on the sale of a $125,900 house if he receives 1.5% of the selling price?

22. Frankie Lopez sold $9638.00 of jewelry this week. Find her commission for the week if she receives a commission rate of 5.6%.

23. A house sold for $85,500 and the real estate agent earned a commission of $2565. Find the rate of commission.

24. A salesperson earned $1750 for selling $25,000 worth of fertilizer. Find the commission rate.

C *Find the amount of discount and the sale price. See Example 5.*

	Original Price	Discount Rate	Amount of Discount	Sale Price
25.	$68.00	10%	_____	_____
26.	$47.00	20%	_____	_____
27.	$96.50	50%	_____	_____
28.	$110.60	40%	_____	_____
29.	$215.00	35%	_____	_____
30.	$370.00	25%	_____	_____
31.	$21,700.00	15%	_____	_____
32.	$17,800.00	12%	_____	_____

33. The price of a $300 fax machine is on sale at 15% off. Find the discount and the sale price.

34. The price of a $2000 designer dress is on sale at 30% off. Find the discount and the sale price.

23. 3%

24. 7%

25. $6.80; $61.20

26. $9.40; $37.60

27. $48.25; $48.25

28. $44.24; $66.36

29. $75.25; $139.75

30. $92.50; $277.50

31. $3255.00; $18,445.00

32. $2136.00; $15,664.00

33. $45; $255

34. $600; $1400

375

Name _____

REVIEW AND PREVIEW

Multiply. See Sections 4.4 and 4.7.

35. $2000 \cdot 0.3 \cdot 2$

36. $500 \cdot 0.08 \cdot 3$

37. $400 \cdot 0.03 \cdot 11$

38. $1000 \cdot 0.05 \cdot 5$

39. $600 \cdot 0.04 \cdot \dfrac{2}{3}$

40. $6000 \cdot 0.06 \cdot \dfrac{3}{4}$

 ## COMBINING CONCEPTS

41. A diamond necklace sells for $24,966. If the tax rate is 7.5%, find the total price.

42. A house recently sold for $562,560. The commission rate on the sale is 5.5%. If a real estate agent is to receive 60% of the commission, find the amount received by the agent.

6.7 PERCENT AND PROBLEM SOLVING: INTEREST

A CALCULATING SIMPLE INTEREST

Interest is money charged for using other people's money. When you borrow money, you pay interest. When you loan or invest money, you earn interest. The money borrowed, loaned, or invested is called the **principal amount**, or simply **principal**. Interest is normally stated in terms of a percent of the principal for a given period of time. The **interest rate** is the percent used in computing the interest. Unless stated otherwise, *the rate is understood to be per year*. When the interest is computed on the original principal, it is called **simple interest**. Simple interest is calculated using the following equation.

> SIMPLE INTEREST
>
>   simple interest = principal · rate · time
>
> where the rate is understood to be per year and time is in years.

Example 1 Finding Simple Interest

Find the simple interest after 2 years on $500 at an interest rate of 12%.

Solution: In this example, the principal is $500, the rate is 12%, and the time is 2 years.

simple interest = principal · rate · time

simple interest = $500 · 12% · 2

= $500 · 0.12 · 2   Write 12% as 0.12.

= $120   Multiply.

The simple interest is $120.

If time is not given in years, we need to convert the given time to years.

Example 2 Finding Simple Interest

Ivan Borski borrowed $2400 at 10% simple interest for 8 months to buy a used Chevy S-10. Find the simple interest he paid.

Solution: Since there are 12 months in a year, we first find what part of a year 8 months is.

$$8 \text{ months} = \frac{8}{12} \text{ year} = \frac{2}{3} \text{ year}$$

Now we find the simple interest.

simple interest = principal · rate · time

simple interest = $2400 · 10% · $\frac{2}{3}$

= $2400 · 0.10 · $\frac{2}{3}$

= $160

The interest on Ivan's loan is $160.

Practice Problem 1

Find the simple interest after 3 years on $750 at an interest rate of 8%.

TEACHING TIP

For Example 1, you may want to have students also calculate the interest earned after 1, 3, and 4 years. Then ask them to predict the amount of simple interest that would be earned in 5 years and 10 years.

Practice Problem 2

Juanita Lopez borrowed $800 for 9 months at a simple interest rate of 20%. How much interest did she pay?

Answers
1. $180, **2.** $120

When money is borrowed, the borrower pays the original amount borrowed, or the principal, as well as the interest. When money is invested, the investor receives the original amount invested, or the principal, as well as the interest. In either case, the **total amount** is the sum of the principal and the interest.

> **FINDING THE TOTAL AMOUNT OF A LOAN OR INVESTMENT**
>
> total amount (paid or received) = principal + interest

Practice Problem 3

If $500 is borrowed at a simple interest rate of 12% for 6 months, find the total amount paid.

Example 3   Finding the Total Amount of an Investment

An accountant invested $2000 at a simple interest rate of 10% for 2 years. What total amount of money will she have from her investment in 2 years?

Solution: First we find her interest.

simple interest	=	principal	·	rate	·	time

simple interest = $2000 · 10% · 2

= $2000 · 0.10 · 2

= $400

The interest is $400.

Next we add the interest to the principal.

total amount	=	principal	+	interest

total amount = $2000 + $400

= $2400

After 2 years, she will have a total amount of $2400.

TRY THE CONCEPT CHECK IN THE MARGIN.

✓ CONCEPT CHECK

Which investment would earn more interest: an amount of money invested at 8% interest for 2 years or the same amount of money invested at 8% for 3 years? Explain.

B CALCULATING COMPOUND INTEREST

Recall that simple interest depends on the original principal only. Another type of interest is compound interest. **Compound interest** is computed on not only the principal, but also on the interest already earned in previous compounding periods. Compound interest is used more often than simple interest.

Let's see how compound interest differs from simple interest. Suppose that $2000 is invested at 7% interest **compounded annually** for 3 years. This means that interest is added to the principal at the end of each year and next year's interest is computed on this new amount. In this section, we will round dollar amounts to the nearest cent.

Answers

3. $530

✓ **Concept Check:** 8% for 3 years. Since the interest rate is the same, the longer you keep the money invested, the more interest you earn.

	Amount at Beginning of Year	Principal · Rate · Time = Interest	Amount at End of Year
1st year	$2000	$2000 · 0.07 · 1 = $140	$2000 + 140 = $2140
2nd year	$2140	$2140 · 0.07 · 1 = $149.80	$2140 + 149.80 = $2289.80
3rd year	$2289.80	$2289.80 · 0.07 · 1 = $160.29	$2289.80 + 160.29 = $2450.09

The compound interest earned can be found by

total amount − original principal = compound interest

$2450.09 − $2000 = $450.09

The simple interest earned would have been

principal · rate · time = interest

$2000 · 0.07 · 3 = $420

Since compound interest earns "interest on interest," compound interest is more than simple interest.

Computing compound interest using the method above can be tedious. We can use a **compound interest table** to compute interest more quickly. The compound interest table is found in Appendix F. This table gives the total compound interest and principal paid on $1 for given rates and number of years. Then we can use the following equation to find the total amount of interest and principal.

FINDING TOTAL AMOUNTS WITH COMPOUND INTEREST

$$\text{total amount} = \text{original principal} \cdot \begin{matrix} \text{compound interest factor} \\ \text{(from table)} \end{matrix}$$

Example 4   Finding Total Amount Received on an Investment

$4000 is invested at 8% compounded semiannually for 10 years. Find the total amount at the end of 10 years.

Solution: Look in Appendix F. The compound interest factor for 10 years at 8% in the compounded semiannually section is 2.19112.

total amount = original principal · compound interest factor

total amount = $4000 · 2.19112

= $8764.48

Therefore, the total amount at the end of 10 years is $8764.48.

Example 5   Finding Compound Interest Earned

In Example 4 we found that the total amount for $4000 invested at 8% compounded semiannually for 10 years is $8764.48. Find the compound interest earned.

Solution: interest earned = total amount − original principal

interest earned = $8764.48 − $4000

= $4764.48

The compound interest earned is $4764.48.

C CALCULATING A MONTHLY PAYMENT

We conclude this section with a method to find the monthly payment on a loan.

Practice Problem 4

$5500 is invested at 7% compounded daily for 5 years. Find the total amount at the end of 5 years.

Practice Problem 5

If the total amount is $9933.14 when $5500 is invested, find the compound interest earned.

Answers
4. $7804.61,   **5.** $4433.14

FINDING THE MONTHLY PAYMENT OF A LOAN	

$$\text{monthly payment} = \frac{\text{principal} + \text{interest}}{\text{total number of payments}}$$

Practice Problem 6

Find the monthly payment on a $3000 3-year loan if the interest on the loan is $1123.58.

Example 6   Finding a Monthly Payment

Find the monthly payment on a $2000 loan for 2 years. The interest on the 2-year loan is $435.88.

Solution: First we determine the total number of monthly payments. The loan is for 2 years. Since there are 12 months per year, the number of payments is 2 · 12, or 24. Now we can calculate the monthly payment.

$$\text{monthly payment} = \frac{\text{principal} + \text{interest}}{\text{total number of payments}}$$

$$\text{monthly payment} = \frac{\$2000 + \$435.88}{24}$$

$$\approx \$101.50$$

The monthly payment is $101.50.

Answer

6. $114.54

CALCULATOR EXPLORATIONS
COMPOUND INTEREST FACTOR

A compound interest factor may be found by using your calculator and evaluating the formula

$$\text{compound interest factor} = \left(1 + \frac{r}{n}\right)^{nt}$$

where r is the interest rate, t is the time in years, and n is the number of times compounded per year. For example, we stated earlier that the compound interest factor for 10 years at 8% compounded semiannually is 2.19112. Let's find this factor by evaluating the compound interest factor formula when $r = 8\%$ or 0.08, $t = 10$, and $n = 2$ (compounded semiannually means 2 times per year). Thus,

$$\text{compound interest factor} = \left(1 + \frac{0.08}{2}\right)^{2 \cdot 10} \quad \text{or} \quad \left(1 + \frac{0.08}{2}\right)^{20}$$

To evaluate, press the keys

$$(\quad 1 \quad + \quad 0.08 \quad \div \quad 2 \quad) \quad y^x \quad 20 \quad =$$

or   or

$$\wedge \quad \boxed{\text{ENTER}}$$

The display will read $\boxed{2.1911231}$. Rounded to 5 decimal places, this is 2.19112.

Find the compound interest factor. Use the table in Appendix F to check your answer.

1. 5 years, 9%, compounded quarterly   1.56051

2. 15 years, 14%, compounded daily   8.16288

3. 20 years, 11%, compounded annually   8.06231

4. 1 year, 7%, compounded semiannually   1.07123

5. Find the total amount after 4 years if $500 is invested at 6% compounded quarterly. $634.50

6. Find the total amount for 19 years if $2500 is invested at 5% compounded daily. $6463.85

EXERCISE SET 6.7

A *Find the simple interest. See Examples 1 and 2.*

Principal	Rate	Time	Principal	Rate	Time
1. $200	8%	2 years	**2.** $800	9%	3 years
3. $160	11.5%	4 years	**4.** $950	12.5%	5 years
5. $5000	10%	$1\frac{1}{2}$ years	**6.** $1500	14%	$2\frac{1}{4}$ years
7. $375	18%	6 months	**8.** $1000	10%	18 months
9. $2500	16%	21 months	**10.** $775	15%	8 months

Solve. See Examples 1 through 3.

11. A company borrows $62,500 for 2 years at a simple interest of 12.5% to buy an airplane. Find the total amount paid on the loan.

12. $65,000 is borrowed to buy a house. If the simple interest rate on the 30-year loan is 10.25%, find the total amount paid on the loan.

13. A money market fund advertises a simple interest rate of 9%. Find the total amount received on an investment of $5000 for 15 months.

14. The Real Service Company takes out a 270-day (9-month) short-term, simple interest loan of $4500 to finance the purchase of some new equipment. If the interest rate is 14%, find the total amount that the company pays back.

15. Marsha Waide borrows $8500 and agrees to pay it back in 4 years. If the simple interest rate is 12%, find the total amount she pays back.

16. Ms. Lapchinski gives her 18-year-old daughter a graduation gift of $2000. If this money is invested at 8% simple interest for 5 years, find the total amount.

B *Find the total amount in each compound interest account. See Example 4.*

17. $6150 is compounded semiannually at a rate of 14% for 15 years.

18. $2060 is compounded annually at a rate of 15% for 10 years.

19. $1560 is compounded daily at a rate of 8% for 5 years.

20. $1450 is compounded quarterly at a rate of 10% for 15 years.

21. $10,000 is compounded semiannually at a rate of 9% for 20 years.

22. $3500 is compounded daily at a rate of 8% for 10 years.

ANSWERS

1. $32
2. $216
3. $73.60
4. $593.75
5. $750
6. $472.50
7. $33.75
8. $150
9. $700
10. $77.50
11. $78,125
12. $264,875
13. $5562.50
14. $4972.50
15. $12,580
16. $2800
17. $46,815.40
18. $8333.85
19. $2327.15
20. $6379.70
21. $58,163.60
22. $7788.73

382

Name _____

Find the total amount of compound interest earned. See Example 5.

23. $2675 is compounded annually at a rate of 9% for 1 year.

24. $6375 is compounded semiannually at a rate of 10% for 1 year.

25. $2000 is compounded annually at a rate of 8% for 5 years.

26. $2000 is compounded semiannually at a rate of 8% for 5 years.

27. $2000 is compounded quarterly at a rate of 8% for 5 years.

28. $2000 is compounded daily at a rate of 8% for 5 years.

C *Solve. See Example 6.*

29. A college student borrows $1500 for 6 months to pay for a semester of school. If the interest is $61.88, find the monthly payment.

30. Jim Tillman borrows $1800 for 9 months. If the interest is $148.90, find his monthly payment.

31. $20,000 is borrowed for 4 years. If the interest on the loan is $10,588.70, find the monthly payment.

32. $105,000 is borrowed for 15 years. If the interest on the loan is $181,125.00, find the monthly payment.

REVIEW AND PREVIEW

Find the perimeter of each figure. See Section 1.2.

33.

Rectangle — 6 yards, 10 yards

34.
18 centimeters, 16 centimeters, 12 centimeters, Triangle

35.
Regular Pentagon— All sides same length — 7 meters

36.
Square — 21 miles

COMBINING CONCEPTS

37. Explain how to look up the compound interest factor in the compound interest table.

38. Explain how to find the amount of interest on a compounded account.

39. Compare the following accounts:
Account 1: $1000 is invested for 10 years at a simple interest rate of 6%.
Account 2: $1000 is compounded semiannually at a rate of 6% for 10 years.

Discuss how the interest is computed for each account. Determine which account earns more interest. Why?

Name ⎯⎯⎯⎯⎯⎯⎯⎯⎯⎯ **Section** ⎯⎯⎯⎯ **Date** ⎯⎯⎯⎯

CHAPTER 6 ACTIVITY
INVESTIGATING COMPOUND INTEREST

MATERIALS:
▲ calculator

This activity may be completed by working in groups or individually.

Suppose you have just moved to a new city and must open new bank accounts with $5000. You would like to open a checking account, a savings account, and a certificate of deposit (CD). After doing some research, you find the following five options for CDs at five local banks.

Bank	Minimum Deposit	Interest Rate	Number of Compoundings per Year
A	$1000	6.50%	26
B	$1000	6.55%	365
C	$1000	6.60%	4
D	$1000	6.65%	2
E	$1000	6.70%	1

1. Without making any calculations, which CD investment option do you think would make the best choice? Explain your reasoning. answers may vary

2. For each CD option, find the values of a CD opened with the minimum deposit after an investment of 2, 5, and 10 years (use the compound interest table on the next page). Create a table of your results.

3. Based on your table from Question 2, which CD option would be the best choice? Explain your reasoning. Does this surprise you? What other factors should you consider when choosing a savings tool such as a CD? answers may vary

4. You may have noticed that it can sometimes be difficult to compare options when different interest rates and numbers of compoundings are involved. When describing investment options, banks often use *annual percentage yield* (APY) instead to help eliminate confusion. The APY of an investment represents the percent increase from the original principal to the ending value of the investment after one year. Find the APY of each CD option. Do you think that the APY makes it easier to compare the options? Explain.

2. Bank	Years	Value of CD
A	2	$1138.64
	5	$1383.47
	10	$1913.99
B	2	$1139.95
	5	$1387.45
	10	$1925.03
C	2	$1139.88
	5	$1387.23
	10	$1924.40
D	2	$1139.78
	5	$1386.93
	10	$1923.57
E	2	$1138.49
	5	$1383.00
	10	$1912.69

Answers may vary

4. Bank	APY
A	6.7%
B	6.8%
C	6.8%
D	6.8%
E	6.7%

Answers may vary

CHAPTER 6 ACTIVITY COMPOUND INTEREST TABLE
Compounded Annually

Years	6.50%	6.55%	6.60%	6.65%	6.70%	6.75%	6.80%	6.85%	6.90%	6.95%	7.00%
2	1.13423	1.13529	1.13636	1.13742	1.13849	1.13956	1.14062	1.14169	1.14276	1.14383	1.14490
5	1.37009	1.37331	1.37653	1.37976	1.38300	1.38624	1.38949	1.39275	1.39601	1.39928	1.40255
10	1.87714	1.88597	1.89484	1.90374	1.91269	1.92167	1.93069	1.93975	1.94884	1.95798	1.96715

Compounded Semiannually

Years	6.50%	6.55%	6.60%	6.65%	6.70%	6.75%	6.80%	6.85%	6.90%	6.95%	7.00%
2	1.13648	1.13758	1.13868	1.13978	1.14089	1.14199	1.14309	1.14420	1.14531	1.14641	1.14752
5	1.37689	1.38023	1.38358	1.38693	1.39029	1.39365	1.39703	1.40041	1.40380	1.40720	1.41060
10	1.89584	1.90504	1.91428	1.92357	1.93290	1.94227	1.95169	1.96115	1.97065	1.98020	1.98979

Compounded Quarterly

Years	6.50%	6.55%	6.60%	6.65%	6.70%	6.75%	6.80%	6.85%	6.90%	6.95%	7.00%
2	1.13764	1.13876	1.13988	1.14100	1.14212	1.14325	1.14437	1.14550	1.14663	1.14775	1.14888
5	1.38042	1.38382	1.38723	1.39064	1.39407	1.39750	1.40094	1.40439	1.40784	1.41131	1.41478
10	1.90556	1.91496	1.92440	1.93389	1.94342	1.95300	1.96263	1.97230	1.98202	1.99179	2.00160

Compounded Monthly

Years	6.50%	6.55%	6.60%	6.65%	6.70%	6.75%	6.80%	6.85%	6.90%	6.95%	7.00%
2	1.13843	1.13956	1.14070	1.14183	1.14297	1.14410	1.14524	1.14638	1.14752	1.14866	1.14981
5	1.38282	1.38626	1.38971	1.39317	1.39664	1.40011	1.40360	1.40709	1.41060	1.41411	1.41763
10	1.91218	1.92172	1.93130	1.94092	1.95060	1.96032	1.97009	1.97991	1.98978	1.99970	2.00966

Compounded Biweekly

Years	6.50%	6.55%	6.60%	6.65%	6.70%	6.75%	6.80%	6.85%	6.90%	6.95%	7.00%
2	1.13864	1.13978	1.14092	1.14206	1.14320	1.14434	1.14548	1.14662	1.14777	1.14891	1.15006
5	1.38347	1.38692	1.39039	1.39386	1.39734	1.40083	1.40432	1.40783	1.41134	1.41487	1.41840
10	1.91399	1.92356	1.93317	1.94284	1.95255	1.96232	1.97213	1.98199	1.99189	2.00185	2.01186

Compounded Daily

Years	6.50%	6.55%	6.60%	6.65%	6.70%	6.75%	6.80%	6.85%	6.90%	6.95%	7.00%
2	1.13882	1.13995	1.14109	1.14224	1.14338	1.14452	1.14567	1.14681	1.14796	1.14911	1.15026
5	1.38399	1.38745	1.39093	1.39441	1.39790	1.40140	1.40490	1.40842	1.41194	1.41548	1.41902
10	1.91543	1.92503	1.93468	1.94437	1.95412	1.96391	1.97375	1.98364	1.99359	2.00358	2.01362

CHAPTER 6 HIGHLIGHTS

DEFINITIONS AND CONCEPTS	EXAMPLES

SECTION 6.1   INTRODUCTION TO PERCENT

Percent means per hundred. The % symbol denotes percent.	$51\% = \dfrac{51}{100}$ 51 per 100 $7\% = \dfrac{7}{100}$ 7 per 100
To write a percent as a decimal, drop the % symbol and move the decimal point two places to the left.	$32\% = 0.32$
To write a decimal as a percent, move the decimal point two places to the right and attach the % symbol.	$0.08 = 08\% = 8\%$

SECTION 6.2   PERCENTS AND FRACTIONS

To write a percent as a fraction, drop the % symbol and write the number over 100.	$25\% = \dfrac{25}{100} = \dfrac{\overset{1}{\cancel{25}}}{4\cdot\cancel{25}} = \dfrac{1}{4}$
To write a fraction as a percent, multiply the fraction by 100%.	$\dfrac{1}{6} = \dfrac{1}{6}\cdot 100\% = \dfrac{1}{6}\cdot\dfrac{100}{1}\% = \dfrac{100}{6}\% = 16\frac{2}{3}\%$

SECTION 6.3   SOLVING PERCENT PROBLEMS WITH EQUATIONS

Three key words in the statement of a percent problem are **of**, which means multiplication ($\cdot$) **is**, which means equals ($=$) **what** (or some equivalent word or phrase), which stands for the unknown	Solve. 6 is 12% of what number? $6 = 12\%\cdot n$ $6 = 0.12\cdot n$ Write 12% as a decimal. $\dfrac{6}{0.12} = n$ Divide 6 by 0.12, the number multiplied by n. $50 = n$ Thus, 6 is 12% of 50.

SECTION 6.4   SOLVING PERCENT PROBLEMS WITH PROPORTIONS

PERCENT PROPORTION

$$\frac{\text{amount}}{\text{base}} = \frac{\text{percent}}{100} \quad \leftarrow \text{always 100}$$

or

$$\text{amount} \rightarrow \frac{a}{b} = \frac{p}{100} \quad \leftarrow \text{percent}$$
$$\text{base} \rightarrow$$

Solve.

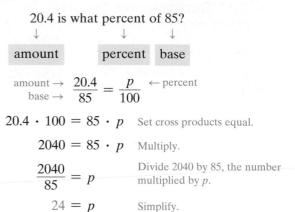

20.4 is what percent of 85?

$$\text{amount} \rightarrow \frac{20.4}{85} = \frac{p}{100} \quad \leftarrow \text{percent}$$
$$\text{base} \rightarrow$$

$20.4 \cdot 100 = 85 \cdot p$ Set cross products equal.

$2040 = 85 \cdot p$ Multiply.

$\dfrac{2040}{85} = p$ Divide 2040 by 85, the number multiplied by p.

$24 = p$ Simplify.

Thus, 20.4 is 24% of 85.

SECTION 6.5   APPLICATIONS OF PERCENT

PERCENT OF INCREASE

$$\text{percent of increase} = \frac{\text{amount of increase}}{\text{original amount}}$$

PERCENT OF DECREASE

$$\text{percent of decrease} = \frac{\text{amount of decrease}}{\text{original amount}}$$

A town with a population of 16,480 decreased to 13,870 over a 12-year period. Find the percent decrease. Round to the nearest whole percent.

$$\text{amount of decrease} = 16,480 - 13,870$$
$$= 2610$$

$$\text{percent of decrease} = \frac{\text{amount of decrease}}{\text{original amount}}$$

$$= \frac{2610}{16,480} \approx 0.16$$

$$= 16\%$$

The town's population decreased by 16%.

SECTION 6.6   PERCENT AND PROBLEM SOLVING: SALES TAX, COMMISSION, AND DISCOUNT

SALES TAX

$$\text{sales tax} = \text{sales tax rate} \cdot \text{purchase price}$$
$$\text{total price} = \text{purchase price} + \text{sales tax}$$

Find the sales tax and the total price of a purchase of $42.00 if the sales tax rate is 9%.

$$\text{sales tax} = \text{sales tax rate} \cdot \text{purchase price}$$

$$\text{sales tax} = 9\% \cdot \$42$$
$$= 0.09 \cdot \$42$$
$$= \$3.78$$

The total price is

$$\text{total price} = \text{purchase price} + \text{sales tax}$$

$$\text{total price} = \$42.00 + \$3.78$$
$$= \$45.78$$

SECTION 6.6 *(CONTINUED)*	
COMMISSION commission = commission rate · sales	A salesperson earns a commission of 3%. Find the commission from sales of $12,500 worth of appliances. commission = commission rate · sales commission = 3% · $12,500 = 0.03 · 12,500 = $375
DISCOUNT AND SALE PRICE amount of discount = discount rate · original price sale price = original price − amount of discount	A suit is priced at $320 and is on sale today for 25% off. What is the sale price? amount of discount = discount rate · original price amount of discount = 25% · $320 = 0.25 · 320 = $80 sale price = original price − amount of discount sale price = $320 − $80 = $240 The sale price is $240.

SECTION 6.7 PERCENT AND PROBLEM SOLVING: INTEREST	
SIMPLE INTEREST interest = principal · rate · time where the rate is understood to be per year.	Find the simple interest after 3 years on $800 at an interest rate of 5%. interest = principal · rate · time interest = $800 · 5% · 3 = $800 · 0.05 · 3 Write 5% as 0.05. = $120 Multiply. The interest is $120.
Compound interest is computed not only on the principal, but also on interest already earned in previous compounding periods. (See Appendix F) total amount = original principal · compound interest factor	$800 is invested at 5% compounded quarterly for 10 years. Find the total amount at the end of 10 years. total amount = original principal · compound interest factor total amount = $800 · 1.64362 ≈ $1314.90

Focus On the Real World

How Much Can You Afford for a House?

When a home buyer takes out a mortgage to buy a house, the loan is generally repaid on a monthly basis with a monthly mortgage payment. (Some banks also offer biweekly payment programs.) An important consideration in choosing a house is the amount of the monthly payment. Usually, the amount that a home buyer can afford to make as a monthly payment will dictate the house purchase price that can be afforded.

The first steps in deciding how much can be afforded for a house is finding out how much income the household has each month before taxes. The Mortgage Bankers Association of America (MBAA) suggests that the monthly mortgage payment be between 25% and 28% of the total monthly income. If other long-term debts exist (such as car or education loans and long-term credit card debt repayment), the MBAA further recommends that the total of housing costs and other monthly debt payments not exceed 36% of the total monthly income.

Once the size of the monthly payment that can be afforded has been found, a mortgage calculator can be used to work backwards to estimate the mortgage amount that will give that monthly payment. For example, the Mortgage Bankers Association of America includes a mortgage calculator on its Website at http://www.mbaa.org/cgi-bin/calculator.pl. (Alternatively, visit the MBAA homepage at http://www.mbaa.org and navigate to Consumer Info. Then look for the Mortgage Calculator option.) With this mortgage calculator, the user can input the initial principal (mortgage amount), annual interest rate, and loan term. This information is then used to calculate the associated monthly payment. To work backwards with a mortgage calculator:

▲ Enter the interest rate that is likely for your loan and the loan term in which you are interested.
▲ Then make a guess for the initial principal that can be afforded.
▲ Have the mortgage calculator calculate the monthly payment.
▲ If the monthly payment calculated is higher than the range that can be afforded, repeat the calculation using the same interest rate and loan term but using a lower value for the initial principal.
▲ If the monthly payment calculated is lower than the range that can be afforded, repeat the calculation using the same interest rate and loan term but using a higher value for the initial principal.
▲ Repeat these calculations methodically until a monthly payment is obtained that is in the range that can be afforded. The initial principal value that gave this monthly payment amount is an estimate of the mortgage amount that can be afforded to buy a home.

Group Activity

1. Research current interest rates on 30-year mortgages.
2. Use the method described above to find the size of mortgage that can be afforded by households with the following total monthly incomes before taxes. Use a loan term of 30 years and a current interest rate on a 30-year mortgage.
    a. $1500   b. $2000   c. $2500   d. $3000
    e. $3500   f. $4000
    Assume in each case that the household has no other debts.
3. Create a table of your results.

CHAPTER 6 REVIEW

(6.1) *Solve.*

1. In a survey of 100 adults, 37 preferred pepperoni on their pizza. What percent preferred pepperoni?
   37%

2. A basketball player made 77 out 100 attempted free throws. What percent of free throws was made?
   77%

Write each percent as a decimal.

3. 83%   0.83   **4.** 75%   0.75   **5.** 73.5%   0.735   **6.** 1.5%   0.015

7. 125%   1.25   **8.** 145%   1.45   **9.** 0.5%   0.005   **10.** 0.7%   0.007

11. 200%   2.00 or 2   **12.** 400%   4.00 or 4   **13.** 26.25%   0.2625   **14.** 85.34%   0.8534

Write each decimal as a percent.

15. 2.6   260%   **16.** 0.055   5.5%   **17.** 0.35   35%   **18.** 1.02   102%

19. 0.725   72.5%   **20.** 0.25   25%   **21.** 0.076   7.6%   **22.** 0.085   8.5%

23. 0.75   75%   **24.** 0.65   65%   **25.** 4.00   400%   **26.** 9.00   900%

(6.2) *Write each percent as a fraction or a mixed number in simplest form.*

27. 1% $\dfrac{1}{100}$ **28.** 10% $\dfrac{1}{10}$ **29.** 25% $\dfrac{1}{4}$ **30.** 8.5% $\dfrac{17}{200}$

31. 10.2% $\dfrac{51}{500}$ **32.** $16\dfrac{2}{3}$% $\dfrac{1}{6}$ **33.** $33\dfrac{1}{3}$% $\dfrac{1}{3}$ **34.** 110% $1\dfrac{1}{10}$

Write each fraction or mixed number as a percent.

35. $\frac{1}{5}$ 20%

36. $\frac{7}{10}$ 70%

37. $\frac{5}{6}$ $83\frac{1}{3}$%

38. $\frac{5}{8}$ 62.5%

39. $1\frac{2}{3}$ $166\frac{2}{3}$%

40. $1\frac{1}{4}$ 125%

41. $\frac{3}{5}$ 60%

42. $\frac{1}{16}$ 6.25%

(6.3) *Translate each to an equation and solve.*

43. 1250 is 1.25% of what number?   100,000

44. What number is $33\frac{1}{3}$% of 24,000?   8000

45. 124.2 is what percent of 540?   23%

46. 22.9 is 20% of what number?   114.5

47. What number is 40% of 7500?   3000

48. 693 is what percent of 462?   150%

(6.4) *Translate each to a proportion and solve.*

49. 104.5 is 25% of what number?   418

50. 16.5 is 5.5% of what number?   300

51. What number is 36% of 180?   64.8

52. 63 is what percent of 35?   180%

53. 93.5 is what percent of 85?   110%

54. What number is 33% of 500?   165

(6.5) *Solve.*

55. In a survey of 2000 people, it was found that 1320 have a microwave oven. Find the percent of people who own microwaves.   66%

56. Of the 12,360 freshmen entering County College, 2000 are enrolled in Basic College Mathematics. Find the percent of entering freshmen who are enrolled in Basic College Mathematics. Round to the nearest whole percent.   16%

57. The current charge for dumping waste in a local landfill is $16 per cubic foot. To cover new environmental costs, the charge will increase to $33 per cubic foot. Find the percent increase.   106.25%

58. The number of violent crimes in a city decreased from 675 to 534. Find the percent decrease. Round to the nearest tenth.   20.9%

59. This year the fund drive for a charity collected $215,000. Next year, a 4% decrease is expected. Find how much is expected to be collected in next year's drive. $206,400

60. A local union negotiated a new contract that increases the hourly pay 15% over last year's pay. The old hourly rate was $11.50. Find the new hourly rate rounded to the nearest cent. $13.23

(6.6) *Solve.*

61. If the sales tax rate is 5.5%, what is the total amount charged for a $250 coat? $263.75

62. Find the sales tax paid on a $25.50 purchase if the sales tax rate is 4.5%. $1.15

63. Russ James is a sales representative for a chemical company and is paid a commission rate of 5% on all sales. Find his commission if he sold $100,000 worth of chemicals last month. $5000

64. Carol Sell is a sales clerk in a clothing store. She receives a commission of 7.5% on all sales. Find her commission for the week if her sales for the week were $4005. Round to the nearest cent. $300.38

65. A $3000 mink coat is on sale for 30% off. Find the discount and the sale price.
discount: $900; sale price: $2100

66. A $90 calculator is on sale for 10% off. Find the discount and the sale price.
discount: $9; sale price: $81

(6.7) *Solve.*

67. Find the simple interest due on $4000 loaned for 3 months at 12% interest. $120

68. Find the total amount due on an 8-month loan of $1200 at a simple interest rate of 15%. $1320

69. Find the total amount in an account if $5500 is compounded annually at 12% for 15 years. $30,104.64

70. Find the total amount in an account if $6000 is compounded semiannually at 11% for 10 years. $17,506.56

71. Find the compound interest earned if $100 is compounded quarterly at 12% for 5 years. $80.61

72. Find the compound interest earned if $1000 is compounded quarterly at 18% for 20 years. $32,830.10

Chapter 6 Test

Write each percent as a decimal.

1. 85%

2. 500%

3. 0.6%

Write each decimal as a percent.

4. 0.056

5. 6.1

6. 0.35

Write each percent as a fraction or a mixed number in simplest form.

7. 120%

8. 38.5%

9. 0.2%

Write each fraction or mixed number as a percent.

10. $\frac{11}{20}$

11. $\frac{3}{8}$

12. $1\frac{3}{4}$

Solve.

13. What number is 42% of 80?

14. 0.6% of what number is 7.5?

15. 567 is what percent of 756?

Solve. Round all dollar amounts to the nearest cent.

16. An alloy is 12% copper. How much copper is contained in 320 pounds of this alloy?

17. A farmer in Nebraska estimates that 20% of his potential crop, or $11,350, has been lost to a hard freeze. Find the total value of his potential crop.

18. If the local sales tax rate is 1.25%, find the total amount charged for a stereo system priced at $354.

19. A town's population increased from 25,200 to 26,460. Find the percent increase.

ANSWERS

1. 0.85

2. 5

3. 0.006

4. 5.6%

5. 610%

6. 35%

7. $\frac{6}{5}$

8. $\frac{77}{200}$

9. $\frac{1}{500}$

10. 55%

11. 37.5%

12. 175%

13. 33.6

14. 1250

15. 75%

16. 38.4 lb

17. $56,750

18. $358.43

19. 5%

Name _____

20. A $120 framed picture is on sale for 15% off. Find the discount and the sale price.

21. Randy Nguyen is paid a commission rate of 4% on all sales. Find Randy's commission if his sales were $9875.

22. A sales tax of $1.53 is added to an item's price of $152.99. Find the sales tax rate. Round to the nearest whole percent.

23. Find the simple interest earned on $2000 saved for $3\frac{1}{2}$ years at an interest rate of 9.25%.

24. $1365 is compounded annually at 8%. Find the total amount in the account after 5 years.

25. A couple borrowed $400 from a bank at 13.5% for 6 months for car repairs. Find the total amount due the bank at the end of the 6-month period.

Name _____ Section _____ Date _____

CUMULATIVE REVIEW

1. How many cases can be filled with 9900 cans of jalapenos if each case holds 48 cans? How many cans will be left over? Will there be enough cases to fill an order for 200 cases?

2. Write each fraction as a mixed number or a whole number.

   a. $\dfrac{30}{7}$

   b. $\dfrac{16}{15}$

   c. $\dfrac{84}{6}$

3. Use a factor tree to find the prime factorization of 24.

4. Write $\dfrac{10}{27}$ in simplest form.

5. Multiply and simplify: $\dfrac{23}{32} \cdot \dfrac{4}{7}$

6. Find the reciprocal of $\dfrac{11}{8}$.

7. Find the perimeter of the rectangle.

   $\frac{2}{15}$ in.

   $\frac{4}{15}$ in.

8. Find the LCM of 12 and 20.

9. Add: $\dfrac{2}{5} + \dfrac{4}{15}$

10. Subtract: $7\dfrac{3}{14} - 3\dfrac{6}{7}$

Perform each indicated operation.

11. $\dfrac{1}{2} \div \dfrac{8}{7}$

12. $\dfrac{2}{9} \cdot \dfrac{3}{11}$

Write each fraction as a decimal.

13. $\dfrac{8}{10}$

14. $\dfrac{87}{10}$

Name _____

15. The price of a gallon of gasoline in Aimsville is currently $1.0279. Round this to the nearest cent.

16. Add: $763.7651 + 22.001 + 43.89$

17. Multiply: $23.6 \cdot 0.78$

Divide.

18. $\dfrac{786}{10,000}$

19. $\dfrac{0.12}{10}$

20. Simplify: $(1.3)^2$

21. Write $\dfrac{1}{4}$ as a decimal.

22. Write the following rate as a fraction in simplest form.

10 nails every 6 feet

23. Is $\dfrac{4.1}{7} = \dfrac{2.9}{5}$ a true proportion?

24. On a Chamber of Commerce map of Abita Springs, 5 miles corresponds to 2 inches. How many miles correspond to 7 inches?

25. Translate to an equation: What number is 25% of 0.008?

Measurement

The use of measurements is common in everyday life. A sales representative records the number of miles she has driven when she submits her travel expense report. A respiratory therapist measures the volume of air exhaled by a patient. A measurement is necessary in each case.

The Statue of Liberty, standing on Liberty Island in New York City's harbor, is one of the largest statues in the world. Originally named the Statue of Liberty Enlightening the World, this symbol of freedom was a gift from the people of France. U.S. citizens raised money for the pedestal on which the statue stands and for the installation of the statue. After the pedestal had been prepared in 1885, the statue arrived dismantled in 214 packing cases of iron framework and copper sheeting. The statue was assembled using rivets, and over a year later it was dedicated by President Grover Cleveland. The statue alone weighs 225 tons and is over 150 feet tall. For its 100th anniversary in 1986, the Statue of Liberty received extensive renovations to its framework and torch. In Exercises 39 and 40 on page 408, we will explore some of the measurements of the Statue of Liberty.

Name _____ **Section** _____ **Date** _____

CHAPTER 7 PRETEST

Convert.

1. 6 ft to inches

2. 98 in. = ____ ft ____ in.

3. 2.8 mi to feet

4. 9.8 m to centimeters

5. 7 lb to ounces

6. 29 dg to grams

7. 0.6 kg to grams

8. 18 qt to gallons

9. 6280 L to kiloliters

10. 45°C to Fahrenheit

11. 77°F to Celsius

12. 15 BTU to foot-pounds

Perform the indicated operation.

13. 9 ft 10 in. + 3 ft 7 in.

14. 6 yd 1 ft − 1 yd 2 ft

15. 5.4 mm × 9

16. 5 tons 1300 lb ÷ 4

17. 20.6 g + 391 mg

18. 9 qt 1 pt × 7

19. 3c 2 fl oz + 5c 6 fl oz

20. Jay Dolesky paid $13 to fill his car with 40 liters of gasoline. Find the price per liter of gasoline to the nearest tenth of a cent.

7.1 LENGTH: U.S. AND METRIC SYSTEMS OF MEASUREMENT

A CONVERTING U.S. SYSTEM UNITS OF LENGTH

In the United States, two systems of measurement are commonly used. They are the **United States (U.S.), or English, measurement system** and the **metric system**. The U.S. measurement system is familiar to most Americans. Units such as feet, miles, ounces, and gallons are used. However, the metric system is also commonly used in fields such as medicine, sports, international marketing, and certain physical sciences. We are used to buying 2-liter bottles of soft drinks, watching televised coverage of the 100-meter dash at the Olympic games, or taking a 200-milligram dose of pain reliever.

The U.S. system of measurement uses the **inch, foot, yard**, and **mile** to measure **length**. The following is a summary of equivalencies between units of length.

U.S. UNITS OF LENGTH	UNIT FRACTIONS
12 inches (in.) = 1 foot (ft)	$\dfrac{12 \text{ in.}}{1 \text{ ft}} = \dfrac{1 \text{ ft}}{12 \text{ in.}} = 1$
3 feet = 1 yard (yd)	$\dfrac{3 \text{ ft}}{1 \text{ yd}} = \dfrac{1 \text{ yd}}{3 \text{ ft}} = 1$
5280 feet = 1 mile (mi)	$\dfrac{5280 \text{ ft}}{1 \text{ mi}} = \dfrac{1 \text{ mi}}{5280 \text{ ft}} = 1$

To convert from one unit of length to another, **unit fractions** may be used. A unit fraction is a fraction that equals 1. For example, since 12 in. = 1 ft, we have the unit fractions

$$\frac{12 \text{ in.}}{1 \text{ ft}} = \frac{1 \text{ ft}}{12 \text{ in.}} = 1$$

For example, to convert 48 inches to feet, we *multiply by a unit fraction that relates feet to inches*. The unit fraction should be written so that *the units we are converting to*, feet, *are in the numerator and the original units*, inches, *are in the denominator*. We do this so that like units will divide out, as shown next.

$$48 \text{ in.} = \frac{48 \text{ in.}}{1} \cdot \overbrace{\frac{1 \text{ ft}}{12 \text{ in.}}}^{\text{Unit fraction}} \quad \begin{array}{l} \leftarrow \text{ Units converting to} \\ \leftarrow \text{ Original units} \end{array}$$

$$= \frac{48 \cdot 1 \text{ ft}}{1 \cdot 12}$$

$$= \frac{48 \text{ ft}}{12}$$

$$= 4 \text{ ft}$$

Therefore, 48 inches equals 4 feet, as seen in the diagram.

12 in.	12 in.	12 in.	12 in.	
1 ft	1 ft	1 ft	1 ft	48 in. = 4 ft

Practice Problem 1

Convert 5 feet to inches.

Practice Problem 2

Convert 7 yards to feet.

Example 1   Convert 8 feet to inches.

Solution: We multiply 8 feet by a unit fraction that compares 12 inches to 1 foot. The unit fraction should be $\frac{\text{units converting to}}{\text{original units}}$ or $\frac{12 \text{ inches}}{1 \text{ foot}}$.

$$8 \text{ ft} = \frac{8 \cancel{\text{ft}}}{1} \cdot \frac{12 \text{ in.}}{1 \cancel{\text{ft}}}$$

$$= 8 \cdot 12 \text{ in.}$$

$$= 96 \text{ in.} \qquad \text{Multiply.}$$

Then 8 ft = 96 in., as shown in the diagram.

1 ft	1 ft	1 ft	1 ft	1 ft	1 ft	1 ft	1 ft
12 in.	12 in.	12 in.	12 in.	12 in.	12 in.	12 in.	12 in.

8 ft = 96 in.

Example 2   Convert 7 feet to yards.

Solution: We multiply by a unit fraction that compares 1 yard to 3 feet.

$$7 \text{ ft} = \frac{7 \cancel{\text{ft}}}{1} \cdot \frac{1 \text{ yd}}{3 \cancel{\text{ft}}} \begin{array}{l} \leftarrow \text{Units converting to} \\ \leftarrow \text{Original units} \end{array}$$

$$= \frac{7 \text{ yd}}{3}$$

$$= 2\frac{1}{3} \text{ yd} \qquad \text{Divide.}$$

Thus 7 ft = $2\frac{1}{3}$ yd.

B Using Mixed U.S. System Units of Length

Sometimes it is more meaningful to express a measurement of length with mixed units such as 1 ft and 5 in. We usually condense this and write 1 ft 5 in.

In Example 2, we found that 7 feet was the same as $2\frac{1}{3}$ yards. The measurement can also be written as a mixture of yards and feet. That is,

7 ft = ____ yd ____ ft

Because 3 ft = 1 yd, we divide 3 into 7 to see how many whole yards are in 7 feet. The quotient is the number of yards, and the remainder is the number of feet.

$$\begin{array}{r} 2 \text{ yd } 1 \text{ ft} \\ 3\overline{)7} \\ \underline{-6} \\ 1 \end{array}$$

Thus 7 ft = 2 yd 1 ft, as seen in the diagram.

Answers

1. 60 inches,   **2.** 21 feet

Example 3

Convert: 134 in. = _____ ft _____ in.

Solution: Because 12 in. = 1 ft, we divide 12 into 134. The quotient is the number of feet. The remainder is the number of inches. To see why we divide 12 into 134, notice that

$$134 \text{ in.} = \frac{134 \text{ in.}}{1} \cdot \frac{1 \text{ ft}}{12 \text{ in.}} = \frac{134}{12} \text{ ft}$$

$$\begin{array}{r} 11 \text{ ft } 2 \text{ in.} \\ 12\overline{)134} \\ -12 \\ \hline 14 \\ -12 \\ \hline 2 \end{array}$$

Thus 134 in. = 11 ft 2 in.

Practice Problem 3

Practice Problem 3

Convert: 68 in. = _____ ft _____ in.

Example 4

Convert 3 feet 7 inches to inches.

Solution: First, we will convert 3 feet to inches. Then we will add 7 inches.

$$3 \text{ ft} = \frac{3 \text{ ft}}{1} \cdot \frac{12 \text{ in.}}{1 \text{ ft}} = 36 \text{ in.}$$

Then

$$3 \text{ ft } 7 \text{ in.} = 36 \text{ in.} + 7 \text{ in.} = 43 \text{ in.}$$

Practice Problem 4

Convert 5 yards 2 feet to feet.

TEACHING TIP

Point out to students that mixed units are just a way of writing mixed numbers. For example, 3 ft 2 in. = $3\frac{2}{12}$ ft. Then have them add the measurements in Example 5 as mixed numbers.

C OPERATIONS ON U.S. SYSTEM UNITS OF LENGTH

Finding sums or differences of measurements often involves converting units as shown in the next example. Just remember that, as usual, only like units can be added or subtracted.

Example 5

Add 3 ft 2 in. and 5 ft 11 in.

Solution: To add, we line up the similar units.

$$\begin{array}{r} 3 \text{ ft } 2 \text{ in.} \\ +5 \text{ ft } 11 \text{ in.} \\ \hline 8 \text{ ft } 13 \text{ in.} \end{array}$$

Since 13 inches is the same as 1 ft 1 in., we have

$$8 \text{ ft } 13 \text{ in.} = 8 \text{ ft} + 1 \text{ ft } 1 \text{ in.}$$
$$= 9 \text{ ft } 1 \text{ in.}$$

TRY THE CONCEPT CHECK IN THE MARGIN.

Practice Problem 5

Add 4 ft 8 in. to 8 ft 11 in.

✓ CONCEPT CHECK

How could you estimate the following sum?

$$\begin{array}{r} 7 \text{ yd } 4 \text{ in.} \\ +3 \text{ yd } 27 \text{ in.} \end{array}$$

Example 6

Multiply 8 ft 9 in. by 3.

Solution: By the distributive property, we multiply 8 ft by 3 and 9 in. by 3.

$$\begin{array}{r} 8 \text{ ft } 9 \text{ in.} \\ \times \phantom{8 \text{ ft } 9 \text{ in}} 3 \\ \hline 24 \text{ ft } 27 \text{ in.} \end{array}$$

Practice Problem 6

Multiply 4 ft 7 in. by 4.

Answers

3. 5 ft 8 in., **4.** 17 ft, **5.** 13 ft 7 in.,
6. 18 ft 4 in.

✓ **Concept Check:** Round each to the nearest yard: 7 yd + 4 yd = 11 yd

Since 27 in. is the same as 2 ft 3 in., we simplify the product as

$$24 \text{ ft } 27 \text{ in.} = 24 \text{ ft } + 2 \text{ ft } 3 \text{ in.}$$
$$= 26 \text{ ft } 3 \text{ in.}$$

Practice Problem 7

Divide 18 ft 6 in. by 2.

TEACHING TIP

After going over Example 7, have students try the problem again after converting 24 yd 6 in. to either yards or inches.

Example 7   Divide 24 yd 6 in. by 3.

Solution: We divide each of the units by 3.

```
         8 yd 2 in.
   3) 24 yd 6 in.
     −24 yd
            6 in.
           −6 in.
              0
```

The quotient is 8 yd 2 in.

To check, see that 8 yd 2 in. multiplied by 3 is 24 yd 6 in.

Practice Problem 8

A carpenter cuts 1 ft 9 in. from a board of length 5 ft 8 in. Find the remaining length of the board.

Example 8   Finding the Length of a Piece of Rope

A rope of length 6 yd 1 ft has 2 yd 2 ft cut from one end. Find the length of the remaining rope.

Solution: Subtract 2 yd 2 ft from 6 yd 1 ft.

```
  beginning length  →    6 yd 1 ft
−     amount cut    →  −2 yd 2 ft
   remaining length
```

We cannot subtract 2 ft from 1 ft, so we borrow 1 yd from the 6 yd. One yard is converted to 3 ft and combined with the 1 ft already there.

Borrow 1 yd = 3 ft   The problem now reads:

```
5 yd + 1 yd 3 ft
      6 yd 1 ft                5 yd 4 ft
     −2 yd 2 ft               −2 yd 2 ft
                               3 yd 2 ft
```

The remaining rope is 3 yd 2 ft long.

D CONVERTING METRIC SYSTEM UNITS OF LENGTH

The basic unit of length in the metric system is the **meter**. A meter is slightly longer than a yard. It is approximately 39.37 inches long, and recall that a yard is 36 inches long.

1 yard
1 meter

All units of length in the metric system are based on the meter. The following is a summary of the prefixes used in the metric system. Also shown are equivalencies between units of length. Like the decimal system, the metric system uses powers of 10 to define units.

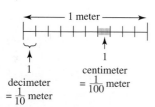

1 decimeter = $\frac{1}{10}$ meter

1 centimeter = $\frac{1}{100}$ meter

Answers

7. 9 ft 3 in.,   **8.** 3 ft 11 in.

Prefix	Meaning	Metric Unit of Length
kilo	1000	1 **kilo**meter (km) = 1000 meters (m)
hecto	100	1 **hecto**meter (hm) = 100 m
deka	10	1 **deka**meter (dam) = 10 m
		1 meter (m) = 1 m
deci	1/10	1 **deci**meter (dm) = 1/10 m or 0.1 m
centi	1/100	1 **centi**meter (cm) = 1/100 m or 0.01 m
milli	1/1000	1 **milli**meter (mm) = 1/1000 m or 0.001 m

These same prefixes are used in the metric system for mass and capacity. The most commonly used measurements of length in the metric system are the **meter, millimeter, centimeter**, and **kilometer**.

Being comfortable with the metric units of length means gaining a "feeling" for metric lengths, just as you have a "feeling" for the length of an inch, a foot, and a mile. To help you accomplish this, study the following examples.

A millimeter is about the thickness of a large paper clip wire.
A centimeter is about the width of a large paper clip.

A meter is slightly longer than a yard.
A kilometer is about two-thirds of a mile.
The length of this workbook is approximately 27.5 centimeters.
The width of this workbook is approximately 21.5 centimeters.

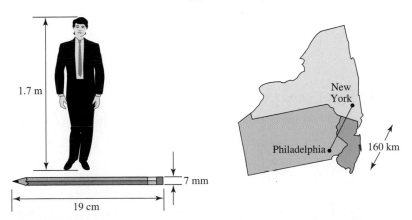

Just as for the U.S. system of measurement, unit fractions may be used to convert from one unit of length to another. The metric system does, however, have a distinct advantage over the U.S. system of measurement: the ease of converting from one unit of length to another. Since all units of length are powers of 10 of the meter, converting from one unit of length to another is as simple as moving the decimal point. Listing units of length in order from largest to smallest helps to keep track of how many places to move the decimal point when converting.

For example, let's convert 1200 meters to kilometers. To convert from meters to kilometers, we move along the chart shown from meters 3 units to the left to kilometers. This means that we move the decimal point 3 places to the left to convert from meters to kilometers.

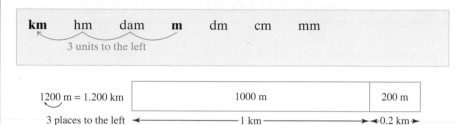

The same conversion can be made using unit fractions.

$$1200 \text{ m} = \frac{1200 \text{ m}}{1} \cdot \overbrace{\frac{1 \text{ km}}{1000 \text{ m}}}^{\text{Unit fraction}} = \frac{1200 \text{ km}}{1000} = 1.2 \text{ km}$$

Practice Problem 9

Convert 3.5 m to kilometers.

Example 9   Convert 2.3 m to centimeters.

Solution: First we will convert by using a unit fraction.

$$2.3 \text{ m} = \frac{2.3 \text{ m}}{1} \cdot \overbrace{\frac{100 \text{ cm}}{1 \text{ m}}}^{\text{Unit fraction}} = 230 \text{ cm}$$

Now we will convert by listing the units of length in a chart and moving from meters to centimeters.

km   hm   dam   m   dm   cm   mm

2 units to the right

2.30 m = 230. cm

2 places to the right

With either method, we get 230 cm.

Practice Problem 10

Convert 2.5 m to millimeters.

✓ CONCEPT CHECK

What is wrong with the following conversion? Convert 150 cm to meters.

150.00 cm = 15,000 m

Example 10   Convert 450,000 mm to meters.

Solution: We list the units of length in a chart and move from millimeters to meters.

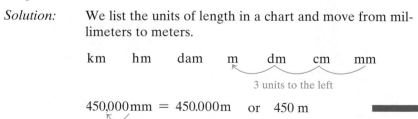

450,000 mm = 450.000 m   or   450 m

TRY THE CONCEPT CHECK IN THE MARGIN.

Answers

9. 0.0035 km,   **10.** 2500 mm

✓ **Concept Check:**   Should move decimal to the left: 1.5 m

E OPERATIONS ON METRIC SYSTEM UNITS OF LENGTH

To add, subtract, multiply, or divide with metric measurements of length, we write all numbers using the same unit of length and then add, subtract, multiply, or divide as with decimals.

Example 11   Subtract 430 m from 1.3 km.

Solution: First we convert both mesurements to kilometers or both to meters.

$$430\,m = 0.43\,km \qquad or \qquad 1.3\,km = 1300\,m$$

$$\begin{array}{r} 1.30\,km \\ -\ 0.43\,km \\ \hline 0.87\,km \end{array} \qquad\qquad \begin{array}{r} 1300\,m \\ -\ \ \ 430\,m \\ \hline 870\,m \end{array}$$

The difference is 0.87 km or 870 m.

Example 12   Multiply 5.7 mm by 4.

Solution: Here we simply multiply the two numbers. Note that the unit of measurement remains the same.

$$\begin{array}{r} 5.7\ mm \\ \times\ \ \ 4 \\ \hline 22.8\ mm \end{array}$$

Example 13   Finding a Person's Height

Fritz Martinson was 1.2 meters tall on his last birthday. Since then he has grown 14 centimeters. Find his current height in meters.

Solution:

$$\begin{array}{ll} \text{original height} & \rightarrow \quad\ 1.20\ m \\ +\ \text{height grown} & \rightarrow \ +0.14\ m \quad \text{(since 14 cm = 0.14 m)} \\ \hline \text{current height} & \qquad\ \ 1.34\ m \end{array}$$

Fritz is now 1.34 meters tall.

Practice Problem 11

Subtract 640 m from 2.1 km.

Practice Problem 12

Multiply 18.3 hm by 5.

Practice Problem 13

Doris Blackwell is knitting a scarf that is currently 0.8 meter long. If she knits an additional 45 centimeters, how long will the scarf be?

Answers

11. 1.46 km or 1460 m,   **12.** 91.5 hm,
13. 125 cm or 1.25 m

CALCULATOR EXPLORATIONS
METRIC TO U.S. SYSTEM CONVERSIONS IN LENGTH

To convert between the two systems of measurement in length, the following **approximations** may be used.

meters $\times$ 3.28 $\approx$ feet	feet $\times$ 0.305 $\approx$ meters
meters $\times$ 1.09 $\approx$ yards	yards $\times$ 0.914 $\approx$ meters
centimeters $\times$ 0.39 $\approx$ inches	inches $\times$ 2.54 $=$ centimeters
kilometers $\times$ 0.62 $\approx$ miles	miles $\times$ 1.609 $\approx$ kilometers

Example   The distance from New Orleans, Louisiana, to Pensacola, Florida, is about 400 miles. How many kilometers is this?

Solution: From the above approximations,

$$\text{miles} \times 1.609 \approx \text{kilometers}$$
$$\downarrow$$
$$400 \times 1.609 \approx \text{kilometers}$$

To multiply on your calculator, press the keys

$$\boxed{400} \;\; \boxed{\times} \;\; \boxed{1.609} \;\; \boxed{=}$$

or

$$\boxed{\text{ENTER}}$$

The display will read $\boxed{643.6}$. Then
$$400 \text{ miles} \approx 643.6 \text{ kilometers.}$$
$$\uparrow$$
(is approximately)

Convert as indicated.

1. 7 meters to feet. $\approx$22.96 ft

2. 11.5 yards to meters. $\approx$10.511 m

3. 8.5 inches to centimeters. $\approx$21.59 cm

4. 15 kilometers to miles. $\approx$9.3 mi

5. A 5-kilometer race is being held today. How many miles is this? $\approx$3.1 mi

6. A 100-meter dash is being held today. How many yards is this? $\approx$109 yd

TEACHING TIP

Ask students to use the approximations to determine which unit is longer:
meter or foot
meter or yard
centimeter or inch
kilometer or mile

Name _____ **Section** _____ **Date** _____

MENTAL MATH

Convert as indicated.

1. 12 inches to feet
2. 6 feet to yards
3. 24 inches to feet
4. 36 inches to feet
5. 36 inches to yards
6. 2 yards to inches

Determine whether the measurement in each statement is reasonable.

7. The screen of a home television set has a 30-meter diagonal.

8. A window measures 1 meter by 0.5 meter.

9. A drinking glass is made of glass 2 millimeters thick.

10. A paper clip is 4 kilometers long.

11. The distance across the Colorado River is 50 kilometers.

12. A model's hair is 30 centimeters long.

EXERCISE SET 7.1

A *Convert each measurement as indicated. See Examples 1 and 2.*

1. 60 in. to feet
2. 84 in. to feet
3. 12 yd to feet

4. 18 yd to feet
5. 42,240 ft to miles
6. 36,960 ft to miles

7. 102 in. to feet
8. 150 in. to feet
9. 10 ft to yards

10. 25 ft to yards
11. 6.4 mi to feet
12. 3.8 mi to feet

B *Convert each measurement as indicated. See Examples 3 and 4.*

13. 40 ft = ____ yd ____ ft
14. 100 ft = ____ yd ____ ft

15. 41 in. = ____ ft ____ in.
16. 75 in. = ____ ft ____ in.

17. 10,000 ft = ____ mi ____ ft
18. 25,000 ft = ____ mi ____ ft

19. 5 ft 2 in. = ____ in.
20. 4 ft 11 in. = ____ in.

21. 17 ft

22. 22 ft

23. 84 in.

24. 60 in.

25. 12 ft 3 in.

26. 18 ft 2 in.

27. 22 yd 1 ft

28. 25 yd

29. 8 ft 5 in.

30. 7 ft 3 in.

31. 5 ft 6 in.

32. 10 ft 9 in.

33. 3 ft 4 in.

34. 13 ft 5 in.

35. 50 yd 2 ft

36. 122 yd 2 ft

37. 10 ft 6 in.

38. 3 mi 120 ft

39. 18 ft 5 in.

40. 11 ft 11 in.

Name _____

21. 5 yd 2 ft = ____ ft

22. 7 yd 1 ft = ____ ft

23. 2 yd 1 ft = ____ in.

24. 1 yd 2 ft = ____ in.

C *Perform each indicated operation. Simplify the result if possible. See Examples 5 through 7.*

25. 5 ft 8 in. + 6 ft 7 in.

26. 9 ft 10 in. + 8 ft 4 in.

27. 12 yd 2 ft + 9 yd 2 ft

28. 16 yd 2 ft + 8 yd 1 ft

29. 24 ft 8 in. − 16 ft 3 in.

30. 15 ft 5 in. − 8 ft 2 in.

31. 16 ft 3 in. − 10 ft 9 in.

32. 14 ft 8 in. − 3 ft 11 in.

33. 6 ft 8 in. ÷ 2

34. 26 ft 10 in. ÷ 2

35. 12 yd 2 ft × 4

36. 15 yd 1 ft × 8

Solve. Remember to insert units when writing your answers. See Example 8.

37. The National Zoo maintains a small patch of bamboo, which it grows as a food supply for its pandas. Two weeks ago, the bamboo was 6 ft 10 in. tall. Since then, the bamboo has grown 3 ft 8 in. taller. How tall is the bamboo now?

38. While exploring in the Marianas Trench, a submarine probe was lowered to a point 1 mile 1400 feet below the ocean's surface. Later it was lowered an additional 1 mile 4000 feet below this point. How far is the probe below the surface of the Pacific?

39. The right arm of the Statue of Liberty in New York harbor is 42 ft long. The tablet held in the Statue's left arm is 23 ft 7 in. long. How much longer is the right arm than the tablet? (*Source: 1998 World Almanac*)

40. The length of one of the Statue of Liberty's hands is 16 ft 5 in. The nose of the Statue of Liberty is 4 ft 6 in. long. How much longer is a hand than the nose? (*Source: 1998 World Almanac*)

41. The Amana Corporation stacks up its microwave ovens in a distribution warehouse. Each stack is 1 ft 9 in. wide. How far from the wall would 9 of these stacks extend?

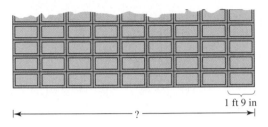

1 ft 9 in

|← ——————— ? ——————— →|

42. The highway commission is installing concrete sound barriers along a highway. Each barrier is 1 yd 2 ft long. How far will 25 barriers in a row reach?

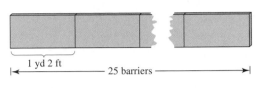

1 yd 2 ft

|← ——————— 25 barriers ——————— →|

43. A carpenter needs to cut a board into thirds. If the board is 9 ft 3 in. long originally, how long will each cut piece be?

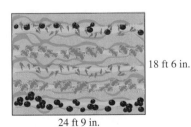

|← ——————— 9 ft 3 in. ——————— →|

44. A wall is erected exactly halfway between two buildings that are 192 ft 8 in. apart. If the wall is 8 in. wide, how far is it from the wall to either of the buildings?

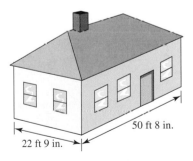

192 ft 8 in.

? 8 in. ?

45. Evelyn Pittman plans to erect a fence around her garden to keep the rabbits out. If the garden is a rectangle 24 ft 9 in. long by 18 ft 6 in. wide, what is the length of the fencing material she must purchase?

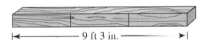

18 ft 6 in.

24 ft 9 in.

46. Ronnie Hall needs new gutters for the front and *both sides* of his home. The front of the house is 50 ft 8 in., and each side is 22 ft 9 in. wide. What length of gutter must he buy?

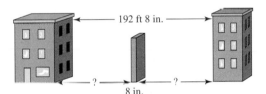

50 ft 8 in.

22 ft 9 in.

47. The world's longest Coca-Cola truck is in Sweden and is 79 feet long. How many *yards* long are 4 of these trucks? (*Source: Coca-Cola Today*)

48. The world's largest Coca-Cola sign is in Arica, Chile. It is in the shape of a rectangle whose length is 400 feet and whose width is 131 feet. Find the area of the sign. (*Source: Coca-Cola Today*) (*Hint:* Recall that area of a rectangle is the product: length · width.)

49. 4000 cm

50. 1800 cm

51. 4.0 cm

52. 1.8 cm

53. 0.3 km

54. 0.4 km

55. 1.4 m

56. 6.4 m

57. 15 m

58. 64 m

59. 83 mm

60. 46 mm

61. 0.201 dm

62. 1.402 dm

63. 40 mm

64. 200 mm

65. 8.94 m

66. 18.06 cm

67. 2.94 m or 2940 mm

68. 9.20 m or 920 cm

69. 1.29 cm or 12.9 mm

70. 23.7 m or 2.37 dam

71. 12.640 km or 12,640 m

72. 12.5 cm or 125 mm

73. 54.9 m

74. 56.4 m

75. 1.55 km

76. 1.92 m

77. 9.12 m

78. 2.13 m

410

Name _____

D *Convert as indicated. See Examples 9 and 10.*

49. 40 m to centimeters   **50.** 18 m to centimeters   **51.** 40 mm to centimeters

52. 18 mm to centimeters   **53.** 300 m to kilometers   **54.** 400 m to kilometers

55. 1400 mm to meters   **56.** 6400 mm to meters   **57.** 1500 cm to meters

58. 6400 cm to meters   **59.** 8.3 cm to millimeters   **60.** 4.6 cm to millimeters

61. 20.1 mm to decimeters   **62.** 140.2 mm to decimeters

63. 0.04 m to millimeters   **64.** 0.2 m to millimeters

E *Perform each indicated operation. See Examples 11 and 12.*

65. 8.6 m + 0.34 m   **66.** 14.1 cm + 3.96 cm   **67.** 2.9 m + 40 mm

68. 30 cm + 8.9 m   **69.** 24.8 mm − 1.19 cm   **70.** 45.3 m − 2.16 dam

71. 15 km − 2360 m   **72.** 14 cm − 15 mm   **73.** 18.3 m × 3

74. 14.1 m × 4   **75.** 6.2 km ÷ 4   **76.** 9.6 m ÷ 5

Solve. Remember to insert units when writing your answers. See Example 13.

77. A 3.4-m rope is attached to a 5.8-m rope. However, when the ropes are tied, 8 cm of length is lost to form the knot. What is the length of the tied ropes?

78. A 2.15-m-long sash cord has become frayed at both ends, so 1 cm is trimmed from each end. How long is the remaining cord?

79. The ice on Doc Miller's pond is 5.33 cm thick. For safe skating, Doc insists that it must be 80 mm thick. How much thicker must the ice be before Doc goes skating?

80. The sediment on the bottom of the Towamencin Creek is normally 14 cm thick, but the recent flood washed away 22 mm of sediment. How thick is it now?

81. An art class is learning how to make kites. The two sticks used for each kite have lengths of 1 m and 65 cm. What total length of wood must be ordered for the sticks if 25 kites are to be built?

82. The total pages of a hard-bound economics text are 3.1 cm thick. The front and back covers are each 2 mm thick. How high would a stack of 10 of these texts be?

83. A logging firm needs to cut a 67-m-long redwood log into 20 equal pieces before loading it onto a truck for shipment. How long will each piece be?

84. An 18.3-m-tall flagpole is mounted on a 65-cm-high pedestal. How far is the top of the flagpole above the ground?

85. At one time it was believed that the fort of Basasi, on the Indian-Tibetan border, was the highest located structure since it was at an elevation of 5.988 km above sea level. However, a settlement has been located that is 21 m higher than the fort. What is the elevation of this settlement?

86. The average American male at age 35 is 1.75 m tall. The average 65-year-old male is 48 mm shorter. How tall is the average 65-year-old male?

87. A floor tile is 22.86 cm wide. How many tiles in a row are needed to cross a room 3.429 m wide?

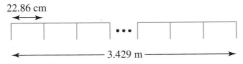

22.86 cm

3.429 m

88. A standard postcard is 1.6 times longer than it is wide. If it is 9.9 cm wide, what is its length?

?

9.9 cm

79. 26.7 mm

80. 11.8 cm

81. 41.25 m or 4125 cm

82. 35 cm

83. 3.35 m

84. 18.95 m

85. 6.009 km or 6009 m

86. 1.702 m

87. 15 tiles

88. 15.84 cm

Name _____

REVIEW AND PREVIEW

Write each decimal or fraction as a percent. See Sections 6.1 and 6.2.

89. 0.21

90. 0.86

91. $\dfrac{13}{100}$

92. $\dfrac{47}{100}$

93. $\dfrac{1}{4}$

94. $\dfrac{3}{20}$

COMBINING CONCEPTS

95. To convert from meters to centimeters, the decimal point is moved two places to the right. Explain how this relates to the fact that the prefix centi-means $\dfrac{1}{100}$.

96. Explain why conversions in the metric system are easier to make than conversions in the U.S. system of measurement.

97. Anoa Longway plans to use 26.3 meters of leftover fencing material to enclose a square garden plot for her daughter. How long will each side of the garden be?

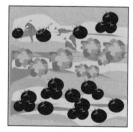

98. A marathon is a running race over a distance of 26 mi 385 yd. If a runner runs five marathons in a year, what is the total distance he or she has run in marathons? (*Source: Microsoft Encarta 97 Encyclopedia*)

7.2   WEIGHT AND MASS: U.S. AND METRIC SYSTEMS OF MEASUREMENT

A   CONVERTING U.S. SYSTEM UNITS OF WEIGHT

Whenever we talk about how heavy an object is, we are concerned with the object's **weight**. We discuss weight when we refer to a 12-ounce box of Rice Krispies, an overweight 19-pound tabby cat, or a barge hauling 24 tons of garbage.

The most common units of weight in the U.S. measurement system are the **ounce**, the **pound**, and the **ton**. The following is a summary of equivalencies between units of weight.

U.S. UNITS OF WEIGHT	UNIT FRACTIONS
16 ounces (oz) = 1 pound (lb)	$\dfrac{16\ oz}{1\ lb} = \dfrac{1\ lb}{16\ oz} = 1$
2000 pounds = 1 ton	$\dfrac{2000\ lb}{1\ ton} = \dfrac{1\ ton}{2000\ lb} = 1$

TRY THE CONCEPT CHECK IN THE MARGIN.

Unit fractions that equal 1 will be used to convert between units of weight in the U.S. system. When converting using unit fractions, recall that the numerator of a unit fraction should contain units we are converting to and the denominator should contain original units.

To convert 40 ounces to pounds, multiply by $\dfrac{1\ lb}{16\ oz}$.   ← Units converting to
                                                                      ← Original units

$$40\ oz = \frac{40\ \cancel{oz}}{1} \cdot \overbrace{\frac{1\ lb}{16\ \cancel{oz}}}^{\text{Unit fraction}}$$

$$= \frac{40\ lb}{16} \qquad \text{Multiply.}$$

$$= \frac{5}{2}\ lb \quad \text{or} \quad 2\frac{1}{2}\ lb, \text{ as a mixed number}$$

Example 1   Convert 9000 pounds to tons.

Solution: We multiply 9000 lb by the unit fraction

$$\frac{1\ ton}{2000\ lb}.$$ ← Units converting to
              ← Original units

TEACHING TIP   Classroom Activity

Have students begin this lesson by working in groups to make a list or poster of items having the following estimated weights: 1 ounce, 10 ounces, 1 pound, 10 pounds, 100 pounds, 1000 pounds, 1 ton, and 10 tons.

✓ CONCEPT CHECK

If you were describing the weight of a semi-truck, which type of unit would you use: ounce, pound, or ton? Why?

Practice Problem 1

Convert 4500 pounds to tons.

Answers

1. $2\frac{1}{4}$ tons

✓ Concept Check: ton

$$9000 \text{ lb} = \frac{9000 \text{ lb}}{1} \cdot \frac{1 \text{ ton}}{2000 \text{ lb}} = \frac{9000 \text{ tons}}{2000} = \frac{9}{2} \text{ tons or } 4\frac{1}{2} \text{ tons}$$

2000 lb	2000 lb	2000 lb	2000 lb	1000 lb	9000 lb
1 ton	1 ton	1 ton	1 ton	$\frac{1}{2}$ ton	$= 4\frac{1}{2}$ tons

---

Practice Problem 2

Convert 56 ounces to pounds.

Example 2   Convert 3 pounds to ounces

Solution: We multiply by the unit fraction $\frac{16 \text{ oz}}{1 \text{ lb}}$ to convert from pounds to ounces.

$$3 \text{ lb} = \frac{3 \text{ lb}}{1} \cdot \frac{16 \text{ oz}}{1 \text{ lb}} = 3 \cdot 16 \text{ oz} = 48 \text{ oz}$$

1 lb	1 lb	1 lb	3 lb
16 oz	16 oz	16 oz	$= 48$ oz

As with length, it is sometimes useful to simplify a measurement of weight by writing it in terms of mixed units.

$$33 \text{ ounces} = \underline{\hspace{2cm}} \text{ lb} \underline{\hspace{2cm}} \text{ oz}$$

Because 16 oz = 1 lb, divide 16 into 33 to see how many pounds are in 33 ounces. The quotient is the number of pounds and the remainder is the number of ounces. To see why we divide 16 into 33, notice that

$$33 \text{ oz} = 33 \text{ oz} \cdot \frac{1 \text{ lb}}{16 \text{ oz}} = \frac{33}{16} \text{ lb}$$

$$\begin{array}{r} 2 \text{ lb } 1 \text{ oz} \\ 16 \overline{)33} \\ -32 \\ \hline 1 \end{array}$$

Thus 33 ounces is the same as 2 lb 1 oz.

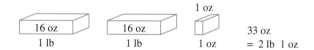

16 oz	16 oz	1 oz	33 oz
1 lb	1 lb	1 oz	$= 2$ lb 1 oz

B   OPERATIONS ON U.S. SYSTEM UNITS OF WEIGHT

Performing arithmetic operations on units of weight works the same way as performing arithmetic operations on units of length.

Practice Problem 3

Subtract 5 tons 1200 lb from 8 tons 100 lb.

Example 3   Subtract 3 tons 1350 lb from 8 tons 1000 lb.

Solution: To subtract, we line up similar units.

$$\begin{array}{r} 8 \text{ tons } 1000 \text{ lb} \\ -3 \text{ tons } 1350 \text{ lb} \\ \hline \end{array}$$

Since we cannot subtract 1350 lb from 1000 lb, we borrow 1 ton from the 8 tons. To do so, we write 1 ton as 2000 lb and combine it with the 1000 lb.

Answers

2. $3\frac{1}{2}$ pounds,   **3.** 2 tons 900 lb

7 tons + (1 ton) 2000 lb

$$
\begin{array}{r}
\cancel{8} \text{ tons } 1000 \text{ lb} \\
-3 \text{ tons } 1350 \text{ lb} \\
\end{array}
\qquad \text{becomes} \qquad
\begin{array}{r}
7 \text{ tons } 3000 \text{ lb} \\
-3 \text{ tons } 1350 \text{ lb} \\
\hline
4 \text{ tons } 1650 \text{ lb} \\
\end{array}
$$

To check, see that the sum of 4 tons 1650 lb and 3 tons 1350 lb is 8 tons 1000 lb.

TEACHING TIP

After doing Example 3 with mixed units, have students repeat the subtraction after converting the units to pounds.

Example 4

Multiply 5 lb 9 oz by 6.

Solution: We multiply 5 lb by 6 and 9 oz by 6.

$$
\begin{array}{r}
5 \text{ lb} \quad 9 \text{ oz} \\
\times \qquad 6 \\
\hline
30 \text{ lb } 54 \text{ oz} \\
\end{array}
$$

To write 54 oz as mixed units, we divide by 16 $\left(1 \text{ lb} = 16 \text{ oz}\right)$:

$$
\begin{array}{r}
3 \text{ lb } 6 \text{ oz} \\
16 \overline{)54} \\
-48 \\
\hline
6 \\
\end{array}
$$

Thus,

$$
30 \text{ lb } 54 \text{ oz} = 30 \text{ lb} + 3 \text{ lb } 6 \text{ oz} = 33 \text{ lb } 6 \text{ oz}
$$

Practice Problem 4

Multiply 4 lb 11 oz by 8.

Example 5

Divide 9 lb 6 oz by 2.

Solution: We divide each of the units by 2.

$$
\begin{array}{r}
4 \text{ lb} \qquad 11 \text{ oz} \\
2 \overline{)9 \text{ lb}} \qquad 6 \text{ oz} \\
-8 \\
\hline
1 \text{ lb} \;=\; 16 \text{ oz} \\
\overline{22 \text{ oz}} \\
\end{array}
$$

Divide 2 into 22 oz to get 11 oz.

To check, multiply 4 pounds 11 ounces by 2. The result will be 9 pounds 6 ounces.

Practice Problem 5

Divide 5 lb 8 oz by 4.

Example 6   Finding the Weight of a Child

Bryan weighed 8 lb 8 oz at birth. By the time he was 1 year old, he had gained 11 lb 14 oz. Find his weight at age 1 year.

Solution:

$$
\begin{array}{rl}
\text{birth weight} \rightarrow & 8 \text{ lb} \quad 8 \text{ oz} \\
+\text{weight gained} \rightarrow & +11 \text{ lb } 14 \text{ oz} \\
\hline
\text{total weight} \rightarrow & 19 \text{ lb } 22 \text{ oz} \\
\end{array}
$$

Since 22 oz equals 1 lb 6 oz.

$$
19 \text{ lb } 22 \text{ oz} = 19 \text{ lb} + 1 \text{ lb } 6 \text{ oz}
$$

$$
= 20 \text{ lb } 6 \text{ oz}
$$

Bryan weighed 20 lb 6 oz on his first birthday.

Practice Problem 6

A 5-lb 14-oz batch of cookies is packed in a 6-oz container before it is mailed. Find the total weight.

Answers

4. 37 lb 8 oz,   **5.** 1 lb 6 oz,   **6.** 6 lb 4 oz

C CONVERTING METRIC SYSTEM UNITS OF MASS

In scientific and technical areas, a careful distinction is made between **weight** and **mass**. **Weight** is really a measure of the pull of gravity. The farther from Earth an object gets, the less it weighs. However, **mass** is a measure of the amount of substance in the object and does not change. Astronauts orbiting Earth weigh much less than they weigh on Earth, but they have the same mass in orbit as they do on Earth. On Earth's surface, weight and mass are the same, so either term may be used.

The basic unit of mass in the metric system is the **gram**. It is defined as the mass of water contained in a cube 1 centimeter (cm) on each side.

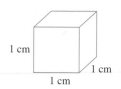

1 cm
1 cm
1 cm

The following examples may help you get a feeling for metric masses.

A tablet contains 200 milligrams of ibuprofen.

A large paper clip weighs approximately 1 gram.

A box of crackers weighs 453 grams.

A kilogram is slightly over 2 pounds. An adult woman may weigh 60 kilograms.

The prefixes for units of mass in the metric system are the same as for units of length, as shown in the following table.

Prefix	Meaning	Metric Unit of Mass
kilo	1000	1 kilogram (kg) = 1000 grams (g)
hecto	100	1 hectogram (hg) = 100 g
deka	10	1 dekagram (dag) = 10 g
		1 gram (g) = 1 g
deci	1/10	decigram (dg) = 1/10 g or 0.1 g
centi	1/100	1 centigram (cg) = 1/100 g or 0.01 g
milli	1/1000	1 milligram (mg) = 1/1000 g or 0.001 g

TRY THE CONCEPT CHECK IN THE MARGIN.

The **milligram**, the **gram**, and the **kilogram** are the three most commonly used units of mass in the metric system.

As with lengths, all units of mass are powers of 10 of the gram, so converting from one unit of mass to another only involves moving the decimal point. To convert from one unit of mass to another in the metric system, list the units of mass in order from largest to smallest.

Let's convert 4300 milligrams to grams. To convert from milligrams to grams, we move along the table 3 units to the left.

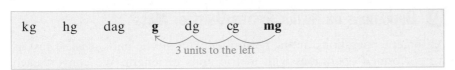

| kg | hg | dag | **g** | dg | cg | **mg** |

3 units to the left

This means that we move the decimal point 3 places to the left to convert from milligrams to grams.

$$4300\,\text{mg} = 4.3\,\text{g}$$

The same conversion can be done with unit fractions.

$$4300\,\text{mg} = \frac{4300\ \cancel{\text{mg}}}{1} \cdot \frac{0.001\ \text{g}}{1\ \cancel{\text{mg}}}$$
$$= 4300 \cdot 0.001\ \text{g}$$
$$= 4.3\ \text{g} \quad \text{To multiply by 0.001, move the decimal point 3 places to the left.}$$

To see that this is reasonable, study the diagram.

| 1000 mg | 1000 mg | 1000 mg | 1000 mg | 300 mg | 4300 mg |
| 1 g | 1 g | 1 g | 1 g | 0.3 g | = 4.3 g |

Thus 4300 mg = 4.3 g.

Example 7   Convert 3.2 kg to grams.

Solution: First we will convert by using a unit fraction.

Unit fraction

$$3.2\ \text{kg} = 3.2\ \cancel{\text{kg}} \cdot \frac{1000\ \text{g}}{1\ \cancel{\text{kg}}} = 3200\ \text{g}$$

Now let's list the units of mass in a chart and move from kilograms to grams.

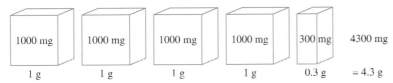

| kg | hg | dag | g | dh | cg | m g |

3 units to the right

$$3.200\ \text{kg} = 3200.\ \text{g}$$

3 places to the right

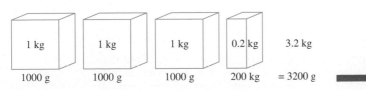

| 1 kg | 1 kg | 1 kg | 0.2 kg | 3.2 kg |
| 1000 g | 1000 g | 1000 g | 200 kg | = 3200 g |

Practice Problem 7

Convert 3.41 g to milligrams.

Answers

7. 3410 mg

✓ **Concept Check:**   False

Practice Problem 8

Convert 56.2 cg to grams.

Example 8   Convert 2.35 cg to grams.

Solution: We list the units of mass in a chart and move from centigrams to grams.

kg	hg	dag	g	dg	cg	m g

2 units to the left

02.35 cg $=$ 0.0235 g

2 places to the left

D OPERATIONS ON METRIC SYSTEM UNITS OF MASS

Arithmetic operations can be performed with metric units of mass just as we performed operations with metric units of length. We convert each number to the same unit of mass and add, subtract, multiply, or divide as with decimals.

Practice Problem 9

Subtract 3.1 dg from 2.5 g.

Example 9   Subtract 5.4 dg from 1.6 g.

Solution: We convert both numbers to decigrams or to grams before subtracting.

5.4 dg $=$ 0.54 g   or   1.6 g $=$ 16 dg

$$\begin{array}{r} 1.60 \text{ g} \\ -0.54 \text{ g} \\ \hline 1.06 \text{ g} \end{array} \qquad \begin{array}{r} 16.0 \text{ dg} \\ -5.4 \text{ dg} \\ \hline 10.6 \text{ dg} \end{array}$$

The difference is 1.06 g or 10.6 dg.

Practice Problem 10

Multiply 12.6 kg by 4.

Example 10   Multiply 15.4 kg by 5.

Solution: We multiply the two numbers together.

$$\begin{array}{r} 15.4 \text{ kg} \\ \times \quad 5 \\ \hline 77.0 \text{ kg} \end{array}$$

The result is 77.0 kg.

Practice Problem 11

Twenty-four bags of cement weigh a total of 550 kg. Find the average weight of 1 bag, rounded to the nearest kilogram.

Example 11   Calculating Allowable Weight in an Elevator

An elevator has a weight limit of 1400 kg. A sign posted in the elevator indicates that the maximum capacity of the elevator is 17 persons. What is the average allowable weight for each passenger, rounded to the nearest kilogram?

Solution: To solve, notice that the total weight of

1400 kilograms $\div$ 17 = average weight

$$\begin{array}{r} 82.3 \text{ kg} \approx 82 \text{ kg} \\ 17\overline{)1400.0 \text{ kg}} \\ \underline{-136} \\ 40 \\ \underline{-34} \\ 60 \\ \underline{-51} \\ 9 \end{array}$$

Answers

8. 0.562 g,   **9.** 2.19 g or 21.9 dg,   **10.** 50.4 kg,
11. 23 kg

Each passenger can weigh an average of 82 kg. (Recall that a kilogram is slightly over 2 pounds, so 82 kilograms is over 164 pounds. For a better approximation, see the Calculator Explorations box.) ▬▬

CALCULATOR EXPLORATIONS
Metric to U.S. System Conversions in Weight

To convert between the two systems of measurement in weight, the following *approximations* can be used.

grams × 0.035 ≈ ounces	ounces × 28.35 ≈ grams
kilograms × 2.20 ≈ pounds	pounds × 0.454 ≈ kilograms
grams × 0.0022 ≈ pounds	pounds × 454 ≈ grams

Example   A bulldog weighs 35 pounds. How many kilograms is this?

Solution: From the above approximations,

pounds × 0.454 ≈ kilograms
↓
35 × 0.454 ≈ kilograms

To multiply on your calculator, press the keys

or

ENTER

The display will read 15.89 . Thus 35 pounds ≈ 15.89 kilograms. ▬▬

Convert as indicated.

1. 15 ounces to grams.   ≈425.25 g

2. 11.2 grams to ounces.   ≈0.392 oz

3. 7 kilograms to pounds.   ≈15.4 lb

4. 23 pounds to kilograms.   ≈10.442 kg

5. A piece of candy weighs 5 grams. How many ounces is this?
   ≈0.175 oz

6. If a person weighs 82 kilograms, how many pounds is this?
   ≈180.4 lb

TEACHING TIP

Ask students to use the approximations to determine which unit is heavier.
grams or ounces
kilograms or pounds
grams or pounds

Focus On History

THE DEVELOPMENT OF UNITS OF MEASURE

The earliest units of measure were based on the human body. The ancient Egyptians, Babylonians, Hebrews, and Mesopotamians used a unit of length called the *cubit*, which represents the distance between a human elbow and fingertips. For instance, in the book of Genesis in the Bible, Noah's ark is described as having a length of "three hundred cubits, its width fifty cubits, and its height thirty cubits." Other commonly used measures found in documents of these ancient cultures include the digit, hand, span, and foot. These measures are shown at the right.

Several thousand years later, the English system of measurement also consisted of a mixed bag of body- and nature-related units of measure. A rod was the combined length of the left feet of 16 men. An inch was the distance spanned by three grains of barley. A foot was the length of the foot of the king currently in power. Around 1100 A.D. King Henry I of England decreed that a yard was the distance between the king's nose and the thumb of his outstretched arm.

Although the English system became somewhat standardized by the 13th century, the problem with it and the ancient systems of measurement was that the relationship between the various units was not necessarily easy to remember. This made it difficult to convert from one type of unit to another.

The metric system grew out of the French Revolution during the 1790s. As a reaction against the lack of consistency and utility of existing systems of measurement, the French Academy of Sciences set out to develop an easy-to-use, internationally standardized system of measurements. Originally, the basic unit of the metric system, the meter, was to be one ten-millionth of the distance between the North Pole and the Equator on a line through Paris. However, it was soon discovered that this distance was nearly impossible to measure accurately. The meter has been redefined several times since the end of the French Revolution. Most recently, the meter was redefined in 1983 as the distance that light travels in a vacuum during $\frac{1}{299,792,458}$ of a second.

Although the definition of the meter has evolved over time, the original idea of developing an easy-to-use system of measurements has endured. The metric system uses a standard set of prefixes for all basic unit types (length, mass, capacity, etc.) and all conversions within a unit type are based on powers of 10.

CRITICAL THINKING

Develop your own unit of measure for (a) length and (b) area. Explain how you chose your units and for what types of relative sizes of measurements they would be most useful. Discuss the advantages and disadvantages of your units of measurement. Then use your units to measure the width of your classroom desk and the area of the front cover of this textbook.

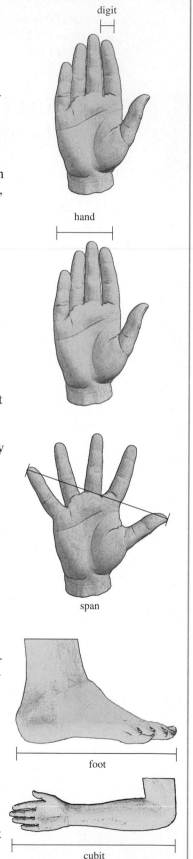

digit

hand

span

foot

cubit

Name _____ **Section** _____ **Date** _____

MENTAL MATH

Convert.

1. 16 ounces to pounds
2. 32 ounces to pounds
3. 1 ton to pounds
4. 3 tons to pounds
5. 1 pound to ounces
6. 3 pounds to ounces
7. 2000 pounds to tons
8. 4000 pounds to tons

Determine whether the measurement in each statement is reasonable.

9. The doctor prescribed a pill containing 2 kg of medication.
10. A full-grown cat weighs approximately 15 g.
11. A bag of flour weighs 4.5 kg.
12. A staple weighs 15 mg.
13. A professor weighs less than 150 g.
14. A car weighs 2000 mg.

EXERCISE SET 7.2

A *Convert as indicated. See Examples 1 and 2.*

1. 2 pounds to ounces
2. 3 pounds to ounces

3. 5 tons to pounds
4. 3 tons to pounds

5. 12,000 pounds to tons
6. 32,000 pounds to tons

7. 60 ounces to pounds
8. 90 ounces to pounds

9. 3500 pounds to tons
10. 9000 pounds to tons

11. 16.25 pounds to ounces
12. 14.5 pounds to ounces

13. 4.9 tons to pounds
14. 8.3 tons to pounds

15. $4\frac{3}{4}$ pounds to ounces
16. $9\frac{1}{8}$ pounds to ounces

MENTAL MATH ANSWERS

1. 1 lb
2. 2 lb
3. 2000 lb
4. 6000 lb
5. 16 oz
6. 48 oz
7. 1 ton
8. 2 tons
9. no
10. no
11. yes
12. yes
13. no
14. no

ANSWERS

1. 32 oz
2. 48 oz
3. 10,000 lb
4. 6000 lb
5. 6 tons
6. 16 tons
7. $3\frac{3}{4}$ lb
8. $5\frac{5}{8}$ lb
9. $1\frac{3}{4}$ tons
10. $4\frac{1}{2}$ tons
11. 260 oz
12. 232 oz
13. 9800 lb
14. 16,600 lb
15. 76 oz
16. 146 oz

17. 1.5 tons

18. 3.2 lb

19. 53 lb 10 oz

20. 17 lb 2 oz

21. 9 tons 390 lb

22. 4 tons 55 lb

23. 3 tons 175 lb

24. 3 tons 590 lb

25. 8 lb 11 oz

26. 18 lb 12 oz

27. 31 lb 2 oz

28. 11 lb 9 oz

29. 1 ton 700 lb

30. 1 ton 600 lb

31. 5 lb 8 oz

32. 6 tons 700 lb

Name _____

17. 2950 pounds to the nearest tenth of a ton

18. 51 ounces to the nearest tenth of a pound

B *Perform each indicated operation. See Examples 3 through 5.*

19. 34 lb 12 oz + 18 lb 14 oz

20. 6 lb 10 oz + 10 lb 8 oz

21. 6 tons 1540 lb + 2 tons 850 lb

22. 2 tons 1575 lb + 1 ton 480 lb

23. 5 tons 1050 lb − 2 tons 875 lb

24. 4 tons 850 lb − 1 ton 260 lb

25. 12 lb 4 oz − 3 lb 9 oz

26. 45 lb 6 oz − 26 lb 10 oz

27. 5 lb 3 oz × 6

28. 2 lb 5 oz × 5

29. 6 tons 1500 lb ÷ 5

30. 5 tons 400 lb ÷ 4

Solve. Remember to insert units when writing your answers. See Example 6.

31. Doris Johnson has two open containers of Uncle Ben's rice. If she combines 1 lb 10 oz from one container with 3 lb 14 oz from the other container, how much total rice does she have?

32. Dru Mizel maintains the records of the amount of coal delivered to his department in the steel mill. In January, 3 tons 1500 lb were delivered. In February, 2 tons 1200 lb were delivered. Find the total amount delivered in these two months.

33. Carla Hamtini was amazed when she grew a 28 lb 10 oz zucchini in her garden, but later she learned that the heaviest zucchini ever grown weighed 64 lb 8 oz. It was grown in Llanharry, Wales by B. Lavery in 1990. How far below the record weight was Carla's zucchini? (*Source: The Guinness Book of Records*, 1996)

34. The heaviest baby born in good health weighed an incredible 22 lb 8 oz. He was born in Italy in September, 1955. How much heavier is this than a 7 lb 12 oz baby? (*Source: The Guinness Book of Records*, 1996)

35. The Shop 'n Bag supermarket chain ships hamburger meat by placing 10 packages of hamburger in a box, with each package weighing 3 lb 4 oz. How much will 4 boxes of hamburger weigh?

36. The Quaker Company ships its 1-lb 2-oz boxes of oatmeal in cartons containing 12 boxes of oatmeal. How much will 3 such cartons weigh?

37. A carton of Del Monte Pineapple weighs 55 lb 4 oz, but 2 lb 8 oz of this weight is due to packaging. Subtract the weight of the packaging to find the actual weight of the pineapple in 4 cartons.

38. The Hormel Corporation ships cartons of canned ham weighing 43 lb 2 oz each. Of this weight, 3 lb 4 oz is due to packaging. Find the actual weight of the ham found in 3 cartons.

39. One bag of Pepperidge Farm Bordeaux cookies weighs $6\frac{3}{4}$ ounces. How many pounds will a dozen bags weigh?

40. One can of Payless Red Beets weighs $8\frac{1}{2}$ ounces. How much will eight cans weigh?

C *Convert as indicated. See Examples 7 and 8.*

41. 500 g to kilograms   **42.** 650 g to kilograms   **43.** 4 g to milligrams

44. 9 g to milligrams   **45.** 25 kg to grams   **46.** 18 kg to grams

---

33. 35 lb 14 oz

34. 14 lb 12 oz

35. 130 lb

36. 40 lb 8 oz

37. 211 lb

38. 119 lb 10 oz

39. 5 lb 1 oz

40. 4 lb 4 oz

41. 0.5 kg

42. 0.65 kg

43. 4000 mg

44. 9000 mg

45. 25,000 g

46. 18,000 g

47. 48 mg to grams   **48.** 112 mg to grams   **49.** 6.3 g to kilograms

50. 4.9 g to kilograms   🔲 **51.** 15.14 g to milligrams   **52.** 16.23 g to milligrams

53. 4.01 kg to grams   **54.** 3.16 kg to grams

D *Perform each indicated operation. See Examples 9 and 10.*

55. 3.8 mg + 9.7 mg   **56.** 41.6 g + 9.8 g   **57.** 205 mg + 5.61 g

58. 2.1 g + 153 mg   **59.** 9 g − 7150 mg   **60.** 4 kg − 2410 g

61. 1.61 kg − 250 g   **62.** 6.13 g − 418 mg   **63.** 5.2 kg × 2.6

64. 4.8 kg × 9.3   **65.** 17 kg ÷ 8   **66.** 8.25 g ÷ 6

Solve. Remember to insert units when writing your answers. See Example 11.

67. A can of 7-Up weighs 336 grams. Find the weight in kilograms of 24 cans.

68. Guy Green normally weighs 73 kg, but he lost 2800 grams after being sick with the flu. Find Guy's new weight.

🔲 **69.** Sudafed is a decongestant that comes in two strengths. The regular strength contains 60 mg of medication. The extra strength contains 0.09 g of medication. How much extra medication is in the extra-strength tablet?

70. A small can of Planters sunflower nuts weighs 177 grams. If each can contains 6 servings, find the weight of one serving.

47. 0.048 g

48. 0.112 g

49. 0.0063 kg

50. 0.0049 kg

51. 15,140 mg

52. 16,230 mg

53. 4010 g

54. 3160 g

55. 13.5 mg

56. 51.4 g

57. 5.815 g or 5815 mg

58. 2.253 g or 2253 mg

59. 1850 mg or 1.850 g

60. 1.59 kg or 1590 g

61. 1360 g or 1.360 kg

62. 5712 mg or 5.712 g

63. 13.52 kg

64. 44.64 kg

65. 2.125 kg

66. 1.375 g

67. 8.064 kg

68. 70.2 kg

69. 30 mg

70. 29.5 g

Name _____

71. Tim Caucutt's doctor recommends that Tim limit his daily intake of sodium to 0.6 gram. A one-ounce serving of Cheerios with $\frac{1}{2}$ cup of fortified skim milk contains 350 mg of sodium. How much more sodium can Tim have after he eats a bowl of Cheerios for breakfast, assuming he intends to follow the doctor's orders?

72. A large bottle of Hire's Root Beer weighs 1900 g. If a carton contains 6 large bottles of root beer, find the weight in kilograms of 5 cartons.

73. Three milligrams of preservatives are added to a 0.5-kg box of dried fruit. How many milligrams of preservatives are in 3 cartons of dried fruit if each carton contains 16 boxes?

74. One box of Swiss Miss Cocoa Mix weighs 0.385 kg, but 39 grams of this weight is the packaging. Find the actual weight of the cocoa in 8 boxes.

75. A carton of 12 boxes of Quaker Oats Oatmeal weighs 6.432 kg. Each box includes 26 grams of packaging material. What is the actual weight of the oatmeal in the carton?

76. The supermarket prepares hamburger in 85-gram market packages. When Leo Gonzalas gets home, he divides the package in half before refrigerating the meat. How much will each package weigh?

77. A package of Trailway's Gorp, a high-energy hiking trail mix, contains 0.3 kg of nuts, 0.15 kg of chocolate bits, and 400 g of raisins. Find the total weight of the package.

78. The manufacturer of Anacin wants to reduce the caffeine content of its aspirin by $\frac{1}{4}$. Currently, each regular tablet contains 32 mg of caffeine. How much caffeine should be removed from each tablet?

79. A regular-size bag of Lay's potato chips weighs 198 grams. Find the weight of a dozen bags, rounded to the nearest hundredth of a kilogram.

80. Clarence Patterson's cat weighs a hefty 9 kg. The vet has recommended that the cat lose 1500 grams. How much should the cat weigh?

71. 250 mg

72. 57 kg

73. 144 mg

74. 2.768 kg

75. 6.12 kg

76. 42.5 g

77. 850 g or 0.85 kg

78. 8 mg

79. 2.38 kg

80. 7.5 kg

425

Name _____

REVIEW AND PREVIEW

Write each fraction as a decimal. See Section 4.7.

81. $\frac{1}{4}$ **82.** $\frac{1}{20}$ **83.** $\frac{4}{25}$

84. $\frac{3}{5}$ **85.** $\frac{7}{8}$ **86.** $\frac{3}{16}$

◆ COMBINING CONCEPTS

87. Why is the decimal point moved to the right when grams are converted to milligrams?

88. To change 8 pounds to ounces, multiply by 16. Why is this the correct procedure?

7.3 CAPACITY: U.S. AND METRIC SYSTEMS OF MEASUREMENT

A CONVERTING U.S. SYSTEM UNITS OF CAPACITY

Units of **capacity** are generally used to measure liquids. The number of gallons of gasoline needed to fill a gas tank in a car, the number of cups of water needed in a bread recipe, and the number of quarts of milk sold each day at a supermarket are all examples of using units of capacity. The following summary shows equivalencies between units of capacity.

> ### U.S. UNITS OF CAPACITY
>
> $$8 \text{ fluid ounces (fl oz)} = 1 \text{ cup (c)}$$
> $$2 \text{ cups} = 1 \text{ pint (pt)}$$
> $$2 \text{ pints} = 1 \text{ quart (qt)}$$
> $$4 \text{ quarts} = 1 \text{ gallon (gal)}$$

Just as with units of length and weight, we can form unit fractions to convert between different units of capacity. For instance,

$$\frac{2 \text{ c}}{1 \text{ pt}} = \frac{1 \text{ pt}}{2 \text{ c}} = 1 \quad \text{and} \quad \frac{2 \text{ pt}}{1 \text{ qt}} = \frac{1 \text{ qt}}{2 \text{ pt}} = 1$$

Example 1   Convert 9 quarts to gallons.

Solution: We multiply by the unit fraction $\frac{1 \text{ gal}}{4 \text{ qt}}$.

$$9 \text{ qt} = \frac{9 \text{ qt}}{1} \cdot \frac{1 \text{ gal}}{4 \text{ qt}}$$

$$= \frac{9 \text{ gal}}{4}$$

$$= 2\frac{1}{4} \text{ gal}$$

Thus, 9 quarts is the same as $2\frac{1}{4}$ gallons, as shown in the diagram.

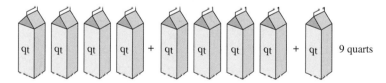

9 quarts

Example 2   Convert 14 cups to quarts.

Solution: Our equivalency table contains no direct conversion from cups to quarts. However, from this table we know that

$$1 \text{ qt} = 2 \text{ pt} = 4 \text{ c}$$

so 1 qt = 4c. Now we have the unit fraction $\frac{1 \text{ qt}}{4 \text{ c}}$. Thus,

$$14 \text{ c} = \frac{14 \text{ c}}{1} \cdot \frac{1 \text{ qt}}{4 \text{ c}} = \frac{7}{2} \text{ qt} \quad \text{or} \quad 3\frac{1}{2} \text{ qt}$$

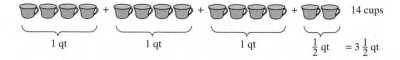

14 cups

$$\underbrace{\qquad}_{1 \text{ qt}} + \underbrace{\qquad}_{1 \text{ qt}} + \underbrace{\qquad}_{1 \text{ qt}} + \underbrace{\qquad}_{\frac{1}{2} \text{ qt}} = 3\frac{1}{2} \text{ qt}$$

TEACHING TIP

Point out to students that ounces which are used to measure weight are not the same as fluid ounces which are used to measure capacity.

Practice Problem 1

Convert 43 pints to quarts.

TEACHING TIP

Ask your students to suggest a method for remembering that there are
8 fluid ounces in 1 cup
2 cups in a pint
2 pints in a quart
4 quarts in a gallon.

Practice Problem 2

Convert 26 quarts to cups.

Answers

1. $21\frac{1}{2}$ quarts,   **2.** 104 cups

✓ **CONCEPT CHECK**

If 50 cups are converted to quarts, will the equivalent number of quarts be less than or greater than 50? Explain.

Practice Problem 3

Subtract 2 qt from 1 gal 1 qt.

TEACHING TIP

After going over Example 3, have students repeat the problem after converting the units to gallons or quarts. Then have them compare their answer to make sure they are equivalent.

Practice Problem 4

Multiply 2 gal 3 qt by 2.

Practice Problem 5

Divide 6 gal 1 qt by 2.

Answers

3. 3 qt, **4.** 5 gal 2 qt, **5.** 3 gal 1 pt
✓ **Concept Check:**   less than 50

TRY THE CONCEPT CHECK IN THE MARGIN.

B **OPERATIONS ON U.S. SYSTEM UNITS OF CAPACITY**

As is true of units of length and weight, units of capacity can be added, subtracted, multiplied, and divided.

Example 3   Subtract 3 qt from 4 gal 2 qt.

Solution: To subtract, we line up similar units.

$$
\begin{array}{r}
4 \text{ gal } 2 \text{ qt} \\
- \quad\quad 3 \text{ qt} \\
\end{array}
$$

We cannot subtract 3 qt from 2 qt. We need to borrow 1 gallon from the 4 gallons, convert it to 4 quarts, and then combine it with the 2 quarts.

Borrow 1 gal = 4 qt

3 gal + (1 gal) 4 qt

$$
\begin{array}{r}
4 \text{ gal } 2 \text{ qt} \\
- \quad\quad 3 \text{ qt} \\
\end{array}
\quad \text{or} \quad
\begin{array}{r}
3 \text{ gal } 6 \text{ qt} \\
- \quad\quad 3 \text{ qt} \\
\hline
3 \text{ gal } 3 \text{ qt}
\end{array}
$$

To check, see that the sum of 3 gal 3 qt and 3 qt is 4 gal 2 qt.

Example 4   Multiply 3 qt 1 pt by 3.

Solution: We multiply each of the units of capacity by 3.

$$
\begin{array}{r}
3 \text{ qt } 1 \text{ pt} \\
\times \quad\quad 3 \\
\hline
9 \text{ qt } 3 \text{ pt}
\end{array}
$$

Since 3 pints is the same as 1 quart and 1 pint, we have

$$9 \text{ qt } 3 \text{ pt} = 9 \text{ qt} + 1 \text{ qt } 1 \text{ pt} = 10 \text{ qt } 1 \text{ pt}$$

The 10 quarts can be changed to gallons by dividing by 4, since there are 4 quarts in a gallon. To see why we divide, notice that

$$10 \text{ qt} = \frac{10 \text{ qt}}{1} \cdot \frac{1 \text{ gal}}{4 \text{ qt}} = \frac{10}{4} \text{ gal}$$

$$
\begin{array}{r}
2 \text{ gal } 2 \text{ qt} \\
4\overline{)10 \text{ qts}} \\
-8 \\
\hline
2
\end{array}
$$

Then the product is 10 qt 1 pt or 2 gal 2 qt 1 pt.

Example 5   Divide 3 gal 2 qt by 2.

Solution: We divide each unit of capacity by 2.

$$
\begin{array}{r}
1 \text{ gal} \quad\quad 3 \text{ qt} \\
2\overline{)3 \text{ gal}} \quad 2 \text{ qt} \\
-2 \\
\hline
1 \text{ gal} = 4 \text{ qt} \\
6 \text{ qt}
\end{array}
$$

6 qt ÷ 2 = 3 qt

Example 6   Finding the Amount of Water in an Aquarium

An aquarium contains 6 gal 3 qt of water. If 2 gal 2 qt of water is added, what is the total amount of water in the aquarium?

Solution:

$$
\begin{array}{rl}
\text{beginning water} \rightarrow & 6 \text{ gal } 3 \text{ qt} \\
+ \quad \text{water added} \rightarrow & +2 \text{ gal } 2 \text{ qt} \\
\hline
\text{total water} \rightarrow & 8 \text{ gal } 5 \text{ qt}
\end{array}
$$

Since 5 qt = 1 gal 1 qt, we have

$$
\overbrace{8 \text{ gal}}^{8 \text{ gal}} \quad \overbrace{5 \text{ qt}}^{5 \text{ qt}}
$$

$$= 8 \text{ gal} + 1 \text{ gal } 1 \text{ qt}$$
$$= 9 \text{ gal } 1 \text{ qt}$$

The total amount of water is 9 gallons 1 quart. ▬▬▬

C   CONVERTING METRIC SYSTEM UNITS OF CAPACITY

Thus far, we know that the basic unit of length in the metric system is the meter and the basic unit of mass in the metric system is the gram. What is the basic unit of capacity? The **liter** is the basic unit of capacity in the metric system. By definition, a **liter** is the capacity or volume of a cube measuring 10 centimeters on each side.

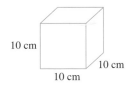

10 cm   10 cm   10 cm

The following examples may help you get a feeling for metric capacities.

One liter of liquid is slightly more than one quart.

1 quart   1 liter

Many soft drinks are packaged in 2-liter bottles.

COLA

The metric system was designed to be a consistent system. Once again, the prefixes for metric units of capacity are the same as for metric units of length and mass, as summarized in the following table.

A large oil drum contains 15 gal 3 qt of oil. How much will be in the drum if an additional 4 gal 3 qt of oil is poured into it?

Answer

6. 20 gal 2 qt

Prefix	Meaning	Metric Unit of Capacity
kilo	1000	1 kiloliter (kl) = 1000 liters (L)
hecto	100	1 hectoliter (hl) = 100 L
deka	10	1 dekaliter (dal) = 10 L
		1 liter (L) = 1 L
deci	1/10	1 deciliter (dl) = 1/10 L or 0.1 L
centi	1/100	1 centiliter (cl) = 1/100 L or 0.01 L
milli	1/1000	1 milliliter (ml) = 1/1000 L or 0.001 L

The **milliliter** and the **liter** are the two most commonly used metric units of capacity.

Converting from one unit of capacity to another involves multiplying by powers of 10 or moving the decimal point to the left or to the right. Listing units of capacity in order from largest to smallest helps to keep track of how many places to move the decimal point when converting.

Let's convert 2.6 liters to milliliters. To convert from liters to milliliters, we move along the table 3 units to the right.

kl   hl   dal   L   dl   cl   ml

3 units to the right

This means that we move the decimal point 3 places to the right to convert from liters to milliliters.

$$2.600 \text{ L} = 2600. \text{ ml}$$

This same conversion can be done with unit fractions.

$$2.6 \text{ L} = \frac{2.6 \text{ L}}{1} \cdot \frac{1000 \text{ ml}}{1 \text{ L}}$$

$$= 2.6 \cdot 1000 \text{ ml}$$

$$= 2600 \text{ ml}$$

To multiply by 1000, move the decimal point 3 places to the right.

To visualize the result, study the diagram below.

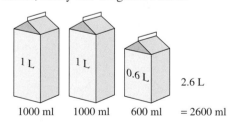

1000 ml   1000 ml   600 ml   = 2600 ml

Thus 2.6 L = 2600 ml.

Practice Problem 7

Convert 2100 ml to liters.

Example 7   Convert 3210 ml to liters.

Solution: Let's use the unit fraction method first.

Unit fraction

$$3210 \text{ ml} = 3210 \text{ ml} \cdot \frac{1 \text{ L}}{1000 \text{ ml}} = 3.21 \text{ L}$$

Now let's list the unit measures in a chart and move from milliliters to liters.

Answer

7. 2.1 L

kl   hl   dal   **L**   dl   cl   ml

3 units to the left

3210 ml = 3.210 L

3 places to the left

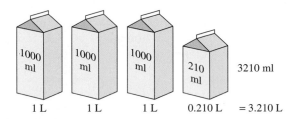

1000 ml   1000 ml   1000 ml   210 ml   3210 ml

1 L   1 L   1 L   0.210 L   = 3.210 L

Example 8

Convert 0.185 dl to milliliters.

Solution: We list the unit measures in a chart and move from deciliters to milliliters.

kl   hl   dal   L   dl   cl   ml

2 units to the right

0.185 dl = 18.5 ml

2 places to the right

Practice Problem 8

Convert 2.13 dal to liters.

D OPERATIONS ON METRIC SYSTEM UNITS OF CAPACITY

As was true for length and weight, arithmetic operations involving metric units of capacity can also be performed. Make sure the metric units of capacity are the same before adding, subtracting, multiplying, or dividing.

Example 9

Add 2400 ml to 8.9 L.

Solution: We must convert both to liters or both to milliliters before adding the capacities together.

2400 ml = 2.4 L   or   8.9 L = 8900 ml

$$\begin{array}{r} 2.4\text{ L} \\ \underline{|\ 8.9\text{ L}} \\ 11.3\text{ L} \end{array}$$

$$\begin{array}{r} 2400\text{ ml} \\ +\ 8900\text{ ml} \\ \hline 11{,}300\text{ ml} \end{array}$$

The total is 11.3 L or 11,300 ml. They both represent the same capacity.

TRY THE CONCEPT CHECK IN THE MARGIN.

Example 10

Divide 18.08 ml by 16.

Solution:

$$\begin{array}{r} 1.13\text{ ml} \\ 16\overline{)18.08\text{ ml}} \\ \underline{-16} \\ 2\ 0 \\ \underline{-1\ 6} \\ 48 \\ \underline{-48} \\ 0 \end{array}$$

The solution is 1.13 ml.

Practice Problem 9

Add 1250 ml to 2.9 L.

✓ CONCEPT CHECK

How could you estimate the following operation? Subtract 950 ml from 7.5 L.

Practice Problem 10

Divide 146.9 L by 13.

Answers

8. 21.3 L,   **9.** 4150 ml or 4.15 L,   **10.** 11.3 L

✓ Concept Check: 950 ml = 0.95 L; round 0.95 to 1; 7.5 − 1 = 6.5 L

Practice Problem 11

If 28.6 L of water can be pumped every minute, how much water can be pumped in 85 minutes?

Example 11   Finding the Amount of Medication a Person Has Received

A patient hooked up to an IV unit in the hospital is to receive 12.5 ml of medication every hour. How much medication does the patient receive in 3.5 hours?

Solution: We multiply 12.5 ml by 3.5.

$$
\begin{array}{r}
\text{medication per hour} \rightarrow \quad 12.5\text{ml} \\
\times \qquad\qquad \text{hours} \rightarrow \quad \times\ 3.5 \\
\hline
\text{total medication} \qquad 625 \\
375\ \ \\
\hline
43.75\text{ml}
\end{array}
$$

The patient receives 43.75 ml of medication.

CALCULATOR EXPLORATIONS
METRIC TO U.S. SYSTEM CONVERSIONS IN CAPACITY

To convert between the two systems of measurement in capacity, the following *approximations* can be used.

liters $\times$ 1.06 $\approx$ quarts   quarts $\times$ 0.946 $\approx$ liters
liters $\times$ 0.264 $\approx$ gallons   gallons $\times$ 3.785 $\approx$ liters

Example   How many quarts are there in a 2-liter bottle of cola?

Solution: From the above approximations,

liters $\times$ 1.06 $\approx$ quarts
$\downarrow$
2 $\times$ 1.06 $\approx$ quarts

To multiply on your calculator, press the keys

$$\boxed{2}\ \boxed{\times}\ \boxed{1.06}\ \boxed{=}$$

or

$$\boxed{\text{ENTER}}$$

The display should read $\boxed{2.12}$.
2 liters $\approx$ 2.12 quarts

Convert as indicated.

1.   5 quarts to liters $\approx$4.73 L

2.   26 gallons to liters $\approx$98.41 L

3.   17.5 liters to gallons $\approx$4.62 gal

4.   7.8 liters to quarts $\approx$8.268 qt

5.   A 1-gallon container holds how many liters? $\approx$3.785 L

6.   How many quarts are contained in a 3-liter bottle of cola?
      $\approx$3.18 qt

Answer
11. 243.1 L

Name _____ **Section** _____ **Date** _____

MENTAL MATH

Convert as indicated.

1. 2 c to pints   **2.** 4 c to pints   **3.** 4 qt to gallons   **4.** 8 qt to gallons

5. 2 pt to quarts   **6.** 6 pt to quarts   **7.** 8 fl oz to cups   **8.** 24 fl oz to cups

9. 1 pt to cups   **10.** 3 pt to cups   **11.** 1 gal to quarts   **12.** 2 gal to quarts

Determine whether the measurement in each statement is reasonable.

13. Clair took a dose of 2 L of cough medicine to cure her cough.

14. John drank 250 ml of milk for lunch.

15. Jeannie likes to relax in a tub filled with 3000 ml of hot water.

16. Sarah pumped 20 L of gasoline into her car yesterday.

EXERCISE SET 7.3

A _Convert each measurement as indicated. See Examples 1 and 2._

1. 32 fluid ounces to cups   **2.** 16 quarts to gallons   **3.** 8 quarts to pints

4. 9 pints to quarts   **5.** 10 quarts to gallons   **6.** 15 cups to pints

7. 80 fluid ounces to pints   **8.** 18 pints to gallons   **9.** 2 quarts to cups

10. 3 pints to fluid ounces   **11.** 120 fluid ounces to quarts   **12.** 20 cups to gallons

13. 6 gallons to fluid ounces   **14.** 5 quarts to cups   **15.** $4\frac{1}{2}$ pints to cups

16. $6\frac{1}{2}$ gallons to quarts   **17.** $2\frac{3}{4}$ gallons to pints   **18.** $3\frac{1}{4}$ quarts to cups

B _Perform each indicated operation. See Examples 3 through 5._

19. 4 gal 3 qt + 5 gal 2 qt   **20.** 2 gal 3 qt + 8 gal 3 qt   **21.** 1 c 5 fl oz + 2 c 7 fl oz

22. 2 c 3 fl oz + 2 c 6 fl oz   **23.** 3 gal − 1 gal 3 qt   **24.** 2 pt − 1 pt 1 c

Name _____

25. 3 gal 1 qt − 1 qt 1 pt   **26.** 3 qt 1 c − 1 c 4 fl oz   **27.** 1 pt 1 c × 3

28. 1 qt 1 pt × 2   **29.** 8 gal 2 qt × 2   **30.** 6 gal 1 pt × 2

31. 9 gal 2 qt ÷ 2   **32.** 5 gal 6 fl oz ÷ 2

Solve. Remember to insert units when writing your answers. See Example 6.

33. A can of Hawaiian Punch holds $1\frac{1}{2}$ quarts of liquid. How many fluid ounces is this?

34. Weight Watchers Double Fudge bars contain 21 fluid ounces of ice cream. How many cups of ice cream is this?

35. Many diet experts advise individuals to drink 64 ounces of water each day. How many quarts of water is this?

36. A recipe for walnut fudge cake calls for $1\frac{1}{4}$ cups of water. How many fluid ounces is this?

37. Can 5 pt 1 c of fruit punch and 2 pt 1 c of ginger ale be poured into a 1-gal container without it overflowing?

38. Three cups of prepared jello are poured into 6 dessert dishes. How many fluid ounces of jello are in each dish?

39. How much punch has been prepared if 1 qt 1 pt of Ocean Spray Cranapple drink is mixed with 1 pt 1 c of ginger ale?

40. Henning's Supermarket sells home-made soup in 1-qt-1-pt containers. How much soup is contained in three such containers?

41. A case of Pepsi Cola holds 24 cans, each of which contains 12 ounces of Pepsi Cola. How many *quarts* are there in a case of Pepsi?

42. Manuela's Service Station has a drum that holds 40 gallons of oil. If 6 gallons and 3 quarts have been used, how much oil remains?

C   *Convert as indicated. See Examples 7 and 8.*

43. 5 L to milliliters   **44.** 8 L to milliliters   **45.** 4500 ml to liters

46. 3100 ml to liters

47. 410 L to kiloliters

48. 250 L to kiloliters

49. 64 ml to liters

50. 39 ml to liters

🖾 51. 0.16 kl to liters

52. 0.48 kl to liters

53. 3.6 L to milliliters

54. 1.9 L to milliliters

55. 0.16 L to kiloliters

56. 0.127 L to kiloliters

D *Perform each indicated operation. See Examples 9 and 10.*

57. 2.9 L + 19.6 L

58. 18.5 L + 4.6 L

59. 2700 ml + 1.8 L

60. 4.6 L + 1600 ml

61. 8.6 L − 190 ml

62. 4.8 L − 283 ml

63. 11,400 ml − 0.8 L

64. 6850 ml − 0.3 L

65. 480 ml × 8

66. 290 ml × 6

67. 81.2 L ÷ 0.5

68. 5.4 L ÷ 3.6

Solve. Remember to insert units when writing your answers. See Example 11.

69. Mike Schaferkotter drank 410 ml of Mountain Dew from a 2-liter bottle. How much Mountain Dew remains in the bottle?

70. The Werners' Volvo has a 54.5-L gas tank. Only 3.8 liters of gasoline still remain in the tank. How much is needed to fill it?

71. Margie Phitts added 354 ml of Prestone dry gas to the 18.6 L of gasoline in her car's tank. Find the total amount of gasoline in the tank.

72. Chris Peckaitis wishes to share a 2-L bottle of Coca Cola equally with 7 of his friends. How much will each person get?

46.	3.1 L
47.	0.41 kl
48.	0.25 kl
49.	0.064 L
50.	0.039 L
51.	160 L
52.	480 L
53.	3600 ml
54.	1900 ml
55.	0.00016 kl
56.	0.000127 kl
57.	22.5 L
58.	23.1 L
59.	4.5 L or 4500 ml
60.	6.2 L or 6200 ml
61.	8410 ml or 8.41 L
62.	4.517 L or 4517 ml
63.	10,600 ml or 10.6 L
64.	6.55 L or 6550 ml
65.	3840 ml
66.	1740 ml
67.	162.4 L
68.	1.5 L
69.	1.59 L
70.	50.7 L
71.	18.954 L
72.	250 ml

73. $0.316

74. 360 ml

75. 474 ml

76. 1730 ml

77. $\dfrac{7}{10}$

78. $\dfrac{9}{10}$

79. $\dfrac{3}{100}$

80. $\dfrac{7}{1000}$

81. $\dfrac{3}{500}$

82. $\dfrac{2}{25}$

83. answers may vary

84. 9.5 bushels

85. 256 fl drams

Name _____

73. Stanley Fisher paid $14.00 to fill his car with 44.3 liters of gasoline. Find the price per liter of gasoline to the nearest tenth of a cent.

74. A student carelessly misread the scale on a cylinder in the chemistry lab and added 40 cl of water to a mixture instead of 40 ml. Find the excess amount of water.

75. A large bottle of Ocean Spray Cranicot drink contains 1.42 L of beverage. The smaller bottle contains only 946 ml. How much more is in the larger bottle?

76. In a lab experiment, Melissa Martin added 400 ml of salt water to 1.65 L of water. Later 320 ml of the solution was drained off. How much of the solution still remains?

REVIEW AND PREVIEW

Write each decimal as a fraction. See Section 4.7.

77. 0.7

78. 0.9

79. 0.03

80. 0.007

81. 0.006

82. 0.08

COMBINING CONCEPTS

83. Explain how to borrow to subtract 1 gal 2 qt from 3 gal 1 qt.

Internet Excursions

Go to http://www.prenhall.com/martin-gay
By going to the World Wide Web address listed above, you will be directed to a site called A Dictionary of Units, or a related site that will help you answer the questions below.

84. There are more units of capacity in use than the ones mentioned in this section. For instance, there are special measures for the capacity of dry items such as grains or fruits. Visit this web site and locate the U.S. System of Measurements area. List the measure equivalencies for dry capacity. Then convert 38 pecks of apples to bushels.

85. There are also special apothecaries' measures for capacity of liquid drugs and medicines. Although metric measures are commonly used in the pharmacy industry today, awareness of these measures is still useful. Within the U.S. System of Measurements area of this web site, find and list the apothecaries' measures equivalencies. Then convert 2 pints to fl drams.

Name _____ **Section** _____ **Date** _____

INTEGRATED REVIEW — LENGTH, WEIGHT, AND CAPACITY

Convert each measurement as indicated.

Length

1. 36 in. = _____ ft

2. 10,560 ft = _____ mi

3. 20 ft = _____ yd

4. 6 yd = _____ ft

5. 2.1 mi = _____ ft

6. 3.2 ft = _____ in.

7. 30 m = _____ cm

8. 24 mm = _____ cm

9. 2000 mm = _____ m

10. 1800 cm = _____ m

11. 7.2 cm = _____ mm

12. 600 m = _____ km

Weight or Mass

13. 5 tons − _____ lb

14. 11,000 lb = _____ tons

15. 8.5 lb = _____ oz

16. 40 oz = _____ lb

17. 56 oz = _____ lb

18. 5 lb = _____ oz

19. 28 kg = _____ g

20. 1400 mg = _____ g

438

Name _____

21. 5.6 g = _____ kg

22. 6 kg = _____ g

23. 670 mg = _____ g

24. 3.6 g = _____ kg

Capacity

25. 6 qt = _____ pt

26. 5 pt = _____ qt

27. 14 qt = _____ gal

28. 17 c = _____ pt

29. $3\frac{1}{2}$ pt = _____ c

30. 26 qt = _____ gal

31. 7 L = _____ ml

32. 350 L = _____ kl

33. 47 ml = _____ L

34. 0.97 kl = _____ L

35. 0.126 kl = _____ L

36. 75 ml = _____ L

7.4 TEMPERATURE: U.S. AND METRIC SYSTEMS OF MEASUREMENT

When Gabriel Fahrenheit and Anders Celsius independently established units for temperature scales, each based his unit on the heat of water the moment it boils compared to the moment it freezes. One degree Celsius is 1/100 of the difference in heat. One degree Fahrenheit is 1/180 of the difference in heat. Celsius arbitrarily labeled the temperature at the freezing point at 0°C, making the boiling point 100°C; Fahrenheit labeled the freezing point 32°F, making the boiling point 212°F. Water boils at 212°F or 100°C.

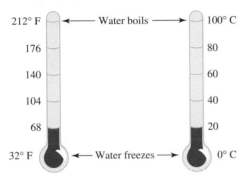

By comparing the two scales in the figure, we see that a 20°C day is as warm as a 68°F day. Similarly, a sweltering 104°F day in the Mojave desert corresponds to a 40°C day.

TRY THE CONCEPT CHECK IN THE MARGIN.

A CONVERTING DEGREES CELSIUS TO DEGREES FAHRENHEIT

To convert from Celsius temperatures to Fahrenheit temperatures, see the box below. In this box, we use the symbol F to represent degrees Fahrenheit and the symbol C to represent degrees Celsius.

CONVERTING CELSIUS TO FAHRENHEIT

$$F = \frac{9}{5} \cdot C + 32 \quad \text{or} \quad F = 1.8 \cdot C + 32$$

(To convert to Fahrenheit temperature, multiply the Celsius temperature by $\frac{9}{5}$ or 1.8, and then add 32.)

Example 1

Convert 15°C to degrees Fahrenheit.

Solution:

$$F = \frac{9}{5} \cdot C + 32$$

$$= \frac{9}{5} \cdot 15 + 32 \quad \text{Replace C with 15.}$$

$$= 27 + 32 \quad \text{Simplify.}$$

$$= 59 \quad \text{Add.}$$

Thus 15°C is equivalent to 59°F.

Objectives

A Convert temperatures from degrees Celsius to degrees Fahrenheit.

B Convert temperatures from degrees Fahrenheit to degrees Celsius.

SSM   CD-ROM   Video 7.4

✓ CONCEPT CHECK

Which of the following statements is correct? Explain.

a. 6°C is below the freezing point of water.

b. 6°F is below the freezing point of water.

Practice Problem 1

Convert 50°C to degrees Fahrenheit.

Answers

1. 122°F

✓ **Concept Check:** **a.** False, **b.** True

Practice Problem 2

Convert 18°C to degrees Fahrenheit.

TEACHING TIP

Ask students to suggest a way to verify the Fahrenheit to Celsius formula.

Practice Problem 3

Convert 86°F to degrees Celsius.

Practice Problem 4

Convert 113°F to degrees Celsius. If necessary, round to the nearest tenth of a degree.

Example 2   Convert 29°C to degrees Fahrenheit.

Solution: $F = 1.8 \cdot C + 32$

$= 1.8 \cdot 29 + 32$ Replace C with 29.

$= 52.2 + 32$ Multiply 1.8 by 29.

$= 84.2$ Add.

Therefore, 29°C is the same as 84.2°F. ▬▬▬

B   CONVERTING DEGREES FAHRENHEIT TO DEGREES CELSIUS

To convert from Fahrenheit temperatures to Celsius temperatures, see the box below. The symbol C represents degrees Celsius and F represents degrees Fahrenheit.

> **CONVERTING FAHRENHEIT TO CELSIUS**
>
> $$C = \frac{5}{9} \cdot (F - 32)$$
>
> (To convert to Celsius temperature, subtract 32 from the Fahrenheit temperature, and then multiply by $\frac{5}{9}$.)

Example 3   Convert 59°F to degrees Celsius.

Solution: We evaluate the formula $C = \frac{5}{9} \cdot (F - 32)$ when F is 59.

$$C = \frac{5}{9} \cdot (F - 32)$$

$$= \frac{5}{9} \cdot (59 - 32)$$ Replace F with 59.

$$= \frac{5}{9} \cdot (27)$$ Subtract in parentheses.

$$= 15$$ Multiply.

Therefore, 59°F is the same temperature as 15°C. ▬▬▬

Example 4   Convert 114°F to degrees Celsius. If necessary, round to the nearest tenth of a degree.

Solution: $$C = \frac{5}{9} \cdot (F - 32)$$

$$= \frac{5}{9} \cdot (114 - 32)$$ Replace F with 114.

$$= \frac{5}{9} \cdot (82)$$ Subtract in parentheses.

$$\approx 45.6$$ Multiply.

Therefore, 114°F is approximately 45.6°C. ▬▬▬

Example 5 The normal body temperature is 98.6°F. What is this temperature in degrees Celsius?

Solution: We evaluate the formula $C = \frac{5}{9} \cdot (F - 32)$ when F is 98.6.

$$C = \frac{5}{9} \cdot (F - 32)$$

$$= \frac{5}{9} \cdot (98.6 - 32) \quad \text{Replace F with 98.6.}$$

$$= \frac{5}{9} \cdot (66.6) \quad \text{Subtract in parentheses.}$$

$$= 37 \quad \text{Multiply.}$$

Therefore, normal body temperature is 37°C. ▬▬▬

TRY THE CONCEPT CHECK IN THE MARGIN.

Practice Problem 5

During a bout with the flu, Albert's temperature reaches 102.8°F. What is his temperature measured in degrees Celsius? Round to the nearest tenth of a degree.

✓ CONCEPT CHECK

Clarissa must convert 40°F to degrees Celsius. What is wrong with her work shown below?

$F = 1.8 \cdot C + 32$
$F = 1.8 \cdot 40 + 32$
$F = 72 + 32$
$F = 104$

Answers

5. 39.3°C

✓ Concept Check: She used the conversion for Celsius to Fahrenheit instead of Fahrenheit to Celsius.

Focus On the Real World

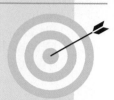

COMPUTER STORAGE CAPACITY

Have you ever heard someone describe a computer as having a 4 gig hard drive and wondered what exactly that meant? In this chapter, we focus on measuring basic physical characteristics such as length, weight/mass, and capacity. However, another type of measurement that we frequently encounter in the real world deals with computer storage capacity, that is, how much information can be stored on a computer device such as a floppy diskette, data cartridge, CD-ROM, hard drive, or computer memory (such as RAM).

To understand computer storage capacities, we must first understand the building blocks of information in the world of computers. The smallest piece of information handled by a computer is called a **bit** (abbreviated b). Bit is short for "binary digit" and consists of either a 0 or a 1. A collection of 8 bits is called a **byte** (abbreviated B). A single byte represents a standard character such as a letter of the alphabet, a number, or a punctuation mark. The following table shows the relationships among various terms used to describe computer storage capacities.

1 nibble = 4 bits
1 byte (B) = 8 bits
1 kilobyte (KB) = 1024 bytes = 2^{10} bytes
1 megabyte (MB) = 1024 kilobytes = 1024^2 bytes = 2^{20} bytes
1 gigabyte (GB) = 1024 megabytes = 1024^3 bytes = 2^{30} bytes
1 terabyte (TB) = 1024 gigabytes = 1024^4 bytes = 2^{40} bytes

Note: Sometimes a megabyte is referred to as a "meg" and a gigabyte as a "gig."

GROUP ACTIVITY

Using advertisements for computer systems, find five different examples of uses of any of the computer storage capacity terms defined above. Then convert each example to both bytes and bits. Make a table to organize your results. Be sure to include an explanation of the original use of each example.

Name _____ **Section** _____ **Date** _____

Mental Math

Determine whether the measurement in each statement is reasonable.

1. A 72°F room feels comfortable.

2. Water heated to 110°F will boil.

3. Josiah has a fever if a thermometer reads his temperature as 40°F.

4. An air temperature of 20°F on a Vermont ski slope can be expected in the winter.

5. When the temperature is 30°C outside an overcoat is needed.

6. An air-conditioned room at 60°C feels quite chilly.

7. Barbara has a fever if a thermometer records her temperature at 40°C.

8. Water cooled to 32°C will freeze.

Exercise Set 7.4

A **B** *Convert as indicated. When necessary, round to the nearest tenth of a degree. See Examples 1 through 5.*

1. 41°F to degrees Celsius

2. 68°F to degrees Celsius

3. 104°F to degrees Celsius

4. 86°F to degrees Celsius

5. 60°C to degrees Fahrenheit

6. 80°C to degrees Fahrenheit

7. 115°C to degrees Fahrenheit

8. 35°C to degrees Fahrenheit

9. 62°F to degrees Celsius

10. 182°F to degrees Celsius

11. 142.1°F to degrees Celsius

12. 43.4°F to degrees Celsius

443

13. 197.6°F

14. 167°F

15. 61.3°F

16. 119.5°F

17. 56.7°C

18. 43.9°C

19. 80.6°F

20. 64.4°F

21. 21.1°C

22. 18.9°C

23. 37.9°C

24. 36.8°C

444

13. 92°C to degrees Fahrenheit

14. 75°C to degrees Fahrenheit

15. 16.3°C to degrees Fahrenheit

16. 48.6°C to degrees Fahrenheit

17. The highest temperature ever recorded in Death Valley was 134°F. Convert this measurement to degrees Celsius. (*Source:* National Climatic Data Center)

18. The hottest day in Pennsylvania reached 111°F. Convert this measurement to degrees Celsius. (*Source:* National Climatic Data Center)

19. A weather forecaster in Caracas predicts a high temperature of 27°C. Find this measurement in degrees Fahrenheit.

20. While driving to work, Alan Olda notices a temperature of 18°C flash on the local bank's temperature display. Find the corresponding temperature in degrees Fahrenheit.

21. At Mack Trucks' headquarters the room temperature is to be set at 70°F, but the thermostat is calibrated in degrees Celsius. Find the temperature to be set.

22. The computer room at Merck, Sharp, and Dohm is normally cooled to 66°F. Find the corresponding temperature in degrees Celsius.

23. Najib Tan is running a fever of 100.2°F. Find his temperature as it would be shown on a Celsius thermometer.

24. William Saylor generally has a subnormal temperature of 98.2°F. Find this temperature on a Celsius thermometer.

Name _____

25. In a European cookbook, the recipe requires the ingredients for caramels to be heated to 118°C, but the cook only has access to a Fahrenheit thermometer. Find the temperature in degrees Fahrenheit that should be used to make the caramels.

26. The ingredients for divinity should be heated to 127°C, but the candy thermometer that Myung Kim has is calibrated to degrees Fahrenheit. Find how hot he should heat the ingredients.

27. Mark Tabbey's recipe for Yorkshire pudding calls for a very hot, 500°F oven. Find the temperature setting he should use with an oven having Celsius controls.

28. The temperature of Earth's core is estimated to be 4000°C. Find the corresponding temperature in degrees Fahrenheit.

29. The surface temperature of Venus can reach 864°F. Find this temperature in degrees Celsius.

REVIEW AND PREVIEW

Find the perimeter of each figure. See Section 1.2.

30.

3 in.
3 in. | Square

31.
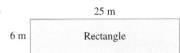
25 m
6 m | Rectangle

32.
4 cm   3 cm   5 cm

33.
3 ft   3 ft
3 ft   3 ft
3 ft

34.
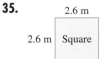
2 ft 8 in.
1 ft 6 in. | Rectangle

35.
2.6 m
2.6 m | Square

---

25. 244.4°F

26. 260.6°F

27. 260°C

28. 7232°F

29. 462.2°C

30. 12 in.

31. 62 m

32. 12 cm

33. 15 ft

34. 8 ft 4 in.

35. 10.4 m

445

36. On February 17, 1995, in the Tokamak Fusion Test Reactor at Princeton University, the highest temperature produced in a laboratory was achieved. This temperature was 918,000,000°F. Convert this temperature to degrees Celsius. (*Source: Guiness Book of Records*, 1998)

37. The hottest-burning substance known is carbon subnitride. Its flame at one atmosphere pressure reaches 9010°F. Convert this temperature to degrees Celsius. (*Source: Guiness Book of Records*, 1998)

37. 4988°C

7.5 ENERGY: U.S. AND METRIC SYSTEMS OF MEASUREMENT

Many people think of energy as a concept that involves movement or activity. However, **energy** is defined as the capacity to do work. Often energy is stored, awaiting use at some point in time.

A USING THE U.S. SYSTEM UNITS OF ENERGY

In the U.S. system of measurement, energy is commonly measured in foot-pounds. One **foot-pound (ft-lb)** is the amount of energy needed to lift a 1-pound object a distance of 1 foot. To determine the amount of energy necessary to move a 50-pound weight a distance of 100 feet, we simply multiply these numbers. That is,

$$(50 \text{ pounds}) \cdot (100 \text{ feet}) = 5000 \text{ ft-lb of energy}$$

Example 1 Finding Energy Needed to Move a Carton

An employee for the Jif Peanut Butter company must lift a carton of peanut butter jars 16 feet to the top of the warehouse. In the carton are 24 jars, each of which weighs 1.125 pounds. How much energy is required to lift the carton?

Solution: First we determine the weight of the carton.

$$\text{weight of carton} = \text{weight of a jar} \cdot \text{number of jars}$$
$$= 1.125 \text{ pounds} \cdot 24$$
$$= 27 \text{ pounds}$$

Thus, the carton weighs 27 pounds.

To find the energy needed to lift the 27-pound carton, we multiply the weight times the distance.

$$\text{energy} = (27 \text{ pounds}) \cdot (16 \text{ feet}) = 432 \text{ ft-lb}$$

Thus 432 ft-lb of energy are required to lift the carton.

TRY THE CONCEPT CHECK IN THE MARGIN.

Another form of energy is heat. In the U.S. system of measurement, heat is measured in **British Thermal Units (BTU)**. A BTU is the amount of heat required to raise the temperature of 1 pound of water 1 degree Fahrenheit. To relate British Thermal Units to foot-pounds, we need to know that

$$1 \text{ BTU} = 778 \text{ ft-lb}$$

Example 2 Converting BTU to Foot-pounds

The Raywall Company produces several different furnace models. Their FC-4 model requires 13,652 BTU every hour to operate. Convert the required energy to foot-pounds.

Solution: To convert BTU to foot-pounds, we multiply by the unit fraction $\frac{778 \text{ ft-lb}}{1 \text{ BTU}}$.

Unit fraction

$$13,652 \text{ BTU} = 13,652 \text{ BTU} \cdot \frac{778 \text{ ft-lb}}{1 \text{ BTU}}$$
$$= 10,621,256 \text{ ft-lb}$$

Objectives

A Define and use U.S. units of energy and convert from one unit to another.

B Define and use metric units of energy.

SSM   CD-ROM   Video 7.5

TEACHING TIP

Ask students to work in groups to find 10 other situations which would require the same amount of energy as needed for moving 50 pounds 100 feet. Then have students try to order them from easiest to hardest. If the energy needed is the same for all of these, why do some seem easier than others?

Practice Problem 1

Three bales of cardboard must be lifted 340 feet. If each bale weighs 63 pounds, find the amount of work required to lift the cardboard.

✓ CONCEPT CHECK

Suppose you would like to find how many foot-pounds of energy are needed to lift an object weighing 12 ounces a total of 14 yards. What adjustments should you make before computing the answer?

Practice Problem 2

The FC-5 model furnace produced by Raywall uses 17,065 BTU every hour. Convert this energy requirement to foot-pounds.

Answers

1. 64,260 ft-lb, **2.** 13,276,570 ft-lb

✓ Concept Check: Convert 12 ounces to .75 pounds and 14 yards to 42 feet; 31.5 ft-lb.

TEACHING TIP

Have students use the definition of BTU and foot-pound to explain in writing what it means for 1 BTU = 778 ft-lb.

70 calories

Practice Problem 3

It takes 30 calories each hour for Alan to fly a kite. How many calories will he use if he flies his kite for 2 hours?

Practice Problem 4

It takes 200 calories for Melanie to play Frisbee for an hour. How many calories will she use playing Frisbee for an hour each day for 5 days?

Practice Problem 5

To play volleyball for an hour requires 300 calories. If Martha plays volleyball 1.25 hours each day for 4 days, how many calories does Martha use?

Thus, 13,652 BTU is equivalent to 10,621,256 ft-lb.

B USING THE METRIC SYSTEM UNITS OF ENERGY

In the metric system, heat is measured in calories. A **calorie (cal)** is the amount of heat required to raise the temperature of 1 kilogram of water 1 degree Celsius.

The fact that an apple contains 70 calories means that 70 calories of heat energy are stored in our bodies whenever we eat an apple. This energy is stored in fat tissue and is burned (or oxidized) by our bodies when we require energy to do work. We need 20 calories each hour just to stand still. This means that 20 calories of heat energy will be burned by our bodies each hour that we spend standing.

Example 3   Finding Number of Calories Needed

It takes 20 calories for Jim to stand for one hour. How many calories does he use to stand for 3 hours at a crowded party?

Solution: We multiply the number of calories used in one hour by the number of hours spent standing.

total calories $= 20 \cdot 3 = 60$ calories

Therefore, Jim uses 60 calories to stand for 3 hours at the party.

Example 4   Finding Number of Calories Needed

It takes 115 calories for Kathy to walk slowly for an hour. How many calories does she use walking slowly for 1 hour a day for 6 days?

Solution: We multiply the total number of calories used in an hour each day by the number of days.

total calories $= 115 \cdot 6 = 690$ calories

Therefore, Kathy uses 690 calories walking slowly for 1 hour for 6 days.

Example 5   Finding Number of Calories Needed

It requires 100 calories to play a game of cards for an hour. If Jason plays poker for 1.5 hours each day for 5 days, how many calories are required?

Solution: We first determine the number of calories Jason uses each day to play poker.

calories used each day $= 100\,(1.5) = 150$ calories

Then we multiply the number of calories used each day by the number of days.

calories used for 5 days $= 150 \cdot 5 = 750$ calories

Thus Jason uses 750 calories to play poker for 1.5 hours each day for 5 days.

Answers

3. 60 calories,   **4.** 1000 calories,
5. 1500 calories

Name _____ **Section** _____ **Date** _____

MENTAL MATH

Solve.

1. How many foot-pounds of energy are needed to lift a 6 pound object 5 feet?

2. How many foot-pounds of energy are needed to lift a 10 pound object 4 feet?

3. How many foot-pounds of energy are need to lift a 3 pound object 20 feet?

4. How many foot-pounds of energy are needed to lift a 5 pound object 9 feet?

5. If 30 calories are burned by the body in 1 hour, how many calories are burned in 3 hours?

6. If 15 calories are burned by the body in 1 hour, how many calories are burned in 2 hours?

7. If 20 calories are burned by the body in 1 hour, how many calories are burned in $\frac{1}{4}$ of an hour?

8. If 50 calories are burned by the body in 1 hour, how many calories are burned in $\frac{1}{2}$ hour?

EXERCISE SET 7.5

A *Solve. See Examples 1 and 2.*

1. How much energy is required to lift a 3-pound math text 380 feet up a hill?

2. How much energy is required to lift a 20-pound sack of potatoes 55 feet?

3. How much energy is required to lift a 168-pound person 22 feet?

4. How much energy is needed to lift a 2250-pound car a distance of 45 feet?

5. How many foot-pounds of energy are needed to lift 2.5 tons of topsoil 85 feet from the pile delivered by the nursery to the garden?

6. How many foot-pounds of energy are needed to lift 4.25 tons of coal 16 feet into a new coal bin?

7. Convert 30 BTU to foot-pounds.

8. Convert 50 BTU to foot-pounds.

9. Convert 1000 BTU to foot-pounds.

10. Convert 10,000 BTU to foot-pounds.

MENTAL MATH ANSWERS

1. 30 ft-lb

2. 40 ft-lb

3. 60 ft-lb

4. 45 ft-lb

5. 90 calories

6. 30 calories

7. 5 calories

8. 25 calories

ANSWERS

1. 1140 ft-lb

2. 1100 ft-lb

3. 3696 ft-lb

4. 101,250 ft-lb

5. 425,000 ft-lb

6. 136,000 ft-lb

7. 23,340 ft-lb

8. 38,900 ft-lb

9. 778,000 ft-lb

10. 7,780,000 ft-lb

11. 15,560,000 ft-lb

12. 18,672,000 ft-lb

13. 26,553,140 ft-lb

14. 31,863,768 ft-lb

15. 10,283 BTU

16. 578 BTU

17. 805 calories

18. 810 calories

19. 750 calories

20. 540 calories

11. A 20,000 BTU air conditioner requires how many foot-pounds of energy to operate?

12. A 24,000 BTU air conditioner requires how many foot-pounds of energy to operate?

13. The Raywall model FC-10 heater uses 34,130 BTU each hour to operate. How many foot-pounds of energy does it use each hour?

14. The Raywall model FC-12 heater uses 40.956 BTU each hour to operate. How many foot-pounds of energy does it use each hour?

15. 8,000,000 ft-lb is equivalent to how many BTU, rounded to the nearest whole number?

16. 450,000 ft-lb is equivalent to how many BTU, rounded to the nearest whole number?

B *Solve. See Examples 3 through 5.*

17. While walking slowly, Janie Gaines burns 115 calories each hour. How many calories does she burn if she walks slowly for an hour every day of the week?

18. Dancing burns 270 calories per hour. How many calories are needed to go dancing an hour a night for 3 nights?

19. Approximately 300 calories are burned each hour when skipping rope. How many calories are required to skip rope $\frac{1}{2}$ hour each day for 5 days?

20. Ebony Jordan burns 360 calories per hour while riding her stationary bike. How many calories does she burn if she rides her bicycle $\frac{1}{4}$ hour each day for 6 days?

450

21. Julius Davenport goes through a rigorous exercise routine each day. He burns calories at a rate of 720 calories per hour. How many calories does he need to exercise 20 minutes per day, 6 days a week?

22. A roller skater can easily use 325 calories per hour while skating. How many calories are needed to roller skate 75 minutes per day for 3 days?

23. A casual stroll burns 165 calories per hour. How long will it take to stroll off the 425 calories contained in a hamburger, to the nearest tenth of an hour?

24. Even when asleep, the body burns 15 calories per hour. How long must a person sleep to burn off the calories in an 80-calorie orange, to the nearest tenth of an hour?

25. One pound of body weight is lost whenever 3500 calories are burned off. If walking briskly burns 200 calories for each mile walked, how far must Sheila Osby walk to lose 1 pound?

26. Bicycling can burn as much as 500 calories per hour. How long must a person ride a bicycle at this rate to use up the 3500 calories needed to lose 1 pound?

REVIEW AND PREVIEW

Write each fraction in simplest form. See Section 2.3.

27. $\frac{20}{25}$

28. $\frac{75}{100}$

29. $\frac{27}{45}$

30. $\frac{56}{60}$

31. $\frac{72}{80}$

32. $\frac{18}{20}$

21. 1440 calories

22. 1218.75 calories

23. 2.6 hours

24. 5.3 hours

25. 17.5 miles

26. 7 hours

27. $\frac{4}{5}$

28. $\frac{3}{4}$

29. $\frac{3}{5}$

30. $\frac{14}{15}$

31. $\frac{9}{10}$

32. $\frac{9}{10}$

Name _____

◆ **COMBINING CONCEPTS**

33. A 123.9-pound pile of prepacked canned goods must be lifted 9 inches to permit a door to close. How much energy is needed to do the job?

34. A 14.3-poound framed picture must be lifted 6 feet 3 inches. How much energy is required to move the picture?

35. 6400 ft-lb of energy was needed to lift an anvil 25 feet. Find the weight of the anvil.

36. 825 ft-lb of energy was needed to lift 40 pounds of apples to a new container. How far were the apples moved?

CHAPTER 7 ACTIVITY
MAP READING

MATERIALS:
▲ ruler
▲ string
▲ calculator

This activity may be completed by working in groups or individually.

Investigate the route you would take from Santa Rosa, New Mexico, to San Antonio, New Mexico. Use the map in the figure to answer the following questions. You may find that fitting string to roads on the map is useful when measuring distances.

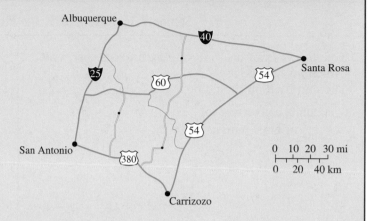

1. How many miles is it from Santa Rosa to San Antonio via Interstate 40 and Interstate 25? Convert this distance to kilometers. approximately 187.5 miles; 301.7 km

2. How many miles is it from Santa Rosa to San Antonio via U.S. 54 and U.S. 380? Convert this distance to kilometers. approximately 180 miles; 289.6 km

3. Assume that the speed limit on Interstates 40 and 25 is 65 miles per hour. How long would the trip take if you took this route? approximately 2 hours 53 minutes

4. At what average speed would you have to travel on the U.S. routes to make the trip from Santa Rosa to San Antonio in the same amount of time that it would take on the interstate routes? Do you think this speed is reasonable on this route? Explain your reasoning. approximately 62 mph

5. Discuss in general the factors that might affect your decision among different routes. answers may vary

6. Explain which route you would choose in this case and why. answers may vary

CHAPTER 7 HIGHLIGHTS

DEFINITIONS AND CONCEPTS	EXAMPLES

SECTION 7.1   LENGTH: U.S. AND METRIC SYSTEMS OF MEASUREMENT

To convert from one unit of length to another, **unit fractions** may be used.

The unit fraction should be in the form

$$\frac{\text{units converting to}}{\text{original units}}$$

$$\frac{12 \text{ inches}}{1 \text{ foot}}, \frac{1 \text{ foot}}{12 \text{ inches}}, \frac{3 \text{ feet}}{1 \text{ yard}}$$

Convert 6 feet to inches.

$$6 \text{ feet} = \frac{6 \cancel{\text{feet}}}{1} \cdot \frac{12 \text{ inches}}{1 \cancel{\text{foot}}} \quad \begin{array}{l} \leftarrow \text{units} \\ \text{converting to} \\ \leftarrow \text{original} \\ \text{units} \end{array}$$

$$= 6 \cdot 12 \text{ inches}$$

$$= 72 \text{ inches}$$

LENGTH: U.S. SYSTEM OF MEASUREMENT

$$12 \text{ inches (in.)} = 1 \text{ foot (ft)}$$
$$3 \text{ feet} = 1 \text{ yard (yd)}$$
$$5280 \text{ feet} = 1 \text{ mile (mi)}$$

The basic unit of length in the metric system is the **meter**. A meter is slightly longer than a yard.

Convert 3650 centimeters to meters.

$$3650 \text{ cm} = \frac{3650 \cancel{\text{cm}}}{1} \cdot \frac{0.01 \text{ m}}{1 \cancel{\text{cm}}} = 36.5 \text{ m}$$

or

km   hm   dam   m   dm   cm   mm

2 units to the left

$$3650 \text{ cm} = 36.5 \text{ m}$$

2 places to the left

LENGTH: METRIC SYSTEM OF MEASUREMENT

Prefix	Meaning	Metric Unit of Length
kilo	1000	1 kilometer (km) = 1000 meters (m)
hecto	100	1 hectometer (hm) = 100 m
deka	10	1 dekameter (dam) = 10 m
		1 meter (m) = 1 m
deci	1/10	1 decimeter (dm) = 1/10 m or 0.1 m
centi	1/100	1 centimeter (cm) = 1/100 m or 0.01 m
milli	1/1000	1 millimeter (mm) = 1/1000 m or 0.001 m

SECTION 7.2   WEIGHT AND MASS: U.S. AND METRIC SYSTEMS OF MEASUREMENT

Weight is really a measure of the pull of gravity.
Mass is a measure of the amount of substance in the object and does not change.

WEIGHT: U.S. SYSTEM OF MEASUREMENT

$$16 \text{ ounces (oz)} = 1 \text{ pound (lb)}$$
$$2000 \text{ pounds} = 1 \text{ ton}$$

Convert 5 pounds to ounces.

$$5 \text{ lb} = \frac{5 \cancel{\text{lb}}}{1} \cdot \frac{16 \text{ oz}}{1 \cancel{\text{lb}}} = 80 \text{ oz}$$

A **gram** is the basic unit of mass in the metric system. It is the mass of water contained in a cube 1 centimeter on each side. A paper clip weighs about 1 gram.

Convert 260 grams to kilograms.

$$260 \text{ g} = \frac{260 \text{ g}}{1} \cdot \frac{1 \text{ kg}}{1000 \text{ g}} = 0.26 \text{ kg}$$

or

kg   hg   dag   g   dg   cg   mg

3 units to the left

$$260 \text{ g} = 0.260 \text{ kg}$$

3 places to the left

MASS: METRIC SYSTEM OF MEASUREMENT

Prefix	Meaning	Metric Unit of Mass
kilo	1000	1 kilogram (kg) = 1000 grams (g)
hecto	100	1 hectogram (hg) = 100 g
deka	10	1 dekagram (dag) = 10 g
		1 gram (g) = 1 g
deci	1/10	1 decigram (dg) = 1/10 g or 0.1 g
centi	1/100	1 centigram (cg) = 1/100 g or 0.01 g
milli	1/1000	1 milligram (mg) = 1/1000 g or 0.001 g

SECTION 7.3 CAPACITY: U.S. AND METRIC SYSTEMS OF MEASUREMENT

CAPACITY: U.S. SYSTEM OF MEASUREMENT

$$8 \text{ fluid ounces (fl oz)} = 1 \text{ cup (c)}$$
$$2 \text{ cups} = 1 \text{ pint (pt)}$$
$$2 \text{ pints} = 1 \text{ quart (qt)}$$
$$4 \text{ quarts} = 1 \text{ gallon (gal)}$$

The **liter** is the basic unit of capacity in the metric system. It is the capacity or volume of a cube measuring 10 centimeters on each side. A liter of liquid is slightly more than 1 quart.

Convert 5 pints to gallons.

$$1 \text{ gal} = 4 \text{ qt} = 8 \text{ pt}$$

$$5 \text{ pt} = \frac{5 \text{ pt}}{1} \cdot \frac{1 \text{ gal}}{8 \text{ pt}} = \frac{5}{8} \text{ gal}$$

Convert 1.5 liters to milliliters.

$$1.5 \text{ L} = \frac{1.5 \text{ L}}{1} \cdot \frac{1000 \text{ ml}}{1 \text{ L}} = 1500 \text{ ml}$$

or

kl   hl   dal   L   dl   cl   ml

3 units to the right

$$1.500 \text{ L} = 1500 \text{ ml}$$

3 places to the right

CAPACITY: METRIC SYSTEM OF MEASUREMENT

Prefix	Meaning	Metric Unit of Capacity
kilo	1000	1 kiloliter (kl) = 1000 liters (L)
hecto	100	1 hectoliter (hl) = 100 L
deka	10	1 dckaliter (dal) = 10 L
		1 liter (L) = 1 L
deci	1/10	1 deciliter (dl) = 1/10 L or 0.1 L
centi	1/100	1 centiliter (cl) = 1/100 L or 0.01 L
milli	1/1000	1 milliliter (ml) = 1/1000 L or 0.001 L

SECTION 7.4 TEMPERATURE: U.S. AND METRIC SYSTEMS OF MEASUREMENT	
CELSIUS TO FAHRENHEIT $$F = \frac{9}{5} \cdot C + 32 \text{ or } F = 1.8 \cdot C + 32$$ FAHRENHEIT TO CELSIUS $$C = \frac{5}{9} \cdot (F - 32)$$	Convert 35°C to degrees Fahrenheit. $$F = \frac{9}{5} \cdot 35 + 32 = 63 + 32 = 95$$ $35°C = 95°F$ Convert 50°F to degrees Celsius. $$C = \frac{5}{9} \cdot (50 - 32) = \frac{5}{9} \cdot 18 = 10$$ $50°F = 10°C$

SECTION 7.5 ENERGY: U.S. AND METRIC SYSTEMS OF MEASUREMENT	
Energy is the capacity to do work. In the U.S. system of measurement, a **foot-pound** (ft-lb) is the amount of energy needed to lift a 1-pound object a distance of 1 foot. In the U.S. system of measurement, a **British Thermal Unit** (BTU) is the amount of heat required to raise the temperature of 1 pound of water 1 degree Fahrenheit. $\left(1 \text{ BTU} = 778 \text{ ft-lb}\right)$ In the metric system, a **calorie** is the amount of heat required to raise the temperature of 1 kilogram of water 1 degree Celsius.	How much energy is needed to lift a 20-pound object 35 feet? $$(20 \text{ pounds}) \cdot (35 \text{ feet}) = \begin{array}{c} 700 \text{ foot-pounds} \\ \text{of energy} \end{array}$$ Convert 50 BTU to foot-pounds. $$50 \text{ BTU} = 50 \text{ B\!T\!U} \cdot \frac{778 \text{ ft-lb}}{1 \text{ B\!T\!U}} = 38,900 \text{ ft-lb}$$ A stationary bicyclist uses 350 calories in one hour. How many calories will the bicyclist use in $\frac{1}{2}$ hour? $$\text{calories used} = 350 \cdot \frac{1}{2} = 175 \text{ calories}$$

CHAPTER 7 REVIEW

(7.1) *Convert.*

1. 108 in. to feet   9 ft

2. 72 ft to yards   24 yd

3. 2.5 mi to feet   13,200 ft

4. 6.25 ft to inches   75 in.

5. 52 ft = ____ yd ____ ft   17 yd 1 ft

6. 46 in. = ____ ft ____ in.   3 ft 10 in.

7. 42 m to centimeters   4200 cm

8. 82 cm to millimeters   820 mm

9. 12.18 mm to meters   0.01218 m

10. 2.31 m to kilometers   0.00231 km

Perform each indicated operation.

11. 4 yd 2 ft + 16 yd 2 ft   21 yd 1 ft

12. 12 ft 1 in. − 4 ft 8 in.   7 ft 5 in.

13. 8 ft 3 in. × 5   41 ft 3 in.

14. 7 ft 4 in. ÷ 2   3 ft 8 in.

15. 8 cm + 15 mm   9.5 cm or 95 mm

16. 4 m + 126 cm   5.26 m or 526 cm

17. 9.3 km − 183 m   9117 m or 9.117 km

18. 4100 mm − 3 m   1.1 m or 1100 mm

Solve.

19. A bolt of cloth contains 333 yd 1 ft of cotton ticking. Find the amount of material that remains after 163 yd 2 ft is removed from the bolt.   169 yd 2 ft

20. The local ambulance corps plans to award 20 framed certificates of valor to some of its outstanding members. If each frame requires 6 ft 4 in. of framing material, how much material is needed for all the frames?   126 ft 8 in.

21. The trip from Philadelphia to Washington, D.C., is 217 km. Four friends agree to share the driving equally. How far must each drive on this round-trip vacation?   108.5 km

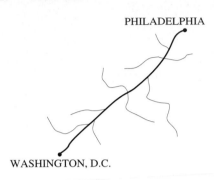

22. The college has ordered that NO SMOKING signs be placed above the doorway to each classroom. Each sign is 0.8 m long and 30 cm wide. Find the area of each sign. (*Hint:* Recall that the area of a rectangle = width · length.)   0.24 sq. m

(7.2) *Convert.*

23. 66 oz to pounds   4.125 lb

24. 2.3 tons to pounds   4600 lb

25. 52 oz = ____ lb ____ oz   3 lb 4 oz

26. 8200 lb = ____ tons ____ lb   4 tons 200 lb

27. 1400 mg to grams   1.4 g

28. 40 kg to grams   40,000 g

29. 2.1 hg to dekagrams   21 dag

30. 0.03 mg to decigrams   0.0003 dg

Perform each indicated operation.

31. 6 lb 5 oz − 2 lb 12 oz   3 lb 9 oz

32. 5 tons 1600 lb + 4 tons 1200 lb   10 tons 800 lb

33. 6 tons 2250 lb ÷ 3   2 tons 750 lb

34. 8 lb 6 oz × 4   33 lb 8 oz

35. 1300 mg + 3.6 g   4.9 g or 4900 mg

36. 4.8 kg + 4200 g   9 kg or 9000 g

37. 9.3 g − 1200 mg   8.1 g or 8100 mg

38. 6.3 kg × 8   50.4 kg

Solve the following.

39. Donshay Berry ordered 1 lb 12 oz of soft-center candies and 2 lb 8 oz of chewy-center candies for his party. Find the total weight of the candy ordered.
4 lb 4 oz

40. Four local townships jointly purchase 38 tons 300 lb of cinders to spread on their roads during an ice storm. Determine the weight of the cinders each township receives if they share the purchase equally.   9 tons 1075 lb

41. Linda Holden ordered 8.3 kg of whole wheat flour from the health store, but she received 450 g less. How much flour did she actually receive?   7.85 kg

42. Eight friends spent a weekend in the Poconos tapping maple trees and preparing 9.3 kg of maple syrup. Find the weight each friend receives if they share the syrup equally.   1.1625 kg

(7.3) *Convert.*

43. 16 pints to quarts   8 qt

44. 40 fluid ounces to cups   5 c

45. 6.75 gallons to quarts   27 qt

46. 8.5 pints to cups   17 c

47. 9 pt = ____ qt ____ pt   4 qt 1 pt

48. 15 qt = ____ gal ____ qt   3 gal 3 qt

49. 3.8 L to milliliters   3800 ml

50. 4.2 ml to deciliters   0.042 dl

51. 14 hl to kiloliters   1.4 kl

52. 30.6 L to centiliters   3060 cl

Perform each indicated operation.

53. 1 qt 1 pt + 3 qt 1 pt   1 gal 1 qt

54. 3 gal 2 qt 1 pt × 2   7 gal 1 qt

55. 0.946 L − 210 ml   736 ml or 0.736 L

56. 6.1 L + 9400 ml   15.5 L or 15,500 mL

Solve.

57. Carlos Perez prepares 4 gal 2 qt of iced tea for a block party. During the first 30 minutes of the party, 1 gal 3 qt of the tea is consumed. How much iced tea remains?   2 gal 3 qt

58. A recipe for soup stock calls for 1 c 4 fl oz of beef broth. How much should be used if the recipe is cut in half?   6 fl oz

59. Each bottle of Kiwi liquid shoe polish holds 85 ml of the polish. Find the number of liters of shoe polish contained in 8 boxes if each box contains 16 bottles.   10.88 L

60. Ivan Miller wants to pour three separate containers of saline solution into a single vat with a capacity of 10 liters. Will 6 liters of solution in the first container combined with 1300 milliliters in the second container and 2.6 liters in the third container fit into the larger vat?   yes, 9.9 L

(7.4) *Convert. Round to the nearest tenth of a degree, if necessary.*

61. 245°C to degrees Fahrenheit   473°F

62. 160°C to degrees Fahrenheit   320°F

63. 42°C to degrees Fahrenheit   107.6°F

64. 86°C to degrees Fahrenheit   186.8°F

65. 93.2°F to degrees Celsius   34°C

66. 51.8°F to degrees Celsius   11°C

67. 41.3°F to degrees Celsius   5.2°C

68. 80°F to degrees Celsius   26.7°C

Solve. Round to the nearest tenth of a degree, if necessary.

69. A sharp dip in the jet stream caused the temperature in New Orleans to drop to 35°F. Find the corresponding temperature in degrees Celsius.   1.7°C

70. The recipe for meat loaf calls for a 165°C oven. Find the setting used if the oven has a Fahrenheit thermometer.   329°F

(7.5) *Solve.*

71. How many foot-pounds of energy are needed to lift a 5.6-pound radio a distance of 12 feet?   67.2 ft-lb

72. How much energy is required to lift a 21-pound carton of Rice-A-Roni a distance of 6.5 feet?   136.5 ft-lb

73. How much energy is used when a 1.2-ton pile of sand is lifted 15 yards?   108,000 ft-lb

74. The energy required to operate a 12,000-BTU air conditioner is equivalent to how many foot-pounds?   9,336,000 ft-lb

75. Convert 2,000,000 foot-pounds to BTU, rounded to the nearest hundred.   2600 BTU

76. Kip Yates burns off 450 calories each hour he plays handball. How many calories does he use to play handball for $2\frac{1}{2}$ hours?   1125 calories

77. Qwanetta Sesson uses 210 calories each hour she spends mowing the grass. Find the number of calories needed to mow the grass if she spends 3 hours mowing each week for 24 weeks.   15,120 calories

78. Four ounces of sirloin steak contain 420 calories. If Edith Lutrell uses 180 calories each hour she walks, how long must she walk to burn off the calories from this steak?   2 hours 20 minutes

CHAPTER 7 TEST

Convert.

1. 280 in. to feet and inches

2. $2\frac{1}{2}$ gal to quarts

3. 30 oz to pounds

4. 2.8 tons to pounds

5. 38 pt to gallons

6. 40 mg to grams

7. 2.4 kg to grams

8. 3.6 cm to millimeters

9. 4.3 dg to grams

10. 0.83 L to milliliters

Perform each indicated operation.

11. 3 qt 1 pt + 2 qt 1 pt

12. 8 lb 6 oz − 4 lb 9 oz

13. 2 ft 9 in. × 3

14. 5 gal 2 qt ÷ 2

15. 8 cm − 14 mm

16. 1.8 km + 456 m

Convert. Round to the nearest tenth of a degree, if necessary.

17. 84°F to degrees Celsius

18. 12.6°C to degrees Fahrenheit

19. The sugar maples in front of Bette MacMillan's house are 8.4 meters tall. Because they interfere with the phone lines, the telephone company plans to remove the top third of the trees. How tall will the maples be after they are cut back?

20. A total of 15 gal 1 qt of oil has been removed from a 20-gallon drum. How much oil still remains in the container?

ANSWERS

1. 23 ft 4 in.

2. 10 qt

3. 1.875 lb

4. 5600 lb

5. $4\frac{3}{4}$ gal

6. 0.04 g

7. 2400 g

8. 36 mm

9. 0.43 g

10. 830 ml

11. 1 gal 2 qt

12. 3 lb 13 oz

13. 8 ft 3 in.

14. 2 gal 3 qt

15. 66 mm or 6.6 cm

16. 2.256 km or 2256 m

17. 28.9°C

18. 54.7°F

19. 5.6 m

20. 4 gal 3 qt

21. 105.8°F

21. The doctors are quite concerned about Lucia Gillespie, who is running a 41°C fever. Find Lucia's temperature in degrees Fahrenheit.

22. Gordan Cooper, the engineer in charge of the bridge construction, said that the span of the bridge would be 88 m. But the actual construction required it to be 340 cm longer. Find the span of the bridge.

22. 91.4 m

23. If 2 ft 9 in. of material is used to manufacture one scarf, how much material is needed for 6 scarves?

24. Phillipe Jordaine must lift a 48.5-pound carton of canned tomatoes a distance of 14 feet. Find the energy required to move the carton.

23. 16 ft 6 in.

25. Energy used by a 26,000-BTU heater is equivalent to how many foot-pounds?

26. Robin Nestle burns 180 calories each hour when she swims. How many calories does she use if she swims 1 hour per day for 5 days?

24. 679 ft-lb

25. 20,228,000 ft-lb

26. 900 calories

Name _____ Section _____ Date _____

CUMULATIVE REVIEW

1. Find the sum: $1647 + 246 + 32 + 85$

2. Find the prime factorization of 80.

3. Find the LCM of 11 and 33.

4. Add: $3\frac{4}{5} + 1\frac{4}{15}$

Write each decimal as a fraction or mixed number in simplest form.

5. 0.125

6. 105.083

7. Insert $<$, $>$, or $=$ to form a true statement.
   0.052   0.236

8. Subtract: $85 - 17.31$. Check your answer.

Multiply.

9. 42.1×0.1

10. 9.2×0.001

11. Divide: $60.24 \div 8$. Check your answer.

12. Estimate the distance in miles between Garden City, Kansas, and Wichita, Kansas, by rounding each given distance to the nearest ten.

13. Write $\frac{2}{3}$ as a decimal.

14. Write the ratio of 12 to 17 using fractional notation.

15. Write "318.5 miles every 13 gallons of gas" as a unit rate.

464

Name _____

16. Find the unknown number, n.

$$\frac{6}{5} = \frac{7}{n}$$

17. A 50-pound bag of fertilizer covers 2400 square feet of lawn. How many bags of fertilizer are needed to cover a town square containing 15,360 square feet of lawn? Round the answer up to the nearest whole bag.

18. Write 23% as a decimal.

19. Write $\dfrac{1}{12}$ as a percent. Round to the nearest hundredth percent.

20. What number is 35% of 40?

21. Translate to a proportion. 75 is what percent of 30?

22. In response to a decrease in sales, a company with 1500 employees reduces the number of employees to 1230. What is the percent decrease?

23. A speaker that normally sells for $65.00 is on sale at 25% off. What is the discount and what is the sale price?

24. Find the simple interest after 2 years on $500 at an interest rate of 12%.

25. Convert 9000 pounds to tons.

Geometry

The word *geometry* is formed from the Greek words *geo*, meaning earth, and *metron*, meaning measure. Geometry literally means to measure the earth. In this chapter we learn about various geometric figures and their properties, such as perimeter, area, and volume. Knowledge of geometry can help us solve practical problems in real-life situations. For instance, knowing certain measures of a circular swimming pool allows us to calculate how much water it can hold.

Just outside of Cairo, Egypt, is a famous plateau called Giza. This is the home of the Great Pyramids, the only surviving entry on the list of the Seven Wonders of the Ancient World. There are three pyramids at Giza. The largest, and oldest, was built as the tomb of the Pharaoh Khufu around 2550 B.C. This pyramid is made from over 2,300,000 blocks of stone weighing a total of 6.5 million tons. It took about 30 years to build this monument. Prior to the 20th century, Khufu's pyramid was the tallest building in the world. Khufu's son Khafre is responsible for the second-largest pyramid at Giza during his rule as pharaoh between 2520 to 2494 B.C. The smallest of the pyramids at Giza is credited to Menkaure, believed to be the son of Khafre and grandson of Khufu. In Exercises 33–36 on page 514, we will use some of the measurements of the Great Pyramids to calculate their volumes.

465

Name _____ **Section** _____ **Date** _____

CHAPTER 8 PRETEST

Classify each angle as acute, right, obtuse, or straight.

1.

2.

3. Find the complement of a 54° angle.

4. Find the measures of angles x, y, and z.

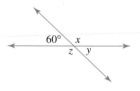

5. Find the measure of $\angle x$.

6. The radius of a sphere is 12.2 in. Find its diameter.

7. Find the perimeter of the given rectangle.

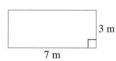

8. Find the circumference of the given circle. Use $\pi \approx 3.14$.

9. Find the area of the given triangle.

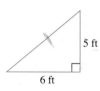

10. Find the volume of a rectangular box 10 in. by 17 in. by 3 in.

11. Find the exact volume of a sphere with a radius of 8 cm.

Simplify.

12. $\sqrt{36}$

13. $\sqrt{\dfrac{81}{49}}$

14. Find the length of the hypotenuse of a right triangle with leg lengths 8 yd and 3 yd. Approximate the length to the nearest thousandth.

15. Given that the pair of triangles are similar, find the length of the side labeled n.

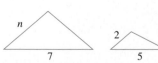

8.1   LINES AND ANGLES

A   IDENTIFY LINES, LINE SEGMENTS, RAYS, AND ANGLES

Let's begin with a review of two important concepts—plane and space.

A **plane** is a flat surface that extends indefinitely. Surfaces like a plane are a classroom floor or a blackboard or whiteboard.

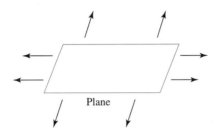

Plane

Space extends in all directions indefinitely. Examples of *objects* in space are houses, grains of salt, bushes, your Basic College Mathematics textbook, and you.

The most basic concept of geometry is the idea of a point in space. A **point** has no length, no width, and no height, but it does have location. We will represent a point by a dot, and we will label points with letters.

Point *P*

A **line** is a set of points extending indefinitely in two directions. A line has no width or height, but it does have length. We can name a line by any two of its points. A **line segment** is a piece of a line with two endpoints.

Line *AB* or $\overleftrightarrow{AB}$ Line segment *AB* or $\overline{AB}$

A **ray** is a part of a line with one endpoint. A ray extends indefinitely in one direction. An **angle** is made up of two rays that share the same endpoint. The common endpoint is called the **vertex**.

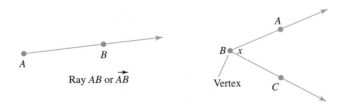

Ray *AB* or $\overrightarrow{AB}$ Vertex

The angle in the figure above can be named

$$\angle ABC \qquad \angle CBA \qquad \angle B \qquad \text{or} \qquad \angle x$$

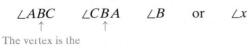

The vertex is the middle point.

Rays *BA* and *BC* are **sides** of the angle.

Practice Problem 1

Identify each figure as a line, a ray, a line segment, or an angle. Then name the figure using the given points.

a.

b.

c.

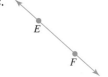

d.

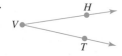

Practice Problem 2

Use the figure for Example 2 to list other ways to name $\angle z$.

Example 1   Identify each figure as a line, a ray, a line segment, or an angle. Then name the figure using the given points.

a.

b.

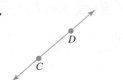

c.

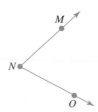

d.

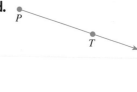

Solution: Figure (a) extends indefinitely in two directions. It is line CD or $\overleftrightarrow{CD}$.

Figure (b) has two endpoints. It is line segment EF or $\overline{EF}$.

Figure (c) has two rays with a common endpoint. It is $\angle MNO$, $\angle ONM$, or $\angle N$.

Figure (d) is part of a line with one endpoint. It is ray PT or $\overrightarrow{PT}$.

Example 2   List other ways to name $\angle y$.

Solution: Two other ways to name $\angle y$ are $\angle QTR$ and $\angle RTQ$. We may not use the vertex alone to name this angle because three different angles have T as their vertex.

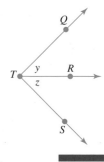

B CLASSIFYING ANGLES

An angle can be measured in **degrees**. The symbol for degrees is a small, raised circle, °. There are 360° in a full revolution, or full circle.

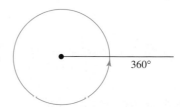

$\frac{1}{2}$ of a revolution measures $\frac{1}{2}\left(360°\right) = 180°$. An angle that measures 180° is called a **straight angle**.

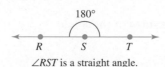

$\angle RST$ is a straight angle.

$\frac{1}{4}$ of a revolution measures $\frac{1}{4}(360°) = 90°$. An angle that measures 90° is called a **right angle**. The symbol ∟ is used to denote a right angle.

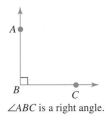

∠ABC is a right angle.

An angle whose measure is between 0° and 90° is called an **acute angle**.

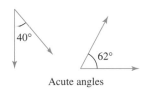

Acute angles

An angle whose measure is between 90° and 180° is called an **obtuse angle**.

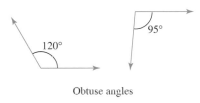

Obtuse angles

Example 3   Classify each angle as acute, right, obtuse, or straight.

a.

b.

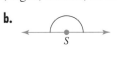

c.

d.

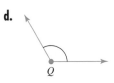

Solution: **a.** ∠R is a right angle, denoted by ∟.
b. ∠S is a straight angle.
c. ∠T is an acute angle. It measures between 0° and 90°.
d. ∠Q is an obtuse angle. It measures between 90° and 180°.

C IDENTIFYING COMPLEMENTARY AND SUPPLEMENTARY ANGLES

Two angles that have a sum of 90° are called **complementary angles**. We say that each angle is the **complement** of the other.

∠R and ∠S are complementary angles because
↓   ↓
60° + 30° = 90°

Practice Problem 3

Classify each angle as acute, right, ob-tuse, or straight.

a.

b.

c.   d.

Answers
3. a. acute, **b.** straight, **c.** obtuse, **d.** right

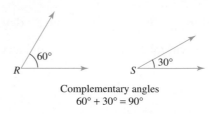

Complementary angles
$60° + 30° = 90°$

Two angles that have a sum of 180° are called **supplementary angles**. We say that each angle is the **supplement** of the other.

$\angle M$ and $\angle N$ are supplementary angles because

$125° + 55° = 180°$

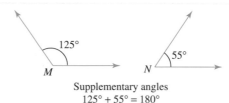

Supplementary angles
$125° + 55° = 180°$

TEACHING TIP

Help students remember that the sum of the measures of *supplementary* angles is 180° by reminding them that an angle of 180° is a *straight* angle and pointing out that both terms begin with an *s*.

Practice Problem 4

Find the complement of a 36° angle.

Practice Problem 5

Find the supplement of an 88° angle.

✓ Concept Check:

True or false? The supplement of a 48° angle is 42°. Explain.

Practice Problem 6

Find the measure of $\angle y$.

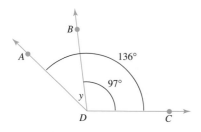

Example 4   Find the complement of a 48° angle.

     Solution: The complement of an angle that measures 48° is an angle that measures $90° - 48° = 42°$.   ▬▬▬

Example 5   Find the supplement of a 107° angle.

     Solution: The supplement of an angle that measures 107° is an angle that measures $180° - 107° = 73°$.   ▬▬▬

TRY THE CONCEPT CHECK IN THE MARGIN.

D   FINDING MEASURES OF ANGLES

Measures of angles can be added or subtracted to find measures of related angles.

Example 6   Find the measure of $\angle x$.

     Solution: $\angle x = 87° - 52° = 35°$

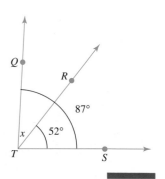

Two lines in a plane can be either parallel or intersecting. **Parallel lines** never meet. **Intersecting lines** meet at a point. The symbol ∥ is used to indicate "is parallel to." For example, in the figure $p \parallel q$.

Answers

4. 54°,   **5.** 92°,   **6.** 39°

✓ **Concept Check:**   False; the complement of a 48° angle is 42°, the supplement of a 48° angle is 132°.

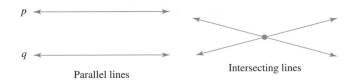

Parallel lines   Intersecting lines

Some intersecting lines are perpendicular. Two lines are **perpendicular** if they form right angles when they intersect. The symbol ⊥ is used to denote "is perpendicular to." For example, in the figure $n \perp m$.

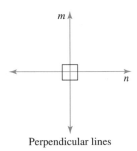

Perpendicular lines

When two lines intersect, four angles are formed. Two of these angles that are opposite each other are called **vertical angles**. Vertical angles have the same measure. Two angles that share a common side are called **adjacent angles**. Adjacent angles formed by intersecting lines are supplementary. That is, they have a sum of 180°.

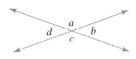

Vertical angles:
∠a and ∠c
∠d and ∠b

Adjacent angles:
∠a and ∠b
∠b and ∠c
∠c and ∠d
∠d and ∠a

Example 7   Find the measure of ∠x, ∠y, and ∠z if the measure of ∠t is 42°.

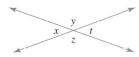

Solution: Since ∠t and ∠x are vertical angles, they have the same measure, so ∠x measures 42°.

Since ∠t and ∠y are adjacent angles, their measures have a sum of 180°. So ∠y measures 180° − 42° = 138°.

Since ∠y and ∠z are vertical angles, they have the same measure. So ∠z measures 138°.

Practice Problem 7

Find the measure of ∠a, ∠b, and ∠c.

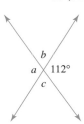

Answer

7. ∠a = 112°; ∠b = 68°; ∠c = 68°

A line that intersects two or more lines at different points is called a **transversal**. Line *l* is a transversal that intersects lines *m* and *n*. The eight angles formed have special names. Some of these names are:

Corresponding Angles: ∠*a* and ∠*e*, ∠*c* and ∠*g*, ∠*b* and ∠*f*, ∠*d* and ∠*h*

Alternate Interior Angles: ∠*c* and ∠*f*, ∠*d* and ∠*e*

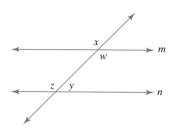

When two lines cut by a transversal are *parallel*, the following are true:

PARALLEL LINES CUT BY A TRANSVERSAL

If two parallel lines are cut by a transversal, then the measures of **corresponding angles are equal** and **alternate interior angles are equal**.

Example 8 Given that $m \parallel n$ and that the measure of ∠*w* is 100°, find the measures of ∠*x*, ∠*y*, and ∠*z*.

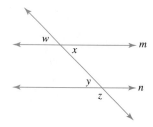

Solution:

The measure of ∠*x* = 100°. ∠*x* and ∠*w* are vertical angles.

The measure of ∠*z* = 100°. ∠*x* and ∠*z* are corresponding angles.

The measure of ∠*y* = 180° − 100° = 80°. ∠*z* and ∠*y* are supplementary angles.

Practice Problem 8

Given that $m \parallel n$ and that the measure of ∠*w* = 40°, find the measures of ∠*x*, ∠*y*, and ∠*z*.

Answer

8. ∠*x* = 40°; ∠*y* = 40°; ∠*z* = 140°

Name _____ **Section** _____ **Date** _____

EXERCISE SET 8.1

A *Identify each figure as a line, a ray, a line segment, or an angle. Then name the figure using the given points. See Example 1.*

1.

2.

3.

4.

5.

6.

7.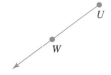

8.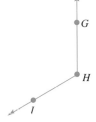

B *Find the measure of each angle in the figure.*

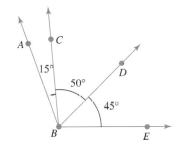

9. ∠ABC

10. ∠EBD

11. ∠CBD

12. ∠CBA

13. ∠DBA

14. ∠EBC

15. ∠CBE

16. ∠ABE

Fill in each blank. See Example 3.

17. A right angle has a measure of _____.

18. A straight angle has a measure of _____.

19. An acute angle measures between _____ and _____.

20. An obtuse angle measures between _____ and _____.

ANSWERS

1. line; line *yz* or $\overleftrightarrow{yz}$

2. angle; ∠*DEF* or ∠*FED* or ∠*E*

3. line segment; line segment *LM* or $\overline{LM}$

4. ray; ray *ST* or $\overrightarrow{ST}$

5. line segment; line segment *PQ* or $\overline{PQ}$

6. line; line *AB* or $\overleftrightarrow{AB}$

7. ray; ray *UW* or $\overrightarrow{UW}$

8. angle; ∠*GHI* or ∠*IHG* or ∠*H*

9. 15°

10. 45°

11. 50°

12. 15°

13. 65°

14. 95°

15. 95°

16. 110°

17. 90°

18. 180°

19. 0°; 90°

20. 90°; 180°

474

Name _____

Classify each angle as acute, right, obtuse, or straight. See Example 3.

21.

22.

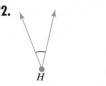

23.

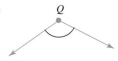

24.

25.

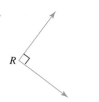

26.

27.

28.

Find each complementary or supplementary angle as indicated. See Examples 4 and 5.

29. Find the complement of a 17° angle.

30. Find the complement of an 87° angle.

31. Find the supplement of a 17° angle.

32. Find the supplement of an 87° angle.

33. Find the complement of a 48° angle.

34. Find the complement of a 22° angle.

35. Find the supplement of a 125° angle.

36. Find the supplement of a 155° angle.

37. Identify the pairs of complementary angles.

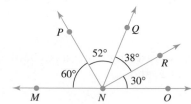

38. Identify the pairs of complementary angles.

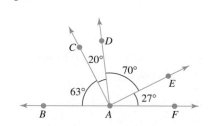

39. Identify the pairs of supplementary angles.

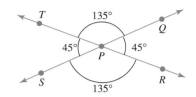

40. Identify the pairs of supplementary angles.

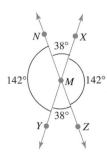

D *Find the measure of ∠x in each figure. See Example 6.*

41.

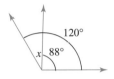

42.

43.

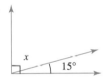

44.

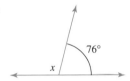

Find the measures of angles x, y, and z in each figure. See Examples 7 and 8.

45.

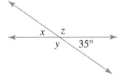

46.

 47.

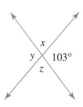

48.

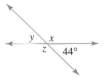

49. *m‖n*

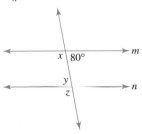

50. *m‖n*

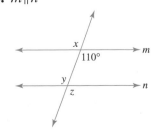

476

Name _____

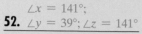 **51.** $m \parallel n$

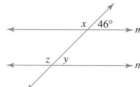

52. $m \parallel n$

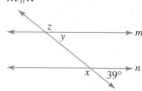

REVIEW AND PREVIEW

Perform each indicated operation. See Sections 2.4, 2.5, 3.3, and 3.4.

53. $\frac{7}{8} + \frac{1}{4}$

54. $\frac{7}{8} - \frac{1}{4}$

55. $\frac{7}{8} \cdot \frac{1}{4}$

56. $\frac{7}{8} \div \frac{1}{4}$

57. $3\frac{1}{3} - 2\frac{1}{2}$

58. $3\frac{1}{3} + 2\frac{1}{2}$

59. $3\frac{1}{3} \div 2\frac{1}{2}$

60. $3\frac{1}{3} \cdot 2\frac{1}{2}$

COMBINING CONCEPTS

61. The angle between the two walls of the Vietnam Veterans Memorial in Washington, D.C., is 125.2°. Find the supplement of this angle. (*Source*: National Park Service)

62. The faces of Khafre's Pyramid at Giza, Egypt, are inclined at an angle of 53.13°. Find the complement of this angle. (*Source*: PBS *NOVA* Online)

63. If lines m and n are parallel, find the measures of angles a through e.

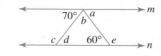

64. In your own words, describe how to find the complement and the supplement of a given angle.

8.2 PLANE FIGURES AND SOLIDS

In order to prepare for the sections ahead in this chapter, we first review plane figures and solids.

A IDENTIFYING PLANE FIGURES

Recall from Section 8.1 that a **plane** is a flat surface that extends indefinitely.

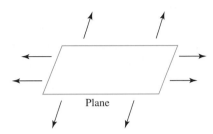

Plane

A **plane figure** is a figure that lies on a plane. Plane figures, like planes, have length and width but no thickness or depth.

A **polygon** is a closed plane figure that basically consists of three or more line segments that meet at their endpoints.

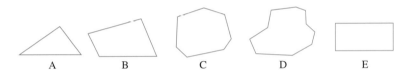

A   B   C   D   E

A **regular polygon** is one whose sides are all the same length and whose angles are the same measure.

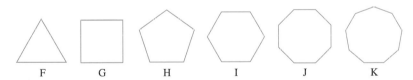

F   G   H   I   J   K

A polygon is named according to the number of its sides.

Some triangles and quadrilaterals are given special names, so let's study these special polygons further. We will begin with triangles. The sum of the measures of the angles of a triangle is 180°.

Example 1

Find the measure of $\angle a$.

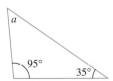

Solution: Since the sum of the measures of the three angles is 180°, we have

measure of $\angle a = 180° - 95° - 35° = 50°$

To check, see that $95° + 35° + 50° = 180°$.   ▬▬▬

We can classify triangles according to the lengths of their sides. (We will use tick marks to denote sides and angles in a figure that are equal.)

Objectives

A Identify plane figures.
B Identify solid figures.

SSM   CD-ROM   Video 8.2

TEACHING TIP

Begin the discussion of planes by reminding students of the relationship between a point and a line. Draw a point on the board and ask how many lines pass through the point. Draw a sample of lines through the point. Then ask if there are any lines which pass through the point which are not part of your drawing surface. Use a meter stick to illustrate such a line. Next consider a line. Ask how many planes pass through the line. Use a model of a plane to illustrate various planes which intersect the line.

Number of Sides	Name	Figure Examples
3	Triangle	A, F
4	Quadrilateral	B, E, G
5	Pentagon	H
6	Hexagon	I
7	Heptagon	C
8	Octagon	J
9	Nonagon	K
10	Decagon	D

Practice Problem 1

Find the measure of $\angle x$.

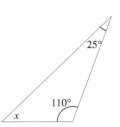

Answer

1. 45°

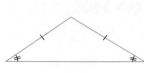
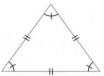
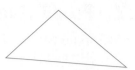

Equilateral triangle
All three sides are the same length. Also, all three angles have the same measure.

Isoceles triangle
Two sides are the same length. Also, the angles opposite the equal sides have equal measure.

Scalene triangle
No sides are the same length. No angles are the same measure.

One other important type of triangle is a right triangle. A **right triangle** is a triangle with a right angle. The side opposite the right angle is called the **hypotenuse** and the other two sides are called **legs**.

Practice Problem 2

Find the measure of ∠y.

HELPFUL HINT

From the previous example, can you see that in a right triangle, the sum of the other two acute angles is 90°? This is because

$$90° \;+\; 90° \;=\; 180°$$

↑ right angle's measure

↑ sum of other two angles' measures

↑ sum of angles' measures

Example 2   Find the measure of ∠b.

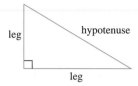

Solution: We know that the measure of the right angle, ⌐, is 90°. Since the sum of the measures of the angles is 180°, we have

measure of ∠b = 180° − 90° − 30° = 60°   ▬▬▬

Now we review some special quadrilaterals. A **parallelogram** is a special quadrilateral with opposite sides parallel and equal in length.

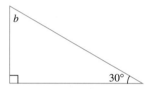

A **rectangle** is a special **parallelogram** that has four right angles.

A **square** is a special **rectangle** that has all four sides equal in length.

Answer

2. 65°

A **rhombus** is a special **parallelogram** that has all four sides equal in length.

A **trapezoid** is a quadrilateral with exactly one pair of opposite sides parallel.

TRY THE CONCEPT CHECK IN THE MARGIN.

In addition to triangles and quadrilaterals, circles are common plane figures. A **circle** is a plane figure that consists of all points that are the same fixed distance from a point *c*. The point *c* is called the **center** of the circle. The **radius** of a circle is the distance from the center of the circle to any point of the circle. The **diameter** of a circle is the distance across the circle passing through the center. The diameter is twice the radius and the radius is half the diameter.

diameter	=	2	·	radius	radius	=	$\dfrac{\text{diameter}}{2}$
↓	↓	↓		↓	↓		↓
d	=	2	·	*r*	*r*	=	$\dfrac{d}{2}$

Example 3   Find the diameter of the circle.

Solution: The diameter is twice the radius.

$$d = 2 \cdot r$$
$$d = 2 \cdot 5 \text{ cm} = 10 \text{ cm}$$

The diameter is 10 centimeters.

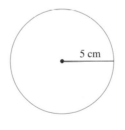

5 cm

B IDENTIFYING SOLID FIGURES

Recall from Section 8.1 that space extends in all directions indefinitely.
    A **solid** is a figure that lies in space. Solids have length, width, and height or depth.
    A **rectangular solid** is a solid that consists of six sides, or faces, all of which are rectangles.

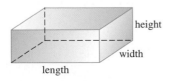

height

width

length

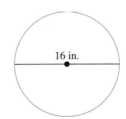

A **cube** is a rectangular solid whose six sides are squares.

A **pyramid** is shown below.

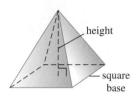

A **sphere** consists of all points in space that are the same distance from a point c. The point c is called the **center** of the sphere. The **radius** of a sphere is the distance from the center to any point of the sphere. The **diameter** of a sphere is the distance across the sphere passing through the center.

The radius and diameter of a sphere are related in the same way as the radius and diameter of a circle.

$$d = 2 \cdot r \qquad \text{or} \qquad r = \frac{d}{2}$$

Practice Problem 4

Find the diameter of the sphere.

Example 4   Find the radius of the sphere.

Solution: The radius is half the diameter.

$$r = \frac{d}{2}$$

$$r = \frac{36 \text{ feet}}{2} = 18 \text{ feet}$$

The radius is 18 feet.

The **cylinders** we will study have bases that are in the shape of circles and are perpendicular to their height.

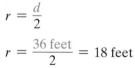

The **cones** we will study have bases that are circles and are perpendicular to their height.

Answer

4. 14 mi

Name	Section _____ Date _____

Exercise Set 8.2

A *Classify each triangle as equilateral, isosceles, or scalene.*

1.

2.

3.

4.

5.

6.

Find the measure of ∠x in each figure. See Examples 1 and 2.

7.

70°
85°
x

8.

x
112°
28°

9.

95°
72°
x

10.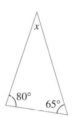

x
80°
65°

11.

x
50°

12.

x
20°

Fill in each blank.

13. Twice the radius of a circle is its _____.

14. A rectangle with all four sides equal is a _____.

15. A parallelogram with four right angles is a _____.

16. Half the diameter of a circle is its _____.

ANSWERS

1. equilateral

2. scalene

3. scalene

4. isosceles

5. isosceles

6. equilateral

7. 25°

8. 40°

9. 13°

10. 35°

11. 40°

12. 70°

13. diameter

14. square

15. rectangle

16. radius

481

Name _____

 17. A quadrilateral with opposite sides parallel is a _____.

18. A quadrilateral with exactly one pair of opposite sides parallel is a _____.

19. The side opposite the right angle of a right triangle is called the _____.

20. A triangle with no equal sides is a _____.

Determine whether each statement is true or false.

21. A square is also a rhombus.

22. A square is also a rectangle.

 23. A rectangle is also a parallelogram.

24. A trapezoid is also a parallelogram.

25. A pentagon is also a quadrilateral.

26. A rhombus is also a parallelogram.

Find the unknown diameter or radius in each figure. See Example 3.

 27.

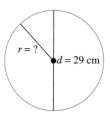

28.

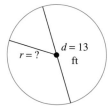

29.

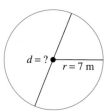

30.

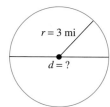

Identify each regular polygon.

31.

32.

33.

34.

Name _____

B *Identify each solid.*

35.

36.

37.

38.

39.

40.

Find each unknown radius or diameter. See Example 4.

41. The radius of a sphere is 7.4 in. Find its diameter.

42. The radius of a sphere is 5.8 m. Find its diameter.

43. The diameter of a sphere is 26 mi. Find its radius.

44. The diameter of a sphere is 78 cm. Find its radius.

Identify the shape of each item.

45.

46.

47.

48.

35. cylinder

36. cube

37. rectangular solid

38. pyramid

39. cone

40. sphere

41. 14.8 in.

42. 11.6 m

43. 13 mi

44. 39 cm

45. cube

46. cone

47. rectangular solid

48. cylinder

484

Name _____

49.

50.

51.

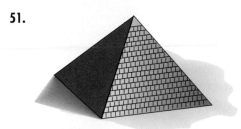

52.

REVIEW AND PREVIEW

Perform each indicated operation. See Sections 4.3 and 4.4.

53. $4(28.6)$ **54.** $2(7.8) + 2(9.6)$ **55.** $2(18) + 2(36)$ **56.** $4(87)$

COMBINING CONCEPTS

57. Saturn has a radius of approximately 36,184 mi. What is its diameter?

58. Is an isosceles right triangle possible? If so, draw one.

59. The following demonstration is credited to the mathematician Pascal, who is said to have developed it as a young boy.

Cut a triangle from a piece of paper. The length of the sides and the size of the angles is unimportant. Tear the points off the triangle.

Place the points of the triangle together. Notice that a straight line is formed. What was Pascal trying to show?

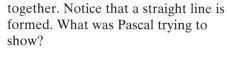

60. In your own words, explain whether a square is also a rhombus.

8.3 Perimeter

A Using Formulas to Find Perimeters

Recall from Section 1.2 that the perimeter of a polygon is the distance around the polygon. This means that the perimeter of a polygon is the sum of the lengths of its sides.

Example 1   Find the perimeter of the rectangle below.

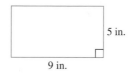

Solution: perimeter = 9 inches + 9 inches + 5 inches + 5 inches

= 28 inches

Notice that the perimeter of the rectangle in Example 1 can be written as $2 \cdot (9 \text{ inches}) + 2 \cdot (5 \text{ inches})$.

↑   ↑

length   width

In general, we can say that the perimeter of a rectangle is always

$2 \cdot \text{length} + 2 \cdot \text{width}$

As we have just seen, the perimeter of some special figures such as rectangles form patterns. These patterns are given as **formulas**. The formula for the perimeter of a rectangle is shown next.

PERIMETER OF A RECTANGLE

$\text{perimeter} = 2 \cdot \text{length} + 2 \cdot \text{width}$

In symbols, this can be written as

$P = 2 \cdot l + 2 \cdot w$

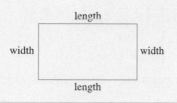

Example 2   Find the perimeter of a rectangle with a length of 11 inches and a width of 3 inches.

Solution: We will use the formula for perimeter and replace the letters by their known lengths.

$P = 2 \cdot l + 2 \cdot w$

$= 2 \cdot 11 \text{ in.} + 2 \cdot 3 \text{ in.}$ Replace l with 11 in. and w with 3 in.

$= 22 \text{ in.} + 6 \text{ in.}$

$= 28 \text{ in.}$

The perimeter is 28 inches.

Objectives

A Use formulas to find the perimeter of special geometric figures.

B Use formulas to find the circumference of a circle.

SSM   CD-ROM   Video 8.3

Practice Problem 1

Find the perimeter of the rectangular lot shown below.

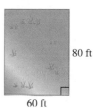

80 ft

60 ft

TEACHING TIP

Point out the word "rim" embedded in pe*rim*eter to help students remember its meaning.

Practice Problem 2

Find the perimeter of a rectangle with a length of 22 cm and a width of 10 cm.

Answers

1. 280 ft, **2.** 64 cm

TEACHING TIP

Consider reminding students that measurement problems require units in the answer. Then point out that in Example 2, 28 is meaningless without inches.

Recall that a square is a special rectangle with all four sides the same length. The formula for the perimeter of a square is shown next.

PERIMETER OF A SQUARE

$$\text{Perimeter} = \text{side} + \text{side} + \text{side} + \text{side}$$
$$= 4 \cdot \text{side}$$

In symbols,

$$P = 4 \cdot s$$

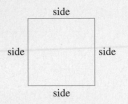

Practice Problem 3

How much fencing is needed to enclose a square field 50 yd on a side?

Example 3   Finding Perimeter of a Table Top

Find the perimeter of a square table top if each side is 5 feet long.

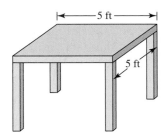

Solution: The formula for the perimeter of a square is $P = 4 \cdot s$. We will use this formula and replace s by 5 feet.

$$P = 4 \cdot s$$
$$= 4 \cdot 5 \text{ feet}$$
$$= 20 \text{ feet}$$

The perimeter of the square table top is 20 feet. ■

The formula for the perimeter of a triangle with sides of length a, b, and c is given next.

PERIMETER OF A TRIANGLE

$$\text{Perimeter} = \text{side } a + \text{side } b + \text{side } c$$

In symbols,

$$P = a + b + c$$

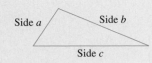

Answer

3. 200 yd

Example 4   Find the perimeter of a triangle if the sides are 3 inches, 7 inches, and 6 inches.

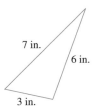

Solution: The formula is $P = a + b + c$, where a, b, and c are the lengths of the sides. Thus,

$$P = a + b + c$$
$$= 3 \text{ in.} + 7 \text{ in.} + 6 \text{ in.}$$
$$= 16 \text{ in.}$$

The perimeter of the triangle is 16 inches.   ▬▬▬▬

Recall that to find the perimeter of other polygons, we find the sum of the lengths of their sides.

Example 5   Find the perimeter of the trapezoid shown.

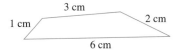

Solution: To find the perimeter, we find the sum of the lengths of its sides.

perimeter $= 3 \text{ cm} + 2 \text{ cm} + 5 \text{ cm} + 2 \text{ cm} = 12 \text{ cm}$

The perimeter is 12 centimeters.   ▬▬▬▬

Example 6   Finding the Perimeter of a Room

Find the perimeter of the room shown below.

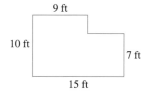

Solution: To find the perimeter of the room, we first need to find the lengths of all sides of the room.

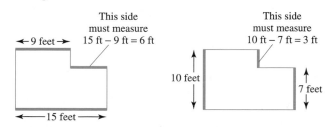

Now that we know the measures of all sides of the room, we can add the measures to find the perimeter.

Practice Problem 4

Find the perimeter of a triangle if the sides are 5 cm, 9 cm, and 7 cm in length.

Practice Problem 5

Find the perimeter of the trapezoid shown.

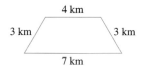

Practice Problem 6

Find the perimeter of the room shown.

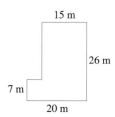

Answers

4. 21 cm,   **5.** 17 km,   **6.** 92 m

9 ft

3 ft

10 ft

6 ft

7 ft

15 ft

perimeter = 10 ft + 9 ft + 3 ft + 6 ft + 7 ft + 15 ft
           = 50 ft

The perimeter of the room is 50 feet.

Practice Problem 7

A rectangular lot measures 60 ft by 120 ft. Find the cost to install fencing around the lot if the cost of fencing is $1.90 per foot.

Example 7   Calculating the Cost of Baseboard

A rectangular room measures 10 feet by 12 feet. Find the cost to install new baseboard around the room if the cost of the baseboard is $0.66 per foot.

Solution: First we find the perimeter of the room.

$P = 2 \cdot l + 2 \cdot w$

   $= 2 \cdot 12 \text{ feet} + 2 \cdot 10 \text{ feet}$ Replace *l* with 12 feet and *w* with 10 feet.

   $= 24 \text{ feet} + 20 \text{ feet}$

   $= 44 \text{ feet}$

The cost for the baseboard is

cost = 0.66 · 44 = 29.04

The cost of the baseboard is $29.04.

B USING FORMULAS TO FIND CIRCUMFERENCES

Recall from Section 4.4 that the distance around a circle is given a special name called the **circumference**. This distance depends on the radius or the diameter of the circle.

The formulas for circumference are shown next.

CIRCUMFERENCE OF A CIRCLE

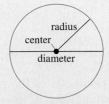

radius

center

diameter

   Circumference = $2 \cdot \pi \cdot$ radius or Circumference = $\pi \cdot$ diameter
In symbols,

   $C = 2 \cdot \pi \cdot r$ or $C = \pi \cdot d$

where $\pi \approx 3.14$ or $\pi \approx \dfrac{22}{7}$.

Answer

7. $684

To better understand circumference and π(pi), try the following experiment. Take any can and measure its circumference and its diameter.

The can in the figure above has a circumference of 23.5 centimeters and a diameter of 7.5 centimeters. Now divide the circumference by the diameter.

$$\frac{\text{circumference}}{\text{diameter}} = \frac{23.5 \text{ cm}}{7.5 \text{ cm}} \approx 3.1$$

Try this with other sizes of cylinders and circles—you should always get a number close to 3.1. The exact ratio of circumference to diameter is π. (Recall that $\pi \approx 3.14$ or $\pi \approx \frac{22}{7}$).

Example 8   Mary Catherine Dooley plans to install a border of new tiling around the circumference of her circular spa. If her spa has a diameter of 14 feet, find its circumference.

Solution: Because we are given the diameter, we use the formula $C = \pi \cdot d$.

$C = \pi \cdot d$

$\quad = \pi \cdot 14 \text{ ft}$ Replace d with 14 feet.

$\quad = 14\,\pi \text{ ft}$

The circumference of the spa is *exactly* $14\,\pi$ feet. By replacing π with the *approximation* 3.14, we find that the circumference is *approximately* 14 feet $\cdot$ 3.14 $= 43.96$ feet.

TRY THE CONCEPT CHECK IN THE MARGIN.

Practice Problem 8

An irrigation device waters a circular region with a diameter of 20 yd. What is the circumference of the watered region?

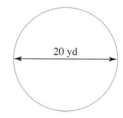

✓ CONCEPT CHECK

The distance around which figure is greater: a square with side length 5 in. or a circle with radius 3 in.?

Answers

8. 62.8 yd

✓ Concept Check: a square with length 5 in.

Focus On History

THE PYTHAGOREAN THEOREM

Pythagoras was a Greek teacher who lived from about 580 B.C. to 501 B.C. He founded his school at Croton in southern Italy. The school was a tightly knit community of eager young students with the motto "All is number." Pythagoras himself taught by lecturing on the subjects of arithmetic, music, geometry, and astronomy—all based on mathematics. In fact, Pythagoras and his students held numbers to be sacred and tried to find mathematical order in all parts of life and nature. Pythagoras' followers, known as Pythagoreans, were sworn to not reveal any of the Master's teachings to outsiders. It is for this reason that very little of his life or teachings are known today—Pythagoreans were forbidden from recording any of the Master's teachings in written form, and Pythagoras left none of his own writings.

Traditionally, the so-called Pythagorean theorem is attributed to Pythagoras himself. However, some scholars question whether Pythagoras ever gave any rigorous proof of the theorem at all! There is ample evidence that several ancient cultures had knowledge of this important theorem and used its results in practical matters well before the time of Pythagoras.

▲ A Babylonian clay tablet (#322 in the Plimpton collection at Columbia University) dating from 1900 B.C. gives numerical evidence that the Babylonians were well aware of what we today call the Pythagorean theorem.

▲ The ancient Chinese knew of a very simple geometric proof of the Pythagorean theorem, which can be seen in the *Arithmetic Classics of the Gnomon and the Circular Paths of Heaven*, dating from about 600 B.C.

▲ The ancient Egyptians used the Pythagorean theorem to form right angles when they needed to measure a plot of land. They put equally spaced knots in pieces of rope so they could form a triangle that had a right angle by stretching the rope out on the ground. They used equally spaced knots to ensure that the lengths of the measuring ropes were in the ratio 3:4:5.

CRITICAL THINKING

In the Egyptian use of the Pythagorean theorem, if three pieces of rope were used and one piece had a total of 10 knots equally spaced and another piece had a total of 13 knots with the same equal spacing, how many knots with the same equal spacing would a third piece of rope need so that the three pieces form a right triangle with the first two pieces as the legs of the triangle? Draw a figure and explain your answer. (Assume that each piece of rope was knotted at both ends.)

EXERCISE SET 8.3

A *Find the perimeter of each figure. See Examples 1 through 6.*

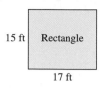 **1.**

15 ft | Rectangle

17 ft

2.
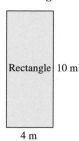

Rectangle | 10 m

4 m

3.
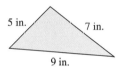

Square

9 cm

4.
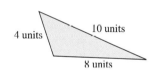

Square

46 mi

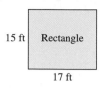 **5.**

5 in.   7 in.

9 in.

6.

4 units   10 units

8 units

7.
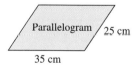

Parallelogram / 25 cm

35 cm

8.
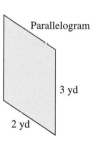

Parallelogram

3 yd

2 yd

9.
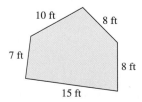

10 ft   8 ft

7 ft

8 ft

15 ft

10.
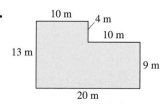

10 m   4 m

10 m

13 m

9 m

20 m

1. 64 ft

2. 28 m

3. 36 cm

4. 184 mi

5. 21 in.

6. 22 units

7. 120 cm

8. 10 yd

9. 48 ft

10. 66 m

11. 66 in.

12. 183 cm

13. 21 ft

14. 30 in.

15. 60 ft

16. 200 ft

17. 346 yd

18. 96 in.

19. 22 ft

20. 182 ft

21. $66

22. $364

492

Name _____

11.

12 in.

3 in.

12 in.

15 in.

24 in.

12.

30 cm   45 cm

25 cm   30 cm

53 cm

Solve. See Example 7.

13. A polygon has sides of length 5 ft, 3 ft, 2 ft, 7 ft, and 4 ft. Find its perimeter.

14. A triangle has sides of length 8 in., 12 in., and 10 in. Find its perimeter.

15. Baseboard is to be installed in a square room that measures 15 ft on one side. Find how much baseboard is needed.

16. Find how much fencing is needed to enclose a rectangular rose garden 85 ft by 15 ft.

17. If a football field is 53 yd wide and 120 yd long, what is the perimeter?

120 yd

53 yd

18. A stop sign has eight equal sides of length 12 in. Find its perimeter.

STOP

19. A metal strip is being installed around a workbench that is 8 ft long and 3 ft wide. Find how much stripping is needed.

20. Find how much fencing is needed to enclose a rectangular garden 70 ft by 21 ft.

21. If the stripping in Exercise 19 costs $3 per foot, find the total cost of the stripping.

22. If the fencing in Exercise 20 costs $2 per foot, find the total cost of the fencing.

Name _____

23. A regular hexagon has a side length of 6 in. Find its perimeter.

24. A regular pentagon has a side length of 14 m. Find its perimeter.

25. Find the perimeter of the top of a square compact disc case if the length of one side is 7 in.

26. Find the perimeter of a square ceramic tile with a side of length 5 in.

27. A rectangular room measures 6 ft by 8 ft. Find the cost of installing a strip of wallpaper around the room if the wallpaper costs $0.86 per foot.

28. A rectangular house measures 75 ft by 60 ft. Find the cost of installing gutters around the house if the cost is $2.36 per foot.

Find the perimeter of each figure. See Example 6.

29.

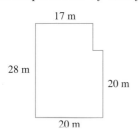

17 m
28 m
20 m
20 m

30.

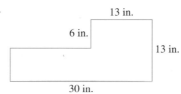

13 in.
6 in.
13 in.
30 in.

31.

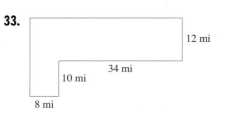

3 ft
4 ft
5 ft
6 ft
15 ft

32.

16 cm
2 cm
11 cm
4 cm
3 cm
9 cm

33.

12 mi
34 mi
10 mi
8 mi

34.

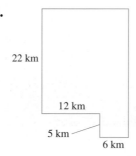

22 km
12 km
5 km
6 km

24. 70 m

25. 28 in.

26. 20 in.

27. $24.08

28. $637.20

29. 96 m

30. 86 in.

31. 66 ft

32. 58 cm

33. 128 mi

34. 90 km

35. 17π cm; 53.38 cm

36. 12π in.; 37.68 in.

37. 16π mi.; 50.24 mi

38. 50π ft.; 157 ft

39. 26π m; 81.64 m

40. 20π yd.; 62.8 yd

41. 31.43 ft

42. 251.2 m

43. 12,560 ft

44. 17.29 in.

494

B *Find the circumference of each circle. Give the exact circumference and then an approximation. Use $\pi \approx 3.14$. See Example 8.*

35.

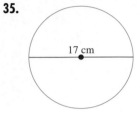
17 cm

36.

6 in.

37.

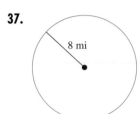
8 mi

38.

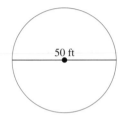
50 ft

39.

26 m

40.

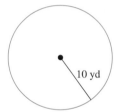
10 yd

41. A circular fountain has a radius of 5 ft. Approximate the distance around the fountain. Use $\frac{22}{7}$ for π.

42. A circular walkway has a radius of 40 m. Approximate the distance around the walkway. Use 3.14 for π.

43. Meteor Crater, near Winslow, Arizona is 4000 ft in diameter. Approximate the distance around the crater. Use 3.14 for π. (*Source: The Handy Science Answer Book*, 1994)

44. The largest pearl, the *Pearl of Lao-tze*, has a diameter of $5\frac{1}{2}$ in. Approximate the distance around the pearl. Use $\frac{22}{7}$ for π. (*Source: The Guinness Book of Records*, 1996, 1998)

Name _____

REVIEW AND PREVIEW

Simplify. See Section 1.8.

45. $5 + 6 \cdot 3$

46. $25 - 3 \cdot 7$

47. $(20 - 16) \div 4$

48. $6 \cdot (8 + 2)$

49. $(18 + 8) - (12 + 4)$

50. $72 \div (2 \cdot 6)$

51. $(72 \div 2) \cdot 6$

52. $4^1 \cdot (2^3 - 8)$

 COMBINING CONCEPTS

Recall from Section 1.5 that area measures the amount of surface of a region. Given the following situations, tell whether you are more likely to be concerned with area or perimeter.

53. ordering fencing to fence a yard

54. ordering grass seed to plant in a yard

55. buying carpet to install in a room

56. buying gutters to install on a house

57. ordering paint to paint a wall

58. ordering baseboards to install in a room

59. buying a wallpaper border to go on the walls around a room

60. buying fertilizer for your yard

45. 23

46. 4

47. 1

48. 60

49. 10

50. 6

51. 216

52. 0

53. perimeter

54. area

55. area

56. perimeter

57. area

58. perimeter

59. perimeter

60. area

61. a. 62.8 m; 125.6 m

61. a. Find the circumference of each circle. Approximate the circumference by using 3.14 for π.

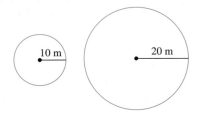

b. If the radius of a circle is doubled, is its corresponding circumference doubled?

62. a. Find the circumference of each circle. Approximate the circumference by using 3.14 for π.

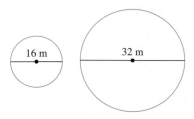

b. If the diameter of a circle is doubled, is its corresponding circumference doubled?

b. yes

62. a. 50.24 m; 100.48 m

63. Find the perimeter of the skating rink.

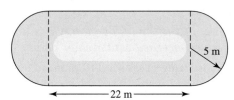

b. yes

63. $44 + 10\pi \approx 75.4$ m

8.4 AREA

A FINDING AREA

Recall that area measures the amount of surface of a region. Thus far, we know how to find the area of a rectangle and a square. These formulas, as well as formulas for finding the areas of other common geometric figures, are given next.

AREA FORMULAS OF COMMON GEOMETRIC FIGURES

Geometric Figure	Area Formula
RECTANGLE 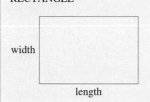	Area of a rectangle: **Area = length · width** $A = lw$
SQUARE 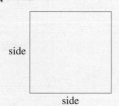	Area of a square: **Area = side · side** $A = s \cdot s = s^2$
TRIANGLE 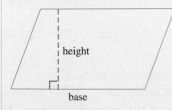	Area of a triangle: **Area $= \dfrac{1}{2} \cdot$ base $\cdot$ height** $A = \dfrac{1}{2} \cdot b \cdot h$
PARALLELOGRAM 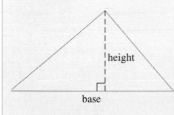	Area of a parallelogram: **Area = base · height** $A = b \cdot h$
TRAPEZOID 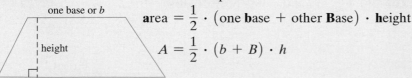	Area of a trapezoid: **area $= \dfrac{1}{2} \cdot ($ one base $+$ other Base$) \cdot$ height** $A = \dfrac{1}{2} \cdot (b + B) \cdot h$

TEACHING TIP

Begin this lesson by asking students to list some everyday situations in which they would need to calculate the area. (Painting a room, carpeting a room, tiling a patio)

Use these formulas for the following examples.

HELPFUL HINT

Area is always measured in square units.

Practice Problem 1

Find the area of the triangle.

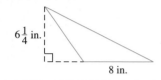

Example 1   Find the area of the triangle.

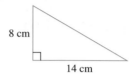

Solution: $A = \dfrac{1}{2} \cdot b \cdot h$

$= \dfrac{1}{2} \cdot 14\,\text{cm} \cdot 8\,\text{cm}$

$= \dfrac{\overset{1}{\cancel{2}} \cdot 7 \cdot 8}{\underset{1}{\cancel{2}}}$ square centimeters

$= 56$ square centimeters

The area is 56 square centimeters.

Practice Problem 2

Find the area of the square.

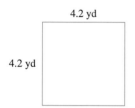

Example 2   Find the area of the parallelogram.

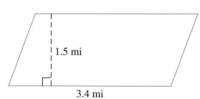

Solution: $A = b \cdot h$

$= 3.4\ \text{miles} \cdot 1.5\ \text{miles}$

$= 5.1$ square miles

The area is 5.1 square miles.

HELPFUL HINT

When finding the area of figures, be sure all measurements are changed to the same unit before calculations are made.

Practice Problem 3

Find the area of the figure.

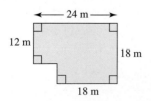

Example 3   Find the area of the figure.

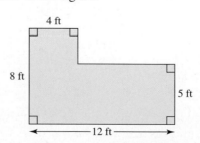

Answers

1. 25 sq. in.,   **2.** 17.64 sq. yd,   **3.** 396 sq. m

Solution: Split the figure into two rectangles. To find the area of the figure, we find the sum of the areas of the two rectangles.

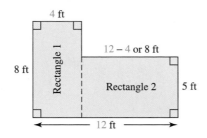

Area of Rectangle 1 $= l \cdot w$

$\qquad\qquad\qquad = 8 \text{ feet} \cdot 4 \text{ feet}$

$\qquad\qquad\qquad = 32 \text{ square feet}$

Notice that the length of Rectangle 2 is 12 feet $-$ 4 feet or 8 feet.

Area of Rectangle 2 $= l \cdot w$

$\qquad\qquad\qquad = 8 \text{ feet} \cdot 5 \text{ feet}$

$\qquad\qquad\qquad = 40 \text{ square feet}$

Area of the Figure $=$ Area of Rectangle 1 $+$ Area of Rectangle 2

$\qquad\qquad\qquad = 32 \text{ square feet} + 40 \text{ square feet}$

$\qquad\qquad\qquad = 72 \text{ square feet}$

HELPFUL HINT

The figure in Example 3 can also be split into two rectangles as shown.

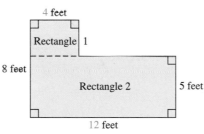

To better understand the formula for area of a circle, try the following. Cut a circle into many pieces as shown.

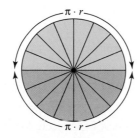

The circumference of a circle is $2 \cdot \pi \cdot r$. This means that the circumference of half a circle is half of $2 \cdot \pi \cdot r$, or $\pi \cdot r$.

Then unfold the two halves of the circle and place them together as shown.

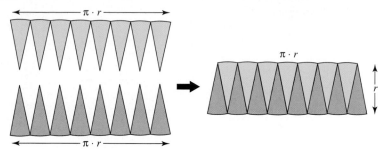

The figure on the right is almost a parallelogram with a base of $\pi \cdot r$ and a height of r. The area is

$$A = \boxed{\text{base}} \cdot \boxed{\text{height}}$$
$$= (\pi \cdot r) \cdot r$$
$$= \pi \cdot r^2$$

This is the formula for area of a circle.

TEACHING TIP   Classroom Activity

Before discussing Example 4, have students work in groups to estimate the number of square feet in a circle with radius 3 feet. You may want to hand out a scaled drawing of the circle on 1/4 inch graph paper where 1 foot is scaled to 1 inch. Have them first use the drawing to give an upper bound finding the area of the 6′ × 6′ square which circumscribes the circle. Then have them give a better estimate of the number of square feet (not the number of 1/4 foot squares!) contained in the circle. Have each group share its estimate and the way it arrived at its estimate. Then compare their results with the answer.

> **AREA FORMULA OF A CIRCLE**
>
> Circle
>
>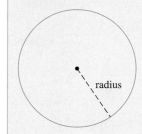
>
> Area of a circle:
> **Area** $= \pi \cdot (\text{radius})^2$
> $A = \pi \cdot r^2$
> (A fraction approximation for π is $\frac{22}{7}$.)
> (A decimal approximation for π is 3.14.)

Practice Problem 4

Find the area of the given circle. Find the exact area and an approximation. Use 3.14 as an approximation for π.

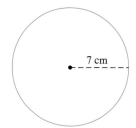

7 cm

✓ CONCEPT CHECK

Use estimation to decide which figure would have a larger area: a circle of diameter 10 in. or a square 10 in. long on each side.

Answers

4. 49π sq. cm $\approx$ 153.86 sq. cm

✓ **Concept Check:** a square 10 in. long on each side

Example 4   Find the area of a circle with a radius of 3 feet. Find the exact area and an approximation. Use 3.14 as an approximation for π.

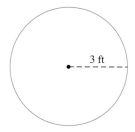

3 ft

Solution: We let $r = 3$ feet and use the formula

$$A = \pi \cdot r^2$$
$$= \pi \cdot (3 \text{ feet})^2$$
$$= 9 \cdot \pi \text{ square feet}$$

To approximate this area, we substitute 3.14 for π.

$$9 \cdot \pi \text{ square feet} \approx 9 \cdot 3.14 \text{ square feet}$$
$$= 28.26 \text{ square feet}$$

The *exact* area of the circle is 9π square feet, which is *approximately* 28.26 square feet.

TRY THE CONCEPT CHECK IN THE MARGIN.

Name _____ **Section** _____ **Date** _____

EXERCISE SET 8.4

A *Find the area of the geometric figure. If the figure is a circle, give an exact area and then use the given **approximation** for π to approximate the area. See Examples 1 through 4.*

 1.
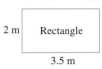
2 m | Rectangle
3.5 m

2.
2.75 ft | Rectangle
7 ft

 3.
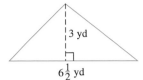
3 yd
6½ yd

4.
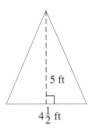
5 ft
4½ ft

5.
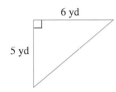
6 yd
5 yd

6.
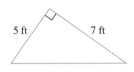
5 ft   7 ft

 7. Use 3.14 for π.
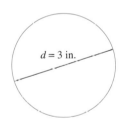
d = 3 in.

8. Use $\frac{22}{7}$ for π.

r = 2 cm

9.
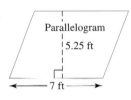
Parallelogram
5.25 ft
7 ft

10.
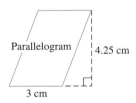
Parallelogram
4.25 cm
3 cm

11.
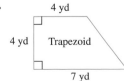
5 m
Trapezoid
4 m
9 m

12.
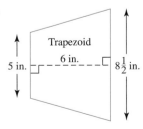
Trapezoid
6 in.
5 in.   8½ in.

13.
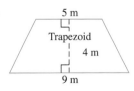
4 yd
4 yd   Trapezoid
7 yd

ANSWERS

1. 7 sq. m

2. 19.25 sq. ft

3. $9\frac{3}{4}$ sq. yd

4. $11\frac{1}{4}$ sq. ft

5. 15 sq. yd

6. 17.5 sq. ft

7. 2.25π sq. in. ≈ 7.065 sq. in.

8. 4π sq. cm ≈ $12\frac{4}{7}$ sq. cm

9. 36.75 sq. ft

10. 12.75 sq. cm

11. 28 sq. m

12. $40\frac{1}{2}$ sq. in.

13. 22 sq. yd

14. 22.5 sq. ft

15. $36\frac{3}{4}$ sq. ft

16. $12\frac{3}{4}$ sq. cm

17. $22\frac{1}{2}$ sq. in.

18. 24 sq. m

19. 25 sq. cm

20. 82 sq. km

21. 86 sq. mi

22. 360 sq. cm

502

14.

10 ft

3 ft Trapezoid

5 ft

15.

7 ft

Parallelogram

$5\frac{1}{4}$ ft

16.

Parallelogram $4\frac{1}{4}$ cm

3 cm

17.

$4\frac{1}{2}$ in. Parallelogram

5 in.

18.

4 m 6 m

Parallelogram

19.

2 cm

$1\frac{1}{2}$ cm $1\frac{1}{2}$ cm

3 cm

7 cm

20.

6 km

4 km

5 km

10 km

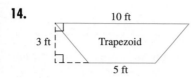

21.

5 mi

10 mi

3 mi

17 mi

22.

25 cm

15 cm

12 cm

5 cm

23.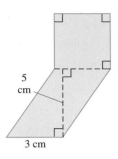

5 cm

3 cm

24.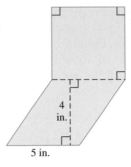

4 in.

5 in.

25. Use $\frac{22}{7}$ for π.

$r = 6$ in.

26. Use 3.14 for π.

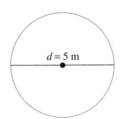

$d = 5$ m

Solve. See Examples 1 through 4.

27. A $10\frac{1}{2}$-foot by 16-foot concrete wall is to be built using concrete blocks. Find the area of the wall.

28 The floor of Terry's attic is 24 ft by 35 ft. Find how many square feet of insulation are needed to cover the attic floor.

29. Reebok International Ltd. of Massachusetts flew the largest banner from a single-seater plane in 1990. Use the diagram to find its area. (*Source: The Guinness Book of Records*, 1988)

← 100 ft →

Reebok Totally Beachin' 50 ft

30. The longest illuminated sign is in Ramat Gan, Israel and measures 197 ft by 66 ft. Find its area. (*Source: The Guinness Book of Records*, 1998)

66 ft

← 197 ft →

23. 24 sq. cm

24. 45 sq. in.

25. 36π sq. in. ≈ 113.1 sq. in.

26. 6.25π sq. m ≈ 19.625 sq. m

27. 168 sq. ft

28. 840 sq. ft

29. 5000 sq. ft

30. 13,002 sq. ft

31. 128 sq. in.; $\frac{8}{9}$ sq. ft

31. One side of a concrete block measures 8 in. by 16 in. Find the area of the side in square inches. Find the area in square feet (144 sq. in. = 1 sq. ft).

32. A standard *double* roll of wall paper is $6\frac{5}{6}$ ft wide and 33 ft long. Find the area of the *double* roll.

32. $225\frac{1}{2}$ sq. ft

33. 510 sq. in.

33. A picture frame measures 20 in. by $25\frac{1}{2}$ in. Find how many square inches of glass the frame requires.

34. A mat to go under a table cloth is made to fit a round dining table with a 4-foot diameter. Approximate how many square feet of mat there are. Use 3.14 as an approximation for π.

34. 12.56 sq. ft

35. A drapery panel measures 6 ft by 7 ft. Find how many square feet of material are needed for *four* panels.

36. A page in this book measures 27.5 cm by 20.5 cm. Find its area.

35. 168 sq. ft

37. Find how many square feet of land are in the following plot.

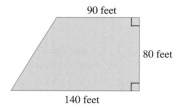

36. 563.75 sq. cm

38. For Gerald Gomez to determine how much grass seed he needs to buy, he must know the size of his yard. Use the drawing to determine how many square feet are in his yard.

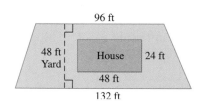

37. 9200 sq. ft

39. The shaded part of the roof shown is in the shape of a trapezoid and needs to be shingled. The number of packages of shingles to buy depends on the area. Use the dimensions given to find the area of the shaded part of the roof to the nearest whole square foot.

38. 4320 sq. ft

40. The end of the building shaded in the drawing is to be bricked. The number of bricks to buy depends on the area. Find the area.

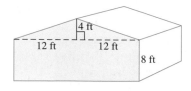

39. 381 sq. ft

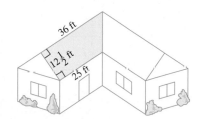

40. 240 sq. ft

504

Name _____

REVIEW AND PREVIEW

Find the perimeter or circumference of each geometric figure. See Section 8.3.

41.

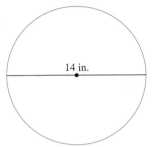

Use 3.14 for π.

42.

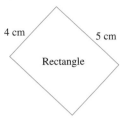

4 cm   5 cm

Rectangle

43.

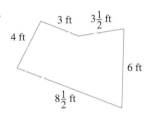

3 ft $3\frac{1}{2}$ ft

4 ft

6 ft

$8\frac{1}{2}$ ft

44.

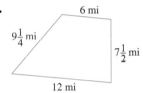

6 mi

$9\frac{1}{4}$ mi

$7\frac{1}{2}$ mi

12 mi

45.

$2\frac{1}{8}$ ft

Regular hexagon

46.

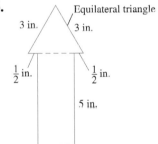

Equilateral triangle

3 in.   3 in.

$\frac{1}{2}$ in. $\frac{1}{2}$ in.

5 in.

COMBINING CONCEPTS

47. A pizza restaurant recently advertised two specials. The first special was a 12-inch pizza for $10. The second special was two 8-inch pizzas for $9. Determine the better buy. (*Hint:* First compare the areas of the two specials and then find a price per square inch for both specials.)

48. Find the approximate area of the state of Utah.

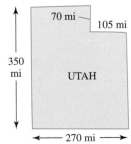

70 mi   105 mi

350 mi

UTAH

270 mi

49. Find the area of a rectangle that measures 2 *feet* by 8 *inches*. Give the area in square feet and in square inches.

Internet Excursions

Go to http://www.prenhall.com/martin-gay
The World Wide Web address listed above will direct you to the official Web site of the U.S. Department of Defense and the Pentagon in Washington, D.C., or a related site. You will find information that helps you complete the questions below.

50. Visit this Website and locate information on the length of each outer wall of the Pentagon. Then calculate the outer perimeter of the Pentagon.

?

51. Notice that each of the five outer walls of the Pentagon is in the shape of a rectangle. Visit this Website and locate information on the height of the building. Then use this height information along with the length information found in Exercise 50 to calculate the area of each outer wall of the Pentagon.

8.5 VOLUME

A FINDING VOLUME

Volume is a measure of the space of a region. The volume of a box or can, for example, is the amount of space inside. Volume can be used to describe the amount of juice in a pitcher or the amount of concrete needed to pour a foundation for a house.

The volume of a solid is the number of **cubic units** in the solid. A cubic centimeter and a cubic inch are illustrated.

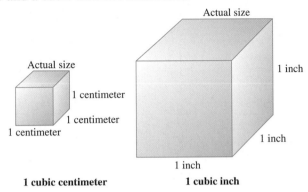

Actual size

1 centimeter
1 centimeter
1 centimeter

1 cubic centimeter

Actual size

1 inch
1 inch
1 inch

1 cubic inch

Formulas for finding the volumes of some common solids are given next.

VOLUME FORMULAS OF COMMON SOLIDS

Solid

Rectangular Solid

height
width
length

Volume Formulas

Volume of a rectangular solid:
Volume = length · width · height
$$V = l \cdot w \cdot h$$

CUBE

side
side
side

Volume of a cube:
Volume = side · side · side
$$V = s^3$$

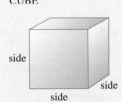

SPHERE

radius

Volume of a sphere:
Volume = $\frac{4}{3} \cdot \pi \cdot (\textbf{radius})^3$

$$V = \frac{4}{3} \cdot \pi \cdot r^3$$

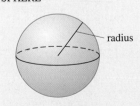

CIRCULAR CYLINDER

height
radius

Volume of a circular cylinder:
Volume = $\pi \cdot (\textbf{radius})^2 \cdot \textbf{height}$
$$V = \pi \cdot r^2 \cdot h$$

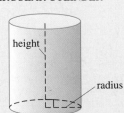

Ojective

A Find the volume of solids.

SSM   CD-ROM   Video 8.5

TEACHING TIP

Point out that the volume of a rectangular solid, a cube and a circular cylinder are easy to remember because their volumes are just the product of the area of their bases and their height.

VOLUME FORMULAS OF COMMON SOLIDS

Solid	Volume Formulas
CONE	Volume of a cone:

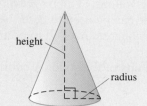

$$\textbf{Volume} = \frac{1}{3} \cdot \pi \cdot (\textbf{radius})^2 \cdot \textbf{height}$$

$$V = \frac{1}{3} \cdot \pi \cdot r^2 \cdot h$$

SQUARE-BASED PYRAMID

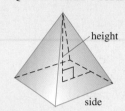

Volume of a square-based pyramid:

$$\textbf{Volume} = \frac{1}{3} \cdot (\textbf{side})^2 \cdot \textbf{height}$$

$$V = \frac{1}{3} \cdot s^2 \cdot h$$

Practice Problem 1

Draw a diagram and find the volume of a rectangular box that is 5 ft long, 2 ft wide, and 4 ft deep.

✓ CONCEPT CHECK

Juan is calculating the volume of the following rectangular solid. Find the error in his calculation.

$$\text{Volume} = l + w + h$$
$$= 14 + 8 + 5$$
$$= 27 \text{ cubic cm}$$

Practice Problem 2

Draw a diagram and approximate the volume of a ball of radius $\frac{1}{2}$ cm. Use $\frac{22}{7}$ for π.

Answers

1. 40 cu. ft,   **2.** $\frac{11}{21}$ cu. cm

✓ Concept Check

$$\text{Volume} = l \cdot w \cdot h$$
$$= 14 \cdot 8 \cdot 5$$
$$= 560 \text{ cu. cm}$$

HELPFUL HINT

Volume is always measured in cubic units.

Example 1   Find the volume of a rectangular box that is 12 inches long, 6 inches wide, and 3 inches high.

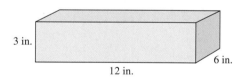

Solution: $V = l \cdot w \cdot h$

$V = 12$ inches $\cdot$ 6 inches $\cdot$ 3 inches $= 216$ cubic inches

The volume of the rectangular box is 216 cubic inches.

TRY THE CONCEPT CHECK IN THE MARGIN.

Example 2   Approximate the volume of a ball of radius 3 inches.

Use the approximation $\frac{22}{7}$ for π.

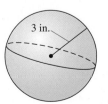

Solution: $V = \dfrac{4}{3} \cdot \pi \cdot r^3$

$\approx \dfrac{4}{3} \cdot \dfrac{22}{7} (3 \text{ inches})^3$

$= \dfrac{4}{3} \cdot \dfrac{22}{7} \cdot 27 \text{ cubic inches}$

$= \dfrac{4 \cdot 22 \cdot \overset{1}{\cancel{3}} \cdot 9}{\underset{1}{\cancel{3}} \cdot 7} \text{ cubic inches}$

$= \dfrac{792}{7} \quad \text{ or } \quad 113\dfrac{1}{7} \text{ cubic inches}$

The volume is *approximately* $113\dfrac{1}{7}$ cubic inches. ▬▬▬

Example 3   Approximate the volume of a can that has a $3\dfrac{1}{2}$-inch radius and a height of 6 inches. Use $\dfrac{22}{7}$ for π.

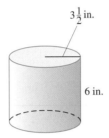

$3\dfrac{1}{2}$ in.

6 in.

Solution: Using the formula for a circular cylinder, we have

$V = \pi \cdot r^2 \cdot h$

$\qquad \qquad 3\dfrac{1}{2} = \dfrac{7}{2}$

$= \pi \cdot \left(\dfrac{7}{2} \text{ inches} \right)^2 \cdot 6 \text{ inches}$

or approximately

$\approx \dfrac{22}{7} \cdot \dfrac{49}{4} \cdot 6 \text{ cubic inches}$

$= 231 \text{ cubic inches}$

The volume is approximately 231 cubic inches. ▬▬▬

Practice Problem 3

Approximate the volume of a cylinder of radius 5 in. and height 7 in. Use 3.14 for π.

TEACHING TIP   Classroom Activity

Have students work in groups to create a sketch of a solid which combines 2 or more solids. Then have them find the volume of their solid. Finally have them exchange the sketch of their solid with another group to find the volume.

Answer

3. 549.5 cu. in.

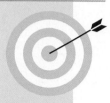

Focus On the Real World

THE COST OF ROAD SIGNS

There are nearly 4 million miles of streets and roads in the United States. With streets, roads, and highways comes the need for traffic control, guidance, warning, and regulation. Road signs perform many of these tasks. Just in our routine travels, we see a wide variety of road signs every day. Think how many road signs must exist on the 4 million miles of roads in the United States. Have you ever wondered how much signs like these cost?

The cost of a road sign generally depends on the type of sign. Costs for several types of signs and sign posts are listed in the table. Examples of various types of signs are shown below.

ROAD SIGN COSTS	
Type of Sign	Cost
Regulatory, warning, marker	$15–$18 per square foot
Large guide	$20–$25 per square foot
Type of Post	Cost
U-channel	$125–$200 each
Square tube	$10–$15 per foot
Steel breakaway posts	$15–$25 per foot

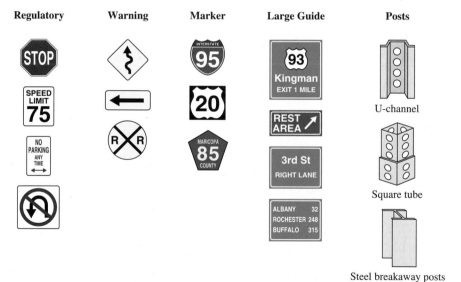

The cost of a sign is based on its area. For diamond, square, or rectangular signs, the area is found by multiplying the length (in feet) times the width (in feet). Then the area is multiplied by the cost per square foot. For signs with irregular shapes, costs are generally figured *as if* the sign were a rectangle, multiplying the height and width at the tallest and widest parts of the sign.

GROUP ACTIVITY

Locate four different kinds of road signs on or near your campus. Measure the dimensions of each sign, including the height of the post on which it is mounted. Using the cost data given in the table, find the minimum and maximum cost of each sign, including its post. Summarize your results in a table, and include a sketch of each sign.

Name _____ **Section** _____ **Date** _____

EXERCISE SET 8.5

A *Find the volume of each solid. See Examples 1 through 3. Use* $\frac{22}{7}$ *for* π.

1.

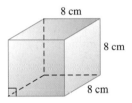

4 in.

3 in.

6 in.

2.

3 mi

3.

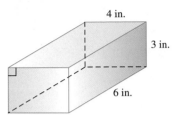

8 cm

8 cm

8 cm

4.

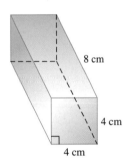

8 cm

4 cm

4 cm

5.

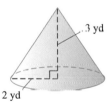

.3 yd

2 yd

6.

10 ft

6 ft

7.

10 in.

8.

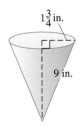

$1\frac{3}{4}$ in.

9 in.

9. 75 cu. cm

10. 18 cu. ft

11. $2\frac{10}{27}$ cu. in.

12. 3080 cu. ft

13. 8.4 cu. ft

14. 125 cu. ft

15. $10\frac{5}{6}$ cu. in.

16. 33.49 cu. cm

9.

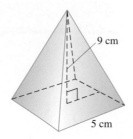

9 cm

5 cm

10.

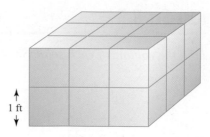

1 ft

Solve.

11. Find the volume of a cube with edges of $1\frac{1}{3}$ inches.

$1\frac{1}{3}$ in.

12. A water storage tank is in the shape of a cone with the pointed end down. If the radius is 14 ft and the depth of the tank is 15 ft, approximate the volume of the tank in cubic feet. Use $\frac{22}{7}$ for π.

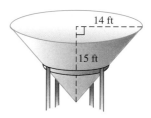

14 ft

15 ft

13. Find the volume of a rectangular box 2 ft by 1.4 ft by 3 ft.

14. Find the volume of a box in the shape of a cube that is 5 ft on each side.

15. Find the volume of a pyramid with a square base 5 in. on a side and a height of 1.3 in.

16. Approximate to the nearest hundredth the volume of a sphere with a radius of 2 cm. Use 3.14 for π.

17. A paperweight is in the shape of a square-based pyramid 20 cm tall. If an edge of the base is 12 cm, find the volume of the paperweight.

18. A bird bath is made in the shape of a hemisphere (half-sphere). If its radius is 10 in., approximate the volume. Use $\frac{22}{7}$ for π.

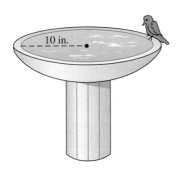

19. Find the exact volume of a sphere with a radius of 7 in.

20. A tank is in the shape of a cylinder 8 ft tall and 3 ft in radius. Find the exact volume of the tank.

21. Find the volume of a rectangular block of ice 2 ft by $2\frac{1}{2}$ ft by $1\frac{1}{2}$ feet.

22. Find the capacity (volume in cubic feet) of a rectangular ice chest with inside measurements of 3 ft by $1\frac{1}{2}$ ft by $1\frac{3}{4}$ ft.

23. An ice cream cone with a 4-cm diameter and 3-cm depth is filled exactly level with the top of the cone. Approximate how much ice cream (in cubic centimeters) is in the cone. Use $\frac{22}{7}$ for π.

24. A child's toy is in the shape of a square-based pyramid 10 in. tall. If an edge of the base is 7 in., find the volume of the toy.

REVIEW AND PREVIEW

Evaluate. See Section 1.8.

25. 5^2 **26.** 7^2 **27.** 3^2 **28.** 20^2

29. $1^2 + 2^2$ **30.** $5^2 + 3^2$ **31.** $4^2 + 2^2$ **32.** $1^2 + 6^2$

---

17. 960 cu. cm

18. $2095\frac{5}{21}$ cu. in.

19. $\frac{1372}{3}\pi$ cu. in. or $\left(457\frac{1}{3}\right)\pi$ cu. in.

20. 72π cu. ft

21. $7\frac{1}{2}$ cu. ft

22. $7\frac{7}{8}$ cu. ft

23. $12\frac{4}{7}$ cu. cm

24. $163\frac{1}{3}$ cu. in.

25. 25

26. 49

27. 9

28. 400

29. 5

30. 34

31. 20

32. 37

◢ **COMBINING CONCEPTS**

33. 2,583,283 cu. m

33. The Great Pyramid of Khufu at Giza is the largest of the ancient Egyptian pyramids. Its original height was 146.5 meters. The length of each side of its square base was originally 230 meters. Find the volume of the Great Pyramid of Khufu as it was origiinally built. Round to the nearest whole cubic meter. (*Source*: PBS *NOVA* Online)

34. 77,811,712 cu. ft

34. The second largest pyramid at Giza is Khafre's Pyramid. Its original height was 471 feet. The length of each side of its square base was originally 704 feet. Find the volume of Khafre's Pyramid as it was originally built. (*Source*: PBS *NOVA* Online)

35. 2,583,669 cu. m

35. Menkaure's Pyramid, the smallest of the three great pyramids at Giza, was originally 65.5 meters tall. Each of the sides of its square base was originally 344 meters long. What was the volume of Menkaure's Pyramid as it was originally built? Round to the nearest whole cubic meter. (*Source*: PBS *NOVA* Online)

36. 230,125 cu. m

36. Due to factors such as weathering and loss of outer stones, the Great Pyramid of Khufu now stands only 137 meters tall. Its square base is now only 227 meters on a side. Find the current volume of the Great Pyramid of Khufu to the nearest whole cubic meter. How much has its volume decreased since it was built? See Exercise 33 for comparison. (*Source*: PBS *NOVA* Online)

37. 26,696.5 cu. ft

37. The centerpiece of the New England Aquarium in Boston is its Giant Ocean Tank. This exhibit is a four-story cylindrical saltwater tank containing sharks, sea turtles, stingrays, and tropical fish. The radius of the tank is 16.3 feet and its height is 32 feet (assuming that a story is 8 feet). What is the volume of the Giant Ocean Tank? Use $\pi \approx 3.14$ and round to the nearest tenth of a cubic foot. (*Source*: New England Aquarium)

38. a. 5 cu. ft

b. 1 cu. ft; (a) is larger

38. Except for service dogs for guests with disabilities, Walt Disney World does not allow pets in its parks or hotels. However, the resort does make pet-boarding services available to guests. The pet-care kennels at Walt Disney World offer three different sizes of indoor kennels. Of these, the smaller two kennels measure (a) 2'1″ × 1'8″ × 1'7″ and (b) 1'1″ × 2' × 8″. What is the volume of each kennel rounded to the nearest cubic foot? Which is larger? (*Source*: Walt Disney World Resort)

Name _____ **Section** _____ **Date** _____

INTEGRATED REVIEW — GEOMETRY CONCEPTS

1. Find the supplement and the complement of a 27° angle.

Find the measures of angles x, y, and z in each figure.

2.

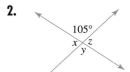

105°

x z
y

3. $m \parallel n$.

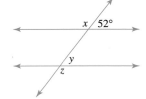

x / 52°

y
z

2. $\angle x = 75°$;
$\angle y = 105°$; $\angle z = 75°$

3. $\angle x = 128°$;
$\angle y = 52°$; $\angle z = 128°$

4. Find the measure of $\angle x$.

x

38°

4. $\angle x = 52°$

*Find the perimeter (or circumference) and area of each figure. For the circle give an exact cir-
cumference and area. Then use $\pi \approx 3.14$ to approximate each. Don't forget to attach correct
units.*

5.

Square | 5 m

5. 20 m; 25 sq. m

6.

4 ft

3 ft

5 ft

6. 12 ft; 6 sq. ft

7.

3 cm

8.

11 mi

Parallelogram / 5 mi
4 mi

7. 6π cm ≈ 18.84 cm;
9π sq. cm ≈ 28.26
sq. cm

8. 32 mi; 44 sq. mi

9. The smallest cathedral is in High-landville, Missouri, and measures 14 feet by 17 feet. Find its perimeter and its area. (*Source: The Guinness Book of Records*, 1998)

Find the volume of each solid. Don't forget to attach correct units.

10. A cube with edges of 4 inches each.

11. A rectangular box 2 feet by 3 feet by 5.1 feet.

12. A pyramid with a square base 10 centimeters on a side and a height of 12 centimeters.

13. A sphere with a diameter of 16 miles. Give the exact volume and then use $\pi \approx \frac{22}{7}$ to approximate.

8.6   SQUARE ROOTS AND THE PYTHAGOREAN THEOREM

A FINDING SQUARE ROOTS

The square of a number is the number times itself. For example,

The square of 5 is 25 because 5^2 or $5 \cdot 5 = 25$.
The square of 3 is 9 because 3^2 or $3 \cdot 3 = 9$.
The square of 10 is 10^2 or $10 \cdot 10 = 100$.

The reverse process of squaring is finding a **square root**. For example,

A square root of 9 is 3 because $3^2 = 9$.
A square root of 25 is 5 because $5^2 = 25$.
A square root of 100 is 10 because $10^2 = 100$.

We use the symbol $\sqrt{}$, called a **radical sign**, to name square roots. For example,

$\sqrt{9} = 3$ because $3^2 = 9$.
$\sqrt{25} = 5$ because $5^2 = 25$.

SQUARE ROOT OF A NUMBER

A square root of a number a is a number b whose square is a. We use the radical sign $\sqrt{}$ to name square roots.

Example 1   Find each square root.

   a. $\sqrt{49}$ **b.** $\sqrt{36}$ **c.** $\sqrt{1}$ **d.** $\sqrt{81}$

Solution: **a.** $\sqrt{49} = 7$ because $7^2 = 49$.
   b. $\sqrt{36} = 6$ because $6^2 = 36$.
   c. $\sqrt{1} = 1$ because $1^2 = 1$.
   d. $\sqrt{81} = 9$ because $9^2 = 81$.

Example 2   Find: $\sqrt{\dfrac{1}{36}}$

Solution: $\sqrt{\dfrac{1}{36}} = \dfrac{1}{6}$ because $\dfrac{1}{6} \cdot \dfrac{1}{6} = \dfrac{1}{36}$.

Example 3   Find: $\sqrt{\dfrac{4}{25}}$

Solution: $\sqrt{\dfrac{4}{25}} = \dfrac{2}{5}$ because $\dfrac{2}{5} \cdot \dfrac{2}{5} = \dfrac{4}{25}$.

B APPROXIMATING SQUARE ROOTS

Thus far, we have found square roots of perfect squares. Numbers like $\dfrac{1}{4}$, 36, $\dfrac{4}{25}$, and 1 are called **perfect squares** because their square root is a whole number or a fraction. A square root such as $\sqrt{5}$ cannot be written as a whole number or a fraction since 5 is not a perfect square.

   Although $\sqrt{5}$ cannot be written as a whole number or a fraction, it can be approximated by estimating, by using a table (as in Appendix E), or by using a calculator.

Objectives

A Find the square root of a number.
B Approximate square roots.
C Use the Pythagorean theorem.

SSM   CD-ROM   Video
8.6

Practice Problem 1

Find each square root.

a. $\sqrt{100}$

b. $\sqrt{64}$

c. $\sqrt{121}$

d. $\sqrt{0}$

Practice Problem 2

Find: $\sqrt{\dfrac{1}{4}}$

Practice Problem 3

Find: $\sqrt{\dfrac{9}{16}}$

Answers
1. a. 10, **b.** 8, **c.** 11, **d.** 0, **2.** $\dfrac{1}{2}$, **3.** $\dfrac{3}{4}$

Practice Problem 4

Use Appendix E or a calculator to approximate the square root of 11 to the nearest thousandth.

Practice Problem 5

Approximate $\sqrt{29}$ to the nearest thousandth.

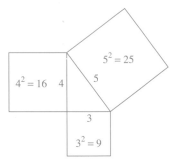
Example 4   Use Appendix E or a calculator to approximate the square root of 43 to the nearest thousandth.

Solution: $\sqrt{43} \approx 6.557$ ▬▬▬

> **HELPFUL HINT**
>
> $\sqrt{43}$ is *approximately* 6.557. This means that if we multiply 6.557 by 6.557, the product is *close* to 43.
>
> $6.557 \times 6.557 = 42.994249$

Example 5   Approximate $\sqrt{32}$ to the nearest thousandth.

Solution: $\sqrt{32} \approx 5.657$ ▬▬▬

C USING THE PYTHAGOREAN THEOREM

One important application of square roots has to do with right triangles. Recall that a **right triangle** is a triangle in which one of the angles is a right angle, or measures 90° (degrees). The **hypotenuse** of a right triangle is the side opposite the right angle. The **legs** of a right triangle are the other two sides. These are shown in the following figure. The right angle in the triangle is indicated by the small square drawn in that angle.

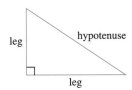

The following theorem is true for all right triangles.

> **PYTHAGOREAN THEOREM**
>
> In any **right triangle**,
>
> $$(\text{leg})^2 + (\text{other leg})^2 = (\text{hypotenuse})^2$$
>
>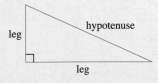

Using the Pythagorean theorem, we can use one of the following formulas to find an unknown length of a right triangle.

> **FINDING AN UNKNOWN LENGTH OF A RIGHT TRIANGLE**
>
> $$\text{hypotenuse} = \sqrt{(\text{leg})^2 + (\text{other leg})^2}$$
>
> or
>
> $$\text{leg} = \sqrt{(\text{hypotenuse})^2 - (\text{other leg})^2}$$

Answers
4. 3.317, **5.** 5.385

Example 6   Find the length of the hypotenuse of the given right triangle.

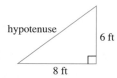

Practice Problem 6

Find the length of the hypotenuse of the given right triangle.

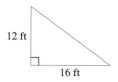

Solution: Since we are finding the hypotenuse, we use the formula

$$\text{hypotenuse} = \sqrt{(\text{leg})^2 + (\text{other leg})^2}$$

Putting the known values into the formula, we have

$$\text{hypotenuse} = \sqrt{(6)^2 + (8)^2}$$ Legs are 6 feet and 8 feet.
$$= \sqrt{36 + 64}$$
$$= \sqrt{100}$$
$$= 10$$

The hypotenuse is 10 feet long.

Example 7   Approximate the length of the hypotenuse of the given right triangle. Round the length to the nearest whole unit.

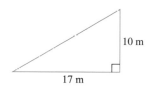

Practice Problem 7

Approximate the length of the hypotenuse of the given right triangle. Round to the nearest whole unit.

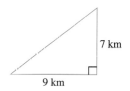

Solution: $$\text{hypotenuse} = \sqrt{(\text{leg})^2 + (\text{other leg})^2}$$
$$= \sqrt{(17)^2 + (10)^2}$$ The legs are 10 meters and 17 meters.
$$= \sqrt{289 + 100}$$
$$= \sqrt{389}$$ From Appendix E or a calculator
$$\approx 20$$

The hypotenuse is exactly $\sqrt{389}$ meters, which is approximately 20 meters.

Example 8   Find the length of the leg in the given right triangle. Give the exact length and a two-decimal-place approximation.

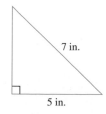

Practice Problem 8

Find the length of the leg in the given right triangle. Give the exact length and a two-decimal-place approximation.

Answers
6. 20 ft,   **7.** 11 km,   **8.** $\sqrt{72}$ ft $\approx$ 8.49 ft

✓ **CONCEPT CHECK**

The following lists are the lengths of the sides of two triangles. Which set forms a right triangle?
a. 8, 15, 17
b. 24, 30, 40

Practice Problem 9

A football field is a rectangle measuring 100 yards by 53 yards. Draw a diagram and find the length of the diagonal of a football field to the nearest yard.

Solution: Notice that the hypotenuse measures 7 inches and that the length of one leg measures 5 inches. Since we are looking for the length of the other leg, we use the formula

$$\text{leg} = \sqrt{(\text{hypotenuse})^2 - (\text{other leg})^2}$$

Putting the known values into the formula, we have

$$\text{leg} = \sqrt{(7)^2 - (5)^2}$$ The hypotenuse is 7 inches and the other leg is 5 inches.
$$= \sqrt{49 - 25}$$
$$= \sqrt{24}$$
$$\approx 4.90$$ From Appendix C or a calculator

The length of the leg is exactly $\sqrt{24}$ inches, which is approximately 4.90 inches.

TRY THE CONCEPT CHECK IN THE MARGIN.

Example 9   Finding the Diagonal Length of a City Block

A standard city block is a square that measures 300 feet on a side. Find the length of the diagonal of a city block rounded to the nearest whole foot.

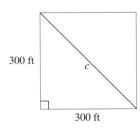

Solution: The diagonal is the hypotenuse of a right triangle, so we use the formula

$$\text{hypotenuse} = \sqrt{(\text{leg})^2 + (\text{other leg})^2}$$

Putting the known values into the formula, we have

$$\text{hypotenuse} = \sqrt{(300)^2 + (300)^2}$$ The legs are both 300 feet.
$$= \sqrt{90{,}000 + 90{,}000}$$
$$= \sqrt{180{,}000}$$
$$\approx 424$$ From Appendix E or a calculator

The length of the diagonal is approximately 424 feet.

Answers

9. 113 yd

✓ **Concept Check:**   **a.** $8^2 + 15^2 = 17^2$

CALCULATOR EXPLORATIONS
FINDING SQUARE ROOTS

To simplify or approximate square roots using a calculator, locate the key marked $\boxed{\sqrt{}}$.

To simplify $\sqrt{64}$, for example, press the keys

$\boxed{64}$ $\boxed{\sqrt{}}$ or $\boxed{\sqrt{}}$ $\boxed{64}$

The display should read $\boxed{8}$. Then

$\sqrt{64} = 8$

To *approximate* $\sqrt{10}$, press the keys

$\boxed{10}$ $\boxed{\sqrt{}}$ or $\boxed{\sqrt{}}$ $\boxed{10}$

The display should read $\boxed{3.16227766}$. This is an *approximation* for $\sqrt{10}$. A three-decimal-place approximation is:

$\sqrt{10} \approx 3.162$

Is this answer reasonable? Since 10 is between perfect squares 9 and 16, $\sqrt{10}$ is between $\sqrt{9} = 3$ and $\sqrt{16} = 4$. Our answer is reasonable since 3.162 is between 3 and 4.

Simplify.

1. $\sqrt{1024}$ 32

2. $\sqrt{676}$ 26

Approximate each square root. Round each answer to the nearest thousandth.

3. $\sqrt{15}$ 3.873

4. $\sqrt{19}$ 4.359

5. $\sqrt{97}$ 9.849

6. $\sqrt{56}$ 7.483

Focus On Business and Career

PRODUCT PACKAGING

Suppose you have just developed a new product that you would like to market. You will need to think about who would like to buy it, where and how it should be sold, for how much it will sell, how to package it, and other pressing concerns. Although all of these items are important to think through, many package designers believe that the packaging in which a product is sold is at least as important as the product itself.

Product packaging contains the product, keeping it from leaking out, keeping it fresh if perishable, providing protective cushioning against breakage, and keeping all the pieces together as a bundle. Product packaging also provides a way to give information about the product: what it is, how to use it, for whom it is designed, how it is beneficial or advantageous, who to contact if more information is needed, what other products are necessary to use with the product, etc. Product packaging must be pleasing and eyecatching to the product's audience. It must be capable of selling its contents without further assistance.

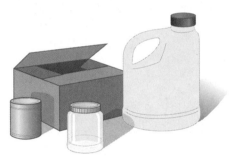

CRITICAL THINKING

1. How can a knowledge of geometry be helpful in the packaging design process?

2. Design two different packages for the same product that have roughly the same volume. Does one package "look" larger than the other? How could this be useful to a package designer?

Name _____ Section _____ Date _____

EXERCISE SET 8.6

A *Find each square root. See Examples 1 through 3.*

 1. $\sqrt{4}$ **2.** $\sqrt{9}$ **3.** $\sqrt{625}$ **4.** $\sqrt{16}$

 5. $\sqrt{\dfrac{1}{81}}$ **6.** $\sqrt{\dfrac{1}{64}}$ **7.** $\sqrt{\dfrac{144}{64}}$ **8.** $\sqrt{\dfrac{36}{81}}$

9. $\sqrt{256}$ **10.** $\sqrt{144}$ **11.** $\sqrt{\dfrac{9}{4}}$ **12.** $\sqrt{\dfrac{121}{169}}$

B *Use Appendix E or a calculator to approximate each square root. Round the square root to the nearest thousandth. See Examples 4 and 5.*

13. $\sqrt{3}$ **14.** $\sqrt{5}$ **15.** $\sqrt{15}$ **16.** $\sqrt{17}$

17. $\sqrt{14}$ **18.** $\sqrt{18}$ **19.** $\sqrt{47}$ **20.** $\sqrt{85}$

21. $\sqrt{8}$ **22.** $\sqrt{10}$ **23.** $\sqrt{26}$ **24.** $\sqrt{35}$

25. $\sqrt{71}$ **26.** $\sqrt{62}$ **27.** $\sqrt{7}$ **28.** $\sqrt{2}$

C *Find the unknown length of each right triangle. If necessary, approximate the length to the nearest thousandth. See Examples 6 through 8.*

29.
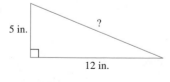
5 in.
?
12 in.

30.
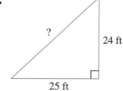
?
24 ft
25 ft

523

Name _____

31.

32.

Sketch each right triangle and find the length of the side not given. If necessary, approximate the length to the nearest thousandth. See Examples 6 through 8.

33. leg = 3, leg = 4

34. leg = 9, leg = 12

35. leg = 6, hypotenuse = 10

36. leg = 48, hypotenuse = 53

37. leg = 10, leg = 14

38. leg = 32, leg = 19

39. leg = 2, leg = 16

40. leg = 27, leg = 36

41. leg = 5, hypotenuse = 13

42. leg = 45, hypotenuse = 117

43. leg = 35, leg = 28

44. leg = 30, leg = 15

45. leg = 30, leg = 30

46. leg = 110, leg = 132

47. hypotenuse = 2, leg = 1

48. hypotenuse = 7, leg = 6

Name _____

Solve. See Example 9

49. A standard city block is a square with each side measuring 100 yards. Find the length of the diagonal of a city block to the nearest hundredth yard.

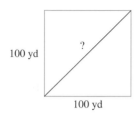

100 yd

? 100 yd

50. A section of land is a square with each side measuring 1 mile. Find the length of the diagonal of a section of land to the nearest thousandth mile.

? 1 mi

51. Find the height of the tree. Round the height to one decimal place.

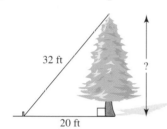

32 ft

?

20 ft

52. Find the height of the antenna. Round the height to one decimal place.

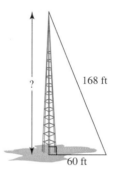

? 168 ft

60 ft

53. A football field is a rectangle that is 300 feet long by 160 feet wide. Find, to the nearest foot, the length of a straight-line run that started at one corner and went diagonally to end at the opposite corner.

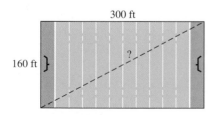

300 ft

160 ft

?

54. A baseball diamond is the shape of a square that has sides of length 90 feet. Find the distance across the diamond from third base to first base to the nearest tenth of a foot.

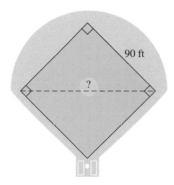

90 ft

?

Name _____

REVIEW AND PREVIEW

Find the value of n in each proportion. See Section 5.3.

55. $\dfrac{n}{6} = \dfrac{2}{3}$

56. $\dfrac{8}{n} = \dfrac{4}{8}$

57. $\dfrac{9}{11} = \dfrac{n}{55}$

58. $\dfrac{5}{6} = \dfrac{35}{n}$

59. $\dfrac{3}{n} = \dfrac{7}{14}$

60. $\dfrac{n}{9} = \dfrac{4}{6}$

◢ COMBINING CONCEPTS

Determine what two whole numbers each square root is between without using a calculator or table. Then use a calculator or table to check.

61. $\sqrt{38}$

62. $\sqrt{27}$

63. $\sqrt{101}$

64. $\sqrt{85}$

8.7 CONGRUENT AND SIMILAR TRIANGLES

A DECIDING WHETHER TRIANGLES ARE CONGRUENT

Two triangles are **congruent** if they have the same shape and the same size. In congruent triangles, the measures of corresponding angles are equal and the lengths of corresponding sides are equal. The following triangles are congruent.

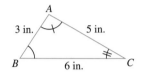 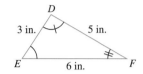

Since these triangles are congruent, the measures of corresponding angles are equal.

Angles with Equal Measure: $\angle A$ and $\angle D$, $\angle B$ and $\angle E$, $\angle C$ and $\angle F$

Also, the lengths of corresponding sides are equal.

Equal Corresponding Sides: $\overline{AB}$ and $\overline{DE}$, $\overline{BC}$ and $\overline{EF}$, $\overline{CA}$ and $\overline{FD}$

Any one of the following may be used to determine whether two triangles are congruent.

CONGRUENT TRIANGLES

Angle-Side-Angle (ASA)

If the measures of two angles of a triangle equal the measures of two angles of another triangle and the lengths of the sides between each pair of angles are equal, the triangles are congruent.

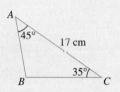 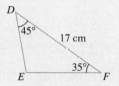

For example, these two triangles are congruent by Angle-Side-Angle.

Side-Side-Side (SSS)

If the lengths of the three sides of a triangle equal the lengths of the corresponding sides of another triangle, the triangles are congruent.

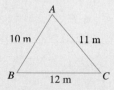 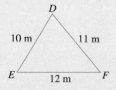

For example, these two triangles are congruent by Side-Side-Side.

TEACHING TIP   Classroom Activity

After discussing congruent triangles, have students work individually to draw 5 triangles with the following criteria:

1. Angles measuring 40°, 65°, 75°
2. Sides measuring 8 cm, 10 cm, 15 cm (Hint: After drawing a segment 15 cm long, draw a circle with 10 cm radius from one end point and a circle with 8 cm radius from the other endpoint.)
3. Two angles measuring 50° and 70° with the length between them measuring 7 cm
4. Two sides measuring 5 cm and 8 cm with the angle between them measuring 45°
5. One angle measuring 30°, with the side opposite the angle measuring 5 cm and one of the other sides measuring 7 cm

Then have students compare their triangles to determine which criteria appears to create congruent triangles, that is, identical triangles except for their orientation or measurement errors. Have the groups share their conjectures.

CONGRUENT TRIANGLES, CONTINUED

Side-Angle-Side (SAS)

If the lengths of two sides of a triangle equal the lengths of corresponding sides of another triangle, and the measures of the angles between each pair of sides are equal, the triangles are congruent.

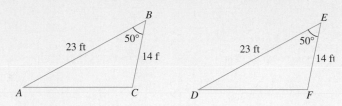

For example, these two triangles are congruent by Side-Angle-Side.

Practice Problem 1

Determine whether triangle *MNO* is congruent to triangle *RTS*.

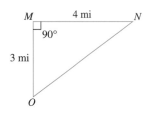

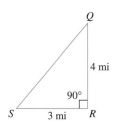

Example 1   Determine whether triangle *ABC* is congruent to triangle *DEF*.

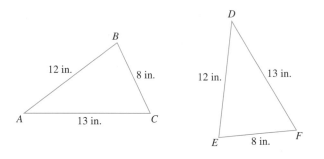

Solution: Since the lengths of all three sides of triangle *ABC* equal the lengths of all three sides of triangle *DEF*, the triangles are congruent.

In Example 1, notice that once we know that the two triangles are congruent, we know that all three corresponding angles are congruent.

B FINDING THE RATIOS OF CORRESPONDING SIDES IN SIMILAR TRIANGLES

Two triangles are **similar** if they have the same shape but not necessarily the same size. In similar triangles, the measures of corresponding angles are equal and corresponding sides are in proportion. The following triangles are similar.

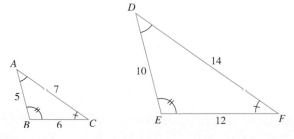

Since these triangles are similar, the measures of corresponding angles are equal.

Answer

1. congruent

Angles with Equal Measure: $\angle A$ and $\angle D$, $\angle B$ and $\angle E$, $\angle C$ and $\angle F$

Also, the lengths of corresponding sides are in proportion.

Sides in Proportion: $\dfrac{AB}{DE} = \dfrac{5}{10} = \dfrac{1}{2}, \dfrac{BC}{EF} = \dfrac{6}{12} = \dfrac{1}{2}, \dfrac{CA}{FD} = \dfrac{7}{14} = \dfrac{1}{2}$

The ratio of corresponding sides is $\dfrac{1}{2}$.

Example 2   Find the ratio of corresponding sides for the similar triangles ABC and DEF.

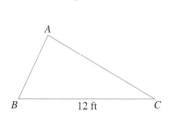

Solution: We are given the lengths of two corresponding sides. Their ratio is

$$\frac{12 \text{ feet}}{17 \text{ feet}} = \frac{12}{17}$$

C FINDING UNKNOWN LENGTHS OF SIDES IN SIMILAR TRIANGLES

Because the ratios of lengths of corresponding sides are equal, we can use proportions to find unknown lengths in similar triangles.

Example 3   Given that the triangles are similar, find the missing length n.

Solution: Since the triangles are similar, corresponding sides are in proportion. Thus, the ratio of 2 to 3 is the same as the ratio of 10 to n, or

$$\frac{2}{3} = \frac{10}{n}$$

To find the unknown length n, we set cross products equal.

$2 \cdot n = 30$ Set cross products equal.

$n = \dfrac{30}{2}$

$n = 15$

The missing length is 15 units.

Practice Problem 2

Find the ratio of corresponding sides for the similar triangles QRS and XYZ.

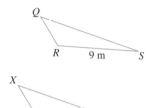

Practice Problem 3

Given that the triangles are similar, find the missing length n.

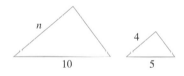

Answers

2. $\dfrac{9}{13}$, **3.** $n = 8$

✓ CONCEPT CHECK

The following two triangles are similar. Which vertices of the first triangle appear to correspond to which vertices of the second triangle.

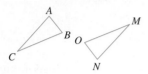

Practice Problem 4

Tammy Shultz, a firefighter, needs to estimate the height of a burning building. She estimates the length of her shadow to be 8 feet long and the length of the building's shadow to be 60 feet long. Find the height of the building if she is 5 feet tall.

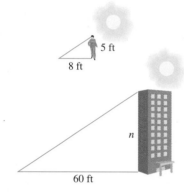

TEACHING TIP

Conclude this section by asking the following: Are congruent triangles similar triangles? Are similar triangles congruent triangles?

TRY THE CONCEPT CHECK IN THE MARGIN.

Many applications involve a diagram containing similar triangles. Surveyors, astronomers, and many other professionals use ratios of similar triangles continually in their work.

Example 4   Finding the Height of a Tree

Mel Rose is a 6-foot-tall park ranger who needs to know the height of a particular tree. He notices that when the shadow of the tree is 69 feet long, his own shadow is 9 feet long. Find the height of the tree.

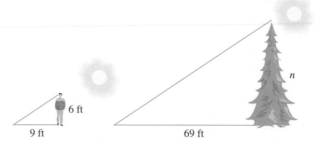

Solution:

1. UNDERSTAND. Read and reread the problem. Notice that the triangle formed by the sun's rays, Mel, and his shadow is similar to the triangle formed by the sun's rays, the tree, and its shadow.

2. TRANSLATE. Write a proportion from the similar triangles formed.

Mel's height → $\dfrac{6}{n}$ = $\dfrac{9}{69}$ ← length of Mel's shadow
height of tree → $\dfrac{6}{n}$ = $\dfrac{9}{69}$ ← length of tree's shadow

or $\dfrac{6}{n} = \dfrac{3}{23}$ (ratio in lowest terms)

3. SOLVE for n.

$$\frac{6}{n} = \frac{3}{23}$$

$6 \cdot 23 = n \cdot 3$ Set cross products equal.

$138 = n \cdot 3$

$\dfrac{138}{3} = n$

$46 = n$

4. INTERPRET. *Check* to see that replacing n with 46 in the proportion makes the proportion true. *State* your conclusion: The height of the tree is 46 feet. ▬▬▬

Answers

4. 37.5 ft

✓ Concept Check: *A* corresponds to *O*; *B* corresponds to *N*; *C* corresponds to *M*

EXERCISE SET 8.7

A *Determine whether each pair of triangles is congruent. See Example 1.*

1.

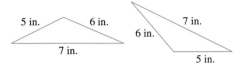

2.

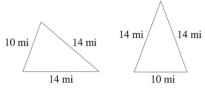

3.

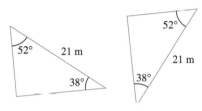

4.

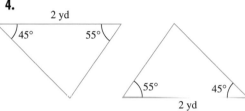

B *Find each ratio of the corresponding sides of the given similar triangles. See Example 2.*

 5.

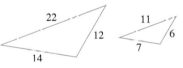

6.

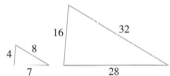

7.

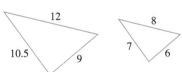

8.

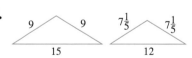

C *Given that the pairs of triangles are similar, find the length of the side labeled n. See Example 3.*

9.

10.

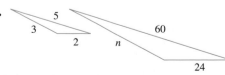

1. congruent

2. congruent

3. congruent

4. congruent

5. $\frac{2}{1}$

6. $\frac{1}{4}$

7. $\frac{3}{2}$

8. $\frac{5}{4}$

9. 4.5

10. 36

11. 6

12. 8

13. 5

14. 13.5

15. 13.5

16. 15.75

17. 17.5

18. 28.8

19. 8

20. 13.5

21. 21.25

22. 3.5

532

11.

12.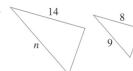

13.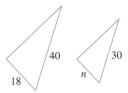

14.

15.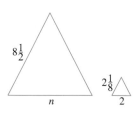

16.
14
8
n
9

17.

18.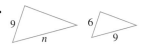
21.6 n 7.2 9.6

19.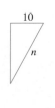
$8\frac{1}{2}$ $2\frac{1}{8}$ n 2

20.
9 6
n 9

21.
16 10
34 n

22.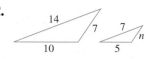
14 7 7 n
10 5

Name _____

Solve. See Example 4.

23. Lloyd White, a firefighter, needs to estimate the height of a burning building. He estimates the length of his shadow to be 9 feet long and the length of the building's shadow to be 75 feet long. Find the height of the building if he is 6 feet tall.

24. Samantha Black, a 5-foot-tall park ranger, needs to know the height of a tree. She notices that when the shadow of the tree is 48 feet long her shadow is 4 feet long. Find the height of the tree.

25. Gepetto the toy maker wants to make a triangular mainsail for a toy sailboat in the same shape as a full-size sailboat's mainsail. Use the diagram to find the lengths of the toy mainsail's sides.

26. A parent wants to make a doll's triangular diaper in the same shape as a full-size diaper. Use the diagram to find the lengths of the doll diaper's sides.

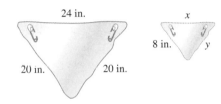

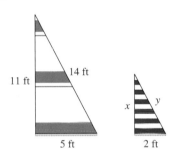

27. If a 30-foot tree casts an 18-foot shadow, find the length of the shadow cast by a 24-foot tree.

28. If a 24-foot flagpole casts a 32-foot shadow, find the length of the shadow cast by a 44-foot antenna. Round to the nearest tenth.

534

Name _____

REVIEW AND PREVIEW

Find the average of each list of numbers. See Section 1.6.

29. 14, 17, 21, 18

30. 87, 84, 93

31. 76, 79, 88

32. 7, 8, 4, 6, 3, 8

 ## COMBINING CONCEPTS

Given that the pairs of triangles are similar. Find the length of the side labeled n. Round your result to 1 decimal place.

33.

34.

35. In your own words, describe any differences in similar triangles and congruent triangles.

CHAPTER 8 ACTIVITY
FINDING PATTERNS

MATERIALS:
▲ Cardboard cut into 1-inch squares

This activity may be completed by working in groups or individually. You will explore the perimeters, areas, and diagonal lengths of patterns formed by flooring tiles and sidewalk tiles. Complete each of the following tables. You will analyze the data in your tables and identify the numerical patterns that are present. Note: A diagonal is a line that extends from the upper-left corner of the figure to the lower-right corner.

Figure 1 Sidewalk Tile Patterns

Sidewalk Pattern Number	1	2	3	4	5	6
Perimeter	4	6	8	10	12	14
Area	1	2	3	4	5	6
Length of diagonal	$\sqrt{2}$	$\sqrt{5}$	$\sqrt{10}$	$\sqrt{17}$	$\sqrt{26}$	$\sqrt{37}$

1. In the table, record the perimeter, area, and length of diagonal for each of the three sidewalk tile patterns shown in Figure 1. (*Hint*: For the length of the diagonal, you can use the Pythagorean theorem.)

2. Form the appropriate sidewalk tile patterns for 4, 5, and 6 sidewalk tiles with cardboard squares. Record the perimeters, areas, and diagonal lengths in the table.

3. Study the results for perimeter, area, and length of diagonal. What patterns do you notice? Describe each pattern.   answers may vary

4. Use the patterns you observed in Question 3 to predict the perimeter, area, and diagonal length for a sidewalk made of 7 tiles. Use the cardboard tiles to form this pattern and verify your predictions.   perimeter: 16; area: 7; length of diagonal: $\sqrt{50}$

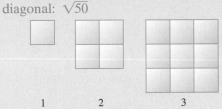

Figure 2 Floor Tile Patterns

Floor Pattern Number	1	2	3	4	5	6
Perimeter	4	8	12	16	20	24
Area	1	4	9	16	25	36
Length of diagonal	$\sqrt{2}$	$\sqrt{8}$	$\sqrt{18}$	$\sqrt{32}$	$\sqrt{50}$	$\sqrt{72}$

5. In the table, record the perimeter, area, and length of diagonal for each of the three floor tile patterns shown in Figure 2.

6. Form the appropriate floor tile patterns with cardboard squares for 4, 5, and 6. Record the perimeters, areas, and diagonal lengths in the table.

7. Study the results for perimeter, area, and length of diagonal. What patterns do you notice? Describe each pattern.   answers may vary

8. Use the patterns you observed in Question 7 to predict the perimeter, area, and diagonal length for the floor tile pattern corresponding to the number 7. Use the cardboard tiles to form this pattern and verify your predictions.
perimeter: 28; area: 49; length of diagonal: $\sqrt{98}$

CHAPTER 8 HIGHLIGHTS

DEFINITIONS AND CONCEPTS	EXAMPLES

A **line** is a set of points extending indefinitely in two directions. A line has no width or height, but it does have length. We can name a line by any two of its points.

Line AB or $\overleftrightarrow{AB}$

A **line segment** is a piece of a line with two endpoints.

Line segment AB or $\overline{AB}$

A **ray** is a part of a line with one endpoint. A ray extends indefinitely in one direction.

Ray AB or $\overrightarrow{AB}$

An **angle** is made up of two rays that share the same endpoint. The common endpoint is called the **vertex**.

Vertex

An angle that measures 180° is called a **straight angle**.

$\angle RST$ is a straight angle.

An angle that measures 90° is called a **right angle**. The symbol ⌐ is used to denote a right angle.

$\angle ABC$ is a right angle.

An angle whose measure is between 0° and 90° is called an **acute angle**.

Acute angles

An angle whose measure is between 90° and 180° is called an **obtuse angle**.

Obtuse angles

Two angles that have a sum of 90° are called **complementary angles**. We say that each angle is the **complement** of the other.

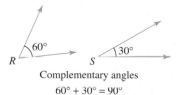

Complementary angles
$60° + 30° = 90°$

SECTION 8.1 (CONTINUED)

Two angles that have a sum of 180° are called **supplementary angles**. We say that each angle is the **supplement** of the other

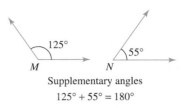

Supplementary angles
125° + 55° = 180°

When two lines intersect, four angles are formed. Two of these angles that are opposite each other are called **vertical angles**. Vertical angles have the same measure.

Two of these angles that share a common side are called **adjacent angles**. Adjacent angles formed in intersecting lines are supplementary.

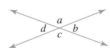

Vertical angles:
∠a and ∠c
∠d and ∠b
Adjacent angles:
∠a and ∠b
∠b and ∠c
∠c and ∠d
∠d and ∠a

A line that intersects two or more lines at different points is called a **transversal**. Line l is a transversal that intersects lines m and n. The eight angles formed have special names. Some of these names are:

Corresponding Angles: ∠a and ∠e, ∠c and ∠g, ∠b and ∠f, ∠d and ∠h

Alternate Interior Angles: ∠c and ∠f, ∠d and ∠e

PARALLEL LINES CUT BY A TRANSVERSAL

If two parallel lines are cut by a transversal, then the measures of **corresponding angles are equal** and the measures of **alternate interior angles are equal**.

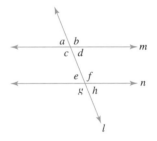

SECTION 8.2 PLANE FIGURES AND SOLIDS

The **sum of the measures** of the angles of a triangle is 180°.

Find the measure of ∠x.

The measure of ∠x = 180° − 85° − 45° = 50°

A **right triangle** is a triangle with a right angle. The side opposite the right angle is called the **hypotenuse** and the other two sides are called **legs**.

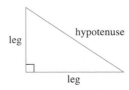

SECTION 8.2   (*CONTINUED*)

For a circle or a sphere:

$$d = 2 \cdot r$$

$$r = \frac{d}{2}$$

Find the diameter of the circle.

6 ft

$$d = 2 \cdot r$$
$$= 2 \cdot 6 \text{ feet} = 12 \text{ feet}$$

SECTION 8.3   PERIMETER

PERIMETER FORMULAS

Rectangle:

$$P = 2 \cdot l + 2 \cdot w$$

Square:

$$P = 4 \cdot s$$

Triangle:

$$P = a + b + c$$

Circumference of a Circle:

$$C = 2 \cdot \pi \cdot r \quad \text{or} \quad C = \pi \cdot d$$

where $\pi \approx 3.14$ or $\pi \approx \dfrac{22}{7}$

Find the perimeter of the rectangle.

28 m

15 m

$$P = 2 \cdot l + 2 \cdot w$$
$$= 2 \cdot 28 \text{ m} + 2 \cdot 15 \text{ m}$$
$$= 56 \text{ m} + 30 \text{ m}$$
$$= 86 \text{ m}$$

The perimeter is 86 meters.

SECTION 8.4   AREA

AREA FORMULAS

Rectangle:

$$A = l \cdot w$$

Square:

$$A = s^2$$

Triangle:

$$A = \frac{1}{2} \cdot b \cdot h$$

Parallelogram:

$$A = b \cdot h$$

Trapezoid:

$$A = \frac{1}{2} \cdot (b + B) \cdot h$$

Circle:

$$A = \pi \cdot r^2$$

Find the area of the square.

8 cm

$$A = s^2$$
$$= (8 \text{ cm})^2$$
$$= 64 \text{ square centimeters}$$

The area of the square is 64 square centimeters.

SECTION 8.5 VOLUME	
VOLUME FORMULAS	Find the volume of the sphere. Use $\dfrac{22}{7}$ for π.
Rectangular Solid:	
$V = l \cdot w \cdot h$	
Cube:	
$V = s^3$	$V = \dfrac{4}{3} \cdot \pi \cdot r^3$
Sphere:	$\approx \dfrac{4}{3} \cdot \dfrac{22}{7} \cdot (2 \text{ inches})^3$
$V = \dfrac{4}{3} \cdot \pi \cdot r^3$	$= \dfrac{4 \cdot 22 \cdot 8}{3 \cdot 7}$ cubic inches
Right Circular Cylinder:	$- \dfrac{704}{21}$ or $33\dfrac{11}{21}$ cubic inches
$V = \pi \cdot r^2 \cdot h$	
Cone:	
$V = \dfrac{1}{3} \cdot \pi \cdot r^2 \cdot h$	
Square-Based Pyramid:	
$V = \dfrac{1}{3} \cdot s^2 \cdot h$	

SECTION 8.6 SQUARE ROOTS AND THE PYTHAGOREAN THEOREM	
SQUARE ROOT OF A NUMBER	$\sqrt{9} = 3$, $\sqrt{100} = 10$, $\sqrt{1} = 1$
A **square root** of a number a is a number b whose square is a. We use the radical sign $\sqrt{}$ to name square roots.	
PYTHAGOREAN THEOREM	Find the hypotenuse of the given triangle.
$(\text{leg})^2 + (\text{other leg})^2 = (\text{hypotenuse})^2$	hypotenuse $= \sqrt{(\text{leg})^2 + (\text{other leg})^2}$
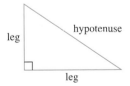	$= \sqrt{(3)^2 + (8)^2}$ The legs are 3 and 8 inches.
	$= \sqrt{9 + 64}$
	$= \sqrt{73}$ inches
	≈ 8.5 inches
TO FIND AN UNKNOWN LENGTH OF A RIGHT TRIANGLE	
hypotenuse $= \sqrt{(\text{leg})^2 + (\text{other leg})^2}$	
leg $= \sqrt{(\text{hypotenuse})^2 - (\text{other leg})^2}$	

SECTION 8.7   CONGRUENT AND SIMILAR TRIANGLES

Congruent triangles have the same shape and the same size. Corresponding angles are equal and corresponding sides are equal.

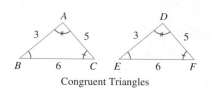

Congruent Triangles

Similar triangles have exactly the same shape but not necessarily the same size. Corresponding angles are equal and the ratios of the lengths of corresponding sides are equal.

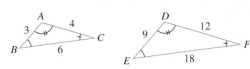

$$\frac{AB}{DE} = \frac{3}{9} = \frac{1}{3}, \frac{BC}{EF} = \frac{6}{18} = \frac{1}{3},$$

$$\frac{CA}{FD} = \frac{4}{12} = \frac{1}{3}$$

CHAPTER 8 REVIEW

(8.1) *Classify each angle as acute, right, obtuse, or straight.*

1.

A
right

2.

B
straight

3.

C
acute

4.
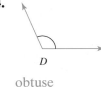
D
obtuse

5. Find the complement of a 25° angle.   65°

6. Find the supplement of a 105° angle.   75°

7. Find the supplement of a 72° angle.   108°

8. Find the complement of a 1° angle.   89°

Find the measure of x in each figure.

9.
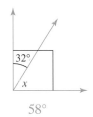
32°
x
58°

10.
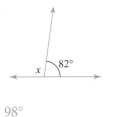
x 82°
98°

11.
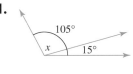
105°
x 15°
90°

12.

20° 45°
x
25°

13. Identify the pairs of supplementary angles.
133° and 47°
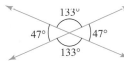
133°
47°   47°
133°

14. Identify the pairs of complementary angles.
43° and 47°; 58° and 32°

58° 32°
47°   43°

Find the measures of angles x, y, and z in each figure.

15.

y
x 80°
z
∠*x* = 80°; ∠*y* = 100°; and ∠*z* = 100°

16.
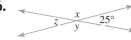
x
z 25°
y
∠*x* = 155°; ∠*y* = 155°; ∠*z* = 25°

17.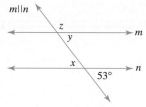

$\angle x = 53°; \angle y = 53°; \angle z = 127°$

18.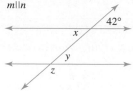

$\angle x = 42°; \angle y = 42°; \angle z = 138°$

(8.2) *Find the measure of ∠x in each figure.*

19.    103°

20. 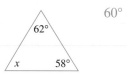   60°

21. 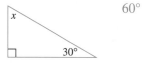   60°

22. 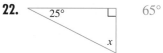   65°

Find the unknown diameter or radius as indicated.

23.

4.2 m

24.

7 ft

25.

9.5 m

26.

15.2 cm

Identify each solid.

27.

cube

28.

cylinder

29.

pyramid

30.

rectangular solid

Find the unknown radius or diameter as indicated.

31. The radius of a sphere is 9 inches. Find its diameter.
18 in.

32. The diameter of a sphere is 4.7 meters. Find its radius.   2.35 m

Identify each regular polygon.

33.    pentagon

34.    hexagon

(8.3) *Find the perimeter of each figure.*

35. 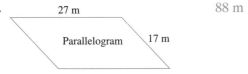   88 m

27 m

Parallelogram   17 m

36. 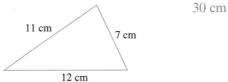   30 cm

11 cm   7 cm

12 cm

37. 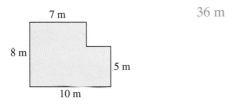   36 m

7 m

8 m

5 m

10 m

38. 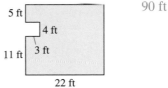   90 ft

5 ft   4 ft

11 ft   3 ft

22 ft

Solve.

39. Find the perimeter of a rectangular sign that measures 6 feet by 10 feet.   32 ft

40. Find the perimeter of a town square that measures 110 feet on a side.   440 ft

Find the circumference of each circle. Use $\pi \approx 3.14$.

41.   5.338 in.

1.7 in.

42. 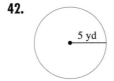   31.4 yd

5 yd

(8.4) *Find the area of each figure. For the circles, find the exact area and then use $\pi \approx 3.14$ to approximate the area.*

43.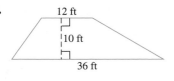

12 ft

10 ft

36 ft

240 sq. ft

44.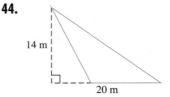

14 m

20 m

140 sq. m

45.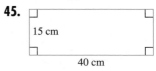

15 cm

40 cm

600 sq. cm

46.

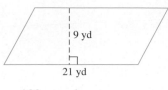

189 sq. yd

47.

49π sq. ft $\approx$ 153.86 sq. ft

48.

4π sq. in. $\approx$ 12.56 sq. in.

49.

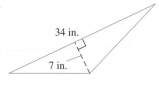

119 sq. in.

50.

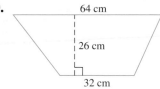

1248 sq. cm

51.

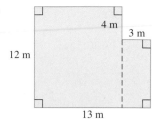

144 sq. m

52. The amount of sealer necessary to seal a driveway depends on the area. Find the area of a rectangular driveway 36 feet by 12 feet.   432 sq. ft

53. Find how much carpet is necessary to cover the floor of the room shown.   130 sq. ft

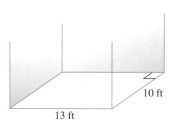

(8.5) *Find the volume of each solid. For Exercises 56 and 57, use* $\pi \approx \dfrac{22}{7}$.

54. $15\dfrac{5}{8}$ cu. in.

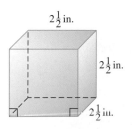

55.   84 cu. ft

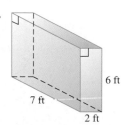

56.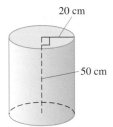

20 cm

50 cm

$62,857\frac{1}{7}$ cu. cm

57.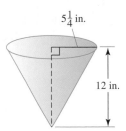

$5\frac{1}{4}$ in.

12 in.

$346\frac{1}{2}$ cu. in.

58. Find the volume of a pyramid with a square base 2 feet on a side and a height of 2 feet. $2\frac{2}{3}$ cu. ft

59. Approximate the volume of a tin can 8 inches high and 3.5 inches in radius. Use 3.14 for π. 307.72 cu. in.

60. A chest has 3 drawers. If each drawer has inside measurements of $2\frac{1}{2}$ feet by $1\frac{1}{2}$ feet by $\frac{2}{3}$ feet, find the total volume of the 3 drawers. $7\frac{1}{2}$ cu. ft

61. A cylindrical canister for a shop vacuum is 2 feet tall and 1 foot in *diameter*. Find its exact volume. 0.5π cu. ft

62. Find the volume of air in a rectangular room 15 feet by 12 feet with a 7-foot ceiling. 1260 cu. ft

63. A mover has two boxes left for packing. Both are cubical, one 3 feet on a side and the other 1.2 feet on a side. Find their combined volume. 28.728 cu. ft

(8.6) *Simplify.*
64. $\sqrt{64}$ 8

65. $\sqrt{144}$ 12

66. $\sqrt{36}$ 6

67. $\sqrt{1}$ 1

68. $\sqrt{\frac{4}{25}}$ $\frac{2}{5}$

69. $\sqrt{\frac{1}{100}}$ $\frac{1}{10}$

Find the unknown length of each given right triangle. If necessary, round to the nearest tenth.
70. leg = 12, leg = 5 13

71. leg = 20, leg = 21 29

72. leg = 9, hypotenuse = 14 10.7

73. leg = 124, hypotenuse = 155 93

74. leg = 66, leg = 56 86.6

75. Find the length to the nearest hundredth of the diagonal of a square that has a side of length 20 centimeters.   28.28 cm

76. Find the height of the building rounded to the nearest tenth.   88.2 ft

(8.7) *Given that the pairs of triangles are similar, find the unknown length n.*

77.

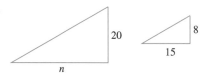

$37\frac{1}{2}$

78.

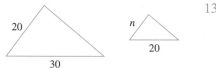

$13\frac{1}{3}$

79.

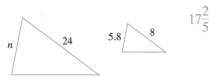

$17\frac{2}{5}$

80.

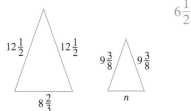

$6\frac{1}{2}$

Solve.

81. A housepainter needs to estimate the height of a condominium. He estimates the length of his shadow to be 7 feet long and the length of the building's shadow to be 42 feet long. Find the height of the building if the housepainter is $5\frac{1}{2}$ feet tall.
33 ft

82. Santa's elves are making a triangular sail for a toy sailboat. The toy sail is to be the same shape as a real sailboat's sail. Use the following diagram to find the unknown lengths x and y.
$x = \frac{5}{6}$ in.; $y = 2\frac{1}{6}$ in.

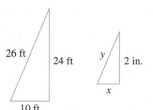

Name _____ Section _____ Date _____

CHAPTER 8 TEST

1. Find the complement of a 78° angle.

2. Find the supplement of a 124° angle.

3. Find the measure of ∠x.

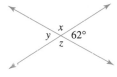

Find the measure of x, y, and z in each figure.

4.

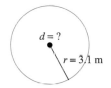

5.

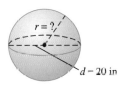

Find the unknown diameter or radius as indicated.

6.

7.

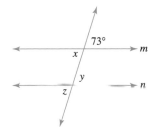

8. Find the measure of ∠x.

Find the perimeter (or circumference) and area of each figure. For the circle, give the exact value and then use π ≈ 3.14 for an approximation.

9.

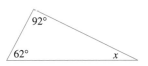

10.

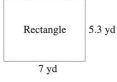

Rectangle   5.3 yd

7 yd

11.

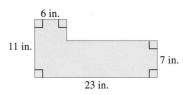

6 in.

11 in.

7 in.

23 in.

1. 12°

2. 56°

3. 50°

4. ∠x = 118°; ∠y = 62°; ∠z = 118°

5. ∠x = 73°; ∠y = 73°; ∠z = 73°

6. 6.2 meters

7. 10 inches

8. 26°

9. circumference = 18π ≈ 56.52 in. area = 81π ≈ 254.34 sq. in.

10. perimeter = 24.6 yd.; area = 37.1 sq. yd

11. perimeter = 68 in.; area = 185 sq. in.

548

Name _____

Find the volume of each solid. For the cylinder, use $\pi \approx \dfrac{22}{7}$.

12.

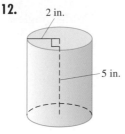

2 in.

5 in.

13.

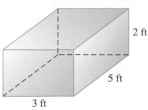

2 ft

5 ft

3 ft

Find each square root and simplify. Round the square root to the nearest thousandth if necessary.

14. $\sqrt{49}$

15. $\sqrt{157}$

16. $\sqrt{\dfrac{64}{100}}$

Solve.

17. Find the perimeter of a square photo with a side length of 4 inches.

4 in.

18. How much soil is needed to fill a rectangular hole 3 feet by 3 feet by 2 feet?

19. Find how much baseboard is needed to go around a rectangular room that measures 18 feet by 13 feet.

20. Approximate to the nearest hundredth of a centimeter the length of the missing side of a right triangle with legs of 4 centimeters each.

21. Vivian Thomas is going to put insecticide on her lawn to control grubworms. The lawn is a rectangle measuring 123.8 feet by 80 feet. The amount of insecticide required is 0.02 ounces per square foot. Find how much insecticide Vivian needs to purchase.

22. Given that the following triangles are similar, find the missing length, *n*.

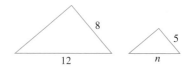

8

12

5

n

23. Tamara Watford, a surveyor, needs to estimate the height of a tower. She estimates the length of her shadow to be 4 feet long and the length of the tower's shadow to be 48 feet long. Find the height of the tower if she is $5\frac{3}{4}$ feet tall.

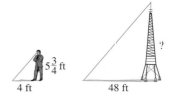

$5\frac{3}{4}$ ft

4 ft

48 ft

Name _____ **Section** _____ **Date** _____

CUMULATIVE REVIEW

1. Write the decimal 19.5023 in words.

2. Round 736.2359 to the nearest hundredth.

3. Add: $45 + 2.06$

Multiply.

4. 7.68×10

5. 76.3×1000

6. Divide: $270.2 \div 7$

7. Simplify: $\dfrac{0.7 + 1.84}{0.4}$

8. Insert $<$, $>$, or $=$ to form a true statement. $\dfrac{1}{8}$ _____ 0.12

9. Write the ratio of 2.6 to 3.1 as a fraction in simplest form.

10. Is $\dfrac{2}{3} = \dfrac{4}{6}$ a true proportion?

11. In a survey of 100 people, 17 people drive blue cars. What percent drive blue cars?

Write each percent as a fraction in simplest form.

12. 1.9%

13. 125%

14. 85% of 300 is what number?

15. 20.8 is 40% of what?

16. Mr. Buccaran, the principal at Slidell High School, counted 31 freshmen absent during a particular day. If this is 4% of the total number of freshmen, how many freshmen are there at Slidell High School?

17. Sherry Souter, a real estate broker for Wealth Investments, sold a house for $114,000 last week. If her commission is 1.5% of the selling price of the home, find the amount of her commission.

ANSWERS

1. nineteen and five thousand twenty-three ten-thousandths (Sec. 4.1, Ex. 3)

2. 736.24 (Sec. 4.2, Ex. 4)

3. 47.06 (Sec. 4.3, Ex. 3)

4. 76.8 (Sec 4.4, Ex. 4)

5. 76,300 (Sec. 4.4, Ex. 6)

6. 38.6 (Sec. 4.5, Ex. 1)

7. 6.35 (Sec. 4.6, Ex. 7)

8. $>$ (Sec. 4.7, Ex. 4)

9. $\dfrac{26}{31}$ (Sec. 5.1, Ex. 5)

10. yes (Sec. 5.3, Ex. 2)

11. 17% (Sec. 6.1, Ex. 1)

12. $\dfrac{19}{1000}$ (Sec. 6.2, Ex. 2)

13. $\dfrac{5}{4}$ or $1\dfrac{1}{4}$ (Sec. 6.2, Ex. 3)

14. 255 (Sec. 6.3, Ex. 8)

15. 52 (Sec. 6.4, Ex. 9)

16. 775 freshmen (Sec. 6.5, Ex. 1)

17. $1710 (Sec. 6.6, Ex. 3)

Name _____

18. Convert 8 feet to inches.

19. Convert 3.2 kg to grams.

20. Subtract 3 qt from 4 gal 2 qt.

21. Convert 29°C to degrees Fahrenheit.

22. Find the measure of $\angle a$.

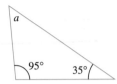

23. Find the perimeter of the rectangle below.

24. Find $\sqrt{\dfrac{4}{25}}$.

25. Find the ratio of corresponding sides for triangles *ABC* and *DEF*.

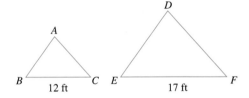

Copyright 1999 Prentice Hall Inc.

Statistics and Probability

We often need to make decisions based on known statistics or the probability of an event occurring. For example, we decide whether or not to bring an umbrella to work based on the probability of rain. We choose an investment based on its mean, or average, return. We can predict which football team will win based on the trend in its previous wins and losses. This chapter reviews presenting data in a usable form on a graph and the basic ideas of statistics and probability.

9.1 Reading Pictographs, Bar Graphs, and Line Graphs

9.2 Reading Circle Graphs

9.3 Reading Histograms

Integrated Review—Reading Graphs

9.4 Mean, Median, and Mode

9.5 Counting and Introduction to Probability

A tornado is a violent whirling column of air that is often spawned by the unstable weather conditions occurring during thunderstorms. Although tornadoes are capable of sustaining wind speeds of 250 to more than 300 mph, most tornadoes have wind speeds under 110 mph. The average forward speed of a tornado is 30 mph, but some tornadoes have been known to travel over land at speeds up to 70 mph. The path of a tornado can extend anywhere from a few feet up to 100 miles long. Each year in the United States, an average of 800 tornadoes occur, causing an average of 80 deaths. The deadliest tornado in the United States was the Tri-State Tornado Outbreak on March 18, 1925, which killed 689 people and injured over 2000 more in Missouri, Illinois, and Indiana. In Exercises 17–22 on page 560 and the Chapter Highlights on page 596, we will see how graphs can be used to summarize data about tornadoes.

Name _____ Section _____ Date _____

CHAPTER 9 PRETEST

The line graph below shows the number of burglaries in a town during the months of March through September.

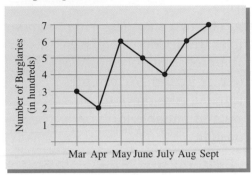

1. During which month, between March and September, did the least number of burglaries occur?

2. During which month, between March and September, were there 400 burglaries?

3. How many burglaries were there in September?

4. The following table shows a break-down of an average day for Dawn Miller.

Attending college classes	4 hours
Studying	3 hours
Working	5 hours
Sleeping	8 hours
Driving	1 hour
Other	3 hours

Draw a circle graph showing this data.

Below is a list of scores from the final exam given in Mrs. Maxwell's Basic College Mathematics class. Use this list to complete the table below.

| 76 | 71 | 94 | 73 | 81 | 78 | 96 | 65 |
| 95 | 80 | 90 | 86 | 98 | 88 | 62 | 91 |

Class Intervals (scores)	Tally	Class Frequency (number of exams)
5. 60–69		
6. 70–79		
7. 80–89		
8. 90–99		

9. Use the table from Exercises 5–8 to draw a histogram.

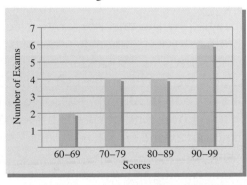

10. Find the grade point average if the following grades were earned in one semester.

Grade	Credit Hours
B	4
B	3
A	3
C	5
D	2

11. Find the median of the following list of numbers.   28, 36, 81, 64, 73, 31, 25, 92, 74

A single die is tossed. Find the probability that the die is each of the following.

12. a 4   13. a number greater than 3   14. a 3 or a 5

9.1 READING PICTOGRAPHS, BAR GRAPHS, AND LINE GRAPHS

Often data is presented visually in a graph. In this section, we practice reading several kinds of graphs including pictographs, bar graphs, and line graphs.

A READING PICTOGRAPHS

A **pictograph** such as the one below is a graph in which pictures or symbols are used. This type of graph will contain a key that explains the meaning of the symbol used. An advantage of using a pictograph to display information is that comparisons can easily be made. A disadvantage of using a pictograph is that it is often hard to tell what fractional part of a symbol is shown. For example, in the pictograph below, Ukraine shows a part of a symbol, but it's hard to read what fractional part of a symbol is shown with any accuracy.

Example 1 The following pictograph shows the approximate nuclear energy generated by selected countries. Use this pictograph to answer the questions.

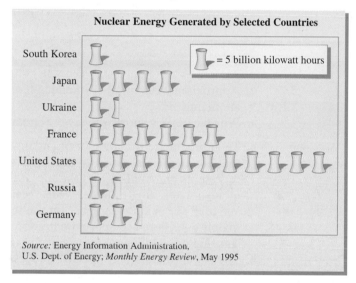

Nuclear Energy Generated by Selected Countries

= 5 billion kilowatt hours

Source: Energy Information Administration, U.S. Dept. of Energy; *Monthly Energy Review,* May 1995

a. Approximate the amount of nuclear energy that is generated in Japan.

b. Approximate how much more nuclear energy is generated in France than in Japan.

Solution: **a.** Japan corresponds to 4 symbols and each symbol represents 5 billion kilowatt hours of energy. This means that Japan generates approximately 4 · (5 billion) or 20 billion kilowatt hours of energy.

b. France shows 2 more symbols than Japan. This means that France generates 2 · (5 billion) or 10 billion more kilowatt hours of nuclear energy than Japan. ▬▬

B READING BAR GRAPHS

Another way to present data by a graph is with a **bar graph**. Bar graphs can appear with vertical bars or horizontal bars. Although we have studied bar graphs in previous sections, we now practice reading the height of the bars contained in a bar graph. An advantage to using bar graphs is that a scale is usually included for greater accuracy. Care must be taken when reading

---

Practice Problem 1

Use the pictograph shown in Example 1 to answer the questions.

a. Approximate the amount of nuclear energy that is generated in the United States.

b. Approximate the total nuclear energy generated in Japan and France.

TEACHING TIP

Ask students to bring a ruler, compass, and protractor to class during this chapter.

Answer

1. a. 55 billion kilowatt hours, **b.** 50 billion kilowatt hours

bar graphs, as well as other types of graphs—they may be misleading, as shown later in this section.

Practice Problem 2

Use the bar graph in Example 2 to answer the questions.

a. Approximate the number of endangered species that are insects.

b. Which category shows the most endangered species?

TEACHING TIP

Have students compare and contrast bar graphs and pictographs.

TEACHING TIP

Consider having students make a bar graph of the data in the pictograph on the previous page.

Example 2 The following bar graph shows the number of endangered species in different categories. Use this graph to answer the questions.

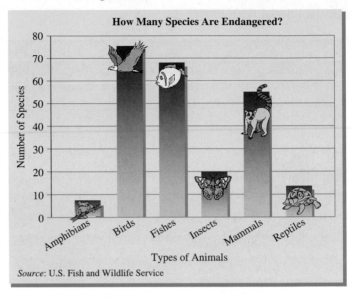

Source: U.S. Fish and Wildlife Service

a. Approximate the number of endangered species that are mammals.
b. Which category shows the fewest endangered species?

Solution: **a.** To approximate the number of endangered species that are mammals, we go to the top of the bar that represents mammals. From the top of this bar, we move horizontally to the left until the scale is reached. We read the height of the bar on the scale as approximately 55. There are approximately 55 mammal species that are endangered, as shown in the next figure.
b. The fewest endangered species is represented by the shortest bar. The shortest bar corresponds to amphibians.

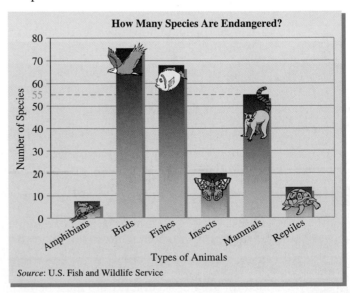

Source: U.S. Fish and Wildlife Service

Answer

2. a. 20, **b.** birds

As mentioned previously, graphs can be misleading. Both graphs below show the same information, but with different scales. Special care should be taken when forming conclusions from the appearance of a graph.

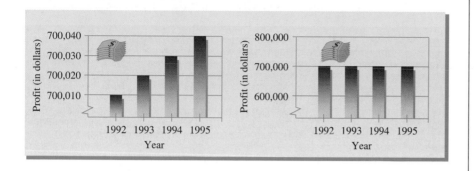

Are profits shown in the graphs above greatly increasing or are they remaining about the same?

C CONSTRUCTING BAR GRAPHS

Next, we practice constructing a bar graph.

Example 3 Draw a vertical bar graph using the information in the table about the caffeine content of selected foods.

Average Caffeine Content of Selected Foods			
Food	Milligrams	Food	Milligrams
Brewed coffee (percolator, 8 ounces)	124	Instant coffee (8 ounces)	104
Brewed decaffeinated coffee (8 ounces)	3	Brewed tea (U.S. brands, 8 ounces)	64
Coca-Cola classic (8 ounces)	31	Mr. Pibb (8 ounces)	27
Dark chocolate (semi-sweet, $1\frac{1}{2}$ ounces)	30	Milk chocolate (8 ounces)	9

Source: International Food Information Council and the Coca-Cola Company

Solution: We draw and label a vertical line and a horizontal line as shown on the next page. We place the different food categories along the horizontal line. Along the vertical line, we place a scale. The scale shown below starts at 0 and then shows multiples of 20 placed at equally distant intervals. It may also be helpful to draw horizontal lines along the scale markings to help draw the vertical bars at the correct height. The finished bar graph is shown on the next page on the right.

Practice Problem 3

Draw a vertical bar graph using the information in the table about electoral votes for selected states.

Electoral Votes for President by Selected States	
State	Electoral Votes
Texas	31
California	54
Florida	25
Nebraska	5
Indiana	12
Georgia	13

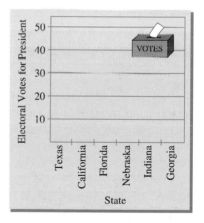

Answer

3.

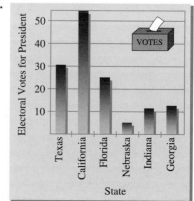

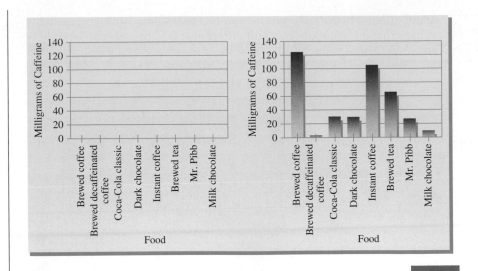

D READING LINE GRAPHS

Another common way to display information with a graph is by using a **line graph**. An advantage of a line graph is that it can be used to visualize relationships between two quantities. A line graph can also be very useful in showing a change over time.

Example 4   The following line graph shows the average daily temperature for each month for Omaha, Nebraska. Use this graph to answer the questions.

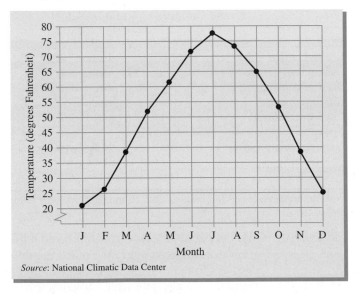

Source: National Climatic Data Center

a. During what month is the average daily temperature the highest?
b. During what month is the average daily temperature 65°F?
c. During what months is the average daily temperature less than 30°F?

Solution: **a.** The month with the highest temperature corresponds to the highest point. We follow the highest point downward to the horizontal month scale and see that this point corresponds to July.

Practice Problem 4

Use the temperature graph in Example 4 to answer the questions.

a. During what month is the average daily temperature the lowest?

b. During what month is the average daily temperature 25°F?

c. During what months is the average daily temperature greater than 70°F?

Answer

4. a. January,   **b.** December,
c. June, July, and August

b. We find the 65°F mark on the vertical scale and move to the right until a darkened point on the graph is reached. From that point, we move downward to the month scale and read the corresponding month. During the month of September, the average daily temperature was 65°F.

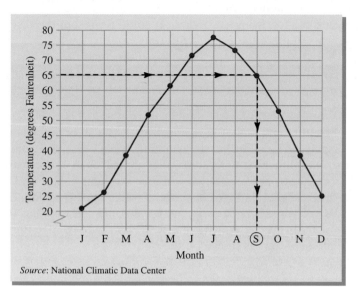

Source: National Climatic Data Center

c. To see what months the temperature is less than 30°F, we find what months correspond to darkened points that fall below the 30°F mark on the vertical scale. These months are January, February, and December.

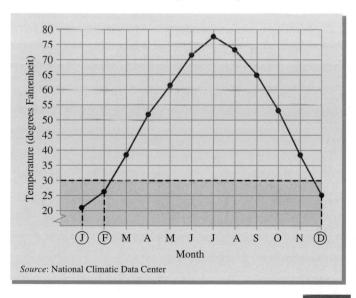

Source: National Climatic Data Center

Focus On Mathematical Connections

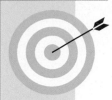

STEM-AND-LEAF DISPLAYS

Stem-and-leaf displays are another way to organize data. Once data is logically organized, it can be much easier to draw conclusions from it.

Suppose we have collected the following set of data. It could represent the test scores for an algebra class or the pulse rates of a group of small children.

90	73	93	99	79	95	69	78	93	80
89	85	97	78	75	79	72	76	97	88
83	98	72	94	92	79	70	98	85	99

TEACHING TIP for Stem and Leaf Displays
Have students compare and contrast the Stem and Leaf Displays to Histograms.

In a stem-and-leaf display, the last digit of each number forms the *leaf* and the remaining digits to the left form the *stem*. For the first number in the list 90, 9 is the stem and 0 is the leaf. To make the stem-and-leaf display, write all of the stems in numerical order in a column. Then write each leaf on the horizontal line next to its stem, aligning leaves in vertical columns. In this case, because the data ranges from 69 to 99, we will use the stems 6, 7, 8, and 9. Each line of the display represents an interval of data; for instance, the line corresponding to the stem **7** represents all data that falls in the interval **70** to **79**, inclusive. Once the data has been divided into its stems and leaves on the display as shown on the left, simply rearrange the leaves on each line to appear in numerical order, as shown on the right.

Stem	Leaf		Stem	Leaf
6	9		6	9
7	3 9 8 8 5 9 2 6 2 9 0	⟶	7	0 2 2 3 5 6 8 8 9 9 9
8	0 9 5 8 3 5		8	0 3 5 5 8 9
9	0 3 9 5 3 7 7 8 4 2 8 9		9	0 2 3 3 4 5 7 7 8 8 9 9

Now that the data has been organized in a stem-and-leaf display, it is easy to answer questions about the data such as: What are the least and greatest values in the set of data? Which data interval contains the most items from the data set? How many values fall between 74 and 84? Which data value occurs most frequently in the data set? What patterns or trends do you see in the data?

CRITICAL THINKING

Make a stem-and-leaf display of the weekend emergency room admission data shown at the right. Then answer the following questions.

1. What was the difference between the least number and greatest number of weekend ER admissions?   86

2. How many weekends had between 125 and 165 ER admissions?   16

3. What number of weekend ER admissions occurred most frequently?   198

4. Which interval contains the most weekend ER admissions?   160–169

Number of Emergency Room Admissions on Weekends				
198	168	117	185	159
160	177	169	112	175
170	188	137	117	145
198	169	154	163	192
167	179	155	133	121
162	188	124	145	146
128	181	198	149	140
122	162	161	180	177

Name _____ **Section** _____ **Date** _____

EXERCISE SET 9.1

A *The following pictograph shows the annual automobile production by one plant for the years 1992–1998. Use this graph to answer Exercises 1–8. See Example 1.*

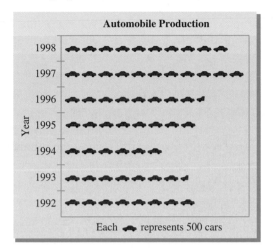

Automobile Production

Each 🚗 represents 500 cars

1. In what year was the greatest number of cars manufactured?

2. In what year was the least number of cars manufactured?

3. Approximate the number of cars manufactured in the year 1995.

4. Approximate the number of cars manufactured in the year 1996.

5. In what year(s) did the production of cars decrease from the previous year?

6. In what year(s) did the production of cars increase from the previous year?

7. In what year(s) were 4000 cars manufactured?

8. In what year(s) were 5500 cars manufactured?

The following pictograph shows the average number of ounces of chicken consumed per person per week in the United States. Use this graph to answer Exercises 9–16. See Example 1.

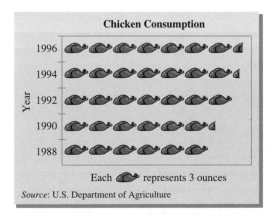

Chicken Consumption

Each 🍗 represents 3 ounces

Source: U.S. Department of Agriculture

9. Approximate the number of ounces of chicken consumed per week in 1992.

10. Approximate the number of ounces of chicken consumed per week in 1996.

559

560

Name _____

 **11.** In what year(s) was the number of ounces of chicken consumed per week greater than 21 ounces?

12. In what year(s) was the number of ounces of chicken consumed per week 21 ounces or less?

13. What was the increase in average chicken consumption from 1988 to 1992?

14. What was the increase in average chicken consumption from 1992 to 1996?

 **15.** Describe a trend in eating habits shown by this graph.

16. In 1996, did you eat less than, greater than, or about the same as the U.S. average number of ounces consumed per week?

B *The following bar graph shows the average number of people killed by tornadoes during the months of the year. Use this graph to answer Exercises 17–22. See Example 2.*

Tornado Deaths

Source: Storm Prediction Center

17. In which month(s) did the most tornado-related deaths occur?

18. In which month(s) did the fewest tornado-related deaths occur?

19. Approximate the number of tornado-related deaths that occurred in May.

20. Approximate the number of tornado-related deaths that occurred in April.

21. In which month(s) did over 5 deaths occur?

22. In which month(s) did over 10 deaths occur?

Name _____

The following horizontal bar graph shows the population (in millions) of the world's largest cities. Use this graph to answer Exercises 23–28. See Example 2.

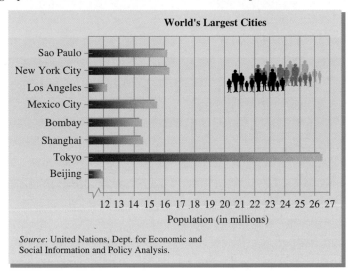

World's Largest Cities

Population (in millions)

Source: United Nations, Dept. for Economic and Social Information and Policy Analysis.

23. Name the city with the largest population and estimate its population.

24. Name the city whose population is between 15 and 16 million and estimate its population.

25. Name the United States city with the largest population and estimate its population.

26. Name the city shown with the smallest population and estimate its population.

27. How much larger is Tokyo than Bombay?

28. How much larger is Mexico City than Beijing?

C *Use the information given to draw a vertical bar graph. Clearly label the bars in each graph. See Example 3*

29.

Best-Selling Magazines in 1996 (in millions)	
Magazine	Circulation
Modern Maturity	21
Reader's Digest	15
TV Guide	13
National Geographic	9
Better Homes and Gardens	8
Family Circle	5

Source: Audit Bureau of Circulations

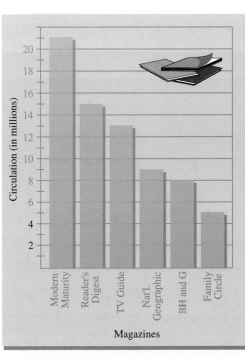

30.

Leading U.S. Passenger Airlines in 1996 (in millions)	
Airline	Passengers
Delta	97
United	82
American	79
US Airways	57
Southwest	55
Northwest	53

Source: Air Transport Association of America

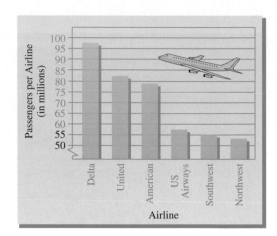

D *The following line graph shows NCAA Division I average number of field goals per game. Use this graph to answer Exercises 31–36. See Example 4.*

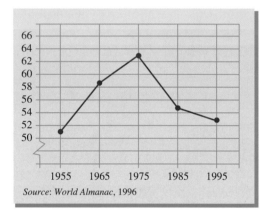

Source: World Almanac, 1996

31. Approximate the average number of field goals per game in 1985.

32. Approximate the average number of field goals per game in 1975.

33. During what year shown was the average number of field goals per game the highest?

34. During what year shown was the average number of field goals per game the lowest?

35. Between 1955 and 1965 did the average number of field goals per game increase or decrease?

36. Between 1985 and 1995 did the average number of field goals per game increase or decrease?

Name _____

Review and Preview

Find each percent. See Sections 6.3 and 6.4.

37. 30% of 12 **38.** 45% of 120 **39.** 10% of 62 **40.** 95% of 50

Write each fraction as a percent. See Section 6.2

41. $\dfrac{1}{4}$ **42.** $\dfrac{2}{5}$ **43.** $\dfrac{17}{50}$ **44.** $\dfrac{9}{10}$

Combining Concepts

The following double line graph shows temperature highs and lows for a week. Use this graph to answer Exercises 45–50.

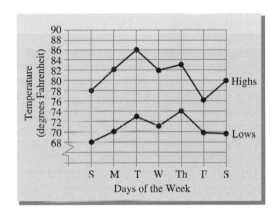

45. What was the high temperature reading on Thursday?

46. What was the low temperature reading on Thursday?

47. What day was the temperature the lowest? What was this low temperature?

48. What day of the week was the temperature the highest? What was this high temperature?

49. Tuesday, 13°F

49. What day of the week was the difference between the high temperature and the low temperature the greatest? What was this difference in temperature?

50. What day of the week was the difference between the high temperature and the low temperature the least? What was this difference in temperature?

50. Friday, 6°F

Internet Excursions

Go to http://www.prenhall.com/martin-gay

The Toy Manufacturers of America (TMA) is a trade association for U.S. toy makers and importers. Among its many goals, TMA supports the production of safe toys and compiles industry statistics. The World Wide Web address listed here will provide you with access to the web site of the Toy Manufacturers of America (TMA), or a related site. You will find statistical information that will help you answer the questions below.

51. Visit this Website and view the graph of Annual Sales. What type of graph is this? For the most recent year shown on the graph, report that year's retail sales. Describe any trends you see in the graph.

52. Visit this Website and view the graph of Imports and Exports. For the most recent year shown on the graph, report both the value of imports and exports. Describe any trends you see in the graph.

51. bar graph; answers may vary

52. answers may vary

9.2 Reading Circle Graphs

A Reading Circle Graphs

In Section 6.1, the following circle graph was shown. This particular graph shows the favorite cookie for every 100 people.

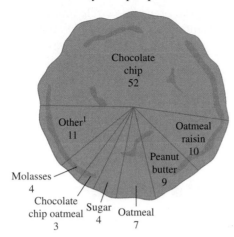

Chocolate chip 52

Other[1] 11

Oatmeal raisin 10

Peanut butter 9

Molasses 4

Chocolate chip oatmeal 3

Sugar 4

Oatmeal 7

[1]–all 1% or less
Source: USA Today, 2/23/96

TEACHING TIP

Have students add the numbers in each sector. Ask what the sum represents.

Each sector of the graph (shaped like a piece of pie) shows a category and the relative size of the category. In other words, the most popular cookie is the chocolate chip cookie, and it is represented by the largest sector.

Example 1 Find the ratio of people preferring chocolate chip cookies to total people. Write the ratio as a fraction in simplest form.

Solution: The ratio is

$$\frac{52 \text{ people preferring chocolate chip}}{100 \text{ people}} = \frac{52}{100} = \frac{13}{25}$$

Practice Problem 1

Find the ratio of people preferring oatmeal raisin cookies to total people. Write the ratio as a fraction in simplest form.

A circle graph is often used to show percents in different categories, with the whole circle representing 100%. For example, in 1997, 45,900 public schools had CD-ROM technology. The next circle graph shows the percent of these schools that are elementary, junior high, or senior high. Notice that the percents in each category sum to 100%.

Type of U.S. Public Schools with
CD-ROM Technology in 1997

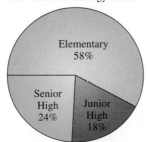

Elementary 58%

Senior High 24%

Junior High 18%

Source: Quality Education Data, Inc.

Example 2 Using the circle graph shown, determine the percent of schools with CD-ROM technology that are either junior or senior high schools.

Practice Problem 2

Using the circle graph shown, determine the percent of schools with CD-ROM technology that are either elementary or junior high schools.

Answers

1. $\frac{1}{10}$, **2.** 76%

Solution: To find this percent, we add the junior high and senior high percents. The percent of schools with CD-ROM technology that are either junior or senior high schools is

24% + 18% = 42%.

Practice Problem 3

Using the circle graph for Example 2, find the number of junior high schools with CD-ROM technology in 1997.

TEACHING TIP
Have students find factors of 360.
360: 1, 2, 3, 4, 5, 6, 8, 9, 10, 12, 15, 18, 20, 24, 30, 36, 40, 45, 60, 72, 90, 120, 180, 360
Then ask them to find the number of degrees and the corresponding percent for the following fractional parts of a circle:
$\frac{1}{2}, \frac{1}{3}, \frac{1}{4}, \frac{1}{5}, \frac{1}{6}, \frac{1}{8}, \frac{1}{9}, \frac{1}{10},$ and $\frac{1}{12}.$

✓ CONCEPT CHECK

Can the following data be represented by a circle graph? Why or why not?

Responses to the question "In which activities are you involved?"	
Intramural sports	60%
On-campus job	42%
Fraternity/sorority	27%
Academic clubs	21%
Music programs	14%

Practice Problem 4

Use the data shown to draw a circle graph.

Freshmen	30%
Sophomores	27%
Juniors	25%
Seniors	18%

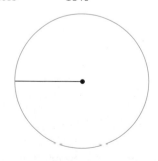

Answers

3. 8262, **4.** See page 567.

✓ Concept Check: No, since the percents add up to more than 100%.

Example 3 Using the preceding circle graph, find the number of elementary schools with CD-ROM technology in 1997.

Solution: We use the percent equation:

$$\boxed{amount} = \boxed{percent} \cdot \boxed{base}$$
$$\downarrow \qquad\qquad \downarrow \qquad\qquad \downarrow$$
$$amount = 58\% \cdot 45{,}900 = 0.58(45{,}900) = 26{,}622$$

Thus 26,622 elementary schools had CD-ROM technology in 1997.

TRY THE CONCEPT CHECK IN THE MARGIN.

B DRAWING CIRCLE GRAPHS

To draw a circle graph, we will use the fact that a whole circle contains 360° (degrees).

Example 4 The following table shows the percent of U.S. armed forces personnel that are in each branch of the service. (*Source:* U.S. Department of Defense)

Branch of Service	Percent
Army	33%
Navy	27%
Marine Corps	12%
Air Force	26%
Coast Guard	2%

Draw a circle graph showing this data.

Solution: First we find the number of degrees in each sector representing each branch of service. Remember that the whole circle contains 360°. (We will round degrees to the nearest whole.)

Sector	Degrees in Each Sector
Army	33% × 360° = 118.8° ≈ 119°
Navy	27% × 360° = 97.2° ≈ 97°
Marine Corps	12% × 360° = 43.2° ≈ 43°
Air Force	26% × 360° = 93.6° ≈ 94°
Coast Guard	2% × 360° = 7.2° ≈ 7°

Next we draw a circle and mark its center. Then we draw a line from the center of the circle to the circle itself.

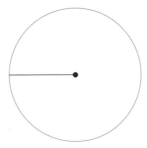

To construct the sectors, we will use a **protractor**. We place the hole in the protractor over the center of the circle. Then we adjust the protractor so that 0° on the protractor is aligned with the line that we drew.

It makes no difference which sector we draw first. To construct the "Army" sector, we find 119° on the protractor and mark our circle. Then we remove the protractor and use this mark to draw a second line from the center to the circle itself.

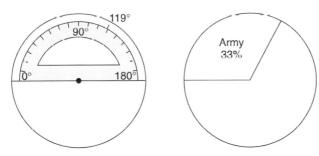

To construct the "Navy" sector next, we follow the same procedure as above except that we line 0° up with the second line we drew and mark the protractor this time at 97°.

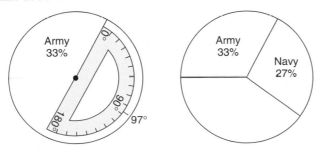

TEACHING TIP

Before making an accurate circle graph of the U.S. Armed Forces Personnel, consider having students make a quick rough sketch. When they complete the accurate graph, have them compare their accurate graph with the sketch to see where their estimates were off.

Answer

4.

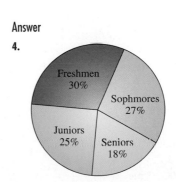

We continue in this manner until the circle graph is complete.

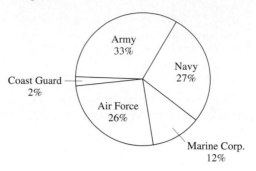

✓ CONCEPT CHECK

True or false? The larger a sector in a circle graph, the larger the percent of the total it represents. Explain your answer.

TRY THE CONCEPT CHECK IN THE MARGIN.

Answer

✓ **Concept Check:** True

Name _____ **Section** _____ **Date** _____

EXERCISE SET 9.2

A *The following circle graph is a result of surveying 700 college students. They were asked where they live while attending college. Use this graph to answer Exercises 1–6. Write all ratios as fractions in simplest form. See Example 1.*

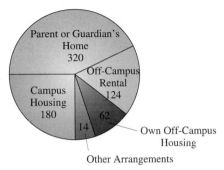

Parent or Guardian's Home 320
Off-Campus Rental 124
Campus Housing 180
62
14
Own Off-Campus Housing
Other Arrangements

1. Where do most of these college students live?

2. Besides the category "Other Arrangements," where do least of these college students live?

3. Find the ratio of students living in campus housing to total students.

4. Find the ratio of students living in off-campus rentals to total students.

5. Find the ratio of students living in campus housing to students living at home.

6. Find the ratio of students living in off-campus rentals to students living at home.

The following circle graph shows how much money teenagers spend on a backpack. Use this graph for Exercises 7–14. See Examples 2 and 3.

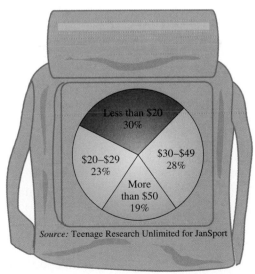

Less than $20 30%
$20–$29 23%
$30–$49 28%
More than $50 19%
Source: Teenage Research Unlimited for JanSport

7. Backpacks in which price range are purchased most often?

8. Backpacks in which price range are purchased least often?

1. Parent or guardian's home

2. Own off-campus housing

3. $\frac{9}{35}$

4. $\frac{31}{175}$

5. $\frac{9}{16}$

6. $\frac{31}{80}$

7. less than $20

8. more than $50

569

9. 53%

10. 47%

11. 1410 teenagers

12. 1081 teenagers

13. 1316 teenagers

14. 893 teenagers

15. 55%

16. 42%

17. nonfiction

18. children's fiction

19. 31,400 books

20. 21,352 books

21. 27,632 books

22. 41,448 books

23. 25,120 books

24. 35,168 books

Name ______________________________

9. What percent of backpacks purchased cost less than $30?

10. What percent of backpacks purchased cost $30 or more?

If Central High School has 4700 teenagers, use the graph to find how many Central High teenagers spent the dollar amounts given in Exercises 7–10 on backpacks.

11. Less than $20

12. $20–$29

13. $30–$49

14. More than $50

The following circle graph shows the percent of types of books available at Midway Memorial Library. Use this graph for Exercises 15–24. See Examples 2 and 3.

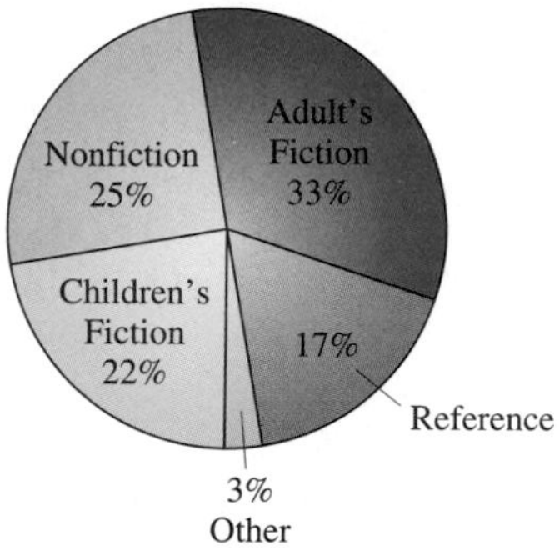

15. What percent of books are classified as some type of fiction?

16. What percent of books are nonfiction or reference?

17. What is the second-largest category of books?

18. What is the third-largest category of books?

If this library has 125,600 books, find how many books are in each category given in Exercises 19–24.

19. Nonfiction

20. Reference

21. Children's fiction

22. Adult's fiction

23. Reference or other

24. Nonfiction or other

Name _____

25. see graph

B *Draw a circle graph to represent the information given in each table. See Example 4.*

25.

Absence from Work in 1995 Due to Carpal Tunnel Syndrome	
Under 3 days	7%
3–10 days	18%
11–20 days	14%
21 or more days	61%

(*Note:* Carpal tunnel syndrome is a nerve disorder causing pain mostly in the wrist and hand.)

Source: Bureau of Labor Statistics

26.

Most Important Reason for Picking a Favorite Restaurant Among Wendy's, Burger King, and McDonalds	
Convenience	13%
Taste of Food	70%
Price	9%
Atmosphere	4%
Others	4%

Source: TeleNation

26. see graph

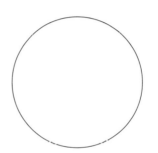

27. $2^2 \times 5$

28. 5^2

29. $2^3 \times 5$

REVIEW AND PREVIEW

Write the prime factorization of each number. See Section 2.2.

27. 20   **28.** 25   **29.** 40

30. 2^4

31. 5×17

30. 16   **31.** 85   **32.** 105

32. $3 \times 5 \times 7$

◆ **COMBINING CONCEPTS**

The following circle graph shows the relative sizes of the great oceans. These oceans together make up 264,489,800 square kilometers of the Earth's surface. Find the square kilometers for each ocean.

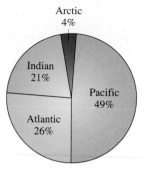

Source: Philip's World Atlas, 1994

33. Pacific Ocean

34. Atlantic Ocean

35. Indian Ocean

36. Arctic Ocean

9.3 READING HISTOGRAMS

A READING HISTOGRAMS

Suppose that the test scores of 36 students are summarized in the table below.

Student Scores	Frequency (Number of Students)
40–49	1
50–59	3
60–69	2
70–79	10
80–89	12
90–99	8

The results in the table can be displayed in a histogram. A **histogram** is a special bar graph. The width of each bar represents a range of numbers called a **class interval**. The height of each bar corresponds to how many times a number in the class interval occurred and is called the **class frequency**.

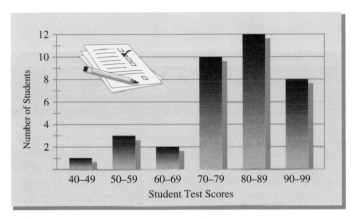

Example 1 Use the preceding histogram to determine how many students scored 50–59 on the test.

Solution: We find the bar representing 50–59. The height of this bar is 3, which means 3 students scored 50–59 on the test.

Example 2 Use the preceding histogram to determine how many students scored 80 or above on the test.

Solution: We see that two different bars fit this description. There are 12 students who scored 80–89 and 8 students who scored 90–99. The sum of these two categories is 12 + 8 or 20 students. Thus, 20 students scored 80 or above on the test.

B CONSTRUCTING HISTOGRAMS

The daily high temperatures for one month in New Orleans, Louisiana, are recorded in the following list.

Objectives

A Read histograms.
B Construct histograms.

SSM   CD-ROM   Video 9.3

TEACHING TIP Classroom Activity

Before beginning this lesson, ask students to discuss "frequency." Then use responses from the class to make a table of the frequency for some quantity such as age, height, or bedtime.

TEACHING TIP

Have students compare and contrast bar graphs and histograms.

TEACHING TIP

Point out to students that the vertical axes for a histogram is a frequency, that is the number of times something occurs.

Practice Problem 1

Use the histogram for Examples 1 and 2 to determine how many students scored 70–79 on the test.

Practice Problem 2

Use the histogram for Examples 1 and 2 to determine how many students scored less than 60 on the test.

Answers

1. 10, **2.** 4

Copyright 1999 Prentice-Hall, Inc.

TEACHING TIP
Ask students to determine how the class intervals might have been selected in Practice Problem 3.

Practice Problem 3

Complete the frequency distribution table for the data below. Each number represents a credit card owner's unpaid balance each month.

0	53	89	125
265	161	37	76
62	201	136	42

Class Intervals (Credit Card Balances)	Tally	Class Frequency (Number of Months)
$0–$49	———	———
$50–$99	———	———
$100–$149	———	———
$150–$199	———	———
$200–$249	———	———
$250–$299	———	———

Practice Problem 4

Construct a histogram from the frequency distribution table above.

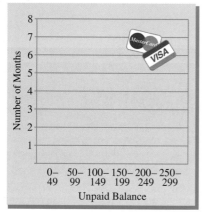

✓ CONCEPT CHECK

Which of the following sets of data is better suited to representation by a histogram? Explain.

Grade on Final	# of Students	Section Number	Avg. Grade on Final
51–60	12	150	78
61–70	18	151	83
71–80	29	152	87
81–90	23	153	73
91–100	25		

Answers

3.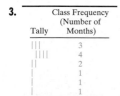

4.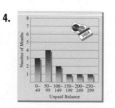

✓ Concept Check: first set of data

85°	90°	95°	89°	88°	94°
87°	90°	95°	92°	95°	94°
82°	92°	96°	91°	94°	92°
89°	89°	90°	93°	95°	91°
88°	90°	88°	86°	93°	89°

The data in this list has not been organized and can be hard to interpret. One way to organize the data is to place it in a **frequency distribution table**. We will do this in Example 3.

Example 3

Complete the frequency distribution table for the preceding temperature data.

Solution: Go through the data and place a tally mark next to its class interval (in the second table column). Then count the tally marks and write each total in the third table column.

Class Intervals (Temperatures)	Tally	Class Frequency (Number of Days)
82°–84°	I	1
85°–87°	III	3
88°–90°	ⅢⅢ I	11
91°–93°	Ⅲ II	7
94°–96°	Ⅲ III	8

Example 4

Construct a histogram from the frequency distribution table in Example 3.

Solution:

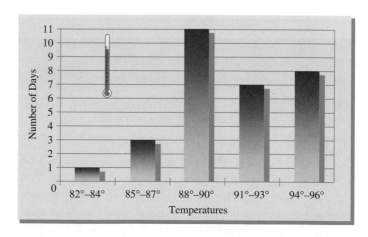

TRY THE CONCEPT CHECK IN THE MARGIN.

TEACHING TIP
While discussing Example 3, ask students to determine how the class intervals might have been selected.

Name _____ **Section** _____ **Date** _____

Exercise Set 9.3

A *The following histogram shows the number of miles each adult from a survey of 100 adults drives per week. Use this histogram to answer Exercises 1–10. See Examples 1 and 2.*

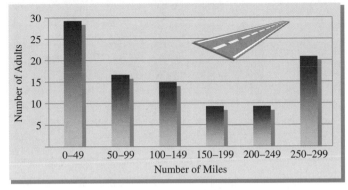

1. How many adults drive 100–149 miles per week?

2. How many adults drive 200–249 miles per week?

3. How many adults drive less than 150 miles per week?

4. How many adults drive 200 miles or more per week?

5. How many adults drive 100–199 miles per week?

6. How many adults drive 0–149 miles per week?

7. How many more adults drive 250–299 miles per week than 200–249 miles per week?

8. How many more adults drive 0–49 miles per week than 50–99 miles per week?

9. What is the ratio of adults who drive 150–199 miles per week to total adults surveyed?

10. What is the ratio of adults who drive 50–99 miles per week to total adults surveyed?

ANSWERS

1. 15 adults

2. 9 adults

3. 61 adults

4. 30 adults

5. 24 adults

6. 61 adults

7. 12 adults

8. 12 adults

9. $\dfrac{9}{100}$

10. $\dfrac{17}{100}$

Name _____

The following histogram shows the projected ages of householders for the year 2005. Use this histogram to answer Exercises 11–18. See Examples 1 and 2.

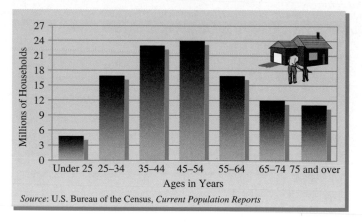

Source: U.S. Bureau of the Census, *Current Population Reports*

11. The most householders will be in what age range?

12. The least householders will be in what age range?

13. How many householders will be 55–64 years old?

14. How many householders will be 35–44 years old?

15. How many householders will be 44 years old or younger?

16. How many householders will be 55 years old or older?

17. Which bar represents the household you expect to be in in the year 2005?

18. How many more householders will be 45–54 years old than 55–64 years old?

B *The following list shows the golf scores for an amateur golfer. Use this list to complete the frequency distribution table below. See Example 3.*

78	84	91	93	97
97	95	85	95	96
101	89	92	89	100

Class Intervals (Scores)	Tally	Class Frequency (Number of Games)
19. 70–79	I	1
20. 80–89	IIII	4
21. 90 99	ⅣⅢ III	8
22. 100–109	II	2

Name _____

Twenty-five people in a survey were asked to give their current checking account balances. Use the balances shown in the following list to complete the frequency distribution table below. See Example 3.

$ 53	$105	$162	$443	$109
$468	$ 47	$259	$316	$228
$207	$357	$ 15	$301	$ 75
$ 86	$ 77	$512	$219	$100
$192	$288	$352	$166	$292

	Class Intervals (Account Balances)	Tally	Class Frequency (Number of People)
23.	$0–$99	⊤⊣⊣ I	6
24.	$100–$199	⊤⊣⊣ I	6
25.	$200 $299	⊤⊣⊣ I	6
26.	$300–$399	IIII	4
27.	$400–$499	II	2
28.	$500–$599	I	1

29. Use the table from Exercises 19 22 to construct a histogram. See Example 4.

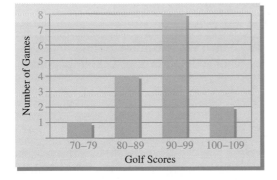

30. Usc the table from Exercises 23–28 to construct a histogram. See Example 4.

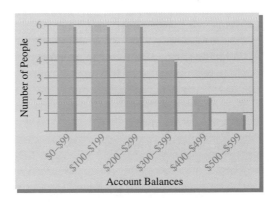

578

Name _____

REVIEW AND PREVIEW

Find the average of each list of numbers. See Section 1.6. (Recall that the average of a list of numbers is the sum of the numbers divided by the number of numbers.)

31. 86, 94

32. 75, 87

33. 12, 28, 20

34. 19, 10, 22

35. 30, 22, 23, 33

36. 39, 25, 31, 37

COMBINING CONCEPTS

The following graph is called a double bar graph. Study this graph and use it to answer Exercises 37–39.

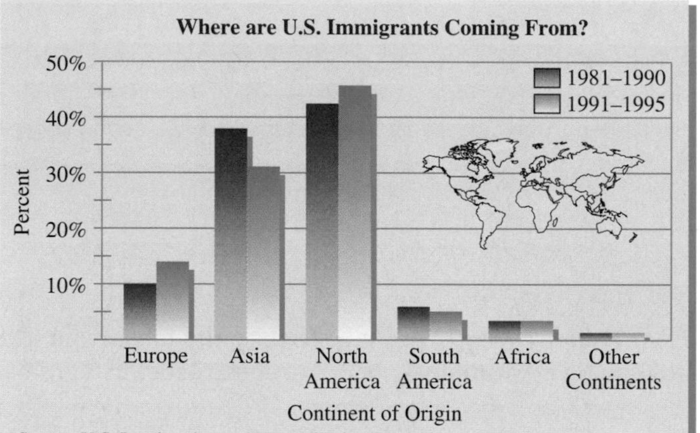

Where are U.S. Immigrants Coming From?

Legend: 1981–1990, 1991–1995

Source: U.S. Immigration and Naturalization Service

37. Which continent shows the greatest change in percent of U.S. immigrants?

38. Which single continent shows no change in percent of U.S. immigrants?

39. How might a graph of this type be helpful?

Name _____ **Section** _____ **Date** _____

1. ___65 pounds___

CHAPTER 9 INTEGRATED REVIEW — READING GRAPHS

The following pictograph shows the average number of pounds of beef and veal consumed per person per year in the United States. Use this graph to answer Exercises 1–4.

2. ___73 pounds___

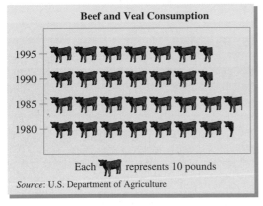

Beef and Veal Consumption

1995
1990
1985
1980

Each 🐄 represents 10 pounds

Source: U.S. Department of Agriculture

1. Approximate the number of pounds of beef and veal consumed per person for the year 1995.

2. Approximate the number of pounds of beef and veal consumed per person for the year 1980.

3. In what year(s) was the number of pounds consumed the greatest?

4. In what year(s) was the number of pounds consumed the least?

3. ___1985___

4. ___1990 and 1995___

The following bar graph shows the highest United States dams. Use this graph to answer Exercises 5–8.

5. ___Oroville Dam, 755 feet___

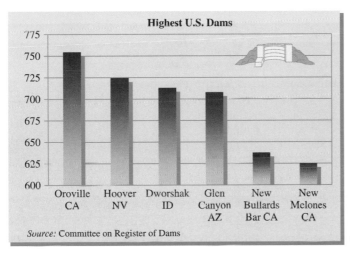

Highest U.S. Dams

775
750
725
700
675
650
625
600

Oroville CA | Hoover NV | Dworshak ID | Glen Canyon AZ | New Bullards Bar CA | New Melones CA

Source: Committee on Register of Dams

6. ___New Bullards Bar Dam, 635 feet___

5. Name the United States dam with the greatest height and estimate this height.

6. Name the United States dam whose height is between 625 and 650 feet and estimate its height.

7. Estimate how much higher the Hoover Dam is than the Glen Canyon Dam.

8. How many United States dams have heights over 700 feet?

7. ___15 feet___

8. ___4 dams___

The following line graph shows the daily high temperatures for one week in Annapolis, Maryland. Use this graph to answer Exercises 9–12.

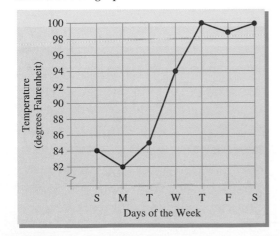

100
98
96
94
92
90
88
86
84
82

Temperature (degrees Fahrenheit)

S M T W T F S

Days of the Week

9. Name the day(s) of the week with the highest temperature and give that high temperature.

10. Name the day(s) of the week with the lowest temperature and give that low temperature.

11. What days of the week was the temperature less than 90° Fahrenheit?

12. What days of the week was the temperature greater than 90° Fahrenheit?

9. ___Thursday and Saturday, 100°F___

10. ___Monday, 82°F___

11. ___Sunday, Monday, and Tuesday___

12. ___Wednesday, Thursday, Friday, and Saturday___

13. 70 quart containers

14. 52 quart containers

15. 2 quart containers

16. 6 quart containers

The following circle graph shows the type of beverage milk consumed in the United States. Use this graph for Exercises 13–16. If a store in Kerrville, Texas sells 200 quart containers of milk per week, estimate how many quart containers are sold in each category below.

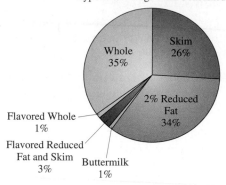

Type of Beverage Milk Consumed

Source: U.S. Department of Agriculture

13. Whole milk

14. Skim milk

15. Buttermilk

16. Flavored Reduced Fat and Skim milk.

The following list shows weekly quiz scores for a student in Basic College Mathematics. Use this list to complete the frequency distribution table.

50	80	71	83	86
67	89	93	88	97
	53	90		
75	80	78	93	99

17. see table

18. see table

19. see table

20. see table

21. see table

	Class Intervals (Scores)	Tally	Class Frequency (Number of Quizzes)
17.	50–59	||	2
18.	60–69	|	1
19.	70–79	|||	3
20.	80–89	⊪|	6
21.	90–99	⊪	5

22. Use the table from Exercises 17–21 to construct a histogram.

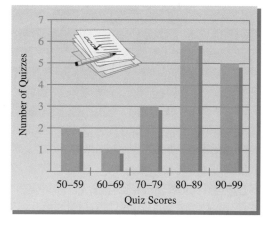

22. see graph

580

9.4   MEAN, MEDIAN, AND MODE

A   FINDING THE MEAN

Sometimes we want to summarize data by displaying them in a graph, but sometimes it is also desirable to be able to describe a set of data, or a set of numbers by a single "middle" number. Three such **measures of central tendency** are the mean, the median, and the mode.

The most common measure of central tendency is the mean (sometimes called the arithmetic mean or the average). Recall that we first introduced finding the average of a list of numbers in Section 1.6.

> The **mean (average)** of a set of number items is the sum of the items divided by the number of items.

Example 1   Seven students in a psychology class conducted an experiment on mazes. Each student was given a pencil and asked to successfully complete the same maze. The timed results are below.

Student	Ann	Thanh	Carlos	Jesse	Melinda	Ramzi	Dayni
Time (seconds)	13.2	11.8	10.7	16.2	15.9	13.8	18.5

  a. Who completed the maze in the shortest time? Who completed the maze in the longest time?
  b. Find the mean time.
  c. How many students took longer than the mean time? How many students took shorter than the mean time?

Solution: **a.** Carlos completed the maze in 10.7 seconds, the shortest time. Dayni completed the maze in 18.5 seconds, the longest time.

  b. To find the mean (or average), we find the sum of the number items and divide by 7, the number of items.

$$\text{mean} = \frac{13.2 + 11.8 + 10.7 + 16.2 + 15.9 + 13.8 + 18.5}{7}$$

$$= \frac{100.1}{7} = 14.3$$

  c. Three students, Jesse, Melinda, and Dayni, had times longer than the mean time. Four students, Ann, Thanh, Carlos, and Ramzi had times shorter than the mean time.

TRY THE CONCEPT CHECK IN THE MARGIN.

Often in college, the calculation of a **grade point average** (GPA) is a **weighted mean** and is calculated as shown in Example 2.

Objectives

A Find the mean of a list of numbers.
B Find the median of a list of numbers.
C Find the mode of a list of numbers.

SSM   CD-ROM   Video 9.4

TEACHING TIP

While discussing Example 1, ask if any student completed the maze in the mean time.

Practice Problem 1

Find the mean of the following test scores: 77, 85, 86, 91, and 88.

✓ CONCEPT CHECK

Estimate the mean of the following set of data:

5, 10, 10, 10, 10, 15

Answers

1. 85.4

✓ Concept Check: 10

Practice Problem 2

Find the grade point average if the following grades were earned in one semester.

Grade	Credit Hours
A	2
C	4
B	5
D	2
A	2

TEACHING TIP

Point out that in the weighted mean the grade is weighted so a grade for a 3 credit hour class counts 3 times as much as a grade for a 1 credit hour class. An A in a 3 credit hour class is the same as A's in 3 one credit hour classes. Then have the class calculate the GPA without weighting the grades. Which method gives a better estimate of the student's performance?

Example 2

The following grades were earned by a student during one semester. Find the student's grade point average.

Course	Grade	Credit Hours
College Mathematics	A	3
Biology	B	3
English	A	3
PE	C	1
Social Studies	D	2

Solution: To calculate the grade point average, we need to know the point values for the different possible grades. The point values of grades commonly used in colleges and universities are given below.

A: 4,   B: 3,   C: 2,   D: 1,   F: 0

Now, to find the grade point average, we multiply the number of credit hours for each course by the point value of each grade. The grade point average is the sum of these products divided by the sum of the credit hours.

Course	Grade	Point Value of Grade	Credit Hours	(Point Value) · (Credit Hours)
College Mathematics	A	4	3	12
Biology	B	3	3	9
English	A	4	3	12
PE	C	2	1	2
Social Studies	D	1	2	2
			Totals: 12	37

$$\text{grade point average} = \frac{37}{12} \approx 3.08 \text{ rounded to two decimal places}$$

The student earned a grade point average of 3.08.

B FINDING THE MEDIAN

You may have noticed that a very low number or a very high number can affect the mean of a list of numbers. Because of this, you may sometimes want to use another measure of central tendency. A second measure of central tendency is called the **median**. The median of a list of numbers is not affected by a low or high number in the list.

> The **median** of an ordered set of numbers is the middle number. If the number of items is even, the median is the mean of the two middle numbers.

Practice Problem 3

Find the median of the list of numbers:
7, 9, 13, 23, 24, 35, 38, 41, 43

Example 3

Find the median of the list of numbers:
25, 54, 56, 57, 60, 71, 98

Solution: Because this list is in numerical order, the median is the middle number, 57.

Answers

2. 2.73,   **3.** 24

Example 4   Find the median of the list of scores: 67, 91, 75, 86, 55, 91

Solution: First we list the scores in numerical order and then find the middle number.

55, 67, 75, 86, 91, 91

Since there is an even number of scores, there are two middle numbers. The median is the mean of the two middle numbers.

$$\text{median} = \frac{75 + 86}{2} = 80.5$$

The median is 80.5.

C FINDING THE MODE

The last common measure of central tendency is called the **mode**.

> The **mode** of a set of numbers is the number that occurs most often. (It is possible for a set of numbers to have more than one mode or to have no mode.)

Example 5   Find the mode of the list of numbers:
11, 14, 14, 16, 31, 56, 65, 77, 77, 78, 79

Solution: There are two numbers that occur the most often. They are 14 and 77. This list of numbers has two modes, 14 and 77.

Example 6   Find the median and the mode of the following set of numbers. These numbers were high temperatures for fourteen consecutive days in a city in Montana.

76, 80, 85, 86, 89, 87, 82, 77, 76, 79, 82, 89, 89, 92

Solution: First we write the numbers in numerical order.

76, 76, 77, 79, 80, 82, 82, 85, 86, 87, 89, 89, 89, 92

Since there is an even number of items, the median is the mean of the two middle numbers.

$$\text{median} = \frac{82 + 85}{2} = 83.5$$

The mode is 89, since 89 occurs most often.

TRY THE CONCEPT CHECK IN THE MARGIN.

> **⌈HELPFUL HINT**
>
> Don't forget that it is possible for a list of numbers to have no mode. For example, the list
>
>   2, 4, 5, 6, 8, 9
>
> has no mode. There is no number or numbers that occur more often than the others.

Practice Problem 4

Find the median of the list of scores:
43, 89, 78, 65, 95, 95, 88, 71

TEACHING TIP   Classroom Activity
In groups, have students discuss the following:

If you were the manager of a shoe store, which measure of central tendency would be more useful to you? Explain.

If you were a baseball coach, which measure of central tendency would be most useful to you? Explain.

If you were a real estate agent, which measure of central tendency would be most useful to you? Explain.

Describe a situation in which you would use the mean as a measure of central tendency. Describe a situation in which you would use the median as a measure of central tendency. Describe a situation in which you would use the mode as a measure of central tendency.

Then have the groups share their conclusions with the class.

Practice Problem 5

Find the mode of the list of numbers:
9, 10, 10, 13, 15, 15, 15, 17, 18, 18, 20

Practice Problem 6

Find the median and the mode of the list of numbers:
26, 31, 15, 15, 26, 30, 16, 18, 15, 35

✓ CONCEPT CHECK

True or False? Every set of numbers *must* have a mean, median, and mode. Explain your answer.

Answers

4. 83,   **5.** 15,   **6.** median: 22; mode: 15

✓ **Concept Check:**   False, a set of numbers may have no mode.

Focus On the Real World

MISLEADING GRAPHS

Graphs are very common in magazines and in newspapers such as *USA Today*. Graphs can be a convenient way to get an idea across because, as the old saying goes, "a picture is worth a thousand words." However, some graphs can be deceptive, which may or may not be intentional. It is important to know some of the ways that graphs can be misleading.

Beware of graphs like the one at the right. Notice that the graph shows a company's profit for various months. It appears that profit is growing quite rapidly. However, this impressive picture tells us little without knowing what units of profit are being graphed. Does the graph show profit in dollars or millions of dollars? An unethical company with profit increases of only a few pennies could use a graph like this one to make the profit increase seem much more substantial than it really is. A truthful graph describes the size of the units used along the vertical axis.

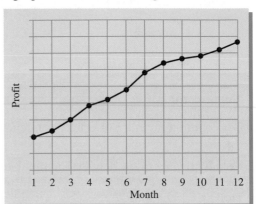

Another type of graph to watch for is one that misrepresents relationships. This can occur in both bar graphs and circle graphs. For example, the bar graph at the right shows the number of men and women employees in the accounting and shipping departments of a certain company. In the accounting department, the bar representing the number of women is shown twice as tall as the bar representing the number of men. However, the number of women (13) is not twice the number of men (10). This set of bars misrepresents the relationship between the number of men and women. Do you see how the relationship between the number of men and women in the shipping department is distorted by the heights of the bars used? A truthful graph will use bar heights or circle sectors that are proportional in size to the numbers the represent.

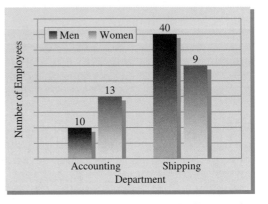

We have already seen that the impression a graph can give also depends on its vertical scale. Here is another example. The two graphs below represent exactly the same data. The only difference between the two graphs is the vertical scale—one shows enrollments from 246 to 260 students and the other shows enrollments between 0 and 300 students. If you were trying to convince readers that algebra enrollment at UPH had changed drastically over the period 1996–2000, which graph would you use? Which graph do you think gives the more honest representation?

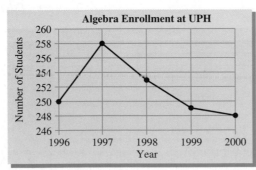

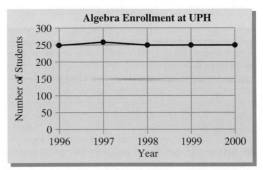

Name _____ Section _____ Date _____

MENTAL MATH

State the mean for each list of numbers.

1. 3, 5   **2.** 10, 20   **3.** 1, 3, 5   **4.** 7, 7, 7

EXERCISE SET 9.4

A **B** **C** *For each set of numbers, find the mean, the median, and the mode. If necessary, round the mean to one decimal place. See Examples 1 and 3 through 6.*

1. 21, 28, 16, 42, 38

2. 42, 35, 36, 40, 50

3. 7.6, 8.2, 8.2, 9.6, 5.7, 9.1

4. 4.9, 7.1, 6.8, 6.8, 5.3, 4.9

5. 0.2, 0.3, 0.5, 0.6, 0.6, 0.9, 0.2, 0.7, 1.1

6. 0.6, 0.6, 0.8, 0.4, 0.5, 0.3, 0.7, 0.8, 0.1

7. 231, 543, 601, 293, 588, 109, 334, 268

8. 451, 356, 478, 776, 892, 500, 467, 780

The ten tallest buildings in the United States are listed in the following table. Use this table to answer Exercises 9–12. If necessary, round results to one decimal place. See Examples 1 and 3 through 6.

Building	Height (in feet)
Sears Tower, Chicago	1450
One World Trade Center (1972), New York	1368
Two World Trade Center (1973), New York	1362
Empire State Building, New York	1250
Amoco, Chicago	1136
John Hancock Center, Chicago	1127
Stratosphere Tower, Las Vegas	1049
Chrysler Building, New York	1046
NationsBank Tower, Atlanta	1023
First Interstate World Center, Los Angles	1018

Source: World Almanac, 1998

9. Find the mean height for the five tallest buildings.

10. Find the median height for the five tallest buildings.

11. Find the median height for the ten tallest buildings.

12. Find the mean height for the ten tallest buildings.

For Exercises 13–16, the grades are given for a student for a particular semester. Find the grade point average. If necessary, round the grade point average to the nearest hundredth. See Example 2.

13.

Grade	Credit Hours
B	3
C	3
A	4
C	4

14.

Grade	Credit Hours
D	1
F	1
C	4
B	5

15.

Grade	Credit Hours
A	3
A	3
B	4
B	1
B	2

16.

Grade	Credit Hours
B	2
B	2
A	3
C	3
B	3

ANSWERS

1. mean: 29, median: 28, no mode

2. mean: 40.6, median: 40, no mode

3. mean: 8.1, median: 8.2, mode: 8.2

4. mean: 6.0, median: 6.05, mode: 6.8 and 4.9

5. mean: 0.6, median: 0.6, mode: 0.2 and 0.6

6. mean: 0.5, median: 0.6, mode: 0.6 and 0.8

7. mean: 370.9, median: 313.5, no mode

8. mean: 587.5, median: 489, no mode

9. 1313.2 feet

10. 1362 feet

11. 1131.5 feet

12. 1182.9 feet

13. 2.79

14. 2.18

15. 3.46

16. 3.0

585

17. 6.8

18. 6.95

19. 6.9

20. 84.67

21. 85.5

22. no mode

23. 73

24. 71

25. 70 and 71

26. 6

27. 9

28. $\dfrac{3}{5}$

29. $\dfrac{1}{3}$

30. $\dfrac{1}{9}$

31. $\dfrac{3}{5}$

32. $\dfrac{7}{20}$

33. $\dfrac{11}{15}$

34. 20, 21, 21

35. 35, 35, 37, 43

36. answers may vary

586

During an experiment, the following times (in seconds) were recorded:
7.8, 6.9, 7.5, 4.7, 6.9, 7.0.

17. Find the mean. Round to the nearest tenth.

18. Find the median.

19. Find the mode.

In a mathematics class, the following test scores were recorded for a student:
86, 95, 91, 74, 77, 85.

20. Find the mean. Round to the nearest hundredth.

21. Find the median.

22. Find the mode.

The following pulse rates were recorded for a group of 15 students:
78, 80, 66, 68, 71, 64, 82, 71, 70, 65, 70, 75, 77, 86, 72.

23. Find the mean.

24. Find the median.

25. Find the mode.

26. How many rates were higher than the mean?

27. How many rates were lower than the mean?

REVIEW AND PREVIEW

Write each fraction in simplest form. See Section 2.3.

28. $\dfrac{12}{20}$

29. $\dfrac{6}{18}$

30. $\dfrac{4}{36}$

31. $\dfrac{18}{30}$

32. $\dfrac{35}{100}$

33. $\dfrac{55}{75}$

COMBINING CONCEPTS

Find the missing numbers in each set of numbers.

34. 16, 18, —, —, —.
The mode is 21.
The median is 20.

35. —, —, —, 40, —.
The mode is 35.
The median is 37.
The mean is 38.

36. Write a list of numbers that you feel the median would be a better measure of central tendency than the mean.

9.5 COUNTING AND INTRODUCTION TO PROBABILITY

A USING A TREE DIAGRAM

In our daily conversations, we often talk about the likelihood or the probability of a given result occurring. For example:

The *chance* of thundershowers is 70 percent.
What are the *odds* that the Saints will go to the Super Bowl?
What is the *probability* that you will finish cleaning your room today?

Each of these chance happenings—thundershowers, the Saints playing in the Super Bowl, and cleaning your room today—is called an **experiment**. The possible results of an experiment are called **outcomes**. For example, flipping a coin is an experiment and the possible outcomes are heads (H) or tails (T).

One way to picture the outcomes of an experiment is to draw a tree diagram. Each outcome is shown on a separate branch. For example, the outcomes of flipping a coin are

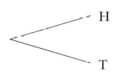

Head   Tail

Example 1   Draw a tree diagram for tossing a coin twice. Then use the diagram to find the number of possible outcomes.

Solution: There are 4 possible outcomes when tossing a coin twice.

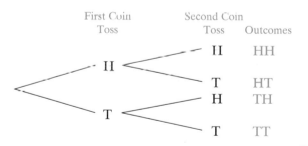

First Coin Toss | Second Coin Toss | Outcomes
H < H → HH
H < T → HT
T < H → TH
T < T → TT

Example 2   Draw a tree diagram for an experiment consisting of rolling a die and then tossing a coin. Then use the diagram to find the number of possible outcomes.

Die

Solution: Recall that a die has six sides and each side represents a number, 1 through 6.

Objectives

A Use a tree diagram to count outcomes.
B Find the probability of an event.

SSM   CD-ROM   Video 9.5

TEACHING TIP

Have students make a row at the bottom of their tree diagram with the following information:

	First Coin Toss	Second Coin Toss	Both Coin Tosses Together
Number of Possible Outcomes	2	2	4

Practice Problem 1

Draw a tree diagram for tossing a coin three times. Then use the diagram to find the number of possible outcomes.

Practice Problem 2

Draw a tree diagram for an experiment consisting of tossing a coin and then rolling a die. Then use the diagram to find the number of possible outcomes.

Answers

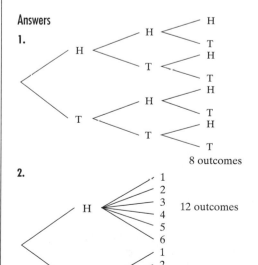

1.

8 outcomes

2.

12 outcomes

TEACHING TIP

Have students make a row at the bottom of their tree diagram with the following information:

	Die Roll	Coin Toss	Die Roll & Coin Toss Together
Number of Possible Outcomes	6	2	12

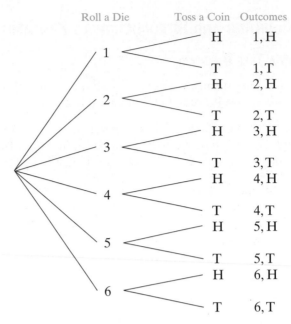

There are 12 possible outcomes for rolling a die and then tossing a coin.

Any number of outcomes considered together is called an **event**. For example, when tossing a coin twice, HH is an event. The event is tossing a head first and then tossing a head second. Another event would be tossing a tail first and then a head (TH), and so on.

B FINDING THE PROBABILITY OF AN EVENT

As we mentioned earlier, the probability of an event is a measure of the chance or likelihood of it occuring. For example, if a coin is tossed, what is the probability that a head occurs? Since one of two equally likely possible outcomes is heads, the probability is $\frac{1}{2}$.

THE PROBABILITY OF AN EVENT

$$\text{probability of an event} = \frac{\text{number of ways that the event can occur}}{\text{number of possible outcomes}}$$

TEACHING TIP

Ask students to describe an event using the roll of a die having a probability of 0. If they are stumped, give them a hint such as, "Is there any number you are certain you will not roll?" Then ask students to describe an event using the roll of a die with a probability of 1.

┌ HELPFUL HINT

Note from the definition of probability that the probability of an event is always between 0 and 1, inclusive (including 0 and 1). A probability of 0 means an event won't occur and a probability of 1 means that an event is certain to occur.

Example 3   If a coin is tossed twice, find the probability of tossing a head and then a head (HH).

Solution:

1 way the event can occur

$$\underbrace{\text{HT, HH, TH, TT}}_{\text{4 possible outcomes}}$$

$$\text{probability} = \frac{1}{4} \quad \begin{matrix} \text{Number of ways the event can occur} \\ \text{Number of possible outcomes} \end{matrix}$$

The probability of tossing a head and then a head is $\frac{1}{4}$.

Practice Problem 3

If a coin is tossed three times, find the probability of tossing a head, then a tail, then a tail (HTT).

Example 4   If a die is rolled one time, find the probability of rolling a 3 or a 4.

Solution: Recall that there are 6 possible outcomes when rolling a die.

2 ways that the event can occur

$$\text{possible outcomes: } 1, \quad 2, \quad \underbrace{3, \quad 4,}_{} \quad 5, \quad 6$$

6 possible outcomes

$$\text{probability of a 3 or a 4} = \frac{2}{6} \quad \begin{matrix} \text{Number of ways the event} \\ \text{can occur} \\ \text{Number possible outcomes} \end{matrix}$$

$$= \frac{1}{3} \quad \text{Simplest form.}$$

Practice Problem 4

If a die is rolled one time, find the probability of rolling a 1 or a 2.

TRY THE CONCEPT CHECK IN THE MARGIN.

Example 5   Find the probability of choosing a red marble from a box containing 1 red, 1 yellow, and 2 blue marbles.

Solution:

1 way that event can occur

$$\underbrace{\text{yellow} \quad \text{blue} \quad \text{blue} \quad \text{red}}_{\text{4 possible outcomes}}$$

$$\text{probability} = \frac{1}{4}$$

✓ CONCEPT CHECK

Suppose you have calculated a probability of $\frac{11}{9}$. How do you know that you have made an error in your calculation.

Practice Problem 5

Use the diagram from Example 5 and find the probability of choosing a blue marble from the box.

Answers

3. $\frac{1}{8}$,   **4.** $\frac{1}{3}$,   **5.** $\frac{1}{2}$

✓ Concept Check: Number of ways an event can occur can't be larger than possible outcomes.

Focus On Business and Career

SURVEYS

How often have you read an article in a newspaper or a magazine that included results from a survey or a poll? Surveys have become very popular ways for businesses and organizations to get feedback on a variety of topics. A political organization may hire a polling group to gauge the public's response to a political candidate. A food company may send out surveys to customers to gather market research on a new food product. A health club may ask its patrons to fill out a brief comment card to report their views on the services offered at the club. Surveys are useful for collecting information needed to make a decision: Should we do a media blitz to increase public awareness of our candidate? Do we need to change the recipe for our new food product to make it more appealing to a wider audience? Should we add a new service to attract new customers and retain our current clientele?

Once data has been collected from a survey, it must be summarized to be useful in decision making. Survey results can be summarized in any of the types of graphs presented in this chapter. Survey results can also be summarized by reporting average responses or with a combination of basic statistics and graphs.

GROUP ACTIVITY

1. Conduct a survey of 30 students in one of your classes. Ask each student to report his or her age.

2. Find the difference between the ages of the youngest and oldest survey respondent (this difference between the largest and smallest value is called the *range*). Divide the range into five or six equal age categories. Tally the number of your respondents that fall into each category. Make a histogram of your results. What does this graph tell you about the ages of your survey respondents?

3. Find the average age of your survey respondents.

4. Find the median age of your survey respondents.

5. Find the mode of your survey respondents.

6. Compare the mean, median, and mode of your age data. Are these measures similar? Which is the largest? Which is the smallest? If there is a noticeable difference between any of these measures, can you explain why?

Name _____ **Section** _____ **Date** _____

MENTAL MATH

If a coin is tossed once, find the probability of each event.

1. The coin lands heads up.

2. The coin lands tails up.

If the spinner shown is spun once, find the probability of each event.

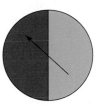

3. The spinner stops on red.

4. The spinner stops on blue.

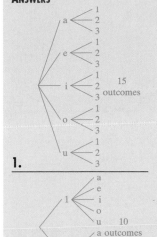

1. _____

2. _____

EXERCISE SET 9.5

A *Draw a tree diagram for each experiment. Then use the diagram to find the number of possible outcomes. See Examples 1 and 2.*

1. Choose a vowel, a, e, i, o, u, and then a number, 1, 2, or 3.

2. Choose a number 1 or 2 and then a vowel, a, e, i, o, u.

3. _____

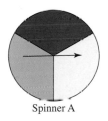 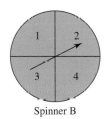

Spinner A   Spinner B

3. Spin Spinner A once.

4. Spin Spinner B once.

4. _____

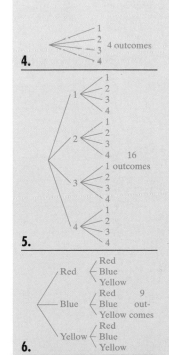

5. Spin Spinner B twice.

6. Spin Spinner A twice.

5. _____

6. _____

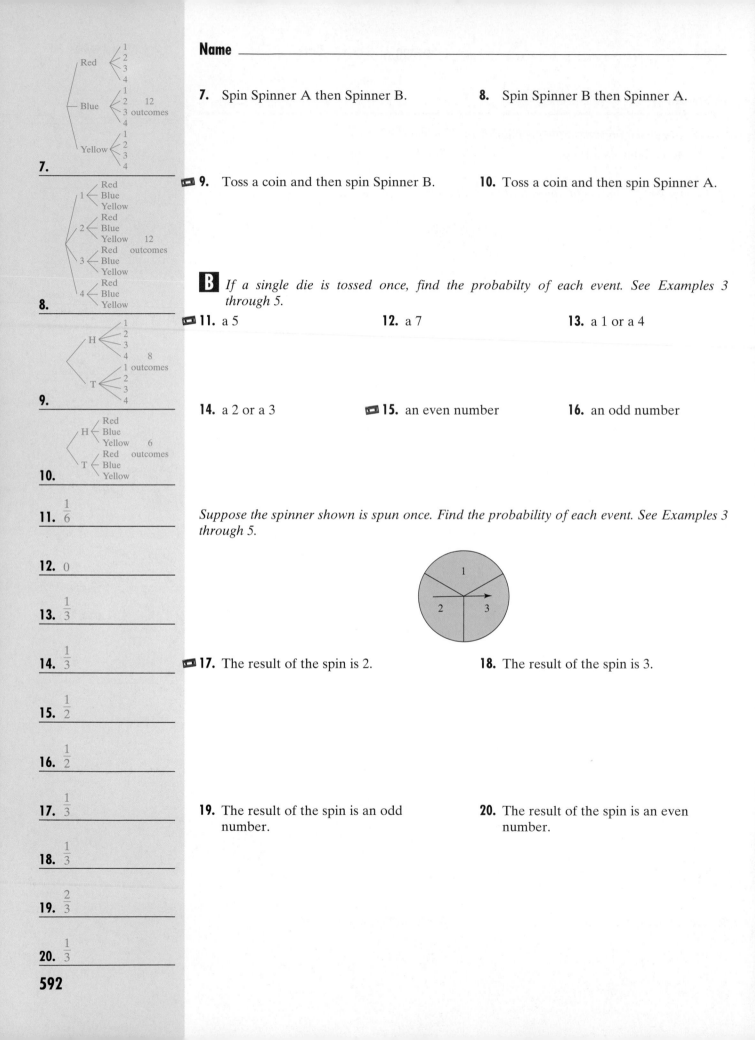

Red
1
2
3
4

Blue
1
2   12
3 outcomes
4

Yellow
1
2
3
4

7. _____

1
Red
Blue
Yellow
2
Red
Blue
Yellow   12
3   outcomes
Red
Blue
Yellow
4
Red
Blue
Yellow

8. _____

H
1
2
3
4   8
1 outcomes
T
2
3
4

9. _____

H
Red
Blue
Yellow   6
Red   outcomes
T
Blue
Yellow

10. _____

11. $\frac{1}{6}$ _____

12. 0 _____

13. $\frac{1}{3}$ _____

14. $\frac{1}{3}$ _____

15. $\frac{1}{2}$ _____

16. $\frac{1}{2}$ _____

17. $\frac{1}{3}$ _____

18. $\frac{1}{3}$ _____

19. $\frac{2}{3}$ _____

20. $\frac{1}{3}$ _____

592

Name _____

7. Spin Spinner A then Spinner B.

8. Spin Spinner B then Spinner A.

9. Toss a coin and then spin Spinner B.

10. Toss a coin and then spin Spinner A.

B _If a single die is tossed once, find the probabilty of each event. See Examples 3 through 5._

11. a 5

12. a 7

13. a 1 or a 4

14. a 2 or a 3

15. an even number

16. an odd number

Suppose the spinner shown is spun once. Find the probability of each event. See Examples 3 through 5.

17. The result of the spin is 2.

18. The result of the spin is 3.

19. The result of the spin is an odd number.

20. The result of the spin is an even number.

Name _____

If a single choice is made from the bag of marbles shown, find the probability of each event. See Examples 3 through 5.

21. A red marble is chosen.

22. A blue marble is chosen.

23. A yellow marble is chosen.

24. A green marble is chosen.

Review and Preview

Perform each indicated operation. See Sections 2.4, 2.5, and 3.3.

25. $\dfrac{1}{2} + \dfrac{1}{3}$

26. $\dfrac{7}{10} - \dfrac{2}{5}$

27. $\dfrac{1}{2} \cdot \dfrac{1}{3}$

28. $\dfrac{7}{10} \div \dfrac{2}{5}$

29. $5 \div \dfrac{3}{4}$

30. $\dfrac{3}{5} \cdot 10$

21. $\dfrac{1}{7}$

22. $\dfrac{1}{7}$

23. $\dfrac{2}{7}$

24. $\dfrac{3}{7}$

25. $\dfrac{5}{6}$

26. $\dfrac{3}{10}$

27. $\dfrac{1}{6}$

28. $\dfrac{7}{4}$ or $1\dfrac{3}{4}$

29. $\dfrac{20}{3}$ or $6\dfrac{2}{3}$

30. 6

593

594

Name _____

◇ **COMBINING CONCEPTS**

Recall that a deck of cards contains 52 cards. These cards consist of 4 suits (hearts, spades, clubs, and diamonds) of each the following: 2, 3, 4, 5, 6, 7, 8, 9, 10, Jack, Queen, and King, and Ace. If a card is chosen from a deck of cards, find the probability of each event.

31. the King of hearts

32. the 10 of spades

33. a King

34. a 10

35. a heart

36. a club

Two die are tossed. Find the probability of each sum of the dice. (Hint: Draw a tree diagram of the possibilities of two tosses of a die and then find the sum of the numbers on each branch.)

37. a sum of 4

38. a sum of 11

39. a sum of 13

40. a sum of 2

Chapter 9 Activity
Investigating Probability

MATERIALS:
▲ paper or foam cup
▲ 30 thumbtacks

This activity may be completed by working in groups or individually. Recall that probability is a measure of how likely it is that an event will occur. We can report probabilities as fractions or percents because any fraction can be written as a percent. In this activity you will investigate one way that probabilities can be estimated.

Number of Tacks	Number of Tacks Landing Point Up	Fraction of Point-Up Tacks	Percent of Point-Up Tacks
30			
60			
90			
120			
150			

1. Place the thumbtacks in the cup. Shake the cup and toss out the thumbtacks onto a flat surface. Count the number of tacks that land point up, and record this number in the table. Complete the experiment for 30, 60, 90, 120, and 150 tacks. (*Hint:* For 60 thumbtacks, count the number of tacks landing point up in two tosses of the 30 thumbtacks, etc.)

2. For each row of the table, find the fraction of tacks that landed point up. Then express each fraction as a percent. Add these values to the table in the columns labeled "Fraction of Point-Up Tacks" and "Percent of Point-Up Tacks."

3. Each of the percents you computed in Question 2 is an *estimate* of the probability that a single thumbtack will land point up when tossed. When you estimate a probability experimentally, the larger the number of trials used (i.e., number of tacks tossed), the better the estimate of the actual probability. What do you suppose the value of the actual probability is? Explain your reasoning.

4. Combine your results for all 450 of your tack tosses recorded in the table with the other students' results. Of this total number of tacks, compute the percent of tacks that landed point up. This is your best estimate of the probability that a tack will land point up when tossed.

5. If you tossed 200 thumbtacks, what percent would you expect to land point up? How many tacks would you expect to land point up? Use the percent (probability) you computed in Question 4 to make this calculation. What if you tossed 300 thumbtacks?

CHAPTER 9 HIGHLIGHTS

DEFINITIONS AND CONCEPTS	EXAMPLES

SECTION 9.1 READING PICTOGRAPHS, BAR GRAPHS, AND LINE GRAPHS

A **pictograph** is a graph in which pictures or symbols are used to visually present data.

A **line graph** displays information with a line that connects data points.

A **bar graph** presents data using vertical or horizontal bars.

The following bar graph shows the number of acres of wheat harvested in 1996 for leading states.

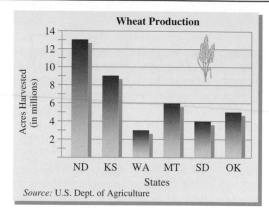

1. Approximately how many acres of wheat were harvested in Kansas?

   9,000,000 acres

2. About how many more acres of wheat were harvested in North Dakota than South Dakota?

   $$\begin{array}{r} 13 \text{ million} \\ -\ 4 \text{ million} \\ \hline 11 \text{ million} \end{array}$$ or 11,000,000 acres

SECTION 9.2 READING CIRCLE GRAPHS

In a **circle graph**, each section (shaped like a piece of pie) shows a category and the relative size of the category.

The following circle graph classifies tornadoes by wind speed.

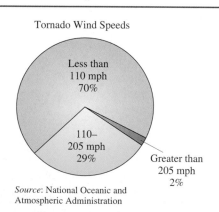

1. What percent of tornadoes have wind speeds of 110 mph or greater?

   $29\% + 2\% = 31\%$

2. If there were 1235 tornadoes in the United States in 1995, how many of these might we expect to have had wind speeds less than 110 mph? Find 70% of 1235.

   $70\%(1235) = 0.70(1235) = 864.5 \approx 865$

   Around 865 tornadoes would be expected to have had wind speeds of less than 110 mph.

SECTION 9.3 READING HISTOGRAMS

A **histogram** is a special bar graph in which the width of each bar represents a **class interval** and the height of each bar represents the **class frequency**. The following histogram shows student quiz scores.

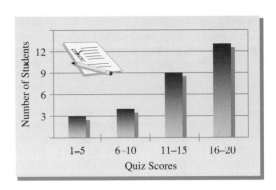

1. How many students received a score of 6–10?

4 students

2. How many students received a score of 11–20?

9 + 13 = 22 students

SECTION 9.4 MEAN, MEDIAN, AND MODE

The **mean** (or **average**) of a set of number items is

$$\text{mean} = \frac{\text{sum of items}}{\text{number of items}}$$

The **median** of an ordered set of numbers is the middle number. If the number of items is even, the median is the mean of the two middle numbers.

The **mode** of a set of numbers is the number that occurs most often. (A set of numbers may have no mode or more than one mode.)

Find the mean, median, and mode of the set of numbers: 33, 35, 35, 43, 68, 68

$$\text{mean} = \frac{33 + 35 + 35 + 43 + 68 + 68}{6} = 47$$

The median is the mean of the two middle numbers:

$$\text{median} = \frac{35 + 43}{2} = 39$$

There are two modes because there are two numbers that both occur twice:

modes: 35 and 68

SECTION 9.5 COUNTING AND INTRODUCTION TO PROBABILITY

An **experiment** is an activity being considered, such as tossing a coin or rolling a die. The possible results of an experiment are the **outcomes**. A **tree diagram** is one way to picture and count outcomes.

Draw a tree diagram for tossing a coin and then choosing a number from 1 to 4.

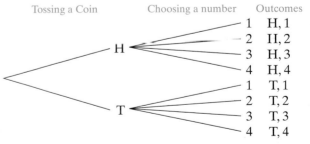

SECTION 9.5 (CONTINUED)	
Any number of outcomes considered together is called an **event**. The **probability** of an event is a measure of the chance or likelihood of it occurring. probability of an event $=\dfrac{\text{number of ways that the event can occur}}{\text{number of possible outcomes}}$	Find the probability of tossing a coin twice and a tail occurring each time. 1 way the event can occur HH HT TH TT 4 possible outcomes probability $=\dfrac{1}{4}$

CHAPTER 9 REVIEW

(9.1) *The following pictograph shows the number of new homes constructed, by state. Use this graph to answer Exercises 1–6.*

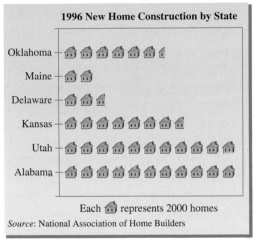

1. How many new homes were constructed in Oklahoma in 1996?   13,000

2. How many new homes were constructed in Kansas in 1996?   15,000

3. Which state shown had the most new homes constructed?   Utah and Alabama

4. Which state shown had the fewest new homes constructed?   Maine

5. Which state(s) shown had more than 13,000 new homes constructed?   Kansas, Utah, and Alabama

6. Which state(s) shown had fewer than 8000 new homes constructed?   Maine and Delaware

The following bar graph shows percent of persons age 25 or more who completed four or more years of college. Use this graph to answer Exercises 7–10.

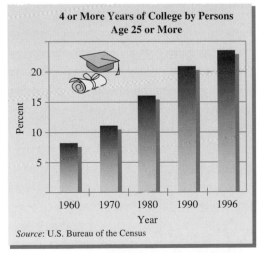

7. Approximate the percent of persons who completed four or more years of college in 1960.   8%

8. What year shown had the greatest percent of persons completing four or more years of college?   1996

9. What years shown had 15% or more of persons completing four or more years of college? 1980, 1990, 1996

10. Describe any patterns you notice in this graph.
answers may vary

The following double line graph shows the number of deaths per 100,000 persons from motor vehicle accidents or firearms. Use this graph to answer Exercises 11–15.

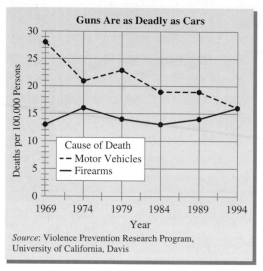

Guns Are as Deadly as Cars

Source: Violence Prevention Research Program, University of California, Davis

11. Approximate the number of deaths per 100,000 persons caused by firearms in 1969. 13

12. Approximate the number of deaths per 100,000 persons caused by motor vehicle accidents in 1969. 28

13. In what year(s) is the number of deaths by firearms decreasing? 1979, 1984

14. In what year(s) is the number of deaths by motor vehicle accidents increasing? 1979

15. Describe the meaning of the single point for 1994.
answers may vary

(9.2) *The following circle graph shows a family's $4000 monthly budget. Use this graph to answer Exercises 16–22. Write all ratios as fractions in simplest form.*

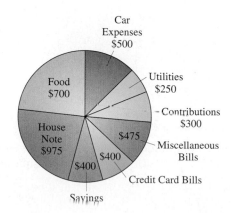

Car Expenses $500
Utilities $250
Contributions $300
Miscellaneous Bills
Credit Card Bills
$475
$400
Savings
$400
House Note $975
Food $700

16. What is the largest budget item?   House Note

17. What is the smallest budget item?   Utilities

18. How much money is budgeted for either the house note or utilities? $1225

19. How much money is budgeted for either savings or contributions? $700

20. Find the ratio of house note to the total monthly budget. $\frac{39}{160}$

21. Find the ratio of food to the total monthly budget. $\frac{7}{40}$

22. Find the ratio of car expenses to food. $\frac{5}{7}$

The following circle graph shows the percent of states with various interstate highway speed limits in 1997. Use this graph to determine the number of states with each speed limit in Exercises 23–26.

Percent of States with
Interstate Highway Speed Limit

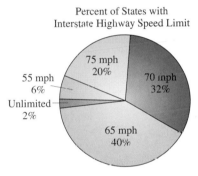

75 mph 20%

55 mph 6%

Unlimited 2%

70 mph 32%

65 mph 40%

Source: National Motorist Association

23. How many states have an interstate highway speed limit of 65 mph?   20 states

24. How many states have an interstate highway speed limit of 75 mph?   10 states

25. How many states have an interstate highway speed limit of either 70 mph or 75 mph?   26 states

26. How many states have an interstate highway speed limit of either 55 mph or 65 mph?   23 states

(9.3) *The following histogram shows the hours worked per week by the employees of Southern Star Furniture. Use this histogram to answer Exercises 27–30.*

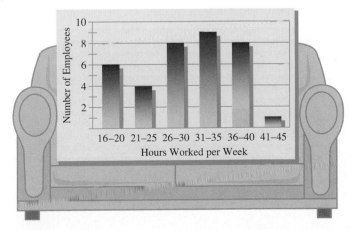

27. How many employees work 21–25 hours per week?
4 employees

28. How many employees work 41–45 hours per week?
1 employee

29. How many employees work 36 hours or more per week?   9 employees

30. How many employees work 30 hours or less per week?   18 employees

Following is a list of monthly record high temperatures for New Orleans, Louisiana. Use this list to complete the frequency distribution table below.

83	96	101	92
85	100	92	102
89	101	87	84

	Class Intervals (temperatures)	Tally	Class Frequency (number of months)
31.	80°–89°	ℍℍ	5
32.	90°–99°	|||	3
33.	100°–109°	||||	4

34. Use the table from Exercises 31, 32, and 33 to draw a histogram.

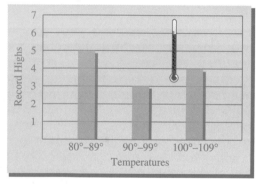

(9.4) *Find the mean, the median, and any mode(s) for each list of numbers.*

35. 13, 23, 33, 14, 6   mean: 17.8; median: 14; no mode

36. 45, 21, 60, 86, 64   mean: 55.2; median: 60; no mode

37. $14,000, $20,000, $12,000, $20,000, $36,000, $45,000
mean: $24,500; median: $20,000; mode: $20,000

38. 560, 620, 123, 400, 410, 300, 400, 780, 430, 450
mean: 447.3; median: 420; mode: 400

For Exercises 39 and 40, the grades are given for a student for a particular semester. Find each grade point average. If necessary, round the grade point average to the nearest hundredth.

39.

Grade	Credit Hours
A	3
A	3
C	2
B	3
C	1

3.25

40.

Grade	Credit Hours
B	3
B	4
C	2
D	2
B	3

2.57

(9.5) *Draw a tree diagram for each experiment. Then use the diagram to determine the number of outcomes.*

Spinner 1   Spinner 2

41. Toss a coin and then spin Spinner 1.

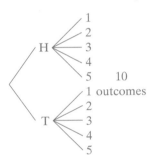

10 outcomes

42. Spin Spinner 2 and then toss a coin.

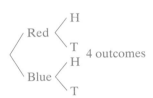

4 outcomes

43. Spin Spinner 1 twice.

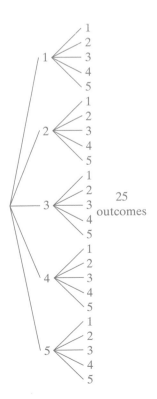

25 outcomes

44. Spin Spinner 2 twice.

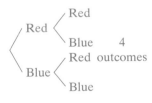

4 outcomes

45. Spin Spinner 1 and then Spinner 2.

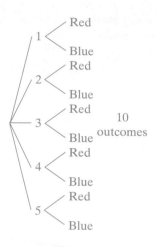

Find the probability of each event.

Die

46. Roll a 4 on a die. $\frac{1}{6}$

47. Roll a 3 on a die. $\frac{1}{6}$

48. Spin a 4 on Spinner 1. $\frac{1}{5}$

49. Spin a 3 on Spinner 1. $\frac{1}{5}$

50. Spin either a 1, 3, or 5 on Spinner 1. $\frac{3}{5}$

51. Spin either a 2 or 4 on Spinner 1. $\frac{2}{5}$

Name _____ **Section** _____ **Date** _____

CHAPTER 9 TEST

The following pictograph shows the money collected each week from a wrapping paper fundraiser. Use this graph to answer Exercises 1–3.

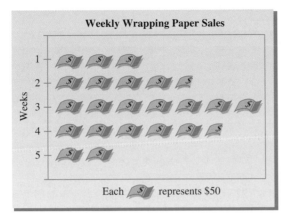

Weekly Wrapping Paper Sales

Each $ represents $50

1. How much money was collected during the second week?

2. During which week was the most money collected? How much money was collected during that week?

3. What was the total money collected for the fundraiser?

The following bar graph shows the normal monthly precipitation in centimeters for Chicago, Illinois. Use this graph to answer Exercises 4–6.

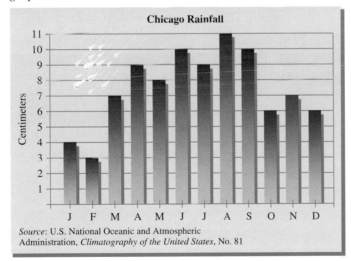

Chicago Rainfall

Source: U.S. National Oceanic and Atmospheric Administration, *Climatography of the United States*, No. 81

4. During which months does Chicago normally have greater than 9 centimeters of rainfall?

5. During which month does Chicago normally have the least amount of rainfall? How much rain falls during that month?

6. During which month(s) does 7 centimeters of rain normally fall?

ANSWERS

1. $225

2. 3rd week, $350

3. $1100

4. June, August, September

5. February, 3 centimeters

6. March and November

605

Name _____

7. Use the information in the table to draw a bar graph. Clearly label each bar.

Selected 1997 Average Math SAT Scores	
State	Score
Florida	499
Alabama	555
California	514
Oklahoma	560
New Jersey	508

Source: The College Board

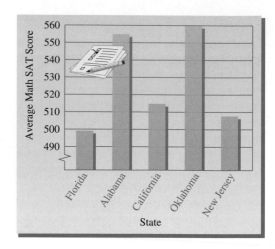

The following line graph shows online sales of recorded music (CDs, albums, and cassettes) and their projected growth.

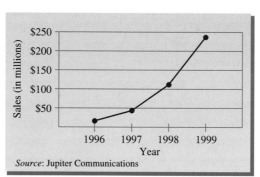

Source: Jupiter Communications

8. Approximate the 1998 online sales of recorded music.

9. What year has shown the greatest increase in online sales?

10. During what years were sales less than $50 million?

Name _____

The result of a survey of 200 people is shown in the following circle graph. Each person was asked to tell his or her favorite type of music. Use this graph to answer Exercises 11–12.

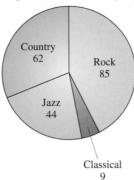

11. Find the ratio of those who prefer rock music to total number surveyed.

12. Find the ratio of those who prefer country music to those who prefer jazz.

The following circle graph shows U.S. car sales by type in 1997. If there were approximately 8,272,000 cars sold in the U.S in 1997, find how many cars of the type given were sold.

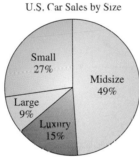

U.S. Car Sales by Size

Source: Ward's Automotive Reports

13. Small cars

14. Luxury cars

13. 2,233,440 small cars

14. 1,240,800 luxury cars

A professor measures the heights of the students in her class. The results are shown on the following histogram. Use this histogram to answer Exercises 15–16.

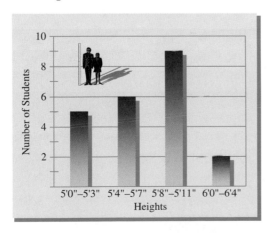

15. How many students are 5′8″–5′11″ tall?

16. How many students are 5′7″ or shorter?

15. 9 students

16. 11 students

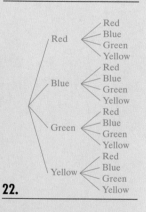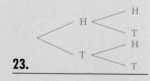
17. The history test scores of 25 students are shown below. Use these scores to complete the frequency distribution table.

70 86 81 65 92
43 72 85 69 97
82 51 75 50 68
88 83 85 77 99
77 63 59 84 90

Class Intervals (scores)	Tally	Class Frequency (number of students)
40–49	|	1
50–59	|||	3
60–69	||||	4
70–79	⊬⊬	5
80–89	⊬⊬ |||	8
90–99	||||	4

18. Use the results of Exercise 17 to draw a histogram.

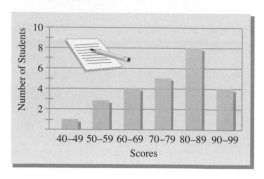

Find the mean, median, and mode of each list of numbers.

19. 26, 32, 42, 43, 49

20. 8, 10, 16, 16, 14, 12, 12, 13

Find the grade point average. If necessary, round to the nearest hundredth.

21.

Grade	Credit Hours
A	3
B	3
C	3
B	4
A	1

22. Draw a tree diagram for the experiment of spinning the spinner twice.

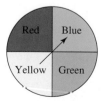

23. Draw a tree diagram for the experiment of tossing a coin twice.

Suppose that the numbers 1 to 10 are each written on a scrap of paper and placed in a bag. You then select one number from the bag.

24. What is the probability of choosing a 6 from the bag?

25. What is the probability of choosing a 3 or a 4 from the bag?

CUMULATIVE REVIEW

1. Simplify: $(8 - 6)^2 + 2^3 \cdot 3$

2. Write $\dfrac{30}{108}$ in simplest form.

3. Add: $1\dfrac{4}{5} + 4 + 2\dfrac{1}{2}$

4. The formula for finding the area of a triangle is Area $= \dfrac{1}{2} \cdot$ base $\cdot$ height. Find the area of the triangle shown.

5. Write the ratio of $10 to $15 as a fraction in simplest form.

Write each rate as a fraction in simplest form.

6. $2160 for 12 weeks

7. 360 miles on 16 gallons of gasoline

8. Is $\dfrac{1\frac{1}{6}}{10\frac{1}{2}} = \dfrac{\frac{1}{2}}{4\frac{1}{2}}$ a true proportion?

9. The standard dose of an antibiotic is 4 cc (cubic centimeters) for every 25 pounds (lb) of body weight. At this rate, find the standard dose for a 140-lb woman.

Write each percent as a decimal.

10. 4.6%

11. 190%

Write each percent as a fraction in simplest form.

12. 40%

13. $33\dfrac{1}{3}\%$

610

Name _____

14. Translate the following to an equation: Five is what percent of 20?

15. Find the sales tax and the total price on a purchase of an $85.50 trench coat in a city where the sales tax rate is 7.5%.

16. An accountant invested $2000 at a simple interest rate of 10% for 2 years. What total amount of money will she have from her investment in 2 years?

17. Convert 7 feet to yards.

18. Divide 9 lb 6 oz by 2.

19. Convert 2.35 cg to grams.

20. Convert 3210 ml to liters.

21. Convert 15°C to degrees Fahrenheit.

22. Find the complement of a 48° angle.

23. Find $\sqrt{\dfrac{1}{36}}$.

24. Find the mode of the following list of numbers:
11, 14, 14, 16, 31, 56, 65, 77, 77, 78, 79

25. If a coin is tossed twice, find the probability of tossing a head and then a head.

Signed Numbers

Thus far, we have studied whole numbers, fractions, and decimals. However, these numbers are not sufficient for representing many situations in real life. For example, to express 5° below zero or $100 in debt, numbers less than zero are needed. This chapter is devoted to signed numbers, which include numbers less than zero, and to operations on these numbers.

10.1 Signed Numbers

10.2 Adding Signed Numbers

10.3 Subtracting Signed Numbers

Integrated Review—Signed Numbers

10.4 Multiplying and Dividing Signed Numbers

10.5 Order of Operations

Golf continues to increase in popularity. This game has its roots in Scotland in the 14th and 15th centuries. It became so popular that in 1457 the Scottish Parliament had to outlaw playing golf because it was keeping people from practicing their archery, which was extremely important from a military standpoint. The game of golf eventually spread across the European continent and then to North America and around the world. A variety of golf governing bodies and golfers associations now exist. One of these is the Ladies Professional Golf Association (LPGA). The LPGA Tour was founded in 1950, offering 21 events by 1952 and prize money totaling $200,000 by 1959. In 1996 LPGA rookie Karrie Webb became the first rookie in all of the golf world to earn $1 million in a single season of play. By 1997, the annual LPGA Tour prize money totaled $30.2 million. In Exercise 58 on page 627 and Exercises 65 and 66 on page 643, we will see how negative numbers are used in scoring in the game of golf.

612

Name _____ **Section** _____ **Date** _____

CHAPTER 10 PRETEST

1. Graph each signed number in the list on the same number line.

$$-4, 1, 2\frac{1}{2}, -1\frac{1}{4}$$

Insert < *or* > *between each pair of numbers to make a true statement.*

2. −15 ⟷ −17

3. $\dfrac{3}{5}$ ⟷ $-\dfrac{1}{5}$

Find the absolute value.

4. $|-22|$

5. $|9.8|$

6. Find the opposite of −12.

Perform indicated operations.

7. −31 + 50

8. −8 − 16

9. 14 − 29

10. 4 − 9 + 10

11. 7 − (−41) + (−9)

12. −6(−9)

13. −3(−4)(10)

14. $(-3)^4$

15. $\dfrac{-92}{-2}$

16. $\dfrac{1.44}{-1.2}$

17. 5 + 8(−3)

18. (−2 ÷ 2) − 6 · 3 − 9

19. Juanita has $275 in her checking account. She writes a check for $102, makes a deposit of $29, and then writes another check for $210. Find the amount left in her account. (Write the amount as an integer.)

20. A card player had a score of −24 for each of 3 games. Find his total score.

10.1 SIGNED NUMBERS

A REPRESENTING REAL-LIFE SITUATIONS

Thus far in this text, all numbers have been 0 or greater than 0. Numbers greater than 0 are called **positive numbers**. However, sometimes situations exist that cannot be represented by a number greater than 0. For example,

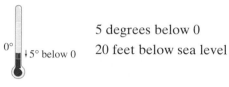

5 degrees below 0

20 feet below sea level

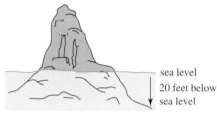

sea level
20 feet below
sea level

To represent these situations, we need numbers less than 0.

Extending the number line to the left of 0 allows us to picture **negative numbers**, numbers that are less than 0.

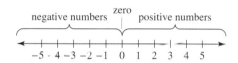

When a single + sign or no sign is in front of a number, the number is a positive number. When a single − sign is in front of a number, the number is a negative number. Together, we call positive numbers, negative numbers, and zero the **signed numbers**.

−5 indicates "negative five."
5 and +5 both indicate "positive five."
The number 0 is neither positive nor negative.

Some signed numbers are integers. The **integers** consist of the positive numbers, zero, and the negative numbers labeled on the number line above. The integers are

$$\ldots, -3, -2, -1, 0, 1, 2, 3, \ldots$$

Now we have numbers to represent the situations previously mentioned.

5 degrees below 0   −5°
twenty feet below sea level   −20 feet

Example 1   Representing Depth with a Signed Number

Jack Mayfield, a miner for the Molly Kathleen Gold Mine, is presently 150 feet below the surface of the earth. Represent this position using a signed number.

Solution: If 0 represents the surface of the earth, then 150 feet below the surface can be represented by −150.   ■

TEACHING TIP   Classroom Activity

Ask students to help create a list of real-life situations in which negative numbers are used or occur.

Practice Problem 1

a. A deep-sea diver is 800 feet below the surface of the ocean. Represent this position using a signed number.

b. A company reports a $2 million loss for the year. Represent this amount using a signed number.

Answers

1. a. −800, **b.** −2 million

Practice Problem 2

Graph the signed numbers -5, 3, -3, $-1\frac{3}{4}$, and -4.5 on a number line.

✓ CONCEPT CHECK

Is there a smallest negative number? Is there a largest positive number? Explain.

Practice Problems 3–9

Insert $<$ or $>$ between each pair of numbers to make a true statement.

3. 8 -8

4. -11 0

5. -15 -14

6. 3 -4.6

7. 0 -2

8. -7.9 7.9

9. $-\frac{3}{8}$ $-1\frac{1}{7}$

Answers

2.

3. $>$, 4. $<$, 5. $<$, 6. $>$, 7. $>$,
8. $<$, 9. $>$

✓ **Concept Check:** No, there are no smallest negative or largest positive numbers.

B GRAPHING SIGNED NUMBERS

Example 2   Graph the signed numbers -3, 2, -2, $-\frac{1}{2}$, and -3.8 on a number line.

Solution:

C COMPARING SIGNED NUMBERS

For any two numbers graphed on a number line, the number to the **right** is the **greater number** and the number to the **left** is the **smaller number**. Recall that the symbol $>$ means "is greater than" and the symbol $<$ means "is less than."

To illustrate, both -5 and -7 are graphed on the number line shown.

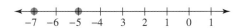

The graph of -7 is **to the left of** -5, so -7 **is less than** -5. We can write this as

$$-7 < -5$$

We can also write

$$-5 > -7$$

since -5 is **to the right** of -7, so -5 **is greater than** -7.

TRY THE CONCEPT CHECK IN THE MARGIN.

Examples   Insert $<$ or $>$ between each pair of numbers to make a true statement.

3. -7 7 -7 is to the left of 7, so $-7 < 7$.
4. 0 -4 0 is to the right of -4, so $0 > -4$.
5. -9 -11 -9 is to the right of -11, so $-9 > -11$.
6. -2.9 -1 -2.9 is to the left of -1, so $-2.9 < -1$.
7. -6 0 -6 is to the left of 0, so $-6 < 0$.
8. 8.6 -8.6 8.6 is to the right of -8.6, so $8.6 > -8.6$.
9. $-\frac{1}{4}$ $-2\frac{1}{2}$ $-\frac{1}{4}$ is to the right of $-2\frac{1}{2}$, so $-\frac{1}{4} > -2\frac{1}{2}$.

HELPFUL HINT

If you think of $<$ and $>$ as arrowheads, notice that in a true statement the arrow always points to the smaller number.

$5 > -4$ $-3 < -1$

↑   ↑

smaller   smaller
number   number

D Finding the Absolute Value of a Number

The **absolute value** of a number is the number's distance from 0 on the number line. The symbol for absolute value is | |. For example, |3| is read as "the absolute value of 3."

|3| = 3 because 3 is 3 units from 0.

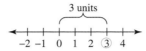

|−3| = 3 because −3 is 3 units from 0.

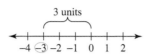

Examples   Find each absolute value.

10. $|-2| = 2$ because −2 is 2 units from 0.
11. $|5| = 5$ because 5 is 5 units from 0.
12. $|0| = 0$ because 0 is 0 units from 0.
13. $\left|-\dfrac{3}{4}\right| = \dfrac{3}{4}$ because $-\dfrac{3}{4}$ is $\dfrac{3}{4}$ unit from 0.
14. $|1.2| = 1.2$ because 1.2 is 1.2 units from 0.

HELPFUL HINT

Since the absolute value of a number is that number's *distance* from 0, the absolute value of a number is always 0 or positive. It is never negative.

$$|0| = 0 \qquad |-6| = 6$$
  ↑   ↑
 zero   a positive number

E Finding Opposites

Two numbers that are the same distance from 0 on the number line but are on opposite sides of 0 are called **opposites**.

4 and −4 are opposites.

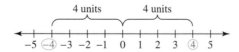

When two numbers are opposites, we say that each is the opposite of the other. Thus **4 is the opposite of −4** and **−4 is the opposite of 4.**

The phrase "the opposite of" is written in symbols as " − ". For example,

the opposite of   5   is   −5
      ↓   ↓   ↓   ↓
      −   (5) =   −5

the opposite of   −3   is   3
      ↓   ↓   ↓   ↓
      −   (−3) =   3

Notice we just stated that

$$-(-3) = 3$$

Practice Problems 10–14

Find each absolute value.

10. $|6|$

11. $|-4|$

12. $|0|$

13. $\left|\dfrac{7}{8}\right|$

14. $|-3.4|$

Answers

10. 6, **11.** 4, **12.** 0, **13.** $\dfrac{7}{8}$, **14.** 3.4

In general, we have the following:

> **OPPOSITES**
>
> If a is a number, then $-(-a) = a$.

Notice that because "the opposite of" is written as " $-$ ", to find the opposite of a number we place a " $-$ " sign in front of the number.

Practice Problems 15–19

Find the opposite of each number.

15. -7 16. 0

17. $\dfrac{11}{15}$ 18. -9.6

19. 4

Examples   Find the opposite of each number.

15. 12   The opposite of 12 is -12.

16. 1.7   The opposite of 1.7 is -1.7.

17. -3 The opposite of -3 is $-(-3)$ or 3.

18. 0   The opposite of 0 is -0 or 0.

19. $-\dfrac{10}{13}$ The opposite of $-\dfrac{10}{13}$ is $-\left(-\dfrac{10}{13}\right)$ or $\dfrac{10}{13}$.

> **HELPFUL HINT**
>
> Remember that 0 is neither positive or negative.

Practice Problems 20–22

Simplify.

20. $-(-11)$

21. $-|7|$

22. $-|-2|$

Examples   Simplify.

20. $-(-4) = 4$ The opposite of negative 4 is 4.

21. $-|6| = -6$ The opposite of the absolute value of 6 is the opposite of 6, which is -6.

22. $-|-5| = -5$ The opposite of the absolute value of -5 is the opposite of 5, which is -5.

Answers

15. 7, **16.** 0, **17.** $-\dfrac{11}{15}$, **18.** 9.6, **19.** -4,

20. 11, **21.** -7, **22.** -2

EXERCISE SET 10.1

A *Represent each quantity by a signed number. See Example 1.*

1. A worker in a silver mine in Nevada works 1445 feet underground.

2. A scuba diver is swimming 35 feet below the surface of the water in the Gulf of Mexico.

3. The peak of Mount Whitney in California is 14,494 feet above sea level. (*Source:* U.S. Geological Survey)

4. The average depth of the Atlantic Ocean is 11,730 feet below the surface of the ocean. (*Source: 1998 World Almanac*)

5. The Virginia Cavaliers football team lost 15 yards on a play.

6. The record high temperature in Arizona is 128 degrees Fahrenheit above zero. (*Source:* National Climatic Data Center)

7. The Dow Jones stock market average fell 317 points in one day.

8. The lowest elevation in the United States is found at Death Valley, California at an elevation of 282 feet below sea level. (*Source:* U.S. Geological Survey)

9. Converse, Inc. manufactures a wide variety of footwear. In fiscal year 1998, Converse posted a net loss of $5.049 million. (*Source:* Converse, Inc.)

10. For fiscal year 1997, Apple Computer, Inc., reported a net loss of $1.045 billion. (*Source:* Apple Computer, Inc.)

11. The temperature on one January day in Chicago was −10 degrees Celsius. Tell whether this temperature is cooler or warmer than −5 degrees Celsius.

12. Two divers are exploring the bottom of a trench in the Pacific Ocean. Joe is at 135 feet below the surface of the ocean and Sara is at 157 feet below the surface. Determine who is deeper in the water.

1. −1445

2. −35

3. +14,494

4. −11,730

5. −15

6. +128

7. −317

8. −282

9. −5.049 million

10. −1.045 billion

11. cooler

12. Sara

13. −3.3%

14. −23.4%

15. see number line

16. see number line

17. see number line

18. see number line

19. <

20. >

21. >

22. >

23. >

24. <

25. >

26. >

27. <

28. >

29. <

30. >

31. <

32. <

33. >

34. <

35. 5

36. 7

37. 8

38. 19

618

Name _____

13. In 1997, the number of music CDs shipped to retailers reflected a 3.3 percent loss from the previous year. Write a signed number to represent the percent loss in CDs shipped. (*Source:* Recording Industry Association of America)

14. In 1997, the number of music cassettes shipped to retailers reflected a 23.4 percent loss from the previous year. Write a signed number to represent the percent loss of cassettes shipped. (*Source:* Recording Industry Association of America)

B *Graph the signed numbers in each list on a number line. See Example 2.*

15. $-3, 0, 4, -1\frac{1}{2}$

16. $-4, 0, 2, -3\frac{1}{4}$

17. $5, -2, -4.7$

18. $3, -1, -4, -2.1$

C *Insert* < *or* > *between each pair of numbers to make a true statement. See Examples 3 through 9.*

19. 5   7

20. 16   10

21. 4   0

22. 8   0

23. −5   −7

24. −12   −10

25. 0   −3

26. 0   −7

27. −26   26

28. 13   −13

29. −4.6   −2.7

30. 0 $-\frac{7}{8}$

31. $-1\frac{3}{4}$ 0

32. −8.4   −1.6

33. $\frac{1}{4}$ $-\frac{8}{11}$

34. −0.2   6

D *Find each absolute value. See Examples 10 through 14.*

35. $|5|$

36. $|7|$

37. $|-8|$

38. $|-19|$

39. $|0|$ **40.** $|100|$ **41.** $|-5|$ **42.** $|-10|$

43. $|-8.1|$ **44.** $\left|-\dfrac{1}{2}\right|$ **45.** $\left|\dfrac{9}{10}\right|$ **46.** $|-31.6|$

47. $\left|-\dfrac{3}{8}\right|$ **48.** $\left|\dfrac{20}{23}\right|$ **49.** $|7.6|$ **50.** $|-0.6|$

E *Find the opposite of each number. See Examples 15 through 19.*

51. 5   **52.** 8   **53.** -4 **54.** -6

55. 23.6   **56.** 123.9   **57.** $-\dfrac{9}{16}$ **58.** $-\dfrac{4}{9}$

59. -0.7 **60.** -4.4 **61.** $\dfrac{17}{18}$ **62.** $\dfrac{2}{3}$

Simplify. See Examples 20 through 22.

63. $|-7|$ **64.** $|-11|$ **65.** $-|20|$ **66.** $-|43|$

67. $-|-3|$ **68.** $-|-18|$ **69.** $-(-8)$ **70.** $-(-7)$

71. $|-14|$ **72.** $-(-14)$ **73.** $-(-29)$ **74.** $-|-29|$

#	Answer
39.	0
40.	100
41.	5
42.	10
43.	8.1
44.	$\dfrac{1}{2}$
45.	$\dfrac{9}{10}$
46.	31.6
47.	$\dfrac{3}{8}$
48.	$\dfrac{20}{23}$
49.	7.6
50.	0.6
51.	-5
52.	-8
53.	4
54.	6
55.	-23.6
56.	-123.9
57.	$\dfrac{9}{16}$
58.	$\dfrac{4}{9}$
59.	0.7
60.	4.4
61.	$-\dfrac{17}{18}$
62.	$-\dfrac{2}{3}$
63.	7
64.	11
65.	-20
66.	-43
67.	-3
68.	-18
69.	8
70.	7
71.	14
72.	14
73.	29
74.	-29

75. 13

76. 9

77. 35

78. 35

79. 360

80. 489

81. true

82. false

83. true

84. answers may vary

85. answers may vary

620

Name _____

REVIEW AND PREVIEW

Add. See Section 1.2.

75. $0 + 13$ **76.** $9 + 0$ **77.** $15 + 20$

78. $20 + 15$ **79.** $47 + 236 + 77$ **80.** $362 + 37 + 90$

◆ COMBINING CONCEPTS

For Exercises 81–83, determine whether each statement is true or false.

81. A positive number is always greater than a negative number.

82. The absolute value of a number is *always* a positive number.

83. Zero is always less than a positive number.

84. Explain how to determine which of two signed numbers is larger.

85. Write in your own words how to find the absolute value of a signed number.

10.2 ADDING SIGNED NUMBERS

A ADDING SIGNED NUMBERS

Adding signed numbers can be visualized by using a number line. A positive number can be represented on the number line by an arrow of appropriate length pointing to the right, and a negative number by an arrow of appropriate length pointing to the left.

Both arrows represent 2 or +2. They both point to the right and they are both 2 units long.

Both arrows represent −3. They both point to the left and they are both 3 units long.

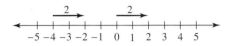
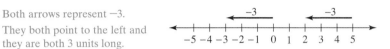

To add signed numbers on a number line, such as $5 + (-2)$, we start at 0 on the number line and draw an arrow representing 5. From the tip of this arrow, we draw another arrow representing −2. The tip of the second arrow ends at their sum, 3.

$$5 + (-2) = 3$$

To add $-1 + (-4)$ on the number line, we start at 0 and draw an arrow representing −1. From the tip of this arrow, we draw another arrow representing −4. The tip of the second arrow ends at their sum, −5.

$$-1 + (-4) = -5$$

Example 1   Add using a number line: $-3 + (-4)$

Solution:

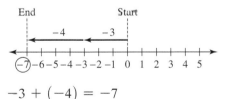

$$-3 + (-4) = -7$$

Example 2   Add using a number line: $-7 + 3$

Solution:

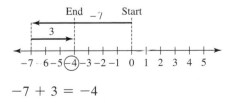

$$-7 + 3 = -4$$

Using a number line each time we add two numbers can be time consuming. Instead, we can notice patterns in the previous examples and write rules for adding signed numbers.
Rules for adding signed numbers depend on whether we are adding

Ojectives

A Add signed numbers.
B Solve problems by adding signed numbers.

SSM   CD-ROM   Video
10.2

TEACHING TIP

Before discussing the rules for adding signed numbers, have students list the possible combinations of signs when adding 2 numbers and the results which may occur for these sums. Get them started by saying, "One possibility when adding two signed numbers is that both numbers are positive. What results are possible when adding two positive numbers? What other possibilities exist when adding two signed numbers?" When they get to a possibility which can give a positive or negative result, ask them to describe the subset of this group which gives only positive results.

Practice Problem 1

Add using a number line: $5 + (-1)$

Practice Problem 2

Add using a number line: $-6 + (-2)$

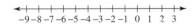

Answers

1.

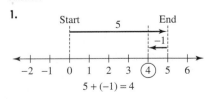

$5 + (-1) = 4$

2.

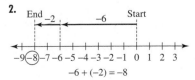

$-6 + (-2) = -8$

numbers with the same sign or different signs. When adding two numbers with the same sign, notice that the sign of the sum is the same as the sign of the addends.

ADDING TWO NUMBERS WITH THE SAME SIGN
Step 1. Add their absolute values.
Step 2. Use their common sign as the sign of the sum.

Practice Problem 3

Add: $(-3) + (-9)$

Example 3 Add: $-2 + (-21)$

Solution: **Step 1.** $|-2| = 2$, $|-21| = 21$, and $2 + 21 = 23$.

Step 2. Their common sign is negative, so the sum is negative:

$$-2 + (-21) = -23$$

Practice Problems 4–7

Add.

4. $-12 + (-3)$

5. $9 + 5$

6. $-\dfrac{3}{7} + \left(-\dfrac{2}{7}\right)$

7. $-8.3 + (-5.7)$

Examples Add.

4. $-5 + (-1) = -6$

5. $2 + 6 = 8$

6. $-\dfrac{3}{8} + \left(-\dfrac{1}{8}\right) = -\dfrac{\overset{1}{\cancel{4}}}{\underset{2}{\cancel{8}}} = -\dfrac{1}{2}$

7. $-1.2 + (-7.1) = -8.3$

The rule for adding two numbers with different signs follows.

ADDING TWO NUMBERS WITH DIFFERENT SIGNS
Step 1. Find the larger absolute value minus the smaller absolute value.
Step 2. Use the sign of the number with the larger absolute value as the sign of the sum.

Practice Problem 8

Add: $-3 + 9$

Example 8 Add: $-2 + 5$

Solution: **Step 1.** $|-2| = 2$, $|5| = 5$, and $5 - 2 = 3$.

Step 2. 5 has the larger absolute value and its sign is an understood $+$:

$$-2 + 5 = +3 \text{ or } 3$$

Practice Problem 9

Add: $2 + (-8)$

Example 9 Add: $3 + (-7)$

Solution: **Step 1.** $|3| = 3$, $|-7| = 7$, and $7 - 3 = 4$.

Step 2. -7 has the larger absolute value and its sign is $-$:

$$3 + (-7) = -4$$

Practice Problems 10–13

Add.

10. $-46 + 20$ 11. $8.6 + (-6.2)$

12. $-\dfrac{3}{4} + \dfrac{1}{8}$ 13. $-2 + 0$

Examples Add.

10. $-18 + 10 = -8$

11. $12.9 + (-8.6) = 4.3$

12. $-\dfrac{1}{2} + \dfrac{1}{6} = -\dfrac{3}{6} + \dfrac{1}{6} = -\dfrac{2}{6} = -\dfrac{1}{3}$

13. $0 + (-5) = -5$ The sum of 0 and any number is the number.

Answers

3. -12, 4. -15, 5. 14, 6. $-\dfrac{5}{7}$, 7. -14,

8. 6, 9. -6, 10. -26, 11. 2.4, 12. $-\dfrac{5}{8}$,

13. -2

Copyright 1999 Prentice-Hall, Inc.

TRY THE CONCEPT CHECK IN THE MARGIN.

When we add three or more numbers, we follow the order of operations and add from left to right. That is, we start at the left and add the first two numbers. Then we add their sum to the next number. Continue this process until the addition is completed.

Example 14   Add: $(-3) + 4 + (-11)$

Solution: $(-3) + 4 + (-11) = 1 + (-11)$
$$= -10$$

Example 15   Add: $1 + (-10) + (-8) + 9$

Solution: $1 + (-10) + (-8) + 9 = -9 + (-8) + 9$
$$= -17 + 9$$
$$= -8$$

TRY THE CONCEPT CHECK IN THE MARGIN.

B   SOLVING PROBLEMS BY ADDING SIGNED NUMBERS

Next, we practice solving problems that require adding signed numbers.

Example 16   Calculating Temperature

On January 6th the temperature in Caribou, Maine at 8 A.M. was $-12°$ Fahrenheit. By 9 A.M., the temperature had risen by 4 degrees, and by 10 A.M. it had risen 6 degrees from the 9 A.M. temperature. What was the temperature at 10 A.M.?

Solution: **1.** UNDERSTAND. Read and reread the problem.
2. TRANSLATE.

In words:

temperature at 10 A.M.	=	8 A.M. temperature	+	rise of 4°	+	rise of 6°

Translate:
$$\text{temperature at 10 A.M.} = -12 + (+4) + (+6)$$

3. SOLVE.

$$\text{temperature at 10 A.M.} = -12 + (+4) + (+6)$$
$$= -8 + (+6)$$
$$= -2$$

4. INTERPRET. Check and state your conclusion: The temperature was $-2°F$ at 10 A.M.

CALCULATOR EXPLORATIONS
ENTERING NEGATIVE NUMBERS

To enter a negative number on a calculator, find the key marked
+/− . (Some calculators have a key marked CHS and some calculators have a special key (−) for entering a negative sign.) To
enter the number −2, for example, press the keys 2 +/− . The
display will read $\boxed{\qquad -2}$.

To find $-32 + (-131)$, press the keys

32 +/− + 131 +/− =

or

ENTER

The display will read $\boxed{\qquad -163}$. Thus $-32 + (-131) = -163$.

Use a calculator to perform each indicated operation.

1. $-256 + 97$ -159
2. $811 + (-1058)$ -247
3. $6(15) + (-46)$ 44
4. $-129 + 10(48)$ 351
5. $-108.65 + (-786.205)$
   -894.855
6. $-196.662 + (-129.856)$
   -326.518

Name _____ **Section** _____ **Date** _____

MENTAL MATH

Add.
1. $5 + 0$ 2. $(-2) + 0$ 🔲 3. $0 + (-35)$ 4. $0 + 3$

EXERCISE SET 10.2

A *Add using a number line. See Examples 1 and 2.*

1. $8 + 2$

$$+8 \qquad +2$$
$$\leftarrow\!+\!+\!+\!+\!+\!+\!+\!+\!+\!+\!+\!+\!+\!+\!+\!+\!+\!\rightarrow$$
$$-7\,-6\,-5\,-4\,-3\,-2\,-1\;0\;1\;2\;3\;4\;5\;6\;7\;8\;9\;(10)$$

2. $9 + (-4)$

$$+9$$
$$-4$$
$$\leftarrow\!+\!+\!+\!+\!+\!+\!+\!+\!+\!+\!+\!+\!+\!+\!+\!+\!+\!\rightarrow$$
$$-7\,-6\,-5\,-4\,-3\,-2\,-1\;0\;1\;2\;3\;4\;(5)\;6\;7\;8\;9\;10$$

3. $-4 + 7$

$$-4$$
$$+7$$
$$\leftarrow\!+\!+\!+\!+\!+\!+\!+\!+\!+\!+\!+\!+\!+\!+\!+\!+\!+\!\rightarrow$$
$$-7\,-6\,-5\,-4\,-3\,-2\,-1\;0\;1\;2\;(3)\;4\;5\;6\;7\;8\;9\;10$$

4. $10 + (-3)$

$$+10$$
$$-3$$
$$\leftarrow\!+\!+\!+\!+\!+\!+\!+\!+\!+\!+\!+\!+\!+\!+\!+\!+\!+\!\rightarrow$$
$$-7\,-6\,-5\,-4\;3\,-2\,-1\;0\;1\;2\;3\;4\;5\;6\;(7)\;8\;9\;10$$

5. $-13 + 7$

$$-13$$
$$+7$$
$$\leftarrow\!+\!+\!+\!+\!+\!+\!+\!+\!+\!+\!+\!+\!+\!+\!+\!+\!\rightarrow$$
$$-13\,-12\,-11\,-10\,-9\,-8\,-7\,(6)\,-5\,-4\,-3\,-2\,-1\;0\;1\;2\;3\;4$$

6. $(-6) + (-5)$

$$-5 \qquad -6$$
$$\leftarrow\!+\!+\!+\!+\!+\!+\!+\!+\!+\!+\!+\!+\!+\!+\!+\!+\!\rightarrow$$
$$-13\,-12\,(11)\,-10\,-9\,-8\,-7\,-6\,-5\,-4\,-3\,-2\,-1\;0\;1\;2\;3\;4$$

1. see number line

2. see number line

3. see number line

4. see number line

5. see number line

6. see number line

Add. See Examples 3 through 13.
7. $23 + 12$ 8. $15 + 42$ 🔲 9. $-6 + (-2)$ 10. $-5 + (-4)$

11. $-43 + 43$ 12. $-62 + 62$ 🔲 13. $6 + (-2)$ 14. $8 + (-3)$

15. $-6 + 8$ 16. $-8 + 12$ 17. $3 + (-5)$ 🔲 18. $5 + (-9)$

7. 35

8. 57

9. −8

10. −9

11. 0

12. 0

13. 4

14. 5

15. 2

16. 4

17. −2

18. −4

625

626

Name _____

19. $-2 + (-9)$ **20.** $-6 + (-1)$ **21.** $-12 + (-12)$ **22.** $-23 + (-23)$

23. $-25 + (-32)$ **24.** $-45 + (-90)$ **25.** $-123 + (-100)$

26. $-500 + (-230)$ **27.** $-7 + 7$ **28.** $-10 + 10$

29. $12 + (-5)$ **30.** $24 + (-10)$ **31.** $-6 + 3$

32. $-8 + 2$ **33.** $-12 + 3$ **34.** $-15 + 5$

35. $56 + (-26)$ **36.** $89 + (-37)$ **37.** $-37 + 57$

38. $-25 + 65$ **39.** $-42 + 93$ **40.** $-64 + 164$

41. $-6.3 + (-2.2)$ **42.** $-9.1 + (-4.6)$ **43.** $-10.7 + 15.3$ **44.** $-19.6 + 23.1$

45. $-\dfrac{2}{3} + \left(-\dfrac{1}{6}\right)$ **46.** $-\dfrac{3}{8} + \left(-\dfrac{1}{4}\right)$ **47.** $-\dfrac{4}{5} + \dfrac{1}{10}$ **48.** $-\dfrac{9}{16} + \dfrac{3}{8}$

Add. See Examples 14 and 15.

49. $-4 + 2 + (-5)$ **50.** $-1 + 5 + (-8)$

51. $-5.2 + (-7.7) + (-11.7)$ **52.** $-10.3 + (-3.2) + (-2.7)$

53. $12 + (-4) + (-4) + 12$ **54.** $18 + (-9) + 5 + (-2)$

55. $(-10) + 14 + 25 + (-16)$ **56.** $34 + (-12) + (-11) + 213$

B *Solve. See Example 16.*

57. The temperature at 4 P.M. on February 2nd was −10° Celsius. By 11 P.M. the temperature had risen 12 degrees. Find the temperature at 11 P.M.

58. Scores in golf can be positive or negative integers. For example, a score of 3 *over* par can be represented by +3 and a score of 5 *under* par can be represented by −5. If Fred Couples had scores of 3 over par, 6 under par, and 7 under par for three games of golf, what was his total score?

59. Suppose a deep-sea diver dives from the surface to 165 feet below the surface. He then dives down 16 more feet. Use positive and negative numbers to represent this situation. Then find the diver's present depth.

60. Suppose a diver dives from the surface to 248 meters below the surface and then swims up 6 meters, down 17 meters, down another 24 meters, and then up 23 meters. Use positive and negative numbers to represent this situation. Then find the diver's depth after these movements.

In some card games, it is possible to have positive and negative scores. The table shows the scores for two teams playing a series of four card games. Use this table to answer Exercises 61 and 62.

	Game 1	Game 2	Game 3	Game 4
Team 1	−2	−13	20	2
Team 2	5	11	−7	−3

61. Find each team's total score after four games. If the winner is the team with the greater score, find the winning team.

62. Find each team's total score after three games. If the winner is the team with the greater score, which team was winning after three games?

63. The all-time record low temperature for Utah is −69°F, which was recorded on February 1, 1985. Maine's all-time record low temperature is 21°F higher than Utah's record low. What is Maine's record low temperature? (*Source*: National Climatic Data Center)

64. The all-time record low temperature for Louisiana is −16°F, which occurred on February 13, 1899. In Hawaii, the lowest temperature ever recorded is 28°F more than Louisiana's all-time low temperature. What is the all-time record low temperature for Hawaii? (*Source*: National Climatic Data Center)

57. 2°C

58. −10 or 10 under par

59. −165 + (−16) = −181; 181 feet below the surface

60. −248 + 6 + (−17) + (−24) + 23 = −260 260 feet below the surface

61. Team 1: 7; Team 2: 6; winning team: Team 1

62. Team 1: 5; Team 2: 9; winning team: Team 2

63. −48°F

64. 12°F

65. −7679 meters

66. −7535 meters

67. 44

68. 91

69. 0

70. 0

71. 28

72. 14

73. true

74. true

75. false

76. true

77. answers may vary

78. answers may vary

Name _____

65. The deepest spot in the Pacific Ocean is the Mariana Trench, which has an elevation of 10,924 meters below sea level. The bottom of the Pacific's Aleutian Trench has an elevation 3245 meters higher than that of the Mariana Trench. Use a negative number to represent the depth of the Aleutian Trench. (*Source*: Defense Mapping Agency)

66. The deepest spot in the Atlantic Ocean is the Puerto Rico Trench, which has an elevation of 8605 meters below sea level. The bottom of the Atlantic's Cayman Trench has an elevation 1070 meters above the level of the Puerto Rico Trench. Use a negative number to represent the depth of the Cayman Trench. (*Source*: Defense Mapping Agency)

REVIEW AND PREVIEW

Subtract. See Section 1.3.

67. $44 - 0$

68. $91 - 0$

69. $52 - 52$

70. $103 - 103$

71. $87 - 59$

72. $32 - 18$

COMBINING CONCEPTS

For Exercises 73–76, determine whether each statement is true or false.

73. The sum of two negative numbers is always a negative number.

74. The sum of two positive numbers is always a positive number.

75. The sum of a positive number and a negative number is always a negative number.

76. The sum of zero and a negative number is always a negative number.

77. In your own words, explain how to add two negative numbers.

78. In your own words, explain how to add a positive number and a negative number.

10.3 SUBTRACTING SIGNED NUMBERS

In Section 10.1 we discussed the opposite of a number.

The opposite of 3 is -3.
The opposite of -6.7 is 6.7.

Now let's find the sum of these opposites.

$$3 + (-3) = 0 \quad \text{and} \quad -6.7 + 6.7 = 0$$

From this, we see that the sum of a number and its opposite is 0. In this section, we use opposites to subtract signed numbers.

A SUBTRACTING SIGNED NUMBERS

To subtract signed numbers, we will write the subtraction problem as an addition problem. To see how we do this, study the examples below.

$$10 - 4 = 6$$
$$10 + (-4) = 6$$

Since both expressions simplify to 6, this means that

$$10 - 4 = 10 + (-4) = 6$$

Also,

$$3 - 2 = 3 + (-2) = 1$$
$$15 - 1 = 15 + (-1) = 14$$

Thus, to subtract two numbers, we add the first number to the opposite (also called the additive inverse) of the second number.

Examples   Subtract.

subtraction	=	first number	+	opposite of the second number		
1. $8 - 5$	=	8	+	(-5)	=	3
2. $-4 - 10$	=	-4	+	(-10)	=	-14
3. $6 - (-5)$	=	6	+	5	=	11
4. $-11 - (-7)$	=	-11	+	7	=	-4

Examples   Subtract.

5. $-10 - 5 = -10 + (-5) = -15$

6. $8 - 15 = 8 + (-15) = -7$

7. $-4 - (-5) = -4 + 5 = 1$

8. $-1.7 - 6.2 = -1.7 + (-6.2) = -7.9$

9. $-\dfrac{10}{11} - \left(-\dfrac{3}{11}\right) = -\dfrac{10}{11} + \dfrac{3}{11} = -\dfrac{7}{11}$

TRY THE CONCEPT CHECK IN THE MARGIN.

Objectives

A Subtract signed numbers.
B Add and subtract signed numbers.
C Solve problems by subtracting signed numbers.

SSM   CD-ROM   Video
                                       10.3

SUBTRACTING TWO NUMBERS

If a and b are numbers, then
$a - b = a + (-b)$.

Practice Problems 1–4

Subtract.

1. $12 - 7$ 2. $-6 - 4$
3. $11 - (-14)$ 4. $-9 - (-1)$

Practice Problems 5–9

Subtract

5. $5 - 9$ 6. $-12 - 4$
7. $-2 - (-7)$ 8. $-10.5 - 14.3$
9. $\dfrac{5}{13} - \dfrac{12}{13}$

✓ CONCEPT CHECK

What is wrong with the following calculation?

$$-7 - (-3) = -10$$

Answers

1. 5, **2.** -10, **3.** 25, **4.** -8, **5.** -4,
6. -16, **7.** 5, **8.** -24.8, **9.** $-\dfrac{7}{13}$

✓ Concept Check: $-7 - (-3) = -4$

Practice Problem 10

Subtract 5 from −10.

Practice Problem 11

Simplify: $-4 - 3 - 7 - (-5)$

TEACHING TIP
Suggest that students rewrite the expression followed by an equal sign and first simplification. Then write each additional simplification below the previous one making sure to align equal signs. This format will help eliminate careless errors.

Practice Problem 12

Simplify: $3 + (-5) - 6 - (-4)$

TEACHING TIP   Classroom Activity
Have students work in groups to write their own word problems. Then they can exchange their problem with another group to solve.

Practice Problem 13

The highest point in Asia is the top of Mount Everest, at a height of 29,028 feet above sea level. The lowest point is the Dead Sea, which is 1312 feet below sea level. How much higher is Mount Everest than the Dead Sea? (*Source*: National Geographic Society)

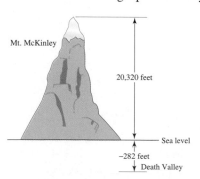

Mt. McKinley

20,320 feet

Sea level
−282 feet
Death Valley

Answers

10. −15, **11.** −9, **12.** −4, **13.** 30,340 feet

Example 10   Subtract 7 from −3.

Solution: To subtract 7 *from* −3, we find
$$-3 - 7 = -3 + (-7) = -10$$

B ADDING AND SUBTRACTING SIGNED NUMBERS

If a problem involves adding or subtracting more than two signed numbers, we rewrite differences as sums and add from left to right.

Example 11   Simplify: $7 - 8 - (-5) - 1$

Solution:
$$7 - 8 - (-5) - 1 = 7 + (-8) + 5 + (-1)$$
$$= -1 + 5 + (-1)$$
$$= 4 + (-1)$$
$$= 3$$

Example 12   Simplify: $7 + (-12) - 3 - (-8)$

Solution:
$$7 + (-12) - 3 - (-8) = 7 + (-12) + (-3) + 8$$
$$= -5 + (-3) + 8$$
$$= -8 + 8$$
$$= 0$$

C SOLVING PROBLEMS BY SUBTRACTING SIGNED NUMBERS

Solving problems often requires subtraction of signed numbers.

Example 13   Finding a Change in Elevation

The highest point in the United States is the top of Mount McKinley, at a height of 20,320 feet above sea level. The lowest point is Death Valley, California, which is 282 feet below sea level. How much higher is Mount McKinley than Death Valley? (*Source*: U.S. Geological Survey)

Solution:
1. UNDERSTAND. Read and reread the problem. To find how much higher, we subtract. Don't forget that if Death Valley is 282 feet *below* sea level, we represent its height by −282. Draw a diagram to help visualize the problem.

2. TRANSLATE.

how much higher is Mt. McKinley	=	height of Mt. McKinley	minus	height of Death Valley
How much higher is Mt. McKinley	=	20,320	−	(−282)

In words / Translate:

3. SOLVE.
$$20,320 - (-282) = 20,320 + 282 = 20,602$$

4. INTERPRET. Check and state your conclusion: Mount McKinley is 20,602 feet higher than Death Valley.

Name _____ Section _____ Date _____

MENTAL MATH

Subtract.
1. $5 - 5$ 2. $7 - 7$ 3. $6.2 - 6.2$ 4. $1.9 - 1.9$

EXERCISE SET 10.3

A *Subtract. See Examples 1 through 9.*
1. $-5 - (-5)$ 2. $-6 - (-6)$ 3. $8 - 3$ 4. $5 - 2$

5. $3 - 8$ 6. $2 - 5$ 7. $7 - (-7)$ 8. $12 - (-12)$

9. $-5 - (-8)$ 10. $-25 - (-25)$ 11. $-14 - 4$ 12. $-2 - 42$

13. $2 - 16$ 14. $8 - 9$ 15. $2.2 - 5.5$ 16. $1.7 - 6.3$

17. $3.62 - (0.4)$ 18. $8.44 - (-0.2)$ 19. $-\dfrac{3}{10} - \left(-\dfrac{7}{10}\right)$ 20. $-\dfrac{2}{11} - \left(-\dfrac{5}{11}\right)$

21. $\dfrac{2}{5} - \dfrac{7}{10}$ 22. $\dfrac{3}{16} - \dfrac{7}{8}$ 23. $\dfrac{1}{2} - \left(-\dfrac{1}{3}\right)$ 24. $\dfrac{1}{7} - \left(-\dfrac{1}{4}\right)$

Solve. See Example 10.
25. Subtract 18 from -20.

26. Subtract 10 from -22.

27. Find the difference of -20 and -3.

28. Find the difference of -8 and -13.

29. Subtract -11 from 2.

30. Subtract -50 from -50.

MENTAL MATH ANSWERS
1. 0
2. 0
3. 0
4. 0

ANSWERS
1. 0
2. 0
3. 5
4. 3
5. -5
6. -3
7. 14
8. 24
9. 3
10. 0
11. -18
12. -44
13. -14
14. -1
15. -3.3
16. -4.6
17. 4.02
18. 8.64
19. $\dfrac{2}{5}$
20. $\dfrac{3}{11}$
21. $-\dfrac{3}{10}$
22. $-\dfrac{11}{16}$
23. $\dfrac{5}{6}$
24. $\dfrac{11}{28}$
25. -38
26. -32
27. -17
28. 5
29. 13
30. 0

31. 2

32. 3

33. 0

34. 11

35. −1

36. −7

37. −27

38. −27

39. 40

40. 31

41. 262°F

42. 214°F

43. −$16

44. −5 points

45. −12°C

46. 36,290 feet

632

B *Simplify: See Examples 11 and 12.*

31. $7 - 3 - 2$

32. $8 - 4 - 1$

33. $12 - 5 - 7$

34. $30 - 7 - 12$

35. $-5 - 8 - (-12)$

36. $-10 - 6 - (-9)$

37. $-10 + (-5) - 12$

38. $-15 + (-8) - 4$

39. $12 - (-34) + (-6)$

40. $23 - (-17) + (-9)$

C *Solve. See Example 13.*

41. The coldest temperature ever recorded on earth was −126°F in Antarctica. The warmest temperature ever recorded was 136°F in the Sahara Desert. How many degrees warmer is 136°F than −126°F? (*Source: Questions Kids Ask*, Grolier Limited, 1991)

42. The coldest temperature ever recorded in the United States was −80°F in Alaska. The warmest temperature ever recorded was 134°F in California. How many degrees warmer is 134°F than −80°F? (*Source: The World Almanac and Book of Facts, 1996*)

43. Aaron Aiken has $125 in his checking account. He writes a check for $117, makes a deposit of $45, and then writes another check for $69. Find the amount left in his account. (Write the amount as an integer.)

44. In canasta, it is possile to have a negative score. If Juan Santanilla's score is 15, what is his new score if he loses 20 points?

45. The temperature on a February morning is −6° Celsius at 6 A.M. If the temperature drops 3 degrees by 7 A.M., rises 4 degrees between 7 A.M. and 8 A.M., and then drops 7 degrees between 8 A.M. and 9 A.M., find the temperature at 9 A.M.

46. Mt. Elbert in Colorado has an elevation of 14,433 feet above sea level. The Mid-America Trench in the Pacific Ocean has an elevation of 21,857 feet below sea level. Find the difference in elevation between those two points. (*Source:* National Geographic Society and Defense Mapping Agency)

Name _____

The bar graph shows heights of selected lakes. For Exercises 47–50, find the difference in elevation for the lakes listed. (Source: U.S. Geological Survey)

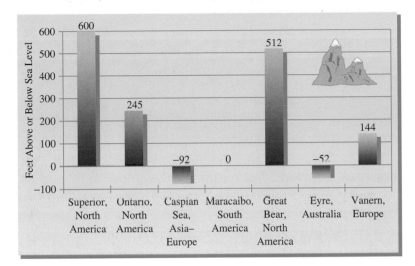

47. Lake Superior and Lake Eyre

48. Great Bear Lake and Caspian Sea

49. Lake Maracaibo and Lake Vanern

50. Lake Eyre and Caspian Sea

51. The average temperature on the surface of Mercury is 166.85°C. The average temperature on the surface of Neptune is −215.15°C. How many degrees warmer is the surface of Mercury than the surface of Neptune? (*Source*: National Space Science Data Center)

52. The average temperature on the surface of Earth is 14.85°C. The average temperature on the surface of Mars is −63.15°C. How many degrees warmer is the surface of Earth than the surface of Mars? (*Source*: National Space Science Data Center)

53. The average temperature on the surface of Saturn is −176°C. The average temperature on the surface of Uranus is −215°C. Which planet has the warmer average temperature? How much warmer is it? (*Source*: National Space Science Data Center)

54. The average temperature on the surface of Pluto is −223.15°C. The average temperature on the surface of Jupiter is −144°C. What is the difference between the average temperature on Jupiter and the average temperature on Pluto? (*Source*: National Space Science Data Center)

REVIEW AND PREVIEW

Multiply. See Section 1.5.

55. $8 \cdot 0$

56. $0 \cdot 8$

57. $1 \cdot 8$

58. $8 \cdot 1$

59.
$$\begin{array}{r} 23 \\ \times\ 46 \\ \hline \end{array}$$

60.
$$\begin{array}{r} 51 \\ \times\ 89 \\ \hline \end{array}$$

47. 652 feet

48. 604 feet

49. 144 feet

50. 40 feet

51. 382°C

52. 78°C

53. Saturn; 39°C

54. 79.15°C

55. 0

56. 0

57. 8

58. 8

59. 1058

60. 4539

◢ **COMBINING CONCEPTS**

Simplify. (Hint: Find the absolute values first.)

61. $|-3| - |-7|$ **62.** $|-12| - |-5|$ **63.** $|-6| - |6|$

64. $|-9| - |9|$ **65.** $|-17| - |-29|$ **66.** $|-23| - |-42|$

For Exercises 67 and 68, determine whether each statement is true or false.

67. $|-8 - 3| = 8 - 3$ **68.** $|-2 - (-6)| = |-2| - |-6|$

69. In your own words, explain how to subtract one signed number from another.

Internet Excursions

Go to http://www.prenhall.com/martin-gay
The World Wide Web address listed here will direct you to the web site of the CNN Financial Network, or a related site. You will be able to collect data on prices of stocks traded on the New York Stock Exchange and answer the questions below.

70. Pick stocks for any four companies. You can use the Website to look up the stock market ticker symbol for the companies. Then get a stock quote for each company. Complete the table below. For each company in the table, show that the sum of the previous closing price and the change in price give the last price.

71. For the same four companies you used in Exercise 70, show how to use the last price and the change in price to calculate the previous closing price.

Date of stock quotes:_____   Time of stock quotes: _____

Company Name	Ticker Symbol	Previous Close	Change	Last

Name _____ **Section** _____ **Date** _____

CHAPTER 10 INTEGRATED REVIEW — SIGNED NUMBERS

Represent each quantity by a signed number.

1. The peak of Mount Everest in Asia is 29,028 feet above sea level. (*Source:* U.S. Geological Survey)

2. The Mariana Trench in the Pacific Ocean is 35,840 feet below sea level. (*Source: The World Almanac,* 1998)

3. Graph the signed numbers on the given number line. $-4, 0, -1.5, 3\frac{1}{4}$

Insert $<$ *or* $>$ *between each pair of numbers to make a true statement.*

4. 0 -3

5. -15 -5

6. $-1\frac{1}{2}$ $1\frac{1}{4}$

7. -2 -7.6

Simplify.

8. $|-1|$

9. $-|-4|$

10. $|-8.6|$

11. $-\left(-\frac{3}{4}\right)$

Find the opposite of each number.

12. 6

13. -3

14. 89.1

15. $-\frac{2}{9}$

Add or subtract as indicated.

16. $-7 + 12$

17. $-9.2 + (-11.6)$

18. $\frac{5}{9} + \left(-\frac{1}{3}\right)$

19. $1 - 3$

20. $\frac{3}{8} - \left(-\frac{3}{8}\right)$

21. $-2.6 - 1.4$

ANSWERS

1. $+29{,}028$

2. $-35{,}840$

3. see number line

4. $>$

5. $<$

6. $<$

7. $>$

8. 1

9. -4

10. 8.6

11. $\frac{3}{4}$

12. -6

13. 3

14. -89.1

15. $\frac{2}{9}$

16. 5

17. -20.8

18. $\frac{2}{9}$

19. -2

20. $\frac{3}{4}$

21. -4

22. $-18 - (-102)$

23. $-8 + (-6) + 20$

24. $-\dfrac{4}{5} - \dfrac{3}{10}$

25. $-11 - 7 - (-19)$

26. $8.65 - 12.09$

27. $-4 + (-8) - 16 - (-9)$

28. Subtract 14 from 26.

29. Subtract -8 from -12.

Side column (answers):

22. 84

23. 6

24. $-\dfrac{11}{10} = -1\dfrac{1}{10}$

25. 1

26. -3.44

27. -19

28. 12

29. -4

10.4 MULTIPLYING AND DIVIDING SIGNED NUMBERS

Multiplying and dividing signed numbers is similar to multiplying and dividing whole numbers. One difference is that we need to determine whether the result is a positive number or a negative number.

A MULTIPLYING SIGNED NUMBERS

Consider the following pattern of products.

First factor decreases by 1 each time.

$3 \cdot 2 = 6$
$2 \cdot 2 = 4$ Product decreases by 2 each time.
$1 \cdot 2 = 2$
$0 \cdot 2 = 0$

This pattern can be continued, as follows.

$-1 \cdot 2 = -2$
$-2 \cdot 2 = -4$
$-3 \cdot 2 = -6$

This suggests that the product of a negative number and a positive number is a negative number.

What is the sign of the product of two negative numbers? To find out, we form another pattern of products. Again, we decrease the first factor by 1 each time, but this time the second factor is negative.

$2 \cdot (-3) = -6$
$1 \cdot (-3) = -3$ Product increases by 3 each time.
$0 \cdot (-3) = 0$

This pattern continues as:

$-1 \cdot (-3) = 3$
$-2 \cdot (-3) = 6$
$-3 \cdot (-3) = 9$

This suggests that the product of two negative numbers is a positive number. Thus we can determine the sign of a product when we know the signs of the factors.

> **MULTIPLYING SIGNED NUMBERS**
>
> The product of two numbers having the same sign is a positive number. The product of two numbers having different signs is a negative number.

Examples   Multiply.

1. $-7 \cdot 3 = -21$
2. $-2(-5) = 10$
3. $0 \cdot (-4) = 0$
4. $\left(-\dfrac{1}{2}\right)\left(-\dfrac{2}{3}\right) = \dfrac{1 \cdot \overset{1}{\cancel{2}}}{\underset{1}{\cancel{2}} \cdot 3} = \dfrac{1}{3}$
5. $-3(1.2) = -3.6$

Objectives

A Multiply signed numbers.
B Divide signed numbers.
C Solve problems by multiplying or dividing signed numbers.

SSM   CD-ROM   Video
                       10.4

TEACHING TIP

Before discussing the rules for multiplying signed numbers, have students list the possible combinations of signs when multiplying 2 numbers and the results which may occur for these products. Get them started by saying, "One possibility when multiplying two signed numbers is that both numbers are positive. What results are possible when multiplying two positive numbers? What other possibilities exist when multiplying two signed numbers?"

Practice Problems 1–5

Multiply.

1. $-2 \cdot 6$ 2. $-4(-3)$

3. $0 \cdot (-10)$ 4. $\left(\dfrac{3}{7}\right)\left(-\dfrac{1}{3}\right)$

5. $-4\left(-3.2\right)$

Answers

1. -12, **2.** 12, **3.** 0, **4.** $-\dfrac{1}{7}$, **5.** 12.8

✓ CONCEPT CHECK

Estimate the product

$$(3.92) \cdot \left(-9\frac{1}{8}\right)$$

Practice Problem 6

Multiply.

a. $(-2)(-1)(-2)(-1)$

b. $7(0)(-4)$

✓ CONCEPT CHECK

What is the sign of the product of five negative numbers? Explain.

Practice Problem 7

Evaluate $(-3)^4$.

TEACHING TIP

Have students develop another strategy for multiplying more than two signed numbers by asking them to use a list to find a relationship between the number of negative numbers and the sign of the product. Then ask them to use this relationship to describe a strategy for multiplying more than two signed numbers.

Practice Problems 8–11

Divide.

8. $\dfrac{28}{-7}$ 9. $-18 \div (-2)$

10. $\dfrac{-4.6}{0.2}$ 11. $-\dfrac{3}{5} \div \dfrac{2}{7}$

Answers

6. a. 4,   **b.** 0,   **7.** 81,   **8.** −4,   **9.** 9,

10. −23,   **11.** $-\dfrac{21}{10}$

✓ Concept Check: $4 \cdot (-9) = -36$

✓ Concept Check: negative

TRY THE CONCEPT CHECK IN THE MARGIN.

To find the product of more than two numbers, we multiply from left to right.

Example 6   Multiply.

a. $7(-6)(-2)$ b. $(-2)(-3)(-4)$

Solution:

a. $7(-6)(-2) = -42(-2) = 84$

b. $(-2)(-3)(-4) = 6(-4) = -24$

TRY THE CONCEPT CHECK IN THE MARGIN.

Recall from our study of exponents that $2^3 = 2 \cdot 2 \cdot 2 = 8$. We can now work with bases that are negative numbers. For example,

$$(-2)^3 = (-2)(-2)(-2) = -8$$

Example 7   Evaluate $(-5)^2$.

Solution: Remember that $(-5)^2$ means 2 factors of -5.

$$(-5)^2 = (-5)(-5) = 25$$

B DIVIDING SIGNED NUMBERS

Division of signed numbers is related to multiplication of signed numbers. The sign rules for division can be discovered by writing a related multiplication problem. For example,

$$\frac{6}{2} = 3 \qquad \text{because } 3 \cdot 2 = 6$$

$$\frac{-6}{2} = -3 \qquad \text{because } -3 \cdot 2 = -6$$

$$\frac{6}{-2} = -3 \qquad \text{because } -3 \cdot (-2) = 6$$

$$\frac{-6}{-2} = 3 \qquad \text{because } 3 \cdot (-2) = -6$$

> **DIVIDING SIGNED NUMBERS**
>
> The quotient of two numbers having the same sign is a positive number. The quotient of two numbers having different signs is a negative number.

Examples   Divide.

8. $\dfrac{-12}{6} = -2$

9. $-20 \div (-4) = 5$

10. $\dfrac{1.2}{-0.6} = -2$ $0.6\overline{)1.2}$ → 2

11. $-\dfrac{7}{9} \div \dfrac{2}{5} = -\dfrac{7}{9} \cdot \dfrac{5}{2} = -\dfrac{7 \cdot 5}{9 \cdot 2} = -\dfrac{35}{18}$

TRY THE CONCEPT CHECK IN THE MARGIN.

Examples   Divide, if possible.

12. $\frac{0}{-5} = 0$ because $0 \cdot -5 = 0$.

13. $\frac{-7}{0}$ is undefined because there is no number that gives a product of -7 when multiplied by 0. ▬▬

C SOLVING PROBLEMS BY MULTIPLYING AND DIVIDING SIGNED NUMBERS

Many real-life problems involve multiplication and division of signed numbers.

Example 14 Calculating Total Golf Score

A professional golfer finished seven strokes under par (-7) for each of three days of a tournament. What was his total score for the tournament?

Solution: **1.** UNDERSTAND. Read and reread the problem. Although the key word is *total*, since this is repeated addition of the same number, we multiply.
2. TRANSLATE.

In words: | golfer's total score | = | number of days | · | score each day |

Translate: golfer's total score = 3 · (-7)

3. SOLVE. $3 \cdot (-7) = -21$
4. INTERPRET. Check and state your conclusion: The golfer's total score is -21, or 21 strokes under par. ▬▬

Focus On Business and Career

NET INCOME AND NET LOSS

For most businesses, a financial goal is to "make money." But what does this mean from a mathematical point of view? To find out, we must first discuss some common business terms.

▲ **Revenue** is the amount of money a business takes in. A company's annual revenue is the amount of money it collects during its fiscal, or business, year. For most companies, the largest source of revenue is from the sales of their products or services. For instance, a grocery store's annual revenue is the amount of money it collects during the year from selling groceries to customers. Large companies may also have revenues from interest or rentals.

▲ **Expenses** are the costs of doing business. For instance, a large part of a grocery store's expenses include the cost of the food items it buys from wholesalers to resell to customers. Other expenses include salaries, mortgage payments, equipment, taxes, advertising, and so on.

▲ **Net income/loss** is the difference between a company's annual revenues and expenses. If the company's revenues are larger than its expenses, the difference is a positive number and the company posts a net income for the year. Posting a net income can be interpreted as "making money." If the company's revenues are smaller than its expenses, the difference is a negative number and the company posts a net loss for the year. Posting a net loss can be interpreted as "losing money."

Net income can also be thought of as profit. A negative profit is a net loss.

GROUP ACTIVITY

Search for corporate annual reports or articles in financial newspapers and magazines that report a net income or net loss. Describe what the income or loss means for the company. What were some of the factors that contributed to the net income or loss?

Name _____ **Section** _____ **Date** _____

Exercise Set 10.4

A *Multiply. See Examples 1 through 5.*

1. $-2(-3)$ **2.** $5(-3)$ **3.** $-4(9)$ **4.** $-7(-2)$

5. $(2.6)(-1.2)$ **6.** $-0.3(5.6)$ **7.** $0(-14)$ **8.** $-6(0)$

9. $-\dfrac{3}{5}\left(-\dfrac{2}{7}\right)$ **10.** $-\dfrac{3}{4}\left(\dfrac{9}{10}\right)$

Multiply. See Example 6.

11. $6(-4)(2)$ **12.** $-2(3)(-7)$ **13.** $-1(-2)(-4)$ **14.** $8(-3)(3)$

15. $-4(4)(-5)$ **16.** $-2(-5)(-4)$ **17.** $10(-5)(0)$ **18.** $2(-1)(3)(-2)$

Evaluate. See Example 7.

19. $(-2)^2$ **20.** $(-2)^4$ **21.** $(-3)^3$ **22.** $(-1)^4$

23. $(-5)^2$ **24.** $(-4)^3$ **25.** $(-5)^3$ **26.** $(-3)^2$

Multiply. See Examples 1 through 7.

27. $-12(0)$ **28.** $0(-100)$ **29.** $\dfrac{3}{4}\left(-\dfrac{7}{8}\right)$ **30.** $-\dfrac{3}{11}\cdot\dfrac{1}{2}$

31. $-1\cdot(-1)\cdot(-1)\cdot(-1)$ **32.** $-1\cdot(-1)\cdot(-1)$

33. $-1(2)(7)(-3.1)$ **34.** $-2(3)(5)(-6.2)$

35. $(-2)^3$ **36.** $(-3)^5$

Answers	
1.	6
2.	-15
3.	-36
4.	14
5.	-3.12
6.	-1.68
7.	0
8.	0
9.	$\dfrac{6}{35}$
10.	$\dfrac{-27}{40}$
11.	-48
12.	42
13.	-8
14.	-72
15.	80
16.	-40
17.	0
18.	12
19.	4
20.	16
21.	-27
22.	1
23.	25
24.	-64
25.	-125
26.	9
27.	0
28.	0
29.	$\dfrac{-21}{32}$
30.	$\dfrac{3}{22}$
31.	1
32.	-1
33.	43.4
34.	186
35.	-8
36.	-243

37. −4

38. −10

39. −5

40. −7

41. 8

42. −8

43. 0

44. undefined

45. −13

46. 2

47. −26

48. −1.1

49. $\dfrac{7}{2}$

50. $\dfrac{21}{16}$

51. −5

52. −5

53. −6

54. −60

55. 3

56. 12

57. −300

58. −80

59. $-\dfrac{4}{5}$

60. $-\dfrac{3}{7}$

61. $3 \cdot (-4) =$ −12 yards

62. $7 \cdot (-400) =$ −$2800

63. $5 \cdot (-20) =$ −100 feet

64. $6 \cdot (-5) = -30°$

642

B *Divide. See Examples 8 through 13.*

37. $-24 \div 6$ **38.** $90 \div (-9)$ **39.** $\dfrac{-30}{6}$ **40.** $\dfrac{56}{-8}$

41. $\dfrac{-88}{-11}$ **42.** $\dfrac{-32}{4}$ **43.** $\dfrac{0}{14}$ **44.** $\dfrac{-13}{0}$

45. $\dfrac{39}{-3}$ **46.** $\dfrac{-24}{-12}$ **47.** $\dfrac{7.8}{-0.3}$ **48.** $\dfrac{1.21}{-1.1}$

49. $-\dfrac{7}{12} \div \left(-\dfrac{1}{6}\right)$ **50.** $-\dfrac{3}{8} \div \left(-\dfrac{2}{7}\right)$ **51.** $\dfrac{100}{-20}$ **52.** $\dfrac{45}{-9}$

53. $240 \div (-40)$ **54.** $480 \div (-8)$ **55.** $\dfrac{-12}{-4}$ **56.** $\dfrac{-36}{-3}$

57. $\dfrac{-120}{0.4}$ **58.** $\dfrac{-200}{2.5}$ **59.** $-\dfrac{8}{15} \div \dfrac{2}{3}$ **60.** $-\dfrac{1}{6} \div \dfrac{7}{18}$

C *Solve. See Example 14.*

61. A football team lost four yards on each of three consecutive plays. Represent the total loss as a product of signed numbers and find the total loss.

62. Joe Norstrom lost $400 on each of seven consecutive days in the stock market. Represent his total loss as a product of signed numbers and find his total loss.

63. A deep-sea diver must move up or down in the water in short steps in order to keep from getting a physical condition called the bends. Suppose the diver moves down from the surface in five steps of 20 feet each. Represent his total movement as a product of signed numbers and find the product.

64. A weather forecaster predicts that the temperature will drop five degrees each hour for the next six hours. Represent this drop as a product of signed numbers and find the total drop in temperature.

65. During the 1998 LPGA Corning Classic golf tournament, the winner Tammie Green had scores of $-5, -2, -6,$ and -7 in four rounds of golf. Find her average score per round. (*Source:* Ladies Professional Golf Association)

66. During the 1998 LPGA Rochester International golf tournament, the winner Rosie Jones had scores of $+2, -3, -8,$ and 0 in four rounds of golf. Find her average score per round. (*Source:* Ladies Professional Golf Association)

67. The Scotts Company produces a full line of lawn and garden care products. In 1996, Scotts posted an income of $-\$2.5$ million. If this continued, what would Scotts income be after three years? (*Source:* The Scotts Company)

68. In 1997, the earnings for a share of Apple Computer stock were $-\$8.29$ per share. If an investor owned 15 shares of Apple Computer stock in 1997, what were his total earnings for these shares? (*Source:* Apple Computer Inc.)

69. In 1979, there were 35 California Condors in the entire world. By 1987, there were only 27 California Condors remaining. (*Source:* California Department of Fish and Game)
  a. Find the change in the number of California Condors from 1979 to 1987.
  b. Find the average change per year in the California Condor population over this period.

70. In 1993, music cassettes with a total value of $2915.8 million were produced by American music manufacturers. In 1997, this value had dropped to $1522.7 million. (*Source:* Recording Industry Association of America)
  a. Find the change in the value of music cassettes produced from 1993 to 1997.
  b. Find the average change per year in the value of music cassettes produced over this period.

REVIEW AND PREVIEW

Perform each indicated operation. See Section 1.8.

71. $(3 \cdot 5)^2$

72. $(12 - 3)^2(18 - 10)$

73. $90 + 12^2 - 5^3$

74. $3 \cdot (7 - 4) + 2 \cdot 5^2$

75. $12 \div 4 - 2 + 7$

76. $12 \div (4 - 2) + 7$

---

65. -5

66. -2.25

67. $-\$7.5$ million

68. $-\$124.35$

69. a. -8 condors

  b. -1 condor per year

70. a. $-\$1393.1$ million

  b. $-\$348.275$ per year

71. 225

72. 648

73. 109

74. 59

75. 8

76. 13

Name _____

COMBINING CONCEPTS

In Exercises 77 through 79, determine whether each statement is true or false.

77. The product of two negative numbers is always a negative number.

78. The product of a positive number and a negative number is always a negative number.

79. The quotient of two negative numbers is always a positive number.

80. In 1996 there were 2558 commercial country music radio stations in the United States. By 1997, that number had declined to 2502. (*Source:* M Street Corporation)
  a. Find the change in the number of country music radio stations from 1996 to 1997.
  b. If this change continues, what will be the total change in the number of country music stations after another four years?
  c. Based on your answer to part (b), how many country music radio stations will there be in 2001?

81. In 1995 Saturn sold a total of 285,674 cars. By 1996, the number of Saturns sold had decreased to 278,574. (*Source:* American Automobile Manufacturers Association)
  a. Find the change in the number of Saturns sold from 1995 to 1996.
  b. If this change continues, what will be the total change in the number of Saturns sold after another two years?
  c. Based on your answer to part (b), how many Saturns would have been sold in 1998?

82. In your own words, explain how to multiply two signed numbers.

83. In your own words, explain how to divide two signed numbers.

10.5 ORDER OF OPERATIONS

A SIMPLIFYING EXPRESSIONS

We first discussed the order of operations in Chapter 1. In this section, you are given an opportunity to practice using the order of operations when expressions contain signed numbers. The rules for order of operations from Section 1.8 are repeated here.

   If there are no other grouping symbols such as fraction bars or absolute values, perform operations in the following order.

> ### ORDER OF OPERATIONS
>
> 1. Do all operations within parentheses or brackets.
> 2. Evaluate any expressions with exponents and find any square roots.
> 3. Multiply or divide in order from left to right.
> 4. Add or subtract in order from left to right.

Examples   Find the value of each expression.

1. $(-3)^2 = (-3)(-3) = 9$

2. $-3^2 = -(3)(3) = -9$

3. $\left(-\frac{1}{2}\right)^3 = \left(-\frac{1}{2}\right)\left(-\frac{1}{2}\right)\left(-\frac{1}{2}\right) = \frac{1}{4}\left(-\frac{1}{2}\right) = -\frac{1}{8}$

HELPFUL HINT

When simplifying expressions with exponents, notice that parentheses make an important difference.

$(-3)^2$ and -3^2 do not mean the same thing.
$(-3)^2$ means $(-3)(-3) = 9$.
-3^2 means the opposite of $3 \cdot 3$, or -9.

Only with parentheses is the -3 squared.

Example 4   Simplify: $\dfrac{-6(2)}{-3}$

Solution: First we multiply -6 and 2. Then we divide.

$$\frac{-6(2)}{-3} = \frac{-12}{-3}$$
$$= 4$$

Example 5   Simplify: $\dfrac{12 - 16}{-1 + 3}$

Solution: We simplify above and below the fraction bar separately. Then we divide.

$$\frac{12 - 16}{-1 + 3} = \frac{-4}{2}$$
$$= -2$$

TEACHING TIP

Begin this lesson by asking students to recall and write the rules for order of operations. Then have them compare their rules with the rules in the book.

Practice Problems 1–3

Find the value of each expression.

1. $(-2)^4$

2. -2^4

3. $\left(-\frac{1}{3}\right)^3$

Practice Problem 4

Simplify: $\dfrac{25}{5(-1)}$

Practice Problem 5

Simplify: $\dfrac{-18 + 6}{-3 - 1}$

Answers

1. 16, **2.** -16, **3.** $-\frac{1}{27}$, **4.** -5, **5.** 3

Practice Problem 6

Simplify: $20 + 50 + (-4)^3$

Practice Problem 7

Simplify: $-2^3 + (-4)^2 + 1^5$

Practice Problem 8

Simplify: $2(2 - 8) + (-12) - 3$

TEACHING TIP   Classroom Activity

Have students work in groups to form 4 expressions which are identical except for the placement of parentheses. Have them simplify each expression. Then have them give the expression without parentheses and the 4 answers to another group to write the expression with parentheses which matches each answer.

Practice Problem 9

Simplify: $(-5) \cdot |-4| + (-3) + 2^3$

Practice Problem 10

Simplify: $4(-6) \div [3(5 - 7)^2]$

✓ CONCEPT CHECK

True or false? Explain your answer. The result of

$$-4 \cdot (3 - 7) - 8 \cdot (9 - 6)$$

is positive because there are four negative signs.

Practice Problem 11

$$\frac{5}{8} \div \left(\frac{1}{5} + \frac{3}{4} \right)$$

Answers

6. 6, **7.** 9, **8.** −27, **9.** −15, **10.** −2,

11. $\frac{25}{38}$

✓ Concept Check:

False; $-4 \cdot (3 - 7) - 8 \cdot (9 - 6) = -8$

Example 6   Simplify: $60 + 30 + (-2)^3$

Solution:

$$60 + 30 + (-2)^3 = 60 + 30 + (-8)$$ Write $(-2)^3$ as -8. Add from left to right.
$$= 90 + (-8)$$
$$= 82$$

Example 7   Simplify: $-4^2 + (-3)^2 - 1^3$

Solution: $-4^2 + (-3)^2 - 1^3 = -16 + 9 - 1$ Simplify expressions with exponents. Add or subtract from left to right.
$$= -7 - 1$$
$$= -8$$

Example 8   Simplify: $3(4 - 7) + (-2) - 5$

Solution:

$3(4 - 7) + (-2) - 5 = 3(-3) + (-2) - 5$ Simplify inside parentheses.
$$= -9 + (-2) - 5$$ Multiply.
$$= -11 - 5$$ Add or subtract from left to right.
$$= -16$$

Example 9   Simplify: $(-3) \cdot |-5| - (-2) + 4^2$

Solution:

$(-3) \cdot |-5| - (-2) + 4^2 = (-3) \cdot 5 - (-2) + 4^2$ Write $|-5|$ as 5.
$$= (-3) \cdot 5 - (-2) + 16$$ Write 4^2 as 16.
$$= -15 - (-2) + 16$$ Multiply.
$$= -13 + 16$$ Add or subtract from left to right.
$$= 3$$

Example 10   Simplify: $-2[-3 + 2(-1 + 6)] - 5$

Solution: Here we begin with the innermost set of parentheses.

$-2[-3 + 2(-1 + 6)] - 5 = -2[-3 + 2(5)] - 5$ Write $-1 + 6$ as 5.
$$= -2[-3 + 10] - 5$$ Multiply.
$$= -2(7) - 5$$ Add.
$$= -14 - 5$$ Multiply.
$$= -19$$ Subtract.

TRY THE CONCEPT CHECK IN THE MARGIN.

Example 11   Simplify: $\left(\frac{1}{6} - \frac{5}{6} \right) \cdot \frac{3}{7}$

Solution: $\left(\frac{1}{6} - \frac{5}{6} \right) \cdot \frac{3}{7} = -\frac{2}{3} \cdot \frac{3}{7}$ $\frac{1}{6} - \frac{5}{6} = -\frac{4}{6} = -\frac{2}{3}$

$$= -\frac{2 \cdot \overset{1}{\cancel{3}}}{\underset{1}{\cancel{3}} \cdot 7}$$ Multiply.

$$= -\frac{2}{7}$$ Simplify.

CALCULATOR EXPLORATIONS
SIMPLIFYING AN EXPRESSION CONTAINING A FRACTION BAR

Even though most calculators follow the order of operations, parentheses must sometimes be inserted. For example, to simplify $\frac{-8+6}{-2}$ on a calculator, enter parentheses about the expression above the fraction bar so that it is simplified separately.

To simplify $\frac{-8+6}{-2}$, press the keys

$$(\quad 8 \quad +/- \quad + \quad 6 \quad) \quad \div \quad 2 \quad +/- \quad =$$

or

ENTER

The display will read $\boxed{\qquad 1}$.

Thus $\frac{-8+6}{-2} = 1$.

Use a calculator to simplify.

1. $\dfrac{-12 - 36}{-10}$ 4.8

2. $\dfrac{475}{-0.2 + (-1.7)}$ -250

3. $\dfrac{-316 + (-458)}{28 + (-25)}$ -258

4. $\dfrac{-234 + 86}{-18 + 16}$ 74

Focus On History

MAGIC SQUARES

A magic square is a set of numbers arranged in a square table so that the sum of the numbers in each column, row, and diagonal is the same. For instance, in the magic square at the right, the sum of each column, row, and diagonal is 15. Notice that no number is used more than once in the magic square.

2	9	4
7	5	3
6	1	8

The properties of magic squares have been known for a very long time and once were thought to be good-luck charms. The ancient Egyptians and Greeks understood their patterns. A magic square even made it into a famous work of art. The engraving titled *Melencolia I*, created by German artist Albrecht Dürer in 1514, features the following four-by-four magic square on the building behind the central figure.

16	3	2	13
5	10	11	8
9	6	7	12
4	15	14	1

(continued)

CRITICAL THINKING

1. Verify that what is shown in the Dürer engraving is, in fact, a magic square. What is the common sum of the columns, rows, and diagonals?   34

2. Negative numbers can also be used in magic squares. Complete the following magic square.

2	-3	-2
-5	-1	3
0	1	-4

3. Use the numbers $-12, -9, -6, -3, 0, 3, 6, 9,$ and 12 to form a magic square.
   answers may vary

9	-6	-3
-12	0	12
3	6	-9

Exercise Set 10.5

A *Simplify. See Examples 1–11.*

1. $-1(-2) + 1$

2. $3 + (-8) \div 2$

3. $3 - 6 + 2$

4. $5 - 9 + 2$

5. $9 - 12 - 4$

6. $10 - 23 - 12$

7. $4 + 3(-6)$

8. $8 + 4(-3)$

9. $\dfrac{4}{9}\left(\dfrac{2}{10} - \dfrac{7}{10}\right)$

10. $\dfrac{2}{5}\left(\dfrac{3}{8} - \dfrac{4}{8}\right)$

11. $(-10) + 4 \div 2$

12. $(-12) + 6 \div 3$

13. $25 \div (-5) + 12$

14. $28 \div (-7) + 10$

15. $\dfrac{16 - 13}{-3}$

16. $\dfrac{20 - 15}{-1}$

17. $\dfrac{24}{10 + (-4)}$

18. $\dfrac{88}{-8 - 3}$

19. $5(-3) - (-12)$

20. $7(-4) - (-6)$

21. $(-19) - 12(3)$

22. $(-24) - 14(2)$

23. $8 + 4^2$

24. $12 + 3^3$

25. $[8 + (-4)]^2$

26. $[9 + (-2)]^3$

27. $3^3 - 12$

ANSWERS

1. 3
2. -1
3. -1
4. -2
5. -7
6. -25
7. -14
8. -4
9. $-\dfrac{2}{9}$
10. $-\dfrac{1}{20}$
11. -8
12. -10
13. 7
14. 6
15. -1
16. -5
17. 4
18. -8
19. -3
20. -22
21. -55
22. -52
23. 24
24. 39
25. 16
26. 343
27. 15

28. -75

29. -3

30. -1

31. -3

32. 21

33. 4

34. 25

35. 16

36. 3

37. -27

38. -20

39. 34

40. 1

41. 0.65

42. 0.72

43. -59

44. -109

45. -7

46. -2

47. $\frac{5}{36}$

48. $-\frac{3}{16}$

49. -11

650

Name _____

28. $5^2 - 100$

29. $(3 - 12) \div 3$

30. $(12 - 19) \div 7$

31. $5 + 2^3 - 4^2$

32. $12 + 5^2 - 2^4$

33. $(5 - 9)^2 \div (4 - 2)^2$

34. $(2 - 7)^2 \div (4 - 3)^4$

35. $|8 - 24| \cdot (-2) \div (-2)$

36. $|3 - 15| \cdot (-4) \div (-16)$

37. $(-12 - 20) \div 16 - 25$

38. $(-20 - 5) \div 5 - 15$

39. $5(5 - 2) + (-5)^2 - 6$

40. $3 \cdot (8 - 3) + (-4) - 10$

41. $(0.2 - 0.7)(0.6 - 1.9)$

42. $(0.4 - 1.2)(0.8 - 1.7)$

43. $2 - 7 \cdot 6 - 19$

44. $4 - 12 \cdot 8 - 17$

45. $(-36 \div 6) - (4 \div 4)$

46. $(-4 \div 4) - (8 \div 8)$

47. $\left(\frac{1}{2}\right)^2 - \left(\frac{1}{3}\right)^2$

48. $\left(\frac{1}{4}\right)^2 - \left(\frac{1}{2}\right)^2$

49. $(-5)^2 - 6^2$

Name _____

50. $(-4)^4 - (5)^4$ **51.** $(10 - 4^2)^2$ **52.** $(11 - 3^2)^3$

53. $2(8 - 10)^2 - 5(1 - 6)^2$ **54.** $-3(4 - 8)^2 + 5(14 - 16)^3$

55. $3(-10) \div [5(-3) - 7(-2)]$ **56.** $12 - [7 - (3 - 6)] + (2 - 3)^3$

57. $\dfrac{(-7)(-3) - (4)(3)}{3[7 \div (3 - 10)]}$ **58.** $\dfrac{10(-1) - (-2)(-3)}{2[-8 \div (-2 - 2)]}$

59. $(0.2)^2 - (1.5)^2$ **60.** $(1.3)^2 - (2.2)^2$

Review and Preview

Perform each indicated operation. See Sections 1.2, 1.3, 1.5, and 1.6.

61. $45 \cdot 90$ **62.** $90 \div 45$

63. $90 - 45$ **64.** $45 + 90$

Find the perimeter of each figure. See Section 1.2.

65. Square

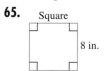

8 in.

66. Parallelogram

5 cm

3 cm

50. -369

51. 36

52. 8

53. -117

54. -88

55. 30

56. 1

57. -3

58. -4

59. -2.21

60. -3.15

61. 4050

62. 2

63. 45

64. 135

65. 32 inches

66. 16 centimeters

651

Name _____

67.
Rectangle

6 ft
9 ft

68.
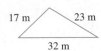
17 m   23 m
32 m

◤ COMBINING CONCEPTS

Insert parentheses where needed so that each expression evaluates to the given number.

69. $2 \cdot 7 - 5 \cdot 3$; evaluates to 12

70. $7 \cdot 3 - 4 \cdot 2$; evaluates to 34

71. $-6 \cdot 10 - 4$; evaluates to -36

72. $2 \cdot 8 \div 4 - 20$; evaluates to -36

Evaluate.

▦ **73.** $(-12)^4$

▦ **74.** $(-17)^6$

69. $2 \cdot (7 - 5) \cdot 3$

70. $(7 \cdot 3 - 4) \cdot 2$

71. $-6 \cdot (10 - 4)$

72. $2 \cdot (8 \div 4 - 20)$

73. 20,736

74. 24,137,569

652

CHAPTER 10 ACTIVITY
INVESTIGATING POSITIVE AND NEGATIVE NUMBERS

MATERIALS:
▲ colored thumbtacks
▲ coin
▲ cardboard
▲ six-sided die
▲ tape

Work with a partner or a small group to try the following activity. The object is to have the largest absolute value at the end of the game.

1. Attach this page to a piece of cardboard with tape. Each person should choose a different colored thumbtack as his or her playing piece. Insert each thumbtack at the starting place 0 on the number line.

2. Each player takes a turn as follows. Roll the die and flip the coin. "Heads" on the coin makes the number that lands face up on the die positive. "Tails" on the coin makes the number that lands face up on the die negative. Record your number along with its sign (positive or negative) in the table below. Move your thumbtack on the number line according to your number.

3. Continue taking turns until each person has taken five turns. Verify your final position on the number line by finding the total of the integers in the table. Do your total and final position agree?

	Positive or Negative Number
Turn 1	
Turn 2	
Turn 3	
Turn 4	
Turn 5	
Total	

4. The winner is the player having the final position with the largest absolute value. Find the absolute value of your final position. Create a table listing the absolute values of each person's final position. Who won? How could you tell who won just by looking at the number line?

5. Many board games include instructions for moving playing pieces forward or backward. Make a list of games that include such instructions. Then explain how these instructions for moving forward or backward are related to positive and negative numbers.

CHAPTER 10 HIGHLIGHTS

DEFINITIONS AND CONCEPTS	EXAMPLES

SECTION 10.1   SIGNED NUMBERS

Together, positive numbers, negative numbers, and 0 are called **signed numbers**.

The **integers** are . . . , −3, −2, −1, 0, 1, 2, 3, ...

$-432, -10, 0, 15$

The **absolute value** of a number is that number's distance from 0 on the number line. The symbol for absolute value is | |.

$|-2| = 2$

$|2| = 2$

Two numbers that are the same distance from 0 on the number line but are on opposite sides of 0 are called **opposites**.

5 and −5
are opposites.

If a is a number, then $-(-a) = a$.

$-(-11) = 11 \qquad -|-3| = -3$

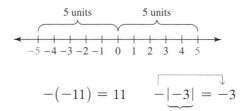

SECTION 10.2   ADDING SIGNED NUMBERS

ADDING TWO NUMBERS WITH THE SAME SIGN

Step 1. Add their absolute values.

Step 2. Use their common sign as the sign of the sum.

ADDING TWO NUMBERS WITH DIFFERENT SIGNS

Step 1. Find the larger absolute value minus the smaller absolute value.

Step 2. Use the sign of the number with the larger absolute value as the sign of the sum.

Add:

$$-3 + (-2) = -5$$
$$-7 + (-15) = -22$$
$$-1.2 + (-5.7) = -6.9$$

$$-6 + 4 = -2$$
$$17 + (-12) = 5$$
$$-\frac{4}{11} + \frac{1}{11} = -\frac{3}{11}$$
$$-32 + (-2) + 14 = -34 + 14$$
$$= -20$$

SECTION 10.3   SUBTRACTING SIGNED NUMBERS

SUBTRACTING TWO NUMBERS

If a and b are numbers, then $a - b = a + (-b)$.

Subtract:

$$-35 - 4 = -35 + (-4) = -39$$
$$3 - 8 = 3 + (-8) = -5$$
$$-7.8 - (-10.2) = -7.8 + 10.2 = 2.4$$
$$7 - 20 - 18 - (-3) = 7 + (-20) + (-18) + 3$$
$$= -13 + (-18) + 3$$
$$= -31 + 3$$
$$= -28$$

SECTION 10.4   MULTIPLYING AND DIVIDING SIGNED NUMBERS

MULTIPLYING SIGNED NUMBERS

The product of two numbers having the same sign is a positive number.
The product of two numbers having different signs is a negative number.

Multiply:

$$(-7)(-6) = 42$$
$$9(-4) = -36$$
$$-3(0.7) = -2.1$$

Evaluate:

$$(-3)^2 = (-3)(-3) = 9$$
$$\left(-\frac{2}{3}\right)^2 = \left(-\frac{2}{3}\right)\left(-\frac{2}{3}\right) = \frac{4}{9}$$

DIVIDING SIGNED NUMBERS

The quotient of two numbers having the same sign is a positive number.
The quotient of two numbers having different signs is a negative number.

Divide:

$$-100 \div (-10) = 10$$
$$\frac{14}{-2} = -7, \frac{-3.6}{0.3} = 12, \frac{0}{-3} = 0, \frac{22}{0} \text{ is undefined.}$$

SECTION 10.5   ORDER OF OPERATIONS

ORDER OF OPERATIONS

1. Do all operations within parentheses or brackets.
2. Evaluate any expressions with exponents.
3. Multiply or divide in order from left to right.
4. Add or subtract in order from left to right.

Simplify:

$$3 + 2 \cdot (-5) = 3 + (-10)$$
$$= -7$$
$$\frac{-2(5 - 7)}{-7 + |-3|} = \frac{-2(-2)}{-7 + 3}$$
$$= \frac{4}{-4}$$
$$= -1$$

Focus On Mathematical Connections

MODELING ADDING AND SUBTRACTING INTEGERS

Just as we used physical models to represent fractions and operations on whole numbers, we can use a physical model to help us add and subtract integers. In this model, we use objects called counters to represent numbers. Black counters ● represent positive numbers and red counters ● represent negative numbers. The key to this model is remembering that taking a ● and ● together creates a neutral or zero pair. Once a neutral pair has been formed, it can be removed from or added to the model without changing the overall value.

Adding Integers

$5 + (-8)$
Begin with 5 black counters and add to that 8 red counters.

● ● ● ● ●
● ● ● ● ● ● ● ●

Next, form and remove neutral pairs.

● ● ● ● ● ● ● ●

Now there are only 3 red counters left, so
$5 + (-8) = -3$

$-4 + (-3)$
Begin with 4 red counters and add to that 3 red counters.

● ● ● ●   ● ● ●

Because this group does not contain a mixture of red and black counters, there is no need to form neutral pairs. Simply find the total number of red counters and remember that red counters represent negative numbers. So,
$-4 + (-3) = -7$

Subtracting Integers

$6 - (-2)$
Begin with 6 black counters.

● ● ● ● ● ●

Subtracting a negative two indicates taking away 2 red counters. However, there are no red counters in the model at this point. Add enough neutral pairs to the model to obtain 2 red counters.

● ● ● ● ● ●   ● ●   ● ●

Now take away 2 red counters.

● ● ● ● ● ● ● ● ●

Because there are 8 black counters remaining, $6 - (-2) = 8$.

$-3 - 9$
Begin with 3 red counters.

● ● ●

Subtracting 9 indicates taking away 9 black counters, However, there are no black counters in the model at this point. Add enough neutral pairs to the model to obtain 9 black counters.

● ● ●   ● ●   ● ●   ● ●
● ● ●   ● ●   ● ●   ● ●

Now take away 9 black counters.

● ●   ● ●   ● ●   ● ●

Because there are 12 red counters remaining, $-3 - 9 = -12$.

CRITICAL THINKING

Use the counter model to perform each indicated operation.

1. $3 + 7$ **2.** $(-5) + 6$ **3.** $(-8) + (-4)$ **4.** $9 + (-11)$ **5.** $(-2) + 2$

6. $8 - 4$ **7.** $(-3) - (-1)$ **8.** $2 - (-5)$ **9.** $(-6) - 4$ **10.** $5 - 7$

Chapter 10 Review

(10.1) *Represent each quantity by a signed number.*

1. A gold miner is working 1435 feet down in a mine.
-1435

2. A mountain peak is 7562 meters above sea level.
$+7562$

Graph each number on a number line.

3. $-2, 4, -3.5, 0$

-3.5
$\longleftarrow$ −7−6−5−4−3−2−1 0 1 2 3 4 5 6 7 $\longrightarrow$

4. $-7, -1.6, 3, -4\frac{1}{3}$

Insert $<$ *or* $>$ *between each pair of numbers to make a true statement.*

5. $-18 \quad -20 \quad >$

6. $-5 \quad 5 \quad <$

7. $12.3 \quad -19.8 \quad >$

Find each absolute value.

8. $|-12| \quad 12$

9. $|0| \quad 0$

10. $\left|-\frac{7}{8}\right| \quad \frac{7}{8}$

Find the opposite of each number.

11. $-12 \quad 12$

12. $\frac{1}{2} \quad -\frac{1}{2}$

Simplify.

13. $-(-7) \quad 7$

14. $-|-7| \quad -7$

Determine whether each statement is true or false.

15. A negative number is always less than a positive number. true

16. The absolute value of a number is always 0 or a positive number. true

(10.2) *Add.*

17. $5 + (-3) \quad 2$

18. $18 + (-4) \quad 14$

19. $-12 + 16 \quad 4$

20. $-23 + 40$ $\quad 17$

21. $-8 + (-15)$ $\quad -23$

22. $-5 + (-17)$ $\quad -22$

23. $-2.4 + 0.3$ $\quad -2.1$

24. $-8.9 + 1.9$ $\quad -7$

25. $\dfrac{2}{3} + \left(-\dfrac{2}{5}\right)$ $\quad \dfrac{4}{15}$

26. $-\dfrac{8}{9} + \dfrac{1}{3}$ $\quad -\dfrac{5}{9}$

27. $-43 + (-108)$ $\quad -151$

28. $-100 + (-506)$ $\quad -606$

29. The temperature at 5 A.M. on a day in January was $-15°$ Celsius. By 6 A.M. the temperature had fallen 5 degrees. Use a signed number to represent the temperature at 6 A.M. $\quad -20°C$

30. A diver starts out at 127 feet below the surface and then swims downward another 23 feet. Use a signed number to represent the diver's current depth. -150 feet

31. During the 1998 Memorial golf tournament, the winner Fred Couples had scores of $-4, -5, -5,$ and -3 over four rounds of golf. What was his total score for the tournament? (*Source:* Professional Golfer's Association) $\quad -17$

32. During the 1998 MCI Classic golf tournament in Hilton Head, South Carolina, the winner Davis Love III had a score of -18. The second-place finisher Glen Day had a score that was 7 points more than the winning score. What was Day's score in the MCI Classic? (*Source:* Professional Golfer's Association) $\quad -11$

(10.3) *Subtract.*

33. $12 - 4$ $\quad 8$

34. $-12 - 4$ $\quad -16$

35. $\dfrac{2}{5} - \dfrac{7}{10}$ $\quad -\dfrac{3}{10}$

36. $-8 - 19$ $\quad -27$

37. $7 - (-13)$ $\quad 20$

38. $-6 - (-14)$ $\quad 8$

39. $16 - 16$ $\quad 0$

40. $-16 - 16$ $\quad -32$

41. $-12 - (-12)$ $\quad 0$

42. $-0.5 - (-1.2)$ $\quad 0.7$

43. $-(-5)$ $\quad 12 - (-3)$ $\quad -4$

44. $\dfrac{3}{7} - \dfrac{5}{7} - \left(-\dfrac{1}{14}\right)$ $\quad -\dfrac{3}{14}$

45. Josh Weidner has $142 in his checking account. He writes a check for $125, makes a deposit for $43, and then writes another check for $85. Represent the balance in his account by a signed number.
−25

46. If the elevation of Lake Superior is 600 feet above sea level and the elevation of the Caspian Sea is 92 feet below sea level, find their difference of elevation. Represent the difference by a signed number.   +692 feet

Determine whether each statement is true or false.

47. $|-5| - |-6| = 5 - 6$ true

48. $|-5 - (-6)| = 5 + 6$ false

(10.4) *Multiply.*

49. $(-3) \cdot (-7)$ 21

50. $(-6) \cdot 3$ −18

51. $-\dfrac{2}{3} \cdot \dfrac{7}{8}$ $-\dfrac{7}{12}$

52. $(-0.5) \cdot (-12)$ 6

53. $-2 \div 0$ undefined

54. $\dfrac{-38}{-1}$ 38

55. $45 \div (-9)$ −5

56. A football team lost five yards on each of two consecutive plays. Represent the total loss by a product of signed numbers and find the product.
$(-5)(2) = -10$ yards

57. A race horse bettor lost $50 on each of four consecutive races. Represent the total loss by a product of signed numbers and find the product.
$(-50)(4) = -\$200$

(10.5) *Simplify.*

58. $5 - 8 + 3$ 0

59. $-3 + 12 + (-7) - 10$ −8

60. $-10 + 3 \cdot (-2)$ −16

61. $5 - 10 \cdot (-3)$ 35

62. $16 \cdot (-2) + 4$ -28

63. $3 \cdot (-12) - 8$ -44

64. $5.2 + 6 \div (-0.3)$ -14.8

65. $-6 + (-10) \div (-2)$ -1

66. $16 + (-3) \cdot 12 \div 4$ 7

67. $\dfrac{3}{4} \cdot \left(-\dfrac{1}{9}\right) - \left(\dfrac{3}{4} - \dfrac{5}{12}\right)$ $-\dfrac{5}{12}$

68. $4^3 - (8 - 3)^2$ 39

69. $\left(-\dfrac{1}{3}\right)^2 - \dfrac{8}{3}$ $-\dfrac{23}{9}$

70. $-(-4) \cdot |-3| - 5$ 7

71. $|5 - 1|^2 \cdot (-5)$ -80

72. $\dfrac{(-4)(-3) - (-2)(-1)}{-10 + 5}$ -2

73. $\dfrac{4(12 - 18)}{-10 \div (-2 - 3)}$ -12

Name _____ **Section** _____ **Date** _____

CHAPTER 10 TEST

Simplify each expression.

1. $-5 + 8$

2. $18 - 24$

3. $0.5 \cdot (-20)$

4. $(-16) \div (-4)$

5. $-\dfrac{3}{11} + \left(-\dfrac{5}{22}\right)$

6. $-7 - (-19)$

7. $(-5) \cdot (-13)$

8. $\dfrac{-2.5}{-0.5}$

9. $|-25| + (-13)$

10. $14 - |-20|$

11. $|5| \cdot |-10|$

12. $\dfrac{|-10|}{-|-5|}$

13. $(-8) + 9 \div (-3)$

14. $-7 + (-32) - 12 + 5$

15. $(-5)^3 - 24 \div (-3)$

16. $(5 - 9)^2 \cdot (8 - 2)^3$

17. $\left(\dfrac{5}{9} - \dfrac{7}{9}\right)^2 + \left(-\dfrac{2}{9}\right)$

18. $3 - (8 - 2)^3$

19. $-6 + (-15) \div (-3)$

20. $\dfrac{4}{2} - \dfrac{8^2}{16}$

21. $\dfrac{-3(-2) + 12}{-1(-4 - 5)}$

22. $\dfrac{|25 - 30|^2}{2(-6) + 7}$

ANSWERS

1. 3

2. -6

3. -10

4. 4

5. $-\dfrac{1}{2}$

6. 12

7. 65

8. 5

9. 12

10. -6

11. 50

12. -2

13. -11

14. -46

15. -117

16. 3456

17. $-\dfrac{14}{81}$

18. 213

19. -1

20. -2

21. 2

22. -5

Name _____

23. Mt. Washington in New Hampshire has an elevation of 6288 feet above sea level. The Romanche Gap in the Atlantic Ocean has an elevation of 25,354 feet below sea level. Represent the difference in elevation between these two points by a signed number. (*Source:* National Geographic Society and Defense Mapping Agency)

24. A mountain climber is at an elevation of 14,893 feet and moves down the mountain a distance of 147 feet. Represent his final elevation as a sum of signed numbers and find the sum.

24. $14,893 + (-147) = 14,746$; +14,746 feet

25. Jane Hathaway has $237 in her checking account. She writes a check for $157, she writes another check for $77, and she deposits $38. Represent the balance in her account by a signed number.

25. +41

Name _____ **Section** _____ **Date** _____

CUMULATIVE REVIEW

1. Multiply: 631×125

2. Divide: $\dfrac{2}{5} \div \dfrac{1}{2}$

3. Add: $\dfrac{2}{3} + \dfrac{1}{7}$

Write each decimal in standard form.

4. Forty-eight and twenty-six hundredths

5. Six and ninety-five thousandths

6. Subtract: $3.5 - 0.068$

7. Simplify: $5.68 + (0.9)^2 \div 100$

8. Write "$27,000 every 6 months" as a unit rate.

9. Find the value of the unknown number, n. $\dfrac{34}{51} = \dfrac{n}{3}$

10. 46 out of every 100 college students live at home. What percent of students live at home? (*Source*: Independent Insurance Agents of America)

Write each fraction or mixed number as a percent.

11. $\dfrac{9}{20}$

12. $1\dfrac{1}{2}$

13. 13 is $6\dfrac{1}{2}\%$ of what number?

14. Translate to a proportion. 101 is what percent of 200?

15. Ivan Borski borrowed $2400 at 10% simple interest for 8 months to buy a used Chevy S-10. Find the simple interest he paid.

ANSWERS

1. 78,875 (Sec. 1.5, Ex. 5)

2. $\dfrac{4}{5}$ (Sec. 2.5, Ex. 7)

3. $\dfrac{17}{21}$ (Sec. 3.3, Ex. 3)

4. 48.26 (Sec. 4.1, Ex. 5)

5. 6.095 (Sec. 4.1, Ex. 6)

6. 3.432 (Sec. 4.3, Ex. 5)

7. 5.6881 (Sec. 4.6, Ex. 8)

8. $\dfrac{4500 \text{ dollars}}{1 \text{ month}}$ or 4500 dollars/month (Sec. 5.2, Ex. 4)

9. $n = 2$ (Sec. 5.3, Ex. 5)

10. 46% (Sec. 6.1, Ex. 2)

11. 45% (Sec. 6.2, Ex. 6)

12. 150% (Sec. 6.2, Ex. 8)

13. 200 (Sec. 6.3, Ex. 10)

14. $\dfrac{101}{200} = \dfrac{p}{100}$ (Sec. 6.4, Ex. 2)

15. $160 (Sec. 6.7, Ex. 2)

663

Name _____

16. Add 3 ft 2 in. and 5 ft 11 in.

17. The normal body temperature is 98.6°F. What is this temperature in degrees Celsius?

18. Find the supplement of a 107° angle.

19. Find the measure of $\angle b$.

20. Find the area of the parallelogram.

1.5 miles

3.4 miles

21. Approximate the volume of a ball of radius 3 inches. Use the approximation $\frac{22}{7}$ for π.

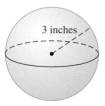

3 inches

22. Find the median of the list of numbers: 25, 54, 56, 57, 60, 71, 98

23. If a die is rolled, find the probability of rolling a 3 or a 4.

24. Add: $-2 + (-21)$

25. Simplify: $60 + 30 + (-2)^3$

Introduction to Algebra

CHAPTER 11

In this chapter we make the transition from arithmetic to algebra. In algebra, letters are used to stand for unknown quantities. Using variables is a very powerful tool for solving problems that cannot be solved with arithmetic alone. This chapter introduces variables, algebraic expressions, and solving variable equations.

Accounting is a system of recording and reporting financial information about a business. Keeping track of numerical data can be traced to prehistoric times when primitive records were kept by scratching marks into branches, bones, or cave walls. The earliest known financial accounting documents are ancient Babylonian payroll records dating from about 3500 B.C. It wasn't until around 1494 A.D. that the first known text on accounting practices was published in Italy. As business has grown and changed and become more complicated, so has financial accounting. However, the basic relationships still remain the same and can be described by very simple equations. In Exercises 47 and 48 on page 682, we will see one way that an equation is used in accounting.

Name _____ Section _____ Date _____

CHAPTER 11 PRETEST

1. Evaluate $3x - 2y$ if $x = -1$ and $y = 3$.

Simplify by combining like terms.

2. $x - 11x$

3. $5x + 4 + 2x - 3y + 7 - y + 9$

4. Multiply: $-2(6m - 3)$

Decide whether the given number is a solution to the given equation.

5. Is -4 a solution to $x + 11 = 15$?

6. Is 7 a solution to $6 - n = -1$?

Solve.

7. $8 = 6 + y$

8. $4 - 9 = x + 3$

9. $-4z = 32$

10. $\dfrac{2}{5}a = 14$

11. $-12x = -4 + 28$

12. $-39 - 3 = -7x + x$

13. $9b - 12 = 6$

14. $4n + 2 = 10n - 16$

15. $3(y + 2) = 9$

16. $20 + 2(w - 7) = 3w + 4$

Translate each sentence into an equation. Use x to represent "a number."

17. The product of 3 and 12 is 36.

18. Twice the sum of 4 and 6 yields 20.

19. The quotient of 10 and three times a number equals 40.

Solve.

20. A number less 7 is 8. Find the number.

11.1   INTRODUCTION TO VARIABLES

A   EVALUATING ALGEBRAIC EXPRESSIONS

Perhaps the most important quality of mathematics is that it is a science of pattern. Communicating about patterns is made possible by using a letter to represent all the numbers fitting a pattern. We call such a letter a **variable**. For example, in Section 1.2 we presented the addition property of 0, which states that the sum of 0 and any whole number is that number. We might write

$$0 + 1 = 1$$
$$0 + 2 = 2$$
$$0 + 3 = 3$$
$$0 + 4 = 4$$
$$0 + 5 = 5$$
$$0 + 6 = 6$$

.

.

.

continuing indefinitely. This is a pattern, and all whole numbers fit the pattern. We can communicate this pattern for all whole numbers by letting a letter, such as a, represent all whole numbers. We can then write

$$0 + a = a$$

HELPFUL HINT

Variables have been used in previous chapters, although we have not called them such. For example, in the ratio and proportion chapter, we wrote equations such as

$$\frac{4}{n} = \frac{6}{12}$$

Here, the letter n is a variable.

Learning to use variable notation is a primary goal of learning **algebra**. We now take some important beginning steps in learning to use variable notation.

A combination of operations on letters (variables) and numbers is called an **algebraic expression** or simply an **expression**.

$$3 + x \qquad 5 \cdot y \qquad 2 \cdot z - 1 + x$$

If two variables or a number and a variable are next to each other, with no operation sign between them, the operation is multiplication. For example,

$$2x \qquad \text{means} \qquad 2 \cdot x$$

and

$$xy \text{ or } x(y) \qquad \text{means} \qquad x \cdot y$$

Also, the meaning of an exponent remains the same when the base is a variable. For example,

TEACHING TIP   Classroom Activity

Have students work in groups to write an algebraic equation to generalize the following patterns:

$1 \cdot 1 = 1$	$1 + 1 = 2$
$2 \cdot 1 = 2$	$2 + 2 = 4$
$3 \cdot 1 = 3$	$3 + 3 = 6$
$4 \cdot 1 = 4$	$4 + 4 = 8$

Copyright 1999 Prentice-Hall, Inc.

TEACHING TIP

Point out that the arithmetic expression $2 \cdot 8 + 7$ is one of the infinite number of arithmetic expressions which can be formed from the algebraic expression $2x + y$. Ask students to identify the value of x and the value of y in this arithmetic expression. Then ask students to write another arithmetic expression which fits the algebraic pattern.

Practice Problem 1

Evaluate: $5x - y$ if $x = 2$ and $y = 3$

Practice Problem 2

Evaluate: $x - y$ if $x = -12$ and $y = -2$

Practice Problem 3

Evaluate: $\dfrac{5r - 2s}{3q}$ if $r = 3$, $s = 3$, and $q = 1$

Practice Problem 4

Evaluate: $b^2 - (3a + c)$ if $a = 2$, $b = -3$, and $c = -1$

Practice Problem 5

The formula for finding the perimeter of a rectangle is $P = 2l + 2w$. Find the perimeter of a rectangular garden that is 25 meters wide and 40 meters long.

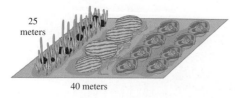

25 meters

40 meters

Answers

1. 7, **2.** −10, **3.** 3, **4.** 4, **5.** 130 meters

$$x^2 = \underbrace{x \cdot x}_{\text{2 factors of } x} \qquad \text{and} \qquad y^5 = \underbrace{y \cdot y \cdot y \cdot y \cdot y}_{\text{5 factors of } y}$$

Algebraic expressions have different values depending on replacement values for x. Replacing a variable in an expression by a number and then finding the value of the expression is called **evaluating the expression** for the variable. When finding the value of an expression, remember to follow the order of operations.

Example 1   Evaluate: $2x + y$ if $x = 8$ and $y = 7$

Solution: Replace x with 8 and y with 7 in $2x + y$.

$$2x + y = 2 \cdot 8 + 7 \quad \text{Replace } x \text{ with 8 and } y \text{ with 7.}$$
$$= 16 + 7 \quad \text{Multiply first because of order of operations.}$$
$$= 23 \quad \text{Add.}$$

Example 2   Evaluate: $x - y$ if $x = 14$ and $y = -3$

Solution: $x - y = 14 - (-3) \quad \text{Replace } x \text{ with 14 and } y \text{ with } -3.$
$$= 14 + (3) \quad \text{To subtract, add the opposite of } -3.$$
$$= 17 \quad \text{Add.}$$

Example 3   Evaluate: $\dfrac{3m - 2n}{2q}$ if $m = 8$, $n = 4$, and $q = 1$

Solution: $\dfrac{3m - 2n}{2q} = \dfrac{3 \cdot 8 - 2 \cdot 4}{2 \cdot 1} \quad \text{Replace } m \text{ with 8, } n \text{ with 4, and } q \text{ with 1.}$
$$= \dfrac{24 - 8}{2} \quad \text{Multiply.}$$
$$= \dfrac{16}{2} \quad \text{Subtract in the numerator.}$$
$$= 8 \quad \text{Divide.}$$

Example 4   Evaluate: $x^3 - (a - b)$ if $x = 2$, $a = -3$, and $b = 5$

Solution:

$$x^3 - (a - b) = 2^3 - (-3 - 5) \quad \text{Replace } x \text{ with 2, } a \text{ with } -3, \text{ and } b \text{ with 5.}$$
$$= 2^3 - (-8) \quad \text{Simplify inside the parentheses.}$$
$$= 8 - (-8) \quad \text{Evaluate the exponential expression.}$$
$$= 8 + 8$$
$$= 16 \quad \text{Add.}$$

Example 5   Finding Area of a Rectangle

The formula for finding the area of a rectangle is $A = lw$, where l is the length of the rectangle and w is the width. Find the area of a rectangular floor that is 45 feet long and 30 feet wide.

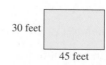

30 feet

45 feet

Solution:　$A = l \cdot w$

$= (45 \text{ feet}) \cdot (30 \text{ feet})$　Replace l with 45 feet and w with 30 feet.

$= 1350 \text{ square feet}$

The area of the floor is 1350 square feet.

B　COMBINING LIKE TERMS

The addends of an algebraic expression are called the **terms** of the expression.

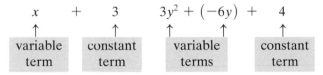

$x + 3$　2 terms

$3y^2 + (-6y) + 4$　3 terms

A term that is only a number has a special name. It is called a **constant term**, or simply a **constant**. A term that contains a variable is called a **variable term**.

x	$+$	3		$3y^2 + (-6y)$	$+$	4
↑		↑		↑		↑
variable term		constant term		variable terms		constant term

The number factor of a variable term is called the **numerical coefficient**. A numerical coefficient of 1 is usually not written.

$5x$	x or $1x$	$3y^2$	$-6y$
↑	↑	↑	↑
Numerical coefficient is 5.	Understood numerical coefficient is 1.	Numerical coefficient is 3.	Numerical coefficient is -6.

Terms that are exactly the same, except that they may have different numerical coefficients, are called **like terms**.

Like Terms	Unlike Terms
$3x, \frac{1}{2}x$	$5x, x^2$
$-6y, 2y, y$	$7x, 7y$

A sum or difference of like terms can be simplified using the **distributive property**. Recall from Chapter 1 that the distributive property says that multiplication distributes over addition (and subtraction). Using variables, we can write the distributive property as follows.

DISTRIBUTIVE PROPERTY

If a, b, and c are numbers, then

$$ac + bc = (a + b)c$$

Also,

$$ac - bc = (a - b)c$$

The distributive property guarantees that, no matter what number x is, $7x + 5x$ (for example) has the same value as $(7 + 5)x$ or $12x$. We have then that

$$7x + 5x = (7 + 5)x = 12x$$

This is an example of **combining like terms**. An algebraic expression is **simplified** when all like terms have been combined.

Practice Problem 6

Simplify each expression by combining like terms.

a. $8m - 11m$

b. $5a + a$

TEACHING TIP

Have students check their simplifications using a specific number for x. For instance, in Example 6a have students substitute 4 for x and show that

$3 \cdot 4 + 2 \cdot 4 = 5 \cdot 4$

$\quad\quad 12 + 8 = 20$

Example 6   Simplify each expression by combining like terms.

a. $3x + 2x$ **b.** $y - 7y$

Solution: We add or subtract like terms.

a. $3x + 2x = (3 + 2)x$

$\quad\quad\quad\quad = 5x$

Understood 1

b. $y - 7y = 1y - 7y$

$\quad\quad\quad\quad = (1 - 7)y$

$\quad\quad\quad\quad = -6y$

The commutative and associative properties of addition and multiplication can also help us simplify expressions. We presented these properties in Sections 1.2 and 1.5 and state them again using variables.

PROPERTIES OF ADDITION AND MULTIPLICATION

If a, b, and c are numbers, then

$a + b = b + a$ Commutative property of addition

$a \cdot b = b \cdot a$ Commutative property of multiplication

That is, the **order** of adding or multiplying two numbers can be changed without changing their sum or product.

$(a + b) + c = a + (b + c)$ Associative property of addition

$(a \cdot b) \cdot c = a \cdot (b \cdot c)$ Associative property of multiplication

That is, the **grouping** of numbers in addition or multiplication can be changed without changing their sum or product.

Practice Problem 7

Simplify: $8m + 5 + m - 4$

Example 7   Simplify: $2y - 6 + 4y + 8$

Solution: We begin by writing subtraction as the opposite of addition.

$2y - 6 + 4y + 8 = 2y + (-6) + 4y + 8$

$\quad\quad\quad\quad = 2y + 4y + (-6) + 8$ Apply the commutative property of addition.

$\quad\quad\quad\quad = (2 + 4)y + (-6) + 8$ Apply the distributive property.

$\quad\quad\quad\quad = 6y + 2$ Simplify.

Answers

6. a. $-3m$,   **b.** $6a$,   **7.** $9m + 1$

Examples

Simplify each expression by combining like terms.

8. $6x + 2x - 5 = 8x - 5$

9. $4x + 2 - 5x + 3 = 4x - 5x + 2 + 3$
$= -1x + 5$ or $-x + 5$

10. $1.2y + 10 - 5.7y - 9 = 1.2y - 5.7y + 10 - 9$
$= -4.5y + 1$

11. $2x - 5 + 3y + 4x - 10y + 11 = 6x - 7y + 6$

C MULTIPLYING EXPRESSIONS

We can also use properties of numbers to multiply expressions such as $3(2x)$. By the associative property of multiplication, we can write the product $3(2x)$ as $(3 \cdot 2)x$, which simplifies to $6x$.

Examples

Multiply:

12. $5(3y) = (5 \cdot 3)y$ Apply the associative property of multiplication.
$= 15y$ Multiply.

13. $-2(4x) = (-2 \cdot 4)x$ Apply the associative property of multiplication.
$= -8x$ Multiply.

We can use the distributive property to combine like terms, which we have done, and also to multiply expressions such as $2(3 + x)$. By the distributive property, we have that

$2(3 + x) = 2 \cdot 3 + 2 \cdot x$ Apply the distributive property.

$= 6 + 2x$ Multiply.

Example 14

Use the distributive property to multiply: $6(x + 4)$

Solution: By the distributive property,

$6(x + 4) = 6 \cdot x + 6 \cdot 4$ Apply the distributive property.

$= 6x + 24$ Multiply.

TRY THE CONCEPT CHECK IN THE MARGIN.

Example 15

Multiply: $-3(5a + 2)$

Solution: By the distributive property,

$-3(5a + 2) = -3(5a) + (-3)(2)$ Apply the distributive property.

$= (-3 \cdot 5)a + (-6)$ Multiply.

$= -15a - 6$ Multiply.

To simplify expressions containing parentheses, we first use the distributive property and multiply.

Practice Problems 8–11

Simplify each expression by combining like terms.

8. $7y + 11y - 8$

9. $2y - 6 + y + 7$

10. $3.7x + 5 - 4.2x + 15$

11. $-9y + 2 - 4y - 8x + 12 - x$

Practice Problems 12–13

Multiply:

12. $7(8a)$

13. $-5(9x)$

Practice Problem 14

Use the distributive property to multiply: $7(y + 2)$

✓ CONCEPT CHECK

What's wrong with the following?
$8(a - b) = -8ab$

Practice Problem 15

Multiply: $4(7a - 5)$

Answers

8. $18y - 8$,   **9.** $3y + 1$,   **10.** $-0.5x + 20$,
11. $-13y - 9x + 14$,   **12.** $56a$,   **13.** $-45x$,
14. $7y + 14$,   **15.** $28a - 20$
✓ **Concept Check:**   Did not distribute the 8.

Practice Problem 16

Simplify: $5(y - 3) - 8 + y$

Example 16   Simplify: $2(3 + x) - 15$

Solution: First we use the distributive property to remove parentheses.

$$2(3 + x) - 15 = 2(3) + 2(x) - 15 \quad \text{Apply the distributive property.}$$
$$= 6 + 2x - 15 \quad \text{Multiply.}$$
$$= 2x + (-9) \quad \text{or} \quad 2x - 9 \quad \text{Combine like terms.}$$

> **HELPFUL HINT**
>
> 2 is *not* distributed to the -15 since it is not within the parentheses.

Practice Problem 17

Simplify: $5(2x - 3) + 7(x - 1)$

Example 17   Simplify: $-2(x - 5) + 4(2x + 2)$

Solution: First we use the distributive property to remove parentheses.

$$-2(x - 5) + 4(2x + 2) = -2(x) + (-2)(-5)$$
$$+ 4(2x) + 4(2) \quad \text{Apply the distributive property.}$$
$$= -2x + 10 + 8x + 8 \quad \text{Multiply.}$$
$$= 6x + 18 \quad \text{Combine like terms.}$$

Practice Problem 18

Find the area of the rectangular garden.

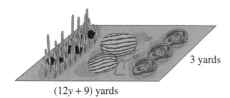

3 yards

$(12y + 9)$ yards

Example 18   Finding the Area of a Deck

Find the area of the rectangular deck.

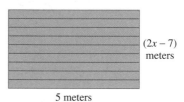

$(2x - 7)$ meters

5 meters

Solution: Recall how to find the area of a rectangle.

$$A = l \cdot w$$
$$= 5(2x - 7) \quad \text{Let length} = 5 \text{ and width} = (2x - 7).$$
$$= 10x - 35 \quad \text{Multiply.}$$

The area is $(10x - 35)$ *square* meters.

Answers

16. $6y - 23$,   **17.** $17x - 22$,   **18.** $(36y + 27)$ square yards

Name _____ **Section** _____ **Date** _____

MENTAL MATH

Identify each pair of terms as like terms or unlike terms.

1. $5x$ and $5y$ **2.** $-3a$ and $-3b$ **3.** x and $-2x$ **4.** $7y$ and y

5. $-5n$ and $6n^2$ **6.** $4m^2$ and $2m$ **7.** $8b$ and $-6b$ **8.** $12a$ and $-11a$

EXERCISE SET 11.1

A *Evaluate each expression if $x = -2$, $y = 5$, and $z = -3$. See Examples 1 through 4.*

1. $3 + 2z$ **2.** $7 + 3z$ **3.** $y - z$ **4.** $z - x$

5. $z - x + y$ **6.** $x + 5y - z$ **7.** $3x - z$ **8.** $2y + 5z$

9. $8 - (y - x)$ **10.** $5 + (2x - 1)$ **11.** $y^3 - 4x$ **12.** $y^3 - z$

13. $\dfrac{6xy}{z}$ **14.** $\dfrac{8yz}{15}$ **15.** $\dfrac{2y - 2}{x}$ **16.** $\dfrac{6 + 3x}{z}$

17. $\dfrac{x + 2y}{2z}$ **18.** $\dfrac{2z + 5}{3x}$ **19.** $\dfrac{5x}{y} - \dfrac{10}{y}$ **20.** $\dfrac{70}{2y} - \dfrac{15}{z}$

Solve. See Example 5.

21. Find the area of a rectangular movie screen that is 50 feet long and 40 feet high. Use $A = lw$.

22. Find the area of a 60-meter by 25-meter rectangular swimming pool. Use $A = lw$.

23. A decorator wishes to put a wallpaper border around a rectangular room that measures 14 feet by 18 feet. Find the room's perimeter. Use $P = 2l + 2w$.

24. How much fencing will a rancher need for a rectangular cattle lot that measures 80 feet by 120 feet? Use $P = 2l + 2w$.

25. How much interest will $3000 in a passbook savings account earn in 2 years at Money Bank, which pays 6% simple interest? Use $I = prt$.

26. How much interest will a $12,000 certificate of deposit earn in 1 year at a rate of 8% simple interest? Use $I = prt$.

MENTAL MATH ANSWERS
1. unlike
2. unlike
3. like
4. like
5. unlike
6. unlike
7. like
8. like

ANSWERS
1. -3
2. -2
3. 8
4. -1
5. 4
6. 26
7. -3
8. -5
9. 1
10. 0
11. 133
12. 128
13. 20
14. -8
15. -4
16. 0
17. $-\dfrac{4}{3}$
18. $\dfrac{1}{6}$
19. -4
20. 12
21. 2000 square feet
22. 1500 square meters
23. 64 feet
24. 400 feet
25. $360
26. $960

27. 78.5 square feet

28. 16″ pizza

29. 23°F

30. −7.78°C

31. 288 cubic inches

32. 96 cubic meters

33. 8x

34. 11y

35. −4n

36. −3z

37. −2c

38. −4b

39. −4x

40. 6y

41. 13a − 8

42. 2b − 15

43. −0.9x + 11.2

44. −11.5y − 13.4

45. 2x − 7

46. 10x − 16

47. −5x + 4y − 5

48. −6a − b − 7

674

27. Find the area of a circular braided rug with a radius of 5 feet. Use $A = \pi r^2$ and $\pi \approx 3.14$.

28. Mario's Pizza sells one 16″ cheese pizza or two 10″ cheese pizzas for $9.99. Which deal gives you more pizza? Use $A = \pi r^2$ with $\pi \approx 3.14$.

29. Convert Paris, France's low temperature of −5°C to Fahrenheit. Use $F = \dfrac{9}{5}C + 32$.

30. Convert Nome, Alaska's 18°F high temperature to Celsius. Use $C = \dfrac{5}{9}(F - 32)$.

31. Find the volume of a box that measures 12″ × 6″ × 4″. Use $V = lwh$.

32. How many cubic meters does a space shuttle cargo hatch have if its dimensions are 8 meters long by 4 meters wide by 3 meters high? Use $V = lwh$.

B *Simplify each expression by combining like terms. See Examples 6 through 11.*

33. $3x + 5x$

34. $8y + 3y$

35. $5n - 9n$

36. $7z - 10z$

37. $4c + c - 7c$

38. $5b - 8b - b$

39. $5x - 7x + x - 3x$

40. $8y + y - 2y - y$

41. $4a + 3a + 6a - 8$

42. $5b - 4b + b - 15$

43. $1.7x + 3.4 - 2.6x + 7.8$

44. $-8.6y + 1.3 - 2.9y - 14.7$

45. $3x + 7 - x - 14$

46. $9x - 6 + x - 10$

47. $4x + 5y + 2 - y - 9x - 7$

48. $a + 4b + 3 - 7a - 5b - 10$

Name _____

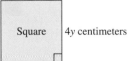 *Multiply. See Examples 12 through 15.*

49. $6(5x)$ **50.** $4(4x)$ **51.** $-2(11y)$ **52.** $-3(21z)$

53. $2(y + 2)$ **54.** $3(x + 1)$ **55.** $5(a - 8)$ **56.** $4(y - 6)$

57. $-4(3x + 7)$ **58.** $-8(8y + 10)$

Simplify each expression. Use the distributive property to remove parentheses first. See Examples 16 and 17.

59. $2(x + 4) - 17$ **60.** $5(6 - y) - 2$ **61.** $4(6n - 5) + 3n$

62. $3(5 - 2b) - 4b$ **63.** $3 + 6(w + 2) + w$ **64.** $8z + 5(6 + z) + 20$

65. $-2(3x + 1) + 5(x - 2)$ **66.** $-3(5x - 2) + 2(3x + 1)$

Find the area of each figure. See Example 18.

67.

Square $4y$ centimeters

68.

12 feet

$(x + 3)$ feet   Rectangle

REVIEW AND PREVIEW

Perform each indicated operation. See Sections 10.2 and 10.3.

69. $-13 + 10$ **70.** $-7 - (-4)$ **71.** $-4 - (-12)$

72. $-15 + 23$ **73.** $-4 + 4$ **74.** $8 + (-8)$

49. $30x$

50. $16x$

51. $-22y$

52. $-63z$

53. $2y + 4$

54. $3x + 3$

55. $5a - 40$

56. $4y - 24$

57. $-12x - 28$

58. $-64y - 80$

59. $2x - 9$

60. $28 - 5y$

61. $27n - 20$

62. $15 - 10b$

63. $7w + 15$

64. $13z + 50$

65. $-x - 12$

66. $-9x + 8$

67. $16y^2$ square centimeters

68. $(12x + 36)$ square feet

69. -3

70. -3

71. 8

72. 8

73. 0

74. 0

675

Name _____

◆ **COMBINING CONCEPTS**

In landscaping, to appraise the value of a large tree, the trunk area is used. The trunk area A is calculated with the formula $A = 0.7854d^2$, where d is the diameter of the tree.

75. The national champion pinyon pine tree is located in Cuba, New Mexico. Its trunk has a diameter of 67.8 inches. Use the formula to find the trunk area of this tree. Round your result to the nearest tenth. (*Source:* American Forests)

76. The national champion white oak tree is located in Wye Mills State Park, Maryland. Its trunk has a diameter of 119 inches. Use the formula to find the trunk area of this tree. Round your result to the nearest tenth. (*Source:* American Forests)

Simplify.

77. $9684q - 686 - 4860q + 12{,}960$

78. $76(268x + 592) - 2960$

79. If x is a whole number, which expression is the largest: $2x$, $5x$ or $\frac{1}{3}x$? Explain your answer.

80. If x is a whole number, which expression is the smallest: $2x$, $5x$, or $\frac{1}{3}x$? Explain your answer.

Find the area of each figure.

81.

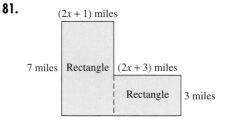

82.

11.2 Solving Equations: The Addition Property

Frequently in this book, we have written statements like $7 + 4 = 11$ or Area = length · width. Each of these statements is called an **equation**. An equation is of the form

expression = expression

An equation can be labeled as

equals sign
↓
$$x + 7 \;=\; 10$$
↑   ↑
left side   right side

Objectives

A Determine whether a given number is a solution of an equation.

B Use the addition property of equality to solve equations.

SSM   CD-ROM   Video 11.2

A Determining Whether a Number is a Solution

When an equation contains a variable, deciding which values of the variable make an equation a true statement is called **solving** an equation for the variable. A **solution** of an equation is a value for the variable that makes an equation a true statement. For example, 2 is a solution of the equation $x + 5 = 7$, since replacing x with 2 results in the *true* statement $2 + 5 = 7$. Similarly, 3 is not a solution of $x + 5 = 7$, since replacing x with 3 results in the *false* statement $3 + 5 = 7$.

Example 1   Determine whether 6 is a solution of the equation $4(x - 3) = 12$.

Solution: We replace x with 6 in the equation.

$$4(\underset{\downarrow}{x} \;\; 3) = 12$$

$$4(6 - 3) \overset{?}{=} 12 \qquad \text{Replace } x \text{ with 6.}$$

$$4(3) \overset{?}{=} 12$$

$$12 \overset{?}{=} 12 \qquad \text{True.}$$

Since $12 = 12$ is a true statement, 6 *is* a solution of the equation.

Example 2   Determine whether -1 is a solution of the equation $3y + 1 = 3$.

Solution:
$$3y + 1 = 3$$
$$3(-1) + 1 \overset{?}{=} 3$$
$$-3 + 1 \overset{?}{=} 3$$
$$-2 \overset{?}{=} 3 \qquad \text{False.}$$

Since $-2 = 3$ is false, -1 is *not* a solution of the equation.

B Using the Addition Property to Solve Equations

To solve an equation, we will use properties of equality to write simpler equations, all equivalent to the original equation, until the final equation has the form

$$x = \textbf{number} \quad \text{or} \quad \textbf{number} = x$$

Practice Problem 1

Determine whether 4 is a solution of the equation $3(y - 6) = 6$.

Practice Problem 2

Determine whether -2 is a solution of the equation $-4x - 3 = 5$.

Answers

1. no, **2.** yes

Equivalent equations have the same solution, so the word "number" on the previous page represents the solution of the orignal equation. The first property of equality to help us write simpler equations is the **addition property of equality**.

ADDITION PROPERTY OF EQUALITY

Let a, b, and c represent numbers.
    If $a = b$, then
        $a + c = b + c$ and $a - c = b - c$

In other words, the same number may be added to or subtracted from both sides of an equation without changing the solution of the equation.

A good way to visualize a true equation is to picture a balanced scale. Since it is balanced, each side of the scale weighs the same amount. Similarly, in a true equation the expressions on each side have the same value. Picturing our balanced scale, if we add the same weight to each side, the scale remains balanced.

Example 3   Solve the equation for x: $x - 2 = 1$

Solution: To solve the equation for x, we need to rewrite the equation in the form $x = $ number. In other words, our goal is to get x alone on one side of the equation. To do so, we add 2 to both sides of the equation.

$$x - 2 = 1$$
$$x - 2 + 2 = 1 + 2 \qquad \text{Add 2 to both sides of the equation.}$$
$$x = 3 \qquad \text{Simplify.}$$

To check, we replace x with 3 in the *original* equation.

$$x - 2 = 1 \qquad \text{Original equation}$$
$$3 - 2 \stackrel{?}{=} 1 \qquad \text{Replace } x \text{ with 3.}$$
$$1 \stackrel{?}{=} 1 \qquad \text{True.}$$

Since $1 - 1$ is a true statement, 3 is the solution of the equation.

Practice Problem 3

Solve the equation for y: $y - 5 = -3$

TEACHING TIP

When discussing Example 3, ask students why 2 was chosen to be added to both sides.

┌─ **HELPFUL HINT**

Note that it is always a good idea to check the solution in the *original* equation to see that it makes the equation a true statement.

Answer

3. 2

Let's visualize how we used the addition property of equality to solve the equation in Example 3. Picture the original equation $x - 2 = 1$ as a balanced scale. The left side of the equation has the same value as the right side.

$x - 2$ 1

If the same weight is added to each side of a scale, the scale remains balanced. Likewise, if the same number is added to each side of an equation, the left side continues to have the same value as the right side.

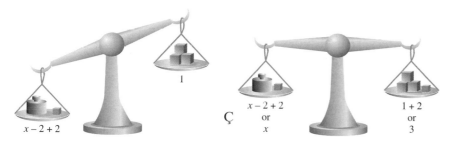

$x - 2 + 2$

$\begin{array}{c} x - 2 + 2 \\ \text{or} \\ x \end{array}$ $\begin{array}{c} 1 + 2 \\ \text{or} \\ 3 \end{array}$

Example 4

Solve: $-8 = x + 1$

Solution: To get x alone on one side of the equation, we subtract 1 from both sides of the equation.

$$-8 = x + 1$$
$$-8 - 1 = x + 1 - 1 \quad \text{Subtract 1 from both sides.}$$
$$-9 = x \quad\quad\quad \text{Simplify.}$$

Check:
$$-8 = x + 1$$
$$-8 \stackrel{?}{=} -9 + 1 \quad \text{Replace } x \text{ with } -9.$$
$$-8 \stackrel{?}{=} -8 \quad\quad \text{True.}$$

The solution is -9.

Practice Problem 4

Solve: $z + 9 = 1$

> **HELPFUL HINT**
>
> Remember that we can get the variable alone on either side of the equation. For example, the equations $x = 2$ and $2 = x$ both have the solution of 2.

Example 5

Solve: $y - 1.2 = -3.2 - 6.6$

Solution: First we simplify the right side of the equation.

$$y - 1.2 = -3.2 - 6.6$$
$$y - 1.2 = -9.8$$

Next, we get y alone on the left side by adding 1.2 to both sides of the equation.

Practice Problem 5

Solve: $x + 6 = 1 - 3$

$$y - 1.2 + 1.2 = -9.8 + 1.2 \quad \text{Add 1.2 to both sides.}$$
$$y = -8.6 \quad \text{Simplify.}$$

Check to see that -8.6 is the solution.   ▬

TRY THE CONCEPT CHECK IN THE MARGIN.

✓ CONCEPT CHECK

What number should be added to or subtracted from both sides of the equation in order to solve the equation $-3.75 = y + 2.1$?

Practice Problem 6

Solve: $-6y - 1 + 7y = 17$

Practice Problem 7

Solve: $\dfrac{2}{3} = x - \dfrac{1}{6}.$

Example 6   Solve: $5x + 2 - 4x = 7 - 9$

     Solution: First we simplify each side of the equation separately.

$$5x + 2 - 4x = 7 - 9$$
$$\underbrace{5x - 4x} + 2 = \underbrace{7 - 9}$$
$$1x + 2 = -2$$

To get x alone on the left side, we subtract 2 from both sides.

$$1x + 2 - 2 = -2 - 2$$
$$1x = -4 \text{ or } x = -4$$

Check to verify that -4 is the solution.   ▬

Example 7   Solve: $\dfrac{7}{8} = y - \dfrac{1}{2}$

     Solution: We use the addition property of equality to add $\dfrac{1}{2}$ to both sides.

$$\frac{7}{8} = y - \frac{1}{2}$$
$$\frac{7}{8} + \frac{1}{2} = y - \frac{1}{2} + \frac{1}{2} \quad \text{Add } \tfrac{1}{2} \text{ to both sides.}$$
$$\frac{7}{8} + \frac{4}{8} = y \quad \text{Simplify.}$$
$$\frac{11}{8} = y \quad \text{Simplify.}$$

Check to see that $\dfrac{11}{8}$ is the solution.   ▬

Answers

6. 18,   **7.** $\dfrac{5}{6}$

✓ **Concept Check:**   subtract 2.1 from both sides

MENTAL MATH

Solve each equation.

1. $x - 2 = 0$ **2.** $x - 5 = 0$ **3.** $x + 1 = 0$ **4.** $x + 6 = 0$

EXERCISE SET 11.2

A *Decide whether the given number is a solution of the given equation. See Examples 1 and 2.*

1. Is 10 a solution of $x - 8 = 2$?

2. Is 9 a solution of $y - 2 = 7$?

3. Is -5 a solution of $x + 12 = 17$?

4. Is -7 a solution of $a + 23 = -16$?

5. Is 8 a solution of $7f = 64 - f$?

6. Is 3 a solution of $12 - k = 9$?

7. Is 3 a solution of
$4c + 2 - 3c = -1 + 6$?

8. Is 1 a solution of $2(b - 3) = 10$?

B *Solve. Check each solution. See Examples 3 through 7.*

9. $a + 5 = 23$ **10.** $s - 7 = 15$ **11.** $d - 9 = -17$ **12.** $f + 4 = -6$

13. $7 = y - 2$ **14.** $-10 = z - 15$ **15.** $-12 = x + 4$ **16.** $1 = y + 7$

17. $x + \dfrac{1}{2} = \dfrac{7}{2}$

18. $x + \dfrac{1}{3} = \dfrac{4}{3}$

19. $y - \dfrac{3}{4} = -\dfrac{5}{8}$

20. $y - \dfrac{5}{6} = -\dfrac{11}{12}$

21. $x - 3 = -1 + 4$

22. $y - 8 = -5 - 1$

23. $-7 + 10 = m - 5$

24. $1 - 8 = n + 2$

25. $x - 0.6 = 4.7$

26. $y - 1.2 = 7.5$

27. $-2 - 3 = -4 + x$

28. $7 - (-10) = x - 5$

29. $y + 2.3 = -9.2 - 8.6$

30. $x + 4.7 = -7.5 + 3.4$

31. $-8x + 4 + 9x = -1 + 7$

32. $3x - 2x + 5 = 5$

33. $2 - 2 = 5x - 4x$

34. $11 + (-15) = 6x - 4 - 5x$

35. $7x + 14 - 6x = -4 + (-10)$

36. $-10x + 11x + 5 = 9 + (-5)$

MENTAL MATH ANSWERS
1. 2
2. 5
3. -1
4. -6
ANSWERS
1. yes
2. yes
3. no
4. no
5. yes
6. yes
7. yes
8. no
9. 18
10. 22
11. -8
12. -10
13. 9
14. 5
15. -16
16. -6
17. 3
18. 1
19. $\dfrac{1}{8}$
20. $-\dfrac{1}{12}$
21. 6
22. 2
23. 8
24. -9
25. 5.3
26. 8.7
27. -1
28. 22
29. -20.1
30. -8.8
31. 2
32. 0
33. 0
34. 0
35. -28
36. -1

681

37. 1

38. 1

39. 1

40. 1

41. 1

42. 1

43. 162,964

44. −28

45. 3705 yards

46. 2378 yards

47. $106,146 million

48. $19,552,709 million

49. answers may vary

50. answers may vary

682

Name _____

REVIEW AND PREVIEW

Perform each indicated operation. See Section 10.4.

37. $\dfrac{-7}{-7}$

38. $\dfrac{4.2}{4.2}$

39. $\dfrac{1}{3} \cdot 3$

40. $\dfrac{1}{5} \cdot 5$

41. $-\dfrac{2}{3} \cdot -\dfrac{3}{2}$

42. $-\dfrac{7}{2} \cdot -\dfrac{2}{7}$

◆ COMBINING CONCEPTS

Solve.

▣ 43. $x - 76{,}862 = 86{,}102$

▣ 44. $-968 + 432 = 86y - 508 - 85y$

A football team's total offense T is found by adding the total passing yardage P to the total rushing yardage R: T = P + R.

45. During the 1997 football season, the Green Bay Packers' total offense was 5614 yards. The Packers' rushing yardage for the season was 1909 yards. How many yards did the Packers gain by passing during the season? (*Source: National Football League*)

46. During the 1997 football season, the Denver Broncos' total offense was 5872 yards. The Broncos' passing yardage for the season was 3494 yards. How many yards did the Broncos gain by rushing during the season? (*Source: National Football League*)

In accounting, a company's annual net income I can be computed using the relation I = R − E, where R is the company's total revenues for a year and E is the company's total expenses for the year.

47. At the end of fiscal year 1997, Wal-Mart had a net income of $3056 million. During the year, Wal-Mart incurred a total of $103,090 million in expenses. What was Wal-Mart's total revenues for the year? (*Source: Wal-Mart Stores, Inc.*)

48. At the end of fiscal year 1997, Home Depot had a net income of $937.739 million. During the year, Home Depot had a total of $18,614.970 million in expenses. What was Home Depot's total revenues for the year? (*Source: Home Depot, Inc.*)

 Internet Excursions

Go to http://www.prenhall.com/martin-gay
The World Wide Web address listed here will provide you with access to the Web site of the National Football League, or a related site. Team and individual player statistics are available that will help you complete the questions below.

49. Choose any NFL team. On its page of statistics, look for the listing of Season Stats. Look for statistics on total net yards (this is the team's total offense yardage), net yards rushing, and net yards passing. Use these statistics to write a problem similar to those in Exercises 45 and 46 about the team's total offense.

50. Choose another NFL team and write a similar problem about the team's total offense. Then trade your problems with another student in your class to solve.

11.3 SOLVING EQUATIONS: THE MULTIPLICATION PROPERTY

A USING THE MULTIPLICATION PROPERTY TO SOLVE EQUATIONS

Although the addition property of equality is a powerful tool for helping us solve equations, it cannot help us solve all types of equations. For example, it cannot help us solve an equation such as $2x = 6$. To solve this equation, we use a second property of equality called the **multiplication property of equality**.

MULTIPLICATION PROPERTY OF EQUALITY

Let a, b, and c represent numbers and let $c \neq 0$.
If $a = b$, then

$$a \cdot c = b \cdot c \qquad \text{and} \qquad \frac{a}{c} = \frac{b}{c}$$

In other words, both sides of an equation may be multiplied or divided by the same nonzero number without changing the solution of the equation.

Picturing again our balanced scale, if we multiply or divide the weight on each side by the same nonzero number, the scale (or equation) remains balanced.

To solve $2x = 6$ for x, we use the multiplication property of equality to divide both sides of the equation by 2, and simplify as follows:

$$2x = 6$$

$$\frac{\overset{1}{\cancel{2}} \cdot x}{\underset{1}{\cancel{2}}} = \frac{6}{2} \qquad \text{Divide both sides by 2.}$$

$$1 \cdot x = 3 \quad \text{or} \quad x = 3$$

Example 1   Solve: $-5x = 15$

Solution: To get x by itself, we divide both sides by -5.

$$-5x = 15 \qquad \text{Original equation}$$

$$\frac{\overset{1}{\cancel{-5}x}}{\underset{1}{\cancel{-5}}} = \frac{15}{-5} \qquad \text{Divide both sides by } -5.$$

$$1x = -3 \quad \text{or} \quad x = -3 \qquad \text{Simplify.}$$

Check: To check, we replace x with -3 in the original equation.

$$-5x = 15 \qquad \text{Original equation}$$

$$-5(-3) \overset{?}{=} 15 \qquad \text{Let } x = -3.$$

$$15 \overset{?}{=} 15 \qquad \text{True.}$$

The solution is -3.

Objective

A Use the multiplication property to solve equations.

SSM   CD-ROM   Video
11.3

Practice Problem 1

Solve: $3y = -18$

Answer
1. -6

Practice Problem 2

Solve: $-16 = 8x$

TEACHING TIP

While discussing Example 2, ask students to try solving the equation by multiplying both sides by $\frac{1}{2}$. Ask if the result is the same. Then discuss why this works.

Practice Problem 3

Solve: $-3y = -27$

Practice Problem 4

Solve: $\frac{5}{7}b = 25$

TEACHING TIP

Remind students that the reciprocal of $\frac{a}{b}$ is $\frac{b}{a}$. Then ask what is the product of reciprocals?

Example 2   Solve: $-8 = 2y$

Solution: To get y alone, we divide both sides of the equation by 2.

$$-8 = 2y$$

$$\frac{-8}{2} = \frac{\overset{1}{\cancel{2}}y}{\cancel{2}}\underset{1}{} \quad \text{Divide both sides by 2.}$$

$$-4 = 1y \quad \text{or} \quad y = -4$$

Check to see that -4 is the solution. ▬

Example 3   Solve: $-1.2x = -36$

Solution: We divide both sides of the equation by the coefficient of x, which is -1.2.

$$-1.2x = -36$$

$$\frac{\overset{1}{\cancel{-1.2}}x}{\underset{1}{\cancel{-1.2}}} = \frac{-36}{-1.2}$$

$$x = 30$$

Check to see that 30 is the solution. ▬

Example 4   Solve: $\frac{3}{5}a = 9$

Solution: Recall that the product of a number and its reciprocal is 1. To get a alone then, we multiply both sides by $\frac{5}{3}$, the reciprocal of $\frac{3}{5}$.

$$\frac{3}{5}a = 9$$

$$\frac{\overset{1}{\cancel{5}}}{\underset{1}{\cancel{3}}} \cdot \frac{\overset{1}{\cancel{3}}}{\underset{1}{\cancel{5}}}a = \frac{5}{3} \cdot 9 \quad \text{Multiply both sides by } \frac{5}{3}.$$

$$1a = \frac{5 \cdot \overset{3}{\cancel{9}}}{\underset{1}{\cancel{3}} \cdot 1} \quad \text{Multiply.}$$

$$a = 15 \quad \text{Simplify.}$$

Check: To check, we replace a with 15 in the original equation.

$$\frac{3}{5}a = 9 \quad \text{Original equation}$$

$$\frac{3}{5} \cdot 15 \overset{?}{=} 9 \quad \text{Replace } a \text{ with 15.}$$

$$\frac{3}{\underset{1}{\cancel{5}}} \cdot \frac{\overset{3}{\cancel{15}}}{1} \overset{?}{=} 9 \quad \text{Multiply.}$$

$$9 \overset{?}{=} 9 \quad \text{True.}$$

Since $9 = 9$ is true, 15 is the solution of $\frac{3}{5}a = 9$. ▬

Answers

2. -2,   **3.** 9,   **4.** 35

Example 5   Solve:

Solution: We multiply both sides of the equation by $\frac{4}{1}$, the reciprocal of $\frac{1}{4}$.

$$\frac{1}{4}x = -\frac{1}{8}$$

$$\frac{\overset{1}{\cancel{4}}}{1} \cdot \frac{1}{\cancel{4}}x = \frac{4}{1} \cdot -\frac{1}{8} \quad \text{Multiply both sides by } \frac{4}{1}.$$

$$1x = -\frac{\overset{1}{\cancel{4}} \cdot 1}{1 \cdot \underset{2}{\cancel{8}}} \quad \text{Multiply.}$$

$$x = -\frac{1}{2} \quad \text{Simplify.}$$

Check to see that $-\frac{1}{2}$ is the solution.

TRY THE CONCEPT CHECK IN THE MARGIN.

We often need to simplify one or both sides of an equation before applying the properties of equality to get the variable alone.

Example 6   Solve: $3y - 7y = 12$

Solution: First we combine like terms.

$$3y - 7y = 12$$

$$-4y = 12 \quad \text{Combine like terms.}$$

$$\frac{\overset{1}{\cancel{-4}}y}{\underset{1}{\cancel{-4}}} = \frac{12}{-4} \quad \text{Divide both sides by } -4.$$

$$y = -3 \quad \text{Simplify.}$$

Check: We replace y with -3.

$$3y - 7y = 12$$

$$3(-3) - 7(-3) \overset{?}{=} 12$$

$$-9 + 21 \overset{?}{=} 12$$

$$12 \overset{?}{=} 12 \quad \text{True.}$$

The solution is -3.

Practice Problem 5

Solve: $-\frac{7}{10}x = \frac{2}{5}$

✓ CONCEPT CHECK

Which operation is appropriate for solving each of the following equations: addition or multiplication?
a. $6 = -4x$

b. $6 = x - 4$

Practice Problem 6

Solve: $10 = 2m - 4m$

Answers

5. $-\frac{4}{7}$,   **6.** -5

✓ Concept Check

a. multiplication,   **b.** addition

Practice Problem 7

Solve: $-8 + 6 = -3a + 2a$

Example 7

Solve: $-z - z = 11 - 5$

Solution: We simplify both sides of the equation first.

$$-z - z = 11 - 5$$

$$-2z = 6 \qquad \text{Combine like terms.}$$

$$\frac{-2z}{-2} = \frac{6}{-2} \qquad \text{Divide both sides by } -2.$$

$$z = -3 \qquad \text{Simplify.}$$

Check to see that -3 is the solution.

Answer

7. 2

Name _____ **Section** _____ **Date** _____

Exercise Set 11.3

A *Solve. See Examples 1 through 5.*

1. $5x = 20$ **2.** $6y = 48$ **3.** $-3z = 12$ **4.** $-2x = 26$

5. $0.4y = 0$ **6.** $0.8x = -8$ **7.** $2z = -34$ **8.** $7y = -21$

9. $-0.3x - -15$ **10.** $-0.4z = -12$ **11.** $\frac{2}{5}x = 10$ **12.** $\frac{3}{7}x = 27$

13. $\frac{1}{6}y = -5$ **14.** $\frac{1}{8}y = -3$ **15.** $\frac{5}{6}x = \frac{5}{18}$ **16.** $\frac{4}{7}y = \frac{8}{21}$

17. $-\frac{2}{9}z = \frac{4}{27}$ **18.** $-\frac{3}{4}v - \frac{9}{14}$ **19.** $\frac{8}{5}t = -\frac{3}{8}$ **20.** $\frac{4}{7}r = -\frac{7}{2}$

21. $-\frac{3}{5}x = -\frac{6}{15}$ **22.** $\frac{6}{7}y = \frac{1}{14}$

Solve. First combine any like terms on each side of the equation. See Examples 6 and 7.

23. $2w - 12w = 40$ **24.** $8y + y = 45$ **25.** $16 = 10t - 8t$

26. $100 = 15y + 5y$ **27.** $2z = 1.2 - 1.4$ **28.** $-3x = 1.1 - 0.2$

29. $4 - 10 = -3z$ **30.** $20 - 12 = -4x$

31. $-3x - 3x = 50 - 2$ **32.** $5y - 9y = -14 + (-14)$

33. $-36 = 9u + 3u$ **34.** $-50 = 4y - 14y$

35. $23x - 25x = 7 - 9$ **36.** $8x - 6x = 12 - 22$

Name _____

37. $5 - 5 = 2x + 7x$ $\qquad$ **38.** $7x + 8x = 12 + (-12)$

39. $-42 + 20 = -2x + 13x$ $\qquad$ **40.** $4y - 9y = -20 + 15$

Review and Preview

Evaluate each expression when $x = 5$. See Section 11.1.

41. $3x + 10$ $\qquad$ **42.** $40x$ $\qquad$ **43.** $\dfrac{x - 3}{2}$

44. $7x - 20$ $\qquad$ **45.** $\dfrac{3x + 5}{x - 7}$ $\qquad$ **46.** $\dfrac{2x - 1}{x - 8}$

Combining Concepts

47. Why does the multiplication property of equality not allow us to divide both sides of an equation by zero?

48. Is the equation $-x = 6$ solved for the variable? Explain why or why not.

49. Solve: $-0.025x = 91.2$

50. Solve: $3.6y = -1.259 - 3.277$

The equation $d = r \cdot t$ describes the relationship between distance d in miles, rate r in miles per hour, and time t in hours.

51. The distance between New Orleans, Louisiana, and Kansas City, Missouri, by road, is 806 miles. How long will it take to drive from New Orleans to Kansas City if the driver maintains a speed of 65 miles per hour? (*Source: 1998 World Almanac*)

52. The distance between Boston, Massachusetts, and Memphis, Tennessee, by road, is 1296 miles. How long will it take to drive from Boston to Memphis if the driver maintains a speed of 60 miles per hour? (*Source: 1998 World Almanac*)

53. The distance between Toledo, Ohio, and Chicago, Illinois, by road, is 232 miles. At what speed should a driver drive if he or she would like to make the trip in 4 hours? (*Source: 1998 World Almanac*)

54. The distance between St. Louis, Missouri, and Minneapolis, Minnesota, by road, is 552 miles. If it took 9 hours to drive from St. Louis to Minneapolis, what was the driver's average speed? (*Source: 1998 World Almanac*)

INTEGRATED REVIEW — EXPRESSIONS AND EQUATIONS

Evaluate each expression when $x = -1$ and $y = 3$.

1. $y - x$ **2.** $\dfrac{y}{x}$ **3.** $5x + 2y$ **4.** $\dfrac{y^2 + x}{2x}$

Simplify each expression by combining like terms.

5. $7x + x$ **6.** $6y - 10y$ **7.** $2a + 5a - 9a - 2$

Multiply.

8. $-2(4x)$ **9.** $5(y + 2)$

10. Find the area

Rectangle	3 meters

$(4x - 2)$ meters

Solve and check.

11. $x + 7 = 20$ **12.** $-11 = x - 2$ **13.** $11x = 55$

14. $-7y = 0$ **15.** $12 = 11x - 14x$ **16.** $\dfrac{3}{5}x = 15$

17. $x - 1.2 = -4.5 + 2.3$ **18.** $8y + 7y = -45$ **19.** $6 - (-5) = x + 5$

20. $-0.2m = -1.6$ **21.** $-\dfrac{2}{3}n = \dfrac{6}{11}$ **22.** $n - \dfrac{2}{5} = \dfrac{3}{10}$

1. 4
2. -3
3. 1
4. -4
5. $8x$
6. $-4y$
7. $-2a - 2$
8. $-8x$
9. $5y + 10$
10. $(12x - 6)$ square meters
11. 13
12. -9
13. 5
14. 0
15. -4
16. 25
17. -1
18. -3
19. 6
20. 8
21. $-\dfrac{9}{11}$
22. $\dfrac{7}{10}$

Focus On Mathematical Connections

MODELING EQUATION SOLVING WITH ADDITION AND SUBTRACTION

We can use positive counters ● and negative counters ● to help us model the equation-solving process. We will also need to use an object that represents a variable. We will use small slips of paper with the variable name written on them.

Recall that taking a ● and ● together creates a neutral or zero pair. Once a neutral pair has been formed, it can be removed from or added to an equation model without changing the overall value. We also need to remember that we can add or remove the same number of positive or negative counters from both sides of an equation without changing the overall value.

We can represent the equation $x + 5 = 2$ as follows

To get the variable by itself, we must remove 5 black counters from both sides of the model. Because there are only 2 counters on the right side, we must add 5 negative counters to both sides of the model. Then we can remove neutral pairs: 5 from the left side and 2 from the right side.

We are left with the following model, which represents the solution, $x = -3$.

Similarly, we can represent the equation $x - 4 = -6$ as follows.

To get the variable by itself, we must remove 4 red counters from both sides of the model

We are left with the following model, which represents the solution, $x = -2$.

CRITICAL THINKING

Use the counter model to solve each equation.

1. $x - 3 = -7$ -4
2. $x - 1 = -9$ -8
3. $x + 2 = 8$ 6
4. $x + 4 = 5$ 1
5. $x + 8 = 3$ -5
6. $x - 5 = -1$ 4
7. $x - 2 = 1$ 3
8. $x - 5 = 10$ 15
9. $x + 3 = -7$ -10
10. $x + 8 = -2$ -10

11.4 SOLVING EQUATIONS USING ADDITION AND MULTIPLICATION PROPERTIES

A SOLVING EQUATIONS USING ADDITION AND MULTIPLICATION PROPERTIES

We will now solve equations using more than one property of equality. To solve an equation such as $2x - 6 = 18$, we must first get the variable term $2x$ alone on one side of the equation.

Example 1   Solve: $2x - 6 = 18$

Solution: We start by adding 6 to both sides to get the variable term $2x$ alone.

$$2x - 6 = 18$$
$$2x - 6 + 6 = 18 + 6 \qquad \text{Add 6 to both sides.}$$
$$2x = 24 \qquad \text{Simplify.}$$

To finish solving, we divide both sides by 2.

$$\frac{\overset{1}{\cancel{2}}x}{\cancel{2}_1} = \frac{24}{2} \qquad \text{Divide both sides by 2.}$$

$$x = 12 \qquad \text{Simplify.}$$

Check:
$$2x - 6 = 18$$
$$2(12) - 6 \overset{?}{=} 18 \qquad \text{Replace } x \text{ with 12 and simplify.}$$
$$24 - 6 \overset{?}{=} 18$$
$$18 \overset{?}{=} 18 \qquad \text{True.}$$

The solution is 12.

Example 2   Solve: $20 - x = 21$

Solution: First we get the variable term alone on one side of the equation.

$$20 - x = 21$$
$$20 - x - 20 = 21 - 20 \qquad \text{Subtract 20 from both sides.}$$
$$-1x = 1 \qquad \text{Simplify. Recall that } -x \text{ means } -1x.$$

$$\frac{\overset{1}{\cancel{-1}}x}{\cancel{-1}_1} = \frac{1}{-1} \qquad \text{Divide both sides by } -1.$$

$$x = -1 \qquad \text{Simplify.}$$

Check:
$$20 - x = 21$$
$$20 - (-1) \overset{?}{=} 21$$
$$21 \overset{?}{=} 21 \qquad \text{True.}$$

The solution is -1.

Example 3   Solve: $1 = \frac{2}{3}x + 7$

Solution: Subtract 7 from both sides to get the variable term alone.

Objectives

A Solve equations using addition and multiplication properties.

B Solve equations containing parentheses.

C Write sentences as equations.

SSM   CD-ROM   Video
11-4

Practice Problem 1

Solve: $5y + 2 = 17$

Practice Problem 2

Solve: $45 = -10 - y$

TEACHING TIP

Consider pointing out to students that their work will be easier to follow and they will make fewer errors if they model their work on the solutions in the text. In particular, they should show each step and align their equal signs.

Practice Problem 3

Solve: $\frac{3}{4}y + 19 = 11$

Answers

1. 3, **2.** -55, **3.** $-\frac{32}{3}$

$$1 - 7 = \frac{2}{3}x + 7 - 7 \quad \text{Subtract 7 from both sides.}$$

$$-6 = \frac{2}{3}x \quad \text{Simplify.}$$

$$\frac{3}{2} \cdot -6 = \frac{\overset{1}{\cancel{3}}}{\cancel{2}} \cdot \frac{\overset{1}{\cancel{2}}}{\cancel{3}}x \quad \text{Multiply both sides by } \frac{3}{2}.$$

$$\frac{3}{\cancel{2}} \cdot \frac{\overset{-3}{\cancel{-6}}}{1} = 1x \quad \text{Simplify.}$$

$$-9 = x \quad \text{Simplify.}$$

Check to see that the solution is -9. ▬▬▬

If an equation contains variable terms on both sides, we use the addition property of equality to get all the variable terms on one side and all the constants or numbers on the other side.

Practice Problem 4

Solve: $7x + 12 = 3x - 4$

Example 4   Solve: $3a - 6 = a + 4$

Solution:

$$3a - 6 = a + 4$$
$$3a - 6 + 6 = a + 4 + 6 \quad \text{Add 6 to both sides.}$$
$$3a = a + 10 \quad \text{Simplify.}$$
$$3a - a = a + 10 - a \quad \text{Subtract } a \text{ from both sides.}$$
$$2a = 10 \quad \text{Simplify.}$$
$$\frac{\overset{1}{\cancel{2}}a}{\cancel{2}} = \frac{10}{2} \quad \text{Divide both sides by 2.}$$
$$a = 5 \quad \text{Simplify.}$$

Check to see that the solution is 5. ▬▬▬

Practice Problem 5

Solve: $8x + 4.2 = 10x + 11.6$

Example 5   Solve: $7x + 3.2 = 4x - 1.6$

Solution:

$$7x + 3.2 = 4x - 1.6$$
$$7x + 3.2 - 3.2 = 4x - 1.6 - 3.2 \quad \text{Subtract 3.2 from both sides.}$$
$$7x = 4x - 4.8 \quad \text{Simplify.}$$
$$7x - 4x = 4x - 4.8 - 4x \quad \text{Subtract } 4x \text{ from both sides.}$$
$$3x = -4.8 \quad \text{Simplify.}$$
$$\frac{\overset{1}{\cancel{3}}x}{\cancel{3}} = -\frac{4.8}{3} \quad \text{Divide both sides by 3.}$$
$$x = -1.6 \quad \text{Simplify.}$$

Check to see that -1.6 is the solution. ▬▬▬

B SOLVING EQUATIONS CONTAINING PARENTHESES

If an equation contains parentheses, we must first use the distributive property to remove them.

Answers

4. -4,   **5.** -3.7

Example 6   Solve: $7(x - 2) = 9x - 6$

Solution: First we apply the distributive property.

$$7(x - 2) = 9x - 6$$
$$7x - 14 = 9x - 6 \quad \text{Apply the distributive property.}$$

Next we move variable terms to one side of the equation and constants to the other side.

$$7x - 14 - 9x = 9x - 6 - 9x \quad \text{Subtract } 9x \text{ from both sides.}$$
$$-2x - 14 = -6 \quad \text{Simplify.}$$
$$-2x - 14 + 14 = -6 + 14 \quad \text{Add 14 to both sides.}$$
$$-2x = 8 \quad \text{Simplify.}$$
$$\frac{-2x}{-2} = \frac{8}{-2} \quad \text{Divide both sides by } -2.$$
$$x = -4 \quad \text{Simplify.}$$

Check to see that -4 is the solution.

You may want to use the steps in the margin to solve equations.

Example 7   Solve: $3(2x - 6) + 6 = 0$

Solution:

$$3(2x - 6) + 6 = 0$$

Step 1. $6x - 18 + 6 = 0$ Apply the distributive property.

Step 2. $6x - 12 = 0$ Combine like terms on the left side of the equation.

Step 3. $6x - 12 + 12 = 0 + 12$ Add 12 to both sides.

$$6x = 12 \quad \text{Simplify.}$$

Step 4. $\dfrac{6x}{6} = \dfrac{12}{6}$ Divide both sides by 6.

$$x = 2 \quad \text{Simplify.}$$

Check: **Step 5.**
$$3(2x - 6) + 6 = 0$$
$$3(2 \cdot 2 - 6) + 6 \stackrel{?}{=} 0$$
$$3(4 - 6) + 6 \stackrel{?}{=} 0$$
$$3(-2) + 6 \stackrel{?}{=} 0$$
$$-6 + 6 \stackrel{?}{=} 0$$
$$0 \stackrel{?}{=} 0 \quad \text{True.}$$

The solution is 2.

C WRITING SENTENCES AS EQUATIONS

Next, we practice translating sentences into equations. Below are key words and phrases that translate to an equal sign.

Practice Problem 6

Solve: $6(a - 5) = 4(a + 1)$

STEPS FOR SOLVING AN EQUATION

Step 1.   If parentheses are present, use the distributive property.

Step 2.   Combine any like terms on each side of the equation.

Step 3.   Use the addition property of equality to rewrite the equation so that variable terms are on one side of the equation and constant terms are on the other side.

Step 4.   Use the multiplication property of equality to divide both sides by the numerical coefficient of the variable to solve.

Step 5.   Check the solution in the *original equation*.

Practice Problem 7

Solve: $4(x + 3) - 12$

TEACHING TIP

Before going over the steps for solving an equation, have students work in groups to make their own steps. Then have them compare their steps to the ones in the book.

Answers

6. 17,   **7.** 0

Key Words or Phrases	Examples	Symbols
equals	3 equals 2 plus 1	$3 = 2 + 1$
gives	the quotient of 10 and -5 gives -2	$\dfrac{10}{-5} = -2$
is/was	x is 5	$x = 5$
yields	y plus 2 yields 13	$y + 2 = 13$
amounts to	twice x amounts to -30	$2x = -30$
is equal to	-24 is equal to 2 times -12	$-24 = 2(-12)$

Practice Problem 8

Translate each sentence into an equation.

a. The difference of 110 and 80 is 30.

b. The product of 3 and the sum of -9 and 11 amounts to 6.

c. The quotient of twice 12 and -6 yields -4.

TEACHING TIP   Classroom Activity

Have students work in groups to translate each of the following equations into a sentence.

$14 \cdot 6 - 3 = 81$

$7(12 - 4) = 84 - 28$

$\dfrac{45}{360} = 0.125$

Example 8

Translate each sentence into an equation.

a. The product of 7 and 6 is 42.
b. Twice the sum of 3 and 5 is equal to 16.
c. The quotient of -45 and 5 yields -9.

Solution:

a. In words:

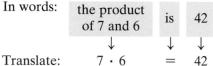

Translate: $7 \cdot 6 \qquad = \qquad 42$

b. In words:

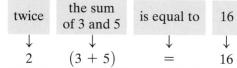

Translate: $2 \qquad (3 + 5) \qquad = \qquad 16$

c. In words:

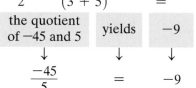

Translate: $\dfrac{-45}{5} \qquad = \qquad -9$

CALCULATOR EXPLORATIONS
CHECKING EQUATIONS

A calculator can be used to check possible solutions of equations. To do this, replace the variable by the possible solution and evaluate each side of the equation separately. For example, to see whether 7 is a solution of the equation $52x = 15x + 259$, replace x with 7 and use your calculator to evaluate each side separately.

Equation: $52x = 15x + 259$

$52 \cdot 7 \stackrel{?}{=} 15 \cdot 7 + 259$ Replace x with 7.

Evaluate left side: $\boxed{52}\ \boxed{\times}\ \boxed{7}\ \boxed{=}$ Display: $\boxed{364}$.

or

$\boxed{\text{ENTER}}$

Evaluate right side: $\boxed{15}\ \boxed{\times}\ \boxed{7}\ \boxed{+}\ \boxed{259}\ \boxed{=}$ Display: $\boxed{364}$.

or

$\boxed{\text{ENTER}}$

Since the left side equals the right side, 7 is a solution of the equation $52x = 15x + 259$.

Use a calculator to determine whether the numbers given are solutions of each equation.

1. $76(x - 25) = -988$;   12   yes

2. $-47x + 862 = -783$;   35   yes

3. $x + 562 = 3x + 900$; -170 no

4. $55(x + 10) = 75x + 910$; -18 yes

5. $29x - 1034 = 61x - 362$; -21 yes

6. $-38x + 205 = 25x + 120$;   25   no

Answers

8. a. $110 - 80 = 30$,   **b.** $3(-9 + 11) = 6$,

c. $\dfrac{2(12)}{-6} = -4$

Name _____ **Section** _____ **Date** _____

ANSWERS

1. 3
2. 4
3. 4
4. 8
5. −4
6. −2
7. −3
8. −5
9. −12
10. 1
11. 100
12. −49
13. −3.9
14. −2.5
15. −4
16. −4
17. 5
18. 2
19. 1
20. −1
21. −4
22. 5
23. 2
24. 7
25. 5
26. 9
27. −3
28. −3
29. 2
30. 4
31. −2
32. −3
33. 3
34. 6
35. −1
36. −1

EXERCISE SET 11.4

A *Solve each equation. See Examples 1 through 5.*

1. $2x - 6 = 0$

2. $3y - 12 = 0$

3. $5n + 10 = 30$

4. $4z + 8 = 40$

5. $6 - n = 10$

6. $7 - y = 9$

7. $10x + 15 = 6x + 3$

8. $5x - 3 = 2x - 18$

9. $3x - 7 = 4x + 5$

10. $3x + 1 = 8x - 4$

11. $-\dfrac{2}{5}x + 19 = -21$

12. $-\dfrac{3}{7}y - 14 = 7$

13. $1.7 = 2y + 9.5$

14. $-5.1 = 3x + 2.4$

15. $9a + 29 = -7$

16. $10 + 4v = -6$

17. $8 - t = 3$

18. $6 - x = 4$

19. $0 = 4x - 4$

20. $0 = 5y + 5$

21. $2n + 8 = 0$

22. $8w - 40 = 0$

23. $7 = 4c - 1$

24. $9 = 2b - 5$

25. $3r + 4 = 19$

26. $5m + 1 = 46$

27. $2x - 1 = -7$

28. $3t - 2 = -11$

29. $2 = 3z - 4$

30. $4 = 4p - 12$

31. $5x - 2 = -12$

32. $7y - 3 = -24$

33. $-7c + 1 = -20$

34. $-2b + 5 = -7$

35. $-5 = -13 - 8k$

36. $-7 = -17 - 10d$

695

37. $4x + 3 = 2x + 11$ **38.** $6y - 8 = 3y + 7$ **39.** $-2y - 10 = 5y + 18$

40. $7n + 5 = 12n - 10$ **41.** $-8n + 1 = -6n - 5$ **42.** $10w + 8 = w - 10$

43. $9 - 3x = 14 + 2x$ **44.** $4 - 7m = -3m + 4$

45. $\dfrac{3}{8}x + 14 = \dfrac{5}{8}x - 2$ **46.** $\dfrac{2}{7}x - 9 = \dfrac{5}{7}x - 15$

47. $-1.4x - 2 = -1.2x + 7$ **48.** $5.7y + 14 = 5.4y - 10$

B *Solve each equation. See Examples 6 and 7.*

49. $3(x - 1) = 12$ **50.** $2(x + 5) = -8$ **51.** $-2(y + 4) = 2$

52. $-1(y + 3) = 10$ **53.** $35 = 17 + 3(x - 2)$ **54.** $22 - 42 = 4(x - 1)$

55. $2(y - 3) = y - 6$ **56.** $3(z + 2) = 5z + 6$ **57.** $2t - 1 = 3(t + 7)$

58. $4 + 3c = 2(c + 2)$ **59.** $3(5c - 1) - 2 = 13c + 3$

60. $4(3t + 4) - 20 = 3 + 5t$ **61.** $10 + 5(z - 2) = 4z + 1$

62. $14 + 4(w - 5) = 6 - 2w$ **63.** $7(6 + w) = 6(2 + w)$

64. $6(5 + c) = 5(c - 4)$

---

37. 4

38. 5

39. -4

40. 3

41. 3

42. -2

43. -1

44. 0

45. 64

46. 14

47. -45

48. -80

49. 5

50. -9

51. -5

52. -13

53. 8

54. -4

55. 0

56. 0

57. -22

58. 0

59. 4

60. 1

61. 1

62. 2

63. -30

64. -50

696

Name _____

65. $-42 + 16 = -26$

C *Write each sentence as an equation. See Example 8.*

65. The sum of -42 and 16 is -26.

66. The difference of -30 and 10 equals -40.

66. $-30 - 10 = -40$

67. The product of -5 and -29 gives 145.

68. The quotient of -16 and 2 yields -8.

67. $-5(-29) = 145$

68. $\dfrac{-16}{2} = -8$

69. Three times the difference of -14 and 2 amounts to -48.

70. The product of -2 and the sum of 3 and 12 is -30.

69. $3(-14 - 2) = -48$

71. The quotient of 100 and twice 50 is equal to 1.

72. Seventeen subtracted from -12 equals -29.

70. $-2(3 + 12) = -30$

REVIEW AND PREVIEW

71. $\dfrac{100}{2(50)} = 1$

The following bar graph shows the average monthly payments when financing a new car. Use this graph to answer Exercises 73–76. See Section 9.1.

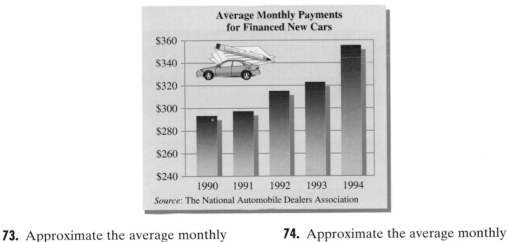

Average Monthly Payments for Financed New Cars

Source: The National Automobile Dealers Association

72. $-12 - 17 = -29$

73. Approximate the average monthly payment for new cars in 1993.

74. Approximate the average monthly payment for new cars in 1994.

73. $322

74. $355

75. How much more is the average monthly payment for financed new cars in 1994 than in 1990?

76. Describe any trends shown in this graph.

75. $62

76. answers may vary

697

COMBINING CONCEPTS

The equation $C = \dfrac{5}{9}(F - 32)$ *gives the relationship between equivalent Celsius temperatures*
C and Fahrenheit temperatures F.

77. The average high temperature during July in Toronto, Canada, is 26.8°C. Use the given equation to convert this temperature to degrees Fahrenheit. (*Source:* World Meteorological Organization)

78. The average high temperature during July in Cairo, Egypt, is 34.4°C. Use the given equation to convert this temperature to degrees Fahrenheit. (*Source:* World Meteorological Organization)

79. The average low temperature in January in Montreal, Canada, is −14.9°C. Use the given equation to convert this temperature to degrees Fahrenheit. (*Source:* World Meteorological Organization)

80. The average low temperature in January in Stockholm, Sweden, is −5°C. Use the given equation to convert this temperature to degrees Fahrenheit. (*Source:* World Meteorological Organization)

11.5   EQUATIONS AND PROBLEM SOLVING

A   WRITING PHRASES AS ALGEBRAIC EXPRESSIONS

Now that we have practiced solving equations for a variable, we can extend considerably our problem-solving skills. We begin by writing phrases as algebraic expressions using the following key words and phrases as a guide.

Addition	Subtraction	Multiplication	Division	Equal Sign
sum	difference	product	quotient	equals
plus	minus	times	divided by	gives
added to	subtracted from	multiply	into	is/was
more than	less than	twice	per	yields
increased by	decreased by	of		amounts to
total	less	double		is equal to

Example 1   Write each phrase as an algebraic expression. Use x to represent "a number."

    a. 7 increased by a number
    b. 15 decreased by a number
    c. the product of 2 and a number
    d. the quotient of a number and 5
    e. 2 subtracted from a number

Solution:

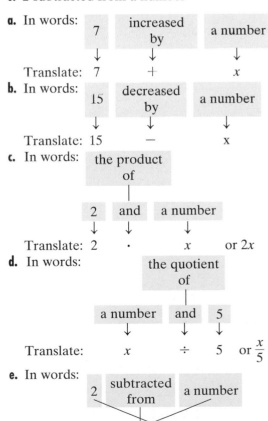

a. In words:

7	increased by	a number
↓	↓	↓

Translate: 7   + x

b. In words:

15	decreased by	a number
↓	↓	↓

Translate: 15   −   x

c. In words:

the product of

2	and	a number
↓	↓	↓

Translate: 2   · x or $2x$

d. In words:

the quotient of

a number	and	5
↓	↓	↓

Translate: x ÷   5   or $\frac{x}{5}$

e. In words:

2	subtracted from	a number

Translate: x −   2

Objectives

A Write phrases as algebraic expressions.

B Write sentences as equations.

C Use problem-solving steps to solve problems.

SSM   CD-ROM   Video
                             11.5

Practice Problem 1

Write each phrase as an algebraic expression. Use x to represent "a number."

a. twice a number

b. 8 increased by a number

c. 10 minus a number

d. 10 subtracted from a number

e. the quotient of 6 and a number

TEACHING TIP   Classroom Activity

Have students work in groups to write a different phrase to represent each of the algebraic expressions found in Example 1. Then have the groups share their phrases and explain why each is equivalent to the original phrase and algebraic expression.

Answers

1. a. $2x$,   **b.** $8 + x$,   **c.** $10 - x$,   **d.** $x - 10$,
e. $\frac{6}{x}$

B WRITING SENTENCES AS EQUATIONS

Now that we have practiced writing phrases as algebraic expressions, let's write sentences as equations. You may want to first study the key words and phrases chart to review some key words and phrases that translate to an equal sign.

Example 2   Write each sentence as an equation. Use x to represent "a number."

a. Nine increased by a number is 5.
b. Twice a number equals -10.
c. A number minus 6 amounts to 168.
d. Three times the sum of a number and 5 is -30.
e. The quotient of 8 and twice a number is equal to 2.

Solution: **a.** In words:

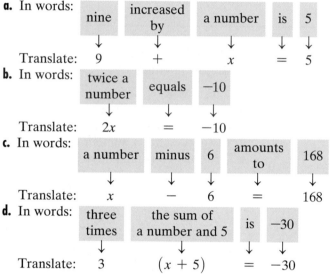

nine	increased by	a number	is	5
↓	↓	↓	↓	↓

Translate: 9 $+$ x $=$ 5

b. In words:

twice a number	equals	-10
↓	↓	↓

Translate: $2x$ $=$ -10

c. In words:

a number	minus	6	amounts to	168
↓	↓	↓	↓	↓

Translate: x $-$ 6 $=$ 168

d. In words:

three times	the sum of a number and 5	is	-30
↓	↓	↓	↓

Translate: 3 $(x + 5)$ $=$ -30

e. In words:

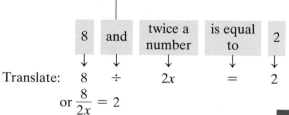

the quotient of

8	and	twice a number	is equal to	2
↓	↓	↓	↓	↓

Translate: 8 $\div$ $2x$ $=$ 2

or $\dfrac{8}{2x} = 2$

C USING PROBLEM-SOLVING STEPS TO SOLVE PROBLEMS

Our main purpose for studying arithmetic and algebra is to solve problems. The same problem-solving steps that have been used throughout this text will be used in this section also. Those steps are repeated here.

The first problem that we will solve consists of finding an unknown number.

Example 3   Finding an Unknown Number

Twice a number added to 3 is the same as the number minus 6. Find the number.

Solution: **1.** UNDERSTAND the problem. To do so, we read and reread the problem. Then we assign a variable to the unknown. We will let $x =$ the unknown number.

Practice Problem 2

Write each sentence as an equation. Use x to represent "a number."

a. Five times a number is 20.

b. The sum of a number and -5 yields 14.

c. Ten subtracted from a number amounts to -23.

d. Five times a number added to 7 is equal to -8.

e. The quotient of 6 and the sum of a number and 4 gives 1.

Practice Problem 3

Translate "the sum of a number and 2 equals 6 added to three times the number" into an equation and solve.

Answers

2. a. $5x = 20$, **b.** $x + (-5) = 14$,

c. $x - 10 = -23$, **d.** $5x + 7 = -8$,

e. $\dfrac{6}{x + 4} = 1$, **3.** -2

2. TRANSLATE the problem into an equation.

In words:

twice a number	added to 3	is the same as	the number minus 6
↓	↓	↓	↓

Translate: $2x$ $+\ 3$ $=$ $x - 6$

3. SOLVE the equation. To solve the equation, we first subtract x from both sides.

$$2x + 3 = x - 6$$
$$2x + 3 - x = x - 6 - x$$
$$x + 3 = -6 \qquad \text{Simplify.}$$
$$x + 3 - 3 = -6 - 3 \qquad \text{Subtract 3 from both sides.}$$
$$x = -9 \qquad \text{Simplify.}$$

4. INTERPRET the results. First, *check* the proposed solution in the stated problem. Twice "-9" is -18 and $-18 + 3$ is -15. This is equal to the number minus 6 or "-9" $- 6$ or -15. Then *state* your conclusion: The unknown number is -9.

Try the Concept Check in the margin.

Example 4   Determining Voter Counts

In an election, the incumbent city council member, Laura Hartley, received 275 *more* votes than her challenger. If a total of 7857 votes were cast in the election, find how many votes the incumbent, Laura Hartley, received.

Solution:

1. UNDERSTAND the problem. We read and reread the problem. Then we assign a variable to an unknown. We use this variable to represent any other unknown quantities. We let

$x =$ the number of challenger votes

Then

$x + 275 =$ the number of incumbent votes
since she received 275 more votes.

2. TRANSLATE the problem into an equation.

In words:

challenger votes	+	incumbent votes	=	total votes
↓		↓		↓

Translate: x $+$ $x + 275$ $=$ 7857

3. SOLVE the equation.

$$x + x + 275 = 7857$$
$$2x + 275 = 7857 \qquad \text{Combine like terms.}$$
$$2x + 275 - 275 = 7857 - 275 \qquad \text{Subtract 275 from both sides.}$$
$$2x = 7582 \qquad \text{Simplify.}$$
$$\frac{2x}{2} = \frac{7582}{2} \qquad \text{Divide both sides by 2.}$$
$$x = 3791 \qquad \text{Simplify.}$$

Problem-Solving Steps

1. UNDERSTAND the problem. During this step, become comfortable with the problem. Some ways of doing this are: Read and reread the problem. Choose a variable to represent the unknown. Construct a drawing. Propose a solution and check. Pay careful attention to how you check your proposed solution. This will help when writing an equation to model the problem.
2. TRANSLATE the problem into an equation.
3. SOLVE the equation.
4. INTERPRET the results: *Check* the proposed solution in the stated problem and *state* your conclusion.

✓ **Concept Check**

Suppose you have solved an equation involving perimeter to find the length of a rectangular table. Explain why you would want to recheck your math if you obtain the result of -5.

Practice Problem 4

At a recent United States/Japan summit meeting, 121 delegates attended. If the United States sent 19 more delegates than Japan, find how many the United States sent.

4. INTERPRET the results. First *check* the proposed solution in the stated problem. Since x represents the number of votes the challenger received, the challenger received 3791 votes. The incumbent received $x + 275 = 3791 + 275 = 4066$ votes. To check, notice that the total number of challenger votes and incumbent votes is $3791 + 4066 = 7857$ votes, the given total of votes cast. Also, 4066 is 275 more votes than 3791, so the solution checks. Then, *state* your conclusion: The incumbent, Laura Hartley, received 4066 votes.

Practice Problem 5

A woman's $21,000 estate is to be divided so that her husband receives twice as much as her son. How much will each receive?

TEACHING TIP   Classroom Activity

Have students work in groups to write and solve word problems similar to Example 3, Example 4 and Example 5. Then have them exchange their word problems with another group to solve.

Example 5   Calculating Separate Costs

Leo Leal sold a used computer system and software for $2100, receiving four times as much money for the computer system as for the software. Find the price of each.

Solution:

1. UNDERSTAND the problem. We read and reread the problem. Then we assign a variable to an unknown. We use this variable to represent any other unknown quantities. We let

$x =$ the software price
$4x =$ the computer system price

2. TRANSLATE the problem into an equation.

In words: software price **and** computer price **is** 2100

Translate: $x + 4x = 2100$

3. SOLVE the equation

$x + 4x = 2100$
$5x = 2100$ Combine like terms.
$\frac{5x}{5} = \frac{2100}{5}$ Divide both sides by 5.
$x = 420$ Simplify.

4. INTERPRET the results. *Check* the proposed solution in the stated problem. The software sold for $420. The computer system sold for $4x = 4(\$420) = \$1680$. Since $420 + $1680 = 2100, the total price, and $1680 is four times $420, the solution checks. *State* your conclusion: The software sold for $420 and the computer system sold for $1680.

Answer

5. husband: $14,000; son: $7,000

EXERCISE SET 11.5

A _Write each phrase as a variable expression. Use x to represent "a number." See Example 1._

1. The sum of a number and five

2. Ten plus a number

3. The total of a number and eight

4. The difference of a number and five hundred

5. Twenty decreased by a number

6. A number less thirty

7. The product of 512 and a number

8. A number times twenty

9. A number divided by 2

10. The quotient of six and a number

11. The sum of seventeen and a number added to the product of five and the number

12. The difference of twice a number, and four

13. The product of five and a number

14. The quotient of twenty and a number, decreased by three

15. A number subtracted from 11

16. Twelve subtracted from a number

17. Fifty decreased by eight times a number

18. Twenty decreased by twice a number

1. $x + 5$

2. $10 + x$

3. $x + 8$

4. $x - 500$

5. $20 - x$

6. $x - 30$

7. $512x$

8. $20x$

9. $\dfrac{x}{2}$

10. $\dfrac{6}{x}$

11. $(17 + x) + 5x$

12. $2x - 4$

13. $5x$

14. $\dfrac{20}{x} - 3$

15. $11 - x$

16. $x - 12$

17. $50 - 8x$

18. $20 - 2x$

19. $-5 + x = -7$

20. $x - 5 = 10$

21. $3x = 27$

22. $\frac{8}{x} = -2$

23. $-20 - x = 104$

24. $2 + 2x = -14$

25. 8

26. 20

27. 9

28. 16

29. 5

30. 7

31. 12

32. 5

33. 8

34. 3

35. 5

36. 10

704

Name _____

B *Write each sentence as an equation. Use x to represent "a number." See Example 2.*

19. A number added to -5 is -7.

20. Five subtracted from a number equals 10.

21. Three times a number yields 27.

22. The quotient of 8 and a number is -2.

23. A number subtracted from -20 amounts to 104.

24. Two added to twice a number gives -14.

C *Solve. See Example 3.*

25. Three times a number added to 9 is 33. Find the number.

26. Twice a number subtracted from 60 is 20. Find the number.

27. The sum of 3, 4, and a number amounts to 16. Find the number.

28. A number less 5 is 11. Find the number.

29. Eight decreased by some number equals the quotient of 15 and 5. Find the number.

30. The product of some number plus 2 and 5 is 11 less than the number times 8. Find the number.

31. Five times a number less 40 is 8 more than the number.

32. The product of 4 and a number is the same as 30 less twice that same number.

33. Three times the difference of some number and 5 amounts to the quotient of 108 and 12. Find the number.

34. Thirty less a number is equal to the product of 3 and the sum of the number and 6.

35. The difference of a number and 3 is equal to the quotient of 10 and 5.

36. The product of a number and 3 is twice the sum of that number and 5.

Name _____

Solve. See Examples 4 and 5.

37. In the 1996 presidential election, Bill Clinton received 220 more electoral votes than Bob Dole. If a total of 538 electoral votes were cast for the two candidates, find how many votes each candidate received. (*Source:* Voter News Service)

38. California has 22 more electoral votes for president than Texas. If the total number of electoral votes for these two states is 86, find the number for each state. (*Source: The World Almanac 1998*)

39. Mark and Stuart Martin collect comic books. Mark has twice the number of books Stuart has. Together they have 120 comic books. Find how many books Mark has.

40. Heather and Mary Gamber collect baseball cards. Heather's collection is three times as large as Mary's. Together they have 400 baseball cards. Find the number of baseball cards in each collection.

41. A Toyota Camry is traveling twice as fast as a Dodge truck. If their combined speed is 105 miles per hour, find the speed of the car and find the speed of the truck.

42. A crow will eat five more ounces of food a day than a finch. If together they eat 13 ounces of food, find how many ounces of food the crow consumes and how many ounces of food the finch consumes.

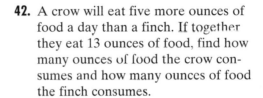

43. Anthony Tedesco sold his used mountain bike and accessories for $270. If he received five times as much money for the bike as he did for the accessories, find how much money he received for the bike.

44. A tractor and a plow attachment are worth $1200. The tractor is worth seven times as much money as the plow. Find the value of the tractor and the value of the plow.

45. 93 points

46. 78 points

47. 590

48. 80

49. 1000

50. 52,000

51. 3000

52. 101,600

53. answers may vary

Name _____

45. During the 1998 Women's NCAA Division I basketball championship game, the Tennessee Volunteers scored 18 more points than the Louisiana Tech Bulldogs. Together, both teams scored a total of 168 points. How many points did the 1998 Champion Tennessee Lady Volunteers score during this game? (*Source:* National Collegiate Athletic Association)

46. During the 1998 Men's NCAA Division I basketball championship game, the Utah Runnin' Utes scored 9 fewer points than the Kentucky Wildcats. The Utes scored 69 points. How many points did the 1998 Champion Kentucky Wildcats score during this game? (*Source:* National Collegiate Athletic Association)

REVIEW AND PREVIEW

Round each number to the given place value. See Section 1.4.

47. 586 to the nearest ten

48. 82 to the nearest ten

49. 1026 to the nearest hundred

50. 52,333 to the nearest thousand

51. 2986 to the nearest thousand

52. 101,552 to the nearest hundred

COMBINING CONCEPTS

53. Solve Example 4 again, but this time let x be the number of incumbent votes. Did you get the same results? Explain why or why not.

Name _____ **Section** _____ **Date** _____

CHAPTER 11 ACTIVITY
ANALYZING OPTIONS

This activity may be completed by working in groups or individually.

A couple is getting married and they have decided to hold their wedding reception at a local hotel. The hotel charges a flat fee of $1200 for renting their large reception hall. This cost does not include food, wedding cake, or drinks. The costs of the food items will depend on the number of guests. The hotel offers the following buffet and cake packages.

Option	Description	Cost of Buffet Per Person	Cost of Cake and Beverages Per Person
A	Cold cuts, white cake, punch	$16	$2
B	Hot buffet, white cake, punch	$22	$2
C	Hot buffet, salad bar, white cake, mints, nuts, punch, coffee	$28	$3
D	Hot buffet, salad bar, white cake; mints, nuts, punch, coffee, open bar	$28	$10

1. The couple decides that they can afford to spend a total of $7000 on the reception. Translate each option into an algebraic equation that describes the total cost of the package in terms of the flat fee and the costs that depend on the number of guests. Let x represent the number of guests and assume that the couple will spend the entire $7000. For instance, the cost of Option A can be represented by the equation
$7000 = 1200 + 16x + 2x$ or
$7000 = 1200 + 18x$.
A: $7000 = 1200 + 18x$;
B: $7000 = 1200 + 24x$;
C: $7000 = 1200 + 31x$;
D: $7000 = 1200 + 38x$

2. If the couple can spend only a total of $7000 on the reception, how many guests can the couple invite under each option? (Be sure to round your answers so that the couple would not spend over $7000.)   A: 322 guests; B: 241 guests; C: 187 guests; D: 152 guests

4. (Optional) Suppose as a member of a student organization, you have been asked to organize an awards banquet. Investigate the cost of a catered meal with three different caterers in your area. If the organization has a fixed budget of $2000 for the banquet, how many people could be invited in each scenario? Be sure to take any hall rentals or other fixed fees into consideration.   answers may vary

3. The couple prefers Option C but they want to invite a total of 240 people. How much additional money would they need to cover the cost of the reception? $1640

CHAPTER 11 HIGHLIGHTS

DEFINITIONS AND CONCEPTS	EXAMPLES

SECTION 11.1 INTRODUCTION TO VARIABLES

A letter used to represent a number is called a **variable**.

A combination of operations on variables and numbers is called an **algebraic expression**.

Replacing a variable in an expression by a number and then finding the value of the expression is called **evaluating the expression** for the variable.

x, y, z, a, b

$3 + x, 7y, x^3 + y - 10$

Evaluate: $2x + y$ if $x = 22$ and $y = 4$

$2x + y = 2 \cdot 22 + 4$ Replace x with 22 and y with 4.

$\qquad\quad = 44 + 4$ Multiply.

$\qquad\quad = 48$ Add.

The addends of an algebraic expression are called the **terms** of the expression.

The number factor of a variable term is called the **numerical coefficient**.

Terms that are exactly the same, except that they may have different numerical coefficients, are called **like terms**.

$$5x^2 + (-4x) + (-2)$$
$$\uparrow \qquad \uparrow \qquad\quad \uparrow$$
3 terms

Term	Numerical Coefficient
$7x$	7
$-6y$	-6
x or $1x$	1

$5x + 11x = (5 + 11)x = 16x$

like terms

$y - 6y = (1 - 6)y = -5y$

An algebraic expression is **simplified** when all like terms have been **combined**.

Use the distributive property to multiply an algebraic expression by a term.

Simplify:

$$-4(x + 2) + 3(5x - 7)$$
$$= -4(x) + (-4)(2) + 3(5x) + 3(-7)$$
$$= -4x + (-8) + 15x + (-21)$$
$$= 11x + (-29) \quad \text{or} \quad 11x - 29$$

SECTION 11.2 SOLVING EQUATIONS: THE ADDITION PROPERTY

ADDITION PROPERTY OF EQUALITY

Let a, b, and c represent numbers.

    If $a - b$, then

$a + c = b + c$ and $a - c = b - c$

In other words, the same number may be added to or subtracted from both sides of an equation without changing the solution of the equation.

Solve for x:

$$x + 8 = 2 + (-1)$$
$$x + 8 = 1$$
$$x + 8 - 8 = 1 - 8 \quad \text{Subtract 8 from both sides.}$$
$$x = -7 \quad \text{Simplify.}$$

The solution is -7.

Section 11.3   Solving Equations: The Multiplication Property

Multiplication Property of Equality

Let a, b, and c represent numbers and let $c \neq 0$.

If $a = b$, then

$$a \cdot c = b \cdot c \quad \text{and} \quad \frac{a}{c} = \frac{b}{c}$$

In other words, both sides of an equation may be multiplied or divided by the same nonzero number without changing the solution of the equation.

Solve: $-7x = 42$

$$\frac{-7x}{-7} = \frac{42}{-7} \quad \text{Divide both sides by } -7.$$

$$x = -6 \quad \text{Simplify.}$$

Solve: $\frac{2}{3}x = -10$

$$\frac{3}{2} \cdot \frac{2}{3}x = \frac{3}{2} \cdot -10 \quad \text{Multiply both sides by } \frac{3}{2}.$$

$$x = -15 \quad \text{Simplify.}$$

Section 11.4   Solving Equations Using Addition and Multiplication Properties

Steps for Solving an Equation

Step 1. If parentheses are present, use the distributive property.

Step 2. Combine any like terms on each side of the equation.

Step 3. Use the addition property of equality to rewrite the equation so that variable terms are on one side of the equation and constant terms are on the other side.

Step 4. Use the multiplication property of equality to divide both sides by the numerical coefficient of the variable to solve.

Step 5. Check the solution in the *original equation*.

Solve for x: $5(3x - 1) + 15 = -5$

Step 1. $\quad 15x - 5 + 15 = -5 \quad$ Apply the distributive property.

Step 2. $\quad 15x + 10 = -5 \quad$ Combine like terms.

Step 3. $\quad 15x + 10 - 10 = -5 - 10 \quad$ Subtract 10 from both sides.
$$15x = -15$$

Step 4. $\quad \frac{15x}{15} = \frac{-15}{15} \quad$ Divide by 15.
$$x = -1$$

Step 5. Check to see that -1 is the solution.

Section 11.5   Equations and Problem Solving

Problem-Solving Steps

1. UNDERSTAND the problem. Some ways of doing this are:
   *Read and reread the problem.
   *Construct a drawing.
   *Assign a variable to an unknown in the problem.

The incubation period for a golden eagle is three times the incubation period for a hummingbird. If the total of their incubation periods is 60 days, find the incubation period for each bird. (*Source: Wildlife Fact File*, International Masters Publishers)

1. UNDERSTAND the problem. Then assign a variable. Let

$$x = \text{incubation period of a hummingbird}$$
$$3x = \text{incubation period of a golden eagle}$$

SECTION 11.5 (*CONTINUED*)

---

2. TRANSLATE the problem into an equation.

2. TRANSLATE.

incubation of hummingbird	+	incubation of golden eagle	is	60
↓		↓	↓	↓
x	+	$3x$	=	60

3. SOLVE the equation.

3. SOLVE.

$$x + 3x = 60$$
$$4x = 60$$
$$\frac{\overset{1}{\cancel{4}}x}{\underset{1}{\cancel{4}}} = \frac{60}{4}$$
$$x = 15$$

4. INTERPRET the results. *Check* the proposed solution in the stated problem and *state* your conclusion.

4. INTERPRET the solution in the stated problem. The incubation period for the hummingbird is 15 days. The incubation period for the golden eagle is $3x = 3 \cdot 15 = 45$ days.
Since 15 days + 45 days = 60 days and 45 is 3(15), the solution checks.

State your conclusion: The incubation period for the hummingbird is 15 days. The incubation period for the golden eagle is 45 days.

CHAPTER 11 REVIEW

(11.1) *Evaluate each expression when $x = 5$, $y = 0$, and $z = -2$.*

1. $\dfrac{2x}{z}$ -5

2. $4x - 3$ 17

3. $\dfrac{x + 7}{y}$ undefined

4. $\dfrac{y}{5x}$ 0

5. $x^3 - 2z$ 129

6. $\dfrac{7 + x}{3z}$ -2

7. $(y + z)^2$ 4

8. $\dfrac{100}{x} + \dfrac{y}{3}$ 20

9. Find the volume of a storage cube whose sides measure 2 feet. Use $V = s^3$.   8 cubic feet

10. Find the volume of a wooden crate in the shape of a cube 4 feet on each side. Use $V = s^3$.
64 cubic feet

11. Lamar deposited his $5000 bonus into an account paying 6% annual interest. How much interest will he earn in 6 years? Use $I = prt$. $1800

12. Jennifer Lewis borrowed $2000 from her grandmother and agreed to pay her 5% simple interest. How much interest will she owe after 3 years? Use $I = prt$. $300

Simplify each expression by combining like terms.

13. $3y + 7y - 15$ $10y - 15$

14. $2y - 10 - 8y$ $-6y - 10$

15. $8a + a - 7 - 15a$ $-6a - 7$

16. $y + 3 - 9y - 1$ $-8y + 2$

17. $1.7x - 3.2 + 2.9x - 8.7$ $4.6x - 11.9$

18. $3.6x - 10.2 - 5.7x - 9.8$ $-2.1x - 20$

Multiply.

19. $-2(x + 5)$ $-2x - 10$

20. $-3(y + 8)$ $-3y - 24$

Simplify.

21. $7x + 3(x - 4) + x$ $11x - 12$

22. $10 - 2(m - 3) - m$ $-3m + 16$

23. $3(5a - 2) - 20a + 10$ $-5a + 4$

24. $6y + 3 + 2(3y - 6)$ $12y - 9$

Name _____

Find the area of each figure.

25.
(2x − 1) yards
3 yards — Rectangle
$(6x − 3)$ square yards

26.
5y meters
Square
$25y^2$ square meters

(11.2)

27. Is 4 a solution of $5(2 − x) = −10$?   yes

28. Is 0 a solution of $6y + 2 = 23 + 4y$?   no

Solve.

29. $z − 5 = −7$ $−2$

30. $x + 1 = 8$ 7

31. $x + \dfrac{7}{8} = \dfrac{3}{8}$ $−\dfrac{1}{2}$

32. $y + \dfrac{4}{11} = −\dfrac{2}{11}$ $−\dfrac{6}{11}$

33. $c − 5 = −13 + 7$ $−1$

34. $7x + 5 − 6x = −20$ $−25$

35. $n + 18 = 10 − (−2)$ $−6$

36. $15 = 8x + 35 − 7x$ $−20$

37. $m − 3.9 = −2.6$ 1.3

38. $z − 4.6 = −2.2$ 2.4

(11.3) *Solve.*

39. $−3y = −21$ 7

40. $−8x = 72$ $−9$

41. $−5n = −5$ 1

42. $−3a = 15$ $−5$

43. $\dfrac{2}{3}x = −\dfrac{8}{15}$ $−\dfrac{4}{5}$

44. $−\dfrac{7}{8}y = 21$ $−24$

45. $−1.2x = 144$ $−120$

46. $−0.8y = −10.4$ 13

47. $−5x = 100 − 120$ 4

712

48. $18 - 30 = -4x$ 3

49. $-7x + 3x = -50 - 2$ 13

50. $-x + 8x = -38 - 4$ -6

(11.4) *Solve.*

51. $3x - 4 = 11$ 5

52. $6y + 1 = 73$ 12

53. $14 - y = -3$ 17

54. $7 - z = 0$ 7

55. $-\dfrac{5}{9}x + 23 = -12$ 63

56. $-\dfrac{4}{3}x - 11 = -55$ 33

57. $6.8 + 4y = -2.2$ -2.25

58. $-9.6 + 5y = -3.1$ 1.3

59. $5(n - 3) = 7 + 3n$ 11

60. $7(2 + x) = 4x - 1$ -5

61. $2x + 7 = 6x - 1$ 2

62. $5x - 18 = -4x + 36$ 6

Write each sentence as an equation.

63. The difference of 20 and -8 is 28.
$20 - (-8) = 28$

64. The product of 5 and the sum of 2 and -6 yields -20. $5(2 + (-6)) = -20$

65. The quotient of -75 and the sum of 5 and 20 is equal to -3. $\dfrac{-75}{5 + 20} = -3$

66. Nineteen subtracted from -2 amounts to -21.
$-2 - 19 = -21$

(11.5) *Write each phrase as an algebraic expression. Use x to represent "a number."*

67. Eleven added to twice a number $2x + 11$

68. The product of -5 and a number, decreased by 50
$-5x - 50$

69. The quotient of 70 and the sum of a number and 6
$\dfrac{70}{x + 6}$

70. Twice the difference of a number and 13
$2(x - 13)$

Write each sentence as an equation using x as the variable.

71. Twice a number minus 8 is 40. $2x - 8 = 40$

72. Twelve subtracted from the quotient of a number and 2 is 10. $\dfrac{x}{2} - 12 = 10$

73. The difference of a number and 3 is the quotient of the number and 4.

$$x - 3 = \frac{x}{4}$$

74. The product of some number and 6 is equal to the sum of the number and 2. $6x = x + 2$

Solve.

75. Five times a number subtracted from 40 is the same as three times the number. Find the number.   5

76. The product of a number and 3 is twice the difference of that number and 8. Find the number.
-16

77. In an election the incumbent received 14,000 votes of the 18,500 votes cast. Of the remaining votes, the Democratic candidate received 272 more than the Independent candidate. Find how many votes the Democratic candidate received.   2386 votes

78. Rajiv Puri has twice as many cassette tapes as he has compact discs. Find the number of CDs if he has a total of 126 music recordings.   42 CDs

CHAPTER 11 TEST

1. Evaluate $\dfrac{3x - 5}{2y}$ if $x = 7$ and $y = -8$.

2. Simplify $7x - 5 - 12x + 10$ by combining like terms.

3. Multiply: $-2(3y + 7)$

4. Simplify: $5(3z + 2) - z - 18$

5. Find the area.

3 meters

| Rectangle | $(3x - 1)$ meters |

Solve.

6. $x - 17 = -10$

7. $y + \dfrac{3}{4} = \dfrac{1}{4}$

8. $-4x = 48$

9. $-\dfrac{5}{8}x = -25$

10. $5x + 12 - 4x - 14 = 22$

11. $2 - c + 2c = 5$

12. $3x - 5 = -11$

13. $-4x + 7 = 15$

14. $3.6 - 2x = -5.4$

15. $3(4 + 2y) = 12$

16. $5x - 2 = x - 10$

17. $10y - 1 = 7y + 20$

18. $6 + 2(3n - 1) = 28$

19. $4(5x + 3) = 2(7x + 6)$

ANSWERS

1. -1

2. $-5x + 5$

3. $-6y - 14$

4. $14z - 8$

5. $(9x - 3)$ square meters

6. 7

7. $-\dfrac{1}{2}$

8. -12

9. 40

10. 24

11. 3

12. -2

13. -2

14. 4.5

15. 0

16. -2

17. 7

18. 4

19. 0

Name _____

Solve.

20. A lawn is in the shape of a trapezoid with a height of 60 feet and bases of 70 feet and 130 feet. Find the area of the lawn. Use
$$A = \frac{1}{2} \cdot h \cdot (B + b).$$

21. If the height of a triangularly shaped jib sail is 12 feet and its base is 5 feet, find the area of the sail. Use
$$A = \frac{1}{2} \cdot b \cdot h.$$

22. Translate the following phrases into mathematical expressions. Use x to represent "a number."
  a. The product of a number and 17
  b. Twice a number subtracted from 20

23. The difference of three times a number and five times the same number is 4. Find the number.

24. In a championship basketball game, Paula Zimmerman made twice as many free throws as Maria Kaminsky. If the total number of free throws made by both women was 12, find how many free throws Paula made.

25. In a 10-kilometer race, there are 112 more men entered than women. Find the number of women runners if the total number of runners in the race is 600.

Name _____ **Section** _____ **Date** _____

CUMULATIVE REVIEW

1. Multiply 0.0531×16

2. Given the rectangle shown:

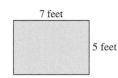

7 feet

5 feet

    a. Find the ratio of its width to its length.

    b. Find the ratio of its length to its perimeter.

3. 12% of what number is 0.6?

4. What percent of 12 is 9?

5. Convert 3 pounds to ounces.

6. Divide 18.08 ml by 16.

7. Identify each figure as a line, a ray, a line segment, or an angle.

    a.

    b.

E   F

    c.

M   N   O

    d. P

T

8. Find the diameter of the circle.

5 cm

9. Find the perimeter of the room shown below.

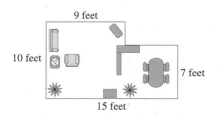

9 feet

10 feet

7 feet

15 feet

10. Find the area of the triangle.

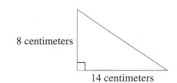

8 centimeters

14 centimeters

ANSWERS

1. 0.8496 (Sec. 4.4, Ex. 3)

2. a. $\frac{5}{7}$

   b. $\frac{7}{24}$ (Sec. 5.1, Ex. 7)

3. 5 (Sec. 6.3, Ex. 9)

4. 75% (Sec. 6.3, Ex. 11)

5. 48 oz (Sec. 7.2, Ex. 2)

6. 1.13 ml (Sec. 7.3, Ex.10)

7. a. line

   b. line segment

   c. angle

   d. ray (Sec. 8.1, Ex. 1)

8. 10 cm (Sec. 8.2, Ex. 3)

9. 50 ft (Sec. 8.3, Ex. 6)

10. 56 square centimeters (Sec. 8.4, Ex. 1)

718

Name _____

11. Simplify: $3\sqrt{81} - \sqrt{4}$

12. Mel is a 6-foot-tall park ranger who needs to know the height of a particular tree. He notices that when the shadow of the tree is 69 feet long, his own shadow is 9 feet long. Find the height of the tree.

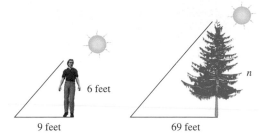

6 feet

9 feet   69 feet n

13. The following bar graph shows the number of endangered species in different categories.

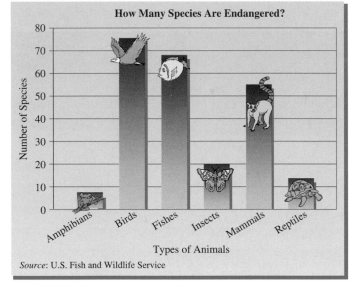

How Many Species Are Endangered?

Number of Species

Amphibians   Birds   Fishes   Insects   Mammals   Reptiles

Types of Animals

Source: U.S. Fish and Wildlife Service

a. Approximate the number of endangered species that are mammals.
b. Which category shows the fewest endangered species?

Insert < *or* > *between each pair of numbers to make a true statement.*

14. −2.9   1

15. −9   −11

Find the absolute value.

16. $|-2|$

17. $|1.2|$

Add.

18. $5 + (-1)$

19. $-\frac{3}{8} + \left(-\frac{1}{8}\right)$

Subtract.

20. $8 - 15$

21. $-4 - (-5)$

Multiply.

22. $-2(-5)$

23. $\left(-\frac{1}{2}\right)\left(-\frac{2}{3}\right)$

24. Simplify: $(-3) \cdot |-5| - (-2) + 4^2$

25. Solve: $3(2x - 6) + 6 = 0$

APPENDIX A

Addition Table and One Hundred Addition Facts

+	0	1	2	3	4	5	6	7	8	9
0	0	1	2	3	4	5	6	7	8	9
1	1	2	3	4	5	6	7	8	9	10
2	2	3	4	5	6	7	8	9	10	11
3	3	4	5	6	7	8	9	10	11	12
4	4	5	6	7	8	9	10	11	12	13
5	5	6	7	8	9	10	11	12	13	14
6	6	7	8	9	10	11	12	13	14	15
7	7	8	9	10	11	12	13	14	15	16
8	8	9	10	11	12	13	14	15	16	17
9	9	10	11	12	13	14	15	16	17	18

Appendix A: One Hundred Addition Facts

Knowledge of the basic addition facts found above is an important prerequisite for a course in prealgebra. Study the table above and then perform the additions. Check your answers either by comparing them with those found in the back-of-the-book answer section or by using the table. Review any facts that you missed.

1. $\dfrac{1\ +4}{5}$ 2. $\dfrac{5\ +6}{11}$ 3. $\dfrac{2\ +3}{5}$ 4. $\dfrac{7\ +8}{15}$ 5. $\dfrac{3\ +9}{12}$ 6. $\dfrac{6\ +1}{7}$ 7. $\dfrac{4\ +4}{8}$ 8. $\dfrac{0\ +6}{6}$ 9. $\dfrac{9\ +5}{14}$ 10. $\dfrac{8\ +2}{10}$

11. $\dfrac{5\ +7}{12}$ 12. $\dfrac{3\ +2}{5}$ 13. $\dfrac{5\ +5}{10}$ 14. $\dfrac{1\ +1}{2}$ 15. $\dfrac{8\ +1}{9}$ 16. $\dfrac{6\ +6}{12}$ 17. $\dfrac{2\ +9}{11}$ 18. $\dfrac{3\ +5}{8}$ 19. $\dfrac{9\ +9}{18}$ 20. $\dfrac{5\ +2}{7}$

21. 6
+4
10

22. 0
+0
0

23. 1
+9
10

24. 3
+7
10

25. 9
+8
17

26. 0
+8
8

27. 4
+9
13

28. 3
+0
3

29. 7
+5
12

30. 8
+9
17

31. 9
+7
16

32. 2
+6
8

33. 4
+3
7

34. 8
+5
13

35. 3
+1
4

36. 0
+3
3

37. 7
+1
8

38. 3
+4
7

39. 8
+0
8

40. 6
+3
9

41. 2
+4
6

42. 0
+9
9

43. 8
+8
16

44. 5
+3
8

45. 3
+6
9

46. 6
+9
15

47. 4
+8
12

48. 0
+1
1

49. 2
+5
7

50. 6
+0
6

51. 2
+0
2

52. 4
+2
6

53. 8
+3
11

54. 7
+4
11

55. 1
+7
8

56. 4
+6
10

57. 0
+5
5

58. 9
+1
10

59. 8
+6
14

60. 5
+1
6

61. 6
+7
13

62. 4
+0
4

63. 1
+6
7

64. 4
+5
9

65. 0
+7
7

66. 5
+8
13

67. 7
+6
13

68. 7
+0
7

69. 4
+1
5

70. 5
+4
9

71. 0
+4
4

72. 1
+2
3

73. 7
+9
16

74. 3
+8
11

75. 7
+7
14

76. 9
+4
13

77. 1
+0
1

78. 4
+7
11

79. 2
+2
4

80. 1
+3
4

81. 2
+8
10

82. 5
+9
14

83. 6
+2
8

84. 9
+6
15

85. 5
+0
5

86. 8
+7
15

87. 7
+3
10

88. 0
+2
2

89. 9
+2
11

90. 3
+3
6

91. 9
+3
12

92. 1
+5
6

93. 2
+7
9

94. 6
+5
11

95. 7
+2
9

96. 1
+8
9

97. 6
+8
14

98. 8
+4
12

99. 9
+0
9

100. 2
+1
3

APPENDIX B

Multiplication Table and One Hundred Multiplication Facts

×	1	2	3	4	5	6	7	8	9
1	1	2	3	4	5	6	7	8	9
2	2	4	6	8	10	12	14	16	18
3	3	6	9	12	15	18	21	24	27
4	4	8	12	16	20	24	28	32	36
5	5	10	15	20	25	30	35	40	45
6	6	12	18	24	30	36	42	48	54
7	7	14	21	28	35	42	49	56	63
8	8	16	24	32	40	48	56	64	72
9	9	18	27	36	45	54	63	72	81

Appendix B: One Hundred Multiplication Facts

Knowledge of the basic multiplication facts found above is an important prerequisite for a course in prealgebra. Study the table above and then perform the multiplications. Check your answers either by comparing them with those found in the back-of-the-book answer section or by using the table. Review any facts that you missed.

1. $\begin{array}{r} 1 \\ \times\,1 \\ \hline 1 \end{array}$
2. $\begin{array}{r} 5 \\ \times\,7 \\ \hline 35 \end{array}$
3. $\begin{array}{r} 7 \\ \times\,8 \\ \hline 56 \end{array}$
4. $\begin{array}{r} 3 \\ \times\,3 \\ \hline 9 \end{array}$
5. $\begin{array}{r} 8 \\ \times\,4 \\ \hline 32 \end{array}$
6. $\begin{array}{r} 9 \\ \times\,5 \\ \hline 45 \end{array}$
7. $\begin{array}{r} 4 \\ \times\,7 \\ \hline 28 \end{array}$
8. $\begin{array}{r} 7 \\ \times\,1 \\ \hline 7 \end{array}$
9. $\begin{array}{r} 2 \\ \times\,2 \\ \hline 4 \end{array}$
10. $\begin{array}{r} 0 \\ \times\,5 \\ \hline 0 \end{array}$

11. $\begin{array}{r} 9 \\ \times\,7 \\ \hline 63 \end{array}$
12. $\begin{array}{r} 8 \\ \times\,8 \\ \hline 64 \end{array}$
13. $\begin{array}{r} 3 \\ \times\,2 \\ \hline 6 \end{array}$
14. $\begin{array}{r} 6 \\ \times\,0 \\ \hline 0 \end{array}$
15. $\begin{array}{r} 5 \\ \times\,6 \\ \hline 30 \end{array}$
16. $\begin{array}{r} 2 \\ \times\,5 \\ \hline 10 \end{array}$
17. $\begin{array}{r} 4 \\ \times\,6 \\ \hline 24 \end{array}$
18. $\begin{array}{r} 0 \\ \times\,7 \\ \hline 0 \end{array}$
19. $\begin{array}{r} 6 \\ \times\,3 \\ \hline 18 \end{array}$
20. $\begin{array}{r} 8 \\ \times\,9 \\ \hline 72 \end{array}$

21. 5
   ×8
   ‾‾
   40

22. 7
   ×2
   ‾‾
   14

23. 4
   ×8
   ‾‾
   32

24. 1
   ×2
   ‾‾
   2

25. 9
   ×6
   ‾‾
   54

26. 3
   ×1
   ‾‾
   3

27. 8
   ×7
   ‾‾
   56

28. 2
   ×8
   ‾‾
   16

29. 6
   ×9
   ‾‾
   54

30. 5
   ×5
   ‾‾
   25

31. 2
   ×1
   ‾‾
   2

32. 8
   ×0
   ‾‾
   0

33. 4
   ×9
   ‾‾
   36

34. 8
   ×3
   ‾‾
   24

35. 6
   ×2
   ‾‾
   12

36. 4
   ×5
   ‾‾
   20

37. 9
   ×4
   ‾‾
   36

38. 2
   ×9
   ‾‾
   18

39. 3
   ×4
   ‾‾
   12

40. 1
   ×6
   ‾‾
   6

41. 8
   ×6
   ‾‾
   48

42. 9
   ×8
   ‾‾
   72

43. 1
   ×8
   ‾‾
   8

44. 5
   ×1
   ‾‾
   5

45. 9
   ×0
   ‾‾
   0

46. 7
   ×4
   ‾‾
   28

47. 9
   ×3
   ‾‾
   27

48. 0
   ×3
   ‾‾
   0

49. 3
   ×5
   ‾‾
   15

50. 6
   ×8
   ‾‾
   48

51. 5
   ×9
   ‾‾
   45

52. 2
   ×6
   ‾‾
   12

53. 1
   ×0
   ‾‾
   0

54. 3
   ×9
   ‾‾
   27

55. 9
   ×9
   ‾‾
   81

56. 5
   ×4
   ‾‾
   20

57. 0
   ×6
   ‾‾
   0

58. 1
   ×9
   ‾‾
   9

59. 5
   ×0
   ‾‾
   0

60. 6
   ×1
   ‾‾
   6

61. 9
   ×2
   ‾‾
   18

62. 1
   ×7
   ‾‾
   7

63. 1
   ×3
   ‾‾
   3

64. 7
   ×3
   ‾‾
   21

65. 6
   ×6
   ‾‾
   36

66. 4
   ×0
   ‾‾
   0

67. 7
   ×9
   ‾‾
   63

68. 4
   ×3
   ‾‾
   12

69. 7
   ×5
   ‾‾
   35

70. 2
   ×0
   ‾‾
   0

71. 6
   ×7
   ‾‾
   42

72. 0
   ×8
   ‾‾
   0

73. 8
   ×5
   ‾‾
   40

74. 2
   ×4
   ‾‾
   8

75. 0
   ×1
   ‾‾
   0

76. 3
   ×8
   ‾‾
   24

77. 9
   ×1
   ‾‾
   9

78. 7
   ×0
   ‾‾
   0

79. 5
   ×3
   ‾‾
   15

80. 4
   ×4
   ‾‾
   16

81. 1
   ×5
   ‾‾
   5

82. 6
   ×5
   ‾‾
   30

83. 3
   ×0
   ‾‾
   0

84. 1
   ×4
   ‾‾
   4

85. 3
   ×7
   ‾‾
   21

86. 4
   ×2
   ‾‾
   8

87. 0
   ×2
   ‾‾
   0

88. 7
   ×7
   ‾‾
   49

89. 8
   ×2
   ‾‾
   16

90. 6
   ×4
   ‾‾
   24

91. 0
   ×0
   ‾‾
   0

92. 2
   ×7
   ‾‾
   14

93. 4
   ×1
   ‾‾
   4

94. 0
   ×4
   ‾‾
   0

95. 2
   ×3
   ‾‾
   6

96. 8
   ×1
   ‾‾
   8

97. 3
   ×6
   ‾‾
   18

98. 5
   ×2
   ‾‾
   10

99. 0
   ×9
   ‾‾
   0

100. 7
    ×6
    ‾‾
    42

APPENDIX C

Review of Geometric Figures

	PLANE FIGURES HAVE LENGTH AND WIDTH BUT NO THICKNESS OR DEPTH.	
Name	Description	Figure
POLYGON	Union of three or more coplanar line segments that intersect with each other only at each end point, with each end point shared by two segments.	
TRIANGLE	Polygon with three sides (sum of measures of three angles is 180°).	
SCALENE TRIANGLE	Triangle with no sides of equal length.	
ISOSCELES TRIANGLE	Triangle with two sides of equal length.	
EQUILATERAL TRIANGLE	Triangle with all sides of equal length.	
RIGHT TRIANGLE	Triangle that contains a right angle.	leg, hypotenuse, leg
QUADRILATERAL	Polygon with four sides (sum of measures of four angles is 360°).	
TRAPEZOID	Quadrilateral with exactly one pair of opposite sides parallel.	base, leg, parallel sides, leg, base
ISOSCELES TRAPEZOID	Trapezoid with legs of equal length.	
PARALLELOGRAM	Quadrilateral with both pairs of opposite sides parallel.	
RHOMBUS	Parallelogram with all sides of equal length.	

A-5

Name	Description	Figure
RECTANGLE	Parallelogram with four right angles.	
SQUARE	Rectangle with all sides of equal length.	
CIRCLE	All points in a plane the same distance from a fixed point called the **center**.	

SOLID FIGURES HAVE LENGTH, WIDTH, AND HEIGHT OR DEPTH.

Name	Description	Figure
RECTANGULAR SOLID	A solid with six sides, all of which are rectangles.	
CUBE	A rectangular solid whose six sides are squares.	
SPHERE	All points the same distance from a fixed point, called the **center**.	
RIGHT CIRCULAR CYLINDER	A cylinder having two circular bases that are perpendicular to its altitude.	
RIGHT CIRCULAR CONE	A cone with a circular base that is perpendicular to its altitude.	

PERCENT, DECIMAL, AND FRACTION EQUIVALENTS		
Percent	Decimal	Fraction
1%	0.01	$\frac{1}{100}$
5%	0.05	$\frac{1}{20}$
10%	0.1	$\frac{1}{10}$
12.5% or $12\frac{1}{2}\%$	0.125	$\frac{1}{8}$
$16.\overline{6}\%$ or $16\frac{2}{3}\%$	$0.1\overline{6}$	$\frac{1}{6}$
20%	0.2	$\frac{1}{5}$
25%	0.25	$\frac{1}{4}$
30%	0.3	$\frac{3}{10}$
$33.\overline{3}\%$ or $33\frac{1}{3}\%$	$0.\overline{3}$	$\frac{1}{3}$
37.5% or $37\frac{1}{2}\%$	0.375	$\frac{3}{8}$
40%	0.4	$\frac{2}{5}$
50%	0.5	$\frac{1}{2}$
60%	0.6	$\frac{3}{5}$
62.5% or $62\frac{1}{2}\%$	0.625	$\frac{5}{8}$
$66.\overline{6}\%$ or $66\frac{2}{3}\%$	$0.\overline{6}$	$\frac{2}{3}$
70%	0.7	$\frac{7}{10}$
75%	0.75	$\frac{3}{4}$
80%	0.8	$\frac{4}{5}$
$83.\overline{3}\%$ or $83\frac{1}{3}\%$	$0.8\overline{3}$	$\frac{5}{6}$
87.5% or $87\frac{1}{2}\%$	0.875	$\frac{7}{8}$
90%	0.9	$\frac{9}{10}$
100%	1.0	1
110%	1.1	$1\frac{1}{10}$
125%	1.25	$1\frac{1}{4}$
$133.\overline{3}\%$ or $133\frac{1}{3}\%$	$1.\overline{3}$	$1\frac{1}{3}$
150%	1.5	$1\frac{1}{2}$
$166.\overline{6}\%$ or $166\frac{2}{3}\%$	$1.\overline{6}$	$1\frac{2}{3}$
175%	1.75	$1\frac{3}{4}$
200%	2.0	2

APPENDIX E

TABLE OF SQUARES AND SQUARE ROOTS

n	n^2	$\sqrt{n}$	n	n^2	$\sqrt{n}$
1	1	1.000	51	2601	7.141
2	4	1.414	52	2704	7.211
3	9	1.732	53	2809	7.280
4	16	2.000	54	2916	7.348
5	25	2.236	55	3025	7.416
6	36	2.449	56	3136	7.483
7	49	2.646	57	3249	7.550
8	64	2.828	58	3364	7.616
9	81	3.000	59	3481	7.681
10	100	3.162	60	3600	7.746
11	121	3.317	61	3721	7.810
12	144	3.464	62	3844	7.874
13	169	3.606	63	3969	7.937
14	196	3.742	64	4096	8.000
15	225	3.873	65	4225	8.062
16	256	4.000	66	4356	8.124
17	289	4.123	67	4489	8.185
18	324	4.243	68	4624	8.246
19	361	4.359	69	4761	8.307
20	400	4.472	70	4900	8.367
21	441	4.583	71	5041	8.426
22	484	4.690	72	5184	8.485
23	529	4.796	73	5329	8.544
24	576	4.899	74	5476	8.602
25	625	5.000	75	5625	8.660
26	676	5.099	76	5776	8.718
27	729	5.196	77	5929	8.775
28	784	5.292	78	6084	8.832
29	841	5.385	79	6241	8.888
30	900	5.477	80	6400	8.944
31	961	5.568	81	6561	9.000
32	1024	5.657	82	6724	9.055
33	1089	5.745	83	6889	9.110
34	1156	5.831	84	7056	9.165
35	1225	5.916	85	7225	9.220
36	1296	6.000	86	7396	9.274
37	1369	6.083	87	7569	9.327
38	1444	6.164	88	7744	9.381
39	1521	6.245	89	7921	9.434
40	1600	6.325	90	8100	9.487
41	1681	6.403	91	8281	9.539
42	1764	6.481	92	8464	9.592
43	1849	6.557	93	8649	9.644
44	1936	6.633	94	8836	9.695
45	2025	6.708	95	9025	9.747
46	2116	6.782	96	9216	9.798
47	2209	6.856	97	9409	9.849
48	2304	6.928	98	9604	9.899
49	2401	7.000	99	9801	9.950
50	2500	7.071	100	10,000	10.000

Compound Interest Table

Compounded Annually

	5%	6%	7%	8%	9%	10%	11%	12%	13%	14%	15%	16%	17%	18%
1 year	1.05000	1.06000	1.07000	1.08000	1.09000	1.10000	1.11000	1.12000	1.13000	1.14000	1.15000	1.16000	1.17000	1.18000
5 years	1.27628	1.33823	1.40255	1.46933	1.53862	1.61051	1.68506	1.76234	1.84244	1.92541	2.01136	2.10034	2.19245	2.28776
10 years	1.62889	1.79085	1.96715	2.15892	2.36736	2.59374	2.83942	3.10585	3.39457	3.70722	4.04556	4.41144	4.80683	5.23384
15 years	2.07893	2.39656	2.75903	3.17217	3.64248	4.17725	4.73459	5.47357	6.25427	7.13794	8.13706	9.26552	10.53872	11.97375
20 years	2.65330	3.20714	3.86968	4.66096	5.60441	6.72750	8.06231	9.64623	11.52309	13.74349	16.36654	19.46076	23.10560	27.39303

Compounded Semiannually

	5%	6%	7%	8%	9%	10%	11%	12%	13%	14%	15%	16%	17%	18%
1 year	1.05063	1.06090	1.07123	1.08160	1.09203	1.10250	1.11303	1.12360	1.13423	1.14490	1.15563	1.16640	1.17723	1.18810
5 years	1.28008	1.34392	1.41060	1.48024	1.55297	1.62889	1.70814	1.79085	1.87714	1.96715	2.06103	2.15892	2.26098	2.36736
10 years	1.63862	1.80611	1.98979	2.19112	2.41171	2.65330	2.91776	3.20714	3.52365	3.86968	4.24785	4.66096	5.11205	5.60441
15 years	2.09757	2.42726	2.80679	3.24340	3.74532	4.32194	4.98395	5.74349	6.61437	7.61226	8.75496	10.06266	11.55825	13.26768
20 years	2.68506	3.26204	3.95926	4.80102	5.81636	7.03999	8.51331	10.28572	12.41607	14.97446	18.04424	21.72452	26.13302	31.40942

Compounded Quarterly

	5%	6%	7%	8%	9%	10%	11%	12%	13%	14%	15%	16%	17%	18%
1 year	1.05095	1.06136	1.07185	1.08243	1.09308	1.10381	1.11462	1.12551	1.13648	1.14752	1.15865	1.16986	1.18115	1.19252
5 years	1.28204	1.34686	1.41478	1.48595	1.56051	1.63862	1.72043	1.80611	1.89584	1.98979	2.08815	2.19112	2.29891	2.41171
10 years	1.64362	1.81402	2.00160	2.20804	2.43519	2.68506	2.95987	3.26204	3.59420	3.95926	4.36038	4.80102	5.28497	5.81636
15 years	2.10718	2.44322	2.83182	3.28103	3.80013	4.39979	5.09225	5.89160	6.81402	7.87809	9.10513	10.51963	12.14965	14.02741
20 years	2.70148	3.29066	4.00639	4.87544	5.93015	7.20957	8.76085	10.64089	12.91828	15.67574	19.01290	23.04980	27.93091	33.83010

Compounded Daily

	5%	6%	7%	8%	9%	10%	11%	12%	13%	14%	15%	16%	17%	18%
1 year	1.05127	1.06183	1.07250	1.08328	1.09416	1.10516	1.11626	1.12747	1.13880	1.15024	1.16180	1.17347	1.18526	1.19716
5 years	1.28400	1.34983	1.41902	1.49176	1.56823	1.64861	1.73311	1.82194	1.91532	2.01348	2.11667	2.22515	2.33918	2.45906
10 years	1.64866	1.82203	2.01362	2.22535	2.45933	2.71791	3.00367	3.31946	3.66845	4.05411	4.48031	4.95130	5.47178	6.04696
15 years	2.11689	2.45942	2.85736	3.31968	3.85678	4.48077	5.20569	6.04786	7.02625	8.16288	9.48335	11.01738	12.79950	14.86983
20 years	2.71810	3.31979	4.05466	4.95216	6.04831	7.38703	9.02202	11.01883	13.45751	16.43582	20.07316	24.51533	29.94039	36.56577

Answers to Selected Exercises

Chapter 1 The Whole Numbers

Chapter 1 Pretest

1. hundreds; 1.1A **2.** twenty-three thousand, four hundred ninety; 1.1B **3.** 87; 1.2A **4.** 3717; 1.5B **5.** 626; 1.3A **6.** 32; 1.3B
7. 136 pages; 1.3C **8.** 9050; 1.4A **9.** 3100; 1.4B **10.** $9 \cdot 3 + 9 \cdot 11$; 1.5A **11.** 25 in.; 1.2B **12.** 184 sq yd; 1.5C **13.** 576 seats; 1.5D
14. 243; 1.6A **15.** 446 R9; 1.6B **16.** 39; 1.6D **17.** $9880; 1.7A **18.** 9^7; 1.8A **19.** 2401; 1.8B **20.** 39; 1.8C

Exercise Set 1.1

1. tens **3.** thousands **5.** hundred-thousands **7.** millions **9.** five thousand, four hundred twenty
11. twenty-six thousand, nine hundred ninety **13.** one million, six hundred twenty thousand
15. fifty-three million, five hundred twenty thousand, one hundred seventy
17. five million, six hundred forty-eight thousand, three hundred fifty-nine
19. six hundred twenty thousand **21.** three thousand, eight hundred ninety-three **23.** 6508 **25.** 29,900 **27.** 6,504,019
29. 3,000,014 **31.** 1821 **33.** 63,100,000 **35.** 500,000,000 **37.** $400 + 6$ **39.** $5000 + 200 + 90$
41. $60,000 + 2000 + 400 + 7$ **43.** $30,000 + 600 + 80$ **45.** $30,000,000 + 9,000,000 + 600,000 + 80,000$ **47.** four thousand
49. $4000 + 100 + 40 + 5$ **51.** Nile **53.** Pomeranian; thirty-nine thousand, seven hundred twelve **55.** German Shepherd
57. 55,543 **59.** answers may vary

Calculator Explorations

1. 134 **3.** 340 **5.** 2834

Mental Math

1. 12 **3.** 9000 **5.** 1620

Exercise Set 1.2

1. 36 **3.** 92 **5.** 49 **7.** 5399 **9.** 117 **11.** 71 **13.** 117 **15.** 25 **17.** 62 **19.** 212 **21.** 94 **23.** 910 **25.** 8273
27. 11,926 **29.** 1884 **31.** 16,717 **33.** 1110 **35.** 8999 **37.** 35,901 **39.** 612,389 **41.** 29 in. **43.** 25 ft **45.** 24 in.
47. 8 yd **49.** 383 mi **51.** 340 ft. **53.** 123,100 people **55.** 6117 performances **57.** 11,949 **59.** Texas **61.** 586
63. 1641 **65.** answers may vary **67.** 166,510,192

Calculator Explorations

1. 770 **3.** 109 **5.** 8978

Mental Math

1. 7 **3.** 5 **5.** 0 **7.** 400 **9.** 500

Exercise Set 1.3

1. 44 **3.** 60 **5.** 265 **7.** 254 **9.** 545 **11.** 600 **13.** 25 **15.** 45 **17.** 146 **19.** 288 **21.** 168 **23.** 6 **25.** 447
27. 5723 **29.** 504 **31.** 89 **33.** 79 **35.** 39,914 **37.** 32,711 **39.** 5041 **41.** 31,213 **43.** 4 **45.** 20 **47.** 7
49. 56 crawfish **51.** 264 pages **53.** 6065 ft **55.** $175 **57.** 358 mi **59.** $389 **61.** $448 **63.** 173 men
65. Jo with 1018 votes **67.** 5920 sq. ft **69.** 497,000 soldiers **71.** Dallas/Ft. Worth International **73.** 148 thousand or 148,000
75. Procter & Gamble, General Motors, Philip Morris, Chrysler **77.** $461,357,000 **79.** $10,251,205,000 **81.** $5269 - 2385 = 2884$

Exercise Set 1.4

1. 630 **3.** 640 **5.** 790 **7.** 400 **9.** 1100 **11.** 43,000 **13.** 248,700 **15.** 36,000 **17.** 100,000 **19.** 60,000,000
21. 5280; 5300; 5000 **23.** 9440; 9400; 9000 **25.** 14,880; 14,900; 15,000 **27.** 12,000 **29.** 69,000 **31.** 160,000,000
33. 305,700,000 **35.** 130 **37.** 380 **39.** 5500 **41.** 300 **43.** 8500 **45.** correct **47.** incorrect **49.** correct **51.** correct
53. $3100 **55.** 80 miles **57.** 25,000 ft **59.** 5,700,000 **61.** 14,000,000 votes **63.** 211,000 children **65.** 800,000,000
67. $1,120,000,000 **69.** 8550 **71.** 1,549,999

Calculator Explorations

1. 3456 **3.** 15,322 **5.** 272,291

Mental Math

1. 24 **3.** 0 **5.** 0 **7.** 87

Exercise Set 1.5

1. $4 \cdot 3 + 4 \cdot 9$ **3.** $2 \cdot 4 + 2 \cdot 6$ **5.** $10 \cdot 11 + 10 \cdot 7$ **7.** 252 **9.** 1872 **11.** 1662 **13.** 5310 **15.** 4172 **17.** 10,857 **19.** 11,326 **21.** 24,800 **23.** 0 **25.** 5900 **27.** 59,232 **29.** 142,506 **31.** 1,821,204 **33.** 456,135 **35.** 64,790 **37.** 240,000 **39.** 300,000 **41.** 63 sq. m **43.** 390 sq. ft **45.** 375 calories **47.** $1295 **49.** 192 cans **51.** 9900 sq. ft **53.** 495,864 sq. m **55.** 5828 pixels **57.** 1500 characters **59.** 1280 calories **61.** 71,343 miles **63.** 22,464 acres **65.** 21,700,000 qts **67.** $3,760,000,000 **69.** 50 students **71.** apple and orange **73.** 2, 9 **75.** answers may vary **77.** 621 points

Calculator Explorations

1. 53 **3.** 62 **5.** 261 **7.** 0

Mental Math

1. 5 **3.** 9 **5.** 0 **7.** 9 **9.** 1 **11.** 5 **13.** undefined **15.** 7 **17.** 0 **19.** 8

Exercise Set 1.6

1. 12 **3.** 37 **5.** 338 **7.** 16 R 2 **9.** 563 R 1 **11.** 37 R 1 **13.** 265 R 1 **15.** 49 **17.** 13 **19.** 97 R 40 **21.** 206 **23.** 506 **25.** 202 R 7 **27.** 45 **29.** 98 R 100 **31.** 202 R 15 **33.** 202 **35.** 58 students **37.** $252,000 **39.** 415 bushels **41.** 88 bridges **43.** Yes, she needs 176 ft; she has 9 ft left over. **45.** 16 touchdowns **47.** 1760 yd **49.** 26 **51.** 498 **53.** 79 **55.** 16° **57.** $1,390,596,750 **59.** increase **61.** no

Integrated Review

1. 148 **2.** 6555 **3.** 1620 **4.** 562 **5.** 79 **6.** undefined **7.** 9 **8.** 1 **9.** 0 **10.** 0 **11.** 0 **12.** 3 **13.** 2433 **14.** 9826 **15.** 213 R 3 **16.** 79,317 **17.** 27 **18.** 9 **19.** 138 **20.** 276 **21.** 663 R 6 **22.** 1076 R 60

Exercise Set 1.7

1. 49 **3.** 237 **5.** 42 **7.** 600 **9.** 9600 sq. ft **11.** $15,500 **13.** 168 hr **15.** $800 **17.** $1,149,000,000 **19.** $40,992 **21.** $35,568,000 **23.** $18/hr **25.** 55 calories **27.** 2,338,560 tickets **29.** 39,412,000 students **31.** 13 paychecks **33.** $239 **35.** $1045 **37.** $12 **39.** yes, by $4 **41.** 91 guest rooms **43.** AT&T **45.** 296 patents **47.** Toshiba by 377 patents

Calculator Explorations

1. 729 **3.** 1024 **5.** 2048 **7.** 2526 **9.** 4295 **11.** 8

Exercise Set 1.8

1. 3^4 **3.** 7^8 **5.** 12^3 **7.** $6^2 \cdot 5^3$ **9.** $9^3 \cdot 8$ **11.** $3 \cdot 2^5$ **13.** $3 \cdot 2^2 \cdot 5^3$ **15.** 25 **17.** 125 **19.** 64 **21.** 1024 **23.** 7 **25.** 243 **27.** 256 **29.** 64 **31.** 81 **33.** 729 **35.** 100 **37.** 10,000 **39.** 10 **41.** 1920 **43.** 729 **45.** 21 **47.** 8 **49.** 29 **51.** 4 **53.** 17 **55.** 46 **57.** 28 **59.** 10 **61.** 7 **63.** 4 **65.** 14 **67.** 72 **69.** 2 **71.** 35 **73.** 4 **75.** undefined **77.** 52 **79.** 44 **81.** 12 **83.** 13 **85.** 400 sq. mi **87.** 64 sq. cm **89.** 10,000 sq. m **91.** $(2 + 3) \cdot 6 - 2$ **93.** $24 \div (3 \cdot 2) + 2 \cdot 5$ **95.** 1260 ft **97.** 6,384,814

Chapter 1 Review

1. hundreds **2.** ten millions **3.** five thousand, four hundred eighty **4.** forty-six million, two hundred thousand, one hundred twenty **5.** $6000 + 200 + 70 + 9$ **6.** $400,000,000 + 3,000,000 + 200,000 + 20,000 + 5000$ **7.** 59,800 **8.** 6,304,000,000 **9.** 1,595,138 **10.** 2,811,801 **11.** 193,699 **12.** 175,394 **13.** 13 **14.** 17 **15.** 3 **16.** 10 **17.** 38 **18.** 68 **19.** 56 **20.** 40 **21.** 110 **22.** 120 **23.** 950 **24.** 1250 **25.** 1711 **26.** 9867 **27.** $197,699 **28.** 8032 mi **29.** 276 ft **30.** 66 km **31.** 33 **32.** 43 **33.** 14 **34.** 33 **35.** 362 **36.** 65 **37.** 304 **38.** 476 **39.** 2114 **40.** 321 **41.** $15,626 **42.** 397 pages **43.** May **44.** August **45.** July, August, September **46.** April, May **47.** 90 **48.** 50 **49.** 470 **50.** 500 **51.** 4800 **52.** 58,000 **53.** 50,000,000 **54.** 800,000 **55.** 7400 **56.** 4100 **57.** 10,200 **58.** 1,780,000 **59.** 42 **60.** 24 **61.** 0 **62.** 0 **63.** 1410 **64.** 2898 **65.** 800 **66.** 900 **67.** 3696 **68.** 1694 **69.** 0 **70.** 0 **71.** 16,994 **72.** 8954 **73.** 113,634 **74.** 44,763 **75.** 411,426 **76.** 636,314 **77.** 1500 **78.** 240,000 **79.** 1,040,000 **80.** 7,020,000 **81.** $4,897,341 **82.** 3920 calories **83.** 60 sq. mi **84.** 500 sq. cm **85.** 3 **86.** 4 **87.** 6 **88.** 5 **89.** 5 R 2 **90.** 4 R 2 **91.** undefined **92.** 0 **93.** 1 **94.** 10 **95.** undefined **96.** 0 **97.** 15 **98.** 19 R 7 **99.** 24 R 2 **100.** 56 **101.** 1 R 17 **102.** 35 R 15 **103.** 500 **104.** 21 R 6 **105.** 506 **106.** 16 **107.** 199 R 8 **108.** 200 **109.** 458 ft **110.** 51 **111.** 25 **112.** 15 **113.** 100 **114.** 4 **115.** 27 boxes **116.** $192 **117.** 75¢ **118.** 7 billion tablets **119.** 7^4 **120.** $6^2 \cdot 3^3$ **121.** $4 \cdot 2^3 \cdot 3^2$ **122.** $5^2 \cdot 7^3 \cdot 2^2$ **123.** 49 **124.** 64 **125.** 1125 **126.** 19,600 **127.** 13 **128.** 10 **129.** 3 **130.** 1 **131.** 32 **132.** 33 **133.** 49 sq. m **134.** 9 sq. in.

Chapter 1 Test

1. 141 **2.** 113 **3.** 14,880 **4.** 766 R 42 **5.** 200 **6.** 48 **7.** 98 **8.** 0 **9.** undefined **10.** 33 **11.** 21 **12.** 36 **13.** 52,000 **14.** 13,700 **15.** 1600 **16.** $17 **17.** $119 **18.** $126 **19.** 20 cm, 25 sq. cm **20.** 60 yd, 200 sq. yd **21.** 508 **22.** 7

Chapter 2 MULTIPLYING AND DIVIDING FRACTIONS

CHAPTER 2 PRETEST

1. $\frac{3}{8}$; 2.1B **2.** (a) $\frac{25}{7}$ (b) $\frac{77}{8}$; 2.1D **3.** (a) 4 (b) $11\frac{5}{6}$; 2.1E **4.** 1, 2, 3, 4, 6, 9, 12, 18, 36; 2.2A **5.** prime; 2.2B

6. composite; 2.2B **7.** $2 \cdot 2 \cdot 2 \cdot 2 \cdot 5$ or $2^4 \cdot 5$; 2.2C **8.** $\frac{7}{12}$; 2.3A **9.** $\frac{1}{7}$; 2.3A **10.** equal; 2.3B **11.** not equal; 2.3B

12. $\frac{2}{3}$; 2.3C **13.** $\frac{6}{35}$; 2.4A **14.** $\frac{1}{21}$; 2.4A **15.** 2; 2.4B **16.** $320; 2.4C **17.** $\frac{19}{12}$; 2.5A **18.** 6; 2.5B **19.** $\frac{15}{8}$ or $1\frac{7}{8}$; 2.5C

20. 23 mi; 2.5 D

MENTAL MATH

1. numerator = 1; denominator = 2 **3.** numerator = 10; denominator = 3 **5.** numerator = 3; denominator = 7

EXERCISE SET 2.1

1. $\frac{1}{3}$ **3.** $\frac{4}{7}$ **5.** $\frac{9}{16}$ **7.** $\frac{3}{7}$ **9.** $\frac{4}{9}$ **11.** $\frac{42}{131}$ **13.** 89; $\frac{89}{131}$ **15.** $\frac{4}{10}$ **17.** $\frac{8}{42}$ **19.** (a) $2\frac{3}{4}$ (b) $\frac{11}{4}$ **21.** (a) $3\frac{2}{3}$

(b) $\frac{11}{3}$ **23.** (a) $1\frac{1}{2}$ (b) $\frac{3}{2}$ **25.** (a) $1\frac{1}{3}$ (b) $\frac{4}{3}$ **27.** (a) $2\frac{5}{6}$ (b) $\frac{17}{6}$ **29.** (a) $1\frac{5}{9}$ (b) $\frac{14}{9}$ **31.** $\frac{7}{3}$ **33.** $\frac{11}{3}$

35. $\frac{21}{8}$ **37.** $\frac{41}{15}$ **39.** $\frac{83}{7}$ **41.** $\frac{53}{8}$ **43.** $\frac{18}{5}$ **45.** $\frac{109}{24}$ **47.** $\frac{84}{13}$ **49.** $\frac{187}{20}$ **51.** $3\frac{2}{5}$ **53.** $3\frac{3}{13}$ **55.** $3\frac{2}{15}$ **57.** 33

59. 15 **61.** $4\frac{5}{8}$ **63.** $1\frac{1}{17}$ **65.** $10\frac{17}{23}$ **67.** $4\frac{2}{11}$ **69.** 9 **71.** 125 **73.** 49 **75.** 24 **77.** $\frac{3043}{79}$ **79.** $\frac{1300}{1585}$ **81.** $\frac{56}{172}$

EXERCISE SET 2.2

1. 1, 2, 4, 8 **3.** 1, 5, 25 **5.** 1, 2, 4 **7.** 1, 2, 3, 6, 9, 18 **9.** 1, 7 **11.** 1, 2, 4, 5, 8, 10, 16, 20, 40, 80 **13.** 1, 2, 3, 4, 6, 12
15. 1, 2, 17, 34 **17.** prime **19.** composite **21.** composite **23.** prime **25.** composite **27.** composite **29.** prime
31. composite **33.** $2^2 \cdot 3$ **35.** $3 \cdot 5$ **37.** $2^3 \cdot 5$ **39.** $2^2 \cdot 3^2$ **41.** $3 \cdot 13$ **43.** $2^4 \cdot 3$ **45.** $2 \cdot 3^3$ **47.** $2^2 \cdot 3 \cdot 5$
49. $2 \cdot 5 \cdot 11$ **51.** $2^3 \cdot 11$ **53.** 2^7 **55.** $2 \cdot 3 \cdot 5^2$ **57.** $2^2 \cdot 3 \cdot 5^2$ **59.** $2^1 \cdot 3 \cdot 5$ **61.** $3^3 \cdot 5 \cdot 7$ **63.** $2^2 \cdot 5^2 \cdot 7$
65. 4300 **67.** 4,286,340 **69.** 55,300 **71.** 3500 **73.** 1000 **75.** (a) $2^4 \cdot 3^2 \cdot 5^2$ (b) $2 \cdot 3^3 \cdot 5^2$ **77.** answers may vary

EXERCISE SET 2.3

1. $\frac{1}{4}$ **3.** $\frac{1}{5}$ **5.** $\frac{7}{8}$ **7.** $\frac{4}{5}$ **9.** $\frac{5}{6}$ **11.** $\frac{7}{8}$ **13.** $\frac{3}{7}$ **15.** $\frac{3}{5}$ **17.** $\frac{4}{7}$ **19.** $\frac{4}{5}$ **21.** $\frac{5}{8}$ **23.** $\frac{3}{10}$ **25.** $\frac{3}{2}$ or $1\frac{1}{2}$ **27.** $\frac{5}{8}$

29. $\frac{5}{14}$ **31.** $\frac{3}{14}$ **33.** equivalent **35.** not equivalent **37.** not equivalent **39.** equivalent **41.** equivalent

43. not equivalent **45.** $\frac{3}{4}$ of a shift **47.** $\frac{1}{2}$ mi **49.** $\frac{47}{74}$ of individuals **51.** (a) $\frac{3}{10}$ (b) 35 states (c) $\frac{7}{10}$

53. 364 **55.** 2322 **57.** 2520 **59.** false; $1 \cdot 1 = 1$ **61.** $\frac{3}{5}$ **63.** $\frac{9}{25}$ **65.** $\frac{1}{25}$ **67.** $\frac{3}{20}$

INTEGRATED REVIEW

1. $\frac{1}{4}$ **2.** $\frac{3}{6}$ **3.** $\frac{7}{4}$ **4.** $\frac{73}{85}$ **5.** $\frac{25}{8}$ **6.** $\frac{28}{5}$ **7.** $\frac{69}{7}$ **8.** $2\frac{6}{7}$ **9.** 5 **10.** $4\frac{7}{8}$ **11.** 1, 5, 7, 35 **12.** 1, 2, 4, 5, 8, 10, 20, 40

13. 1, 2, 3, 4, 6, 8, 9, 12, 18, 24, 36, 72 **14.** $2 \cdot 3$ **15.** $2 \cdot 5 \cdot 7$ **16.** $2^2 \cdot 3^2 \cdot 7$ **17.** $\frac{1}{7}$ **18.** $\frac{5}{6}$ **19.** $\frac{9}{19}$ **20.** $\frac{21}{55}$ **21.** $\frac{1}{2}$

22. $\frac{9}{10}$ **23.** not equivalent **24.** equivalent **25.** $\frac{1}{25}$

EXERCISE SET 2.4

1. $\frac{2}{15}$ **3.** $\frac{6}{35}$ **5.** $\frac{9}{80}$ **7.** $\frac{7}{12}$ **9.** $\frac{5}{28}$ **11.** $\frac{1}{15}$ **13.** $\frac{45}{32}$ or $1\frac{13}{32}$ **15.** $\frac{1}{70}$ **17.** $\frac{18}{55}$ **19.** $\frac{7}{2}$ or $3\frac{1}{2}$ **21.** $\frac{5}{8}$ **23.** $\frac{1}{110}$

25. $\frac{27}{80}$ **27.** 1 **29.** $\frac{1}{56}$ **31.** $\frac{3}{4}$ **33.** $\frac{5}{2}$ or $2\frac{1}{2}$ **35.** $\frac{1}{5}$ **37.** $\frac{5}{3}$ or $1\frac{2}{3}$ **39.** $\frac{2}{3}$ **41.** $\frac{77}{10}$ or $7\frac{7}{10}$ **43.** $\frac{836}{35}$ or $23\frac{31}{35}$

45. 12 **47.** $\frac{25}{2}$ or $12\frac{1}{2}$ **49.** 15 **51.** $\frac{3}{2}$ or $1\frac{1}{2}$ in. **53.** $\frac{55}{2}$ or $27\frac{1}{2}$ mi **55.** $\frac{17}{2}$ or $8\frac{1}{2}$ in. **57.** $\frac{8}{5}$ or $1\frac{3}{5}$ ft **59.** $\frac{3}{16}$ of an in.

61. $1838 **63.** $\frac{39}{2}$ or $19\frac{1}{2}$ in. **65.** $\frac{1}{14}$ sq. ft **67.** $\frac{7}{2}$ or $3\frac{1}{2}$ sq. yd **69.** $\frac{121}{4}$ or $30\frac{1}{4}$ sq. in. **71.** 3840 mi **73.** 2400 mi

75. 206 **77.** 56 R 12 **79.** 13,544,400 **81.** answers may vary

Exercise Set 2.5

1. $\frac{7}{4}$　**3.** 11　**5.** $\frac{1}{15}$　**7.** $\frac{7}{12}$　**9.** $\frac{4}{5}$　**11.** $\frac{1}{6}$　**13.** $\frac{16}{9}$ or $1\frac{7}{9}$　**15.** $\frac{18}{35}$　**17.** $\frac{3}{4}$　**19.** $\frac{121}{60}$ or $2\frac{1}{60}$　**21.** $\frac{1}{100}$　**23.** $\frac{1}{3}$

25. $\frac{3}{4}$　**27.** $\frac{21}{20}$ or $1\frac{1}{20}$　**29.** $\frac{35}{36}$　**31.** $\frac{14}{37}$　**33.** $\frac{8}{45}$　**35.** $\frac{1}{6}$　**37.** $\frac{40}{3}$ or $13\frac{1}{3}$　**39.** 5　**41.** $\frac{5}{28}$　**43.** $\frac{36}{35}$ or $1\frac{1}{35}$　**45.** 96

47. $\frac{35}{11}$ or $3\frac{2}{11}$　**49.** $\frac{5}{6}$ Tbsp　**51.** $\frac{235}{34}$ or $6\frac{31}{34}$ gal　**53.** $\frac{19}{30}$ in.　**55.** $\frac{625}{13}$ or $48\frac{1}{13}$ hr　**57.** 201　**59.** 196　**61.** 1569

63. 15,432 species　**65.** 16 games

Chapter 2 Review

1. proper　**2.** improper　**3.** proper　**4.** mixed number　**5.** proper　**6.** mixed number　**7.** $\frac{2}{6}$　**8.** $\frac{4}{7}$　**9.** $\frac{7}{3}$　**10.** $\frac{9}{4}$

11. $\frac{11}{12}$　**12.** $\frac{23}{131}$　**13.** $3\frac{3}{4}$　**14.** 3　**15.** $45\frac{5}{6}$　**16.** $31\frac{1}{4}$　**17.** $\frac{6}{5}$　**18.** $\frac{26}{9}$　**19.** $\frac{47}{12}$　**20.** $\frac{95}{17}$　**21.** composite　**22.** prime

23. composite　**24.** composite　**25.** 1, 2, 3, 6, 7, 14, 21, 42　**26.** 1, 2, 3, 5, 6, 10, 15, 30　**27.** $2^2 \cdot 17$　**28.** $2 \cdot 3^2 \cdot 5$　**29.** $5 \cdot 157$

30. $3 \cdot 5 \cdot 17$　**31.** $\frac{3}{7}$　**32.** $\frac{5}{9}$　**33.** $\frac{1}{3}$　**34.** $\frac{1}{2}$　**35.** $\frac{29}{32}$　**36.** $\frac{18}{23}$　**37.** $\frac{5}{3}$ or $1\frac{2}{3}$　**38.** $\frac{7}{5}$ or $1\frac{2}{5}$　**39.** 8　**40.** 6　**41.** $\frac{14}{15}$

42. $\frac{3}{5}$　**43.** $\frac{3}{10}$　**44.** $\frac{5}{14}$　**45.** $\frac{7}{12}$　**46.** $\frac{1}{4}$　**47.** 9　**48.** $\frac{1}{2}$　**49.** $\frac{35}{8}$ or $4\frac{3}{8}$　**50.** $\frac{5}{2}$ or $2\frac{1}{2}$　**51.** 7　**52.** 21　**53.** $\frac{13}{12}$ or $1\frac{1}{12}$

54. $\frac{15}{11}$ or $1\frac{4}{11}$　**55.** 10　**56.** $\frac{51}{4}$ or $12\frac{3}{4}$　**57.** 203 calories　**58.** $\frac{40}{3}$ or $13\frac{1}{3}$ g　**59.** $\frac{119}{80}$ or $1\frac{39}{80}$ sq. in.　**60.** $\frac{1}{7}$　**61.** 8　**62.** $\frac{23}{14}$

63. $\frac{5}{17}$　**64.** 2　**65.** $\frac{15}{4}$ or $3\frac{3}{4}$　**66.** 9　**67.** $\frac{27}{2}$ or $13\frac{1}{2}$　**68.** $\frac{5}{6}$　**69.** $\frac{8}{3}$ or $2\frac{2}{3}$　**70.** $\frac{21}{4}$ or $5\frac{1}{4}$　**71.** $\frac{121}{46}$ or $2\frac{29}{46}$　**72.** $\frac{7}{3}$ or $2\frac{1}{3}$

73. $\frac{32}{5}$ or $6\frac{2}{5}$　**74.** $200　**75.** $\frac{21}{20}$ or $1\frac{1}{20}$ mi

Chapter 2 Test

1. $\frac{4}{3}$ or $1\frac{1}{3}$　**2.** $\frac{4}{3}$ or $1\frac{1}{3}$　**3.** $\frac{1}{4}$　**4.** $\frac{16}{45}$　**5.** 16　**6.** $\frac{9}{2}$ or $4\frac{1}{2}$　**7.** 1　**8.** 9　**9.** $\frac{64}{3}$ or $21\frac{1}{3}$　**10.** $\frac{45}{2}$ or $22\frac{1}{2}$　**11.** $\frac{18}{5}$ or $3\frac{3}{5}$

12. $\frac{10}{3}$ or $3\frac{1}{3}$　**13.** $\frac{23}{3}$　**14.** $\frac{39}{11}$　**15.** $4\frac{3}{5}$　**16.** $18\frac{3}{4}$　**17.** $\frac{34}{27}$ or $1\frac{7}{27}$ sq. mi　**18.** 24 mi　**19.** $\frac{4}{35}$　**20.** $\frac{3}{5}$　**21.** 8800 sq. yd

22. $2^3 \cdot 5 \cdot 7$　**23.** $2^2 \cdot 3 \cdot 7$　**24.** $90 per share

Cumulative Review

1. ten-thousands; Sec. 1.1, Ex. 1　**2.** 805; Sec. 1.1, Ex. 8　**3.** 184,046; Sec. 1.2, Ex. 2　**4.** 13 in.; Sec. 1.2, Ex. 5
5. $3887; Sec. 1.2, Ex. 7　**6.** 7321; Sec. 1.3, Ex. 2　**7. (a)** walking　**(b)** 10 people; Sec. 1.3, Ex. 7　**8.** 570; Sec. 1.4, Ex. 1
9. 1800; Sec 1.4, Ex. 5　**10. (a)** 6　**(b)** 0　**(c)** 45　**(d)** 0; Sec. 1.5, Ex. 1　**11. (a)** $3 \cdot 4 + 3 \cdot 5$　**(b)** $10 \cdot 6 + 10 \cdot 8$
(c) $2 \cdot 7 + 2 \cdot 3$; Sec. 1.5, Ex. 2　**12. (a)** 0　**(b)** 0　**(c)** 0　**(d)** undefined; Sec. 1.6, Ex. 3　**13.** 208; Sec. 1.6, Ex. 5
14. 7 boxes; Sec. 1.6, Ex. 11　**15.** 40 ft; Sec. 1.7, Ex. 5　**16.** 4^3; Sec. 1.8, Ex. 1　**17.** $6^3 \cdot 8^5$; Sec. 1.8, Ex. 4　**18.** 7; Sec. 1.8, Ex. 9

19. $\frac{3}{4}$; Sec. 2.1 Ex. 4　**20. (a)** $\frac{38}{9}$　**(b)** $\frac{19}{11}$; Sec. 2.1, Ex. 8　**21.** 1, 2, 4, 5, 10, 20; Sec. 2.2, Ex. 1　**22.** $\frac{7}{11}$; Sec. 2.3, Ex. 2

23. $\frac{35}{12}$ or $2\frac{11}{12}$; Sec. 2.4, Ex. 7　**24.** $\frac{3}{1}$ or 3; Sec. 2.5, Ex. 3　**25.** $\frac{5}{12}$; Sec. 2.5, Ex. 6

Chapter 3　Adding and Subtracting Fractions

Chapter 3 Pretest

1. $\frac{8}{9}$; 3.1A　**2.** 1; 3.1 A　**3.** $\frac{11}{12}$; 3.3A　**4.** $\frac{5}{12}$; 3.3A　**5.** $9\frac{7}{12}$; 3.4A　**6.** $\frac{5}{36}$; 3.1B　**7.** $\frac{1}{9}$; 3.1B　**8.** $\frac{29}{42}$; 3.3B　**9.** $\frac{2}{15}$; 3.3B

10. $5\frac{17}{30}$; 3.4B　**11.** 30; 3.2A, B　**12.** 360; 3.2A, B　**13.** $\frac{14}{35}$; 3.2C　**14.** $\frac{39}{27}$; 3.2C　**15.** <; 3.5A　**16.** $\frac{27}{125}$; 3.5B　**17.** $\frac{7}{18}$; 3.5D

18. $\frac{2}{5}$; 3.5D　**19.** $\frac{7}{8}$ lb; 3.3C　**20.** $5\frac{3}{8}$ ft; 3.4C

Mental Math

1. unlike　**3.** like　**5.** like　**7.** unlike

Exercise Set 3.1

1. $\frac{3}{7}$ **3.** $\frac{1}{5}$ **5.** $\frac{2}{3}$ **7.** $\frac{7}{20}$ **9.** $\frac{1}{2}$ **11.** $\frac{13}{11}=1\frac{2}{11}$ **13.** $\frac{7}{13}$ **15.** $\frac{2}{3}$ **17.** $\frac{6}{11}$ **19.** $\frac{3}{5}$ **21.** 1 **23.** $\frac{3}{4}$ **25.** $\frac{5}{6}$

27. $\frac{4}{5}$ **29.** $\frac{19}{33}$ **31.** 1 in. **33.** 2 m **35.** $1\frac{1}{2}$ hr **37.** $\frac{7}{10}$ of a mi **39.** $\frac{61}{100}$ **41.** $\frac{23}{50}$ **43.** $2 \cdot 5$ **45.** 2^3 **47.** $5 \cdot 11$

49. $\frac{5}{8}$ **51.** $\frac{8}{11}$ **53.** $\frac{1}{4}$ of a mi **55.** answers may vary.

Exercise Set 3.2

1. 12 **3.** 45 **5.** 36 **7.** 72 **9.** 126 **11.** 75 **13.** 24 **15.** 42 **17.** 150 **19.** 68 **21.** 588 **23.** 900 **25.** 357

27. 363 **29.** 216 **31.** 60 **33.** $\frac{20}{35}$ **35.** $\frac{14}{21}$ **37.** $\frac{10}{25}$ **39.** $\frac{15}{30}$ **41.** $\frac{30}{21}$ **43.** $\frac{21}{28}$ **45.** $\frac{30}{45}$ **47.** $\frac{36}{81}$ **49.** $\frac{12}{9}$

51. $\frac{18}{10}$ **53.** $\frac{5}{10}=\frac{1}{2}$ **55.** $\frac{2}{5}$ **57.** $\frac{8}{18}=\frac{4}{9}$ **59.** $\frac{9}{9}=1$ **61.** $\frac{814}{3630}$ **63.** answers may vary

Exercise Set 3.3

1. $\frac{5}{6}$ **3.** $\frac{5}{6}$ **5.** $\frac{8}{33}$ **7.** $\frac{9}{14}$ **9.** $\frac{3}{5}$ **11.** $\frac{19}{36}$ **13.** $\frac{53}{60}$ **15.** $\frac{1}{6}$ **17.** $\frac{75}{56}=1\frac{19}{56}$ **19.** $\frac{16}{11}=1\frac{5}{11}$ **21.** $\frac{11}{16}$ **23.** $\frac{17}{42}$

25. $\frac{33}{56}$ **27.** $\frac{4}{33}$ **29.** $\frac{1}{35}$ **31.** $\frac{11}{36}$ **33.** $\frac{1}{20}$ **35.** $\frac{1}{84}$ **37.** $\frac{34}{15}=2\frac{4}{15}$ cm **39.** $\frac{17}{10}=1\frac{7}{10}$ m **41.** $1\frac{3}{8}$ lb **43.** $\frac{49}{100}$ of students

45. $\frac{77}{100}$ of Americans **47.** 5 **49.** $\frac{16}{29}$ **51.** $\frac{19}{3}=6\frac{1}{3}$ **53.** $\frac{49}{44}=1\frac{5}{44}$ **55.** answers may vary **57.** $\frac{212}{579}$ **59.** 21,200,000 sq. mi

Integrated Review

1. 30 **2.** 21 **3.** 14 **4.** 25 **5.** 100 **6.** 90 **7.** $\frac{9}{24}$ **8.** $\frac{28}{36}$ **9.** $\frac{10}{40}$ **10.** $\frac{12}{30}$ **11.** $\frac{55}{75}$ **12.** $\frac{40}{48}$ **13.** $\frac{1}{2}$ **14.** $\frac{2}{5}$

15. $\frac{13}{15}$ **16.** $\frac{7}{12}$ **17.** $\frac{3}{4}$ **18.** $\frac{2}{15}$ **19.** $\frac{17}{45}$ **20.** $\frac{19}{50}$ **21.** $\frac{37}{40}$ **22.** $\frac{11}{36}$ **23.** $\frac{2}{11}$ **24.** $\frac{1}{17}$ **25.** $\frac{5}{33}$ **26.** $\frac{1}{42}$ **27.** $\frac{5}{18}$

28. $\frac{5}{13}$ **29.** $\frac{11}{18}$ **30.** $\frac{37}{50}$

Exercise Set 3.4

1. $6\frac{4}{5}$ **3.** $13\frac{11}{14}$ **5.** $17\frac{7}{25}$ **7.** $7\frac{5}{24}$ **9.** $7\frac{5}{8}$ **11.** $20\frac{1}{15}$ **13.** $28\frac{7}{12}$ **15.** $13\frac{13}{24}$ **17.** $2\frac{3}{5}$ **19.** $7\frac{5}{14}$ **21.** $\frac{24}{25}$ **23.** $5\frac{11}{14}$

25. $2\frac{1}{2}$ **27.** $23\frac{5}{8}$ **29.** $1\frac{4}{5}$ **31.** $1\frac{13}{15}$ **33.** $3\frac{5}{9}$ **35.** $15\frac{3}{4}$ **37.** $2\frac{3}{8}$ hr **39.** $10\frac{1}{4}$ hr **41.** $7\frac{13}{20}$ in.

43. no, she will be $\frac{1}{12}$ of a ft short **45.** $10\frac{5}{8}$ lb **47.** $352\frac{1}{3}$ yd **49.** $306\frac{2}{3}$ ft **51.** $\frac{2}{5}$ of a min **53.** 7 mi **55.** $21\frac{5}{24}$ m **57.** 8

59. 25 **61.** 81 **63.** 64 **65.** Supreme is heavier by $\frac{1}{8}$ lb **67.** answers may vary

Exercise Set 3.5

1. > **3.** < **5.** < **7.** > **9.** > **11.** < **13.** < **15.** > **17.** $\frac{1}{16}$ **19.** $\frac{8}{125}$ **21.** $\frac{64}{343}$ **23.** $\frac{4}{81}$ **25.** $\frac{1}{6}$

27. $\frac{11}{15}$ **29.** $\frac{3}{35}$ **31.** $\frac{1}{12}$ **33.** $\frac{9}{11}$ **35.** $\frac{2}{5}$ **37.** $\frac{2}{77}$ **39.** $\frac{17}{60}$ **41.** $\frac{5}{8}$ **43.** $\frac{1}{2}$ **45.** $\frac{29}{10}=2\frac{9}{10}$ **47.** $\frac{27}{32}$ **49.** $\frac{1}{81}$

51. $\frac{3}{4}$ **53.** $\frac{1}{4}$ **55.** $\frac{5}{6}$ **57.** $\frac{5}{6}$ **59.** $\frac{2}{3}$ **61.** periodicals **63.** savings account

Chapter 3 Review

1. $\frac{10}{11}$ **2.** $\frac{2}{3}$ **3.** $\frac{1}{6}$ **4.** $\frac{1}{5}$ **5.** $\frac{2}{3}$ **6.** $\frac{1}{7}$ **7.** $\frac{3}{5}$ **8.** $\frac{3}{5}$ **9.** 1 **10.** 1 **11.** $\frac{19}{25}$ **12.** $\frac{16}{21}$ **13.** $\frac{1}{2}$ hr **14.** $\frac{3}{8}$ of a gal

15. $\frac{3}{4}$ of his homework. **16.** $\frac{3}{2}=1\frac{1}{2}$ mi **17.** 55 **18.** 60 **19.** 120 **20.** 84 **21.** 252 **22.** 72 **23.** $\frac{56}{64}$ **24.** $\frac{20}{30}$

25. $\frac{21}{33}$ **26.** $\frac{20}{26}$ **27.** $\frac{16}{60}$ **28.** $\frac{25}{60}$ **29.** $\frac{11}{18}$ **30.** $\frac{7}{26}$ **31.** $\frac{7}{12}$ **32.** $\frac{11}{12}$ **33.** $\frac{27}{55}$ **34.** $\frac{7}{15}$ **35.** $\frac{17}{36}$ **36.** $\frac{1}{6}$

37. $\frac{23}{18}=1\frac{5}{18}$ **38.** $\frac{3}{14}$ **39.** $2\frac{1}{9}$ m **40.** $1\frac{1}{2}$ ft **41.** $\frac{1}{4}$ of a yd **42.** $\frac{7}{10}$ has been cleaned. **43.** $45\frac{16}{21}$ **44.** 60 **45.** $32\frac{13}{22}$

46. $3\frac{19}{60}$ **47.** $111\frac{5}{18}$ **48.** $20\frac{7}{24}$ **49.** $5\frac{16}{35}$ **50.** $3\frac{4}{55}$ **51.** $6\frac{7}{20}$ lb **52.** $44\frac{1}{2}$ yd **53.** $7\frac{4}{5}$ in. **54.** $11\frac{1}{6}$ ft **55.** 5 ft

56. $\frac{1}{40}$ oz   **57.** <   **58.** >   **59.** <   **60.** >   **61.** >   **62.** >   **63.** $\frac{4}{49}$ **64.** $\frac{64}{125}$ **65.** $\frac{9}{400}$ **66.** $\frac{9}{100}$ **67.** $\frac{81}{196}$
68. $\frac{1}{7}$ **69.** $\frac{11}{25}$ **70.** $\frac{1}{8}$ **71.** $\frac{1}{27}$ **72.** $3\frac{3}{5}$ **73.** $1\frac{17}{28}$ **74.** $\frac{5}{6}$

CHAPTER 3 TEST

1. 60   **2.** 72   **3.** <   **4.** <   **5.** $\frac{8}{9}$ **6.** $\frac{2}{5}$ **7.** $\frac{13}{10} = 1\frac{3}{10}$ **8.** $\frac{8}{21}$ **9.** $\frac{13}{24}$ **10.** $\frac{1}{7}$ **11.** $\frac{67}{60} = 1\frac{7}{60}$ **12.** $\frac{7}{50}$
13. $\frac{3}{2} = 1\frac{1}{2}$ **14.** $14\frac{1}{40}$ **15.** $1\frac{7}{24}$ **16.** $\frac{5}{3} = 1\frac{2}{3}$ **17.** $\frac{16}{81}$ **18.** $\frac{10}{3} = 3\frac{1}{3}$ **19.** $\frac{153}{200}$ **20.** $\frac{3}{8}$ **21.** $3\frac{3}{4}$ ft   **22.** $\frac{5}{16}$
23. 125,000 backpacks

CUMULATIVE REVIEW

1. eighty-five; Sec. 1.1, Ex. 4   **2.** one hundred twenty-six; Sec. 1.1, Ex. 5   **3.** 159; Sec. 1.2, Ex. 1   **4.** 14; Sec. 1.3, Ex. 3
5. 278,000; Sec.1.4, Ex. 2   **6.** 20,296; Sec. 1.5, Ex. 4   **7. (a)** 8   **(b)** 11   **(c)** 1   **(d)** 1   **(e)** 10   **(f)** 1; Sec.1.6, Ex. 2
8. 1038 mi; Sec. 1.7, Ex. 1   **9.** 64; 1.8, Ex. 5   **10.** 32; Sec. 1.8, Ex. 7   **11.** improper fraction: $\frac{4}{3}$; mixed number: $1\frac{1}{3}$; Sec. 2.1, Ex. 6
12. improper fraction: $\frac{5}{2}$; mixed number: $2\frac{1}{2}$; Sec. 2.1, Ex. 7   **13.** 3, 11, 17; Sec. 2.2, Ex. 2   **14.** $2^2 \cdot 3^2 \cdot 5$; Sec. 2.2, Ex. 4
15. $\frac{36}{13}$ or $2\frac{10}{13}$; Sec. 2.3, Ex. 5   **16.** not equivalent; Sec. 2.3, Ex. 8   **17.** $\frac{10}{33}$; Sec. 2.4, Ex. 1   **18.** $\frac{1}{8}$; Sec. 2.4, Ex. 2   **19.** $\frac{11}{51}$; Sec. 2.5, Ex. 9
20. $\frac{51}{23}$ or $2\frac{5}{23}$; Sec. 2.5, Ex. 10   **21.** $\frac{5}{8}$; Sec. 3.1, Ex. 2   **22.** 24; Sec. 3.2, Ex. 1   **23.** $\frac{6}{6}$ or 1; Sec. 3.3, Ex. 4   **24.** $4\frac{5}{21}$; Sec. 3.4, Ex. 4
25. $\frac{6}{13}$; Sec. 3.5, Ex. 11

Chapter 4 DECIMALS

CHAPTER 4 PRETEST

1. 5.04; 4.1B   **2.** $\frac{17}{25}$; 4.1C   **3.** 0.081; 4.1D   **4.** <; 4.2A   **5.** 364.7; 4.2B   **6.** 72.245; 4.3A   **7.** 47.717; 4.3B   **8.** 1.404; 4.4A
9. 0.02701; 4.4B   **10.** $22\pi \approx 69.08$ in.; 4.4C   **11.** 5000; 4.5A   **12.** 4.5; 4.5A   **13.** 0.01723; 4.5B   **14.** 17.1; 4.6A
15. 516.84; 4.6A   **16.** 4.48; 4.6B   **17.** 1.14; 4.6B   **18.** 0.555; 4.7C   **19.** $\frac{1}{3}$, 0.37, $\frac{7}{18}$; 4.7B   **20.** \$24.70; 4.4D

MENTAL MATH

1. tens   **3.** tenths

EXERCISE SET 4.1

1. six and fifty-two hundredths   **3.** sixteen and twenty-three hundredths   **5.** two hundred five thousandths
7. one hundred sixty-seven and nine thousandths   **9.** thirty-one and four hundredths   **11.** one and eight tenths   **13.** 6.5
15. 9.08   **17.** 5.625   **19.** 0.0064   **21.** 20.33   **23.** 12.9   **25.** $\frac{3}{10}$ **27.** $\frac{27}{100}$ **29.** $5\frac{47}{100}$ **31.** $\frac{6}{125}$ **33.** $7\frac{7}{100}$ **35.** $15\frac{401}{500}$
37. $\frac{601}{2000}$ **39.** $487\frac{8}{25}$ **41.** 0.6   **43.** 0.45   **45.** 3.7   **47.** 0.268   **49.** 0.09   **51.** 4.026   **53.** 0.028   **55.** 56.3   **57.** 47,260
59. 47,000   **61.** answers may vary   **63.** 7.12

EXERCISE SET 4.2

1. <   **3.** >   **5.** <   **7.** =   **9.** <   **11.** >   **13.** 0.6   **15.** 0.23   **17.** 0.594   **19.** 98,210   **21.** 12.3   **23.** 17.67
25. 0.5   **27.** 0.130   **29.** 3830   **31.** \$0.07   **33.** \$42,650   **35.** \$27   **37.** \$0.20   **39.** 0.26499; 0.25786   **41.** 40,000 people
43. 2.44 hr   **45.** 24.623 hr   **47.** 43   **49.** 5766   **51.** 71   **53.** 243   **55.** 225.490; Mike Aulby
57. 225.490, 225.370, 222.980, 222.830, 222.000, 219.702, 218.158, 218.036, 216.081, 215.432   **59.** answers may vary

CALCULATOR EXPLORATION

1. 328.742   **3.** 5.2414   **5.** 865.392

MENTAL MATH

1. 0.5   **3.** 1.26   **5.** 8.9   **7.** 0.6

EXERCISE SET 4.3

1. 3.5 **3.** 6.83 **5.** 0.094 **7.** 622.012 **9.** 583.09 **11.** 465.56 **13.** 115.123 **15.** 27.0578 **17.** 56.432 **19.** 6.5
21. 15.3 **23.** 598.23 **25.** 1.83 **27.** 861.6 **29.** 376.89 **31.** 876.6 **33.** 194.4 **35.** 2.9988 **37.** 16.3 **39.** $454.71
41. $0.06 **43.** $7.52 **45.** 197.8 lb **47.** 15.81 in. **49.** 763.035 mph **51.** 240.8 in. **53.** 67.44 ft **55.** 29.614 mph

57. 715.05 hr **59.** Switzerland **61.** 8.1 lb **63.** **65.** 138 **67.** 960 **69.** $\frac{1}{125}$ **71.** $\frac{5}{12}$

73. answers may vary **75.** 4.59 m

Country	Pounds of Chocolate Per Person
Switzerland	22.0
Norway	16.0
Germany	15.8
United Kingdom	14.5
Belgium	13.9

EXERCISE SET 4.4

1. 0.12 **3.** 0.6 **5.** 1.3 **7.** 22.26 **9.** 43.274 **11.** 8.23854 **13.** 11.2746 **15.** 84.97593 **17.** 65 **19.** 0.65 **21.** 709.3
23. 0.0006 **25.** 9100 **27.** 0.03762 **29.** 5,500,000,000 chocolate bars **31.** 36,400,000 households **33.** 1,600,000 hr
35. $8\pi \approx 25.12$ m **37.** $10\pi \approx 31.4$ cm **39.** $18.2\pi \approx 57.148$ yd **41.** 24.8 g **43.** $1900 **45.** 64.9605 in. **47.** $555.20
49. 26 **51.** 36 **53.** 8 **55.** 9 **57.** 3,831,600 mi **59.** $250\pi \approx 785$ ft **61.** answers may vary

EXERCISE SET 4.5

1. 0.094 **3.** 300 **5.** 5.8 **7.** 6.6 **9.** 0.413 **11.** 7 **13.** 4.8 **15.** 2100 **17.** 30 **19.** 7000 **21.** 9.8 **23.** 9.6
25. 45 **27.** 200 **29.** 23.87 **31.** 110 **33.** 0.54982 **35.** 0.0129 **37.** 8.7 **39.** 24 mo **41.** $855.85 **43.** 202.1 lb
45. 5.1 m **47.** $0.587 per lb **49.** 127.2 mph **51.** 19.43 points/game **53.** 345.22 **55.** 1001.0 **57.** 20 **59.** 18
61. 85 **63.** 14,483,000 CDs **65.** 45.2 cm

INTEGRATED REVIEW

1. 2.57 **2.** 4.05 **3.** 8.9 **4.** 3.5 **5.** 0.16 **6.** 0.24 **7.** 0.27 **8.** 0.52 **9.** 4.8 **10.** 6.09 **11.** 75.56 **12.** 289.12
13. 25.026 **14.** 44.125 **15.** 8.6 **16.** 5.4 **17.** 280 **18.** 1600 **19.** 224.938 **20.** 145.079 **21.** 0.56 **22.** 0.63
23. 27.6092 **24.** 145.6312 **25.** 5.4 **26.** 17.74 **27.** 414.44 **28.** 1295.03 **29.** 34 **30.** 28 **31.** 116.81 **32.** 18.79

EXERCISE SET 4.6

1. 12 **3.** 2.8 **5.** 2898.66 **7.** 4.2 **9.** 149.8 **11.** 22.89 **13.** 36 in. **15.** 39 ft **17.** 43.96 m **19.** 47 gal
21. about $12,000 **23.** about 53 mi **25.** $2109 million **27.** 126,900 people **29.** 0.16 **31.** 3 **33.** 0.28 **35.** 80.52 **37.** 5.5

39. 5.29 **41.** 7.6 **43.** 0.2025 **45.** 129 **47.** $\frac{5}{16}$ **49.** $\frac{3}{4}$ **51.** $\frac{1}{12}$ **53.** 43.388569

EXERCISE SET 4.7

1. 0.2 **3.** 0.5 **5.** 0.75 **7.** 0.08 **9.** 0.375 **11.** $0.91\overline{6}$ **13.** 0.425 **15.** 0.45 **17.** $0.\overline{3}$ **19.** 0.4375 **21.** $0.\overline{2}$
23. $1.\overline{6}$ **25.** 0.33 **27.** 0.44 **29.** 0.2 **31.** 1.7 **33.** 0.194 **35.** 0.44 **37.** 0.8 **39.** < **41.** > **43.** <

45. < **47.** < **49.** > **51.** < **53.** < **55.** 0.32, 0.34, 0.35 **57.** 0.49, 0.491, 0.498 **59.** 0.73, $\frac{3}{4}$, 0.78 **61.** 0.412, 0.453, $\frac{4}{7}$

63. 5.23, $\frac{42}{8}$, 5.34 **65.** $\frac{17}{8}$, 2.37, $\frac{12}{5}$ **67.** 25.65 sq. in. **69.** 9.36 sq. cm **71.** 0.248 sq. yd **73.** 8 **75.** 72 **77.** $\frac{1}{81}$ **79.** $\frac{9}{25}$

81. $\frac{5}{2}$ **83.** 0.243 **85.** 7300 **87.** 0.625 **89.** answers may vary

CHAPTER 4 REVIEW

1. tenths **2.** hundred-thousandths **3.** twenty-three and forty-five hundredths **4.** three hundred forty-five hundred-thousandths
5. one hundred nine and twenty-three hundredths **6.** two hundred and thirty-two millionths **7.** 2.15 **8.** 503.102

9. 16,205.0014 **10.** $\frac{4}{25}$ **11.** $12\frac{23}{1000}$ **12.** $1\frac{9}{2000}$ **13.** $\frac{231}{100,000}$ **14.** $25\frac{1}{4}$ **15.** 0.9 **16.** 0.25 **17.** 0.045 **18.** 0.07

19. > **20.** = **21.** > **22.** < **23.** 0.6 **24.** 0.94 **25.** 42.90 **26.** 16.349 **27.** $0.26 **28.** $12.46 **29.** $123.00

30. $3646.00 **31.** 13,500 people **32.** $10\frac{3}{4}$ tsp **33.** 9.5 **34.** 5.1 **35.** 1.7 **36.** 2.49 **37.** 7.28 **38.** 26 **39.** 320.312
40. 148.74236 **41.** 459.7 **42.** 100.278 **43.** 65.02 **44.** 189.98 **45.** $0.44 **46.** 22.2 in. **47.** 72 **48.** 9345 **49.** 78.246
50. 73,246.446 **51.** 14π m; 43.96 m **52.** 63.8 mi **53.** 70 **54.** 0.21 **55.** 4900 **56.** 23.904 **57.** 8.059 **58.** 15.825
59. 0.0267 **60.** 9.3 **61.** 7.3 m **62.** 45 mo **63.** 18.2 **64.** 50 **65.** 99.05 **66.** 54.1 **67.** 35.5782 **68.** 0.3526
69. 32.7 **70.** 30.4 **71.** 8932 sq. ft **72.** yes, $5 is enough **73.** 16.94 **74.** 3.89 **75.** 0.1024 **76.** 3.6 **77.** 0.8

78. 0.923   **79.** 0.429   **80.** $0.21\overline{6}$ or 0.217 rounded   **81.** 0.1125 or 0.113 rounded   **82.** 51.057   **83.** =   **84.** <   **85.** <

86. <   **87.** 0.832, 0.837, 0.839   **88.** $0.42, \dfrac{3}{7}, 0.43$ **89.** $\dfrac{19}{12}, 1.63, \dfrac{18}{11}$ **90.** $\dfrac{3}{4}, \dfrac{6}{7}, \dfrac{8}{9}$ **91.** 6.9 sq. ft   **92.** 5.46 sq. in.

CHAPTER 4 TEST

1. forty-five and ninety-two thousandths   **2.** 3000.059   **3.** 34.9   **4.** 0.862   **5.** <   **6.** <   **7.** $\dfrac{69}{200}$ **8.** $24\dfrac{73}{100}$ **9.** 0.5

10. 0.941   **11.** 17.583   **12.** 11.4   **13.** 43.86   **14.** 56   **15.** 6.673   **16.** 12,690   **17.** 47.3   **18.** 1.21   **19.** 6.2   **20.** 2.31 sq. mi
21. 198.08 oz   **22.** 18π mi; 56.52 mi   **23.** 86 high-density disks   **24.** 54 mi

CUMULATIVE REVIEW

1. one hundred six million, fifty-two thousand, four hundred forty-seven; Sec. 1.1, Ex. 6   **2.** 445 baseball cards; Sec. 1.2, Ex. 8
3. 726; Sec. 1.3, Ex. 4   **4.** 2300; Sec. 1.4, Ex. 4   **5.** 15,540 thousand bytes; Sec. 1.5, Ex. 7   **6.** 401 R 2; Sec. 1.6, Ex. 8
7. 47; Sec. 1.8, Ex. 11   **8.** numerator; 3; denominator: 7; 2.1, Ex. 1   **9.** $\dfrac{1}{10}$; Sec. 2.3, Ex. 6   **10.** $\dfrac{15}{1}$ or 15; Sec. 2.4, Ex. 9

11. $\dfrac{63}{16}$; Sec. 2.5, Ex. 5   **12.** $\dfrac{15}{4}$ or $3\dfrac{3}{4}$; Sec. 2.4, Ex. 8   **13.** $\dfrac{3}{20}$; Sec. 2.5, Ex. 8   **14.** $\dfrac{7}{9}$; Sec. 3.1, Ex. 4   **15.** $\dfrac{1}{4}$; Sec. 3.1, Ex. 5

16. $\dfrac{15}{20}$; Sec. 3.2, Ex. 8   **17.** $\dfrac{13}{30}$; Sec. 3.3, Ex. 2   **18.** $4\dfrac{7}{40}$ lb; Sec. 3.4, Ex. 7   **19.** $\dfrac{1}{16}$; Sec. 3.5, Ex. 3   **20.** $\dfrac{3}{256}$; Sec. 3.5, Ex. 5

21. $\dfrac{43}{100}$; Sec. 4.1, Ex. 7   **22.** >; Sec. 4.2, Ex. 1   **23.** 11.568; Sec. 4.3, Ex. 4   **24.** 2370.2; Sec. 4.4, Ex. 5   **25.** 768.05; Sec. 4.4, Ex. 8

Chapter 5 RATIO AND PROPORTION

CHAPTER 5 PRETEST

1. $\dfrac{27}{8}$; 5.1A   **2.** $\dfrac{4}{37}$; 5.1A   **3.** $\dfrac{5}{12}$; 5.1B   **4.** $\dfrac{4}{3}$; 5.1B   **5.** $\dfrac{6}{19}$; 5.1B   **6.** $\dfrac{4}{11}$; 5.1B   **7.** $\dfrac{35 \text{ students}}{2 \text{ instructors}}$; 5.2A

8. $\dfrac{7 \text{ c of flour}}{2 \text{ cakes}}$; 5.2A   **9.** $\dfrac{78.75 \text{ km}}{1 \text{ hr}}$ or 78.75 km/hr; 5.2B   **10.** $\dfrac{3 \text{ cookies}}{1 \text{ child}}$ or 3 cookies per child; 5.2B   **11.** 12 oz; 5.2C

12. 12 doughnuts; 5.2C   **13.** $\dfrac{5}{80} = \dfrac{15}{240}$; 5.3A   **14.** false; 5.3B   **15.** true; 5.3B   **16.** $n = 16$; 5.3C   **17.** $n = 4.5$; 5.3C

18. $n = 133$; 5.3C   **19.** \$12,000; 5.4A   **20.** \$472.75; 5.4A

EXERCISE SET 5.1

1. $\dfrac{11}{14}$ **3.** $\dfrac{23}{10}$ **5.** $\dfrac{151}{201}$ **7.** $\dfrac{2.8}{7.6}$ **9.** $\dfrac{5}{7\frac{1}{2}}$ **11.** $\dfrac{3\frac{3}{4}}{1\frac{2}{3}}$ **13.** $\dfrac{2}{3}$ **15.** $\dfrac{77}{100}$ **17.** $\dfrac{463}{821}$ **19.** $\dfrac{3}{4}$ **21.** $\dfrac{5}{12}$ **23.** $\dfrac{8}{25}$ **25.** $\dfrac{12}{7}$

27. $\dfrac{16}{23}$ **29.** $\dfrac{2}{5}$ **31.** $\dfrac{5}{3}$ **33.** $\dfrac{17}{40}$ **35.** $\dfrac{5}{4}$ **37.** $\dfrac{3}{1}$ **39.** $\dfrac{15}{1}$ **41.** $\dfrac{103}{5}$ **43.** $\dfrac{5}{41}$ **45.** 2.3   **47.** 0.15   **49.** $\dfrac{3}{7}$ **51.** $\dfrac{10}{23}$
53. no, the shipment should not be refused

EXERCISE SET 5.2

1. $\dfrac{1 \text{ shrub}}{3 \text{ feet}}$ **3.** $\dfrac{3 \text{ returns}}{20 \text{ sales}}$ **5.** $\dfrac{2 \text{ phone lines}}{9 \text{ employees}}$ **7.** $\dfrac{9 \text{ gal}}{2 \text{ acres}}$ **9.** $\dfrac{3 \text{ flight attendants}}{100 \text{ passengers}}$ **11.** $\dfrac{71 \text{ calories}}{2\text{-fl oz}}$ **13.** 7 riders/car
15. 110 calories/oz   **17.** 6 diapers/baby   **19.** \$50,000/yr   **21.** ≈ 6.67 km/min   **23.** 1,225,000 voters/senator   **25.** 300 good/defective
27. 0.48 dust and dirt/acre   **29.** \$48,000/species   **31.** \$5,134,125/library   **33. (a)** 31.25 computer boards/hr
(b) ≈ 33.3 computer boards/hr   **(c)** Lamont   **35.** \$11.50 per compact disk   **37.** \$0.17 per banana
39. 8 oz: \$0.149 per oz; 12 oz: \$0.133 per oz; 12 oz   **41.** 16 oz: \$0.106 per oz; 6 oz: \$0.115 per oz; 16 oz
43. 12 oz: \$0.191 per oz; 8 oz: \$0.186 per oz; 8 oz   **45.** 100: \$0.006 per napkin; 180: \$0.005 per napkin; 180 napkins   **47.** 10.2   **49.** 4.44
51. 1.9   **53.** Miles driven: 257,352,347; Miles per gallon: 19.2, 22.3, 21.6   **55.** 1.5 steps/ft   **57.** 24 students/teacher

INTEGRATED REVIEW

1. $\dfrac{9}{10}$ **2.** $\dfrac{9}{25}$ **3.** $\dfrac{43}{50}$ **4.** $\dfrac{8}{23}$ **5.** $\dfrac{173}{139}$ **6.** $\dfrac{6}{7}$ **7.** $\dfrac{7}{26}$ **8.** $\dfrac{20}{33}$ **9.** $\dfrac{2}{3}$ **10.** $\dfrac{1}{8}$ **11.** $\dfrac{60}{47}$ **12.** $\dfrac{1015}{202}$

13. $\dfrac{1 \text{ office}}{4 \text{ graduate assistants}}$ **14.** $\dfrac{2 \text{ lights}}{5 \text{ ft}}$ **15.** $\dfrac{2 \text{ Senators}}{1 \text{ state}}$ **16.** $\dfrac{1 \text{ teacher}}{28 \text{ students}}$ **17.** 55 mi/hr   **18.** 140 ft/sec   **19.** 21 employees/fax line
20. 17 phone calls/teenager   **21.** 8 lb: \$0.27 per lb; 18 lb: \$0.28 per lb; 8 lb   **22.** 100: \$0.020 per plate; 500: \$0.018 per plate; 500 paper plates
23. 3 packs: \$0.80 per pack; 8 packs: \$0.75 per pack; 8 packs   **24.** 4: \$0.92 per battery; 10: \$0.99 per battery; 4 batteries

MENTAL MATH

1. true   **3.** false   **5.** true

Exercise Set 5.3

1. $\dfrac{10 \text{ diamonds}}{6 \text{ opals}} = \dfrac{5 \text{ diamonds}}{3 \text{ opals}}$ 3. $\dfrac{3 \text{ printers}}{12 \text{ computers}} = \dfrac{1 \text{ printer}}{4 \text{ computers}}$ 5. $\dfrac{6 \text{ eagles}}{58 \text{ sparrows}} = \dfrac{3 \text{ eagles}}{29 \text{ sparrows}}$ 7. $\dfrac{2\frac{1}{4} \text{ c flour}}{24 \text{ cookies}} = \dfrac{6\frac{3}{4} \text{ c flour}}{72 \text{ cookies}}$

9. $\dfrac{22 \text{ vanilla wafers}}{1 \text{ c cookie crumbs}} = \dfrac{55 \text{ vanilla wafers}}{2.5 \text{ c cookie crumbs}}$ 11. true   13. false   15. true   17. true   19. false   21. true   23. true

25. false   27. 3   29. 5   31. 4   33. 3.2   35. 19.2   37. $\dfrac{9}{20}$ 39. 1   41. 12   43. $\dfrac{3}{4}$ 45. 0.0025   47. 25

49. <   51. >   53. <   55. 0   57. 1400   59. 252.5   61. answers may vary

Exercise Set 5.4

1. 12 passes   3. 165 min   5. 180 students   7. 23 ft   9. 270 sq. ft   11. 56 mi   13. 450 km   15. 24 oz   17. 16 bags

19. $162,000   21. 15 hits   23. 27 people   25. 86 weeks   27. 6 people   29. 112 ft; 11 in. difference   31. $2\frac{2}{3}$ lb   33. 434

emergency room visits   35. $3 \cdot 5$ 37. $2^2 \cdot 5$ 39. $2^3 \cdot 5^2$ 41. 2^5 43. $4\frac{2}{3}$ ft   45. answers may vary

Chapter 5 Review

1. $\dfrac{5}{4}$ 2. $\dfrac{11}{13}$ 3. $\dfrac{9}{40}$ 4. $\dfrac{14}{5}$ 5. $\dfrac{4}{15}$ 6. $\dfrac{1}{2}$ 7. $\dfrac{1}{2}$ 8. $\dfrac{7}{150}$ 9. $\dfrac{1 \text{ stillborn birth}}{125 \text{ live births}}$ 10. $\dfrac{3 \text{ professors}}{10 \text{ assistants}}$ 11. $\dfrac{5 \text{ pages}}{2 \text{ min}}$

12. $\dfrac{4 \text{ computers}}{3 \text{ hr}}$ 13. 52 mi/hr   14. 15 ft/sec   15. $0.31/pear   16. $1.74/diskette   17. 65 km/hr   18. $1\frac{1}{3}$ gal/acre

19. $36.80/course   20. 13 bushels/tree   21. 8-oz size   22. 18-oz size   23. 1-gal size   24. 32-oz size   25. $\dfrac{20 \text{ men}}{14 \text{ women}} = \dfrac{10 \text{ men}}{7 \text{ women}}$

26. $\dfrac{50 \text{ tries}}{4 \text{ successes}} = \dfrac{25 \text{ tries}}{2 \text{ succcsses}}$ 27. $\dfrac{16 \text{ sandwiches}}{8 \text{ players}} = \dfrac{2 \text{ sandwiches}}{1 \text{ player}}$ 28. $\dfrac{12 \text{ tires}}{3 \text{ cars}} = \dfrac{4 \text{ tires}}{1 \text{ car}}$ 29. no   30. yes   31. no

32. yes   33. 5   34. 15   35. 32.5   36. 5.625   37. 32   38. 13.5   39. 60   40. $7\frac{1}{5}$ 41. 0.94   42. 0.36   43. 14

44. 35   45. 8 bags   46. 16 bags   47. no   48. 79 gal   49. $54,600   50. $1023.50   51. $40\frac{1}{2}$ ft   52. $8\frac{1}{4}$ in.

Chapter 5 Test

1. $\dfrac{9}{13}$ 2. $\dfrac{15}{2}$ 3. $\dfrac{7 \text{ men}}{1 \text{ woman}}$ 4. $\dfrac{3 \text{ in.}}{10 \text{ days}}$ 5. 81.25 km/hr   6. $\dfrac{2}{3}$ in./hr   7. 28 students/teacher   8. 8-oz size   9. 16-oz size

10. true   11. false   12. 5   13. $4\frac{4}{11}$ 14. $\dfrac{7}{3}$ 15. 8   16. $49\frac{1}{2}$ ft   17. $3\frac{3}{4}$ hr   18. $53\frac{1}{3}$ g   19. $144\frac{2}{3}$ cartons   20. 14,880 adults

Cumulative Review

1. (a) 3   (b) 5   (c) 0   (d) 7; Sec. 1.3, Ex. 1   2. 249,000; Sec.1.4, Ex. 2   3. 200; Scc. 1.5, Ex. 3   4. $17,820; Sec. 1.7, Ex. 3

5. $3 \cdot 3 \cdot 5$ or $3^2 \cdot 5$; Sec. 2.2, Ex. 3   6. $\dfrac{3}{5}$; Sec. 2.3, Ex. 1   7. $\dfrac{6}{5}$; Sec. 2.4, Ex. 5   8. $\dfrac{2}{5}$; Sec. 2.4, Ex. 6   9. $\dfrac{5}{7}$; Sec. 3.1, Ex. 1

10. 2; Sec. 3.1, Ex. 3   11. 18; Sec. 3.2, Ex. 2   12. $\dfrac{7}{14}$; Sec. 3.2, Ex. 9   13. $\dfrac{8}{33}$; Sec. 3.3, Ex. 6   14. $\dfrac{1}{6}$ of an hr; Sec. 3.3, Ex. 9

15. $7\frac{17}{24}$; Sec. 3.4, Ex. 1   16. >; Sec. 3.5, Ex. 1   17. one and three tenths; Sec. 4.1, Ex. 1   18. 736.2; Sec. 4.2, Ex. 3

19. 25.454; Sec. 4.3, Ex. 1   20. 0.0849; Sec. 4.4, Ex. 2   21. 0.125; Sec. 4.5, Ex. 3   22. 3.7; Sec. 4.6, Ex. 5

23. $\dfrac{4}{9}, \dfrac{9}{20}$, 0.456; Sec. 4.7, Ex. 6   24. $\dfrac{2.6}{3.1}$; Sec. 5.1, Ex. 2   25. $\dfrac{1\frac{1}{2}}{7\frac{3}{4}}$; Sec. 5.1, Ex. 3

Chapter 6 Percent

Chapter 6 Pretest

1. 48%; 6.1A   2. 0.73; 6.1B   3. 0.068; 6.1B   4. 3%; 6.1C   5. 210%; 6.1C   6. $\dfrac{11}{50}$; 6.2A   7. $\dfrac{3}{200}$; 6.2A   8. 90%; 6.2B

9. 220%; 6.2B   10. 0.035; 6.2C   11. $4 = n \cdot 28$; 6.3A   12. $\dfrac{70}{b} = \dfrac{18}{100}$; 6.4A   13. 40; 6.3B, 6.4B   14. 6.4; 6.3B; 6.4B

15. discount: $46; sale price: $184; 6.5A   16. 21%; 6.5B   17. sales tax: $2.52; total price: $44.52; 6.6A   18. 4%; 6.6B   19. $144; 6.7A

20. $103.33; 6.7C

EXERCISE SET 6.1

1. 81%   **3.** 9%   **5.** chocolate chip; 52%   **7.** 12%   **9.** 0.48   **11.** 0.06   **13.** 1   **15.** 0.613   **17.** 0.028   **19.** 0.006
21. 3   **23.** 0.3258   **25.** 0.737   **27.** 0.25   **29.** 0.818   **31.** 98%   **33.** 310%   **35.** 2900%   **37.** 0.3%   **39.** 22%
41. 530%   **43.** 5.6%   **45.** 33.28%   **47.** 300%   **49.** 70%   **51.** 10%   **53.** 70.2%   **55.** 38%   **57.** 0.25   **59.** 0.65
61. 0.9   **63.** home health aides   **65.** 1.3   **67.** answers may vary

MENTAL MATH

1. 13%   **3.** 87%   **5.** 1%

EXERCISE SET 6.2

1. $\frac{3}{25}$ **3.** $\frac{1}{25}$ **5.** $\frac{9}{200}$ **7.** $1\frac{3}{4}$ **9.** $\frac{73}{100}$ **11.** $\frac{1}{8}$ **13.** $\frac{1}{16}$ **15.** $\frac{2}{25}$ **17.** $\frac{31}{300}$ **19.** $\frac{179}{800}$ **21.** 75%   **23.** 70%

25. 40%   **27.** 59%   **29.** 34%   **31.** $37\frac{1}{2}$%   **33.** $31\frac{1}{4}$%   **35.** 160%   **37.** $66\frac{2}{3}$%   **39.** 65%   **41.** 250%   **43.** 190%

45. 63.64%   **47.** 26.67%   **49.** 14.29%   **51.** 91.67%   **53.** $0.35, \frac{7}{20}$; 20%; 0.2; 50%; $\frac{1}{2}$; $0.7, \frac{7}{10}$; 37.5%, 0.375

55. $0.4, \frac{2}{5}$; $23\frac{1}{2}$%, $\frac{47}{200}$; 80%, 0.8; $0.33\overline{3}, \frac{1}{3}$; 87.5%, 0.875; 0.075, $\frac{3}{40}$ **57.** $\frac{47}{250}$ **59.** 0.667   **61.** 8.5%   **63.** 0.0825   **65.** 35%
67. $n = 15$ **69.** $n = 10$ **71.** $n = 12$ **73.** 0.266; 26.6%   **75.** 1.155; 115.5%   **77.** greater   **79.** answers may vary

MENTAL MATH

1. percent: 42; base: 50; amount: 21   **3.** percent: 125; base: 86; amount: 107.5

EXERCISE SET 6.3

1. $15\% \cdot 72 = n$ **3.** $30\% \cdot n = 80$ **5.** $n \cdot 90 = 20$ **7.** $1.9 = 40\% \cdot n$ **9.** $n = 9\% \cdot 43$ **11.** 3.5   **13.** 7.28   **15.** 600
17. 10   **19.** 110%   **21.** 32%   **23.** 1   **25.** 45   **27.** 500   **29.** 400%   **31.** 25.2   **33.** 45%   **35.** 35   **37.** $n = 30$
39. $n = 3\frac{7}{11}$ **41.** $\frac{17}{12} = \frac{n}{20}$ **43.** $\frac{8}{9} = \frac{14}{n}$ **45.** 686.625   **47.** 12,285

MENTAL MATH

1. amount: 12.6; base: 42; percent: 30   **3.** amount: 102; base: 510; percent: 20

EXERCISE SET 6.4

1. $\frac{a}{65} = \frac{32}{100}$ **3.** $\frac{75}{b} = \frac{40}{100}$ **5.** $\frac{70}{200} = \frac{p}{100}$ **7.** $\frac{2.3}{b} = \frac{58}{100}$ **9.** $\frac{a}{130} = \frac{19}{100}$ **11.** 5.5   **13.** 18.9   **15.** 400   **17.** 10

19. 125%   **21.** 28%   **23.** 29   **25.** 1.92   **27.** 1000   **29.** 210%   **31.** 55.18   **33.** 45%   **35.** 85   **37.** $\frac{7}{8}$ **39.** $3\frac{2}{15}$
41. 0.7   **43.** 2.19   **45.** 12,011.226 or 12,011.2 rounded   **47.** 7270.6

INTEGRATED REVIEW

1. 12%   **2.** 68%   **3.** 25%   **4.** 50%   **5.** 520%   **6.** 780%   **7.** 6%   **8.** 44%   **9.** 250%   **10.** 325%   **11.** 3%   **12.** 5%
13. 0.65   **14.** 0.31   **15.** 0.08   **16.** 0.07   **17.** 1.42   **18.** 5.38   **19.** 0.029   **20.** 0.066   **21.** $\frac{3}{100}$ **22.** $\frac{2}{25}$ **23.** $\frac{21}{400}$
24. $\frac{51}{400}$ **25.** $\frac{19}{50}$ **26.** $\frac{9}{20}$ **27.** $\frac{37}{300}$ **28.** $\frac{1}{6}$ **29.** 8.4   **30.** 100   **31.** 250   **32.** 120%   **33.** 28%   **34.** 76   **35.** 11
36. 130%   **37.** 86%   **38.** 37.8   **39.** 150   **40.** 62

EXERCISE SET 6.5

1. 1600 bolts   **3.** 30 hr   **5.** 15%   **7.** 295 components   **9.** 29.2%   **11.** 496 chairs; 6696 chairs   **13.** $136
15. $867.87; $20,153.87   **17.** 8.844 million; 35.644 million   **19.** 10; 25%   **21.** 102; 120%   **23.** 2; 25%   **25.** 120; 75%   **27.** 44%
29. 21.5%   **31.** 18.3%   **33.** 84.8%   **35.** 81.25% or 81.3% rounded   **37.** 4.56   **39.** 11.18   **41.** 58.54   **43.** 8.2%

EXERCISE SET 6.6

1. $7.50   **3.** $858.93   **5.** 9%   **7.** $125.40   **9.** $1917   **11.** $11,500   **13.** $112.35   **15.** 6%   **17.** $49,474.24   **19.** 14%
21. $1888.50   **23.** 3%   **25.** $6.80; $61.20   **27.** $48.25; $48.25   **29.** $75.25; $139.75   **31.** $3255.00; $18,445.00   **33.** $45; $255
35. 1200   **37.** 132   **39.** 16   **41.** $26,838.45

CALCULATOR EXPLORATIONS

1. 1.56051   **3.** 8.06231   **5.** $634.50

EXERCISE SET 6.7

1. $32   **3.** $73.60   **5.** $750   **7.** $33.75   **9.** $700   **11.** $78,125   **13.** $5562.50   **15.** $12,580   **17.** $46,815.40
19. $2327.15   **21.** $58,163.60   **23.** $240.75   **25.** $938.66   **27.** $971.90   **29.** $260.31   **31.** $637.26   **33.** 32 yd   **35.** 35 m
37. answers may vary   **39.** answers may vary

Chapter 6 Review

1. 37% **2.** 77% **3.** 0.83 **4.** 0.75 **5.** 0.735 **6.** 0.015 **7.** 1.25 **8.** 1.45 **9.** 0.005 **10.** 0.007 **11.** 2.00 or 2
12. 4.00 or 4 **13.** 0.2625 **14.** 0.8534 **15.** 260% **16.** 5.5% **17.** 35% **18.** 102% **19.** 72.5% **20.** 25% **21.** 7.6%
22. 8.5% **23.** 75% **24.** 65% **25.** 400% **26.** 900% **27.** $\frac{1}{100}$ **28.** $\frac{1}{10}$ **29.** $\frac{1}{4}$ **30.** $\frac{17}{200}$ **31.** $\frac{51}{500}$ **32.** $\frac{1}{6}$ **33.** $\frac{1}{3}$

34. $1\frac{1}{10}$ **35.** 20% **36.** 70% **37.** $83\frac{1}{3}\%$ **38.** 62.5% **39.** $166\frac{2}{3}\%$ **40.** 125% **41.** 60% **42.** 6.25% **43.** 100,000
44. 8000 **45.** 23% **46.** 114.5 **47.** 3000 **48.** 150% **49.** 418 **50.** 300 **51.** 64.8 **52.** 180% **53.** 110% **54.** 165
55. 66% **56.** 16% **57.** 106.25% **58.** 20.9% **59.** $206,400 **60.** $13.23 **61.** $263.75 **62.** $1.15 **63.** $5000 **64.** $300.38
65. discount: $900; sale price: $2100 **66.** discount: $9; sale price: $81 **67.** $120 **68.** $1320 **69.** $30,104.64 **70.** $17,506.56
71. $80.61 **72.** $32,830.10

Chapter 6 Test

1. 0.85 **2.** 5 **3.** 0.006 **4.** 5.6% **5.** 610% **6.** 35% **7.** $\frac{6}{5}$ **8.** $\frac{77}{200}$ **9.** $\frac{1}{500}$ **10.** 55% **11.** 37.5% **12.** 175%

13. 33.6 **14.** 1250 **15.** 75% **16.** 38.4 lb **17.** $56,750 **18.** $358.43 **19.** 5% **20.** discount: $18; sale price: $102
21. $395 **22.** 1% **23.** $647.50 **24.** $2005.64 **25.** $427

Cumulative Review

1. 206 cases; 12 cans; yes; Sec. 1.7, Ex. 2 **2. (a)** $4\frac{2}{7}$ **(b)** $1\frac{1}{15}$ **(c)** 14; Sec. 2.1, Ex. 9 **3.** $2 \cdot 2 \cdot 2 \cdot 3$ or $2^3 \cdot 3$; Sec. 2.2, Ex. 7

4. $\frac{10}{27}$; Sec. 2.3, Ex. 3 **5.** $\frac{23}{56}$; Sec. 2.4, Ex. 4 **6.** $\frac{8}{11}$; Sec. 2.5, Ex. 2 **7.** $\frac{4}{5}$ in.; Sec. 3.1, Ex. 6 **8.** 60; Sec. 3.2, Ex. 4 **9.** $\frac{2}{3}$; Sec. 3.3, Ex. 1

10. $3\frac{5}{14}$; Sec. 3.4, Ex. 5 **11.** $\frac{7}{16}$; Sec. 3.5, Ex. 6 **12.** $\frac{2}{33}$; Sec. 3.5, Ex. 8 **13.** 0.8; Sec. 4.1, Ex. 12 **14.** 8.7; Sec. 4.1, Ex. 13

15. $1.03; Sec. 4.2; Ex. 5 **16.** 829.6561; Sec. 4.3, Ex. 2 **17.** 18.408; Sec. 4.4, Ex. 1 **18.** 0.0786; Sec. 4.5, Ex. 7

19. 0.012; Sec. 4.5, Ex. 8 **20.** 1.69; Sec. 4.6, Ex. 6 **21.** 0.25; Sec. 4.7, Ex. 1 **22.** $\frac{5 \text{ nails}}{3 \text{ ft}}$; Sec. 5.2, Ex. 1 **23.** no; Sec. 5.3, Ex. 3

24. $17\frac{1}{2}$ mi; Sec. 5.4, Ex. 1 **25.** $n = 25\% \cdot 0.008$; Sec. 6.3, Ex. 3

Chapter 7 Measurement

Pretest

1. 72 in.; 7.1A **2.** 8 ft 2 in.; 7.1B **3.** 14,784 ft; 7.1A **4.** 980 cm; 7.1D **5.** 112 oz; 7.2A **6.** 2.9 g; 7.2C **7.** 600g; 7.2 C

8. $4\frac{1}{2}$ gal; 7.3A **9.** 6.28 kl; 7.3C **10.** 113°F; 7.4A **11.** 25°C; 7.4B **12.** 11,670 ft-lb; 7.5A **13.** 13 ft 5 in.; 7.1C

14. 4 yd 2 ft; 7.1C **15.** 48.6 mm; 7.1E **16.** 1 ton 825 lb; 7.2B **17.** 20.991 g or 20,991 mg; 7.2D **18.** 66 qt 1 pt; 7.3B
19. 9 c; 7.3B **20.** $0.325; 7.3D

Calculator Explorations

1. ≈ 22.96 ft **3.** ≈ 21.59 cm **5.** ≈ 3.1 mi

Mental Math

1. 1 ft **3.** 2 ft **5.** 1 yd **7.** no **9.** yes **11.** no

Exercise Set 7.1

1. 5 ft **3.** 36 ft **5.** 8 mi **7.** $8\frac{1}{2}$ ft **9.** $3\frac{1}{3}$ yd **11.** 33,792 ft **13.** 13 yd 1 ft **15.** 3 ft 5 in. **17.** 1 mi 4720 ft

19. 62 in. **21.** 17 ft **23.** 84 in. **25.** 12 ft 3 in. **27.** 22 yd 1 ft **29.** 8 ft 5 in. **31.** 5 ft 6 in. **33.** 3 ft 4 in. **35.** 50 yd 2 ft

37. 10 ft 6 in. **39.** 18 ft 5 in. **41.** 15 ft 9 in. **43.** 3 ft 1 in. **45.** 86 ft 6 in. **47.** $105\frac{1}{3}$ yd **49.** 4000 cm **51.** 4.0 cm

53. 0.3 km **55.** 1.4 m **57.** 15 m **59.** 83 mm **61.** 0.201 dm **63.** 40 mm **65.** 8.94 m **67.** 2.94 m or 2940 mm
69. 1.29 cm or 12.9 mm **71.** 12.640 km or 12,640 m **73.** 54.9 m **75.** 1.55 km **77.** 9.12 m **79.** 26.7 mm
81. 41.25 m or 4125 cm **83.** 3.35 m **85.** 6.009 km or 6009 m **87.** 15 tiles **89.** 21% **91.** 13% **93.** 25%
95. answers may vary **97.** 6.575 m

Calculator Explorations

1. ≈ 425.25 g **3.** ≈ 15.4 lb **5.** ≈ 0.175 oz

MENTAL MATH

1. 1 lb **3.** 2000 lb **5.** 16 oz **7.** 1 ton **9.** no **11.** yes **13.** no

EXERCISE SET 7.2

1. 32 oz **3.** 10,000 lb **5.** 6 tons **7.** $3\frac{3}{4}$ lb **9.** $1\frac{3}{4}$ tons **11.** 260 oz **13.** 9800 lb **15.** 76 oz **17.** 1.5 tons **19.** 53 lb 10 oz
21. 9 tons 390 lb **23.** 3 tons 175 lb **25.** 8 lb 11 oz **27.** 31 lb 2 oz **29.** 1 ton 700 lb **31.** 5 lb 8 oz **33.** 35 lb 14 oz
35. 130 lb **37.** 211 lb **39.** 5 lb 1 oz **41.** 0.5 kg **43.** 4000 mg **45.** 25,000 g **47.** 0.048 g **49.** 0.0063 kg **51.** 15,140 mg
53. 4010 g **55.** 13.5 mg **57.** 5.815 g or 5815 mg **59.** 1850 mg or 1.850 g **61.** 1360 g or 1.360 kg **63.** 13.52 kg **65.** 2.125 kg
67. 8.064 kg **69.** 30 mg **71.** 250 mg **73.** 144 mg **75.** 6.12 kg **77.** 850 g or 0.85 kg **79.** 2.38 kg **81.** 0.25 **83.** 0.16
85. 0.875 **87.** answers may vary

CALCULATOR EXPLORATIONS

1. ≈ 4.73 L **3.** ≈ 4.62 gal **5.** ≈ 3.785 L

MENTAL MATH

1. 1 pt **3.** 1 gal **5.** 1 qt **7.** 1 c **9.** 2 c **11.** 4 qt **13.** no **15.** no

EXERCISE SET 7.3

1. 4 c **3.** 16 pt **5.** $2\frac{1}{2}$ gal **7.** 5 pt **9.** 8 c **11.** $3\frac{3}{4}$ qt **13.** 768 fl oz **15.** 9 c **17.** 22 pt **19.** 10 gal 1 qt
21. 4 c 4 fl oz **23.** 1 gal 1 qt **25.** 2 gal 3 qt 1 pt **27.** 2 qt 1 c **29.** 17 gal **31.** 4 gal 3 qt **33.** 48 fl oz **35.** 2 qt **37.** yes
39. 2 qt 1 c **41.** 9 qt **43.** 5000 ml **45.** 4.5 L **47.** 0.41 kl **49.** 0.064 L **51.** 160 L **53.** 3600 ml **55.** 0.00016 kl
57. 22.5 L **59.** 4.5 L or 4500 ml **61.** 8410 ml or 8.410 L **63.** 10,600 ml or 10.6 L **65.** 3840 ml **67.** 162.4 L **69.** 1.59 L

71. 18.954 L **73.** $0.316 **75.** 474 ml **77.** $\frac{7}{10}$ **79.** $\frac{3}{100}$ **81.** $\frac{3}{500}$ **83.** answers may vary

INTEGRATED REVIEW

1. 3 ft **2.** 2 mi **3.** $6\frac{2}{3}$ yd **4.** 18 ft **5.** 11,088 ft **6.** 38.4 in. **7.** 3000 cm **8.** 2.4 cm **9.** 2 m **10.** 18 m
11. 72 mm **12.** 0.6 km **13.** 10,000 lb **14.** 5.5 tons **15.** 136 oz **16.** 2.5 lb **17.** 3.5 lb **18.** 80 oz **19.** 28,000 g
20. 1.4 g **21.** 0.0056 kg **22.** 6000 g **23.** 0.6 g **24.** 0.0036 kg **25.** 12 pt **26.** 2.5 qt **27.** 3.5 gal **28.** 8.5 pt **29.** 7 c
30. 6.5 gal **31.** 7000 ml **32.** 0.35 kl **33.** 0.047 L **34.** 970 L **35.** 126 L **36.** 0.075 L

MENTAL MATH

1. yes **3.** no **5.** no **7.** yes

EXERCISE SET 7.4

1. 5°C **3.** 40°C **5.** 140°F **7.** 239°F **9.** 16.7°C **11.** 61.2°C **13.** 197.6°F **15.** 61.3°F **17.** 56.7°C **19.** 80.6°F
21. 21.1°C **23.** 37.9°C **25.** 244.4°F **27.** 260°C **29.** 462.2°C **31.** 62 m **33.** 15 ft **35.** 10.4 m **37.** 4988°C

MENTAL MATH

1. 30 ft-lb **3.** 60 ft-lb **5.** 90 calories **7.** 5 calories

EXERCISE SET 7.5

1. 1140 ft-lb **3.** 3696 ft-lb **5.** 425,000 ft-lb **7.** 23,340 ft-lb **9.** 778,000 ft-lb **11.** 15,560,000 ft-lb **13.** 26,553,140 ft-lb

15. 10,283 BTU **17.** 805 calories **19.** 750 calories **21.** 1440 calories **23.** 2.6 hr **25.** 17.5 mi **27.** $\frac{4}{5}$ **29.** $\frac{3}{5}$ **31.** $\frac{9}{10}$

33. 92.925 ft-lb **35.** 256 lb

CHAPTER 7 REVIEW

1. 9 ft **2.** 24 yd **3.** 13,200 ft **4.** 75 in. **5.** 17 yd 1 ft **6.** 3 ft 10 in. **7.** 4200 cm **8.** 820 mm **9.** 0.01218 m
10. 0.00231 km **11.** 21 yd 1 ft **12.** 7 ft 5 in. **13.** 41 ft 3 in. **14.** 3 ft 8 in. **15.** 9.5 cm or 95 mm **16.** 5.26 m or 526 cm
17. 9117 m or 9.117 km **18.** 1.1 m or 1100 mm **19.** 169 yd 2 ft **20.** 126 ft 8 in. **21.** 108.5 km **22.** 0.24 sq. m **23.** 4.125 lb
24. 4600 lb **25.** 3 lb 4 oz **26.** 4 tons 200 lb **27.** 1.4 g **28.** 40,000 g **29.** 21 dag **30.** 0.0003 dg **31.** 3 lb 9 oz
32. 10 tons 800 lb **33.** 2 tons 750 lb **34.** 33 lb 8 oz **35.** 4.9 g or 4900 mg **36.** 9 kg or 9000 g **37.** 8.1 g or 8100 mg
38. 50.4 kg **39.** 4 lb 4 oz **40.** 9 tons 1075 lb **41.** 7.85 kg **42.** 1.1625 kg **43.** 8 qt **44.** 5 c **45.** 27 qt **46.** 17 c
47. 4 qt 1 pt **48.** 3 gal 3 qt **49.** 3800 ml **50.** 0.042 dl **51.** 1.4 kl **52.** 3060 cl **53.** 1 gal 1 qt **54.** 7 gal 1 qt
55. 736 ml or 0.736 L **56.** 15.5 L or 15,500 ml **57.** 2 gal 3 qt **58.** 6 fl oz **59.** 10.88 L **60.** yes **61.** 473°F **62.** 320°F
63. 107.6°F **64.** 186.8°F **65.** 34°C **66.** 11°C **67.** 5.2°C **68.** 26.7°C **69.** 1.7°C **70.** 329°F **71.** 67.2 ft-lb **72.** 136.5 ft-lb
73. 108,000 ft-lb **74.** 9,336,000 ft-lb **75.** 2600 BTU **76.** 1125 calories **77.** 15,120 calories **78.** 2 hr 20 min

Chapter 7 Test

1. 23 ft 4 in.   **2.** 10 qt   **3.** 1.875 lb   **4.** 5600 lb   **5.** $4\frac{3}{4}$ gal   **6.** 0.04 g   **7.** 2400 g   **8.** 36 mm   **9.** 0.43 g   **10.** 830 ml
11. 1 gal 2 qt   **12.** 3 lb 13 oz   **13.** 8 ft 3 in.   **14.** 2 gal 3 qt   **15.** 66 mm or 6.6 cm   **16.** 2.256 km or 2256 m   **17.** 28.9°C
18. 54.7°F   **19.** 5.6 m   **20.** 4 gal 3 qt   **21.** 105.8°F   **22.** 91.4 m   **23.** 16 ft 6 in.   **24.** 679 ft-lb   **25.** 20,228,000 ft-lb
26. 900 calories

Cumulative Review

1. 2010; Sec. 1.2, Ex. 4   **2.** $2 \cdot 2 \cdot 2 \cdot 2 \cdot 5$ or $2^4 \cdot 5$; Sec. 2.2, Ex. 5   **3.** 33; Sec. 3.2, Ex. 7   **4.** $5\frac{1}{15}$; Sec. 3.4; Ex. 2   **5.** $\frac{1}{8}$; Sec. 4.1, Ex. 9

6. $105\frac{83}{1000}$; Sec. 4.1, Ex. 11   **7.** $<$; Sec. 4.2, Ex. 2   **8.** 67.69; Sec. 4.3, Ex. 6   **9.** 4.21; Sec. 4.4, Ex. 7   **10.** 0.0092; Sec. 4.4, Ex. 9

11. 7.53; Sec. 4.5, Ex. 2   **12.** 200 mi; Sec. 4.6, Ex. 4   **13.** $0.\overline{6}$; Sec. 4.7, Ex. 3   **14.** $\frac{12}{17}$; Sec. 5.1, Ex. 1   **15.** 24.5 mi/gal; Sec. 5.2, Ex. 5

16. $n = \frac{35}{6}$ or $5\frac{5}{6}$; Sec. 5.3, Ex. 6   **17.** 7 bags; Sec. 5.4, Ex. 3   **18.** 0.23; Sec. 6.1, Ex. 3   **19.** 8.33%; Sec. 6.2, Ex. 9   **20.** 14; Sec. 6.3, Ex. 7

21. $\frac{75}{30} = \frac{p}{100}$; Sec. 6.4, Ex. 5   **22.** 18%, Sec. 6.5, Ex. 5   **23.** discount: $16.25; sale price: $48.75, Sec. 6.6, Ex. 5   **24.** $120; Sec. 6.7, Ex. 1

25. $4\frac{1}{2}$ tons; Sec. 7.2, Ex. 1

Chapter 8 Geometry

Chapter 8 Pretest

1. obtuse; 8.1B   **2.** acute; 8.1B   **3.** 36°; 8.1C   **4.** $x = 120°$, $y = 60°$, $z = 120°$; 8.1D   **5.** 70°; 8.2A   **6.** 24.4 in.; 8.2B

7. 20 m; 8.3A   **8.** 56.52 cm; 8.3B   **9.** 15 sq. ft; 8.4A   **10.** 510 cu. in.; 8.5A   **11.** $\frac{2048}{3}\pi$ cu. cm; 8.5A   **12.** 6; 8.6A   **13.** $\frac{9}{7}$; 8.6A

14. 8.544 yd; 8.6C   **15.** $\frac{14}{5}$; 8.7C

Exercise Set 8.1

1. line; line yz or $\overleftrightarrow{yz}$ **3.** line segment; line segment LM or $\overline{LM}$ **5.** line segment; line segment PQ or $\overline{PQ}$ **7.** ray; ray UW or $\overrightarrow{UW}$
9. 15°   **11.** 50°   **13.** 65°   **15.** 95°   **17.** 90°   **19.** 0°; 90°   **21.** straight   **23.** right   **25.** obtuse   **27.** right   **29.** 73°
31. 163°   **33.** 45°   **35.** 55°   **37.** $\angle MNP$ and $\angle RNO$; $\angle PNQ$ and $\angle QNR$ **39.** $\angle SPT$ and $\angle TPQ$; $\angle SPR$ and $\angle RPQ$; $\angle SPT$ and
$\angle SPR$; $\angle TPQ$ and $\angle QPR$ **41.** 32°   **43.** 75°   **45.** $\angle x = 35°$; $\angle y = 145°$; $\angle z = 145°$ **47.** $\angle x = 77°$; $\angle y = 103°$; $\angle z = 77°$
49. $\angle x = 100°$; $\angle y = 80°$; $\angle z = 100°$ **51.** $\angle x = 134°$, $\angle y = 46°$; $\angle z = 134°$ **53.** $\frac{9}{8}$ or $1\frac{1}{8}$ **55.** $\frac{7}{32}$ **57.** $\frac{5}{6}$ **59.** $\frac{4}{3}$ or $1\frac{1}{3}$
61. 54.8°   **63.** $\angle a = 60°$; $\angle b = 50°$; $\angle c = 110°$; $\angle d = 70°$; $\angle e = 120°$

Exercise Set 8.2

1. equilateral   **3.** scalene   **5.** isosceles   **7.** 25°   **9.** 13°   **11.** 40°   **13.** diameter   **15.** rectangle   **17.** parallelogram
19. hypotenuse   **21.** true   **23.** true   **25.** false   **27.** 14 m   **29.** 14.5 cm   **31.** pentagon   **33.** hexagon   **35.** cylinder
37. rectangular solid   **39.** cone   **41.** 14.8 in.   **43.** 13 mi   **45.** cube   **47.** rectangular solid   **49.** sphere   **51.** pyramid
53. 114.4   **55.** 108   **57.** 72,368 mi   **59.** answers may vary

Exercise Set 8.3

1. 64 ft   **3.** 36 cm   **5.** 21 in.   **7.** 120 cm   **9.** 48 ft   **11.** 66 in.   **13.** 21 ft   **15.** 60 ft   **17.** 346 yd   **19.** 22 ft   **21.** $66
23. 36 in.   **25.** 28 in.   **27.** $24.08   **29.** 96 m   **31.** 66 ft   **33.** 128 mi   **35.** 17π cm; 53.38 cm   **37.** 16π mi; 50.24 mi
39. 26π m; 81.64 m   **41.** 31.43 ft   **43.** 12,560 ft   **45.** 23   **47.** 1   **49.** 10   **51.** 216   **53.** perimeter   **55.** area   **57.** area
59. perimeter   **61. a.** 62.8 m; 125.6 m   **b.** yes   **63.** $44 + 10\pi \approx 75.4$ m

Exercise Set 8.4

1. 7 sq. m   **3.** $9\frac{3}{4}$ sq. yd   **5.** 15 sq. yd   **7.** 2.25π sq. in. ≈ 7.065 sq. in.   **9.** 36.75 sq. ft   **11.** 28 sq. m   **13.** 22 sq. yd

15. $36\frac{3}{4}$ sq. ft   **17.** $22\frac{1}{2}$ sq. in.   **19.** 25 sq. cm   **21.** 86 sq. mi   **23.** 24 sq. cm   **25.** 36π sq. in. ≈ 113.1 sq. in.   **27.** 168 sq. ft

29. 5000 sq. ft   **31.** 128 sq. in.; $\frac{8}{9}$ sq. ft   **33.** 510 sq. in.   **35.** 168 sq. ft   **37.** 9200 sq. ft   **39.** 381 sq. ft   **41.** 14π in. ≈ 43.96 in.

43. 25 ft   **45.** $12\frac{3}{4}$ ft   **47.** 12-in. pizza   **49.** $1\frac{1}{3}$ sq. ft; 192 sq. in.

EXERCISE SET 8.5

1. 72 cu. in.   **3.** 512 cu. cm   **5.** $12\frac{4}{7}$ cu. yd   **7.** $523\frac{17}{21}$ cu. in.   **9.** 75 cu. cm   **11.** $2\frac{10}{27}$ cu. in.   **13.** 8.4 cu. ft   **15.** $10\frac{5}{6}$ cu. in.

17. 960 cu. cm   **19.** $\frac{1372}{3}\pi$ cu. in. or $\left(457\frac{1}{3}\right)\pi$ cu. in.   **21.** $7\frac{1}{2}$ cu. ft   **23.** $12\frac{4}{7}$ cu. cm   **25.** 25   **27.** 9   **29.** 5   **31.** 20

33. 2,583,283 cu. m   **35.** 2,583,669 cu. m   **37.** 26,696.5 cu. ft

INTEGRATED REVIEW

1. 153°; 63°   **2.** $\angle x = 75°$; $\angle y = 105°$; $\angle z = 75°$ **3.** $\angle x = 128°$; $\angle y = 52°$; $\angle z = 128°$ **4.** $\angle x = 52°$ **5.** 20 m; 25 sq. m
6. 12 ft; 6 sq. ft   **7.** 6π cm $\approx$ 18.84 cm; 9π sq. cm $\approx$ 28.26 sq. cm   **8.** 32 mi; 44 sq. mi   **9.** 62 ft; 238 sq. ft   **10.** 64 cu. in.

11. 30.6 cu. ft   **12.** 400 cu. cm   **13.** $\frac{2048}{3}\pi$ cu. mi $\approx$ $2145\frac{11}{21}$ cu. mi

CALCULATOR EXPLORATIONS

1. 32   **3.** 3.873   **5.** 9.849

EXERCISE SET 8.6

1. 2   **3.** 25   **5.** $\frac{1}{9}$ **7.** $\frac{12}{8} = \frac{3}{2}$ **9.** 16   **11.** $\frac{3}{2}$ **13.** 1.732   **15.** 3.873   **17.** 3.742   **19.** 6.856   **21.** 2.828   **23.** 5.099
25. 8.426   **27.** 2.646   **29.** 13 in.   **31.** 6.633 cm   **33.** 5   **35.** 8   **37.** 17.205   **39.** 16.125   **41.** 12   **43.** 44.822   **45.** 42.426
47. 1.732   **49.** 141.42 yd   **51.** 25.0 ft   **53.** 340 ft   **55.** $n = 4$ **57.** $n = 45$ **59.** $n = 6$ **61.** 6, 7   **63.** 10, 11

EXERCISE SET 8.7

1. congruent   **3.** congruent   **5.** $\frac{2}{1}$ **7.** $\frac{3}{2}$ **9.** 4.5   **11.** 6   **13.** 5   **15.** 13.5   **17.** 17.5   **19.** 8   **21.** 21.25   **23.** 50 ft
25. $x = 4.4$ ft; $y = 5.6$ ft   **27.** 14.4 ft   **29.** 17.5   **31.** 81   **33.** 8.4   **35.** answers may vary

CHAPTER REVIEW

1. right   **2.** straight   **3.** acute   **4.** obtuse   **5.** 65°   **6.** 75°   **7.** 108°   **8.** 89°   **9.** 58°   **10.** 98°   **11.** 90°   **12.** 25°
13. 133° and 47°   **14.** 43° and 47°; 58° and 32°   **15.** $\angle x = 80°$; $\angle y = 100°$; $\angle z = 100°$ **16.** $\angle x = 155°$; $\angle y = 155°$; $\angle z = 25°$
17. $\angle x = 53°$; $\angle y = 53°$; $\angle z = 127°$ **18.** $\angle x = 42°$; $\angle y = 42°$; $\angle z = 138°$ **19.** 103°   **20.** 60°   **21.** 60°   **22.** 65°   **23.** 4.2 m
24. 7 ft   **25.** 9.5 m   **26.** 15.2 cm   **27.** cube   **28.** cylinder   **29.** pyramid   **30.** rectangular solid   **31.** 18 in.   **32.** 2.35 m
33. pentagon   **34.** hexagon   **35.** 88 m   **36.** 30 cm   **37.** 36 m   **38.** 90 ft   **39.** 32 ft   **40.** 440 ft   **41.** 5.338 in.
42. 31.4 yd   **43.** 240 sq. ft   **44.** 140 sq. m   **45.** 600 sq. cm   **46.** 189 sq. yd   **47.** 49π sq. ft $\approx$ 153.86 sq. ft

48. 4π sq. in. $\approx$ 21.56 sq. in.   **49.** 119 sq. in.   **50.** 1248 sq. cm   **51.** 144 sq. m   **52.** 432 sq. ft   **53.** 130 sq. ft   **54.** $15\frac{5}{8}$ cu. in.

55. 84 cu. ft   **56.** $62,857\frac{1}{7}$ cu. cm   **57.** $346\frac{1}{2}$ cu. in.   **58.** $2\frac{2}{3}$ cu. ft   **59.** 307.72 cu. in.   **60.** $7\frac{1}{2}$ cu. ft   **61.** 0.5π cu. ft

62. 1260 cu. ft   **63.** 28.728 cu. ft   **64.** 8   **65.** 12   **66.** 6   **67.** 1   **68.** $\frac{2}{5}$ **69.** $\frac{1}{10}$ **70.** 13   **71.** 29   **72.** 10.7   **73.** 93

74. 86.6   **75.** 28.28 cm   **76.** 88.2 ft   **77.** $37\frac{1}{2}$ **78.** $13\frac{1}{3}$ **79.** $17\frac{2}{5}$ **80.** $6\frac{1}{2}$ **81.** 33 ft   **82.** $x = \frac{5}{6}$ in.; $y = 2\frac{1}{6}$ in.

CHAPTER 8 TEST

1. 12°   **2.** 56°   **3.** 50°   **4.** $\angle x = 118°$; $\angle y = 62°$; $\angle z = 118°$ **5.** $\angle x = 73°$; $\angle y = 73°$; $\angle z = 73°$ **6.** 6.2 m   **7.** 10 in.
8. 26°   **9.** circumference $= 18\pi \approx 56.52$ in.; area $= 81\pi \approx 254.34$ sq. in.   **10.** perimeter $= 24.6$ yd; area $= 37.1$ sq. yd

11. perimeter $= 68$ in.; area $= 185$ sq. in.   **12.** $62\frac{6}{7}$ cu. in.   **13.** 30 cu. ft   **14.** 7   **15.** 12.530   **16.** $\frac{8}{10} = \frac{4}{5}$ **17.** 16 in.

18. 18 cu. ft   **19.** 62 ft   **20.** 5.66 cm   **21.** 198.08 oz   **22.** 7.5   **23.** 69 ft

CUMULATIVE REVIEW

1. nineteen and five thousand twenty-three ten-thousandths; Sec. 4.1, Ex. 3   **2.** 736.24; Sec. 4.2, Ex. 4   **3.** 47.06; Sec. 4.3, Ex. 3
4. 76.8; Sec. 4.4, Ex. 4   **5.** 76,300; Sec. 4.4, Ex. 6   **6.** 38.6; Sec. 4.5, Ex. 1   **7.** 6.35; Sec. 4.6, Ex. 7   **8.** >; Sec. 4.7, Ex. 4
9. $\frac{26}{31}$; Sec. 5.1, Ex. 5   **10.** yes; Sec. 5.3, Ex. 2   **11.** 17%; Sec. 6.1, Ex. 1   **12.** $\frac{19}{1000}$; Sec. 6.2, Ex. 2   **13.** $\frac{5}{4}$ or $1\frac{1}{4}$; Sec. 6.2, Ex. 3
14. 255; Sec. 6.3, Ex. 8   **15.** 52; Sec. 6.4, Ex. 9   **16.** 775 freshmen; Sec. 6.5, Ex. 1   **17.** $1710; Sec. 6.6, Ex. 3
18. 96 in.; Sec. 7.1, Ex. 1   **19.** 3200 g; Sec. 7.2, Ex. 7   **20.** 3 gal 3 qt; Sec. 7.3, Ex. 3   **21.** 84.2° F; Sec. 7.4, Ex. 2

22. 50°; Sec. 8.2, Ex. 1   **23.** 28 in.; Sec. 8.3, Ex. 1   **24.** $\frac{2}{5}$; Sec. 8.6, Ex. 3   **25.** $\frac{12}{17}$; Sec. 8.7, Ex. 2

Chapter 9 STATISTICS AND PROBABILITY

PRETEST

1. April; 9.1D   **2.** July; 9.1D   **3.** 700; 9.1D   **4.** 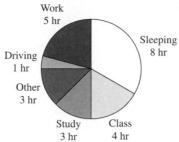 ; 9.2B   **5.** ‖; 2   **6.** ‖‖; 4   **7.** ‖‖; 4
8. ⑷ |; 6

Work
5 hr

Sleeping
8 hr

Driving
1 hr

Other
3 hr

Study
3 hr

Class
4 hr

9. 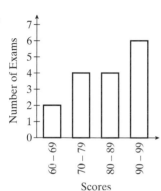 ; 9.3B   **10.** 2.65; 9.4A   **11.** 64; 9.4B   **12.** $\frac{1}{6}$; 9.5B   **13.** $\frac{1}{2}$; 9.5B   **14.** $\frac{1}{3}$; 9.5B

EXERCISE SET 9.1

1. 1997   **3.** 4000   **5.** 1993, 1994, 1998   **7.** 1992, 1995   **9.** 21 oz   **11.** 1994, 1996   **13.** 3 oz per week
15. consumption of chicken is increasing   **17.** April   **19.** 12   **21.** February, March, April, May, June
23. Tokyo, 26.5 million or 26,500,000   **25.** New York City, 16.2 million or 16,200,000   **27.** 12 million or 12,000,000
29. 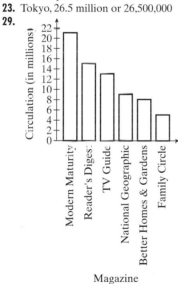   **31.** 54.5   **33.** 1975   **35.** increase   **37.** 3.6   **39.** 6.2   **41.** 25%   **43.** 34%
45. 83°F   **47.** Sunday, 68°F   **49.** Tuesday, 13°F

EXERCISE SET 9.2

1. parent or guardian's home   **3.** $\frac{9}{35}$ **5.** $\frac{9}{16}$ **7.** less than $20   **9.** 53%   **11.** 1410 teenagers   **13.** 1316 teenagers   **15.** 55%
17. nonfiction   **19.** 31,400 books   **21.** 27,632 books   **23.** 25,120 books

25.

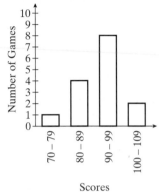

3 – 10 days 18%
Under 3 days 7%
21 or more days 61%
11 – 20 days 14%

27. $2^2 \times 5$ **29.** $2^3 \times 5$ **31.** 5×17 **33.** 129,600,002 sq. km
35. 55,542,858 sq. km

EXERCISE SET 9.3

1. 15 adults   **3.** 61 adults   **5.** 24 adults   **7.** 12 adults   **9.** $\dfrac{9}{100}$ **11.** 45–54   **13.** 17 million householders
15. 45 million householders   **17.** answers may vary   **19.** |; 1   **21.** ⧠⧠; 8   **23.** ⧠|; 6   **25.** ⧠|; 6   **27.** ||; 2
29.

Number of Games (y-axis, 0–10)
Scores (x-axis): 70 – 79, 80 – 89, 90 – 99, 100 – 109

31. 90   **33.** 20   **35.** 27   **37.** Asia   **39.** answers may vary

INTEGRATED REVIEW

1. 65 lb   **2.** 73 lb   **3.** 1985   **4.** 1990 and 1995   **5.** Oroville Dam, 755 ft   **6.** New Bullards Bar Dam, 635 ft   **7.** 15 ft   **8.** 4 dams
9. Thursday and Saturday, 100°F   **10.** Monday, 82°F   **11.** Sunday, Monday, and Tuesday   **12.** Wednesday, Thursday, Friday, and Saturday
13. 70 qt containers   **14.** 52 qt containers   **15.** 2 qt containers   **16.** 6 qt containers   **17.** ||; 2   **18.** |; 1   **19.** |||; 3
20. ⧠|; 6   **21.** ⧠; 5   **22.**

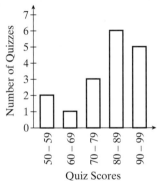

Number of Quizzes (y-axis, 0–7)
Quiz Scores (x-axis): 50 – 59, 60 – 69, 70 – 79, 80 – 89, 90 – 99

MENTAL MATH

1. 4   **3.** 3

EXERCISE SET 9.4

1. mean: 29; median: 28; no mode   **3.** mean: 8.1; median: 8.2; mode: 8.2   **5.** mean: 0.6; median: 0.6; mode: 0.2 and 0.6
7. mean: 370.9; median: 313.5; no mode   **9.** 1313.2 ft   **11.** 1131.5 ft   **13.** 2.79   **15.** 3.46   **17.** 6.8   **19.** 6.9   **21.** 85.5
23. 73   **25.** 70 and 71   **27.** 9   **29.** $\dfrac{1}{3}$ **31.** $\dfrac{3}{5}$ **33.** $\dfrac{11}{15}$ **35.** 35, 35, 37, 43

MENTAL MATH

1. $\dfrac{1}{2}$ **3.** $\dfrac{1}{2}$

ANSWERS

EXERCISE SET 9.5

1.

Outcomes

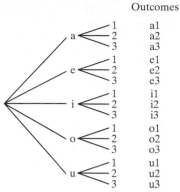

a1
a2
a3
e1
e2
e3
i1
i2
i3
o1
o2
o3
u1
u2
u3

15 outcomes

3.

Outcomes

Red — Red
Blue — Blue
Yellow — Yellow

3 outcomes

5.

Outcomes

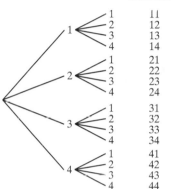

11
12
13
14
21
22
23
24
31
32
33
34
41
42
43
44

16 outcomes

7.

Outcomes

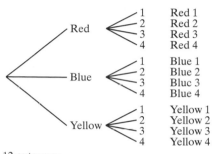

Red 1
Red 2
Red 3
Red 4
Blue 1
Blue 2
Bluc 3
Blue 4
Yellow 1
Yellow 2
Yellow 3
Yellow 4

12 outcomes

9.

Outcomes

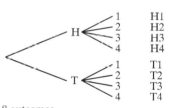

H1
H2
H3
H4
T1
T2
T3
T4

8 outcomes

11. $\frac{1}{6}$ **13.** $\frac{1}{3}$ **15.** $\frac{1}{2}$ **17.** $\frac{1}{3}$ **19.** $\frac{2}{3}$ **21.** $\frac{1}{7}$ **23.** $\frac{2}{7}$ **25.** $\frac{5}{6}$ **27.** $\frac{1}{6}$

29. $\frac{20}{3}$ or $6\frac{2}{3}$ **31.** $\frac{1}{52}$ **33.** $\frac{1}{13}$ **35.** $\frac{1}{4}$ **37.** $\frac{1}{12}$ **39.** 0

CHAPTER 9 REVIEW

1. 13,000   **2.** 15,000   **3.** Utah and Alabama   **4.** Maine   **5.** Kansas, Utah, and Alabama   **6.** Maine and Delaware   **7.** 8%
8. 1996   **9.** 1980, 1990, 1996   **10.** answers may vary   **11.** 13   **12.** 28   **13.** 1979, 1984   **14.** 1979   **15.** answers may vary

16. House Note   **17.** Utilities   **18.** $1225   **19.** $700   **20.** $\frac{39}{160}$ **21.** $\frac{7}{40}$ **22.** $\frac{5}{7}$ **23.** 20 states   **24.** 10 states

25. 26 states   **26.** 23 states   **27.** 4 employees   **28.** 1 employee   **29.** 9 employees   **30.** 18 employees   **31.** ⦀⦀ ; 5   **32.** ⦀ ; 3
33. ⦀⦀ ; 4   **34.**

      35. mean: 17.8; median: 14; no mode   **36.** mean: 55.2; median: 60; no mode
      37. mean: $24,500; median: $20,000; mode: $20,000
      38. mean: 447.3; median: 420; mode: 400   **39.** 3.25   **40.** 2.57

[Bar graph for problem 34 — y-axis: Record Highs (0–6), x-axis: Temperatures, bars at 80°–89° = 5, 90°–99° = 3, 100°–109° = 4]

41.

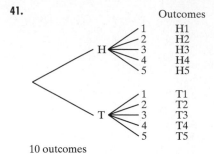

42.

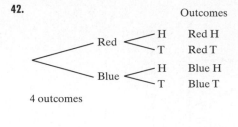

43.

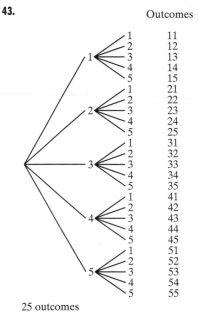

25 outcomes

44.

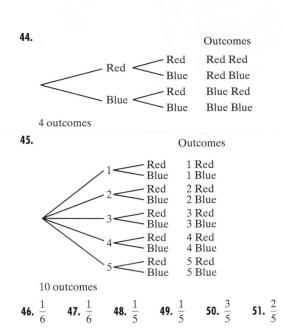

46. $\frac{1}{6}$ **47.** $\frac{1}{6}$ **48.** $\frac{1}{5}$ **49.** $\frac{1}{5}$ **50.** $\frac{3}{5}$ **51.** $\frac{2}{5}$

CHAPTER 9 TEST

1. $225   **2.** 3rd week; $350   **3.** $1100   **4.** June, August, and September   **5.** February, 3 cm   **6.** March and November

7.

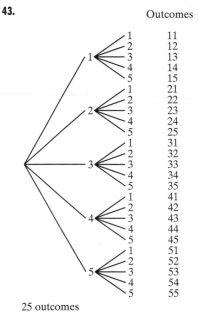

8. 110 million   **9.** 1999   **10.** 1996 and 1997   **11.** $\frac{17}{40}$ **12.** $\frac{31}{22}$ **13.** 2,233,440 small cars

14. 1,240,800 luxury cars   **15.** 9 students   **16.** 11 students

17.

Class Interval (scores)	Tally	Class Frequency (number of students)
40–49			1
50–59	|||	3
60–69	||||	4
70–79	||||	5
80–89	|||| |||	8
90–99	||||	4

18.
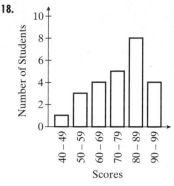

19. mean: 38.4; median: 42; no mode   **20.** mean: 12.625; median: 21.5; mode: 12 and 16   **21.** 3.07

22.
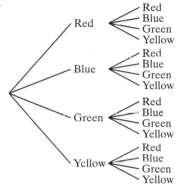

23.
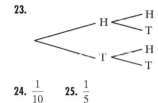

24. $\frac{1}{10}$ **25.** $\frac{1}{5}$

CUMULATIVE REVIEW

1. 28; Sec. 1.8, Ex. 10   **2.** $\frac{5}{18}$; Sec. 2.3, Ex. 4   **3.** $8\frac{3}{10}$; Sec. 3.4, Ex. 3   **4.** 8.4 sq. ft; Sec. 4.7, Ex. 7   **5.** $\frac{2}{3}$; Sec. 5.1, Ex. 4

6. $\frac{\$180}{1\ \text{week}}$; Sec. 5.2, Ex. 2   **7.** $\frac{45\ \text{mi}}{2\ \text{gal}}$; Sec. 5.2, Ex. 3   **8.** yes; Sec. 5.3, Ex. 4   **9.** 22.4 cc; Sec. 5.4, Ex. 2   **10.** 0.046; Sec. 6.1, Ex. 4

11. 1.9; Sec. 6.1, Ex. 5   **12.** $\frac{2}{5}$; Sec. 6.2, Ex. 1   **13.** $\frac{1}{3}$; Sec. 6.2, Ex. 4   **14.** $5 = n \cdot 20$; Sec. 6.3, Ex. 1

15. sales tax: $6.41; total price: $91.91; Sec. 6.6, Ex. 1   **16.** $2400; Sec. 6.7, Ex. 3   **17.** $2\frac{1}{3}$ yd; Sec. 7.1, Ex. 2   **18.** 4 lb 11 oz; Sec. 7.2, Ex. 5

19. 0.0235 g; Sec. 7.2, Ex. 8   **20.** 3.21 L; Sec. 7.3, Ex. 7   **21.** 59°F; Sec. 7.4, Ex. 1   **22.** 42°; Sec. 8.1, Ex. 4   **23.** $\frac{1}{6}$; Sec. 8.6, Ex. 2

24. 14, 77; Sec. 9.4, Ex. 5   **25.** $\frac{1}{4}$; Sec. 9.5, Ex. 3

Chapter 10 SIGNED NUMBERS

CHAPTER 10 PRETEST

1. $-1\frac{1}{4}$ $2\frac{1}{2}$; 10.1B   **2.** > ; 10.1C   **3.** > ; 10.1C   **4.** 22; 10.1D   **5.** 9.8; 10.1D

6. 12; 10.1E   **7.** 19; 10.2A   **8.** −24; 10.3A   **9.** −15; 10.3A   **10.** 5; 10.3B   **11.** 39; 10.3B   **12.** 54; 10.4A   **13.** 120; 10.4A
14. 81; 10.4A   **15.** 46; 10.4B   **16.** −1.2; 10.4B   **17.** −19; 10.5A   **18.** −28; 10.5A   **19.** −$8; 10.3C   **20.** −72; 10.4A

EXERCISE SET 10.1

1. −1445   **3.** + 14,494   **5.** −15   **7.** −317   **9.** −5.049 million   **11.** cooler   **13.** −3.3%
15. $-1\frac{1}{2}$ **17.** −4.7

19. <   **21.** >   **23.** >   **25.** >   **27.** <   **29.** <   **31.** <   **33.** >   **35.** 5   **37.** 8   **39.** 0   **41.** 5   **43.** 8.1
45. $\frac{9}{10}$ **47.** $\frac{3}{8}$ **49.** 7.6   **51.** −5   **53.** 4   **55.** −23.6   **57.** $\frac{9}{16}$ **59.** 0.7   **61.** $-\frac{17}{18}$ **63.** 7   **65.** −20   **67.** −3
69. 8   **71.** 14   **73.** 29   **75.** 13   **77.** 35   **79.** 360   **81.** true   **83.** true   **85.** answers may vary

CALCULATOR EXPLORATIONS

1. 159 **3.** 44 **5.** −894.855

MENTAL MATH

1. 5 **3.** −35

EXERCISE SET 10.2

1.

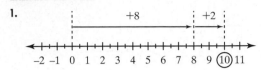

3.

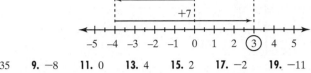

5.

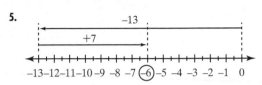

7. 35 **9.** −8 **11.** 0 **13.** 4 **15.** 2 **17.** −2 **19.** −11 **21.** −24
23. −57 **25.** −223 **27.** 0 **29.** 7 **31.** −3 **33.** −9 **35.** 30 **37.** 20
39. 51 **41.** −8.5 **43.** 4.6 **45.** $-\dfrac{5}{6}$ **47.** $-\dfrac{7}{10}$ **49.** −7 **51.** −24.6
53. 16 **55.** 13 **57.** 2°C

59. −165 + (−16) = −181; 181 ft below the surface **61.** Team 1, 7; Team 2, 6; winning team, Team 1 **63.** −48°F **65.** −7679 m
67. 44 **69.** 0 **71.** 28 **73.** true **75.** false **77.** answers may vary

MENTAL MATH

1. 0 **3.** 0

EXERCISE SET 10.3

1. 0 **3.** 5 **5.** −5 **7.** 14 **9.** 3 **11.** −18 **13.** −14 **15.** −3.3 **17.** 4.02 **19.** $\dfrac{2}{5}$ **21.** $-\dfrac{3}{10}$ **23.** $\dfrac{5}{6}$
25. −38 **27.** −17 **29.** 13 **31.** 2 **33.** 0 **35.** −1 **37.** −27 **39.** 40 **41.** 262°F **43.** −$16 **45.** −12°C
47. 652 ft **49.** 144 ft **51.** 382°C **53.** Saturn; 39°C **55.** 0 **57.** 8 **59.** 1058 **61.** −4 **63.** 0 **65.** −12 **67.** false
69. answers may vary **71.** answers may vary

INTEGRATED REVIEW

1. +29,028 **2.** −35,840 **3.** **4.** > **5.** < **6.** < **7.** > **8.** 1

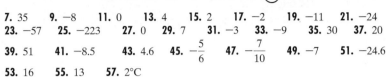

9. −4 **10.** 8.6 **11.** $\dfrac{3}{4}$ **12.** −6 **13.** 3 **14.** −89.1 **15.** $\dfrac{2}{9}$ **16.** 5 **17.** −20.8 **18.** $\dfrac{2}{9}$ **19.** −2 **20.** $\dfrac{3}{4}$ **21.** −4

22. 84 **23.** 6 **24.** $-\dfrac{11}{10} = -1\dfrac{1}{10}$ **25.** 1 **26.** −3.44 **27.** −19 **28.** 12 **29.** −4

EXERCISE SET 10.4

1. 6 **3.** −36 **5.** −3.12 **7.** 0 **9.** $\dfrac{6}{35}$ **11.** −48 **13.** −8 **15.** 80 **17.** 0 **19.** 4 **21.** −27 **23.** 25 **25.** −125

27. 0 **29.** $-\dfrac{21}{32}$ **31.** 1 **33.** 43.4 **35.** −8 **37.** −4 **39.** −5 **41.** 8 **43.** 0 **45.** −13 **47.** −26 **49.** $\dfrac{7}{2}$ **51.** −5

53. −6 **55.** 3 **57.** −300 **59.** $-\dfrac{4}{5}$ **61.** 3·(−4) = −12 yd **63.** 5·(−20) = −100 ft **65.** −5 **67.** $7.5 million
69. a. −8 condors; **b.** −1 condor per year **71.** 225 **73.** 109 **75.** 8 **77.** false **79.** true
81. a. −7100 Saturns **b.** −14,200 Saturns **c.** 264,374 Saturns **83.** answers may vary

CALCULATOR EXPLORATIONS

1. 4.8 **3.** 258

EXERCISE SET 10.5

1. 3 **3.** −1 **5.** 7 **7.** −14 **9.** $-\dfrac{2}{9}$ **11.** −8 **13.** 7 **15.** −1 **17.** 4 **19.** −3 **21.** −55 **23.** 24 **25.** 16

27. 15 **29.** −3 **31.** −3 **33.** 4 **35.** 16 **37.** −27 **39.** 34 **41.** 0.65 **43.** −59 **45.** −7 **47.** $\dfrac{5}{36}$ **49.** −11

51. 36 **53.** −117 **55.** 30 **57.** −3 **59.** −2.21 **61.** 4050 **63.** 45 **65.** 32 in. **67.** 30 ft **69.** 2·(7 − 5)·3 = 12
71. −6·(10 − 4) = −36 **73.** 20,736

CHAPTER 10 REVIEW

1. -1435 **2.** $+7562$ **3.**

$$-3.5$$
$$\longleftarrow \;|\;|\;\bullet\;|\;\bullet\;|\;\bullet\;|\;|\;|\;\bullet\;|\;|\;\longrightarrow$$
$$-5\;-4\;-3\;-2\;-1\;\;0\;\;1\;\;2\;\;3\;\;4\;\;5$$

4.

$$-4\tfrac{1}{3}\qquad -1.6$$
$$\longleftarrow \bullet\;|\;|\;\bullet\;|\;|\;|\;\bullet\;|\;|\;|\;|\;|\;\bullet\;\longrightarrow$$
$$-7\;-6\;-5\;-4\;-3\;-2\;-1\;\;0\;\;1\;\;2\;\;3$$

5. $>$ **6.** $<$ **7.** $>$ **8.** 12   **9.** 0   **10.** $\dfrac{7}{8}$ **11.** 12   **12.** $-\dfrac{1}{2}$ **13.** 7   **14.** -7 **15.** true   **16.** true   **17.** 2   **18.** 14

19. 4   **20.** 17   **21.** -23 **22.** -22 **23.** -2.1 **24.** -7 **25.** $\dfrac{4}{15}$ **26.** $-\dfrac{5}{9}$ **27.** -151 **28.** -606 **29.** $-20°C$

30. -150 ft   **31.** -17 **32.** -11 **33.** 8   **34.** -16 **35.** $-\dfrac{3}{10}$ **36.** -27 **37.** 20   **38.** 8   **39.** 0   **40.** -32 **41.** 0

42. 0.7   **43.** -4 **44.** $-\dfrac{3}{14}$ **45.** $-\$25$ **46.** $+692$ ft   **47.** true   **48.** false   **49.** 21   **50.** -18 **51.** $-\dfrac{7}{12}$ **52.** 6

53. undefined   **54.** 38   **55.** -5 **56.** $(-5 \text{ yds})(2) = -10$ yds   **57.** $(-\$50)(4) = -\$200$ **58.** 0   **59.** -8 **60.** -16

61. 35   **62.** -28 **63.** -44 **64.** -14.8 **65.** -1 **66.** 7   **67.** $-\dfrac{5}{12}$ **68.** 39   **69.** $-\dfrac{23}{9}$ **70.** 7   **71.** -80 **72.** -2

73. -12

CHAPTER 10 TEST

1. 3   **2.** -6 **3.** -10 **4.** 4   **5.** $-\dfrac{1}{2}$ **6.** 12   **7.** 65   **8.** 5   **9.** 12   **10.** -6 **11.** 50   **12.** -2 **13.** -11

14. -46 **15.** -117 **16.** 3456   **17.** $-\dfrac{14}{81}$ **18.** -213 **19.** -1 **20.** -2 **21.** 2   **22.** -5 **23.** $+31,642$ ft

24. $14,893 + (-147) = +14,746$; $14,746$ ft   **25.** $|41$

CUMULATIVE REVIEW

1. 78,875; Sec. 1.5, Ex. 5   **2.** $\dfrac{4}{5}$; Sec. 2.5, Ex. 7   **3.** $\dfrac{17}{21}$; Sec. 3.3, Ex. 3   **4.** 48.26; Sec. 4.1, Ex. 5   **5.** 6.095; Sec. 4.1, Ex. 6

6. 3.432; Sec. 4.3, Ex. 5   **7.** 5.6881; Sec. 4.6, Ex. 8   **8.** $\dfrac{4500 \text{ dollars}}{1 \text{ month}}$ or 4500 dollars/month; Sec. 5.2, Ex. 4   **9.** $n = 2$; Sec. 5.3, Ex. 5

10. 46%; Sec. 6.1, Ex. 2   **11.** 45%; Sec. 6.2, Ex. 6   **12.** 150%; Sec. 6.2, Ex. 8   **13.** 200; Sec. 6.3, Ex. 10   **14.** $\dfrac{101}{200} = \dfrac{p}{100}$; Sec. 6.4, Ex. 2

15. $\$160$; Sec. 6.7, Ex. 2   **16.** 9 ft 1 in.; Sec. 7.1, Ex. 5   **17.** $37°C$; Sec. 7.4, Ex. 5   **18.** $73°$; Sec. 8.1, Ex. 5   **19.** $60°$; Sec. 8.2, Ex. 2

20. 5.1 sq. mi; Sec. 8.4, Ex. 2   **21.** $113\tfrac{1}{7}$ cu. in.; Sec. 8.5, Ex. 2   **22.** 57; Sec. 9.4, Ex. 3   **23.** $\dfrac{1}{3}$; Sec. 9.5, Ex. 4   **24.** -23; Sec. 10.2, Ex. 3

25. 82; Sec. 10.5, Ex. 6

Chapter 11 INTRODUCTION TO ALGEBRA

CHAPTER 11 PRETEST

1. -9; 11.1A   **2.** $-10x$; 11.1B   **3.** $7x - 4y + 20$; 11.1B   **4.** $-12m + 6$; 11.1C   **5.** no; 11.2A   **6.** yes; 11.2A   **7.** $y = 2$; 11.2B
8. $x = -8$; 11.2B   **9.** $z = -8$; 11.3A   **10.** $a = 35$; 11.3A   **11.** $x = -2$; 11.3A   **12.** $x = 7$; 11.3A   **13.** $b = 2$; 11.4A

14. $n = 3$; 11.4A   **15.** $y = 1$; 11.4B   **16.** $w = 2$; 11.4B   **17.** $3 \cdot 12 = 36$; 11.4C   **18.** $2(4 + 6) = 20$; 11.4C   **19.** $\dfrac{10}{3x} = 40$; 11.5B

20. 15; 11.5C

MENTAL MATH

1. unlike   **3.** like   **5.** unlike   **7.** like

EXERCISE SET 11.1

1. -3 **3.** 8   **5.** 4   **7.** -3 **9.** 1   **11.** 133   **13.** 20   **15.** -4 **17.** $-\dfrac{4}{3}$ **19.** -4 **21.** 2000 sq. ft   **23.** 64 ft

25. $\$360$ **27.** 78.5 sq. ft   **29.** $23°F$ **31.** 288 cu. in.   **33.** $8x$ **35.** $-4n$ **37.** $-2c$ **39.** $-4x$ **41.** $13a$ 8
43. $-0.9x + 11.2$ **45.** $2x - 7$ **47.** $-5x + 4y - 5$ **49.** $30x$ **51.** $-22y$ **53.** $2y + 4$ **55.** $5a - 40$ **57.** $-12x - 28$
59. $2x - 9$ **61.** $27n - 20$ **63.** $7w + 15$ **65.** $-x - 12$ **67.** $16y^2$ sq. cm   **69.** -3 **71.** 8   **73.** 0   **75.** 3610.4 sq. in.
77. $4824q + 12,274$ **79.** answers may vary   **81.** $(20x + 16)$ sq. mi

MENTAL MATH

1. 2   **3.** -1

Exercise Set 11.2

1. yes **3.** no **5.** yes **7.** yes **9.** 18 **11.** −8 **13.** 9 **15.** −16 **17.** 3 **19.** $\frac{1}{8}$ **21.** 6 **23.** 8 **25.** 5.3
27. −1 **29.** −20.1 **31.** 2 **33.** 0 **35.** −28 **37.** 1 **39.** 1 **41.** 1 **43.** 162,964 **45.** 3705 yd **47.** $106,146 million

Exercise Set 11.3

1. 4 **3.** −4 **5.** 0 **7.** −17 **9.** 50 **11.** 25 **13.** −30 **15.** $\frac{1}{3}$ **17.** $-\frac{2}{3}$ **19.** $-\frac{15}{64}$ **21.** $\frac{2}{3}$ **23.** −4 **25.** 8
27. −0.1 **29.** 2 **31.** −8 **33.** −3 **35.** 1 **37.** 0 **39.** −2 **41.** 25 **43.** 1 **45.** −10 **47.** answers may vary
49. −3648 **51.** 12.4 hrs **53.** 58 mi per hr

Integrated Review

1. 4 **2.** −3 **3.** 1 **4.** −4 **5.** 8x **6.** −4y **7.** −2a − 2 **8.** −8x **9.** 5y + 10 **10.** (12x − 6) sq. m **11.** 13
12. −9 **13.** 5 **14.** 0 **15.** −4 **16.** 25 **17.** −1 **18.** −3 **19.** 6 **20.** 8 **21.** $-\frac{9}{11}$ **22.** $\frac{7}{10}$

Calculator Explorations

1. yes **3.** no **5.** yes

Exercise Set 11.4

1. 3 **3.** 4 **5.** −4 **7.** −3 **9.** −12 **11.** 100 **13.** −3.9 **15.** −4 **17.** 5 **19.** 1 **21.** −4 **23.** 2 **25.** 5
27. −3 **29.** 2 **31.** −2 **33.** 3 **35.** −1 **37.** 4 **39.** −4 **41.** 3 **43.** −1 **45.** 64 **47.** −45 **49.** 5 **51.** −5
53. 8 **55.** 0 **57.** −22 **59.** 4 **61.** 1 **63.** −30 **65.** −42 + 16 = −26 **67.** −5(−29) = 145 **69.** 3(−14 − 2) = −48
71. $\frac{100}{2(50)} = 1$ **73.** $322 **75.** $62 **77.** 80.24°F **79.** 5.18°F

Exercise Set 11.5

1. x + 5 **3.** x + 8 **5.** 20 − x **7.** 512x **9.** $\frac{x}{2}$ **11.** (17 + x) + 5x **13.** 5x **15.** 11 − x **17.** 50 − 8x
19. −5 + x = −7 **21.** 3x = 27 **23.** −20 − x = 104 **25.** 8 **27.** 9 **29.** 5 **31.** 12 **33.** 8 **35.** 5
37. Dole: 159 votes; Clinton: 379 votes **39.** 80 books **41.** truck, 35 mph; car, 70 mph **43.** $225 **45.** 93 points **47.** 590
49. 1000 **51.** 3000 **53.** answers may vary

Chapter 11 Review

1. −5 **2.** 17 **3.** undefined **4.** 0 **5.** 129 **6.** −2 **7.** 4 **8.** 20 **9.** 8 cu. ft **10.** 64 cu. ft **11.** $1800 **12.** $300
13. 10y − 15 **14.** −6y − 10 **15.** −6a − 7 **16.** −8y + 2 **17.** 4.6x − 11.9 **18.** −2.1x − 20 **19.** −2x − 10
20. −3y − 24 **21.** 11x − 12 **22.** −3m + 16 **23.** −5a + 4 **24.** 12y − 9 **25.** (6x − 3) sq. yd **26.** $25y^2$ sq. m **27.** yes
28. no **29.** −2 **30.** 7 **31.** $-\frac{1}{2}$ **32.** $-\frac{6}{11}$ **33.** −1 **34.** −25 **35.** −6 **36.** −20 **37.** 1.3 **38.** 2.4 **39.** 7
40. −9 **41.** 1 **42.** −5 **43.** $-\frac{4}{5}$ **44.** −24 **45.** −120 **46.** 13 **47.** 4 **48.** 3 **49.** 13 **50.** −6 **51.** 5 **52.** 12
53. 17 **54.** 7 **55.** 63 **56.** 33 **57.** −2.25 **58.** 1.3 **59.** 11 **60.** −5 **61.** 2 **62.** 6 **63.** 20 − (−8) = 28
64. 5(2 + (−6)) = −20 **65.** $\frac{-75}{5 + 20} = -3$ **66.** −2 − 19 = −21 **67.** 2x + 11 **68.** −5x − 50 **69.** $\frac{70}{x + 6}$
70. 2(x − 13) **71.** 2x − 8 = 40 **72.** $\frac{x}{2} - 12 = 10$ **73.** $x - 3 = \frac{x}{4}$ **74.** 6x = x + 2 **75.** 5 **76.** −16
77. 2386 votes **78.** 42 CDs

Chapter 11 test

1. −1 **2.** −5x + 5 **3.** −6y − 14 **4.** 14z − 8 **5.** (9x − 3) sq. m **6.** 7 **7.** $-\frac{1}{2}$ **8.** −12 **9.** 40 **10.** 24 **11.** 3
12. −2 **13.** −2 **14.** 4.5 **15.** 0 **16.** −2 **17.** 7 **18.** 4 **19.** 0 **20.** 6000 sq. ft **21.** 30 sq. ft
22. a. 17x **b.** 20 − 2x **23.** −2 **24.** 8 free throws **25.** 244 women

Cumulative Review

1. 0.8496; Sec. 4.4, Ex. 3 **2. a.** $\frac{5}{7}$ **b.** $\frac{7}{24}$; Sec. 5.1, Ex. 7 **3.** 5; Sec. 6.3, Ex. 9 **4.** 75%; Sec. 6.3, Ex. 11 **5.** 48 oz; Sec. 7.2, Ex. 2
6. 1.13 ml; Sec. 7.3, Ex. 10 **7. a.** line **b.** line segment **c.** angle **d.** ray; Sec. 8.1, Ex. 1 **8.** 10 cm; Sec. 8.2, Ex. 3
9. 50 ft; Sec. 8.3, Ex. 6 **10.** 56 sq. cm; Sec. 8.4, Ex. 1 **11.** 25; Sec. 8.6, Ex. 5 **12.** 46 ft; Sec. 8.7, Ex. 4 **13. a.** 55 mammal species
b. amphibians; Sec. 9.1, Ex. 2 **14.** <; Sec. 10.1, Ex. 6 **15.** >; Sec. 10.1, Ex. 5 **16.** 2; Sec. 10.1, Ex. 10 **17.** 1.2; Sec. 10.1, Ex. 14
18. −6; Sec. 10.2, Ex. 4 **19.** $-\frac{1}{2}$; Sec. 10.2, Ex. 6 **20.** −7; Sec. 10.3, Ex. 6 **21.** 1; Sec. 10.3, Ex. 7 **22.** 10; Sec. 10.4, Ex. 2
23. $\frac{1}{3}$; Sec. 10.4, Ex. 4 **24.** 3; Sec. 10.5, Ex. 7 **25.** 2; Sec. 11.4, Ex. 7

Appendix A

1. 5　2. 11　3. 5　4. 15　5. 12　6. 7　7. 8　8. 6　9. 14　10. 10　11. 12　12. 5　13. 10　14. 2　15. 9
16. 12　17. 11　18. 8　19. 18　20. 7　21. 10　22. 0　23. 10　24. 10　25. 17　26. 8　27. 13　28. 3
29. 12　30. 17　31. 16　32. 8　33. 7　34. 13　35. 4　36. 3　37. 8　38. 7　39. 8　40. 9　41. 6　42. 9
43. 16　44. 8　45. 9　46. 15　47. 12　48. 1　49. 7　50. 6　51. 2　52. 6　53. 11　54. 11　55. 8　56. 10
57. 5　58. 10　59. 14　60. 6　61. 13　62. 4　63. 7　64. 9　65. 7　66. 13　67. 13　68. 7　69. 5　70. 9
71. 4　72. 3　73. 16　74. 11　75. 14　76. 13　77. 1　78. 11　79. 4　80. 4　81. 10　82. 14　83. 8
84. 15　85. 5　86. 15　87. 10　88. 2　89. 11　90. 6　91. 12　92. 6　93. 9　94. 11　95. 9　96. 9　97. 14
98. 12　99. 9　100. 3

Appendix B

1. 1　2. 35　3. 56　4. 9　5. 32　6. 45　7. 28　8. 7　9. 4　10. 0　11. 63　12. 64　13. 6　14. 0　15. 30
16. 10　17. 24　18. 0　19. 18　20. 72　21. 40　22. 14　23. 32　24. 2　25. 54　26. 3　27. 56　28. 16
29. 54　30. 25　31. 2　32. 0　33. 36　34. 24　35. 12　36. 20　37. 36　38. 18　39. 12　40. 6　41. 48
42. 72　43. 8　44. 5　45. 0　46. 28　47. 27　48. 0　49. 15　50. 48　51. 45　52. 12　53. 0　54. 27
55. 81　56. 20　57. 0　58. 9　59. 0　60. 6　61. 18　62. 7　63. 3　64. 21　65. 36　66. 0　67. 63　68. 12
69. 35　70. 0　71. 42　72. 0　73. 40　74. 8　75. 0　76. 24　77. 9　78. 0　79. 15　80. 16　81. 5　82. 30
83. 0　84. 4　85. 21　86. 8　87. 0　88. 49　89. 16　90. 24　91. 0　92. 14　93. 4　94. 0　95. 6　96. 8
97. 18　98. 10　99. 0　100. 42

Solutions to Selected Exercises

Chapter 1

Exercise Set 1.1

1. The place value of the 5 in 352 is tens.

5. The place value of the 5 in 62,500,000 is hundred thousands.

9. 5420 is written as five thousand, four hundred twenty.

13. 1,620,000 is written as one million, six hundred twenty-thousand.

17. 5,648,359 is written as five million, six hundred forty-eight thousand, three hundred fifty-nine.

21. 3893 is written as three thousand, eight hundred ninety-three.

25. Twenty-nine thousand, nine hundred in standard form is 29,900.

29. Three million, fourteen in standard form is 3,000,014.

33. Sixty-three million, one hundred thousand in standard form is 63,100,000.

37. $406 = 400 + 6$

41. $62,407 = 60,000 + 2000 + 400 + 7$

45. $39,680,000 = 30,000,000 + 9,000,000 + 600,000 + 80,000$

49. $4145 = 4000 + 100 + 40 + 5$

53. The Pomeranian has the least AKC registrations, with thirty-nine thousand, seven hundred twelve.

57. The largest number is achieved when the largest number available is used for each place value when reading from left to right. Thus, the largest number possible is 55,543.

Exercise Set 1.2

1.
$$\begin{array}{r} 14 \\ + 22 \\ \hline 36 \end{array}$$

5.
$$\begin{array}{r} 12 \\ 13 \\ + 24 \\ \hline 49 \end{array}$$

9.
$$\begin{array}{r} {\scriptstyle 1} \\ 53 \\ + 64 \\ \hline 117 \end{array}$$

13.
$$\begin{array}{r} {\scriptstyle 1\,1} \\ 38 \\ + 79 \\ \hline 117 \end{array}$$

17.
$$\begin{array}{r} {\scriptstyle 2} \\ 6 \\ 21 \\ 14 \\ 9 \\ + 12 \\ \hline 62 \end{array}$$

21.
$$\begin{array}{r} {\scriptstyle 1} \\ 62 \\ 18 \\ + 14 \\ \hline 94 \end{array}$$

25.
$$\begin{array}{r} {\scriptstyle 1\,1\,1} \\ 7542 \\ 49 \\ + 682 \\ \hline 8273 \end{array}$$

29.
$$\begin{array}{r} {\scriptstyle 1\ \ 2} \\ 627 \\ 628 \\ + 629 \\ \hline 1884 \end{array}$$

33.
$$\begin{array}{r} {\scriptstyle 1\,1\,1} \\ 507 \\ 593 \\ + 10 \\ \hline 1110 \end{array}$$

37.
$$\begin{array}{r} {\scriptstyle 1\,1\,2\,2} \\ 49 \\ 628 \\ 5762 \\ + 29,462 \\ \hline 35,901 \end{array}$$

41. $8 + 3 + 5 + 7 + 5 + 1 = 8 + 1 + 3 + 7 + 5 + 5$
$= 9 + 10 + 10 = 29$
The perimeter is 29 inches.

45. Opposite sides of a rectangle have the same lengths.
$4 + 8 + 4 + 8 = 12 + 12 = 24$
The perimeter is 24 inches.

49.
$$\begin{array}{r} {\scriptstyle 1\,1} \\ 285 \\ + 98 \\ \hline 383 \end{array}$$

It is 383 miles from Kansas City to Colby.

53.
$$\begin{array}{r} {\scriptstyle 1\,1} \\ 105,600 \\ + 17,500 \\ \hline 123,100 \end{array}$$

There were 123,100 people employed in 1997.

57.
$$\begin{array}{r} {\scriptstyle 1} \\ 11,099 \\ + 850 \\ \hline 11,949 \end{array}$$

There were 11,949 transplants involving a kidney performed in 1996.

61. Texas, Florida, and Illinois have the most Wal-Mart stores.

$$\begin{array}{r} {\scriptstyle 1} \\ 163 \\ 130 \\ + 293 \\ \hline 586 \end{array} \qquad \begin{array}{r} {\scriptstyle 5\,9\,10} \\ \cancel{600} \\ - 432 \\ \hline 168 \end{array}$$

Texas, Florida, and Illinois have the most Wal-Mart stores, with a total of 586 stores.

65. Answers may vary.

1.
$$\begin{array}{r} 67 \\ -\ 23 \\ \hline 44 \end{array}$$

Check:
$$\begin{array}{r} 44 \\ +\ 23 \\ \hline 67 \end{array}$$

5.
$$\begin{array}{r} 389 \\ -\ 124 \\ \hline 265 \end{array}$$

Check:
$$\begin{array}{r} 265 \\ +\ 124 \\ \hline 389 \end{array}$$

9.
$$\begin{array}{r} 998 \\ -\ 453 \\ \hline 545 \end{array}$$

Check:
$$\begin{array}{r} 545 \\ +\ 453 \\ \hline 998 \end{array}$$

13.
$$\begin{array}{r} 62 \\ -\ 37 \\ \hline 25 \end{array}$$

Check:
$$\begin{array}{r} 1 \\ 25 \\ +\ 37 \\ \hline 62 \end{array}$$

17.
$$\begin{array}{r} 938 \\ -\ 792 \\ \hline 146 \end{array}$$

Check:
$$\begin{array}{r} 1 \\ 146 \\ +\ 792 \\ \hline 938 \end{array}$$

21.
$$\begin{array}{r} 600 \\ -\ 432 \\ \hline 168 \end{array}$$

Check:
$$\begin{array}{r} 1\ 1 \\ 168 \\ +\ 432 \\ \hline 600 \end{array}$$

25.
$$\begin{array}{r} 923 \\ -\ 476 \\ \hline 447 \end{array}$$

Check:
$$\begin{array}{r} 1\ 1 \\ 447 \\ +\ 476 \\ \hline 923 \end{array}$$

29.
$$\begin{array}{r} 533 \\ -\ 29 \\ \hline 504 \end{array}$$

Check:
$$\begin{array}{r} 1 \\ 504 \\ +\ 29 \\ \hline 533 \end{array}$$

33.
$$\begin{array}{r} 1983 \\ -\ 1904 \\ \hline 79 \end{array}$$

Check:
$$\begin{array}{r} 1 \\ 79 \\ +\ 1904 \\ \hline 1983 \end{array}$$

37.
$$\begin{array}{r} 50,000 \\ -\ 17,289 \\ \hline 32,711 \end{array}$$

Check:
$$\begin{array}{r} 1\ 1\ 1\ 1 \\ 32,711 \\ +\ 17,289 \\ \hline 50,000 \end{array}$$

41.
$$\begin{array}{r} 51,111 \\ -\ 19,898 \\ \hline 31,213 \end{array}$$

Check:
$$\begin{array}{r} 1\ 1\ 1\ 1 \\ 31,213 \\ +\ 19,898 \\ \hline 51,111 \end{array}$$

45.
$$\begin{array}{r} 41 \\ -\ 21 \\ \hline 20 \end{array}$$

Check:
$$\begin{array}{r} 20 \\ +\ 21 \\ \hline 41 \end{array}$$

49.
$$\begin{array}{r} 63 \\ -\ 7 \\ \hline 56 \end{array}$$

Ronnie ate 56 crawfish.

53.
$$\begin{array}{r} 20,320 \\ -\ 14,255 \\ \hline 6065 \end{array}$$

Mt. McKinley is 6065 feet higher than Long's Peak.

57.
$$\begin{array}{r} 645 \\ -\ 287 \\ \hline 358 \end{array}$$

The distance between Hays and Denver is 358 miles.

61.
$$\begin{array}{r} 547 \\ -\ 99 \\ \hline 448 \end{array}$$

The sale price is $448.

65. The total number of votes cast for Jo was:
$$\begin{array}{r} 1\ 2\ 1 \\ 276 \\ 362 \\ 201 \\ +\ 179 \\ \hline 1018 \end{array}$$

The total number of votes cast for Trudy was:
$$\begin{array}{r} 2\ 1 \\ 295 \\ 122 \\ 312 \\ +\ 182 \\ \hline 911 \end{array}$$

Since more votes were cast for Jo than for Trudy, Jo won the election.

69.
$$\begin{array}{r} 1,130,000 \\ -\ 633,000 \\ \hline 497,000 \end{array}$$

North Korea's fighting force was 497,000 soldiers larger than South Korea's fighting force.

73.
$$\begin{array}{r} 382 \\ -\ 234 \\ \hline 148 \end{array}$$

Dallas/Ft. Worth International Airport has 148,000 more departures per year than Los Angeles International Airport.

77.
$$\begin{array}{r} 1,236,978,000 \\ -\ 775,621,000 \\ \hline 461,357,000 \end{array}$$

Philip Morris spent $461,357,000 more on ads than Walt Disney.

81.
$$\begin{array}{r} 5269 \\ -\ 2385 \\ \hline 2884 \end{array}$$

EXERCISE SET 1.4

1. To round 632 to the nearest ten, observe that the digit in the ones place is 2. Since this digit is less than 5, we do not add 1 to the digit in the tens place. The number 632 rounded to the nearest ten is 630.

5. To round 792 to the nearest ten, observe that the digit in the ones place is 2. Since this digit is less than 5, we do not add 1 to the digit in the tens place. The number 792 rounded to the nearest ten is 790.

SOLUTIONS

9. To round 1096 to the nearest ten, observe that the digit in the ones place is 6. Since this digit is at least 5, we need to add 1 to the digit in the tens place. The number 1096 rounded to the nearest ten is 1100.

13. To round 248,695 to the nearest hundred, observe that the digit in the tens place is 9. Since this digit is at least 5, we need to add 1 to the digit in the hundreds place. The number 248,695 rounded to the nearest hundred is 248,700.

17. To round 99,995 to the nearest ten, observe that the digit in the ones place is 5. Since this digit is at least 5, we need to add 1 to the digit in the tens place. The number 99,995 rounded to the nearest ten is 100,000.

	Ten	Hundred	Thousand
21. 5281	5280	5300	5000
25. 14,876	14,880	14,900	15,000

29. To round 68,912 to the nearest thousand, observe that the digit in the hundreds place is 9. Since this digit is at least 5, we need to add 1 to the digit in the thousands place. The number 68,912 rounded to the nearest thousand is 69,000.

33. To round 305,718,800 to the nearest hundred-thousand, observe that the digit in the ten-thousands place is 1. Since this digit is less than 5, we do not add 1 to the digit in the hundred-thousands place. The number 305,718,800 rounded to the nearest hundred-thousand is 305,700,000.

37.
$$\begin{array}{r} 649 \\ -272 \\ \end{array} \quad \begin{array}{c} \text{rounds to} \\ \text{rounds to} \end{array} \quad \begin{array}{r} 650 \\ -270 \\ \hline 380 \end{array}$$

The estimated difference is 380.

41.
$$\begin{array}{r} 1774 \\ -1492 \\ \end{array} \quad \begin{array}{c} \text{rounds to} \\ \text{rounds to} \end{array} \quad \begin{array}{r} 1800 \\ -1500 \\ \hline 300 \end{array}$$

The estimated difference is 300.

45. $362 + 419$ is approximately $360 + 420 = 780$. The answer of 781 is correct.

49. $7806 + 5150$ is approximately $7800 + 5200 = 13,000$. The answer of 12,956 is correct.

53.
$$\begin{array}{r} 799 \\ 1299 \\ +999 \\ \end{array} \quad \begin{array}{c} \text{rounds to} \\ \text{rounds to} \\ \text{rounds to} \end{array} \quad \begin{array}{r} 800 \\ 1300 \\ +1000 \\ \hline 3100 \end{array}$$

The total cost is approximately $3100.

57.
$$\begin{array}{r} 29,028 \\ -4039 \\ \end{array} \quad \begin{array}{c} \text{rounds to} \\ \text{rounds to} \end{array} \quad \begin{array}{r} 29,000 \\ -4000 \\ \hline 25,000 \end{array}$$

The difference in elevation is approximately 25,000 feet.

61.
$$\begin{array}{r} 41,126,333 \\ -27,174,898 \\ \end{array} \quad \begin{array}{c} \text{rounds to} \\ \text{rounds to} \end{array} \quad \begin{array}{r} 41,000,000 \\ -27,000,000 \\ \hline 14,000,000 \end{array}$$

Johnson won the election by approximately 14,000,000 votes.

65. 767,978,000 rounds to 800,000,000. PepsiCo spent approximately $800,000,000.

69. The smallest possible number that rounds to 8600 is 8550.

EXERCISE SET 1.5

1. $4(3 + 9) = 4 \cdot 3 + 4 \cdot 9$

5. $10(11 + 7) = 10 \cdot 11 + 10 \cdot 7$

9.
$$\begin{array}{r} 624 \\ \times 3 \\ \hline 1872 \end{array}$$

13.
$$\begin{array}{r} 1062 \\ \times 5 \\ \hline 5310 \end{array}$$

17.
$$\begin{array}{r} 231 \\ \times 47 \\ \hline 1617 \\ 9240 \\ \hline 10,857 \end{array}$$

21.
$$\begin{array}{r} 620 \\ \times 40 \\ \hline 0 \\ 24,800 \\ \hline 24,800 \end{array}$$

25. $(590)(1)(10) = 5900$

29.
$$\begin{array}{r} 609 \\ \times 234 \\ \hline 2436 \\ 18,270 \\ 121,800 \\ \hline 142,506 \end{array}$$

33.
$$\begin{array}{r} 1941 \\ \times 235 \\ \hline 9705 \\ 58,230 \\ 388,200 \\ \hline 456,135 \end{array}$$

37.
$$\begin{array}{r} 576 \\ \times 354 \\ \end{array} \quad \begin{array}{c} \text{rounds to} \\ \text{rounds to} \end{array} \quad \begin{array}{r} 600 \\ \times 400 \\ \hline 240,000 \end{array}$$

576×354 is approximately 240,000.

41. Area = length · width
= (9 meters)(7 meters)
= 63 square meters
The area is 63 square meters.

45.
$$\begin{array}{r} 125 \\ \times 3 \\ \hline 375 \end{array}$$

There are 375 calories in 3 tablespoons of olive oil.

49.
$$\begin{array}{r} 12 \\ \times 8 \\ \hline 96 \end{array}$$

$2 \times 96 = 192$
There are 192 cans in a case.

53.
$$\begin{array}{r} 776 \\ \times 639 \\ \hline 6984 \\ 23,280 \\ 465,600 \\ \hline 495,864 \end{array}$$

The floor area is 495,864 square meters.

57.

$$\begin{array}{r} 60 \\ \times\ 25 \\ \hline 300 \\ 1200 \\ \hline 1500 \end{array}$$

There are 1500 characters in 25 lines.

61.

$$\begin{array}{r} 7927 \\ \times\ 9 \\ \hline 71{,}343 \end{array}$$

Saturn has a diameter of 71,343 miles.

65.

$$\begin{array}{r} 700{,}000 \\ \times\ 31 \\ \hline 700{,}000 \\ 21{,}000{,}000 \\ \hline 21{,}700{,}000 \end{array}$$

21,700,000 quarts of milk would be used in March.

69. $5 \times 10 = 50$

50 students chose grapes as their favorite fruit.

73. The result of multiplying 3 by the digit in the first blank is a number ending in 6. Only 2 works. The result of multiplying the digit in the second blank by 42 is 378, so the digit in the second blank is 9.

$$\begin{array}{r} 42 \\ \times\ 93 \\ \hline 126 \\ 3780 \\ \hline 3906 \end{array}$$

77. $124 \times 2 = 248$ and $67 \times 3 = 201$.

$$\begin{array}{r} 172 \\ 248 \\ +\ 201 \\ \hline 621 \end{array}$$

Cynthia Cooper scored 621 points in the 1997 season.

Exercise Set 1.6

1.

$$\begin{array}{r} 12 \\ 9)\overline{108} \\ -9 \\ \hline 18 \\ -18 \\ \hline 0 \end{array}$$

Check: $12 \cdot 9 = 108$

5.

$$\begin{array}{r} 338 \\ 3)\overline{1014} \\ -9 \\ \hline 11 \\ -9 \\ \hline 24 \\ -24 \\ \hline 0 \end{array}$$

Check: $338 \cdot 3 = 1014$

9.

$$\begin{array}{r} 563 \quad \text{R } 1 \\ 2)\overline{1127} \\ -10 \\ \hline 12 \\ -12 \\ \hline 07 \\ -6 \\ \hline 1 \end{array}$$

Check:
$563 \cdot 2 + 1 = 1127$

13.

$$\begin{array}{r} 265 \quad \text{R } 1 \\ 8)\overline{2121} \\ -16 \\ \hline 52 \\ -48 \\ \hline 41 \\ -40 \\ \hline 1 \end{array}$$

Check:
$265 \cdot 8 + 1 = 2121$

17.

$$\begin{array}{r} 13 \\ 55)\overline{715} \\ -55 \\ \hline 165 \\ -165 \\ \hline 0 \end{array}$$

Check: $13 \cdot 55 = 715$

21.

$$\begin{array}{r} 206 \\ 18)\overline{3708} \\ -36 \\ \hline 10 \\ -0 \\ \hline 108 \\ -108 \\ \hline 0 \end{array}$$

Check: $206 \cdot 18 = 3708$

25.

$$\begin{array}{r} 202 \quad \text{R } 7 \\ 46)\overline{9299} \\ -92 \\ \hline 09 \\ -0 \\ \hline 99 \\ -92 \\ \hline 7 \end{array}$$

Check: $202 \cdot 46 + 7 = 9299$

29.

$$\begin{array}{r} 98 \quad \text{R } 100 \\ 103)\overline{10194} \\ -927 \\ \hline 924 \\ -824 \\ \hline 100 \end{array}$$

Check: $98 \cdot 103 + 100 = 10{,}194$

33.

$$\begin{array}{r} 202 \\ 223)\overline{45046} \\ -446 \\ \hline 44 \\ -0 \\ \hline 446 \\ -446 \\ \hline 0 \end{array}$$

Check: $202 \cdot 223 = 45{,}046$

37.

$$\begin{array}{r} 252000 \\ 21)\overline{5292000} \\ -42 \\ \hline 109 \\ -105 \\ \hline 42 \\ -42 \\ \hline 00 \\ -0 \\ \hline 00 \\ -0 \\ \hline 00 \\ -0 \\ \hline 0 \end{array}$$

Each person receives $252,000.

41.

$$\begin{array}{r} 88 \quad \text{R } 1 \\ 3)\overline{265} \\ -24 \\ \hline 25 \\ -24 \\ \hline 1 \end{array}$$

There are 88 bridges in 265 miles.

45.

$$\begin{array}{r} 16 \\ 6)\overline{96} \\ -6 \\ \hline 36 \\ -36 \\ \hline 0 \end{array}$$

He scored 16 touchdowns during 1997.

49. There are six numbers.

$$
\begin{array}{r}
14 \\
22 \\
45 \\
18 \\
30 \\
+\,27 \\
\hline
156
\end{array}
\qquad
\begin{array}{r}
26 \\
6)\overline{156} \\
-12 \\
\hline
36 \\
-\,36 \\
\hline
0
\end{array}
$$

Average $= \dfrac{156}{6} = 26$

53. There are five numbers.

$$
\begin{array}{r}
86 \\
79 \\
81 \\
69 \\
+\,80 \\
\hline
395
\end{array}
\qquad
\begin{array}{r}
79 \\
5)\overline{395} \\
-35 \\
\hline
45 \\
-\,45 \\
\hline
0
\end{array}
$$

Average $= \dfrac{395}{5} = 79$

57. Add the four largest amounts, then divide by 4.

$$
\begin{array}{r}
1{,}712{,}102{,}000 \\
1{,}493{,}456{,}000 \\
1{,}236{,}978{,}000 \\
+\,1{,}119{,}851{,}000 \\
\hline
5{,}562{,}387{,}000
\end{array}
\qquad
\begin{array}{r}
1390596750 \\
4)\overline{5562387000} \\
-4 \\
\hline
15 \\
-12 \\
\hline
36 \\
-36 \\
\hline
02 \\
-\,0 \\
\hline
23 \\
-\,20 \\
\hline
38 \\
-\,36 \\
\hline
27 \\
-\,24 \\
\hline
30 \\
-\,28 \\
\hline
20 \\
-\,20 \\
\hline
00 \\
-\,0 \\
\hline
0
\end{array}
$$

The average amount spent by the top four companies is $1,390,596,750.

61. No, because all the numbers are greater than 86.

EXERCISE SET 1.7

1.

41	increased by	8	is	some number
↓	↓	↓	↓	↓
41	+	8	=	some number

$$
\begin{array}{r}
41 \\
+\,8 \\
\hline
49
\end{array}
$$

The number is 49.

5.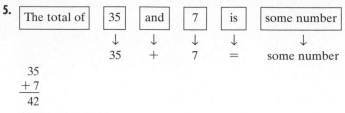

$$35 + 7 = \text{some number}$$

$$\begin{array}{r} 35 \\ +\ 7 \\ \hline 42 \end{array}$$

The number is 42.

9.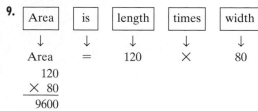

$$\text{Area} = 120 \times 80$$

$$\begin{array}{r} 120 \\ \times\ 80 \\ \hline 9600 \end{array}$$

The area is 9600 square feet.

13.

Hours per week	is	hours per day	times	days per week
↓	↓	↓	↓	↓

$$\text{Hours per week} = 24 \times 7$$

$$\begin{array}{r} 24 \\ \times\ 7 \\ \hline 168 \end{array}$$

There are 168 hours in a week.

17.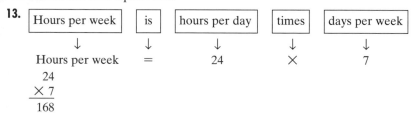

$$\text{1996 income} = \$1,606,000,000 - \$457,000,000$$

$$\begin{array}{r} 1,606,000,000 \\ \times\ 457,000,000 \\ \hline 1,149,000,000 \end{array}$$

PepsiCo's net income in 1996 was $1,149,000,000.

21.

Average	is	total sales	divided by	number of stores
↓	↓	↓	↓	↓

$$\text{Average} = 889,200,000 \div 25$$

$$\begin{array}{r} 35568000 \\ 25)\overline{889200000} \\ -\ 75 \\ \hline 139 \\ -\ 125 \\ \hline 142 \\ -\ 125 \\ \hline 170 \\ -\ 150 \\ \hline 200 \\ -\ 200 \\ \hline 0000 \end{array}$$

The average sales by each store is $35,568,000.

25.

Calories in 1 ounce	is	calories	per	ounces
↓	↓	↓	↓	↓
Calories in 1 ounce	=	165	÷	3

$$
\begin{array}{r}
55 \\
3\overline{)165} \\
-15 \\
\hline
15 \\
-15 \\
\hline
0
\end{array}
$$

There are 55 calories in 1 ounce of canned tuna.

29.

Elementary enrollment	is	total enrollment	minus	secondary enrollment
↓	↓	↓	↓	↓
Elementary enrollment	=	54,807,000	−	15,395,000

$$
\begin{array}{r}
54{,}807{,}000 \\
-\ 15{,}395{,}000 \\
\hline
39{,}412{,}000
\end{array}
$$

There will be 39,412,000 students enrolled in elementary schools in 2001.

33.

Total cost	is	cost of sweaters	plus	cost of shirts
↓	↓	↓	↓	↓
Total cost	=	(3 × 38)	+	(5 × 25)

$$
\begin{array}{ccc}
38 & 25 & 114 \\
\underline{\times\ 3} & \underline{\times\ 5} & \underline{+\ 125} \\
114 & 125 & 239
\end{array}
$$

The total cost is $239.

37. Add costs of the items.
$3 + 4 + 3 + 2(1) = 12$
The order will cost $12.

41. 284 rounds to 300. Divide total rooms by number of properties to find the average number of rooms.

$$
\begin{array}{r}
91 \\
300\overline{)27300} \\
-\ 2700 \\
\hline
300 \\
-\ 300 \\
\hline
0
\end{array}
$$

There are an average of 91 guest rooms per hotel.

45.

$$
\begin{array}{r}
1064 \\
-\ 768 \\
\hline
296
\end{array}
$$

Motorola received 296 more patents than Eastman Kodak.

EXERCISE SET 1.8

1. $3 \cdot 3 \cdot 3 \cdot 3 = 3^4$

5. $12 \cdot 12 \cdot 12 = 12^3$

9. $9 \cdot 9 \cdot 9 \cdot 8 = (9 \cdot 9 \cdot 9)8 = 9^3 \cdot 8$

13. $3 \cdot 2 \cdot 2 \cdot 5 \cdot 5 \cdot 5 = 3(2 \cdot 2)(5 \cdot 5 \cdot 5) = 3 \cdot 2^2 \cdot 5^3$

17. $5^3 = 5 \cdot 5 \cdot 5 = 125$

21. $2^{10} = 2 \cdot 2 \cdot 2 \cdot 2 \cdot 2 \cdot 2 \cdot 2 \cdot 2 \cdot 2 \cdot 2 = 1024$

25. $3^5 = 3 \cdot 3 \cdot 3 \cdot 3 \cdot 3 = 243$

29. $4^3 = 4 \cdot 4 \cdot 4 = 64$

33. $9^3 = 9 \cdot 9 \cdot 9 = 729$

37. $10^4 = 10 \cdot 10 \cdot 10 \cdot 10 = 10{,}000$

41. $1920^1 = 1920$

45. $15 + 3 \cdot 2 = 15 + 6$
$\qquad\qquad = 21$

49. $5 \cdot 9 - 16 = 45 - 16$
$\qquad\qquad = 29$

53. $14 + \dfrac{24}{8} = 14 + 3$
$\qquad\qquad = 17$

57. $0 \div 6 + 4 \cdot 7 = 0 + 4 \cdot 7$
$\qquad\qquad\qquad = 0 + 28$
$\qquad\qquad\qquad = 28$

61. $(6 + 8) \div 2 = 14 \div 2$
$\qquad\qquad\qquad = 7$

65. $(3 + 5^2) \div 2 = (3 + 25) \div 2$
$\qquad\qquad\qquad = 28 \div 2$
$\qquad\qquad\qquad = 14$

69. $\dfrac{18 + 6}{2^4 - 4} = \dfrac{24}{16 - 4}$
$\qquad\qquad = \dfrac{24}{12}$
$\qquad\qquad = 2$

73. $\dfrac{7(9 - 6) + 3}{3^2 - 3} = \dfrac{7 \cdot 3 + 3}{9 - 3}$
$\qquad\qquad = \dfrac{21 + 3}{6}$
$\qquad\qquad = \dfrac{24}{6}$
$\qquad\qquad = 4$

77. $3^4 - [35 - (12 - 6)] = 3^4 - [35 - 6]$
$\qquad\qquad\qquad = 81 - 29$
$\qquad\qquad\qquad = 52$

81. $8 \cdot [4 + (6 - 1) \cdot 2] - 50 \cdot 2 = 8 \cdot [4 + 5 \cdot 2] - 50 \cdot 2$
$\qquad\qquad\qquad = 8 \cdot [4 + 10] - 100$
$\qquad\qquad\qquad = 8 \cdot 14 - 100$
$\qquad\qquad\qquad = 112 - 100$
$\qquad\qquad\qquad = 12$

85. Area of a square $= (\text{side})^2$
$\qquad\qquad\qquad = (20 \text{ miles})^2$
$\qquad\qquad\qquad = 400 \text{ square miles}$

89. Area of base $= (\text{side})^2$
$\qquad\qquad\qquad = (100 \text{ meters})^2$
$\qquad\qquad\qquad = 10,000 \text{ square meters}$

93. $24 \div (3 \cdot 2) + 2 \cdot 5 = 24 \div 6 + 2 \cdot 5$
$\qquad\qquad\qquad = 4 + 10$
$\qquad\qquad\qquad = 14$

97. $(7 + 2^4)^5 - (3^5 - 2^4)^2 = (7 + 16)^5 - (3^5 - 2^4)^2$
$\qquad\qquad\qquad = 23^5 - (243 - 16)^2$
$\qquad\qquad\qquad = 6,436,343 - 227^2$
$\qquad\qquad\qquad = 6,434,343 - 51,529$
$\qquad\qquad\qquad = 6,384,814$

Chapter 1 Test

1.
$\begin{array}{r} 1 \\ 59 \\ + 82 \\ \hline 141 \end{array}$

5. $2^3 \cdot 5^2 = 2 \cdot 2 \cdot 2 \cdot 5 \cdot 5 = 200$

9. $62 \div 0$ is defined

13. To round 52,369 to the nearest thousand observe that the digit in the hundreds place is 3. Since this digit is less than 5, we do not add 1 to the digit in the thousands place. The number 52,369 rounded to the nearest thousand is 52,000.

17.

Total cost	is	17	times	7
↓	↓	↓	↓	↓
Total cost	=	17	×	7

$\begin{array}{r} 17 \\ \times 7 \\ \hline 119 \end{array}$

The total cost of the tickets is $119.

21. For 1997, the average verbal SAT score in Massachusetts is 508.

Chapter 2

Exercise Set 2.1

1. 1 out of 3 equal parts is shaded: $\dfrac{1}{3}$

5. 9 out of 16 equal parts are shaded: $\dfrac{9}{16}$

9. 4 out of 9 equal parts are shaded: $\dfrac{4}{9}$

13. $131 - 42 = 89$ non-freshmen
non-freshmen → 89
$\qquad$ students → 131
$\dfrac{89}{131}$ of the students are not freshmen.

17. born in Virginia → 8
$\quad$ U.S. presidents → 42
$\dfrac{8}{42}$ of U.S. presidents were born in Virginia.

21. Each part is $\dfrac{1}{3}$, and there are 11 parts shaded, or 3 wholes and 2 more parts.

$\quad$ **a.** $3\dfrac{2}{3}$ $\qquad\qquad$ **b.** $\dfrac{11}{3}$

25. Each part is $\dfrac{1}{3}$, and there are 4 parts shaded, or 1 whole and 1 more part.

$\quad$ **a.** $1\dfrac{1}{3}$ $\qquad\qquad$ **b.** $\dfrac{4}{3}$

29. Each part is $\dfrac{1}{9}$, and there are 14 parts shaded, or 1 whole and 5 more parts.

$\quad$ **a.** $1\dfrac{5}{9}$ $\qquad\qquad$ **b.** $\dfrac{14}{9}$

33. $3\dfrac{2}{3} = \dfrac{3 \cdot 3 + 2}{3} = \dfrac{11}{3}$

37. $2\dfrac{11}{15} = \dfrac{15 \cdot 2 + 11}{15} = \dfrac{41}{15}$

41. $6\dfrac{5}{8} = \dfrac{8 \cdot 6 + 5}{8} = \dfrac{53}{8}$

45. $4\dfrac{13}{24} = \dfrac{24 \cdot 4 + 13}{24} = \dfrac{109}{24}$

49. $9\dfrac{7}{20} = \dfrac{20 \cdot 9 + 7}{20} = \dfrac{187}{20}$

53.
$$
\begin{array}{r}
3\ \text{R}\ 3 \\
13\overline{)42} \\
-39 \\
\hline
3
\end{array}
$$

$\dfrac{42}{13} = 3\dfrac{3}{13}$

57.
$$
\begin{array}{r}
33 \\
6\overline{)198} \\
-18 \\
\hline
18 \\
-18 \\
\hline
0
\end{array}
$$

$\dfrac{198}{6} = 33$

61.
$$
\begin{array}{r}
4\ \text{R}5 \\
8\overline{)37} \\
-32 \\
\hline
5
\end{array}
$$

$\dfrac{37}{8} = 4\dfrac{5}{8}$

65.
$$
\begin{array}{r}
10\ \text{R}\ 17 \\
23\overline{)247} \\
-23 \\
\hline
17 \\
-0 \\
\hline
17
\end{array}
$$

$\dfrac{247}{23} = 10\dfrac{17}{23}$

69. $3^2 = 3 \cdot 3 = 9$

73. $7^2 = 7 \cdot 7 = 49$

77. $38\dfrac{41}{79} = \dfrac{79 \cdot 38 + 41}{79} = \dfrac{3043}{79}$

81. Total number of licensees
$= 87 + 56 + 21 + 8$
$= 172$

$\dfrac{56}{172}$ of the licensees are colleges or universities.

Exercise Set 2.2

1. $1 \cdot 8 = 8$
$2 \cdot 4 = 8$
The factors of 8 are 1, 2, 4, and 8.

5. $1 \cdot 4 = 4$
$2 \cdot 2 = 4$
The factors of 4 are 1, 2, and 4.

9. $1 \cdot 7 = 7$
The factors of 7 are 1 and 7.

13. $1 \cdot 12 = 12$
$2 \cdot 6 = 12$
$3 \cdot 4 = 12$
The factors of 12 are 1, 2, 3, 4, 6, and 12.

17. Prime, since its only factors are 1 and 7.

21. Composite, since its factors are 1, 2, 5, and 10.

25. Composite, since its factors are 1, 2, 3, and 6.

29. Prime, since its only factors are 1 and 31.

33.
$$
\begin{array}{r}
3 \\
2\overline{)\ \ 6} \\
2\overline{)\ 12}
\end{array}
$$
$12 = 2^2 \cdot 3$

37.
$$
\begin{array}{r}
5 \\
2\overline{)\ 10} \\
2\overline{)\ 20} \\
2\overline{)\ 40}
\end{array}
$$
$40 = 2^3 \cdot 5$

41.
$$
\begin{array}{r}
13 \\
3\overline{)\ 39}
\end{array}
$$
$39 = 3 \cdot 13$

45.
$$
\begin{array}{r}
3 \\
3\overline{)\ 9} \\
3\overline{)\ 27} \\
2\overline{)\ 54}
\end{array}
$$
$54 = 2 \cdot 3^3$

49.
$$
\begin{array}{r}
11 \\
5\overline{)\ 55} \\
2\overline{)\ 110}
\end{array}
$$
$110 = 2 \cdot 5 \cdot 11$

53.
$$
\begin{array}{r}
2 \\
2\overline{)\ 4} \\
2\overline{)\ 8} \\
2\overline{)\ 16} \\
2\overline{)\ 32} \\
2\overline{)\ 64} \\
2\overline{)\ 128}
\end{array}
$$
$128 = 2^7$

57.
$$
\begin{array}{r}
5 \\
5\overline{)\ 25} \\
3\overline{)\ 75} \\
2\overline{)\ 150} \\
2\overline{)\ 300}
\end{array}
$$
$300 = 2^2 \cdot 3 \cdot 5^2$

61.
$$
\begin{array}{r}
7 \\
5\overline{)\ 35} \\
3\overline{)\ 105} \\
3\overline{)\ 315} \\
3\overline{)\ 945}
\end{array}
$$
$945 = 3^3 \cdot 5 \cdot 7$

65. To round 4267 to the nearest hundred, observe that the digit in the tens place is 6. Since this digit is at least 5, we need to add 1 to the digit in the hundreds place. The number 4267 rounded to the nearest hundred is 4300.

69. To round 55,342 to the nearest hundred, observe that the digit in the tens place is 4. Since this digit is less than 5, we do not add 1 to the digit in the hundreds place. The number 55,342 rounded to the nearest hundred is 55,300.

73. To round 1247 to the nearest thousand, observe that the digit in the hundreds place is 2. Since this digit is less than 5, we do not add 1 to the digit in the thousands place. The number 1247 rounded to the nearest thousand is 1000.

77. Answers may vary.

Exercise Set 2.3

1. $\dfrac{3}{12} = \dfrac{3}{2 \cdot 2 \cdot 3} = \dfrac{1}{2 \cdot 2} = \dfrac{1}{4}$

5. $\dfrac{14}{16} = \dfrac{2 \cdot 7}{2 \cdot 2 \cdot 2 \cdot 2} = \dfrac{7}{2 \cdot 2 \cdot 2} = \dfrac{7}{8}$

9. $\dfrac{35}{42} = \dfrac{5 \cdot 7}{2 \cdot 3 \cdot 7} = \dfrac{5}{2 \cdot 3} = \dfrac{5}{6}$

13. $\dfrac{21}{49} = \dfrac{3 \cdot 7}{7 \cdot 7} = \dfrac{3}{7}$

17. $\dfrac{36}{63} = \dfrac{2 \cdot 2 \cdot 3 \cdot 3}{3 \cdot 3 \cdot 7} = \dfrac{2 \cdot 2}{7} = \dfrac{4}{7}$

21. $\dfrac{25}{40} = \dfrac{5 \cdot 5}{2 \cdot 2 \cdot 2 \cdot 5} = \dfrac{5}{2 \cdot 2 \cdot 2} = \dfrac{5}{8}$

25. $\dfrac{36}{24} = \dfrac{2 \cdot 2 \cdot 3 \cdot 3}{2 \cdot 2 \cdot 2 \cdot 3} = \dfrac{3}{2}$ or $1\dfrac{1}{2}$

29. $\dfrac{70}{196} = \dfrac{2 \cdot 5 \cdot 7}{2 \cdot 2 \cdot 7 \cdot 7} = \dfrac{5}{2 \cdot 7} = \dfrac{5}{14}$

33. Equivalent, since the cross products are equal:
$10 \cdot 9 = 90$ and $15 \cdot 6 = 90$.

37. Not equivalent, since the cross products are not equal:
$10 \cdot 15 = 150$ and $13 \cdot 12 = 156$.

41. Equivalent, since the cross products are equal:
$4 \cdot 15 = 60$ and $10 \cdot 6 = 60$.

45. $\dfrac{6 \text{ hours}}{8 \text{ hours}} = \dfrac{2 \cdot 3}{2 \cdot 2 \cdot 2} = \dfrac{3}{2 \cdot 2} = \dfrac{3}{4}$

6 hours represents $\dfrac{3}{4}$ of a work shift.

49. $\dfrac{235}{370} = \dfrac{5 \cdot 47}{2 \cdot 5 \cdot 37} = \dfrac{47}{2 \cdot 37} = \dfrac{47}{74}$

$\dfrac{47}{74}$ of individuals who have flown in space were Americans.

53. $\begin{array}{r} 91 \\ \times\ 4 \\ \hline 364 \end{array}$

57. $\begin{array}{r} 72 \\ \times\ 35 \\ \hline 360 \\ 2160 \\ \hline 2520 \end{array}$

61. $\dfrac{372}{620} = \dfrac{2 \cdot 2 \cdot 3 \cdot 31}{2 \cdot 2 \cdot 5 \cdot 31} = \dfrac{3}{5}$

65. $3 + 1 = 4$

$\dfrac{4}{100} = \dfrac{2 \cdot 2}{2 \cdot 2 \cdot 5 \cdot 5} = \dfrac{1}{5 \cdot 5} = \dfrac{1}{25}$

$\dfrac{1}{25}$ of donors have type AB (either Rh-positive or Rh-negative).

EXERCISE SET 2.4

1. $\dfrac{1}{3} \cdot \dfrac{2}{5} = \dfrac{1 \cdot 2}{3 \cdot 5} = \dfrac{2}{15}$

5. $\dfrac{3}{10} \cdot \dfrac{3}{8} = \dfrac{3 \cdot 3}{10 \cdot 8} = \dfrac{9}{80}$

9. $\dfrac{2}{7} \cdot \dfrac{5}{8} = \dfrac{2 \cdot 5}{7 \cdot 8} = \dfrac{\overset{1}{\cancel{2}} \cdot 5}{7 \cdot \underset{4}{\cancel{2}} \cdot 4} = \dfrac{5}{28}$

13. $\dfrac{5}{8} \cdot \dfrac{9}{4} = \dfrac{5 \cdot 9}{8 \cdot 4} = \dfrac{45}{32}$ or $1\dfrac{13}{32}$

17. $\dfrac{18}{20} \cdot \dfrac{36}{99} = \dfrac{18 \cdot 36}{20 \cdot 99}$

$= \dfrac{\overset{1}{\cancel{2}} \cdot \overset{1}{\cancel{9}} \cdot \overset{1}{\cancel{2}} \cdot 18}{\underset{1}{\cancel{2}} \cdot \underset{1}{\cancel{2}} \cdot 5 \cdot \underset{1}{\cancel{9}} \cdot 11}$

$= \dfrac{18}{5 \cdot 11}$

$= \dfrac{18}{55}$

21. $\dfrac{14}{21} \cdot \dfrac{15}{16} = \dfrac{14 \cdot 15}{21 \cdot 16}$

$= \dfrac{\overset{1}{\cancel{2}} \cdot \overset{1}{\cancel{7}} \cdot \overset{1}{\cancel{3}} \cdot 5}{\underset{1}{\cancel{3}} \cdot \underset{1}{\cancel{7}} \cdot \underset{1}{\cancel{2}} \cdot 8}$

$= \dfrac{5}{8}$

25. $\dfrac{3}{8} \cdot \dfrac{9}{10} = \dfrac{3 \cdot 9}{8 \cdot 10} = \dfrac{27}{80}$

29. $\dfrac{7}{72} \cdot \dfrac{9}{49} = \dfrac{7 \cdot 9}{72 \cdot 49}$

$= \dfrac{\overset{1}{\cancel{7}} \cdot \overset{1}{\cancel{9}}}{\underset{1}{\cancel{9}} \cdot 8 \cdot \underset{1}{\cancel{7}} \cdot 7}$

$= \dfrac{1}{8 \cdot 7}$

$= \dfrac{1}{56}$

33. $\dfrac{5}{8} \cdot 4 = \dfrac{5}{8} \cdot \dfrac{4}{1}$

$= \dfrac{5 \cdot 4}{8 \cdot 1}$

$= \dfrac{5 \cdot \overset{1}{\cancel{4}}}{\underset{1}{\cancel{4}} \cdot 2 \cdot 1}$

$= \dfrac{5}{2}$ or $2\dfrac{1}{2}$

37. $\dfrac{2}{5} \cdot 4\dfrac{1}{6} = \dfrac{2}{5} \cdot \dfrac{25}{6}$

$= \dfrac{2 \cdot 25}{5 \cdot 6}$

$= \dfrac{\overset{1}{\cancel{2}} \cdot \overset{1}{\cancel{5}} \cdot 5}{\underset{1}{\cancel{5}} \cdot \underset{1}{\cancel{2}} \cdot 3}$

$= \dfrac{5}{3}$ or $1\dfrac{2}{3}$

41. $2\dfrac{1}{5} \cdot 3\dfrac{1}{2} = \dfrac{11}{5} \cdot \dfrac{7}{2} = \dfrac{11 \cdot 7}{5 \cdot 2} = \dfrac{77}{10}$ or $7\dfrac{7}{10}$

45. $\frac{3}{4} \cdot 16 = \frac{3}{4} \cdot \frac{16}{1}$

$= \frac{3 \cdot 16}{4 \cdot 1}$

$= \frac{3 \cdot \overset{1}{\cancel{4}} \cdot 4}{\underset{1}{\cancel{4}} \cdot 1}$

$= \frac{3 \cdot 4}{1}$

$= \frac{12}{1}$

$= 12$

49. $1\frac{1}{5} \cdot 12\frac{1}{2} = \frac{6}{5} \cdot \frac{25}{2}$

$= \frac{6 \cdot 25}{5 \cdot 2}$

$= \frac{\overset{1}{\cancel{2}} \cdot 3 \cdot \overset{1}{\cancel{5}} \cdot 5}{\underset{1}{\cancel{5}} \cdot \underset{1}{\cancel{2}}}$

$= \frac{3 \cdot 5}{1}$

$= 15$

53. $5\frac{1}{2} \cdot 5 = \frac{11}{2} \cdot \frac{5}{1} = \frac{11 \cdot 5}{2 \cdot 1} = \frac{55}{2}$ or $27\frac{1}{2}$

Holly Hanson drives $27\frac{1}{2}$ miles in a week.

57. $\frac{2}{5} \cdot 4 = \frac{2}{5} \cdot \frac{4}{1} = \frac{2 \cdot 4}{5 \cdot 1} = \frac{8}{5}$ or $1\frac{3}{5}$

$1\frac{3}{5}$ feet of the post will be buried.

61. $\frac{2}{3} \cdot 2757 = \frac{2}{3} \cdot \frac{2757}{1}$

$= \frac{2 \cdot 2757}{3 \cdot 1}$

$= \frac{2 \cdot \overset{1}{\cancel{3}} \cdot 919}{\underset{1}{\cancel{3}} \cdot 1}$

$= \frac{2 \cdot 919}{1}$

$= \frac{1838}{1}$

$= 1838$

The cruise sale price is $1838.

65. $\frac{5}{14} \cdot \frac{1}{5} = \frac{\overset{1}{\cancel{5}} \cdot 1}{14 \cdot \underset{1}{\cancel{5}}} = \frac{1}{14}$

The area is $\frac{1}{14}$ square foot.

69. $5\frac{1}{2} \cdot 5\frac{1}{2} = \frac{11}{2} \cdot \frac{11}{2}$

$= \frac{11 \cdot 11}{2 \cdot 2}$

$= \frac{121}{4}$ or $30\frac{1}{4}$

The area of the chip model is $30\frac{1}{4}$ square inches.

73. $\frac{1}{5} \cdot 12{,}000 = \frac{1}{5} \cdot \frac{12{,}000}{1}$

$= \frac{1 \cdot 12{,}000}{5 \cdot 1}$

$= \frac{1 \cdot \overset{1}{\cancel{5}} \cdot 2400}{\underset{1}{\cancel{5}} \cdot 1}$

$= \frac{2400}{1}$

$= 2400$

The Rodriguez family drove 2400 miles on family business.

77.
$$\begin{array}{r} 56 \text{ R } 12 \\ 23\overline{)1300} \\ -115 \\ \hline 150 \\ -138 \\ \hline 12 \end{array}$$

EXERCISE SET 2.5

1. The reciprocal of $\frac{4}{7}$ is $\frac{7}{4}$.

5. The reciprocal of 15 or $\frac{15}{1}$ is $\frac{1}{15}$.

9. $\frac{2}{3} \div \frac{5}{6} = \frac{2}{3} \cdot \frac{6}{5}$

$= \frac{2 \cdot 6}{3 \cdot 5}$

$= \frac{2 \cdot \overset{1}{\cancel{3}} \cdot 2}{\underset{1}{\cancel{3}} \cdot 5}$

$= \frac{4}{5}$

13. $\frac{8}{9} \div \frac{1}{2} = \frac{8}{9} \cdot \frac{2}{1} = \frac{8 \cdot 2}{9 \cdot 1} = \frac{16}{9}$ or $1\frac{7}{9}$

17. $\frac{3}{5} \div \frac{4}{5} = \frac{3}{5} \cdot \frac{5}{4} = \frac{3 \cdot 5}{5 \cdot 4} = \frac{3}{4}$

21. $\frac{1}{10} \div \frac{10}{1} = \frac{1}{10} \cdot \frac{1}{10} = \frac{1 \cdot 1}{10 \cdot 10} = \frac{1}{100}$

25. $\frac{3}{7} \div \frac{4}{7} = \frac{3}{7} \cdot \frac{7}{4} = \frac{3 \cdot 7}{7 \cdot 4} = \frac{3}{4}$

29. $\frac{7}{45} \div \frac{4}{25} = \frac{7}{45} \cdot \frac{25}{4}$

$= \frac{7 \cdot 25}{45 \cdot 4}$

$= \frac{7 \cdot 5 \cdot 5}{5 \cdot 9 \cdot 4}$

$= \frac{7 \cdot 5}{9 \cdot 4}$

$= \frac{35}{36}$

33. $\dfrac{3}{25} \div \dfrac{27}{40} = \dfrac{3}{25} \cdot \dfrac{40}{27}$

$= \dfrac{3 \cdot 40}{25 \cdot 27}$

$= \dfrac{3 \cdot 5 \cdot 8}{5 \cdot 5 \cdot 3 \cdot 9}$

$= \dfrac{8}{5 \cdot 9}$

$= \dfrac{8}{45}$

37. $8 \div \dfrac{3}{5} = \dfrac{8}{1} \div \dfrac{3}{5}$

$= \dfrac{8}{1} \cdot \dfrac{5}{3}$

$= \dfrac{8 \cdot 5}{1 \cdot 3}$

$= \dfrac{40}{3}$ or $13\dfrac{1}{3}$

41. $\dfrac{5}{12} \div 2\dfrac{1}{3} = \dfrac{5}{12} \div \dfrac{7}{3}$

$= \dfrac{5}{12} \cdot \dfrac{3}{7}$

$= \dfrac{5 \cdot 3}{12 \cdot 7}$

$= \dfrac{5 \cdot 3}{3 \cdot 4 \cdot 7}$

$= \dfrac{5}{4 \cdot 7}$

$= \dfrac{5}{28}$

45. $12 \div \dfrac{1}{8} = \dfrac{12}{1} \div \dfrac{1}{8}$

$= \dfrac{12}{1} \cdot \dfrac{8}{1}$

$= \dfrac{12 \cdot 8}{1 \cdot 1}$

$= \dfrac{96}{1}$

$= 96$

49. $3\dfrac{1}{3} \div 4 = \dfrac{10}{3} \div \dfrac{4}{1}$

$= \dfrac{10}{3} \cdot \dfrac{1}{4}$

$= \dfrac{10 \cdot 1}{3 \cdot 4}$

$= \dfrac{2 \cdot 5 \cdot 1}{3 \cdot 2 \cdot 2}$

$= \dfrac{5}{3 \cdot 2}$

$= \dfrac{5}{6}$

Each dose contains $\dfrac{5}{6}$ tablespoon.

53. $15\dfrac{1}{5} \div 24 = \dfrac{76}{5} \div \dfrac{24}{1}$

$= \dfrac{76}{5} \cdot \dfrac{1}{24}$

$= \dfrac{76 \cdot 1}{5 \cdot 24}$

$= \dfrac{4 \cdot 19 \cdot 1}{5 \cdot 4 \cdot 6}$

$= \dfrac{19}{5 \cdot 6}$

$= \dfrac{19}{30}$

An average of $\dfrac{19}{30}$ inch of rain fell each hour.

57.
$$
\begin{array}{r}
\overset{2\,2}{27} \\
76 \\
+\,98 \\
\hline
201
\end{array}
$$

61.
$$
\begin{array}{r}
2000 \\
-\,431 \\
\hline
1569
\end{array}
$$

65. $4 \div \dfrac{1}{4} = \dfrac{4}{1} \div \dfrac{1}{4}$

$= \dfrac{4}{1} \cdot \dfrac{4}{1}$

$= \dfrac{4 \cdot 4}{1 \cdot 1}$

$= \dfrac{16}{1}$

$= 16$

The Broncos played 16 regular season games.

Chapter 2 Test

1. $\dfrac{4}{4} \div \dfrac{3}{4} = \dfrac{4}{4} \cdot \dfrac{4}{3} = \dfrac{\overset{1}{4} \cdot 4}{\underset{1}{4} \cdot 3} = \dfrac{4}{3}$ or $1\dfrac{1}{3}$

5. $8 \div \dfrac{1}{2} = \dfrac{8}{1} \div \dfrac{1}{2} = \dfrac{8}{1} \cdot \dfrac{2}{1} = \dfrac{8 \cdot 2}{1 \cdot 1} = \dfrac{16}{1} = 16$

9. $\dfrac{16}{3} \div \dfrac{3}{12} = \dfrac{16}{3} \cdot \dfrac{12}{3}$

$= \dfrac{16 \cdot 12}{3 \cdot 3}$

$= \dfrac{16 \cdot \overset{1}{3} \cdot 4}{\underset{1}{3} \cdot 3}$

$= \dfrac{64}{3}$ or $21\dfrac{1}{3}$

13. $7\dfrac{2}{3} = \dfrac{3 \cdot 7 + 2}{3} = \dfrac{23}{3}$

17. $1\dfrac{8}{9} \cdot \dfrac{2}{3} = \dfrac{17}{9} \cdot \dfrac{2}{3} = \dfrac{17 \cdot 2}{9 \cdot 3} = \dfrac{34}{27}$ or $1\dfrac{7}{27}$

The area of the figure is $1\dfrac{7}{27}$ square miles.

21. Area = length · width

$$= (100 + 10 + 10) \cdot \left(53\frac{1}{3} + 10 + 10\right)$$

$$= 120 \cdot 73\frac{1}{3}$$

$$= \frac{120}{1} \cdot \frac{220}{3}$$

$$= \frac{120 \cdot 220}{1 \cdot 3}$$

$$= \frac{\overset{1}{\cancel{3}} \cdot 40 \cdot 220}{1 \cdot \underset{1}{\cancel{3}}}$$

$$= \frac{40 \cdot 220}{1}$$

$$= \frac{8800}{1}$$

$$= 8800$$

8800 square yards of turf are needed.

Chapter 3

Exercise Set 3.1

1. $\dfrac{1}{7} + \dfrac{2}{7} = \dfrac{1+2}{7} = \dfrac{3}{7}$

5. $\dfrac{2}{9} + \dfrac{4}{9} = \dfrac{2+4}{9} = \dfrac{6}{9} = \dfrac{2 \cdot 3}{3 \cdot 3} = \dfrac{2}{3}$

9. $\dfrac{3}{14} + \dfrac{4}{14} = \dfrac{3+4}{14} = \dfrac{7}{14} = \dfrac{1 \cdot 7}{2 \cdot 7} = \dfrac{1}{2}$

13. $\dfrac{4}{13} + \dfrac{2}{13} + \dfrac{1}{13} = \dfrac{4+2+1}{13} = \dfrac{7}{13}$

17. $\dfrac{10}{11} - \dfrac{4}{11} = \dfrac{10-4}{11} = \dfrac{6}{11}$

21. $\dfrac{7}{4} - \dfrac{3}{4} = \dfrac{7-3}{4} = \dfrac{4}{4} = \dfrac{1 \cdot 4}{1 \cdot 4} = 1$

25. $\dfrac{25}{12} - \dfrac{15}{12} = \dfrac{25-15}{12} = \dfrac{10}{12} = \dfrac{5 \cdot 2}{6 \cdot 2} = \dfrac{5}{6}$

29. $\dfrac{27}{33} - \dfrac{8}{33} = \dfrac{27-8}{33} = \dfrac{19}{33}$

33. The perimeter is the distance around.
A rectangle has 2 sets of equal sides.
Add the lengths of the sides.

$$\frac{5}{12} + \frac{7}{12} + \frac{5}{12} + \frac{7}{12} = \frac{5+7+5+7}{12}$$

$$= \frac{24}{12}$$

$$= \frac{2 \cdot 12}{1 \cdot 12}$$

$$= 2$$

The perimeter is 2 meters.

37. Find the remaining amount of track to be inspected.
Subtract $\dfrac{5}{20}$ of a mile from $\dfrac{19}{20}$ of a mile.

$$\frac{19}{20} - \frac{5}{20} = \frac{19-5}{20} = \frac{14}{20} = \frac{7 \cdot 2}{10 \cdot 2} = \frac{7}{10}$$

She needs to inspect $\dfrac{7}{10}$ of a mile more.

41. Subtract $\dfrac{16}{50}$ from $\dfrac{39}{50}$.

$$\frac{39}{50} - \frac{16}{50} = \frac{39-16}{50} = \frac{23}{50}$$

The fraction of states that had speed limits that were less than 70 mph was $\dfrac{23}{50}$.

45.
$$
\begin{array}{r}
2 \\
2\overline{)\,4} \\
2\overline{)\,8}
\end{array}
$$

$$8 = 2^3$$

49. $\dfrac{3}{8} + \dfrac{7}{8} - \dfrac{5}{8} = \dfrac{3+7-5}{8}$

$$= \dfrac{5}{8}$$

53. Find the total distance.

$$\frac{3}{8} + \frac{3}{8} - \frac{4}{8} = \frac{3+3-4}{8}$$

$$= \frac{2}{8}$$

$$= \frac{1 \cdot 2}{4 \cdot 2}$$

$$= \frac{1}{4}$$

He is $\dfrac{1}{4}$ of a mile from his starting point.

55. answers may vary

Exercise Set 3.2

1. Multiples of 3:   3, 6, 9, ⑫, 15, ...

Multiples of 4:   4, 8, ⑫, 16, ...

LCM: 12

5. Multiples of 12:   12, 24, ㊱, 48, 60, 72, ...

Multiples of 18:   18, ㊱, 54, ...

LCM: 36

9. $18 = ② \cdot ③ \cdot 3$

$21 = 3 \cdot ⑦$

The LCM is
$2 \cdot 3 \cdot 3 \cdot 7 = 126.$

13. $8 = ②\cdot②\cdot②$

$24 = 2 \cdot 2 \cdot 2 \cdot ③$

The LCM is
$2 \cdot 2 \cdot 2 \cdot 3 = 24.$

17. $25 = \boxed{5 \cdot 5}$

$15 = \boxed{3} \cdot 5$

$6 = \boxed{2} \cdot 3$

The LCM is $2 \cdot 3 \cdot 5 \cdot 5 = 150$.

21. $84 = \boxed{2 \cdot 2} \cdot \boxed{3} \cdot 7$

$294 = 2 \cdot 3 \cdot \boxed{7 \cdot 7}$.

The LCM is $2 \cdot 2 \cdot 3 \cdot 7 \cdot 7 = 588$.

25. $3 = \boxed{3}$

$21 = 3 \cdot \boxed{7}$

$51 = 3 \cdot \boxed{17}$

The LCM is $3 \cdot 7 \cdot 17 = 357$.

29. $8 = \boxed{2 \cdot 2 \cdot 2}$

$6 = 2 \cdot 3$

$27 = \boxed{3 \cdot 3 \cdot 3}$

The LCM is $2 \cdot 2 \cdot 2 \cdot 3 \cdot 3 \cdot 3 = 216$.

33. $\dfrac{4}{7} = \dfrac{4 \cdot 5}{7 \cdot 5} = \dfrac{20}{35}$

37. $\dfrac{2}{5} = \dfrac{2 \cdot 5}{5 \cdot 5} = \dfrac{10}{25}$

41. $\dfrac{10}{7} = \dfrac{10 \cdot 3}{7 \cdot 3} = \dfrac{30}{21}$

45. $\dfrac{2}{3} = \dfrac{2 \cdot 15}{3 \cdot 15} = \dfrac{30}{45}$

49. $\dfrac{4}{3} = \dfrac{4 \cdot 3}{3 \cdot 3} = \dfrac{12}{9}$

53. $\dfrac{7}{10} - \dfrac{2}{10} = \dfrac{7 - 2}{10} = \dfrac{5}{10} = \dfrac{1 \cdot 5}{2 \cdot 5} = \dfrac{1}{2}$

57. $\dfrac{23}{18} - \dfrac{15}{18} = \dfrac{23 - 15}{18} = \dfrac{8}{18} = \dfrac{4 \cdot 2}{9 \cdot 2} = \dfrac{4}{9}$

61. $\dfrac{37}{165} = \dfrac{37 \cdot 22}{165 \cdot 22} = \dfrac{814}{3630}$

63. answers may vary

EXERCISE SET 3.3

1. *Step 1.* The LCD for the denominators 3 and 6 is 6.

Step 2. $\dfrac{2}{3} = \dfrac{2 \cdot 2}{3 \cdot 2} = \dfrac{4}{6}, \dfrac{1}{6}$ already has a denominator of 6.

Step 3. $\dfrac{2}{3} + \dfrac{1}{6} = \dfrac{4}{6} + \dfrac{1}{6} = \dfrac{5}{6}$

Step 4. $\dfrac{5}{6}$ is in simplest form.

5. *Step 1.* The LCD for the denominators 11 and 33 is 33.

Step 2. $\dfrac{2}{11} = \dfrac{2 \cdot 3}{11 \cdot 33} = \dfrac{6}{33}, \dfrac{2}{3}$ already has a denominator of 33.

Step 3. $\dfrac{2}{11} + \dfrac{2}{33} = \dfrac{6}{33} + \dfrac{2}{33} = \dfrac{8}{33}$

Step 4. $\dfrac{8}{33}$ is in simplest form.

9. *Step 1.* The LCD for the denominators 35 and 7 is 35.

Step 2. $\dfrac{11}{35}$ already has a denominator of 35,

$\dfrac{2}{7} = \dfrac{2 \cdot 5}{7 \cdot 5} = \dfrac{10}{35}$

Step 3. $\dfrac{11}{35} + \dfrac{2}{7} = \dfrac{11}{35} + \dfrac{10}{35} = \dfrac{21}{35}$

Step 4. $\dfrac{21}{35} = \dfrac{3 \cdot 7}{5 \cdot 7} = \dfrac{3}{5}$

13. *Step 1.* The LCD for the denominators 15 and 12 is 60.

Step 2. $\dfrac{7}{15} = \dfrac{7 \cdot 4}{15 \cdot 4} = \dfrac{28}{60}, \dfrac{5}{12} = \dfrac{5 \cdot 5}{12 \cdot 5} = \dfrac{25}{60}$

Step 3. $\dfrac{7}{15} + \dfrac{5}{12} = \dfrac{28}{60} + \dfrac{25}{60} = \dfrac{53}{60}$

Step 4. $\dfrac{53}{60}$ is in simplest form.

17. *Step 1.* The LCD for the denominators 7, 8, and 2 is 56.

Step 2. $\dfrac{5}{7} = \dfrac{5 \cdot 8}{7 \cdot 8} = \dfrac{40}{56}$,

$\dfrac{1}{8} = \dfrac{1 \cdot 7}{8 \cdot 7} = \dfrac{7}{56}$,

$\dfrac{1}{2} = \dfrac{1 \cdot 28}{2 \cdot 28} = \dfrac{28}{56}$

Step 3. $\dfrac{5}{7} + \dfrac{1}{8} + \dfrac{1}{2} = \dfrac{40}{56} + \dfrac{7}{56} + \dfrac{28}{56} = \dfrac{75}{56}$

Step 4. $\dfrac{75}{56}$ or $1\dfrac{19}{56}$ is in simplest form.

21. *Step 1.* The LCD for the denominators 8 and 16 is 16.

Step 2. $\dfrac{7}{8} = \dfrac{7 \cdot 2}{8 \cdot 2} = \dfrac{14}{16}, \dfrac{3}{16}$ already has a denominator of 16.

Step 3. $\dfrac{7}{8} - \dfrac{3}{16} = \dfrac{14}{16} - \dfrac{3}{16} = \dfrac{11}{16}$

Step 4. $\dfrac{11}{16}$ is in simplest form.

25. *Step 1.* The LCD for the denominators 7 and 8 is 56.

Step 2. $\dfrac{5}{7} = \dfrac{5 \cdot 8}{7 \cdot 8} = \dfrac{40}{56}, \dfrac{1}{8} = \dfrac{1 \cdot 7}{8 \cdot 7} = \dfrac{7}{56}$

Step 3. $\dfrac{5}{7} - \dfrac{1}{8} = \dfrac{40}{56} - \dfrac{7}{56} = \dfrac{33}{56}$

Step 4. $\dfrac{33}{56}$ is in simplest form.

29. *Step 1.* The LCD for the denominators 35 and 7 is 35.

Step 2. $\dfrac{11}{35}$ already has a denominator of 35,

$\dfrac{2}{7} = \dfrac{2 \cdot 5}{7 \cdot 5} = \dfrac{10}{35}$

Step 3. $\dfrac{11}{35} - \dfrac{2}{7} = \dfrac{11}{35} - \dfrac{10}{35} = \dfrac{1}{35}$

Step 4. $\dfrac{1}{35}$ is in simplest form.

33. *Step 1.* The LCD for the denominators 15 and 12 is 60.

Step 2. $\dfrac{7}{15} = \dfrac{7 \cdot 4}{15 \cdot 4} = \dfrac{28}{60}, \dfrac{5}{12} = \dfrac{5 \cdot 5}{12 \cdot 5} = \dfrac{25}{60}$

Step 3. $\dfrac{7}{15} - \dfrac{5}{12} = \dfrac{28}{60} - \dfrac{25}{60} = \dfrac{3}{60}$

Step 4. $\dfrac{3}{60} = \dfrac{1 \cdot 3}{20 \cdot 3} = \dfrac{1}{20}$

37. Add the lengths of the 4 sides.

Step 1. The LCD for the denominators 3, 5, 3, and 5 is 15.

Step 2. $\dfrac{1}{3} = \dfrac{1 \cdot 5}{3 \cdot 5} = \dfrac{5}{15},$

$\dfrac{4}{5} = \dfrac{4 \cdot 3}{5 \cdot 3} = \dfrac{12}{15},$

$\dfrac{1}{3} = \dfrac{1 \cdot 5}{3 \cdot 5} = \dfrac{5}{15},$

$\dfrac{4}{5} = \dfrac{4 \cdot 3}{5 \cdot 3} = \dfrac{12}{15}$

Step 3. $\dfrac{1}{3} + \dfrac{4}{5} + \dfrac{1}{3} + \dfrac{4}{5} = \dfrac{5}{15} + \dfrac{12}{15} + \dfrac{5}{15} + \dfrac{12}{15}$

$= \dfrac{34}{15}$

Step 4. $\dfrac{34}{15}$ or $2\dfrac{4}{15}$ is in simplest form.

The perimeter is $\dfrac{34}{15}$ or $2\dfrac{4}{15}$ centimeters.

41. Find the total weight by adding the two amounts.

Step 1. The LCD for the denominators 4 and 8 is 8.

Step 2. $\dfrac{3}{4} = \dfrac{3 \cdot 2}{4 \cdot 2} = \dfrac{6}{8}, \dfrac{5}{8}$ already has a denominator of 8.

Step 3. $\dfrac{3}{4} + \dfrac{5}{8} = \dfrac{6}{8} + \dfrac{5}{8} = \dfrac{11}{8}$

Step 4 $\dfrac{11}{8}$ or $1\dfrac{3}{8}$ is in simplest form.

The mixture weighs $\dfrac{11}{8}$ or $1\dfrac{3}{8}$ pounds.

45. Add the fractions for 1 or 2 times per week and 3 times per week.

Step 1. The LCD for the denominators 50 and 100 is 100.

Step 2. $\dfrac{23}{50} = \dfrac{23 \cdot 2}{50 \cdot 2} = \dfrac{46}{100}, \dfrac{31}{100}$ already has a denominator of 100.

Step 3. $\dfrac{23}{50} + \dfrac{31}{100} = \dfrac{46}{100} + \dfrac{31}{100} = \dfrac{77}{100}$

Step 4. $\dfrac{77}{100}$ is in simplest form.

The fraction of Americans that eat pasta 1, 2, or 3 times per week is $\dfrac{77}{100}$.

49. $4 \div 7\dfrac{1}{4} = \dfrac{4}{1} \div \dfrac{29}{4} = \dfrac{4}{1} \cdot \dfrac{4}{29} = \dfrac{4 \cdot 4}{1 \cdot 29} = \dfrac{16}{29}$

53. *Step 1.* The LCD for the denominators 55 and 1760 is 1760.

Step 2. $\dfrac{30}{55} = \dfrac{30 \cdot 32}{55 \cdot 32} = \dfrac{960}{1760}, \dfrac{1000}{1760}$ already has a denominator of 1760.

Step 3. $\dfrac{30}{55} + \dfrac{1000}{1760} = \dfrac{960}{1760} + \dfrac{1000}{1760} = \dfrac{1960}{1760}$

Step 4. $\dfrac{1960}{1760} = \dfrac{49 \cdot 40}{44 \cdot 40} = \dfrac{49}{44}$ or $1\dfrac{5}{44}$

57. Find the sum of the fractions for Asia and Europe.

Step 1. The LCD for the denominators 193 and 579 is 579.

Step 2. $\dfrac{58}{193} = \dfrac{58 \cdot 3}{193 \cdot 3} = \dfrac{174}{579}, \dfrac{38}{579}$ already has a denominator of 579.

Step 3. $\dfrac{58}{193} + \dfrac{38}{579} = \dfrac{174}{579} + \dfrac{38}{579} = \dfrac{212}{579}$

Step 4. $\dfrac{212}{579}$ is in simplest form.

The fraction of the world's land area that is accounted for by Asia and Europe is $\dfrac{212}{579}$.

EXERCISE SET 3.4

1.

$\begin{aligned} 4\dfrac{7}{10} \\ +2\dfrac{1}{10} \\ \hline 6\dfrac{8}{10} \end{aligned} \quad = \quad 6\dfrac{4}{5}$

5.

$\begin{aligned} 9\dfrac{1}{5} \quad &= \quad 9\dfrac{5}{25} \\ +8\dfrac{2}{25} \quad &= \quad +8\dfrac{2}{25} \\ \hline &\qquad\quad 17\dfrac{7}{25} \end{aligned}$

9.
$$3\frac{1}{2} \quad = \quad 3\frac{4}{8}$$
$$+\,4\frac{1}{8} \quad = \quad +\,4\frac{1}{8}$$
$$\overline{} \qquad \overline{7\frac{5}{8}}$$

13.
$$15\frac{1}{6} \quad = \quad 15\frac{2}{12}$$
$$+\,13\frac{5}{12} \quad = \quad +\,13\frac{5}{12}$$
$$\overline{} \qquad \overline{28\frac{7}{12}}$$

17.
$$4\frac{7}{10}$$
$$-\,2\frac{1}{10}$$
$$\overline{2\frac{6}{10}} \quad = \quad 2\frac{3}{5}$$

21.
$$9\frac{1}{5} \quad = \quad 9\frac{5}{25} \quad = \quad 8\frac{30}{25}$$
$$-\,8\frac{6}{25} \quad = \quad -\,8\frac{6}{25} \quad = \quad -\,8\frac{6}{25}$$
$$\overline{} \qquad \qquad \overline{\frac{24}{25}}$$

25.
$$5\frac{2}{3} \quad = \quad 5\frac{4}{6}$$
$$-\,3\frac{1}{6} \quad = \quad -\,3\frac{1}{6}$$
$$\overline{} \qquad \overline{2\frac{3}{6}} \quad = \quad 2\frac{1}{2}$$

29.
$$10 \quad = \quad 9\frac{5}{5}$$
$$-\,8\frac{1}{5} \quad = \quad -\,8\frac{1}{5}$$
$$\overline{} \qquad \overline{1\frac{4}{5}}$$

33.
$$6 \quad = \quad 5\frac{9}{9}$$
$$-\,2\frac{4}{9} \quad = \quad -\,2\frac{4}{9}$$
$$\overline{} \qquad \overline{3\frac{5}{9}}$$

37. The phrase "How much longer" tells us to subtract. Subtract $3\frac{1}{2}$ hours from $5\frac{7}{8}$ hours.
$$5\frac{7}{8} \quad = \quad 5\frac{7}{8}$$
$$-\,3\frac{1}{2} \quad = \quad -\,3\frac{4}{8}$$
$$\overline{} \qquad \overline{2\frac{3}{8}}$$

It takes him $2\frac{3}{8}$ hours longer to prepare the business return.

41. The phrase "how much more" tells us to subtract. Subtract $3\frac{3}{5}$ inches from $11\frac{1}{4}$ inches.
$$11\frac{1}{4} \quad = \quad 11\frac{5}{20} \quad = \quad 10\frac{25}{20}$$
$$-\,3\frac{3}{5} \quad = \quad -\,3\frac{12}{20} \quad = \quad -\,3\frac{12}{20}$$
$$\overline{} \qquad \qquad \overline{7\frac{13}{20}}$$

Tucson gets $7\frac{13}{20}$ inches more, on average, than Yuma.

45. Subtract 50 pounds from $60\frac{5}{8}$ pounds.
$$60\frac{5}{8}$$
$$-\,50$$
$$\overline{10\frac{5}{8}}$$

The traveler will have to pay charges on $10\frac{5}{8}$ pounds.

49. The phrase "overall height" tells us to add. Add the two heights.
$$152\frac{1}{6} \quad = \quad 152\frac{1}{6}$$
$$+\,154\frac{1}{2} \quad = \quad +\,154\frac{3}{6}$$
$$\overline{} \qquad \overline{306\frac{4}{6}} \quad = \quad 306\frac{2}{3}$$

The overall height is $306\frac{2}{3}$ feet.

53. Find the distance around. Add the lengths of the three sides.
$$2\frac{1}{3}$$
$$2\frac{1}{3}$$
$$+\,2\frac{1}{3}$$
$$\overline{6\frac{3}{3}} \quad = \quad 7$$

The perimeter is 7 miles.

57. $2^3 = 2 \cdot 2 \cdot 2 = 8$ **61.** $3^4 = 3 \cdot 3 \cdot 3 \cdot 3 = 81$

65. First, find the total weight of the Supreme box by adding the two weights. Then find the total weight of the Deluxe box by adding the two weights. Finally, subtract the lighter box from the heavier box to see how much more the heavier one weighs.

$$2\frac{1}{4} \quad = \quad 2\frac{1}{4}$$
$$+\ 3\frac{1}{2} \quad = \quad +\ 3\frac{2}{4}$$
$$\overline{\qquad\qquad\qquad 5\frac{3}{4}}$$

The Supreme box weighs $5\frac{3}{4}$ pounds.

$$1\frac{3}{8} \quad = \quad 1\frac{3}{8}$$
$$+\ 4\frac{1}{4} \quad = \quad +\ 4\frac{2}{8}$$
$$\overline{\qquad\qquad\qquad 5\frac{5}{8}}$$

The Deluxe box weighs $5\frac{5}{8}$ pounds.

$$5\frac{3}{4} \quad = \quad 5\frac{6}{8}$$
$$-\ 5\frac{5}{8} \quad = \quad -\ 5\frac{5}{8}$$
$$\overline{\qquad\qquad\qquad \frac{1}{8}}$$

The Supreme box is heavier by $\frac{1}{8}$ pound.

EXERCISE SET 3.5

1. Since $7 > 6$, then $\dfrac{7}{9} > \dfrac{6}{9}$.

5. The LCD is 42.

$\dfrac{9}{42}$ has a denominator of 42 $\dfrac{5}{21} = \dfrac{5 \cdot 2}{21 \cdot 2} = \dfrac{10}{42}$

Since $9 < 10$, then $\dfrac{9}{42} < \dfrac{10}{42}$ or $\dfrac{9}{42} < \dfrac{5}{21}$

9. The LCD is 12.

$\dfrac{3}{4} = \dfrac{3 \cdot 3}{4 \cdot 3} = \dfrac{9}{12}$ $\dfrac{2}{3} = \dfrac{2 \cdot 4}{3 \cdot 4} = \dfrac{8}{12}$

Since $9 > 8$, then $\dfrac{9}{12} > \dfrac{8}{12}$ so $\dfrac{3}{4} > \dfrac{2}{3}$

13. The LCD is 100.

$\dfrac{27}{100}$ has a denominator of 100 $\dfrac{7}{25} = \dfrac{7 \cdot 4}{25 \cdot 4} = \dfrac{28}{100}$

Since $27 < 28$, then $\dfrac{27}{100} < \dfrac{28}{100}$, so $\dfrac{27}{100} < \dfrac{7}{25}$

17. $\left(\dfrac{1}{2}\right)^4 = \dfrac{1}{2} \cdot \dfrac{1}{2} \cdot \dfrac{1}{2} \cdot \dfrac{1}{2} = \dfrac{1}{16}$

21. $\left(\dfrac{4}{7}\right)^3 = \dfrac{4}{7} \cdot \dfrac{4}{7} \cdot \dfrac{4}{7} = \dfrac{64}{343}$

25. $\left(\dfrac{3}{4}\right)^2 \cdot \left(\dfrac{2}{3}\right)^3 = \left(\dfrac{3}{4} \cdot \dfrac{3}{4}\right) \cdot \left(\dfrac{2}{3} \cdot \dfrac{2}{3} \cdot \dfrac{2}{3}\right)$

$\qquad = \dfrac{3 \cdot 3 \cdot 2 \cdot 2 \cdot 2}{4 \cdot 4 \cdot 3 \cdot 3 \cdot 3}$

$\qquad = \dfrac{1}{6}$

29. $\dfrac{3}{7} \cdot \dfrac{1}{5} = \dfrac{3 \cdot 1}{7 \cdot 5} = \dfrac{3}{35}$

33. $\dfrac{6}{11} \div \dfrac{2}{3} = \dfrac{6}{11} \cdot \dfrac{3}{2} = \dfrac{2 \cdot 3 \cdot 3}{11 \cdot 2} = \dfrac{9}{11}$

37. $\dfrac{4}{7} - \dfrac{6}{11} = \dfrac{4 \cdot 11}{7 \cdot 11} - \dfrac{6 \cdot 7}{11 \cdot 7} = \dfrac{44}{77} - \dfrac{42}{77} = \dfrac{2}{77}$

41. $\dfrac{5}{6} \div \dfrac{1}{3} \cdot \dfrac{1}{4} = \dfrac{5}{6} \cdot \dfrac{3}{1} \cdot \dfrac{1}{4}$

$\qquad = \dfrac{5 \cdot 3 \cdot 1}{2 \cdot 3 \cdot 1 \cdot 4}$

$\qquad = \dfrac{5}{8}$

45. $2 \cdot \left(\dfrac{1}{4} + \dfrac{1}{5}\right) + 2 = 2 \cdot \left(\dfrac{5}{20} + \dfrac{4}{20}\right) + 2$

$\qquad = 2 \cdot \dfrac{9}{20} + 2$

$\qquad = \dfrac{9}{10} + \dfrac{20}{10}$

$\qquad = \dfrac{29}{10}$

$\qquad = 2\dfrac{9}{10}$

49. $\left(\dfrac{2}{3} - \dfrac{5}{9}\right)^2 = \left(\dfrac{6}{9} - \dfrac{5}{9}\right)^2$

$\qquad = \left(\dfrac{1}{9}\right)^2$

$\qquad = \dfrac{1}{9} \cdot \dfrac{1}{9}$

$\qquad = \dfrac{1}{81}$

53. $\left(\dfrac{1}{3} + \dfrac{1}{4} + \dfrac{1}{6}\right) \div 3 = \left(\dfrac{4}{12} + \dfrac{3}{12} + \dfrac{2}{12}\right) \div 3$

$\qquad = \dfrac{9}{12} \div 3$

$\qquad = \dfrac{9}{12} \cdot \dfrac{1}{3}$

$\qquad = \dfrac{3 \cdot 3 \cdot 1}{2 \cdot 2 \cdot 3 \cdot 3}$

$\qquad = \dfrac{1}{4}$

57. $\dfrac{20}{24} = \dfrac{2 \cdot 2 \cdot 5}{2 \cdot 2 \cdot 2 \cdot 3} = \dfrac{5}{2 \cdot 3} = \dfrac{5}{6}$

61. Compare the two fractions.
The LCD is 500.
$$\frac{22}{125} = \frac{22 \cdot 4}{125 \cdot 4} = \frac{88}{500}$$

$\frac{93}{500}$ has a denominator of 500.

Since $88 < 93$, then $\frac{88}{500} < \frac{93}{50}$ so $\frac{22}{125} < \frac{93}{500}$.

Periodicals accounted for a greater portion of the mail handled by weight.

Chapter 3 Test

1. $4 = \boxed{2 \cdot 2}$

$15 = \boxed{3} \cdot \boxed{5}$

The LCM is $2 \cdot 2 \cdot 3 \cdot 5 = 60$.

5. $\frac{7}{9} + \frac{1}{9} = \frac{7+1}{9} = \frac{8}{9}$

9. *Step 1.* The LCD for the denominators 8 and 3 is 24.

Step 2. $\frac{7}{8} = \frac{7 \cdot 3}{8 \cdot 3} = \frac{21}{24}, \frac{1}{3} = \frac{1 \cdot 8}{3 \cdot 8} = \frac{8}{24}$

Step 3. $\frac{7}{8} - \frac{1}{3} = \frac{21}{24} - \frac{8}{24} = \frac{13}{24}$

Step 4. $\frac{13}{24}$ is in simplest form.

13. *Step 1.* The LCD for the denominators 12, 8, and 24 is 24.

Step 2. $\frac{11}{12} = \frac{11 \cdot 2}{12 \cdot 2} = \frac{22}{24}, \frac{3}{8} = \frac{3 \cdot 3}{8 \cdot 3} = \frac{9}{24}, \frac{5}{24}$
already has a denominator of 24.

Step 3. $\frac{11}{12} + \frac{3}{8} + \frac{5}{24} = \frac{22}{24} + \frac{9}{24} + \frac{5}{24} = \frac{36}{24}$

Step 4. $\frac{36}{24} = \frac{3 \cdot 12}{2 \cdot 12} = \frac{3}{2}$ or $1\frac{1}{2}$

17. $\left(\frac{2}{3}\right)^4 = \frac{2}{3} \cdot \frac{2}{3} \cdot \frac{2}{3} \cdot \frac{2}{3} = \frac{16}{81}$

21. The phrase "How long is the remaining" tells us to subtract $2\frac{3}{4}$ feet from $6\frac{1}{2}$ feet.

$$
\begin{array}{ccccc}
6\frac{1}{2} & = & 6\frac{2}{4} & = & 5\frac{6}{4} \\
-2\frac{3}{4} & = & -2\frac{3}{4} & = & -2\frac{3}{4} \\
\hline
& & & & 3\frac{3}{4}
\end{array}
$$

The remaining piece is $3\frac{3}{4}$ feet.

Chapter 4

Exercise Set 4.1

1. 6.52 in words is six and fifty-two hundredths.

5. 0.205 in words is two hundred five thousandths.

9. 31.04 in words is thirty-one and four hundredths.

A52

13. Six and five-tenths is 6.5.

17. Five and six hundred twenty-five thousandths is 5.625.

21. Twenty and thirty-three hundredths is 20.33.

25. $0.3 = \frac{3}{10}$

29. $5.47 = 5\frac{47}{100}$

33. $7.07 = 7\frac{7}{100}$

37. $0.3005 = \frac{3005}{10,000} = \frac{601}{2000}$

41. $\frac{6}{10} = 0.6$

45. $\frac{37}{10} = 3.7$

49. $\frac{9}{100} = 0.09$

53. $\frac{28}{1000} = 0.028$

57. To round 47,261 to the nearest ten, observe that the digit in the ones place is 1. Since this digit is less than 5, we do not add 1 to the digit in the tens place. The number 47,261 rounded to the nearest ten is 47,260.

61. answers may vary

Exercise Set 4.2

1. 0.15   0.16
   ↑   ↑
   5 <   6 so
   0.15 < 0.16

5. 0.098   0.1
   ↑   ↑
   0 <   1 so
   0.098 < 0.1

9. 167.908   167.980
      ↑   ↑
      0 <   8 so
   167.908 < 167.980

13. To round 0.57 to the nearest tenth, observe that the digit in the hundredths place is 7. Since this digit is at least 5, we need to add 1 to the digit in the tenths place. The number 0.57 rounded to the nearest tenth is 0.6.

17. To round 0.5942 to the nearest thousandth, observe that the digit in the ten-thousandths place is 2. Since this digit is less than 5, we do not add 1 to the digit in the thousandths place. The number 0.5942 rounded to the nearest thousandth is 0.594.

21. To round 12.342 to the nearest tenth, observe that the digit in the hundredths place is 4. Since this digit is less than 5, we do not add 1 to the digit in the tenths place. The number 12.342 rounded to the nearest tenth is 12.3.

25. To round 0.501 to the nearest tenth, observe that the digit in the hundredths place is 0. Since this digit is less than 5, we do not add 1 to the digit in the tenths place. The number 0.501 rounded to the nearest tenth is 0.5.

29. To round 3829.34 to the nearest ten, observe that the digit in the ones place is 9. Since this digit is at least 5, we need to add 1 to the digit in the tens place. The number 3829.34 rounded to the nearest ten is 3830.

33. To round 42,650.14 to the nearest one, observe that the digit in the tenths place is 1. Since this digit is less than 5, we do not add 1 to the digit in the ones place. The number 42,650.14 rounded to the nearest one is 42,650. The amount is $42,650.

37. To round 0.1992 to the nearest hundredth, observe that the digit in the thousandths place is 9. Since this digit is at least 5, we need to add 1 to the digit in the hundredths place. The number 0.1992 rounded to the nearest hundredth is 0.2. The amount is $0.20.

41. To round 39,867 to the nearest thousand, observe that the digit in the hundreds place is 8. Since this digit is at least 5, we need to add 1 to the digit in the thousands place. The number 39,867 rounded to the nearest thousand is 40,000.

45. To round 24.6229 to the nearest thousandth, observe that the digit in the ten-thousandths place is 9. Since this digit is at least 5, we need to add 1 to the digit in the thousandths place. The number 24.6229 rounded to the nearest thousandth is 24.623. The length is 24.623 hours.

49.
```
  3452
+ 2314
------
  5766
```

53.
```
  482
- 239
-----
  243
```
Check:
```
      1
  243
+ 239
-----
  482
```

57. Compare each place value and list from greatest to least.
225.490, 225.370, 222.980, 222.830, 222.000, 219.702, 218.158, 218.036, 216.081, 215.432

EXERCISE SET 4.3

1.
```
  1.3
+ 2.2
-----
  3.5
```

5.
```
  0.003
+ 0.091
-------
  0.094
```

9.
```
       1
  490.00
+  93.09
--------
  583.09
```

13.
```
    1  11
  100.009
    6.080
+   9.034
---------
  115.123
```

17.
```
   1   1
  45.023
   3.006
+  8.403
--------
  56.432
```

21.
```
  18.0
-  2.7
------
  15.3
```
Check:
```
    1
  15.3
+  2.7
------
  18.0 or 18
```

25.
```
  5.90
- 4.07
------
  1.83
```
Check:
```
    1
  1.83
+ 4.07
------
  5.90 or 5.9
```

29.
```
  500.34
- 123.45
--------
  376.89
```
Check:
```
  1111
  376.89
+ 123.45
--------
  500.34
```

33.
```
  200.0
-   5.6
-------
  194.4
```
Check:
```
    111
  194.4
+   5.6
-------
  200.0 or 200
```

37.
```
  23.0
-  6.7
------
  16.3
```
Check:
```
    11
  16.3
+  6.7
------
  23.0 or 23
```

41. The phrase "By how much did the price change" tells us to subtract. Subtract 0.979 from 1.039.
```
  1.039
- 0.979
-------
  0.060
```
Check:
```
  0.060
+ 0.979
-------
  1.039
```
The price changed by $0.06.

45. The phrase "How much more" tells us to subtract. Subtract 175.9 from 373.7.
```
  373.7
- 175.9
-------
  197.8
```
Check:
```
    111
  197.8
+ 175.9
-------
  373.7
```
The difference in consumption is 197.8 pounds per person.

49. To find Green's speed, we must add the old record to the increase.
```
   11 11
  633.468
+ 129.567
---------
  763.035
```
The new record by Green is 763.035 mph.

53. Find the sum of the lengths of the three sides.
```
    11
  12.40
  29.34
+ 25.70
-------
  67.44
```
The architect needs 67.44 feet of border material.

57. Add the durations of all four missions.
```
  212 11
  330.583
   94.567
  147.000
+ 142.900
---------
  715.050
```
James A. Lovell has spent 715.05 hours in spaceflight.

61. Subtract 13.9 from 22.
```
  22.0
- 13.9
------
   8.1
```
Check:
```
    11
   8.1
+ 13.9
------
  22.0 or 22
```
The difference in consumption is 8.1 pounds.

65.
```
    46
 ×   3
 -----
   138
```

69. $\left(\dfrac{1}{5}\right)^3 = \dfrac{1}{5} \cdot \dfrac{1}{5} \cdot \dfrac{1}{5} = \dfrac{1}{125}$

73. answers may vary

EXERCISE SET 4.4

1.
```
    0.2
 ×  0.6
 ------
   0.12
```

5.
```
   0.26
 ×    5
 ------
  1.30 or 1.3
```

9.
```
    5.62
 ×   7.7
 -------
    3934
   39340
 -------
  43.274
```

13.

$$
\begin{array}{r}
490.2 \\
\times\, 0.023 \\
\hline
14706 \\
98040 \\
000 \\
0000 \\
\hline
11.2746
\end{array}
$$

17. To find 6.5×10, note that 10 has 1 zero. Therefore, we move the decimal point of 6.5 to the right 1 place. The product is 65.

21. To find 7.093×100, note that 100 has 2 zeros. Therefore, we move the decimal point of 7.093 to the right 2 places. The product is 709.3.

25. To find 9.1×1000, note that 1000 has 3 zeros. Therefore, we move the decimal point of 9.1 to the right 3 places. The product is 9100.

29. 5.5 billion $= 5.5 \times 1,000,000,000$
$\quad = 5,500,000,000$
They can make 5,500,000,000 chocolate bars.

33. 1.6 million $= 1.6 \times 1,000,000$
$\quad = 1,600,000$
Americans lose more than 1,600,000 hours in traffic.

37. Circumference $= \pi \cdot$ diameter
$C = \pi \cdot 10 = 10\pi$
$C \approx 10(3.14) = 31.4$
The circumference is 10π centimeters, which is approximately 31.4 centimeters.

41. Multiply the number of ounces by the number of grams of fat in 1 ounce to get the total amount of fat.

$$
\begin{array}{r}
6.2 \\
\times\, 4 \\
\hline
24.8
\end{array}
$$

There are 24.8 grams of fat in a 4-ounce serving of cream cheese.

45. Multiply 39.37 by 1.65.

$$
\begin{array}{r}
39.37 \\
\times\, 1.65 \\
\hline
19685 \\
236220 \\
393700 \\
\hline
64.9605
\end{array}
$$

She is about 64.9605 inches tall.

49.

$$
\begin{array}{r}
26 \\
5\overline{)130} \\
-10 \\
\hline
30 \\
-30 \\
\hline
0
\end{array}
$$

53.

$$
\begin{array}{r}
8 \\
365\overline{)2920} \\
-2920 \\
\hline
0
\end{array}
$$

57. $(20.6)(1.86)(100,000) = 3,831,600$ miles

61. answers may vary

EXERCISE SET 4.5

1.

$$
\begin{array}{r}
0.094 \\
5\overline{)0.470} \\
-45 \\
\hline
20 \\
-20 \\
\hline
0
\end{array}
$$

5. $0.82\overline{)4.756}$ becomes

$$
\begin{array}{r}
5.8 \\
82.\overline{)475.6} \\
-410 \\
\hline
65\,6 \\
-65\,6 \\
\hline
0
\end{array}
$$

9.

$$
\begin{array}{r}
0.413 \\
15\overline{)6.195} \\
-6\,0 \\
\hline
19 \\
-15 \\
\hline
45 \\
-45 \\
\hline
0
\end{array}
$$

13. $0.27\overline{)1.296}$ becomes

$$
\begin{array}{r}
4.8 \\
27.\overline{)129.6} \\
-108 \\
\hline
21\,6 \\
-21\,6 \\
\hline
0
\end{array}
$$

17. $0.6\overline{)18}$ becomes

$$
\begin{array}{r}
30 \\
6.\overline{)180} \\
-18 \\
\hline
00 \\
-0 \\
\hline
0
\end{array}
$$

21. $7.2\overline{)70.56}$ becomes

$$
\begin{array}{r}
9.8 \\
72.\overline{)705.6} \\
-6\,48 \\
\hline
57\,6 \\
-57\,6 \\
\hline
0
\end{array}
$$

25. $0.027\overline{)1.215}$ becomes

$$
\begin{array}{r}
45 \\
27.\overline{)1215.} \\
-108 \\
\hline
135 \\
-135 \\
\hline
0
\end{array}
$$

29. $0.023\overline{)0.549}$ becomes

$$
\begin{array}{r}
23.869 \approx 23.87 \\
23.\overline{)549.000} \\
-46 \\
\hline
89 \\
-69 \\
\hline
20\,0 \\
-18\,4 \\
\hline
1\,60 \\
-1\,38 \\
\hline
220 \\
-207 \\
\hline
13
\end{array}
$$

33. To find $54.982 \div 100$, note that 100 has 2 zeros. Therefore, we move the decimal point of 54.982 to the left 2 places. The quotient is 0.54982.

37. To find $87 \div 10$, note that 10 has 1 zero. Therefore, we move the decimal point of 87 to the left 1 place. The quotient is 8.7.

41. There are 52 weeks per year and 40 hours per week. Therefore, there are $52 \times 40 = 2080$ hours per year.

$$
\begin{array}{r}
855.845 \approx 855.85 \\
2080)\overline{1{,}780{,}159.000} \\
-1\,664\,0 \\
\hline
116\,15 \\
-104\,00 \\
\hline
12\,159 \\
-10\,400 \\
\hline
1\,7590 \\
-1\,6640 \\
\hline
9500 \\
-8320 \\
\hline
11800 \\
-10400 \\
\hline
1400
\end{array}
$$

His hourly wage was \$855.85.

45. $39.37)\overline{200}$ becomes

$$
\begin{array}{r}
5.08 \approx 5.1 \\
3937.)\overline{20{,}000.00} \\
-19\,685 \\
\hline
315\,0 \\
-\quad 0 \\
\hline
315\,00 \\
-314\,96 \\
\hline
4
\end{array}
$$

There are 5.1 meters in 200 inches.

49.
$$
\begin{array}{r}
127.18 \approx 127.2 \\
24.)\overline{3052.45} \\
-24 \\
\hline
65 \\
-48 \\
\hline
172 \\
-168 \\
\hline
4\,4 \\
-2\,4 \\
\hline
2\,05 \\
-1\,92 \\
\hline
13
\end{array}
$$

Their average speed was 127.2 mph.

53. To round 345.219 to the nearest hundredth, observe that the digit in the thousandths place is 9. Since this digit is at least 5, we need to add 1 to the digit in the hundredths place. The number 345.219 rounded to the nearest hundredth is 345.22.

57. $2 + 3 \cdot 6 = 2 + 18 = 20$

61. $(86 + 78 + 91 + 85) \div 4 = 340 \div 4$

$$
\begin{array}{r}
85 \\
4)\overline{340} \\
-32 \\
\hline
20 \\
-20 \\
\hline
0
\end{array}
$$

The average is 85.

65.
$$
\begin{array}{r}
45.2 \\
4.)\overline{180.8} \\
-16 \\
\hline
20 \\
-20 \\
\hline
0\,8 \\
-\,8 \\
\hline
0
\end{array}
$$

The length of a side is 45.2 centimeters.

EXERCISE SET 4.6

1. $2.1 + 5.8 + 4.1 = 12.0$
$2 + 6 + 4 = 12$
Yes, the result and estimate agree.

5.
$$
\begin{array}{rr}
6 & 6 \\
\times\ 483.11 & \times\ 500 \\
\hline
2898.66 & 3000
\end{array}
$$

3000 is close to 2898.66, so the answer is reasonable.

9.
$$
\begin{array}{rr}
69.2 & 70 \\
32.1 & 30 \\
+\ 48.5 & +\ 50 \\
\hline
149.8 & 150
\end{array}
$$

149.8 is close to 150, so the answer is reasonable.

13. $2(12.2) + 2(5.9) \approx 2(12) + 2(6)$
$= 24 + 120 = 36$ inches

17. $3.14(7)(2) = 43.96$ meters

21. 198.79 is approximately 200
$(200)(12)(5) = \$12{,}000$

25.
$$
\begin{array}{rcr}
560.6 & \text{Estimate} & 561 \\
461.0 & & 461 \\
399.8 & & 400 \\
357.1 & & 357 \\
+\ 329.7 & & +\ 330 \\
\hline
& & 2109
\end{array}
$$

The total estimate is \$2109 million.

29. $(0.4)^2 = (0.4)(0.4) = 0.16$

33. $1.4(2 - 1.8) = 1.4(0.2) = 0.28$

37. $7.8 - 4.83 \div 2.1 = 7.8 - 2.3$
$= 5.5$

41. $(3.1 + 0.7)(2.9 - 0.9) = (3.8)(2.0)$
$= 7.6$

45. $\dfrac{7 + 0.74}{0.06} = \dfrac{7.74}{0.06} = 129$

49. $\dfrac{36}{56} \div \dfrac{30}{35} = \dfrac{36}{56} \cdot \dfrac{35}{30}$

$= \dfrac{2 \cdot 2 \cdot 3 \cdot 3 \cdot 5 \cdot 7}{2 \cdot 2 \cdot 2 \cdot 7 \cdot 2 \cdot 3 \cdot 5}$

$= \dfrac{3}{4}$

53. $1.96(7.852 - 3.147)^2 = 1.96(4.705)^2$
$= 1.96(22.137025)$
$= 43.388569$

Estimate: $2(8 - 3)^2 = 2(5)^2 = 2(25) = 50$

1. $\frac{1}{5} = 0.2$

$$5\overline{)1.0}^{\,0.2}$$

5. $\frac{3}{4} = 0.75$

$$\begin{array}{r} 0.75 \\ 4\overline{)3.00} \\ -2\,8 \\ \hline 20 \\ -20 \\ \hline 0 \end{array}$$

9. $\frac{3}{8} = 0.375$

$$\begin{array}{r} 0.375 \\ 8\overline{)3.000} \\ -2\,4 \\ \hline 60 \\ -56 \\ \hline 40 \\ -40 \\ \hline 0 \end{array}$$

13. $\frac{17}{40} = 0.425$

$$\begin{array}{r} 0.425 \\ 40\overline{)17.000} \\ -16\,0 \\ \hline 1\,00 \\ -\ 80 \\ \hline 200 \\ -200 \\ \hline 0 \end{array}$$

17. $\frac{1}{3} = 0.\overline{3}$

$$\begin{array}{r} 0.33... \\ 3\overline{)1.00} \\ -\ 9 \\ \hline 10 \\ -\ 9 \\ \hline 1 \end{array}$$

21. $\frac{2}{9} = 0.\overline{2}$

$$\begin{array}{r} 0.22... \\ 9\overline{)2.00} \\ -1\,8 \\ \hline 20 \\ -18 \\ \hline 2 \end{array}$$

25. $0.\overline{3} = 0.333... \approx 0.33$ **29.** $0.\overline{2} = 0.222... \approx 0.2$

33.
$$\begin{array}{r} 0.1941... \\ 376\overline{)73.0000} \\ -37\,6 \\ \hline 35\,40 \\ -33\,84 \\ \hline 1\,560 \\ -1\,504 \\ \hline 560 \\ -376 \\ \hline 184 \end{array}$$

$\frac{73}{376} \approx 0.194$

37.
$$\begin{array}{r} 0.8 \\ 5\overline{)4.0} \\ -4.0 \\ \hline 0 \end{array}$$

$\frac{4}{5} = 0.8$

41. $2 > 1$, so $0.823 > 0.813$

45. $\frac{2}{3} = \frac{4}{6}$, so $\frac{2}{3} < \frac{5}{6}$

49. $\frac{4}{7} \approx 0.5714$

$0.5714 > 0.14$, so $\frac{4}{7} > 0.14$

53. $\frac{456}{64} = 7.125$

$7.123 < 7.125$, so $7.123 < \frac{456}{64}$

57. $0.49 = 0.490$

$0.49, 0.491, 0.498$

61. $\frac{4}{7} \approx 0.571$

$0.412, 0.453, \frac{4}{7}$

65. $\frac{12}{5} = 2.4$

$\frac{17}{8} = 2.125$

$\frac{17}{8}, 2.37, \frac{12}{5}$

69. Area $= \frac{1}{2} \times$ base $\times$ height

$= \frac{1}{2} \times 5.2 \times 3.6$

$= 0.5 \times 5.2 \times 3.6$

$= 9.36$

Area is 9.36 square centimeters.

73. $2^3 = (2)(2)(2) = 8$

77. $\left(\frac{1}{3}\right)^4 = \frac{1}{3} \cdot \frac{1}{3} \cdot \frac{1}{3} \cdot \frac{1}{3} = \frac{1}{81}$

81. $\left(\frac{2}{5}\right)\left(\frac{5}{2}\right)^2 = \frac{2}{5} \cdot \frac{5}{2} \cdot \frac{5}{2} = \frac{5}{2}$

85.

	Estimate
2502	2500
1521	1500
1313	1300
1054	1100
$+\ 942$	$+\ 900$
	7300

The estimate is 7300 stations.

89. answers may vary

CHAPTER 4 TEST

1. 45.092 in words is forty-five and ninety-two thousandths

5. $25.0909 < 25.9090$

9. $\frac{13}{26} = 0.5$

$$\begin{array}{r} 0.5 \\ 26\overline{)13.0} \end{array}$$

13.
$$\begin{array}{r} 10.2 \\ \times\ 4.3 \\ \hline 306 \\ 4080 \\ \hline 43.86 \end{array}$$

17. $\frac{473}{10} = 47.3$

21. $A = 123.8(80) = 9904$

9904 square feet of lawn

$9904 \times 0.02 = 198.08$

She needs 198.08 ounces of insecticide.

Chapter 5

Exercise Set 5.1

1. 11 to 14 is $\dfrac{11}{14}$

5. 151 to 201 is $\dfrac{151}{201}$

9. 5 to $7\dfrac{1}{2}$ is $\dfrac{5}{7\frac{1}{2}}$

13. $\dfrac{16}{24} = \dfrac{2 \cdot 8}{3 \cdot 8} = \dfrac{2}{3}$

17. $\dfrac{4.63}{8.21} = \dfrac{4.63 \times 100}{8.21 \times 100} = \dfrac{463}{821}$

21. $\dfrac{10 \text{ hours}}{24 \text{ hours}} = \dfrac{5 \cdot 2}{12 \cdot 2} = \dfrac{5}{12}$

25. $\dfrac{24 \text{ days}}{14 \text{ days}} = \dfrac{12 \cdot 2}{7 \cdot 2} = \dfrac{12}{7}$

29. $\dfrac{8 \text{ inches}}{20 \text{ inches}} = \dfrac{2 \cdot 4}{5 \cdot 4} = \dfrac{2}{5}$

33. perimeter $= 8 + 15 + 17 = 40$ feet
$\dfrac{\text{hypotenuse}}{\text{perimeter}} = \dfrac{17 \text{ feet}}{40 \text{ feet}} = \dfrac{17}{40}$

37. $6000 - 4500 = 1500$ married
$\dfrac{4500 \text{ single}}{1500 \text{ married}} = \dfrac{3 \text{ single}}{1 \text{ married}}$

41. $\dfrac{\$5.15}{\$0.25} = \dfrac{103 \cdot 0.05}{5 \cdot 0.05} = \dfrac{103}{5}$

45.
```
     2.3
  9)20.7
     18
     2 7
     2 7
       0
```

49. $\dfrac{15 \text{ states}}{35 \text{ states}} = \dfrac{3 \cdot 5}{7 \cdot 5} = \dfrac{3}{7}$

53. $\dfrac{3 \text{ bruised}}{33 \text{ total}} = \dfrac{1 \cdot 3}{11 \cdot 3} = \dfrac{1}{11}$
$\dfrac{1}{11} < \dfrac{1}{10}$
No, the shipment should not be refused.

Exercise Set 5.2

1. $\dfrac{5 \text{ shrubs}}{15 \text{ feet}} = \dfrac{1 \text{ shrub}}{3 \text{ feet}}$

5. $\dfrac{8 \text{ phone lines}}{36 \text{ employees}} = \dfrac{2 \text{ phone lines}}{9 \text{ employees}}$

9. $\dfrac{6 \text{ flight attendants}}{200 \text{ passengers}} = \dfrac{3 \text{ flight attendants}}{100 \text{ passengers}}$

13. $\dfrac{375 \text{ riders}}{5 \text{ subway cars}} = \dfrac{75 \text{ riders}}{1 \text{ subway car}}$
$\qquad\qquad = 75$ riders/subway car

17. $\dfrac{144 \text{ diapers}}{24 \text{ babies}} = \dfrac{6 \text{ diapers}}{1 \text{ baby}}$
$\qquad\qquad = 6$ diapers/baby

21. $\dfrac{600 \text{ kilometers}}{90 \text{ minutes}} = 6\dfrac{2}{3}$ kilometers/minute

25. $\dfrac{12{,}000 \text{ good}}{40 \text{ defective}} = 300$ good/defective

29. $\dfrac{\$24{,}000{,}000}{500 \text{ species}} = \dfrac{\$48{,}000}{1 \text{ species}}$ or $\$48{,}000$/species

33. a. $\dfrac{250 \text{ boards}}{8 \text{ hours}} = 31.25$ computer boards/hour

b. $\dfrac{400 \text{ boards}}{12 \text{ hours}} \approx 33.3$ computer boards/hour

c. $33.3 > 31.25$ so Lamont can assemble computer boards faster.

37. $\dfrac{\$1.19}{7 \text{ bananas}} = \dfrac{\$0.17}{1 \text{ banana}}$ or $\$0.17$ per banana

41. $\dfrac{\$1.69}{16 \text{ ounces}} \approx \$0.106$ per ounce

$\dfrac{\$0.69}{6 \text{ ounces}} \approx \$0.115$ per ounce

The 16-ounce size is the better buy.

45. $\dfrac{\$0.59}{100 \text{ napkins}} = \$0.006$ per napkin

$\dfrac{\$0.93}{180 \text{ napkins}} \approx \$.005$ per napkin

The package of 180 napkins is the better buy.

49.
```
     3.7
   × 1.2
      74
     370
    4.44
```

53. Fill in the column labeled Miles Driven by subtracting Beginning Odometer Reading from Ending Odometer Reading. Fill in the column labeled Miles per Gallon by dividing Miles driven by Gallons of Gas Used, and round to the nearest tenth.

Miles Driven	Miles Per Gallon
257	19.2
352	22.3
347	21.6

57. $\dfrac{477{,}121 \text{ students}}{20{,}039 \text{ teachers}} \approx \dfrac{24 \text{ students}}{1 \text{ teacher}}$ or 24 students/teacher

Exercise Set 5.3

1. $\dfrac{10 \text{ diamonds}}{6 \text{ opals}} = \dfrac{5 \text{ diamonds}}{3 \text{ opals}}$

5. $\dfrac{6 \text{ eagles}}{58 \text{ sparrows}} = \dfrac{3 \text{ eagles}}{29 \text{ sparrows}}$

9. $\dfrac{22 \text{ vanilla wafers}}{1 \text{ cup cookie crumbs}} = \dfrac{55 \text{ vanilla wafers}}{2.5 \text{ cups cookie crumbs}}$

13. $\dfrac{8}{6} = \dfrac{9}{7}$

$8 \cdot 7 = 9 \cdot 6$

$56 = 54$

false

17. $\dfrac{5}{8} = \dfrac{625}{1000}$

$5 \cdot 1000 = 8 \cdot 625$

$5000 = 5000$

true

21. $\dfrac{4.2}{8.4} = \dfrac{5}{10}$

$(4.2)(10) = 5(8.4)$

$42 = 42$

true

25. $\dfrac{2\frac{2}{5}}{\frac{2}{3}} = \dfrac{\frac{10}{9}}{\frac{1}{4}}$

$2\dfrac{2}{5} \cdot \dfrac{1}{4} = \dfrac{10}{9} \cdot \dfrac{2}{3}$

$\dfrac{12}{5} \cdot \dfrac{1}{4} = \dfrac{10}{9} \cdot \dfrac{2}{3}$

$\dfrac{3}{5} = \dfrac{20}{27}$

false

29. $\dfrac{30}{10} = \dfrac{15}{n}$

$30n = 150$

$n = 5$

33. $\dfrac{n}{6} = \dfrac{8}{15}$

$15n = 48$

$n = 3.2$

37. $\dfrac{\frac{1}{3}}{\frac{3}{8}} = \dfrac{\frac{2}{5}}{x}$

$\dfrac{1}{3}x = \dfrac{3}{8} \cdot \dfrac{2}{5}$

$\dfrac{1}{3}x = \dfrac{3}{20}$

$x = \dfrac{3}{20} \cdot \dfrac{3}{1}$

$x = \dfrac{9}{20}$

41. $\dfrac{\frac{2}{3}}{\frac{6}{9}} = \dfrac{12}{n}$

$\dfrac{2}{3}n = 12 \cdot \dfrac{6}{9}$

$\dfrac{2}{3}n = 8$

$n = 8 \cdot \dfrac{3}{2}$

$n = 12$

45. $\dfrac{n}{0.6} = \dfrac{0.05}{12}$

$12n = (0.05)(0.6)$

$12n = 0.030$

$n = 0.0025$

49. 8.01   8.1

↑   ↑

0 $<$ 1 so

$8.01 < 8.01$

53. $5\dfrac{1}{3} = \dfrac{16}{3}$

$6\dfrac{2}{3} = \dfrac{20}{3}$

$16 < 20$, so $5\dfrac{1}{3} < 6\dfrac{2}{3}$

57. $\dfrac{n}{1150} = \dfrac{588}{483}$

$(588)(1150) = 483n$

$676{,}200 = 483n$

$1400 = n$

61. answers may vary

Exercise Set 5.4

1. Let n = number of completed passes

$\dfrac{4 \text{ completed}}{9 \text{ attempted}} = \dfrac{n \text{ completed}}{27 \text{ attempted}}$

$27 \cdot 4 = 9n$

$108 = 9n$

$12 = n$

12 passes were completed.

5. Let n = number of students accepted

$\dfrac{2 \text{ accepts}}{7 \text{ applicants}} = \dfrac{n \text{ accepts}}{630 \text{ applicants}}$

$1260 = 7n$

$180 = n$

180 students were accepted.

9. Let n = amount of floor space

$\dfrac{9 \text{ square feet}}{1 \text{ student}} = \dfrac{n \text{ square feet}}{30 \text{ students}}$

$270 = n$

270 square feet is required.

13. Let n = distance from Milan to Rome

$\dfrac{1 \text{ centimeter}}{30 \text{ kilometers}} = \dfrac{15 \text{ centimeters}}{n \text{ kilometers}}$

$n = 30 \cdot 15$

$n = 450$

It is 450 km from Milan to Rome.

17. Let n = number of bags of fertilizer

Area of lawn = $260 \times 180 = 46{,}800$ square feet

$\dfrac{1 \text{ bag}}{3000 \text{ square feet}} = \dfrac{n}{46{,}800 \text{ square feet}}$

$46{,}800 = 3000n$

$15.6 = n$

16 bags are needed.

21. Let n = number of hits expected

$\dfrac{3 \text{ hits}}{8 \text{ at bats}} = \dfrac{n \text{ hits}}{40 \text{ at bats}}$

$3 \cdot 40 = 8n$

$120 = 8n$

$15 = n$

He would be expected to make 15 hits.

25. Let n = number of weeks
$$\frac{5 \text{ boxes}}{3 \text{ weeks}} = \frac{144 \text{ boxes}}{n \text{ weeks}}$$
$$5n = 432$$
$$n = 86.4$$
The envelopes will last 86 weeks.

29. Let n = estimated height of Statue of Liberty
$$\frac{\text{statue}}{\text{student}} \quad \frac{42 \text{ feet}}{2 \text{ feet}} = \frac{n \text{ feet}}{5\frac{1}{3} \text{ feet}}$$
$$2n = 42 \cdot 5\frac{1}{3}$$
$$2n = \frac{42}{1} \cdot \frac{16}{3}$$
$$2n = 224$$
$$n = 224 \div 2$$
$$n = 112$$
$$112 - 111\frac{1}{12} = \frac{11}{12}$$
The estimate is 112 feet. The difference is 11 inches.

33. Let n = number of visits including a prescription
$$\frac{7 \text{ prescriptions}}{10 \text{ visits}} = \frac{n \text{ prescriptions}}{620 \text{ visits}}$$
$$10n = 7 \cdot 620$$
$$10n = 4340$$
$$n = 434$$
434 emergency room visits included a prescription.

37.
$$2\overline{)10}^{5}$$
$$2\overline{)20}$$
$$20 = 2^2 \cdot 5$$

41.
$$2\overline{)4}^{2}$$
$$2\overline{)8}$$
$$2\overline{)16}$$
$$2\overline{)32}$$
$$32 = 2^5$$

45. answers may vary

CHAPTER 5 TEST

1. $\dfrac{4500 \text{ trees}}{6500 \text{ trees}} = \dfrac{9 \cdot 500}{13 \cdot 500} = \dfrac{9}{13}$

5. $\dfrac{650 \text{ kilometers}}{8 \text{ hours}} = 81.25 \text{ kilometers per hour}$

$$\begin{array}{r} 81.25 \\ 8\overline{)650.00} \\ -64 \\ \hline 10 \\ -8 \\ \hline 2\,0 \\ -1\,6 \\ \hline 40 \\ -40 \\ \hline 0 \end{array}$$

9. $\dfrac{\$1.49}{16 \text{ ounces}} \approx \dfrac{\$0.09}{1 \text{ ounce}}$

$\dfrac{\$2.39}{24 \text{ ounces}} \approx \dfrac{\$0.10}{1 \text{ ounce}}$

Therefore, the 16-ounce size is the better buy.

13. $\dfrac{8}{n} = \dfrac{11}{6}$
$$11 \cdot n = 8 \cdot 6$$
$$11n = 48$$
$$n = 48 \div 11$$
$$n = 4\frac{4}{11}$$

17. Let n = length of time
$$\frac{3}{80} = \frac{n}{100}$$
$$80n = 300$$
$$n = \frac{15}{4}$$
$$= 3\frac{3}{4}$$
It will take $3\frac{3}{4}$ hours.

Chapter 6

EXERCISE SET 6.1

1. $\dfrac{81}{100} = 81\%$

5. The largest section of the circle graph is chocolate chip. Therefore, chocolate chip was the most preferred cookie.
$$\frac{52}{100} = 52\%$$

9. $48\% = 48.0\% = 0.48$

13. $100\% = 1\,00.\% = 1.00 = 1$

17. $2.8\% = 0.028$

21. $300\% = 3$

25. $73.7\% = 0.737$

29. $81.8\% = 0.818$

33. $3.1 = 3.10 = 310\%$

37. $0.003 = 0.00\,3 = 0.3\%$

41. $5.3 = 5.30 = 530\%$

45. $0.3328 = 33.28\%$

49. $0.7 = 0.70 = 70\%$

53. $0.702 = 70.2\%$

57. $\dfrac{1}{4} = \dfrac{1 \cdot 25}{4 \cdot 25} = \dfrac{25}{100} = 0.25$

61. $\dfrac{9}{10} = 0.9$

65. $130\% = 1.3$

1. $12\% = \dfrac{12}{100} = \dfrac{4\cdot 3}{4\cdot 25} = \dfrac{3}{25}$

5. $4.5\% = \dfrac{4.5}{100} = \dfrac{45}{1000} = \dfrac{5\cdot 9}{5\cdot 200} = \dfrac{9}{200}$

9. $73\% = \dfrac{73}{100}$

13. $6.25\% = \dfrac{6.25}{100} = \dfrac{625}{10{,}000} = \dfrac{625\cdot 1}{625\cdot 16} = \dfrac{1}{16}$

17. $10\frac{1}{3}\% = \dfrac{10\frac{1}{3}}{100} = \dfrac{\frac{31}{3}}{100} = \dfrac{31}{3} \div 100 = \dfrac{31}{3}\cdot\dfrac{1}{100} = \dfrac{31}{300}$

21. $\dfrac{3}{4} = \dfrac{3}{4}\cdot 100\% = \dfrac{300}{4}\% = 75\%$

25. $\dfrac{2}{5} = \dfrac{2}{5}\cdot 100\% = \dfrac{200}{5}\% = 40\%$

29. $\dfrac{17}{50} = \dfrac{17}{50}\cdot 100\% = \dfrac{1700}{50}\% = 34\%$

33. $\dfrac{5}{16} = \dfrac{5}{16}\cdot 100\% = \dfrac{500}{16}\% = 31\frac{1}{4}\%$

37. $\dfrac{2}{3} = \dfrac{2}{3}\cdot 100\% = \dfrac{200}{3}\% = 66\frac{2}{3}\%$

41. $2\frac{1}{2} = \dfrac{5}{2}\cdot 100\% = \dfrac{500}{2}\% = 250\%$

45. $\dfrac{7}{11} = \dfrac{7}{11}\cdot 100\% = \dfrac{700}{11}\% \approx 63.64\%$

$$
\begin{array}{r}
63.636 \approx 63.64 \\
11\overline{)700.000} \\
-66 \\
\hline
40 \\
-33 \\
\hline
70 \\
-66 \\
\hline
40 \\
-33 \\
\hline
70 \\
-66 \\
\hline
4
\end{array}
$$

49. $\dfrac{1}{7} = \dfrac{1}{7}\cdot 100\% = \dfrac{100}{7} \approx 14.29\%$

$$
\begin{array}{r}
14.285 \approx 14.29 \\
7\overline{)100.000} \\
-7 \\
\hline
30 \\
-28 \\
\hline
2\,0 \\
-1\,4 \\
\hline
60 \\
-56 \\
\hline
40 \\
-35 \\
\hline
5
\end{array}
$$

53.

Percent	Decimal	Fraction
35%	0.35	$\frac{7}{20}$
20%	0.2	$\frac{1}{5}$
50%	0.5	$\frac{1}{2}$
70%	0.7	$\frac{7}{10}$
37.5%	0.375	$\frac{3}{8}$

57. $18.8\% = 0.188 = \dfrac{188}{1000} = \dfrac{47}{250}$

61. $\dfrac{17}{200} = 0.085 = 8.5\%$

$$
\begin{array}{r}
0.085 \\
200\overline{)17.000} \\
-16\,00 \\
\hline
10\,00 \\
-10\,00 \\
\hline
0
\end{array}
$$

65. $\dfrac{7}{20} = \dfrac{7}{20}\cdot\dfrac{5}{5} = \dfrac{35}{100} = 35\%$

69. $8\cdot n = 80$

$\dfrac{8\cdot n}{8} = \dfrac{80}{8}$

$n = 10$

73. $\dfrac{21}{79} \approx 0.266 = 26.6\%$

77. A fraction written as a percent is greater than 100% when the numerator is <u>greater</u> than the denominator.

EXERCISE SET 6.3

1. $15\% \cdot 72 = n$

5. $n\cdot 90 = 20$

9. $n = 9\% \cdot 43$

13. $n = 14\% \cdot 52$

$n = 0.14\cdot 52$

$n = 7.28$

17. $1.2 = 12\% \cdot n$

$1.2 = 0.12n$

$\dfrac{1.2}{0.12} = n$

$10 = n$

21. $16 = n\cdot 50$

$\dfrac{16}{50} = n$

$0.32 = n$

$32\% = n$

25. $125\% \cdot 36 = n$

$1.25\cdot 36 = n$

$45 = n$

29. $126 = n\cdot 31.5$

$\dfrac{126}{31.5} = n$

$4 = n$

$n = 400\%$

33. $n \cdot 150 = 67.5$

$$n = \frac{67.5}{150}$$

$$n = 0.45$$

$$n = 45\%$$

37. $\dfrac{27}{n} = \dfrac{9}{10}$

$$9 \cdot n = 27 \cdot 10$$

$$n = \frac{270}{9}$$

$$n = 30$$

41. $\dfrac{17}{12} = \dfrac{n}{20}$

45. $1.5\% \cdot 45,775 = n$

$$0.015 \cdot 45,775 = n$$

$$686.625 = n$$

EXERCISE SET 6.4

1. $\dfrac{a}{65} = \dfrac{32}{100}$

5. $\dfrac{70}{200} = \dfrac{p}{100}$

9. $\dfrac{a}{130} = \dfrac{19}{100}$

13. $\dfrac{a}{105} = \dfrac{18}{100}$

$$\frac{a}{105} = \frac{9}{50}$$

$$a \cdot 50 = 105 \cdot 9$$

$$a \cdot 50 = 945$$

$$a = \frac{945}{50}$$

$$a = 18.9$$

Therefore, 18% of 105 is 18.9.

17. $\dfrac{7.8}{b} = \dfrac{78}{100}$

$$\frac{7.8}{b} = \frac{39}{50}$$

$$7.8 \cdot 50 = b \cdot 39$$

$$390 = b \cdot 39$$

$$\frac{390}{39} = 39$$

$$10 = b$$

Therefore, 78% of 10 is 7.8.

21. $\dfrac{14}{50} = \dfrac{p}{100}$

$$\frac{7}{25} = \frac{p}{100}$$

$$7 \cdot 100 = 25 \cdot p$$

$$700 = 25 \cdot p$$

$$\frac{700}{25} = p$$

$$28 = p$$

Therefore, 28% of 50 is 14.

25. $\dfrac{a}{80} = \dfrac{2.4}{100}$

$$a \cdot 100 = 80 \cdot 2.4$$

$$a \cdot 100 = 192$$

$$a = \frac{192}{100}$$

$$a = 1.92$$

Therefore, 2.4% of 80 is 1.92.

29. $\dfrac{348.6}{166} = \dfrac{p}{100}$

$$348.6 \cdot 100 = 166 \cdot p$$

$$34,860 = 166 \cdot p$$

$$\frac{34,860}{166} = p$$

$$210 = p$$

Therefore, 210% of 166 is 348.6.

33. $\dfrac{3.6}{8} = \dfrac{p}{100}$

$$3.6 \cdot 100 = 8 \cdot p$$

$$360 = 8 \cdot p$$

$$\frac{360}{8} = p$$

$$45 = p$$

Therefore, 45% of 8 is 3.6.

37. $\dfrac{11}{16} + \dfrac{3}{16} = \dfrac{11 + 3}{16} = \dfrac{14}{16} = \dfrac{7 \cdot 2}{8 \cdot 2} = \dfrac{7}{8}$

41.
$$\begin{array}{r} {\scriptstyle 1} \\ 0.41 \\ + 0.29 \\ \hline 0.70 \text{ or } 0.7 \end{array}$$

45. $\dfrac{a}{53,862} = \dfrac{22.3}{100}$

$$a \cdot 100 = 53,862 \cdot 22.3$$

$$a \cdot 100 = 1,201,122.6$$

$$a = \frac{1,201,122.6}{100}$$

$$a = 12,011.226$$

Therefore, 22.3% of 53,862 is 12,011.226.

EXERCISE SET 6.5

1. 1.5% of what number is 24?

$$1.5\% \cdot n = 24$$

$$0.015 \cdot n = 24$$

$$n = \frac{24}{0.015}$$

$$n = 1600$$

1600 bolts were inspected.

5. Let n = percent spent on food

$$\$300 = n \cdot \$2000$$

$$\frac{300}{2000} = \frac{n \cdot 2000}{2000}$$

$$0.15 = n$$

$$0.15 \cdot 100\% = n$$

$$15\% = n$$

She spends 15% of her monthly income on food.

9. Let n = percent of calories from fat

The number of fat calories is what percent of total calories?

$$35 = n \cdot 120$$

$$35 = 120n$$

$$n = \frac{35}{120} \approx 0.292 = 29.2\%$$

29.2% of the food's total calories is from fat.

13. 20% of $170 is what number?

$$20\% \cdot \$170 = n$$

$$0.2 \cdot \$170 = n$$

$$\$34.00 = n$$

$$\text{new bill} = \$170 - \$34$$

$$= \$136$$

Their new bill is $136.

17. 33% of 26.8 million is what number?

$$33\% \cdot 26.8 \text{ million} = n$$

$$0.33 \cdot 26,800,000 = n$$

$$8,844,000 = n$$

Predicted population =

26.8 million + 8.844 million = 35.644 million

The increase is 8.844 million. The predicted population is 35.644 million.

Original Amount	New Amount	Amount of Increase	Percent Increase
21. 85	187	102	120%

	Original Amount	New Amount	Amount of Decrease	Percent Decrease
25.	160	40	120	75%

29. percent increase $= \dfrac{23.7 - 19.5}{19.5}$

$= \dfrac{4.2}{19.5}$

≈ 0.215

$= 21.5\%$

33. percent decrease $= \dfrac{52.1 - 7.9}{52.1}$

$= \dfrac{44.2}{52.1}$

≈ 0.848

$= 84.8\%$

37.
$$\begin{array}{r} 0.12 \\ \times\ 38 \\ \hline 96 \\ 360 \\ \hline 4.56 \end{array}$$

41.
$$\begin{array}{r} 78.00 \\ -\ 19.46 \\ \hline 58.54 \end{array}$$

EXERCISE SET 6.6

1. tax $= 5\% \cdot \$150.00$
$= 0.05 \cdot \$150.00$
$= \$7.50$
The sales tax is $7.50.

5. $\$54 = r \cdot \$600$
$\dfrac{\$54}{\$600} = r$
$0.09 = r$
$9\% = r$
The sales tax rate is 9%.

9. tax $= 6.5\% \cdot \$1800$
$= 0.065 \cdot \$1800$
$= \$117$
total $= \$1800 + \$117 = \$1917$
The total price is $1917.

13. total purchase $= \$90 + \$15 = \$105$
tax $= 7\% \cdot \$105$
$= 0.07 \cdot \$105$
$= \$7.35$
total $= \$105 + \$7.35 = \$112.35$
The total price is $112.35.

17. commission $= 4\% \cdot \$1,236,856$
$= 0.04 \cdot \$1,236,856$
$= \$49,474.24$
Her commission was $49,474.24.

21. commission $= 1.5\% \cdot \$125,900$
$= 0.015 \cdot \$125,900$
$= \$1888.5$
His commission is $1888.5.

	Original Price	Discount Rate	Amount of Discount	Sale Price
25.	$68.00	10%	$6.80	$61.20
29.	$215.00	35%	$75.25	$139.75

33. discount $= 15\% \cdot \$300$
$= 0.15 \cdot \$300$
$= \$45$
sale price $= \$300 - \$45$
$= \$255$

37. $400 \cdot 0.03 \cdot 11 = 12 \cdot 11 = 132$

41. tax $= 7.5\% \cdot \$24,966$
$= 0.075 \cdot \$24,966$
$= \$1872.45$
total price $= \$24,966 + \$1872.45$
$= \$26,838.45$
The total price is $26,838.45.

EXERCISE SET 6.7

1. simple interest $=$ principal $\cdot$ rate $\cdot$ time
$= (\$200)(8\%)(2)$
$= (\$200)(0.08)(2)$
$= \$32$

5. simple interest $=$ principal $\cdot$ rate $\cdot$ time
$= (\$5000)(10\%)\left(1\frac{1}{2}\right)$
$= (\$5000)(0.10)(1.5)$
$= \$750$

9. simple interest $=$ principal $\cdot$ rate $\cdot$ time
$= (\$2500)(16\%)\left(\dfrac{21}{12}\right)$
$= (\$2500)(0.16)(1.75)$
$= \$700$

13. simple interest $=$ principal $\cdot$ rate $\cdot$ time
$= \$5000(9\%)\left(\dfrac{15}{12}\right)$
$= \$5000(0.09)(1.25)$
$= \$562.50$
Total $= \$5000 + \$562.50 = \$5562.50$

17. Total amount $=$ original principal
$\cdot$ compound interest factor
$= \$6150(7.61226)$
$= \$46,815.399$
The total amount is $46,815.40.

21. Total amount $=$ original principal
$\cdot$ compound interest factor
$= \$10,000(5.81636)$
$= \$58,163.60$
The total amount is $58,163.60.

25. Total amount $=$ original principal
$\cdot$ compound interest factor
$= \$2000(1.46933)$
$= \$2938.66$
Compound interest $=$ total amount
$-$ original principal
$= \$2938.66 - \$2000$
$= \$938.66$

29. monthly payment $= \dfrac{\text{principal } + \text{ simple interest}}{\text{number of payments}}$

$$= \frac{\$1500 + \$61.88}{6}$$

$$= \frac{\$1561.88}{6}$$

$$\approx \$260.31$$

The monthly payment is $260.31.

33. perimeter $= 10 + 6 + 10 + 6$
$$= 32$$
The perimeter is 32 yards.

37. answers may vary

Chapter 6 Test

1. $85\% = 85.\% = 0.85$

5. $6.1 = 610\%$

9. $0.2\% = \dfrac{0.2}{100} = \dfrac{2}{1000} = \dfrac{1}{500}$

13. $n = 42\% \cdot 80$
$n = 0.42 \cdot 80$
$n = 33.6$
Therefore, 42% of 80 is 33.6.

17. 20% of what is $11,350?
$0.20n = \$11,350$
$n = \dfrac{\$11,350}{0.20}$
$n = \$56,750$
The value is $56,750.

21. commission $= 4\% \cdot \$9875$
$\qquad\qquad\quad = 0.04 \cdot \$9875$
$\qquad\qquad\quad = \$395$
His commission is $395.

25. simple interest $=$ principal $\cdot$ rate $\cdot$ time
$$= (\$400)(13.5\%)\left(\frac{6}{12}\right)$$
$$= (\$400)(0.135)(0.5)$$
$$= \$27.00$$
Total amount due the bank $= \$400 + \$27$
$\qquad\qquad\qquad\qquad\qquad\quad = \$427$

Chapter 7

Exercise Set 7.1

1. 60 inches $= 60$ inches $\cdot \dfrac{1 \text{ foot}}{12 \text{ inches}} = 5$ feet

5. 42,240 feet $= 42,240$ feet $\cdot \dfrac{1 \text{ mile}}{5280 \text{ feet}} = 8$ miles

9. 10 feet $= 10$ feet $\cdot \dfrac{1 \text{ yard}}{3 \text{ feet}} = 3\dfrac{1}{3}$ yards

13. 40 feet $\div 3 = 13$ with remainder 1
40 feet $= 13$ yards 1 foot

17. 10,000 feet $\div 5280 = 1$ with remainder 4720
10,000 feet $= 1$ mile 4720 feet

21. 5 yards $= 5$ yards $\cdot \dfrac{3 \text{ feet}}{1 \text{ yard}} = 15$ feet
5 yards 2 feet $= 15$ feet $+ 2$ feet $= 17$ feet

25.
$\quad$ 5 feet 8 inches
$+\,$ 6 feet 7 inches
11 feet 15 inches $= 11$ feet $+ 1$ foot 3 inches
$\qquad\qquad\qquad\qquad = 12$ feet 3 inches

29.
$\quad$ 24 feet 8 inches
$-\,$ 16 feet 3 inches
$\quad$ 8 feet 5 inches

33.
$\qquad\quad$ 3 feet 4 inches
2) 6 feet 8 inches
$-\,$ 6 feet
$\quad\; 0 \qquad$ 8 inches
$-\qquad\qquad$ 8 inches
$\qquad\qquad\qquad 0$

37.
$\quad$ 6 feet 10 inches
$+\,$ 3 feet 8 inches
$\quad$ 9 feet 18 inches $= 9$ feet $+ 1$ foot 6 inches
$\qquad\qquad\qquad\qquad = 10$ feet 6 inches
The bamboo is 10 feet 6 inches tall.

41.
$\quad$ 1 foot 9 inches
$\qquad\quad \times\, 9$
9 feet 81 inches $= 9$ feet $+ 6$ feet 9 inches
$\qquad\qquad\qquad\qquad = 15$ feet 9 inches
9 stacks would extend 15 feet 9 inches from the wall.

45. $P = 2L + 2W$
$\quad = 2(24 \text{ feet } 9 \text{ inches}) + 2(18 \text{ feet } 6 \text{ inches})$
$\quad = 48 \text{ feet } 18 \text{ inches} + 36 \text{ feet } 12 \text{ inches}$
$\quad = 84 \text{ feet } 30 \text{ inches}$
$\quad = 84 \text{ feet} + 2 \text{ feet } 6 \text{ inches}$
$\quad = 86 \text{ feet } 6 \text{ inches}$
She must buy 86 feet 6 inches of fencing material.

49. 40 meters to centimeters $\qquad$ *Move the decimal 2 places to the right.*
40 meters $= 4000$ centimeters

53. 300 meters to kilometers $\qquad$ *Move the decimal 3 places to the left.*
300 meters $= 0.300$ kilometer

57. 1500 centimeters to meters $\qquad$ *Move the decimal 2 places to the left.*
1500 centimeters $= 15$ meters

61. 20.1 millimeters to decimeters $\qquad$ *Move the decimal 2 places to the left.*
20.1 millimeters $= 0.201$ decimeter

65.
$\quad$ 8.6$\quad$ meters
$+\,$ 0.34 meters
$\quad$ 8.94 meters

69. 24.8 millimeters $- 1.19$ centimeters
$= 2.48$ centimeters $- 1.19$ centimeters
$= 1.29$ centimeters or 12.9 millimeters

73. 18.3 meters $\times 3 = 54.9$ meters

77. 3.4 meters $+ 5.8$ meters $- 8$ centimeters
$= 3.4$ meters $+ 5.8$ meters $- 0.08$ meter
$= 9.12$ meters
The tied ropes are 9.12 meters long.

81. 25(1 meter + 65 centimeters)
= 25(100 centimeters + 65 centimeters)
= 25(165 centimeters)
= 4125 centimeters or 41.25 meters
4125 centimeters of 41.25 meters of wood must be ordered.

85. 5.988 kilometers + 21 meters
= 5988 meters + 21 meters
= 6009 meters or 6.009 kilometers
The elevation is 6009 meters or 6.009 kilometers.

89. $0.21 = 21\%$

93. $\dfrac{1}{4} = 0.25 = 25\%$

97. A square has 4 equal sides.

$$\begin{array}{r} 6.575 \\ 4\overline{)26.300} \\ \underline{-24} \\ 2\,3 \\ \underline{-2\,0} \\ 30 \\ \underline{-28} \\ 20 \\ \underline{-20} \\ 0 \end{array}$$

Each side will be 6.575 meters long.

EXERCISE SET 7.2

1. $2 \text{ pounds} = 2 \text{ pounds} \cdot \dfrac{16 \text{ ounces}}{1 \text{ pound}} = 32 \text{ ounces}$

5. $12{,}000 \text{ pounds} = 12{,}000 \text{ pounds} \cdot \dfrac{1 \text{ ton}}{2000 \text{ pounds}}$
$= 6 \text{ tons}$

9. $3500 \text{ pounds} = 3500 \text{ pounds} \cdot \dfrac{1 \text{ ton}}{2000 \text{ pounds}}$
$= \dfrac{7}{4} \text{ ton}$
$= 1\dfrac{3}{4} \text{ ton}$

13. $4.9 \text{ tons} = 4.9 \text{ tons} \cdot \dfrac{2000 \text{ pounds}}{1 \text{ ton}}$
$= 9800 \text{ pounds}$

17. $2950 \text{ pounds} = 2950 \text{ pounds} \cdot \dfrac{1 \text{ ton}}{2000 \text{ pounds}}$
$= \dfrac{2950 \text{ tons}}{2000}$
$= 1.475 \text{ tons}$
$\approx 1.5 \text{ tons}$

21.
$$\begin{array}{r} 6 \text{ tons } 1540 \text{ pounds} \\ + 2 \text{ tons } 850 \text{ pounds} \\ \hline 8 \text{ tons } 2390 \text{ pounds} \end{array} = 8 \text{ tons} + 1 \text{ ton } 390 \text{ pounds}$$
$= 9 \text{ tons } 390 \text{ pounds}$

25.
$$\begin{array}{r} 12 \text{ pounds } 4 \text{ ounces} \\ - 3 \text{ pounds } 9 \text{ ounces} \end{array} \qquad \begin{array}{r} 11 \text{ pounds } 20 \text{ ounces} \\ - 3 \text{ pounds } 9 \text{ ounces} \\ \hline 8 \text{ pounds } 11 \text{ ounces} \end{array}$$

29.
$$\begin{array}{r} 1 \text{ ton} \quad 700 \text{ pounds} \\ 5\overline{)6 \text{ tons} \quad 1500 \text{ pounds}} \\ \underline{-5 \text{ tons}} \\ 1 \text{ ton} = 2000 \text{ pounds} \\ \hline 3500 \text{ pounds} \\ \underline{- 3500 \text{ pounds}} \\ 0 \end{array}$$

33.
$$\begin{array}{r} 64 \text{ pounds } 8 \text{ ounces} \\ - 28 \text{ pounds } 10 \text{ ounces} \end{array} \qquad \begin{array}{r} 63 \text{ pounds } 24 \text{ ounces} \\ - 28 \text{ pounds } 10 \text{ ounces} \\ \hline 35 \text{ pounds } 14 \text{ ounces} \end{array}$$
Her zucchini was 35 pounds 14 ounces below the record.

37.
$$\begin{array}{r} 55 \text{ pounds } 4 \text{ ounces} \\ - 2 \text{ pounds } 8 \text{ ounces} \end{array} \qquad \begin{array}{r} 54 \text{ pounds } 20 \text{ ounces} \\ - 2 \text{ pounds } 8 \text{ ounces} \\ \hline 52 \text{ pounds } 12 \text{ ounces} \end{array}$$

$$\begin{array}{r} 52 \text{ pounds } 12 \text{ ounces} \\ \times 4 \\ \hline 208 \text{ pounds } 48 \text{ ounces} \end{array} = 208 \text{ pounds} + 3 \text{ pounds}$$
$= 211 \text{ pounds}$

41. 500 grams to kilograms   *Move the decimal 3 places to the left.*
500 grams = 0.5 kilograms

45. 25 kilograms to grams   *Move the decimal 3 places to the right.*
25 kilograms = 25,000 grams

49. 6.3 grams to kilograms   *Move the decimal 3 places to the left.*
6.3 grams = 0.0063 kilograms

53. 4.01 kilograms to grams   *Move the decimal 3 places to the right.*
4.01 kilograms = 4010 grams

57. $205 \text{ mg} = 0.205 \text{ g}$ or $5.61 \text{ g} = 5610 \text{ mg}$
$$\begin{array}{r} 0.205 \text{ g} \\ 5.610 \text{ g} \\ \hline 5.815 \text{ g} \end{array} \qquad \begin{array}{r} 205 \text{ mg} \\ 5610 \text{ mg} \\ \hline 5815 \text{ mg} \end{array}$$

61. $1.61 \text{ kg} = 1610 \text{ g}$ or $250 \text{ g} = 0.250 \text{ kg}$
$$\begin{array}{r} 1610 \text{ g} \\ - 250 \text{ g} \\ \hline 1360 \text{ g} \end{array} \qquad \begin{array}{r} 1.610 \text{ kg} \\ - 0.250 \text{ kg} \\ \hline 1.360 \text{ kg} \end{array}$$

65. $17 \text{ kilograms} \div 8 = 2.125 \text{ kilograms}$

69. 0.09 grams − 60 milligrams
= 90 milligrams − 60 milligrams = 30 milligrams
The extra-strength tablet contains 30 milligrams more medication.

73. $3 \cdot 16 \cdot 3 \text{ milligrams} = 144 \text{ milligrams}$
3 cartons contain 144 milligrams of preservatives.

77. 0.3 kilograms + 0.15 kilograms + 400 grams
= 0.3 kilograms + 0.15 kilograms + 0.4 kilograms
= 0.85 kilograms or 850 grams
The package weighs 0.85 kilograms or 850 grams.

81. $\dfrac{1}{4} = \dfrac{1 \cdot 25}{4 \cdot 25} = \dfrac{25}{100} = 0.25$

85. $\dfrac{7}{8} = \dfrac{7 \cdot 125}{8 \cdot 125} = \dfrac{875}{1000} = 0.875$

Exercise Set 7.3

1. 32 fluid ounces $= 32$ fluid ounces $\cdot \dfrac{1 \text{ cup}}{8 \text{ fluid ounces}}$

$\qquad\qquad\qquad = 4$ cups

5. 10 quarts $= 10$ quarts $\cdot \dfrac{1 \text{ gallon}}{4 \text{ quarts}} = 2\dfrac{1}{2}$ gallons

9. 2 quarts $= 2$ quarts $\cdot \dfrac{4 \text{ cups}}{1 \text{ quart}} = 8$ cups

13. 6 gallons

$= 6$ gallons $\cdot \dfrac{4 \text{ quarts}}{1 \text{ gallon}} \cdot \dfrac{2 \text{ pints}}{1 \text{ quart}} \cdot \dfrac{2 \text{ cups}}{1 \text{ pint}} \cdot \dfrac{8 \text{ fluid ounces}}{1 \text{ cup}}$

$= 768$ fluid ounces

17. $2\dfrac{3}{4}$ gallons $= \dfrac{11}{4}$ gallons $\cdot \dfrac{4 \text{ quarts}}{1 \text{ gallon}} \cdot \dfrac{2 \text{ pints}}{1 \text{ quart}} = 22$ pints

21.
$$\begin{array}{r} 1 \text{ c } \ 5 \text{ fl oz} \\ + 2 \text{ c } \ 7 \text{ fl oz} \\ \hline 3 \text{ c } 12 \text{ fl oz} \end{array} = 3 \text{ c} + 1 \text{ c } 4 \text{ fl oz} = 4 \text{ c } 4 \text{ fl oz}$$

25.
$$\begin{array}{r} 3 \text{ gal } 1 \text{ qt} \\ - \qquad 1 \text{ qt } 1 \text{ pt} \\ \hline \end{array} \qquad \begin{array}{r} 2 \text{ gal } 4 \text{ qt } 2 \text{ pt} \\ - \qquad 1 \text{ qt } 1 \text{ pt} \\ \hline 2 \text{ gal } 3 \text{ qt } 1 \text{ pt} \end{array}$$

29.
$$\begin{array}{r} 8 \text{ gal } 2 \text{ qt} \\ \times 2 \\ \hline 16 \text{ gal } 4 \text{ qt} = 17 \text{ gal} \end{array}$$

33. $1\dfrac{1}{2}$ quarts $= \dfrac{3}{2}$ quarts $\cdot \dfrac{2 \text{ pints}}{1 \text{ quart}} \cdot \dfrac{2 \text{ cups}}{1 \text{ pint}} \cdot \dfrac{8 \text{ fluid ounces}}{1 \text{ cup}}$

$= 48$ fluid ounces

37.
$$\begin{array}{r} 5 \text{ pints } 1 \text{ cup} \\ + 2 \text{ pints } 1 \text{ cup} \\ \hline 7 \text{ pints } 2 \text{ cups} \end{array}$$

Since 8 pints $= 1$ gallon, the fruit punch can be poured into the container.

41. 12 fluid ounces $\cdot 24 = 288$ fluid ounces

$= 288$ fluid ounces $\cdot \dfrac{1 \text{ cup}}{8 \text{ fluid ounces}} \cdot \dfrac{1 \text{ pint}}{2 \text{ cups}} \cdot \dfrac{1 \text{ quart}}{2 \text{ pints}}$

$= 9$ quarts

45. 4500 ml to liters   *Move the decimal 3 places to the left.*

4500 ml $= 4.5$ liters

49. 64 ml to liters   *Move the decimal 3 places to the left.*

64 ml $= 0.064$ liters

53. 3.6 L to milliliters   *Move the decimal 3 places to the right.*

3.6 L $= 3600$ milliliters

57. 2.9 L $+ 19.6$ L $= 22.5$ L

61. 8.6 L $= 8600$ ml or 190 ml $= 0.190$ L
$$\begin{array}{r} 8600 \text{ ml} \\ - 190 \text{ ml} \\ \hline 8410 \text{ ml} \end{array} \qquad \begin{array}{r} 8.60 \text{ L} \\ - 0.19 \text{ L} \\ \hline 8.41 \text{ L} \end{array}$$

65. 480 ml $\times 8 = 3840$ ml

69. 2 L $- 410$ ml $= 2$ L $- 0.41$ L $= 1.59$ L
1.59 L remains in the bottle.

73. $14.00 \div 44.3$ L $\approx \$0.316$
The cost is about \$0.316 per liter.

77. $0.7 = \dfrac{7}{10}$

81. $0.006 = \dfrac{6}{1000} = \dfrac{3}{500}$

Exercise Set 7.4

1. $C = \dfrac{5}{9} \cdot (41 - 32)$

$= \dfrac{5}{9} \cdot (9)$

$= 5$

$41°F = 5°C$

5. $F = \dfrac{9}{5} \cdot (60) + 32$

$= 9(12) + 32$

$= 108 + 32$

$= 140$

$60°C = 140°F$

9. $C = \dfrac{5}{9} \cdot (62 - 32)$

$= \dfrac{5}{9} \cdot (30)$

$= \dfrac{150}{9}$

≈ 16.7

$62°F \approx 16.7°C$

13. $F = 1.8(92) + 32$

$= 165.6 + 32$

$= 197.6$

$92°C = 197.6°F$

17. $C = \dfrac{5}{9} \cdot (134 - 32)$

$= \dfrac{5}{9} \cdot (102)$

≈ 56.7

$134°F \approx 56.7°C$

21. $C = \dfrac{5}{9} \cdot (70 - 32)$

$= \dfrac{5}{9} \cdot (38)$

$= \dfrac{190}{9}$

≈ 21.1

$70°F \approx 21.1°C$

25. $F = 1.8(118) + 32$

$= 212.4 + 32$

$= 244.4$

$188°C = 244.4°F$

29. $C = \dfrac{5}{9} \cdot (864 - 32)$

$= \dfrac{5}{9} \cdot (832)$

$= \dfrac{4160}{9}$

≈ 462.2

$864°F \approx 462.2°C$

SOLUTIONS

A65

33. $P = 3 \text{ feet} + 3 \text{ feet} + 3 \text{ feet} + 3 \text{ feet} + 3 \text{ feet}$
$= 15 \text{ feet}$

37. $C = \dfrac{5}{9} \cdot (9010 - 32)$
$= \dfrac{5}{9} \cdot (8978)$
≈ 4988
$9010°F$ is about $4988°C$

Exercise Set 7.5

1. energy $= 3 \text{ pounds} \cdot 380 \text{ feet}$
$= 1140 \text{ foot-pounds}$

5. energy $= \dfrac{2.5 \text{ tons}}{1} \cdot \dfrac{2000 \text{ pounds}}{1 \text{ ton}} \cdot 85 \text{ feet}$
$= 425{,}000 \text{ foot-pounds}$

9. $1000 \text{ BTU} = 1000 \text{ BTU} \cdot \dfrac{778 \text{ foot-pounds}}{1 \text{ BTU}}$
$= 778{,}000 \text{ foot-pounds}$

13. $34{,}130 \text{ BTU} = 34{,}130 \text{ BTU} \cdot \dfrac{778 \text{ foot-pounds}}{1 \text{ BTU}}$
$= 26{,}553{,}140 \text{ foot-pounds}$

17. total hours $= 7$ hours
calories $= 7 \cdot 115 = 805$ calories

21. total hours $= \dfrac{1}{3} \cdot 6 = 2$ hours
calories $= 2 \cdot 720 = 1440$ calories

25. miles $= \dfrac{3500}{200} = 17.5$ miles

29. $\dfrac{27}{45} = \dfrac{3 \cdot 3 \cdot 3}{3 \cdot 3 \cdot 5} = \dfrac{3}{5}$

33. energy $= 123.9 \text{ pounds} \cdot \dfrac{9 \text{ inches}}{1} \cdot \dfrac{1 \text{ foot}}{12 \text{ inches}}$
$= 92.925 \text{ foot-pounds}$

Chapter 7 Test

1.
$$\begin{array}{r} 23 \\ 12\overline{)280} \\ -24 \\ \hline 40 \\ -36 \\ \hline 4 \end{array}$$
$280 \text{ inches} = 23 \text{ feet } 4 \text{ inches}$

5. $38 \text{ pints} = 38 \text{ pints} \cdot \dfrac{1 \text{ gallon}}{8 \text{ pints}} = 4\dfrac{3}{4} \text{ gallons}$

9. $4.3 \text{ decigrams} = 0.43 \text{ grams}$

13.
$$\begin{array}{r} 2 \text{ feet } 9 \text{ inches} \\ \times 3 \\ \hline 6 \text{ feet } 27 \text{ inches} \end{array} = 6 \text{ feet} + 2 \text{ feet } 3 \text{ inches}$$
$= 8 \text{ feet } 3 \text{ inches}$

17. $C = \dfrac{5}{9} \cdot (84 - 32)$
$= \dfrac{5}{9} \cdot (52)$
$= \dfrac{260}{9}$
$= 28.\overline{8}$
≈ 28.9
$84°F \approx 28.9°C$

21. $F = 1.8(41) + 32 = 73.8 + 32 = 105.8$
Her fever is $105.8°F$.

25. $26{,}000 \text{ BTU} = 26{,}000 \text{ BTU} \cdot \dfrac{778 \text{ foot-pounds}}{1 \text{ BTU}}$
$= 20{,}228{,}000 \text{ foot-pounds}$

Chapter 8

Exercise Set 8.1

1. The figure extends indefinitely in two directions. It is line yz, or $\overleftrightarrow{yz}$.

5. The figure has two endpoints. It is line segment PQ, or $\overline{PQ}$.

9. $\angle ABC = 15°$

13. $\angle DBA = 50° + 15°$
$= 65°$

17. $90°$

21. $\angle S$ is a straight angle.

25. $\angle Q$ is an obtuse angle. It measures between $90°$ and $180°$.

29. The complement of an angle that measures $17°$ is an angle that measures $90° - 17° = 73°$.

33. The complement of a $45°$ angle is an angle that measures $90° - 45° = 45°$.

37. $\angle MNP$ and $\angle RNO$; are complementary angles since $60° + 30° = 90°$.

41. $\angle x = 120° - 88°$
$= 32°$

45. Since $\angle x$ and the $35°$ angle are vertical angles, $\angle x = 35°$. $\angle x$ and $\angle y$ are adjacent angles, so $\angle y = 180° - 35° = 145°$. $\angle y$ and $\angle z$ are vertical angles, so $\angle z = 145°$.

49. $\angle x$ and the $80°$ angle are adjacent angles, so $\angle x = 180° - 80° = 100°$. $\angle y$ and the $80°$ angle are alternate interior angles, so $\angle y = 80°$. $\angle x$ and $\angle z$ are corresponding angles, so $\angle z = 100°$.

53. $\dfrac{7}{8} + \dfrac{1}{4} = \dfrac{7}{8} + \dfrac{2}{8} = \dfrac{9}{8}$ or $1\dfrac{1}{8}$

57. $3\dfrac{1}{3} - 2\dfrac{1}{2} = \dfrac{10}{3} - \dfrac{5}{2} = \dfrac{20}{6} - \dfrac{15}{6} = \dfrac{5}{6}$

61. The supplement of a $125.2°$ angle is $180° - 125.2° = 54.8°$.

Exercise Set 8.2

1. The triangle is an equilateral triangle because all three sides are the same length.

5. The triangle is an isosceles triangle since two sides are the same length.

9. $\angle x = 180° - 95° - 72° = 13°$

13. diameter **17.** parallelogram

21. true **25.** false

29. $r = \dfrac{d}{2}$

$= \dfrac{29}{2}$

$= 14.5$ centimeters

33. hexagon **37.** rectangular solid

41. $d = 2 \cdot r$

$= 2 \cdot 7.4$

$= 14.8$ inches

45. cube **49.** sphere

53. $4(28.6) = 114.4$

57. $d = 2 \cdot r$

$= 2 \cdot 36,184$

$= 72,368$ miles

Exercise Set 8.3

1. $P = 2l + 2w$

$= 2(17 \text{ feet}) + 2(15 \text{ feet})$

$= 34 \text{ feet} + 30 \text{ feet}$

$= 64 \text{ feet}$

5. $P = a + b + c$

$= 5 \text{ inches} + 7 \text{ inches} + 9 \text{ inches}$

$= 21 \text{ inches}$

9. $P = 10 \text{ feet} + 8 \text{ feet} + 8 \text{ feet} + 15 \text{ feet} + 7 \text{ feet}$

$= 48 \text{ feet}$

13. $P = 5 \text{ feet} + 3 \text{ feet} + 2 \text{ feet} + 7 \text{ feet} + 4 \text{ feet}$

$= 21 \text{ feet}$

17. $P = 2l + 2w$

$= 2(120 \text{ yards}) + 2(53 \text{ yards})$

$= 240 \text{ yards} + 106 \text{ yards}$

$= 346 \text{ yards}$

21. cost $= 3(22)$

$= \$66$

25. $P = 4s$

$= 4(7 \text{ inches})$

$= 28 \text{ inches}$

29. The missing lengths are:

28 meters $-$ 20 meters $=$ 8 meters

20 meters $-$ 17 meters $=$ 3 meters

$P = 8 \text{ meters} + 3 \text{ meters} + 20 \text{ meters} + 20 \text{ meters}$
$+ 28 \text{ meters} + 17 \text{ meters}$

$= 96$ meters

33. The missing lengths are:

12 miles $+$ 10 miles $=$ 22 miles

8 miles $+$ 34 miles $=$ 42 miles

$P = 12 \text{ miles} + 34 \text{ miles} + 10 \text{ miles} + 8 \text{ miles}$
$+ 22 \text{ miles} + 42 \text{ miles}$

$= 128$ miles

37. $C = 2\pi r$

$= 2\pi \cdot 8 \text{ miles}$

$= 16\pi \text{ miles};$

$\approx 16 \cdot 3.14 \text{ miles}$

$= 50.24 \text{ miles}$

41. $C = 2\pi r$

$= 2\pi \cdot 5 \text{ feet}$

$= 10\pi \text{ feet}$

$\approx 10 \cdot \dfrac{22}{7} \text{ feet}$

$= \dfrac{220}{7} \text{ feet}$

$\approx 31.43 \text{ feet}$

45. $5 + 6 \cdot 3 = 5 + 18 = 23$

49. $(18 + 8) - (12 + 4) = 26 - 16 = 10$

53. perimeter **57.** area

61. a. Circumference of the smaller circle:

$C = 2\pi r$

$= 2\pi \cdot 10 \text{ meters}$

$= 20\pi \text{ meters}$

$\approx 20 \cdot 3.14 \text{ meters}$

$= 62.8 \text{ meters,}$

Circumference of the larger circle:

$C = 2\pi r$

$= 2\pi \cdot 20 \text{ meters}$

$= 40\pi \text{ meters}$

$\approx 40 \cdot 3.14 \text{ meters}$

$= 125.6 \text{ meters}$

b. Yes, the circumference is doubled.

Exercise Set 8.4

1. $A = l \cdot w$

$= (3.5 \text{ meters})(2 \text{ meters})$

$= 7 \text{ square meters}$

5. $A = \dfrac{1}{2} \cdot b \cdot h$

$= \dfrac{1}{2}(6 \text{ yards})(5 \text{ yards})$

$= \dfrac{1}{2} \cdot 6 \cdot 5 \text{ square yards}$

$= 15 \text{ square yards}$

9. $A = b \cdot h$

$= (7 \text{ feet})(5.25 \text{ feet})$

$= 36.75 \text{ square feet}$

13. $A = \dfrac{1}{2} \cdot (b + B) \cdot h$

$= \dfrac{1}{2} \cdot (4 \text{ yards} + 7 \text{ yards}) \cdot (4 \text{ yards})$

$= \dfrac{1}{2} \cdot 11 \cdot 4 \text{ square yards}$

$= 22 \text{ square yards}$

17. $A = b \cdot h$

$= (5 \text{ inches})\left(4\dfrac{1}{2} \text{ inches}\right)$

$= 5 \cdot \dfrac{9}{2} \text{ square inches}$

$= \dfrac{45}{2} \text{ or } 22\dfrac{1}{2} \text{ square inches}$

21.

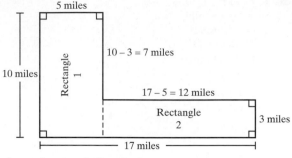

Area of rectangle 1 $= b \cdot h$
$= (5\text{ miles})(10\text{ miles})$
$= 50$ square miles

Area of rectangle 2 $= b \cdot h$
$= (12\text{ miles})(3\text{ miles})$
$= 36$ square miles

Total area $= 50$ square miles $+ 36$ square miles
$= 86$ square miles

25. $A = \pi \cdot r^2$
$= \pi(6\text{ inches})^2$
$= 36\pi$ square inches
$\approx 36 \cdot \dfrac{22}{7}$ square inches
$= \dfrac{792}{7}$ square inches
≈ 113.1 square inches

29. $A = l \cdot w$
$= (100\text{ feet})(50\text{ feet})$
$= 5000$ square feet

33. $A = l \cdot w$
$= (20\text{ inches})\left(25\tfrac{1}{2}\text{ inches}\right)$
$= 20 \cdot \dfrac{51}{2}$ square inches
$= \dfrac{1020}{2}$ or 510 square inches

37. $A = \dfrac{1}{2} \cdot (b + B) \cdot h$
$= \dfrac{1}{2} \cdot (90\text{ feet} + 140\text{ feet}) \cdot (80\text{ feet})$
$= \dfrac{1}{2} \cdot 230 \cdot 80$ square feet
$= 9200$ square feet

41. $C = \pi d$
$= \pi \cdot (14\text{ inches})$
$= 14\pi$ inches
$\approx 14 \cdot 3.14$ inches
$= 43.96$ inches

45. Perimeter $= 6 \cdot s$
$= 6 \cdot \left(2\tfrac{1}{8}\text{ feet}\right)$
$= \left(6 \cdot \dfrac{17}{8}\right)$ feet
$= \dfrac{102}{8}$ or $12\tfrac{3}{4}$ feet

49. 8 inches $= \dfrac{8}{12}$ foot $= \dfrac{2}{3}$ foot

$A = l \cdot w$
$= (2\text{ feet}) \cdot \left(\dfrac{2}{3}\text{ foot}\right)$
$= \dfrac{4}{3}$ or $1\tfrac{1}{3}$ square feet

2 feet $= 24$ inches
$A = l \cdot w$
$= (24\text{ inches}) \cdot (8\text{ inches})$
$= 192$ square inches

EXERCISE SET 8.5

1. $V = l \cdot w \cdot h$
$= (6\text{ inches}) \cdot (4\text{ inches}) \cdot (3\text{ inches})$
$= 72$ cubic inches

5. $V = \dfrac{1}{3} \cdot \pi \cdot r^2 \cdot h$
$= \dfrac{1}{3} \cdot \pi \cdot (2\text{ yards})^2 \cdot (3\text{ yards})$
$= 4\pi$ cubic yards
$\approx 4 \cdot \dfrac{22}{7}$ cubic yards
$= \dfrac{88}{7}$ or $12\tfrac{4}{7}$ cubic yards

9. $V = \dfrac{1}{3} \cdot s^2 \cdot h$
$= \dfrac{1}{3} \cdot (5\text{ centimeters})^2 \cdot (9\text{ centimeters})$
$= 75$ cubic centimeters

13. $V = l \cdot w \cdot h$
$= (2\text{ feet}) \cdot (1.4\text{ feet}) \cdot (3\text{ feet})$
$= 8.4$ cubic feet

17. $V = \dfrac{1}{3} \cdot s^2 \cdot h$
$= \dfrac{1}{3} \cdot (12\text{ centimeters})^2 \cdot (20\text{ centimeters})$
$= \dfrac{2880}{3}$ cubic centimeters
$= 960$ cubic centimeters

21. $V = l \cdot w \cdot h$
$= (2\text{ feet}) \cdot \left(2\tfrac{1}{2}\text{ feet}\right) \cdot \left(1\tfrac{1}{2}\text{ feet}\right)$
$= 2 \cdot \dfrac{5}{2} \cdot \dfrac{3}{2}$ cubic feet
$= \dfrac{30}{4}$ or $7\tfrac{1}{2}$ cubic feet

25. $5^2 = 25$

29. $1^2 + 2^2 = 1 + 4 = 5$

33. $V = \dfrac{1}{3} \cdot s^2 \cdot h$
$= \dfrac{1}{3} \cdot (230\text{ meters})^2 \cdot (146.5\text{ meters})$
$= \dfrac{7{,}749{,}850}{3}$ cubic meters
$\approx 2{,}583{,}283$ cubic meters

37. $V = \pi \cdot r^2 \cdot h$
$= \pi \cdot (16.3 \text{ feet})^2 \cdot (32 \text{ feet})$
$= 8502.08\pi \text{ cubic feet}$
$\approx 8502.08 \cdot 3.14 \text{ cubic feet}$
$\approx 26{,}696.5 \text{ cubic feet}$

Exercise Set 8.6

1. $\sqrt{4} = 2$ because $2^2 = 4$

5. $\sqrt{\dfrac{1}{81}} = \dfrac{1}{9}$ because $\dfrac{1}{9} \cdot \dfrac{1}{9} = \dfrac{1}{81}$

9. $\sqrt{256} = 16$ because $16^2 = 256$

13. $\sqrt{3} \approx 1.732$

17. $\sqrt{14} \approx 3.742$

21. $\sqrt{8} \approx 2.828$

25. $\sqrt{71} \approx 8.426$

29. $a^2 + b^2 = c^2$
$5^2 + 12^2 = c^2$
$25 + 144 = c^2$
$169 = c^2$
$c = \sqrt{169}$
$c = 13$

33.

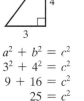

$a^2 + b^2 = c^2$
$3^2 + 4^2 = c^2$
$9 + 16 = c^2$
$25 = c^2$
$c = \sqrt{25}$
$c = 5$

37.

$a^2 + b^2 = c^2$
$10^2 + 14^2 = c^2$
$100 + 196 = c^2$
$296 = c^2$
$c = \sqrt{296}$
$c \approx 17.205$

41.

$a^2 + b^2 = c^2$
$5^2 + b^2 = 13^2$
$25 + b^2 = 169$
$b^2 = 144$
$b = 12$

45.

$a^2 + b^2 = c^2$
$30^2 + 30^2 = c^2$
$900 + 900 = c^2$
$1800 = c^2$
$c = \sqrt{1800}$
$c \approx 42.426$

49. $a^2 + b^2 = c^2$
$100^2 + 100^2 = c^2$
$10{,}000 + 10{,}000 = c^2$
$20{,}000 = c^2$
$c = \sqrt{20{,}000}$
$c \approx 141.42$
Approximately 141.42 yards.

53. $a^2 + b^2 = c^2$
$300^2 + 160^2 = c^2$
$90{,}000 + 25{,}600 = c^2$
$115{,}600 = c^2$
$c = \sqrt{15{,}600}$
$c = 340$

340 feet

57. $\dfrac{9}{11} = \dfrac{n}{55}$
$11 \cdot n = 55 \cdot 9$
$11n = 495$
$\dfrac{11n}{11} = \dfrac{495}{11}$
$n = 45$

61. $\sqrt{38}$ is between 6 and 7.
$\sqrt{38} \approx 6.164$

Exercise Set 8.7

1. The triangles are congruent by Side-Side-Side.

5. Since the triangles are similar, we can compare any of the corresponding sides to find the ratio.
$\dfrac{22}{11} = \dfrac{2}{1}$

9. $\dfrac{3}{n} = \dfrac{6}{9}$
$6n = 27$
$n = 4.5$

13. $\dfrac{n}{3.75} = \dfrac{12}{9}$
$9n = 45$
$n = 5$

17. $\dfrac{n}{3.25} = \dfrac{17.5}{3.25}$
$3.25n = 56.875$
$n = 17.5$

21. $\dfrac{34}{n} = \dfrac{16}{10}$
$16n = 340$
$n = 21.25$

25. $\dfrac{11}{x} = \dfrac{5}{2}$

$5x = 22$

$x = 4.4$

$\dfrac{14}{y} = \dfrac{5}{2}$

$5y = 28$

$y = 5.6$

The lengths are $x = 4.4$ feet and $y = 5.6$ feet.

29. $\dfrac{14 + 17 + 21 + 18}{4} = \dfrac{70}{4} = 17.5$

33. $\dfrac{5.2}{n} = \dfrac{7.8}{12.6}$

$7.8n = 65.52$

$n = 8.4$

CHAPTER 8 TEST

1. The complement of a $78°$ angle is $90° - 78° = 12°$.

5. $\angle x$ and the $73°$ angle are vertical angles, so $\angle x = 73°$. $\angle y$ and the $73°$ angle are corresponding angles, so $\angle y = 73°$. $\angle z$ and $\angle y$ are vertical angles, so $\angle z = 73°$.

9. $C = 2 \cdot \pi \cdot r$

$= 2 \cdot \pi \cdot (9 \text{ inches})$

$= 18\pi$ inches

$\approx 18 \cdot 3.14$ inches

$= 56.52$ inches

$A = \pi \cdot r^2$

$= \pi \cdot (9 \text{ inches})^2$

$= 81\pi$ square inches

$\approx 81 \cdot 3.14$ square inches

$= 254.34$ square inches

13. $V = l \cdot w \cdot h$

$= (3 \text{ feet}) \cdot (5 \text{ feet}) \cdot (2 \text{ feet})$

$= 30$ cubic feet

17. $P = 4 \cdot s$

$= 4 \cdot (4 \text{ inches})$

$= 16$ inches

21. $A = l \cdot w$

$= (123.8 \text{ feet})(80 \text{ feet})$

$= 9904$ square feet

$\dfrac{0.02 \text{ ounces}}{1 \text{ square foot}} = \dfrac{x}{9904 \text{ square feet}}$

$x = 198.08$ ounces

198.08 ounces of insecticide are required.

Chapter 9

EXERCISE SET 9.1

1. The year 1997 has the most cars, so the greatest number of automobiles was manufactured in 1997.

5. Compare each year with the previous year to see that the production of automobiles decreased from the previous year in 1993, 1994, and 1998.

9. The year 1992 has seven chickens and each chicken represents 3 ounces of chicken, so approximately $7 \cdot 3 = 21$ ounces of chicken were consumed per week in 1992.

13. There is one more chicken for 1992 than for 1988, so there was an increase of approximately $1 \cdot 3 = 3$ ounces per week.

17. The tallest bar corresponds to the month of April, so April has the most tornado-related deaths.

21. Look for bars that extend above the horizontal line at 5. The months of February, March, April, May, and June have over 5 tornado-related deaths.

25. The two U.S. cities are New York City and Los Angeles. New York City is the largest, with an estimated population of 16.2 million or 16,200,000 people.

29.

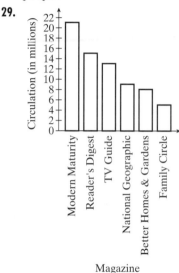

33. 1975 had the highest average number of field goals per game.

37. $30\% \cdot 12 = 0.3 \cdot 12 = 3.6$

41. $\dfrac{1}{4} = 0.25 = 25\%$

45. $83°F$ was the highest temperature reading on Thursday.

49. The greatest difference between high and low temperatures occurred on Tuesday with a difference of $86 - 73 = 13°F$.

EXERCISE SET 9.2

1. The largest sector, or 320 students, corresponds to where most college students live. Most college students live at parent or guardian's home.

5. $\dfrac{\text{students living in campus housing}}{\text{students living at home}} = \dfrac{180}{320} = \dfrac{9}{16}$

9. The sectors which represent "less than \$30" are "less than \$20" and "\$20–29." percent $= 30\% + 23\% = 53\%$

13. amount $= 28\% \cdot 4700 = 1316$ teenagers

17. The second-largest sector is 25% which corresponds to "Nonfiction." Nonfiction is the second-largest category of books.

21. amount $= 22\% \cdot 125{,}600 = 27{,}632$ books

25. Under 3 days: $7\% \cdot 360° = 25.2°$
3–10 days $18\% \cdot 360° = 64.8°$
11–20 days $14\% \cdot 360° = 50.4°$
21 or more days $61\% \cdot 360° = 219.6°$

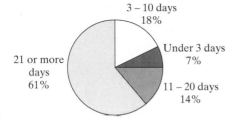

29.
$$
\begin{array}{r}
5 \\
2\overline{)\ 10} \\
2\overline{)\ 20} \\
2\overline{)\ 40}
\end{array}
$$
$40 = 2^3 \cdot 5$

33. Pacific Ocean: $49\% \cdot 264,489,800$
$= 129,600,002$ square kilometers

EXERCISE SET 9.3

1. The height of the bar representing 100–149 miles is 15. Therefore, 15 adults drive 100–149 miles per week.

5. There are two bars that fit this description, namely 100–149 miles and 150–199 miles. Therefore, $15 + 9 = 24$ adults drive 100–199 miles per week.

9. $\dfrac{\text{number of adults who drive 150–199 miles per week}}{\text{total number of adults}}$
$= \dfrac{9}{100}$

13. The height of the bar representing 55–64 years old is 17. Therefore, 17 million householders are 55–64 years old.

17. answers may vary

21.

Class Intervals (Scores)	Tally	Class Frequency (Number of Games)
90–99	ⅲⅲ Ⅲ	8

25.

Class Intervals (Account Balances)	Tally	Class Frequency (Number of People)
$200–$299	ⅲⅲ |	6

29.
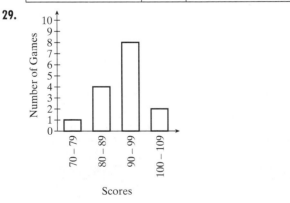

33. $\dfrac{12 + 28 + 20}{3} = \dfrac{60}{3} = 20$

37. Find each difference:

Europe	$14 - 10 = 4\%$
Asia	$38 - 31 = 7\%$
North America	$46 - 42 = 4\%$
South America	$6 - 5 = 1\%$
Africa	$3 - 3 = 0\%$
Other Continents	$1 - 1 = 0\%$

Asia shows the greatest change.

EXERCISE SET 9.4

1. Mean:
$$\frac{21 + 28 + 16 + 42 + 38}{5} = \frac{145}{5} = 29$$

Median: Write the numbers in order.
16, 21, 28, 38, 42
The middle number is 28.
Mode: There is no mode, since each number occurs once.

5. Mean:
$$\frac{0.2 + 0.3 + 0.5 + 0.6 + 0.6 + 0.9 + 0.2 + 0.7 + 1.1}{9}$$
$$= \frac{5.1}{9} \approx 0.6$$

Median: Write the numbers in order.
0.2, 0.2, 0.3, 0.5, 0.6, 0.6, 0.7, 0.9, 1.1
The middle number is 0.6.
Mode: Since 0.2 and 0.6 occur twice, there are two modes, 0.2 and 0.6.

9. Mean:
$$\frac{1450 + 1368 + 1362 + 1250 + 1136}{5}$$
$$= \frac{6566}{5} = 1313.2$$

The mean height of the five tallest buildings is 1313.2 feet.

13.

Grade	Point Value	Credit Hours	$\left(\begin{matrix}\text{Point}\\\text{Value}\end{matrix}\right) \cdot \left(\begin{matrix}\text{Credit}\\\text{Hours}\end{matrix}\right)$
B	3	3	9
C	2	3	6
A	4	4	16
C	2	4	8
Totals	14		39

17. Grade Point Average $= \dfrac{39}{14} \approx 2.76$

Mean:
$$\frac{7.8 + 6.9 + 7.5 + 4.7 + 6.9 + 7.0}{6} = \frac{40.8}{6}$$
$$= 6.8 \text{ seconds}$$

21. Write the numbers in order.
74, 77, 85, 86, 91, 95
Since there is an even number of scores, find the average of the middle two.

median $= \dfrac{85 + 86}{2} = \dfrac{171}{2} = 85.5$

25. 70 and 71; both occur twice.

29. $\dfrac{6}{18} = \dfrac{2 \cdot 3}{2 \cdot 3 \cdot 3} = \dfrac{1}{3}$

33. $\dfrac{55}{75} = \dfrac{5 \cdot 11}{5 \cdot 15} = \dfrac{11}{15}$

EXERCISE SET 9.5

1.

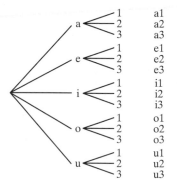

Outcomes

There are 15 outcomes.

5.

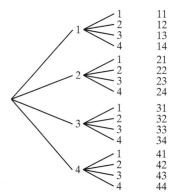

Outcomes

There are 16 outcomes.

9.

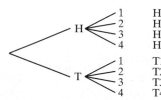

Outcomes

There are 8 outcomes.

13. probability $= \dfrac{2}{6} = \dfrac{1}{3}$

17. probability $= \dfrac{1}{3}$

21. probability $= \dfrac{1}{7}$

25. $\dfrac{1}{2} + \dfrac{1}{3} = \dfrac{1}{2} \cdot \dfrac{3}{3} + \dfrac{1}{3} \cdot \dfrac{2}{2} = \dfrac{3}{6} + \dfrac{2}{6} = \dfrac{5}{6}$

29. $5 \div \dfrac{3}{4} = \dfrac{5}{1} \cdot \dfrac{4}{3} = \dfrac{5 \cdot 4}{1 \cdot 3} = \dfrac{20}{3}$ or $6\dfrac{2}{3}$

33. There are four Kings.

probability $= \dfrac{4}{52} = \dfrac{1}{13}$

37.

Sum

1	2
2	3
3	4
4	5
5	6
6	7
1	3
2	4
3	5
4	6
5	7
6	8
1	4
2	5
3	6
4	7
5	8
6	9
1	5
2	6
3	7
4	8
5	9
6	10
1	6
2	7
3	8
4	9
5	10
6	11
1	7
2	8
3	9
4	10
5	11
6	12

probability $= \dfrac{3}{36} = \dfrac{1}{12}$

CHAPTER 9 TEST

1. The second week has $4\dfrac{1}{2}$ bills and each bill represents $50. $4.5 \cdot \$50 = \$225$ was collected during the second week.

5. The shortest bar is above the month of February. February has the least amount of normal monthly precipitation with 3 centimeters.

9.

Year	Increase
1997	$47 - 18 = 29$
1998	$110 - 47 = 63$
1999	$240 - 110 = 130$

1999 shows the greatest increase.

13. amount $= 27\% \cdot 8,272,000 = 2,233,440$ small cars

17.

Class Interval	Tally	Class Frequency
40–49			1
50–59	|||	3
60–69	||||	4
70–79	![tally]	5
80–89	![tally] |||	8
90–99	||||	4

21.

Grade	Point Value	Credit Hours	$\left(\begin{array}{c}\text{Point}\\\text{Value}\end{array}\right) \cdot \left(\begin{array}{c}\text{Credit}\\\text{Hours}\end{array}\right)$
A	4	3	12
B	3	3	9
C	2	3	6
B	3	4	12
A	4	1	4
	Totals	14	43

25. probability $= \dfrac{2}{10} = \dfrac{1}{5}$

Chapter 10

Exercise Set 10.1

1. If 0 represents ground level, then 1445 feet underground is -1445.

5. If 0 represents the line of scrimmage, a loss of 15 yards is -15.

9. If 0 represents \$0, a loss of \$5.049 million is -5.049 million.

13. If 0 represents 0%, a loss of 3.3% is -3.3.

17.
-4.7
$$\xleftarrow{\hspace{0.5em}\bullet\hspace{2em}\bullet\hspace{4em}\bullet\hspace{0.5em}}\rightarrow$$
$-5 \; -4 \; -3 \; -2 \; -1 \; 0 \; 1 \; 2 \; 3 \; 4 \; 5$

21. $4 > 0$

25. $0 > -3$

29. $-4.6 < -2.7$

33. $\dfrac{1}{4} > -\dfrac{8}{11}$

37. $|-8| = 8$, because -8 is 8 units from 0.

41. $|-5| = 5$, because -5 is 5 units from 0.

45. $\left|\dfrac{9}{10}\right| = \dfrac{9}{10}$ because $\dfrac{9}{10}$ is $\dfrac{9}{10}$ units from 0.

49. $|7.6| = 7.6$ because 7.6 is 7.6 units from 0.

53. The opposite of -4 is $-(-4) = 4$.

57. The opposite of $-\dfrac{9}{16}$ is $-\left(-\dfrac{9}{16}\right) = \dfrac{9}{16}$.

61. The opposite of $\dfrac{17}{18}$ is $-\dfrac{17}{18}$.

65.

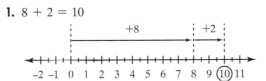

The opposite of the absolute value of 20 is the opposite of 20.

69. $-(-8) = 8$
The opposite of negative 8 is 8.

73. $-(-29) = 29$
The opposite of negative 29 is 29.

77.
$$\begin{array}{r} 15 \\ +\,20 \\ \hline 35 \end{array}$$

81. True. Consider the values on a number line.

85. answers may vary

Exercise Set 10.2

1. $8 + 2 = 10$

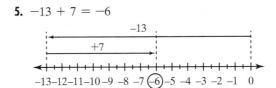

5. $-13 + 7 = -6$

![number line from -13 to 0 with +7 arrow and -13 arrow, -6 circled]

9. $|-6| + |-2| = 8$
Common sign is negative, so -8.

13. $|6| - |-2| = 4$
$6 > 2$, so answer is $+4$.

17. $|-5| - |3| = 2$
$5 > 3$ so answer is -2.

21. $|-12| + |-12| = 24$
The common sign is negative, so -24.

25. $|-123| + |-100| = 223$
The common sign is negative, so -223.

29. $|12| - |-5| = 7$
$12 > 5$, so answer is $+7$.

33. $|-12| - |3| = 9$
$12 > 3$, so answer is -9.

37. $|57| - |-37| = 20$
$57 > 37$, so answer is $+20$.

41. $|-6.3| + |-2.2| = 8.5$
Common sign is negative, so -8.5.

45. $\left|-\dfrac{2}{3}\right| + \left|-\dfrac{1}{6}\right| = \dfrac{4}{6} + \dfrac{1}{6} = \dfrac{5}{6}$
Common sign is negative, so $-\dfrac{5}{6}$.

49. $-4 + 2 + (-5) = -2 + (-5) = -7$

53. $12 + (-4) + (-4) + 12 = 8 + (-4) + 12$
$$= 4 + 12$$
$$= 16$$

57. $-10° + 12° = 2°$
A rise of $12°$ from $-10°$ Celsius puts the 11 P.M. temperature at $2°$C.

61. Team 1: $\quad (-2) + (-13) + 20 + 2$
$$= -15 + 20 + 2$$
$$= 5 + 2$$
$$= 7$$

Team 2: $\quad 5 + 11 + (-7) + (-3)$
$$= 16 + (-7) + (-3)$$
$$= 9 + (-3)$$
$$= 6$$

Team 1, 7; Team 2, 6; Winning team, Team 1

65. $-10,924 + 3245 = -7679$
The depth of the Pacific's Aleutian Trench is -7679 meters.

69. $52 - 52 = 0$

73. True. Add any two negative numbers on a number line to verify.

EXERCISE SET 10.3

1. $-5 - (-5) = -5 + 5 = 0$

5. $3 - 8 = 3 + (-8) = -5$

9. $-5(-8) = -5 + 8 = 3$

13. $2 - 16 = 2 + (-16) = -14$

17. $3.62 - (-0.4) = 3.62 + 0.4 = 4.02$

21. $\dfrac{2}{5} - \dfrac{7}{10} = \dfrac{2}{5} + \left(-\dfrac{7}{10}\right) = \dfrac{4}{10} + \left(-\dfrac{7}{10}\right) = -\dfrac{3}{10}$

25. $-20 - 18 = -20 + (-18) = -38$

29. $2 - (-11) = 2 + 11 = 13$

33. $12 - 5 - 7 = 12 + (-5) + (-7)$
$$= 7 + (-7)$$
$$= 0$$

37. $-10 + (-5) - 12 = -10 + (-5) + (-12)$
$$= -15 + (-12)$$
$$= -27$$

41. To find out how many degrees warmer, subtract the coldest temperature from the warmest temperature.
$136 - (-126) = 136 + 126 = 262$
$136°$F is $262°$F warmer than $-126°$F.

45. $-6 - 3 + 4 - 7 = -9 + 4 - 7$
$$= -5 - 7$$
$$= -12$$
The temperature is $-12°$C.

49. $144 - 0 = 144$ feet

53. $-176 - (-215) = -176 + 215 = 39$
Saturn has the warmest average temperature because $-176 > -215$. It it $39°$C warmer than Uranus.

57. $1 \cdot 8 = 8$

61. $|-3| - |-7| = 3 - 7 = 3 + (-7) = -4$

65. $|-17| - |-29| = 17 - 29 = 17 + (-29) = -12$

69. answers may vary

EXERCISE SET 10.4

1. $-2(-3) = 6$

5. $(2.6)(-1.2) = -3.12$

9. $-\dfrac{3}{5}\left(-\dfrac{2}{7}\right) = \dfrac{(-3) \cdot (-2)}{5 \cdot 7} = \dfrac{6}{35}$

13. $-1(-2)(-4) = 2(-4) = -8$

17. $10(-5)(0) = -50(0) = 0$

21. $(-3)^3 = (-3)(-3)(-3) = 9(-3) = -27$

25. $(-5)^3 = (-5)(-5)(-5) = 25(-5) = -125$

29. $\dfrac{3}{4}\left(-\dfrac{7}{8}\right) = \dfrac{3 \cdot (-7)}{4 \cdot 8} = -\dfrac{21}{32}$

33. $-1(2)(7)(-3.1) = -2(7)(-3.1)$
$$= -14(-3.1)$$
$$= 43.4$$

37. $-24 \div 6 = -4$

41. $\dfrac{-88}{-11} = 8$

45. $\dfrac{39}{-3} = -13$

49. $-\dfrac{7}{12} \div \left(-\dfrac{1}{6}\right) = -\dfrac{7}{12} \cdot \left(-\dfrac{6}{1}\right) = \dfrac{-7 \cdot (-6)}{12 \cdot 1} = \dfrac{7}{2}$

53. $240 \div (-40) = -6$

57. $\dfrac{-120}{0.4} = -300$

61. $(-4)(3) = -12$ yards

65. $\dfrac{-5 + (-2) + (-6) + (-7)}{4} = \dfrac{-20}{4} = -5$
Her average score per round was -5.

69. a. $27 - 35 = -8$
There was a change of -8 condors.

b. 1979 to 1987 is 8 years.
$$\dfrac{-8}{8} = -1$$
There was a change of -1 condor per year.

73. $90 + 12^2 - 5^3 = 90 + 12 \cdot 12 - 5 \cdot 5 \cdot 5$
$$= 90 + 144 - 125$$
$$= 234 - 125$$
$$= 109$$

77. False; The product of two negative numbers is always a positive number.

81. a. $278{,}574 - 285{,}674 = -7100$
There was a change of -7100 Saturns.

b. $2 \cdot (-7100) = -14{,}200$
The change will be $-14{,}200$ Saturns.

c. Two years from 1996 is 1998,
$278{,}574 \quad 14{,}200 - 264{,}374$
Sales would be 264,374 Saturns.

EXERCISE SET 10.5

1. $-1(-2) + 1 = 2 + 1 = 3$

5. $9 - 12 - 4 = 9 + (-12) + (-4)$
$$= -3 + (-4)$$
$$= -7$$

9. $\dfrac{4}{9}\left(\dfrac{2}{10} - \dfrac{7}{10}\right) = \dfrac{4}{9}\left(-\dfrac{5}{10}\right) = \dfrac{4}{9}\left(-\dfrac{1}{2}\right) = -\dfrac{2}{9}$

13. $25 \div (-5) + 12 = -5 + 12 = 7$

17. $\dfrac{24}{10 + (-4)} = \dfrac{24}{6} = 4$

21. $\begin{aligned}(-19) - 12(3) &= (-19) - 36 \\ &= (-19) + (-36) \\ &= -55\end{aligned}$

25. $[8 + (-4)]^2 = [4]^2 = 16$

29. $(3 - 12) \div 3 = (-9) \div 3 = -3$

33. $\begin{aligned}(5 - 9)^2 \div (4 - 2)^2 &= (-4)^2 \div (2)^2 \\ &= 16 \div 4 \\ &= 4\end{aligned}$

37. $\begin{aligned}(-12 - 20) \div 16 - 25 &= (-32) \div 16 - 25 \\ &= -2 - 25 \\ &= -27\end{aligned}$

41. $(0.2 - 0.7)(0.6 - 1.9) = (-0.5)(-1.3) = 0.65$

45. $(-36 \div 6) - (4 \div 4) = -6 - 1 = -7$

49. $(-5)^2 - 6^2 = 25 - 36 = -11$

53. $\begin{aligned}2(8 - 10)^2 - 5(1 - 6)^2 &= 2(-2)^2 - 5(-5)^2 \\ &= 2(-2)(-2) - 5(-5)(-5) \\ &= 2(-2)(-2) - 5(-5)(-5) \\ &= (-4)(-2) + 25(-5) \\ &= 8 + (-125) \\ &= -117\end{aligned}$

57. $\begin{aligned}\dfrac{(-7)(-3) - (4)(3)}{3[7 \div (3 - 10)]} &= \dfrac{(-7)(-3) + (-4)(3)}{3[7 \div (3 + (-10))]} \\[1mm] &= \dfrac{21 + (-12)}{3[7 \div (-7)]} \\[1mm] &= \dfrac{9}{3[-1]} \\[1mm] &= \dfrac{9}{-3} \\[1mm] &= -3\end{aligned}$

61. $45 \cdot 90 = 4050$

65. $P = 4(8 \text{ in.}) = 32 \text{ inches}$

69. $2 \cdot (7 - 5) \cdot 3 = 2 \cdot (2) \cdot 3 = 4 \cdot 3 = 12$

73. $\begin{aligned}(-12)^4 &= (-12)(-12)(-12)(-12) \\ &= 144(-12)(-12) \\ &= -1728(-12) \\ &= 20{,}736\end{aligned}$

CHAPTER 10 TEST

1. $-5 + 8 = 3$

5. $-\dfrac{3}{11} + \left(-\dfrac{5}{22}\right) = -\dfrac{6}{22} + \left(-\dfrac{5}{22}\right) = -\dfrac{11}{22} = -\dfrac{1}{2}$

9. $|-25| + (-13) = 25 + (-13) = 12$

13. $(-8) + 9 \div (-3) = -8 + (-3) = -11$

17. $\left(\dfrac{5}{9} - \dfrac{7}{9}\right)^2 + \left(-\dfrac{2}{9}\right) = \left(-\dfrac{2}{9}\right)^2 + \left(-\dfrac{2}{9}\right)$

$\qquad\qquad\qquad\quad = \dfrac{4}{81} + \left(-\dfrac{2}{9}\right)$

$\qquad\qquad\qquad\quad = \dfrac{4}{81} + \left(-\dfrac{18}{81}\right)$

$\qquad\qquad\qquad\quad = -\dfrac{14}{81}$

21. $\dfrac{-3(-2) + 12}{-1(-4 - 5)} = \dfrac{-3(-2) + 12}{-1[-4 + (-5)]}$

$\qquad\qquad\quad = \dfrac{6 + 12}{-1(-9)}$

$\qquad\qquad\quad = \dfrac{18}{9}$

$\qquad\qquad\quad = 2$

25. $237 - 157 - 17 + 38 = 80 - 77 + 38$

$\qquad\qquad\qquad\qquad\quad = 3 + 38$

$\qquad\qquad\qquad\qquad\quad = +41$

Her balance is +41 dollars.

Chapter 11

Exercise Set 11.1

1. $3 + 2z = 3 + 2(-3)$

$\qquad\quad = 3 - 6$

$\qquad\quad = -3$

5. $z - x + y = -3 - (-2) + 5$

$\qquad\qquad\quad = -3 + 2 + 5$

$\qquad\qquad\quad = 4$

9. $8 - (y - x) = 8 - [5 - (-2)]$

$\qquad\qquad\quad = 8 - (5 + 2)$

$\qquad\qquad\quad = 8 - 7$

$\qquad\qquad\quad = 1$

13. $\dfrac{6xy}{z} = \dfrac{6 \cdot (-2) \cdot 5}{-3}$

$\qquad\quad = \dfrac{-60}{-3}$

$\qquad\quad = 20$

17. $\dfrac{x + 2y}{2z} = \dfrac{-2 + 2 \cdot 5}{2 \cdot (-3)}$

$\qquad\qquad = \dfrac{-2 + 10}{-6}$

$\qquad\qquad = \dfrac{8}{-6}$

$\qquad\qquad = -\dfrac{4}{3}$

21. $A = lw$

$\quad = (50 \text{ feet}) \cdot (40 \text{ feet})$

$\quad = 2000 \text{ square feet}$

The area is 2000 square feet.

25. $I = prt$

$\quad = 3000 \cdot 0.06 \cdot 2$

$\quad = 360$

It will earn $360 in interest.

29. $F = \dfrac{9}{5}C + 32$

$\quad = \dfrac{9}{5} \cdot (-5) + 32$

$\quad = \dfrac{9 \cdot (-5)}{5 \cdot 1} + 32$

$\quad = \dfrac{-45}{5} + 32$

$\quad = -9 + 32$

$\quad = 23$

The temperature is 23°F.

33. $3x + 5x = (3 + 5)x$

$\qquad\qquad = 8x$

37. $4c + c - 7c = (4 + 1 - 7)c$

$\qquad\qquad\qquad = -2c$

41. $4a + 3a + 6a - 8 = (4 + 3 + 6)a - 8$

$\qquad\qquad\qquad\qquad = 13a - 8$

45. $3x + 7 - x - 14 = 3x - x + 7 - 14$

$\qquad\qquad\qquad\qquad = 2x - 7$

49. $6(5x) = (6 \cdot 5)x = 30x$

53. $2(y + 2) = 2 \cdot y + 2 \cdot 2 = 2y + 4$

57. $-4(3x + 7) = -4 \cdot 3x + (-4) \cdot 7$

$\qquad\qquad\qquad = -12x - 28$

61. $4(6n - 5) + 3n = 4 \cdot 6n + 4 \cdot (-5) + 3n$

$\qquad\qquad\qquad\quad = 24n - 20 + 3n$

$\qquad\qquad\qquad\quad = 27n - 20$

65. $-2(3x + 1) + 5(x - 2)$

$\quad = -2 \cdot 3x + (-2) \cdot 1 + 5 \cdot x + 5 \cdot (-2)$

$\quad = -6x - 2 + 5x - 10$

$\quad = -6x + 5x - 2 - 10$

$\quad = -x - 12$

69. $-13 + 10 = -3$

73. $-4 + 4 = 0$

77. $9684q - 686 - 4860q + 12{,}960$

$\quad = (9684 - 4860)q + (12{,}960 - 686)$

$\quad = 4824q + 12{,}274$

81. Add the areas of the two rectangles.

Area = (length) $\cdot$ (width)

$\begin{pmatrix} \text{Area of the} \\ \text{rectangle} \\ \text{on the left} \end{pmatrix} + \begin{pmatrix} \text{Area of the} \\ \text{rectangle} \\ \text{on the right} \end{pmatrix}$

$\quad = 7(2x + 1) + 3(2x + 3)$

$\quad = 7(2x) + 7(1) + 3(2x) + 3(3)$

$\quad = 14x + 7 + 6x + 9$

$\quad = (20x + 16)$

The area is $(20x + 16)$ square miles.

Exercise Set 11.2

1. $x - 8 = 2$

$\quad 10 - 8 \overset{?}{=} 2$

$\qquad\quad 2 \overset{?}{=} 2$ True

Yes, 10 is a solution.

5. $7f = 64 - f$
$7(8) \stackrel{?}{=} 64 - 8$
$56 \stackrel{?}{=} 64 - 8$
$56 \stackrel{?}{=} 56$ True
Yes, 8 is a solution.

9. $a + 5 = 23$
$a + 5 - 5 = 23 - 5$
$a = 18$
Check: $a + 5 = 23$
$18 + 5 \stackrel{?}{=} 23$
$23 \stackrel{?}{=} 23$ True
The solution is 18.

13. $7 = y - 2$
$7 + 2 = y - 2 + 2$
$9 = y$
Check: $7 = y - 2$
$7 \stackrel{?}{=} 9 - 2$
$7 \stackrel{?}{=} 7$ True
The solution is 9.

17. $x + \dfrac{1}{2} = \dfrac{7}{2}$
$x + \dfrac{1}{2} - \dfrac{1}{2} = \dfrac{7}{2} - \dfrac{1}{2}$
$x = \dfrac{6}{2}$
$x = 3$
Check: $x + \dfrac{1}{2} = \dfrac{7}{2}$
$3 + \dfrac{1}{2} \stackrel{?}{=} \dfrac{7}{2}$
$\dfrac{6}{2} + \dfrac{1}{2} \stackrel{?}{=} \dfrac{7}{2}$
$\dfrac{7}{2} \stackrel{?}{=} \dfrac{7}{2}$ True
The solution is 3.

21. $x - 3 = -1 + 4$
$x - 3 = 3$
$x - 3 + 3 = 3 + 3$
$x = 6$
Check: $x - 3 = -1 + 4$
$6 - 3 \stackrel{?}{=} -1 + 4$
$3 \stackrel{?}{=} 3$ True
The solution is 6.

25. $x - 0.6 = 4.7$
$x - 0.6 + 0.6 = 4.7 + 0.6$
$x = 5.3$
Check: $x - 0.6 = 4.7$
$5.3 - 0.6 \stackrel{?}{=} 4.7$
$4.7 \stackrel{?}{=} 4.7$ True
The solution is 5.3.

29. $y + 2.3 = -9.2 - 8.6$
$y + 2.3 = -17.8$
$y + 2.3 - 2.3 = -17.8 - 2.3$
$y = -20.1$
Check: $y + 2.3 = -9.2 - 8.6$
$-20.1 + 2.3 \stackrel{?}{=} -9.2 - 8.6$
$-17.8 \stackrel{?}{=} -17.8$ True
The solution is -20.1.

33. $2 - 2 = 5x - 4x$
$0 = x$
Check: $2 - 2 = 5x - 4x$
$2 - 2 \stackrel{?}{=} 5(0) - 4(0)$
$0 \stackrel{?}{=} 0 - 0$
$0 \stackrel{?}{=} 0$ True
The solution is 0.

37. $\dfrac{-7}{-7} = 1$

41. $-\dfrac{2}{3} \cdot \left(-\dfrac{3}{2}\right) = 1$

45. $T = P + R$
$5614 = P + 1909$
$5614 - 1909 = P + 1909 - 1909$
$3705 = P$
The total passing yardage was 3705 yards.

EXERCISE SET 11.3

1. $5x = 20$
$\dfrac{5x}{5} = \dfrac{20}{5}$
$x = 4$

5. $0.4y = 0$
$\dfrac{0.4y}{0.4} = \dfrac{0}{0.4}$
$y = 0$

9. $-0.3x = -15$
$\dfrac{-0.3x}{-0.3} = \dfrac{-15}{-0.3}$
$x = 50$

13. $\dfrac{1}{6}y = -5$
$6 \cdot \dfrac{1}{6}y = 6 \cdot (-5)$
$y = -30$

17. $-\dfrac{2}{9}z = \dfrac{4}{27}$
$-\dfrac{9}{2} \cdot \left(-\dfrac{2}{9}z\right) = -\dfrac{9}{2} \cdot \dfrac{4}{27}$
$z = -\dfrac{9 \cdot 4}{2 \cdot 27}$
$z = -\dfrac{2}{3}$

21. $-\dfrac{3}{5}x = -\dfrac{6}{15}$
$-\dfrac{5}{3} \cdot \left(-\dfrac{3}{5}x\right) = -\dfrac{5}{3} \cdot \left(-\dfrac{6}{15}\right)$
$x = \dfrac{-5 \cdot (-6)}{3 \cdot 15}$
$x = \dfrac{2}{3}$

25. $16 = 10t - 8t$
$16 = 2t$
$\dfrac{16}{2} = \dfrac{2t}{2}$
$8 = t$

29.
$$4 - 10 = -3z$$
$$-6 = -3z$$
$$\frac{-6}{-3} = \frac{-3z}{-3}$$
$$2 = z$$

33.
$$-36 = 9u + 3u$$
$$-36 = 12u$$
$$\frac{-36}{12} = \frac{12u}{12}$$
$$-3 = u$$

37.
$$5 - 5 = 2x + 7x$$
$$0 = 9x$$
$$\frac{0}{9} = \frac{9x}{9}$$
$$0 = x$$

41.
$$3x + 10 = 3(5) + 10$$
$$= 15 + 10$$
$$= 25$$

45.
$$\frac{3x + 5}{x - 7} = \frac{3(5) + 5}{5 - 7}$$
$$= \frac{15 + 5}{-2}$$
$$= \frac{20}{-2}$$
$$= -10$$

49.
$$-0.025x = 91.2$$
$$\frac{-0.025x}{-0.025} = \frac{91.2}{-0.025}$$
$$x = -3648$$

53.
$$d = rt$$
$$232 = r \cdot 4$$
$$\frac{232}{4} = \frac{r \cdot 4}{4}$$
$$58 = r$$

The driver should maintain a speed of 58 miles per hour.

EXERCISE SET 11.4

1.
$$2x - 6 = 0$$
$$2x - 6 + 6 = 0 + 6$$
$$2x = 6$$
$$\frac{2x}{2} = \frac{6}{2}$$
$$x = 3$$

5.
$$6 - n = 10$$
$$6 - n - 6 = 10 - 6$$
$$-n = 4$$
$$\frac{-n}{-1} = \frac{4}{-1}$$
$$n = -4$$

9.
$$3x - 7 = 4x + 5$$
$$3x - 7 + 7 = 4x + 5 + 7$$
$$3x = 4x + 12$$
$$3x - 4x = 4x + 12 - 4x$$
$$-x = 12$$
$$\frac{-x}{-1} = \frac{12}{-1}$$
$$x = -12$$

13.
$$1.7 = 2y + 9.5$$
$$1.7 - 9.5 = 2y + 9.5 - 9.5$$
$$-7.8 = 2y$$
$$\frac{-7.8}{2} = \frac{2y}{2}$$
$$-3.9 = y$$

17.
$$8 - t = 3$$
$$8 - t - 8 = 3 - 8$$
$$-t = -5$$
$$\frac{-t}{-1} = \frac{-5}{-1}$$
$$t = 5$$

21.
$$2n + 8 = 0$$
$$2n + 8 - 8 = 0 - 8$$
$$2n = -8$$
$$\frac{2n}{2} = \frac{-8}{2}$$
$$n = -4$$

25.
$$3r + 4 = 19$$
$$3r + 4 - 4 = 19 - 4$$
$$3r = 15$$
$$\frac{3r}{3} = \frac{15}{3}$$
$$r = 5$$

29.
$$2 = 3z - 4$$
$$2 + 4 = 3z - 4 + 4$$
$$6 = 3z$$
$$\frac{6}{3} = \frac{3z}{3}$$
$$2 = z$$

33.
$$-7c + 1 = -20$$
$$-7c + 1 - 1 = -20 - 1$$
$$-7c = -21$$
$$\frac{-7c}{-7} = \frac{-21}{-7}$$
$$c = 3$$

37.
$$4x + 3 = 2x + 11$$
$$4x + 3 - 3 = 2x + 11 - 3$$
$$4x = 2x + 8$$
$$4x - 2x = 2x + 8 - 2x$$
$$2x = 8$$
$$\frac{2x}{2} = \frac{8}{2}$$
$$x = 4$$

41.
$$-8n + 1 = -6n - 5$$
$$-8n + 1 - 1 = -6n - 5 - 1$$
$$-8n = -6n - 6$$
$$-8n + 6n = -6n - 6 + 6n$$
$$-2n = -6$$
$$\frac{-2n}{-2} = \frac{-6}{-2}$$
$$n = 3$$

45.
$$\frac{3}{8}x + 14 = \frac{5}{8}x - 2$$
$$\frac{3}{8}x + 14 - 14 = \frac{5}{8}x - 2 - 14$$
$$\frac{3}{8}x = \frac{5}{8}x - 16$$
$$\frac{3}{8}x - \frac{5}{8}x = \frac{5}{8}x - 16 - \frac{5}{8}x$$
$$-\frac{2}{8}x = -16$$
$$-\frac{8}{2} \cdot \left(-\frac{2}{8}x\right) = -\frac{8}{2} \cdot (-16)$$
$$x = 64$$

49.
$$3(x - 1) = 12$$
$$3x - 3 = 12$$
$$3x - 3 + 3 = 12 + 3$$
$$3x = 15$$
$$\frac{3x}{3} = \frac{15}{3}$$
$$x = 5$$

53.
$$35 = 17 + 3(x - 2)$$
$$35 = 17 + 3x - 6$$
$$35 = 11 + 3x$$
$$35 - 11 = 11 + 3x - 11$$
$$24 = 3x$$
$$\frac{24}{3} = \frac{3x}{3}$$
$$8 = x$$

57.
$$2t - 1 = 3(t + 7)$$
$$2t - 1 = 3t + 21$$
$$2t - 1 + 1 = 3t + 21 + 1$$
$$2t = 3t + 22$$
$$2t - 3t = 3t + 22 - 3t$$
$$-1t = 22$$
$$\frac{-t}{-1} = \frac{22}{-1}$$
$$t = -22$$

61.
$$10 + 5(z - 2) = 4z + 1$$
$$10 + 5z - 10 = 4z + 1$$
$$5z = 4z + 1$$
$$5z - 4z = 4z + 1 - 4z$$
$$z = 1$$

65.

the sum of −42 and 16	is	−26

$$-42 + 16 = -26$$
$$-42 + 16 = -26$$

69.

3 times	the difference of −14 and 2	amounts to	−48

$$3 \cdot \qquad (-14 - 2) \qquad = \qquad -48$$
$$3(-14 - 2) = -48$$

73. The height of the bar representing 1993 is approximately 322. Therefore, the average monthly payment for new cars in 1993 was about $322.

77.
$$C = \frac{5}{9}(F - 32)$$
$$26.8 = \frac{5}{9}(F - 32)$$
$$26.8 = \frac{5}{9}F - \frac{160}{9}$$
$$26.8 + \frac{160}{9} = \frac{5}{9}F - \frac{160}{9} + \frac{160}{9}$$
$$\frac{241.2}{9} + \frac{160}{9} = \frac{5}{9}F$$
$$\frac{401.2}{9} = \frac{5}{9}F$$
$$\frac{9}{5} \cdot \frac{401.2}{9} = \frac{9}{5} \cdot \frac{5}{9}F$$
$$80.24 = F$$
The temperature is 80.24°F.

Exercise Set 11.5

1. $x + 5$

5. $20 - x$

9. $\dfrac{x}{2}$

13. $5x$

17. $50 - 8x$

21. $3x = 27$

25. $3x + 9 = 33$
$$3x = 24$$
$$x = 8$$

29. $8 - x = \dfrac{15}{5}$
$$8 - x = 3$$
$$-x = -5$$
$$x = 5$$

33. $3(x - 5) = \dfrac{108}{12}$
$$3x - 15 = 9$$
$$3x = 24$$
$$x = 8$$

37. Dole received x votes. Clinton received $x + 220$ votes.
$$x + (x + 220) = 538$$
$$2x + 220 = 538$$
$$2x + 220 - 220 = 538 - 220$$
$$2x = 318$$
$$\frac{2x}{2} = \frac{318}{2}$$
$$x = 159$$

Dole received 159 votes and Clinton received $159 + 220 = 379$ votes.

41. Truck's speed is t. Car's speed is $2t$.
Combined speed is 105.
$$t + 2t = 105$$
$$3t = 105$$
$$t = 35$$

Truck's speed is 35 mph, and car's speed is $2 \cdot 35 = 70$ mph.

45. Louisiana scored x points. Tennessee scored $x + 18$ points.

$$x + (x + 18) = 168$$
$$2x + 18 = 168$$
$$2x = 150$$
$$x = 75$$
$$x + 18 = 93$$

The Lady Volunteers scored 93 points.

49. 1026 rounded to the nearest hundred is 1000.

53. answers may vary

Chapter 11 Test

1.
$$\frac{3x - 5}{2y} = \frac{3(7) - 5}{2(-8)}$$
$$= \frac{21 - 5}{-16}$$
$$= \frac{16}{-16}$$
$$= -1$$

5. $A = lw$
$$= 3(3x - 1)$$
$$= 3 \cdot 3x + 3 \cdot (-1)$$
$$= 9x - 3$$

The area is $(9x - 3)$ square meters.

9.
$$-\frac{5}{8}x = -25$$
$$-\frac{8}{5} \cdot \left(-\frac{5}{8}x\right) = -\frac{8}{5} \cdot (-25)$$
$$x = 40$$

13.
$$-4x + 7 = 15$$
$$-4x + 7 - 7 = 15 - 7$$
$$-4x = 8$$
$$\frac{-4x}{-4} = \frac{8}{-4}$$
$$x = -2$$

17.
$$10y - 1 = 7y + 20$$
$$10y - 1 + 1 = 7y + 20 + 1$$
$$10y = 7y + 21$$
$$10y - 7y = 7y + 21 - 7y$$
$$3y = 21$$
$$\frac{3y}{3} = \frac{21}{3}$$
$$y = 7$$

21. $A = \frac{1}{2}bh$
$$= \frac{1}{2} \cdot 5 \cdot 12$$
$$= 30$$

The area is 30 square feet.

25. If there are x women entered, then there are $x + 112$ men entered.

$$x + (x + 112) = 600$$
$$2x + 112 = 600$$
$$2x + 112 - 112 = 600 - 112$$
$$2x = 488$$
$$x = 244$$

244 women entered the race.

Index

Photo Credits

Chapter 1 CO The American Kennel Club, **(p. 30)** The Schiller Group, Ltd./Dole Pineapple Plantation, **(p. 30)** Jerry L. Ferrara/Photo Researchers, Inc., **(p. 51)** Hyatt Corporation, **(p. 74)** Tony Freeman/PhotoEdit, **(p. 74)** Morton Beebe-S.F./Corbis-Bettmann, **(p. 85)** Jean-Claude LeJeune/Stock Boston

Chapter 2 CO Gilda Schiff/Photo Researchers, Inc., **(p. 131)** Kevin Horan/Tony Stone Images, **(p. 132)** American Red Cross

Chapter 3 CO Jeff Greenberg/Omni-Photo Communications, Inc., **(p. 183)** Richard Hutchings/PhotoEdit, **(p. 193)** Robert Harbison, **(p. 194)** Rev. Ronald Royer/Photo Researchers, Inc.

Chapter 4 CO Chuck Pefley/Stock Boston, **(p. 220)** AP/Wide World Photos

Chapter 5 CO Henley & Savage/The Stock Market, **(p. 293)** Paul Sakuma/ AP/Wide World Photos, **(p. 298)** Grace Davies/Omni-Photo Communications, Inc., **(p. 298)** British Airways, **(p. 300)** Niesenbahn/Switzerland Tourism, **(p. 311)** Michael Newman/PhotoEdit, **(p. 313)** Stock Boston

Chapter 6 CO Corbis Bettmann, **(p. 333)** C. Borland/PhotoDisc, Inc., **(p. 345)** Telegraph Colour Library/FPG International LLC, **(p. 363)** Jeff Greenberg/Omni-Photo Communications, Inc., **(p. 367)** John Serafin/Simon & Schuster Corporate Digital Archive, **(p. 368)** Michael Heron/Simon & Schuster/PH College

Chapter 7 CO PhotoDisc, Inc.

Chapter 8 CO Bill Gallery/Stock Boston, **(p. 476)** Will & Deni McIntyre/Photo Researchers, Inc., **(p. 476)** Brian K. Diggs/AP/Wide World Photos, **(p. 494)** Tom Bean/DRK Photo, **(p. 506)** UPI/Corbis-Bettmann, **(p. 514)** Will & Deni McIntyre/ Photo Researchers, Inc.

Chapter 9 CO P. Lloyd/Weatherstock

Chapter 10 CO Lennox McLendon/AP/Wide World Photos, **(p. 648)** Melancholia (engraving) by Albrecht Durer (1471–1528). Guildhall Library, Corporation of London, UK/Bridgeman Art Library, London/New York

Chapter 11 CO Donovan Reese/Tony Stone Images